Self-Exciting Fluid Dyn

Exploring the origins and evolution of magnetic fields in planets, stars and galaxies, this book gives a basic introduction to magnetohydrodynamics, and surveys the observational data with particular focus on geomagnetism and solar magnetism. Pioneering laboratory experiments that seek to replicate particular aspects of fluid dynamo action are also described. The authors provide a complete treatment of laminar dynamo theory and of the mean-field electrodynamics that incorporates the effects of random waves and turbulence. Both dynamo theory and its counterpart, the theory of magnetic relaxation, are covered. Topological constraints associated with conservation of magnetic helicity are thoroughly explored, and major challenges are addressed in areas such as fast-dynamo theory, accretion-disc dynamo theory and the theory of magnetostrophic turbulence. The book is aimed at graduate-level students in mathematics, physics, earth sciences and astrophysics, and will be a valuable resource for researchers at all levels.

KEITH MOFFATT FRS is Emeritus Professor of Mathematical Physics at the University of Cambridge. He has served as Head of the Department of Applied Mathematics and Theoretical Physics and as Director of the Isaac Newton Institute for Mathematical Sciences in Cambridge. A former editor of the *Journal of Fluid Mechanics*, he has published papers in fluid dynamics and magnetohydrodynamics and was a pioneer in the development of topological fluid dynamics. He is a Fellow of the Royal Society, a member of Academia Europæa, and a Foreign Member of the Academies of France, Italy, the Netherlands and the USA. He has been awarded numerous prizes, most recently the 2018 Fluid Dynamics Prize of the American Physical Society.

EMMANUEL DORMY is a CNRS Directeur de Recherche in the Department of Mathematics and its Applications at the Ecole Normale Supérieure (ENS) in Paris. He is also Professor at the ENS and at the Ecole Polytechnique, where he teaches different aspects of fluid dynamics. Convinced of the need to embrace all aspects of the dynamo problem, in 2006 he started a research group at the ENS which promotes an interdisciplinary approach and jointly studies all geophysical and astrophysical aspects of dynamo theory. He also founded and directed the Dynamo-GDRE, which promotes exchanges among researchers working on all aspects of dynamo theory throughout Europe and beyond, and he organises widely attended meetings.

Cambridge Texts in Applied Mathematics

The aim of this series is to provide a focus for publishing textbooks in applied mathematics at the advanced undergraduate and beginning graduate level. The books are devoted to covering certain mathematical techniques and theories and exploring their applications.

All titles listed below can be obtained from good booksellers or from Cambridge University Press. For a complete series listing, visit www.cambridge.org/mathematics.

Geometric and Topological Inference
JEAN-DANIEL BOISSONNAT, FRÉDÉRIC CHAZAL & MARIETTE YVINEC

Introduction to Magnetohydrodynamics (2nd Edition)
P. A. DAVIDSON

An Introduction to Stochastic Dynamics
JINQIAO DUAN

Singularities: Formation, Structure and Propagation
J. EGGERS & M. A. FONTELOS

Microhydrodynamics, Brownian Motion and Complex Fluids
MICHAEL D. GRAHAM

Discrete Systems and Integrability
J. HIETARINTA, N. JOSHI & F. W. NIJHOFF

An Introduction to Polynomial and Semi-Algebraic Optimization
JEAN BERNARD LASSERRE

Numerical Linear Algebra
HOLGER WENDLAND

Self-Exciting Fluid Dynamos

KEITH MOFFATT
University of Cambridge

EMMANUEL DORMY
Ecole Normale Supérieure, Paris

CAMBRIDGE
UNIVERSITY PRESS

University Printing House, Cambridge CB2 8BS, United Kingdom

One Liberty Plaza, 20th Floor, New York, NY 10006, USA

477 Williamstown Road, Port Melbourne, VIC 3207, Australia

314–321, 3rd Floor, Plot 3, Splendor Forum, Jasola District Centre,
New Delhi – 110025, India

79 Anson Road, #06–04/06, Singapore 079906

Cambridge University Press is part of the University of Cambridge.

It furthers the University's mission by disseminating knowledge in the pursuit of
education, learning, and research at the highest international levels of excellence.

www.cambridge.org
Information on this title: www.cambridge.org/9781107065871
DOI: 10.1017/9781107588691

First published 2019

Printed in the United Kingdom by TJ International Ltd. Padstow, Cornwall

A catalogue record for this publication is available from the British Library.

Library of Congress Cataloging-in-Publication Data
Names: Moffatt, H. K. (Henry Keith), author. | Dormy, Emmanuel, author.
Title: Self-exciting fluid dynamos / Keith Moffatt (University of Cambridge),
Emmanuel Dormy (Ecole Normale Supérieure, Paris).
Description: Cambridge ; New York, NY : Cambridge University Press, 2019. |
Series: Cambridge texts in applied mathematics | Includes bibliographical references and indexes.
Identifiers: LCCN 2018047454 | ISBN 9781107065871 (hardback : alk. paper)
Subjects: LCSH: Fluid dynamics. | Magnetohydrodynamics. |
Dynamo theory (Cosmic physics) | Geophysics.
Classification: LCC QA912 .M65 2019 | DDC 523.01/886–dc23
LC record available at https://lccn.loc.gov/2018047454

ISBN 978-1-107-06587-1 Hardback
ISBN 978-1-108-71705-2 Paperback

Dedicated to the memory of
George Keith Batchelor
1920–2000

Contents

Preface

In fifty years almost every book begins to require notes either to explain forgotten allusions and obsolete words; or to subjoin those discoveries which have been made by the gradual advancement of knowledge; or to correct those mistakes which time may have discovered.

Samuel Johnson, *Letters*, 1774

This book is an update and amplification of a research monograph *Magnetic Field Generation in Electrically Conducting Fluids* first published 40 years ago (Moffatt 1978a). Despite the passage of time, much of the work described in that monograph remains at the core of dynamo theory, and it has provided a useful text for graduate students approaching the subject for the first time. Nevertheless, as Samuel Johnson so aptly recorded, it is now desirable "to subjoin those discoveries which have been made by the gradual advancement of knowledge" and "to correct those mistakes which time may have discovered".

It is now exactly 100 years since Larmor (1919) enunciated his famous question "How could a rotating body such as the Sun become a magnet?" A reasonably convincing answer can now be given to this question! And "a rotating body such as the Sun" now includes planets, stars and galaxies, which nearly all exhibit internally generated magnetic fields.

Our knowledge of the self-exciting dynamo process has advanced dramatically over the last 40 years, stimulated by a wealth of satellite data, by great advances in computational power and by the extremely challenging laboratory experiments that seek to replicate either laminar or turbulent dynamo action. Mean-field electrodynamics, greatly developed since the late 1960s, remains at the heart of the subject; this theory takes account of a fluctuating velocity field, in the form of either weak random waves or strong 'Kolmogorov-type turbulence'. It is the mean helicity of this type of flow that is known to be particularly conducive to dynamo action in any conducting fluid of sufficiently large extent. Helicity, and its topological interpretation, therefore continues to play a central role in our approach.

The work is presented here in three parts: Part I treats the theoretical and observational background; Part II covers the foundations of the 'kinematic' dynamo theory that pertains to arbitrary velocity fields; and Part III incorporates dynamics governed by the Navier–Stokes equations in a rotating conducting fluid, including the 'back-reaction' of the Lorentz force distribution associated with a dynamo-generated magnetic field. Much of the presentation is completely new; we draw particular attention to the treatment of the VKS experiment (§9.12), fast dynamo action (Chapter 10), low-dimensional models of the geodynamo (Chapter 11), dynamic equilibration and quenching of the α-effect (Chapter 12), magnetorotational instability (§14.7), magnetostrophic turbulence (§15.6) and the final two chapters, 16 and 17, on magnetic relaxation, which can be viewed as 'the other side of the dynamo-theory coin', important in relation to equilibration in a statistically steady state.

Helicity, a topological invariant of the Euler equations of ideal fluid flow, still plays a central role in dynamo theory. Indeed we may confidently assert that turbulence having non-zero mean helicity will always generate a large-scale magnetic field in an electrically conducting fluid of sufficient spatial extent. The mean-field theories on which this assertion is based are presented in Chapters 7–9. These chapters provide a fundamental basis for the dynamical theories presented in subsequent chapters. Magnetic helicity, a similar topological invariant of the equations of ideal magnetohydrodynamics, plays a correspondingly central role in the theory of magnetic relaxation, providing in general a lower bound on the energy of any magnetic field of non-trivial topology.

We have taught much of the content of this book over many years in courses at graduate level in Cambridge, at the Ecole Normale Supérieure (ENS) and at the Ecole Polytechnique ('X'). However, some of the material, particularly in later chapters, records recent work, as yet presentable only at research seminar level.

We acknowledge the invaluable comments and vital input of many collaborators, who have helped to shape our ideas over many years – Philippe Cardin, Atta Chui, Stéphan Fauve, David Gérard-Varet, Andrew Gilbert, David Hughes, Dominique Jault, Yoshifumi Kimura, David Loper, Dionysis Linardatos, Krzysztof Mizerski, Gordon Ogilvie, Michael Proctor, Glyn Roberts, Renzo Ricca, Andrew Soward, Steve Tobias, Juri Toomre, Vladimir Vladimirov and Nigel Weiss all deserve special mention; above all, the late Konrad Bajer (1956–2014), who was involved in initial planning for this volume, but sadly did not live to see its completion.

Finally, we record our deep gratitude to Linty and Ludivine, without whose constant understanding, encouragement and support this work could not have been accomplished!

14 July 2018 HKM, ED

Part I

Basic Theory and Observations

1
Introduction

1.1 What is dynamo theory?

Dynamo theory is concerned with the manner in which magnetic fields are generated and maintained in planets, stars and galaxies. The Earth, Sun and Milky Way provide examples of most immediate interest, for which a huge quantity of observational detail is now available; and yet the fundamental theory applies equally to any sufficiently large mass of electrically conducting fluid, either liquid metal or ionised gas ('plasma' when fully ionised), under the combined effects of global rotation and convective motion, this usually having a turbulent character. This turbulence may be either 'strong turbulence' of a type familiar in aerodynamics and meteorology, or 'weak turbulence' – a field of weakly interacting random waves internal to the fluid. Either way, it is the combination of rotation and convection that turns out to be particularly conducive to the spontaneous growth of magnetic fields in fluid systems of sufficient spatial extent.

It is this latter requirement that has made laboratory realisation of the self-exciting dynamo process such a great challenge for experimentalists. The triple requirements of sufficient conductivity, scale and turbulent intensity have placed huge demands on the design of experiments, and it is only over the last decade that the necessary conditions have been achieved and that self-exciting dynamo action has been convincingly demonstrated. These experimental achievements have run in parallel with great computational achievements in modelling the dynamo process both in planetary liquid cores and in stellar convection zones. Theoretical progress, so essential for a full understanding of the dynamo process, has been much stimulated by the great advances on observational, experimental and numerical fronts.

In this introductory chapter, we first set out some of the historical background, with reference to subsequent chapters where specific issues are treated in detail.

1.2 Historical background

1.2.1 The geodynamo

A very complete history of magnetism over the past millennium may be found in Stern (2002). We content ourselves here with some of the highlights of a fascinating story.[1]

Every child who has played with a magnetic compass knows that the compass needle points North; but he learns as she grows older that magnetic North is not quite the same as 'true' North defined by the Pole star; or to put it differently, that the magnetic dipole axis is slightly inclined to the axis of rotation of the Earth. This worrying mismatch was already known to the Chinese of the Sung dynasty. In his great work *Science and Civilisation in China*, Joseph Needham (1962) quotes the *Mêng Chhi Pi Than* of Shen Kua (c.1088), which he translates thus: "Magicians rub the point of a needle with the lodestone; then it is able to point to the south. But it always inclines slightly to the east, and does not point directly at the south". So the 'declination' of the field was known, at least to the Chinese, 930 years ago. It was rediscovered and charted out by the early navigators of the fifteenth and sixteenth centuries and in particular by Christopher Columbus whose great voyage of discovery in 1492 opened new windows in the Western World. We recognise this declination now as a manifestation of a crucial departure from axisymmetry which is essential for the Earth's internal dynamo to operate.

In his seminal work *De Magnete*, William Gilbert (1600) recognised that 'magnet Earth' could be modelled by a spherical lodestone – his 'terrella' – over whose surface he was able to measure the magnetic field and plot its direction. Figure 1.1 shows a page from the second (1613) edition of this book, describing the sort of measurement that Gilbert was able to make. He spoke scathingly of earlier fanciful speculations concerning magnetism and was a pioneer of the 'scientific approach' based on careful observation and experiment.

The distinction between local magnetic north and 'true' north is often indicated on large-scale maps. The small print usually warns that the angle between the two directions changes irregularly by up to 1° in 6 years. This is the 'secular variation' of the magnetic field which was known to navigators of the seventeenth century and was no doubt a considerable nuisance to them. Edmund Halley considered the possible causes of this secular variation (Halley 1692) and concluded that

> the external parts of the globe may well be reckoned as the shell, and the internal as a nucleus or inner globe included within ours, with a fluid medium between ... only this outer Sphere having its turbinating motion some small matter either swifter or slower than the inner Ball.

[1] Parts of this section are an edited version of the introduction to a Union Lecture (Moffatt 1992) delivered at the IUGG General Assembly (Vienna 1991).

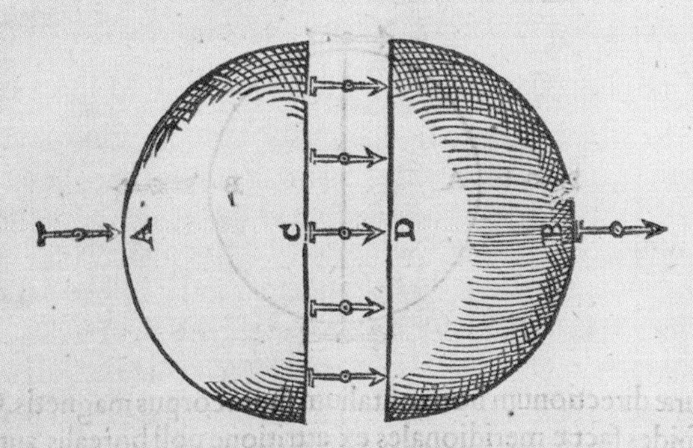

diſſectas, quemadmodùm in præſenti diagrammate. Hoc etiam eodem modo eueniret, ſi per tropici planum lapidis eſſet diſſectio & diſſectarum partium à ſe inuicèm diſiunctio & interuallũ, quemadmodùm priùs per æquinoctialis planũ diuiſo magnete & diſiuncto. Cuſpides enim fugantur à C, alliciuntur à D; & verſoria ſunt parallela, inuicèm imperantibus in finibus vtrinque polis ſeu verticitate.

Dimidium terrellæ per ſe, & eius directiones diſſimiles directionibus duarum partiũ finitimarũ in ſuperiori figurâ oſtenſis. Omnes

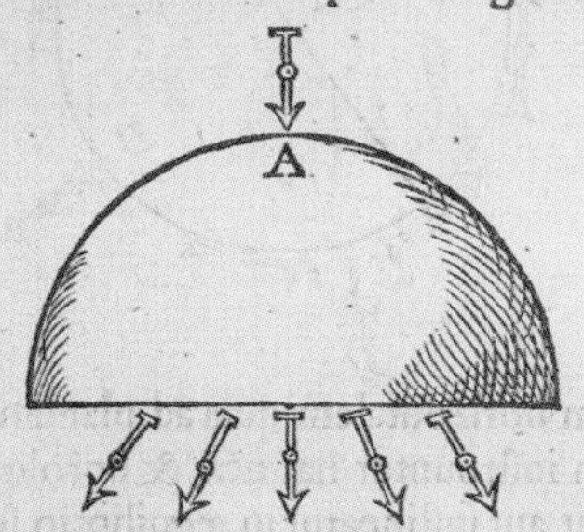

cuſpides tactæ ab A, cruces omnes inferiores præter mediam non rectè ſed obliquè tendunt ad magnetem; quia polus eſt in medio plani, quod anteà fuit æquinoctialis planum. Omnes cuſpides tactæ à locis diſtantibus à polo, mouentur ad polum (haud ſecus ac ſi ſuper ipſum polum fuiſſent attritæ) non ad locum attritionis, vbicunque fuerit in integro lapide inter polum & æquatorem in aliquâ latitudine. Ob eamque cauſam differentiæ regionum ſunt tantùm duæ, ſeptentrionales & meridionales, tam in terrellâ, quàm in generali

Figure 1.1 A page from the 1613 edition of Gilbert's *De Magnete* showing the result of an experiment conducted with his 'terrella', modelling the Earth's magnetic field. [Courtesy of the Wren Library, Trinity College, Cambridge.]

This was a prophetic vision as far as the inner structure of the Earth is concerned, but also remarkable in its perception of the need for *differential* rotation, now recognised as a further key element in the dynamo process.

A discovery of great importance was made by Oersted (1820), namely that current in a wire produces a magnetic field whose field lines embrace the wire. This led Ampère (1822) to propose that an east/west current must flow within the Earth. He wrote[2]

L'idée la plus simple, et celle qui se présenterait immédiatement à celui qui voudrait expliquer cette direction constante de l'aiguille, ne serait-elle pas d'admettre dans la terre un courant électrique, dans une direction telle que le nord se trouvât à gauche d'un homme qui, couché sur sa surface pour avoir la face tournée du côté de l'aiguille, recevrait ce courant dans la direction de ses pieds à sa tête, et d'en conclure qu'il a lieu, de l'est à l'ouest, dans une direction perpendiculaire au méridien magnétique ?

A modern understanding of the origin of these currents is based both on Ampère's law, essentially that electric current is the source of magnetic field, and on Faraday's law of induction (Faraday 1832). By painstaking experiments, Faraday discovered that if a conductor moves across a magnetic field, and if a path is available for the completion of a current circuit, then in general current will flow in that circuit. For this achievement, Faraday was awarded the Copley Medal of the Royal Society of London. The citation records that

he gives indisputable evidence of electric action due to terrestrial magnetism alone. An important addition is thus made to the facts which have long been accumulating for the solution of that most interesting problem, the magnetism of the Earth.

It was in fact more than an important addition; it was the key ingredient of the dynamo process, although this was not recognised till much later.

At about the same time, in two great papers Carl Friedrich Gauss (1832, 1838) established the spherical harmonic decomposition of the Earth's magnetic field and the technique by which secular variation of the field could be quantified. The traditional unit of field intensity in geomagnetism, and equally in astrophysics, is of course the Gauss (G), and it is arguably regrettable that the Système International of units now favours the tesla ($1\mathrm{T} = 10^4\,\mathrm{G}$). Gauss' spherical harmonic decomposition allows us to extrapolate the Earth's field (assumed potential) down to the core–mantle boundary (CMB), to map the contours of constant radial field at the CMB, and to do so at different epochs using all available data (Bloxham et al. 1989, see Chapter 4). In these maps, the dipole ingredient of the field is still quite evident at

[2] This translates, somewhat freely, as follows: "The simplest idea that must occur immediately to anyone attempting to explain the constant direction of the compass needle is this: there must exist an electric current in the Earth, such that, if a man were to lie on the surface of the Earth with the north to his left and his face turned in the direction of the needle, he would sense this current in the direction from his feet to his head; and should we not therefore conclude that this current flows from east to west perpendicular to the magnetic meridian?"

the CMB, but there is also a strong presence of quadrupole, octupole and higher-order ingredients, as is to be expected from the nature of downward extrapolation towards the region where the 'source' currents are confined. The slow evolution of the pattern (i.e. its secular variation) is also evident.

The high point and climax of electromagnetic theory in the nineteenth century came with the publication of James Clerk Maxwell's *Treatise on Electricity and Magnetism* (Maxwell 1873). Maxwell built on Faraday's discoveries and completed the system of equations that bear his name. It is interesting to note however that in a late chapter of the treatise, devoted to *Terrestrial Magnetism*, Maxwell comes nowhere near to any explanation of the real nature of the phenomenon. He confines himself to a description of Gauss' techniques for the determination of the Earth's field and its time variation, and his demonstration that the dominant source for the field is of internal rather than external origin; but as to the root cause of the phenomenon, he writes in sonorous tones:

> The field of investigation into which we are introduced by the study of terrestrial magnetism is as profound as it is extensive. ... What cause [is it], whether exterior to the Earth or in its inner depths, [that] produces such enormous changes in the Earth's magnetism, that its magnetic poles move slowly from one part of the globe to another? ... These immense changes in so large a body force us to conclude that we are not yet acquainted with one of the most powerful agents in nature, the scene of whose activity lies in those inner depths of the Earth, to the knowledge of which we have so few means of access.

It was the science of seismology that was to provide the vital means of access, establishing the existence first of a liquid outer core (Jeffreys 1926, who concluded that "the central core is probably fluid, but its viscosity is uncertain"), and secondly of an inner solid core (Lehmann 1936, Bullen 1946); both inner and outer cores are believed to be important for the operation of the geodynamo.[3]

One of the earliest discussions of possible causes of terrestrial magnetism was given by Arthur Schuster (1911) in his Presidential Address to the Physical Society of London. Schuster discussed the arguments for and against a system of electric currents in the Earth's interior and concluded that "the difficulties which stand in the way of basing terrestrial magnetism on electric currents inside the Earth are insurmountable" – strong words, which have since been invalidated with the passage of time and the birth and advance of magnetohydrodynamics. Nevertheless, even as late as 1940 in their great treatise on *Geomagnetism*, Chapman & Bartels (1940) came to the same defeatist conclusion as Schuster. They discussed Larmor's (1919) suggestion concerning the possibility of self-exciting dynamo action (see below) but stated that "Cowling, however, has shown that such self-excitation

[3] An illuminating discussion of the developments leading to these discoveries is given by Brush (1980).

is not possible. Consequently, Schuster's view still holds, that the difficulties ... are insuperable". Cowling (1934) had not in fact shown that such self-excitation is not possible: he had merely shown that it was not possible for axisymmetric systems (for Cowling's anti-dynamo theorem, see §§6.4 and 6.5), and yet the tilt of the magnetic dipole which had been known for centuries shows that we are dealing with an emphatically non-axisymmetric system. Nevertheless the fact that Chapman & Bartel could be so easily persuaded that Cowling's theorem closed the matter is an indication of the powerful influence that this theorem then had – this no doubt because it was one of the few exact results of the subject. The year 1940 marked a high point in the collection and systematisation of geomagnetic data, but it also marks the nadir as regards real understanding of the origins of terrestrial magnetism,

The post-war years saw a profound transformation in the situation, to the point at which a dynamo theory of the origin of the Earth's magnetic field is now universally accepted among geophysicists. The progress in dynamo theory has been dramatic, and the theory applies with equal force to planets other than the Earth. Statements in textbooks since the 1980s are as vigorously positive as Schuster's (1912) statement was negative. Thus, for example, Jacobs (1994) writes, "There has been much speculation on the origin of the Earth's magnetic field. ... The only possible means seems to be some form of electromagnetic induction, electric currents flowing in the Earth's core"; and Cook (2009) writes, "There is no theory other than a dynamo theory that shows any signs of accounting for the magnetic fields of the planets". It is a dynamo theory based on the principles of magnetohydrodynamics, and ultimately on a suitable exploitation of Faraday's law of induction, that has led to this remarkable revolution in our understanding of Nature.

1.2.2 The solar dynamo

Galileo's celebrated discovery of sunspots dates back to the MDCXIII publication of his *Istoria e Dimostrazioni*. Figure 1.2 shows Galileo's representation of the sunspots that he had by then observed. This apparent 'imperfection' in God's creation caused consternation in the powerful Catholic Church of that epoch; but paradoxically, it is this very imperfection and the manner in which it has evolved over the last four centuries that has provided a prime source of information concerning the physics of the surface layers of the Sun. This will be discussed in detail in Chapter 5; for the moment, we need only note Maunder's discovery of the 11-year sunspot cycle (Maunder 1904), and Hale's discovery of the relatively strong magnetic field in sunspots (Hale 1908) and the polarity laws that govern their behaviour.

Since the 1990s, the science of 'helioseismology' (analysis of the spectrum of solar oscillations, Christensen-Dalsgaard et al. 1996) has provided a wealth of

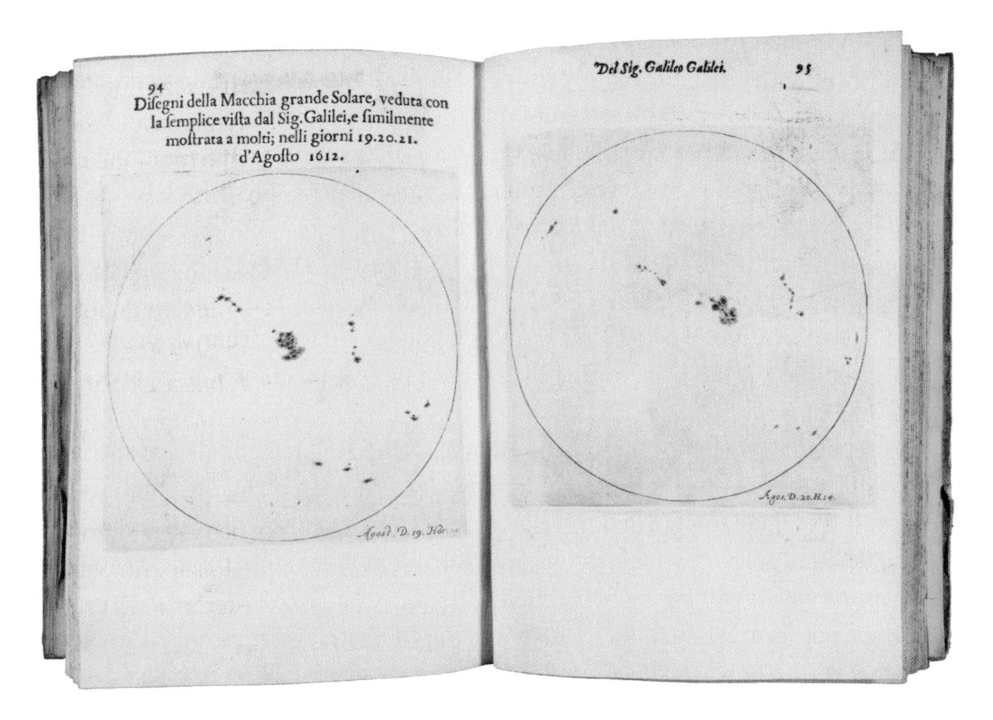

Figure 1.2 Galileo's volume *Istoria e Dimostrazioni*, showing here the record of his observations of sunspots on successive days in August 1612 (Galileo 1613). [Courtesy of the Wren Library, Trinity College, Cambridge.]

information concerning the flow field within the solar interior. In particular, through helioseismology, the differential rotation throughout most of the solar convection zone has been determined, and the presence of the 'tachocline', a layer of rapid shear at the base of the convection zone postulated by Spiegel & Zahn (1992), has been confirmed (Charbonneau et al. 1999). This in turn has led to renewed debate concerning the 'mean-field electrodynamics' applicable to the Sun through the '$\alpha\omega$-mechanism', matters that will be discussed in detail in later chapters.

The birth of solar dynamo theory proper is generally attributed to Joseph Larmor, Lucasian Professor at the University of Cambridge, who, exactly 100 years before publication of this book, posed the question "How could a rotating body such as the Sun become a magnet?" (Larmor 1919); and the question was certainly a natural one since the origin of the magnetic field of the Sun was at that time a total mystery.

And not only the Sun! We now know that a magnetic field is a normal accompaniment of any cosmic body that is both fluid (wholly or in part) and rotating. There appears to be a universal validity about this statement which applies quite irrespective of the length-scales considered. For example, on the planetary length-scale, Jupiter shares with the Earth the property of strong rotation (its rotation period being approximately 10 hours) and it is believed to have a fluid interior composed of an alloy of liquid metallic hydrogen and helium (Hide 1974); it exhibits a surface magnetic field of order 10 G in magnitude (as compared with the Earth's field of order 1 G). On the stellar length-scale, magnetic fields as weak as 1 G are hard to detect in general; there are however numerous examples of stars which rotate with periods ranging from several days to several months, and with detectable surface magnetic fields in the range 10^2 to 3×10^4 G (Preston 1967); and on the galactic length-scale, our own galaxy rotates about the normal to the plane of its disc with a period of order 3×10^8 years and exhibits a galactic-scale magnetic field roughly confined to the plane of the disc whose typical magnitude is of order 3 or 4×10^{-6} G.

The detailed character of these naturally occurring magnetic fields and the manner in which they evolve in time will be described in subsequent chapters; for the moment it is enough to note that it is the mere existence of these fields (irrespective of their detailed properties) which provides the initial motivation for the various investigations which will be described in this book.

Larmor put forward three alternative and very tentative suggestions concerning the origin of the Sun's magnetic field, only one of which has in any sense stood the test of time. This suggestion, which is fundamental to hydromagnetic dynamo theory, was that, just as for the Earth, motion of the electrically conducting fluid within the rotating body, might by its inductive action in flowing across the magnetic field generate just those currents $\mathbf{J}(\mathbf{x})$ required to provide the self-same field $\mathbf{B}(\mathbf{x})$.

1.3 The homopolar disc dynamo

This type of 'bootstrap' effect is most simply illustrated with reference to a system consisting entirely of solid (rather than fluid) conductors. This is the 'homopolar' disc dynamo (Bullard 1955) illustrated in Figure 1.3. A solid copper disc rotates about its axis with angular velocity Ω, and a current path between its rim and its axle is provided by the wire twisted as shown in a loop round the axle. This system can be unstable to the growth of magnetic perturbations. For suppose that a current $I(t)$ flows in the loop; this generates a magnetic flux Φ across the disc, and, provided the conductivity of the disc is not too high,[4] this flux is given by $\Phi = M_0 I$ where M_0 is the mutual inductance between the loop and the rim of the disc. Rotation of the disc leads to an electromotive force $\mathcal{E} = \Omega\Phi/2\pi$ which drives the current I, and the equation for $I(t)$ is then

[4] This proviso is necessary as is evident from the consideration that a superconducting disc would not allow any flux to cross its rim; a highly conducting disc in a time-dependent magnetic field tends to behave in the

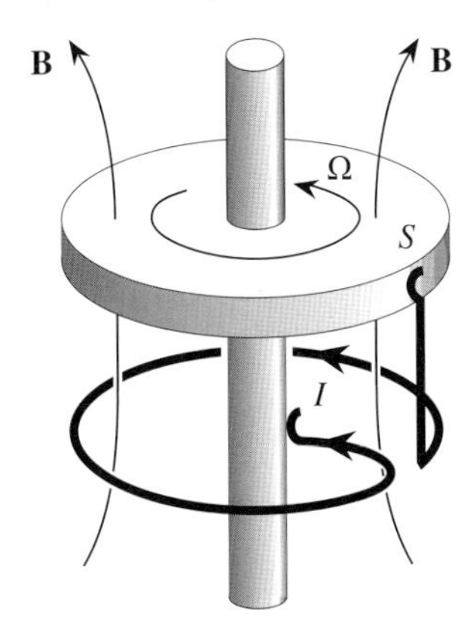

Figure 1.3 The homopolar disc dynamo. Note that the twist in the wire which carries the current $I(t)$ must be in the same sense as the sense of rotation Ω.

$$L\frac{\mathrm{d}I}{\mathrm{d}t} + RI = \mathcal{E} = M\Omega I\,, \tag{1.1}$$

where $M = M_0/2\pi$ and L and R are the self-inductance and resistance of the complete current circuit. The device is evidently unstable to the growth of I (and so of Φ) from an infinitesimal level if

$$\Omega > R/M. \tag{1.2}$$

Under this condition, the current grows exponentially, as does the retarding torque associated with the Lorentz force distribution in the disc. Ultimately the disc angular velocity slows down, and tends to equilibrium at the critical level $\Omega_0 = R/M$ at which the driving torque G just balances the sum of this retarding torque and any frictional torque that may be present. (The system may overshoot this equilibrium state and then oscillate about it, such oscillations being damped by frictional resistance.)

This type of example is certainly suggestive, but it differs from the conducting fluid situation in that the current is constrained by the twisted geometry to follow a very special path that is particularly conducive to dynamo action (i.e. to the conversion of mechanical energy into magnetic energy). No such geometrical constraints are apparent in, say, a spherical body of fluid of uniform electrical conductivity, and the question arises whether fluid motion within such a sphere, or other simply-connected region, can drive a suitably contorted current flow to provide the same sort of homopolar (self-excited) dynamo effect.

There are however two properties of the disc dynamo which reappear in some of

same way; a corrected theory allowing for the associated azimuthal current in the disc will be presented in §10.4.

the hydromagnetic situations to be considered later, and which deserve particular emphasis at this stage. Firstly, there is a discontinuity in angular velocity at the sliding contact S between the rotating disc and the stationary wire, i.e. the system exhibits *differential rotation*. The concentration of this differential rotation at the single point S is by no means essential for the working of the dynamo; we could in principle distribute the differential rotation arbitrarily by dividing the disc into a number of rings, each kept in electrical contact with its neighbours by means of lubricating films of, say, mercury, and by rotating the rings with different angular velocities. If the outermost ring is held fixed (so that there is no longer any sliding at the contact S), then the velocity field is entirely axisymmetric, the differential rotation being distributed across the plane of the disc. The system will still generally work as a dynamo provided the angular velocity of the inner rings is in the sense indicated in Figure 1.3 and sufficiently large.

Secondly, the device *lacks reflectional symmetry*: in Figure 1.3 the disc must rotate in the same sense as the twist in the wire if dynamo action is to occur. Indeed it is clear from equation (1.1) that if $\Omega < 0$ the rotation leads only to an *accelerated decay* of any current that may initially flow in the circuit. Recognition of this essential lack of reflectional symmetry provides the key to understanding the nature of dynamo action as it occurs in conducting fluids undergoing complex motions.

1.4 Axisymmetric and non-axisymmetric systems

It was natural however for early investigators to analyse systems having a maximum degree of symmetry in order to limit the analytical difficulties of the problem. The most natural 'primitive' system to consider in the context of rotating bodies such as the Earth or the Sun is one in which both the velocity field and magnetic field are axisymmetric. As already mentioned, Cowling (1934) considered this idealisation in an investigation of the origin of the much more local and intense magnetic fields of sunspots, but concluded that a steady axisymmetric field could not be maintained by axisymmetric motions. This first 'anti-dynamo' theorem was reinforced by later investigations (Backus & Chandrasekhar 1956; Cowling 1975b) and it was finally shown by Backus (1957) that axisymmetric motions could at most extend the natural decay time of an axisymmetric field in a spherical system by a factor of about 4. In the context of the Earth's magnetic field, whose natural decay time is of the order of 10^4–10^5 years (see Chapter 4), this modest delaying action is totally inadequate to explain the continued existence of the main dipole field for a period of the same order as the age of the Earth itself (3×10^9 years) (the evidence being from studies of rock magnetism), and its relative stability over periods of order 10^6 years and greater (Bullard 1968). It was clear that non-axisymmetric configurations had to be considered if any real progress in dynamo theory were to be made. It is in fact the essentially three-dimensional character of 'the dynamo problem' (as the problem of explaining the origin of the magnetic field of the Earth or of any other cosmic body has come to be called) that provides both its particular difficulty and its peculiar fascination.

Recognition of the three-dimensional nature of the problem led Elsasser (1946) to initiate the study of the interaction of a prescribed non-axisymmetric velocity field with a general non-axisymmetric magnetic field in a conducting fluid contained with a rigid spherical boundary, the medium outside this boundary being assumed non-conducting. Elsasser advocated the technique of expansion of both fields in spherical harmonics, a technique that was greatly developed and extended in the pioneering study of Bullard & Gellman (1954). The discussion of §7(e) of this remarkable paper shows clear recognition of the desirability of two ingredients in the velocity field for effective dynamo action: (i) a differential rotation which would draw out the lines of force of the poloidal magnetic field to generate a toroidal field (for the definition of these terms, see Chapter 2), and (ii) a non-axisymmetric motion capable of distorting a toroidal line of force by an upwelling followed by a twist in such a way as to provide a feedback to the poloidal field.

Interaction of the velocity field $\mathbf{u}(\mathbf{x})$ and the magnetic field $\mathbf{B}(\mathbf{x})$ (through the $\mathbf{u} \wedge \mathbf{B}$ term in Ohm's law) leads to an infinite set of coupled ordinary differential equations for the determination of the various spherical-harmonic ingredients of possible steady magnetic field patterns, and numerical solution of these equations naturally involves truncation of the system and discretisation of radial derivatives. These procedures are of course legitimate in a numerical search for a solution that is known to exist, but they can lead to erroneous conclusions when the existence of an exact steady solution to the problem is in doubt. The dangers were recognised and accepted by Bullard & Gellman, but it was in fact later demonstrated that the velocity field $\mathbf{u}(\mathbf{x})$ that they proposed most forcibly as a candidate for steady dynamo action in a sphere is a failure in this respect under the more searching scrutiny of higher-speed computers (Gibson & Roberts 1969).

The inadequacy of purely computational approaches to the problem intensified the need for theoretical approaches that do not, at the fundamental level, require recourse to the computer. In this respect a breakthrough in understanding was provided by Parker (1955b) who argued that the effect of the non-axisymmetric upwellings (his 'cyclonic events') referred to above might be incorporated by an averaging procedure in equations for the components of the *mean magnetic field* (i.e. the field averaged over the azimuth angle φ about the axis of rotation of the system). Parker's arguments were heuristic rather than deductive, and it was perhaps for this reason that some years elapsed before the power of the approach was generally appreciated. The theory is referred to briefly in Cowling's monograph *Magnetohydrodynamics* (Cowling 1957) with the following conclusions:

The argument is not altogether satisfactory; a more detailed analysis is really needed. Parker does not attempt such an analysis; his mathematical discussion is limited to elucidating the consequences if his picture of what occurs is accepted. But clearly his suggestion deserves a good deal of attention.[5]

This attention was not provided for some years, however, and was finally stimulated by two rather different approaches to the problem, one by Braginskii (1964b) and the other by Steenbeck et al. (1966). The essential idea behind Braginskii's approach was that, while steady axisymmetric solutions to the dynamo problem are ruled out by Cowling's theorem, nevertheless weak departures from axisymmetry might provide a means of regeneration of the mean magnetic field. This approach can succeed only if the fluid conductivity σ is very high (or equivalently if the magnetic diffusivity $\eta = (\mu_0\sigma)^{-1}$ is very weak), and the theory was developed in terms of power series in a small parameter proportional to $\eta^{1/2}$. By this means, Braginskii demonstrated that, as Parker had argued, non-axisymmetric motions could indeed provide an effective mean toroidal electromotive force (emf) in the presence of a predominantly toroidal magnetic field. This emf drives a toroidal current thus generating a poloidal field, and the dynamo cycle anticipated by Bullard & Gellman can be completed. An account of Braginsky's theory, as reformulated by Soward (1972), is presented in Chapter 8.

The approach advocated by Steenbeck, Krause & Rädler is potentially more general, and is applicable when the velocity field consists of a mean and a turbulent (or random) ingredient having widely different length-scales L and ℓ, say ($L \gg \ell$). Attention is then focussed on the evolution of the mean magnetic field on scales large compared with ℓ. The mean-field approach is of course highly developed in the theory of shear-flow turbulence in non-conducting fluids (see, for example, Townsend, 1975) and it had previously been advocated in the hydromagnetic context by, for example, Kovasznay (1960). The power of the approach of Steenbeck et al. (1966) however lay in recognition of the fact that the turbulence can give rise to a mean electromotive force having a component parallel to the prevailing local mean magnetic field (as in Braginskii's model); and these authors succeeded in showing that this effect would certainly occur whenever the statistical properties of the background turbulence *lack reflectional symmetry*. This property of 'chirality' is the random counterpart of the purely geometrical property of the simple disc dynamo discussed above. The theory of 'mean-field electrodynamics' will be presented in Chapter 7, where the central role of chirality will become apparent.

Since 1966, there has been a growing flood of papers developing different aspects of these theories and their applications to the Earth and Sun and other

[5] It is only fair to note that this somewhat guarded assessment is eliminated in the later edition of the book (Cowling 1975a).

celestial systems. It is the aim of this book to provide a coherent account of the most significant of these developments, and reference to specific papers published since 1970 will for the most part be delayed till the appropriate point in the text.

Several other earlier papers are, however, historical landmarks and deserve mention at this stage. The fact that turbulence could be of crucial importance for dynamo action was recognised independently by Batchelor (1950) and Schlüter & Biermann (1950), who considered the effect of a random velocity field on a random magnetic field, both having zero mean. Batchelor recognised that random stretching of magnetic lines of force would lead to exponential increase of magnetic energy in a fluid of infinite conductivity; and, on the basis of the analogy with vorticity (see §3.6), he obtained a criterion for just how large the conductivity must be for this conclusion to remain valid, and an estimate for the ultimate equilibrium level of magnetic energy density that might be expected when Lorentz forces react back upon the velocity field. Schlüter & Biermann, by arguments based on the concept of equipartition of energy, obtained a different criterion for growth and a much greater estimate for the ultimate level of magnetic energy density. Yet a third possibility was advanced by Saffman (1963) who came to the conclusion that, although the magnetic energy might increase for a while from a very weak initial level, ultimately it would always decay to zero due to accelerated ohmic decay associated with persistent decrease in the characteristic length-scale of the magnetic field. It is now known from consideration of the effect of turbulence that lacks reflectional symmetry (see Chapter 7) that none of the conclusions of the above papers can have any general validity, although the question of what happens when the turbulence is reflectionally symmetric remains to some extent open (see §15.4).

The problem as posed by Batchelor has to some extent been bypassed through recognition of the fact that it is the ensemble-average magnetic field that is of real interest and that if this average vanishes, as in the model conceived by Batchelor, the model can have little direct relevance for the Earth and Sun, both of which certainly exhibit a non-zero dipole moment. It is fortunate that the problem has been bypassed, because in a rather pessimistic diagnosis of the various conflicting theories, Kraichnan & Nagarajan (1967) concluded that

> equipartition arguments, the vorticity analogy, and the known turbulence approximations are all found inadequate for predicting whether the magnetic energy eventually dies away or grows exponentially. Lack of bounds on errors makes it impossible to predict reliably the sign of the eventual net growth rate of magnetic energy.

Kraichnan & Nagarajan have not yet been proved wrong as regards the basic problem with homogeneous isotropic reflectionally symmetric turbulence. Parker (1970) comments on the situation in the following terms:

Cyclonic turbulence,[6] together with large-scale shear, generates magnetic field at a very high rate. Therefore we ask whether the possible growth of fields in random turbulence without cyclonic ordering ... is really of paramount physical interest. We suggest that, even if random turbulence could be shown to enhance magnetic field densities, the effect in most astrophysical objects would be obscured by the more rapid generation of fields by the cyclonic turbulence and non-uniform rotation.

The crucial importance of a lack of reflectional symmetry in fluid motions conducive to dynamo action is apparent also in the papers of Herzenberg (1958) and Backus (1958) who provided the first examples of laminar velocity fields inside a sphere which could be shown by rigorous procedures to be capable of sustained dynamo action. Herzenberg's model involved two spherical rotors rotating with angular velocities ω_1 and ω_2 and separated by vector distance $\mathbf{R}$ inside the conducting sphere. The configuration can be described as right-handed or left-handed according as the triple scalar product $[\omega_1, \omega_2, \mathbf{R}]$ is positive or negative. A necessary condition for dynamo action (see §6.9) was that this triple scalar product should be non-zero, and the configuration then certainly lacks reflectional symmetry.

The Backus dynamo followed the pattern of the Bullard & Gellman dynamo, but decomposed temporally into mathematically tractable units. The velocity field considered consisted of three active phases separated by long periods of rest (or 'stasis') to allow unwanted high harmonics of the magnetic field to decay to a negligibly low level. The three phases were: (i) a vigorous differential rotation which generated strong toroidal field from pre-existing poloidal field; (ii) a non-axisymmetric poloidal convection which regenerated poloidal field from toroidal; (iii) a rigid rotation through an angle of $\pi/2$ to bring the newly generated dipole moment into alignment with the direction of the original dipole moment. The lack of reflectional symmetry lies here in the mutual relationship between the phase (i) and phase (ii) velocity fields (see §6.15). These theories, and others relating to laminar dynamo action, will be presented in detail in Chapter 6.

The viewpoint adopted in this book is that random fluctuations in the velocity field and the magnetic field are almost certainly present both in the Earth's core and in the Sun's convection zone, and that a realistic theory of dynamo action should incorporate effects of such fluctuations at the outset. Laminar theories are of course not without value, particularly for the mathematical insight that they provide; but anyone who has conscientiously worked through such papers as those of Bullard & Gellman (1954), Herzenberg (1958) and Backus (1958) will readily admit the enormous complexity of the laminar problem. It is a remarkable fact that acceptance of turbulence (or possibly random wave motions) and appropriate averaging procedures actually leads to a dramatic simplification of the problem. The reason

[6] This is Parker's terminology for turbulence whose statistical properties lack reflectional symmetry.

is that the mean fields satisfy equations to which Cowling's anti-dynamo theorem does not apply, and which are therefore amenable to an axisymmetric analysis with distinctly positive and encouraging results. The equations admit both steady solutions modelling the Earth's quasi-steady dipole field, and, in other circumstances, time-periodic solutions which behave in many respects like the magnetic field of the Sun with its 22-year periodic cycle.

A further crucial advantage of an approach involving random fluctuations is that dynamic, as opposed to purely kinematic, considerations become to some extent amenable to analysis. A kinematic theory is one in which a kinematically possible velocity field $\mathbf{u}(\mathbf{x}, t)$ is assumed known, either in detail or at least statistically when random fluctuations are involved, and its effect on magnetic field evolution is studied. A dynamic theory is one in which $\mathbf{u}(\mathbf{x}, t)$ is constrained to satisfy the relevant equations of motion (generally the Navier–Stokes equations with buoyancy forces, Coriolis forces and Lorentz forces included according to the context); and again the effect of this velocity field on magnetic field evolution is studied. It is only since the advent of the mean-field electrodynamics of Steenbeck, Krause & Rädler that progress on the dynamic aspects of dynamo theory has become possible. These dynamic aspects, which have been increasingly explored over the last 30 years, will be treated in Part III of this book.

The general pattern of the book will therefore be as follows. Chapter 2 will be devoted to simple preliminaries concerning magnetic field structure and diffusion in a stationary conductor. Chapter 3 will be concerned with the interplay of convection and diffusion effects insofar as these influence magnetic field evolution in a moving fluid. In Chapters 4 and 5 we shall digress from the purely mathematical development to provide a necessarily brief survey of the observed properties of the Earth's magnetic field (and other planetary fields) and of the Sun's magnetic field (and other astrophysical fields) and of the relevant physical properties of these systems. This is designed to provide more detailed motivation for the material of subsequent chapters. Some readers may find this motivation superfluous; but it is necessary, particularly when it comes to the study of specific dynamic models, to consider limiting processes in which the various dimensionless numbers characterising the system are either very small or very large; and it is clearly desirable that such limiting processes should at the least be not in contradiction with observation in the particular sphere of relevance claimed for the theory.

Part II (Chapters 6–10) is concerned with the foundations of kinematic dynamo theory, with Chapter 6 focussing on laminar flows and Chapter 7 on both weak and strong turbulence in mean-field electrodynamics, and the theory of the famous α-effect. Chapter 8 will treat nearly axisymmetric systems, and Chapter 9 will survey solutions of the mean-field equations of α^2- or $\alpha\omega$-type. Chapter 10 will treat the concept of the 'fast dynamo' as introduced by Vainshtein & Zel'dovich (1972), i.e. dynamo action for which the growth rate of the magnetic field is independent of resistivity η in the limit $\eta \to 0$, and the pathological structure of magnetic fields that emerge in this situation.

Part III (Chapters 11–17) is concerned with dynamic aspects of the theory. We start in Chapter 11 with low-dimensional models that incorporate dynamic effects, with either mechanical or thermal forcing; such models exhibit the manner in which a dynamo-generated field saturates due to the back reaction of the Lorentz force, frequently in a chaotic state. Chapter 12 treats the phenomenon that has come to be known as α-quenching, which again induces magnetic energy saturation. Chapters 13 and 14 focus respectively on the geodynamo and the solar dynamo, with consideration of the full equations of magnetohydrodynamics in a rotating medium, including the back-reaction of the Lorentz force. Chapter 15 treats specific aspects of turbulence with and without 'helicity' (the simplest measure of chirality) in the statistics of the turbulence.

Finally, Chapters 16 and 17 treat the problem of magnetic relaxation under the topological contraint of a 'frozen-in' magnetic field. In a statistically steady state, the magnetic energy will continue to grow by dynamo action in some regions of the flow, but this growth will be compensated by relaxation of magnetic energy in other regions, a statistically steady state being thereby maintained. Magnetic relaxation is in some respects therefore the counterpart of dynamo action, the two process having comparable 'weight' when statistical equilibrium is attained.

Given the present state of knowledge, it is inevitable that the kinematic theory will occupy a rather greater proportion of the book than it would ideally deserve. It must be remembered however that any results that can be obtained in kinematic theory on the minimal assumption that $\mathbf{u}(\mathbf{x}, t)$ is a kinematically possible but otherwise *arbitrary* velocity field will have a generality that transcends any dynamical model that is subsequently adopted for the determination of $\mathbf{u}$. It is important to seek this generality because, although there is little uncertainty regarding the equations governing magnetic field evolution (i.e. Maxwell's equations and Ohm's law), there are wide areas of uncertainty concerning the relevance of different dynamical models in both terrestrial and solar contexts; for example, it is not yet known what the ultimate source of energy is for core motions that drive the Earth's dynamo. In this situation, any results that do not depend on the details of the governing dynamical equations (whatever these may be) are of particular value. For this reason, the postponement of dynamical considerations to Part III of this work should perhaps be welcomed rather than lamented.

It will be found that the concept of helicity – the spatial average of the scalar product of velocity and vorticity – plays a very central role in dynamo theory and equally in the theory of magnetic relaxation. Figure 1.4 shows a reproduction of Leonardo da Vinci's drawing *The Deluge* held in the Royal Archive at Windsor

Figure 1.4 Leonardo's drawing *The Deluge*, c.1517/18 (pen and black ink with wash): an artist's conception of turbulence with helicity. [RCIN 912380; Royal Collection Trust / © Her Majesty Queen Elizabeth II 2018; reproduced by permission.]

Castle. It would appear that Leonardo was well aware of the helicity of the flows that he imagined, predominantly left-handed in this drawing. The year 2019 is the 500th anniversary of Leonardo's death, when the royal collection of his works will be on open display to the public, an opportunity for helicity to be more widely appreciated! We shall in fact find in Chapter 7 that turbulence with non-zero mean helicity is always capable of generating a large-scale magnetic field in a conducting fluid of sufficient spatial extent, a result that may be conveniently summarised in the couplet

Convection and diffusion in turb'lence with helicity
Yields order from confusion in cosmic electricity,

a fitting note on which to terminate this introductory chapter.

2
Magnetokinematic Preliminaries

2.1 Structural properties of the B-field

2.1.1 Solenoidality

In §§2.1–2.5, we shall be concerned with basic instantaneous properties of magnetic field distributions $\mathbf{B}(\mathbf{x})$; the time dependence of $\mathbf{B}$ is for the moment irrelevant. The first, and perhaps most basic, of these properties is that if S is any closed surface with unit outward normal $\mathbf{n}$, then $\mathbf{B}$ satisfies the 'solenoidal condition'

$$\int_S \mathbf{B}\cdot\mathbf{n}\,\mathrm{d}S = 0\,, \tag{2.1}$$

i.e. magnetic poles do not exist in isolation. This integral statement implies the existence of a single-valued vector potential $\mathbf{A}(\mathbf{x})$ satisfying

$$\mathbf{B} = \nabla\wedge\mathbf{A}\,,\quad \nabla\cdot\mathbf{A} = 0\,. \tag{2.2}$$

$\mathbf{A}$ is not uniquely defined by these equations since we may add to it the gradient of any harmonic function without affecting $\mathbf{B}$; but, in problems involving an infinite domain, $\mathbf{A}$ is made unique by imposing the boundary condition

$$\mathbf{A}\to 0 \qquad \text{as} \quad |\mathbf{x}|\to\infty\,. \tag{2.3}$$

At points where $\mathbf{B}$ is differentiable, (2.1) implies that

$$\nabla\cdot\mathbf{B} = 0\,, \tag{2.4}$$

i.e. $\mathbf{B}$ has zero divergence. Moreover, across any surface of discontinuity S_d of physical properties of the medium (or of other relevant fields such as the velocity field), (2.1) implies that

$$[\![\mathbf{n}.\mathbf{B}]\!] = (\mathbf{n}.\mathbf{B})_+ - (\mathbf{n}.\mathbf{B})_- = 0\,, \tag{2.5}$$

where the + and − refer to the two sides of S_d, and **n** is now the unit normal on S_d directed from the − side to the + side. We shall always use this square bracket notation to denote such surface quantities.

2.1.2 The Biot–Savart integral

We shall describe the field $\mathbf{B}(\mathbf{x})$ as 'localised' in $\mathbb{R}^3$ if $|\mathbf{B}|$ is exponentially decreasing as $|\mathbf{x}| \to \infty$ outside some sphere $|\mathbf{x}| = R$, say. In this situation, the equations (2.2) may be 'solved' for **A** in terms of **B**. First note that **A** itself has a 'vector potential' **P** defined by

$$\mathbf{A} = \nabla \wedge \mathbf{P}, \quad \nabla \cdot \mathbf{P} = 0, \tag{2.6}$$

so that

$$\mathbf{B} = \nabla \wedge (\nabla \wedge \mathbf{P}) = -\nabla^2 \mathbf{P}. \tag{2.7}$$

This Poisson equation has a particular integral,

$$\mathbf{P}(\mathbf{x}) = \frac{1}{4\pi} \int \frac{\mathbf{B}(\mathbf{x}')}{|\mathbf{x} - \mathbf{x}'|} \, dV', \tag{2.8}$$

the integral being over all space. Obviously, for a localised field this integral is convergent,[1] and, as easily verified using (2.4), it satisfies $\nabla \cdot \mathbf{P} = 0$.

It follows from (2.6) that

$$\mathbf{A}(\mathbf{x}) = \frac{1}{4\pi} \int \frac{\mathbf{B}(\mathbf{x}') \wedge (\mathbf{x} - \mathbf{x}')}{|\mathbf{x} - \mathbf{x}'|^3} \, dV', \tag{2.9}$$

a generalised 'Biot–Savart' law that provides the inverse relation, **A** in terms of **B**, as required.

2.1.3 Lines of force ('B-lines')

The lines of force of the **B**-field (or '**B**-lines') are determined as the integral curves of the differential equations

$$d\mathbf{x} \wedge \mathbf{B} = 0. \tag{2.10}$$

A **B**-line may, exceptionally, close on itself. More generally, it may cover a closed surface S in the sense that, if followed far enough, it passes arbitrarily near every point of S. More generally still, a **B**-line may be space-filling in the sense that, if followed far enough, it passes arbitrarily near every point of a three-dimensional subdomain of the region of interest; an explicit example of such a 'chaotic' field is given in §2.9 below. It seems likely that this sort of chaotic behaviour is generic for three-dimensional fields $\mathbf{B}(\mathbf{x})$ having no obvious symmetry, i.e. the **B**-lines of such

[1] Actually $|\mathbf{B}| = \mathcal{O}\left(|\mathbf{x}|^{-3}\right)$ as $|\mathbf{x}| \to \infty$ is sufficient for the convergence of the integral (2.8).

fields are in general space-filling except possibly in some special 'subdomains of regularity' in which the field lines do lie on closed surfaces.

Now let C be any (unknotted) closed curve spanned by an open orientable surface S with normal $\mathbf{n}(\mathbf{x})$. The flux Φ of $\mathbf{B}$ across S is defined by

$$\Phi = \int_S \mathbf{B} \cdot \mathbf{n}\, \mathrm{d}S = \oint_C \mathbf{A} \cdot \mathrm{d}\mathbf{x}, \tag{2.11}$$

where the line integral is described in a right-handed sense relative to the normal on S. A *flux tube* is the aggregate of $\mathbf{B}$-lines passing through a closed curve (usually of small or infinitesimal extent). By virtue of (2.1), Φ is constant along a flux tube. If the field $\mathbf{B}$ is confined to a single flux tube of small cross section centred on a closed curve C', and carrying flux Φ, then, with $\mathbf{x}' \in C'$, the Biot–Savart law reduces to the line integral

$$\mathbf{A}(\mathbf{x}) = \frac{\Phi}{4\pi} \oint_{C'} \frac{\mathrm{d}\mathbf{x}' \wedge (\mathbf{x} - \mathbf{x}')}{|\mathbf{x} - \mathbf{x}'|^3}\,. \tag{2.12}$$

2.1.4 Helicity and flux tube linkage

A measure of the degree of structural complexity of a $\mathbf{B}$-field is provided by a set of 'helicity integrals'

$$\mathcal{H}_m = \int_{V_m} \mathbf{A} \cdot \mathbf{B}\, \mathrm{d}V \quad (m = 1, 2, 3, \ldots), \tag{2.13}$$

where V_m is any volume with surface S_m on which $\mathbf{n} \cdot \mathbf{B} = 0$. Suppose, for example, that $\mathbf{B}$ is identically zero except in two flux tubes occupying volumes V_1 and V_2 of infinitesimal cross section following the closed curves C_1 and C_2 (Figure 2.1a) and let Φ_1 and Φ_2 be the respective fluxes. Suppose further that the curves C_1 and C_2 are unknotted, and that the field lines within each tube are unlinked. The tubes themselves may however be linked; if they are linked as in the figure and the field directions in the tubes are as indicated by the arrows, then each tube has a right-handed orientation relative to the other; if one arrow is reversed the relative orientation becomes left-handed; if both are reversed it remains right-handed.

For the configuration as drawn, evidently $\mathbf{B}\, \mathrm{d}V$ may be replaced by $\Phi_1\, \mathrm{d}\mathbf{x}$ on C_1 and by $\Phi_2\, \mathrm{d}\mathbf{x}$ on C_2 with the result that

$$\mathcal{H}_1 = \Phi_1 \oint_{C_1} \mathbf{A} \cdot \mathrm{d}\mathbf{x} = \Phi_1 \Phi_2, \tag{2.14}$$

and similarly

$$\mathcal{H}_2 = \Phi_2 \oint_{C_2} \mathbf{A} \cdot \mathrm{d}\mathbf{x} = \Phi_2 \Phi_1. \tag{2.15}$$

More generally, if the tubes wind round each other N times (i.e. N is the 'winding

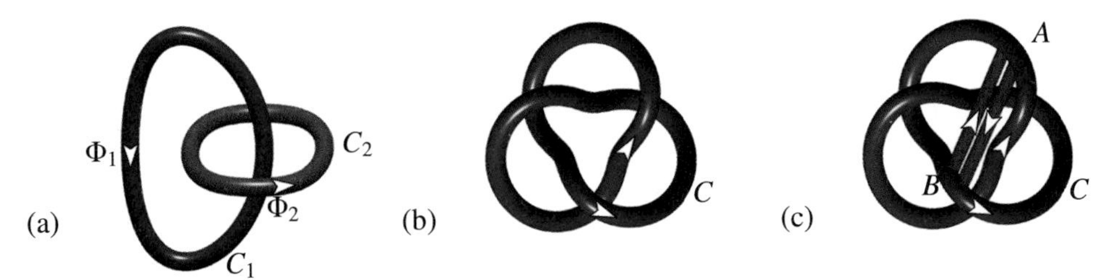

Figure 2.1 (a) The two flux tubes are linked in such a way as to give positive magnetic helicity; (b) a flux tube in the form of a right-handed trefoil knot; (c) insertion of equal and opposite flux tube elements between the points A and B as indicated gives two tubes linked as in (a).

number' of C_1 relative to C_2, taken positive or negative according as the relative orientation of the two flux tubes is right- or left-handed), then

$$\mathcal{H}_1 = \mathcal{H}_2 = \mathcal{N}\,\Phi_1\Phi_2, = \mathcal{H}_{12}, \text{ say.} \tag{2.16}$$

The integral $\mathcal{H}_{12}$ is therefore intimately related to the fundamental topological invariant $\mathcal{N}$ of the pair of curves C_1 and C_2. By virtue of (2.12), $\mathcal{N}$ is given by the Gauss formula

$$\mathcal{N} = \frac{1}{4\pi}\oint_{C_1}\oint_{C_2} \frac{(\mathbf{x}_1 - \mathbf{x}_2)\cdot(d\mathbf{x}_1 \wedge d\mathbf{x}_2)}{|\mathbf{x}_1 - \mathbf{x}_2|^3}\,. \tag{2.17}$$

Note that $\mathcal{N}$, being the integral of a pseudo-scalar, is itself a pseudo-scalar, i.e. it changes sign under change from a right- to a left-handed reference system.

If the **B**-field is localised, then one possible choice of the volume of integration is the whole three-dimensional space V_∞, the corresponding integral being simply denoted $\mathcal{H}$. For the case of the two linked flux tubes, evidently

$$\mathcal{H} = 2\mathcal{H}_{12} = 2\mathcal{N}\,\Phi_1\Phi_2. \tag{2.18}$$

Similarly, if there are m linked flux tubes, then the 'linkage helicity' is given by

$$\mathcal{H} = \sum_{\substack{i,j=1\\ i\neq j}}^{m} \mathcal{H}_{ij}. \tag{2.19}$$

If a single flux tube with flux Φ is knotted, then the integral $\mathcal{H}$ for the associated magnetic field will in general be non-zero. Figure 2.1b shows the simplest non-trivial possibility: the curve C is a right-handed trefoil knot; insertion of the two self-cancelling connecting tubes[2] between the points A and B suggests that this is equivalent to the configuration of Figure 2.1a with $\Phi_1 = \Phi_2 = \Phi$; however, care is needed, because twist may be introduced in either or both tubes in the reconnection

[2] 'self-cancelling' because $\mathbf{B}\cdot d\mathbf{x} = -(\mathbf{B}\cdot d\mathbf{x})'$ on adjacent elements of the two connecting tubes

process, and this will contribute to the helicity. A more detailed consideration of knotted flux tubes requires some elements of differential geometry, and is deferred to §2.10 below.

It may of course happen that $\mathbf{A}\cdot\mathbf{B}\equiv 0$; it is well known that this is the necessary and sufficient condition for the existence of scalar functions $\varphi(\mathbf{x})$ and $\psi(\mathbf{x})$ such that

$$\mathbf{A}=\psi\nabla\phi\,,\quad \mathbf{B}=\nabla\psi\wedge\nabla\phi\,. \tag{2.20}$$

In this situation, the **B**-lines are the intersections of the surfaces $\phi=$ const., $\psi=$ const., and the **A**-lines are everywhere orthogonal to the surfaces $\phi=$ const. It is clear from the above discussion that **B**-fields having linked or knotted **B**-lines cannot admit such a global representation.

The same limitation applies to the use of Clebsch variables $\{\phi,\ \psi,\ \chi\}$, defined (if they exist) by the equations

$$\mathbf{A}=\psi\nabla\phi+\nabla\chi\,,\quad \mathbf{B}=\nabla\psi\wedge\nabla\phi\,. \tag{2.21}$$

For example, if **B** is a field admitting such a representation, with ϕ, ψ and χ single-valued differentiable functions of **x**, then

$$\mathbf{A}\cdot\mathbf{B}=\nabla\chi\cdot\,(\nabla\psi\wedge\nabla\phi)\,, \tag{2.22}$$

and

$$\mathcal{H}_m=\int_{V_m}\nabla\chi\cdot(\nabla\psi\wedge\nabla\phi)\,\mathrm{d}V=\int_{V_m}\nabla\cdot(\chi\nabla\psi\wedge\nabla\phi)\,\mathrm{d}V=\int_{S_m}\chi\,\mathbf{n}\cdot\mathbf{B}\,\mathrm{d}S=0\,, \tag{2.23}$$

since $\mathbf{n}\cdot\mathbf{B}=0$ on S_m. Conversely, if $\mathcal{H}_m\neq 0$ (as will usually happen if the **B**-lines are knotted or linked), then (2.21) is not a possible global representation for **A** and **B** (although it may be useful in a purely local analysis).

The characteristic structure of a field for which the average value $\langle\mathbf{A}\cdot\mathbf{B}\rangle$ is non-zero may be illustrated with reference to the example (in Cartesian coordinates)

$$\mathbf{A}=(0,\ A_0\cos kx,\ B_0\,y-A_0\sin kx)\,, \tag{2.24}$$

$$\mathbf{B}=\nabla\wedge\mathbf{A}=(B_0,\ A_0k\cos kx,\ -A_0k\sin kx)\,, \tag{2.25}$$

where A_0, B_0 and k are constants; here $\langle\mathbf{A}\cdot\mathbf{B}\rangle=kA_0^2$, and the **B**-line through the point (x_0, y_0, z_0) is the helix

$$y-y_0-\frac{A_0}{B_0}\sin kx_0=-\frac{A_0}{B_0}\sin kx,\quad z-z_0-\frac{A_0}{B_0}\cos kx_0=-\frac{A_0}{B_0}\cos kx\,. \tag{2.26}$$

Thus all the **B**-lines of the field (2.25) are helices, right-handed or left-handed according as $kA_0/B_0 <$ or > 0 .

For *any* solenoidal (i.e. divergence-free) vector field, $\mathbf{B}(\mathbf{x})=\nabla\wedge\mathbf{A}(\mathbf{x})$, the quantity $\mathbf{A}\cdot\mathbf{B}$ is the *helicity density* of **B**; its integral $\mathcal{H}$ over V_∞ is then the (total) helicity of **B**; the integrals $\mathcal{H}_m$ over V_m can be described as 'partial' helicities. Helicity is

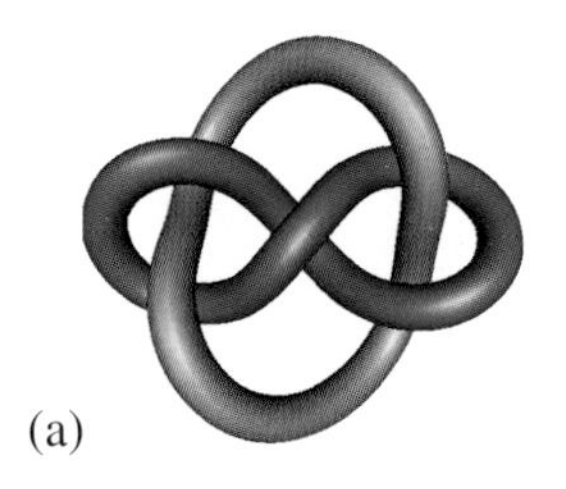

(a)

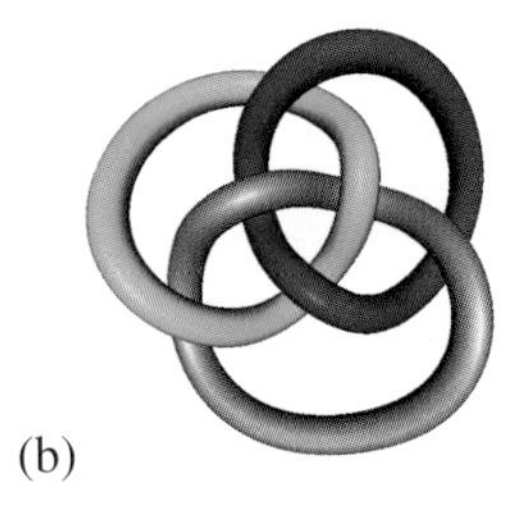

(b)

Figure 2.2 (a) The Whitehead link. Note the right-handed twist in the figure-of-eight tube, a symptom of chirality of the link. If these are magnetic flux tubes, the link helicity is zero. (b) The Borromean rings; each pair of tubes in this configuration is unlinked, so the link helicity is zero; the structure is achiral. [From Scharein: knotplot.]

a pseudo-scalar quantity, being the integral of the scalar product of a polar vector and an axial vector; its sign therefore changes under change from a right-handed to a left-handed frame of reference. A field $\mathbf{B}$ that is 'reflexionally symmetric' (i.e. invariant under the change from a right-handed to a left-handed frame of reference represented by the reflexion $\mathbf{x}' = -\mathbf{x}$) must therefore have zero total helicity. The converse is not true, since of course other pseudo-scalar integral quantities such as

$$\int \mathbf{B} \cdot (\nabla \wedge \mathbf{B})\, \mathrm{d}V \quad \text{or} \quad \int (\nabla \wedge \mathbf{B}) \cdot \nabla \wedge (\nabla \wedge \mathbf{B})\, \mathrm{d}V \tag{2.27}$$

may be non-zero even if $\mathcal{H} = \int \mathbf{A} \cdot \mathbf{B}\, \mathrm{d}V = 0$.

2.2 Chirality

A structure or field that lacks reflexional symmetry is said to have the property of 'chirality'. For example, the trefoil knot shown in Figure 2.1b is right-handed, with or without the orientation arrow, and therefore chiral: its mirror image is a left-handed trefoil knot. Non-zero total helicity of a solenoidal vector field is an indication of chirality; but as will be seen, care is needed since the helicity of a field confined to a flux tube in the form of a trefoil knot may be set to zero through the 'injection' of an appropriate amount of internal twist by the twist-surgery procedure to be described in §2.10.1 below.

The linkage of Figure 2.1a is chiral only when the orientation arrows are included. As drawn, this linkage is right-handed; the mirror image of this oriented link is left-handed. The contribution $\mathcal{N}\Phi_1\Phi_2$ to helicity given by (2.17) and (2.18) may be described as the 'link helicity' of the configuration.

Some exceptional non-trivial links may have zero link helicity. The prototypical example is the 'Whitehead link' shown in Figure 2.2a, for which the net flux in

Figure 2.3 A toy rattleback seen from the side.

either tube across a surface spanning the other is zero. The link helicity of this configuration is thus zero. Note however that, viewed simply as a geometric structure of two linked tubes (without orientation arrows), this structure, like the trefoil knot, is chiral; this is evident from the fact that the figure-of-eight tube in Figure 2.2a exhibits a right-handed twist; the structure cannot therefore be 'isotoped', i.e. continuously deformed (maintaining the linkage topology), to coincide with its mirror image.

Higher-order links like the famous Borromean link of three tubes (Figure 2.2b) are of interest from a purely mathematical point of view. Any two tubes of this linkage are unlinked, and so the helicity is zero. There is nevertheless an indisputable 'triple linkage' in that the tubes cannot be separated by any continuous deformation. This structure *can* be isotoped into coincidence with its mirror image (as may be verified by playing with three coloured linked strings); it is therefore 'achiral' (i.e. without chirality; sometimes the word 'amphichiral' is used with the same meaning[3]).

2.2.1 The rattleback: a prototype of dynamic chirality

The rattleback[4] is a canoe-shaped object (Figure 2.3) with a curious property that is puzzling to understand: when spun on a table in one sense, say clockwise, it spins with a slight wobble and gently comes to rest; when spun in the opposite sense, it wobbles more violently to such an extent that its sense of spin reverses, before again coming slowly to rest. The phenomenon was first described by Walker (1896).

How is such startling behaviour to be understood? It is in fact attributable to a very slight chirality in the rattleback: looked at from above, its centreline may be seen to be very slightly S-shaped – it is not the same as its mirror image; this has the effect of rotating the principal axes of inertia in the horizontal plane through a very

[3] James Clerk Maxwell's *Paradoxical Ode to Hermann Stopffkraft,PhD* opens with the lines "My soul's an amphicheiral knot / Upon a liquid vortex wrought / By Intellect in the Unseen residing". For the historical background to this extraordinary composition, see Silver (2008).

[4] The more technical name for the rattleback is 'celt', pronounced *kelt*, although nothing to do with Scotland!

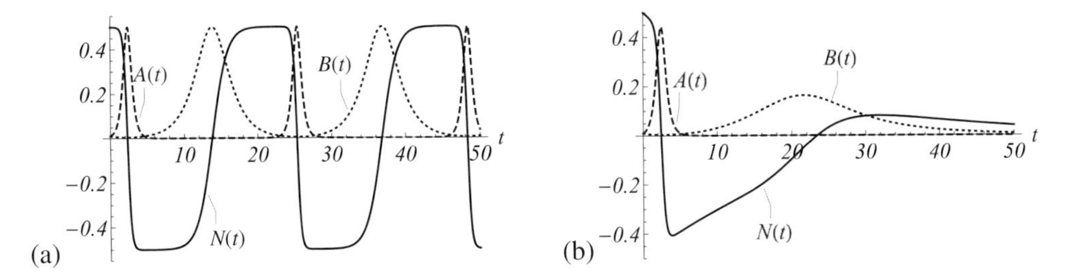

Figure 2.4 (a) Periodic solution of (2.28) when $\chi > 0$, $\lambda = 4$, $\mu = 0$; the pitching mode (A, dashed) is unstable when the spin $N > 0$, and the rolling mode (B, dotted) is unstable when $N < 0$. (b) Solution showing two spin reversals when $\mu = 0.05$.

small angle χ relative to the principal axes of curvature at the point of contact with the table. The result of this slight 'imperfection' is that the state of steady rotation about a vertical axis is susceptible to two modes of instability (Bondi 1986): a weak rolling mode about the axis of the rattleback when the rattleback rotates in the clockwise direction, and a much stronger pitching mode about a horizontal axis perpendicular to this when it rotates in the anti-clockwise direction.[5] These instabilities extract energy from the spin, and in conjunction with inertia, can cause spin reversals.

It turns out that the growth rate of the instabilities is $\mathcal{O}(\chi)$ relative to their frequencies. This allows for the development of a two-timing analysis (Moffatt & Tokieda 2008) that leads to a non-linear dynamical system relating the spin $N(t)$ and the amplitudes $A(t)$ and $B(t)$ of the instability modes. This system may be written in vector form

$$\frac{\mathrm{d}}{\mathrm{d}t}\begin{pmatrix} A \\ B \\ N \end{pmatrix} = \mathrm{sign}[\chi]\begin{pmatrix} B \\ \lambda A \\ 0 \end{pmatrix} \wedge \begin{pmatrix} A \\ B \\ N \end{pmatrix} - \mu \begin{pmatrix} A \\ B \\ N \end{pmatrix}, \tag{2.28}$$

where λ (>1) is determined by the geometry of the rattleback, and μ (> 0) is an empirical parameter introduced here to represent unavoidable effects of rolling and/or slipping friction. When $\mu = 0$, there are two integral invariants of this dynamical system, $I_1 = A^2 + B^2 + N^2$ and $I_2 = AB^\lambda$ and the system is integrable, with periodic solutions. An example is shown in Figure 2.4a: when $N > 0$, the A-mode is strongly unstable, causing a first reversal; when $N < 0$, the B-mode is unstable, with a slower growth rate, but also eventually causing a second reversal; the process then continues periodically. In practice, even weak friction severely limits the number of reversals. Figure 2.4b shows the situation when $\mu = 0.05$; here there is one clear reversal, and a second much weaker reversal, which can actually be observed with the toy rattleback.

[5] These directions are reversed if the sense of chirality of the rattleback is reversed.

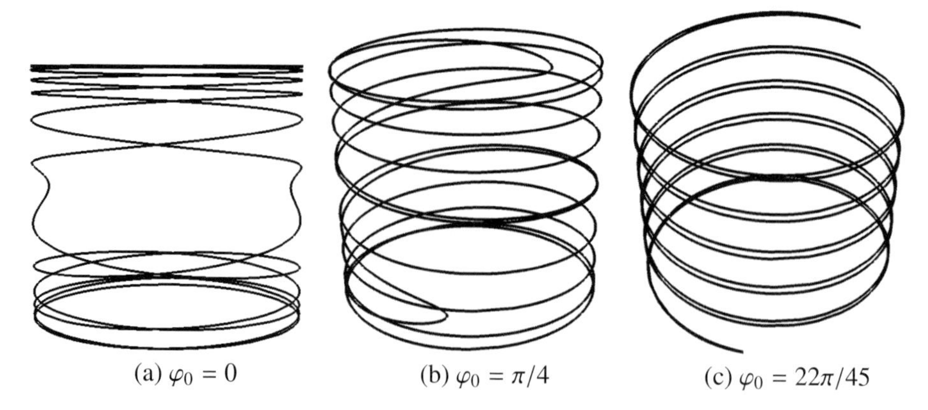

Figure 2.5 Trajectory of a point of the disc $z = Z(t)$, $s < a$, moving with periodic velocity (2.29), here with parameter values $Z_0 = 1, \sigma = 1, \lambda = 0.9, \omega_0 = 5.6\,\pi$ and for three choices of φ_0; in each case, this is a closed curve on a cylinder $s =$ const., which is traversed periodically. As φ_0 approaches $\pi/2$, the trajectory collapses to a helix, traversed up and down periodically, and in this limit the correlation $\mathcal{X} = \langle Z(t)\omega(t)\rangle$ is zero.

A moral may be drawn from this toy example: very weak chirality can have a dramatic dynamical effect. In the dynamo context, helicity is the obvious manifestation of chirality, and, as we shall find in later chapters, even very weak helicity can have a dramatic effect in relation to self-exciting dynamo action.

2.2.2 Mean response provoked by chiral excitation

As a further example of the surprising effects that can result from chirality, consider the following problem.[6] Suppose that viscous fluid fills the half-space $z > 0$ above a rigid plane boundary $z = 0$, and that a circular disc of radius a and negligible thickness is immersed in the fluid parallel to the boundary and at distance $Z_0 \ll a$ from it. We adopt cylindrical polar coordinates (s, φ, z), where s is the radial distance from the axis of the disc.

Suppose now that the disc is caused to oscillate so that its distance from the plane is $Z(t) = Z_0(1+\lambda\cos(\sigma t+\varphi_0)$ where $0 < \lambda < 1$, and that it is simultaneously caused to rotate about its axis with angular velocity $\omega(t) = \omega_0\cos\sigma t$; thus any point of the disc moves with velocity

$$\mathbf{U} = (U, V, W) = (0, s\,\omega_0\cos\sigma t, -\lambda Z_0\sigma\sin(\sigma t + \varphi_0))\,. \qquad (2.29)$$

The trajectory of any such point has parametric equation $s = s_0$, $\varphi = \int_0^t \omega(t)\,dt$, $z = Z(t)$, and, being periodic in t, is a closed curve on the cylinder $s = s_0$; this is illustrated in Figure 2.5 for particular values of λ and ω_0, and for three choices of

[6] This section contains part of work in progress (Vladimirov & Moffatt 2019).

the phase difference φ_0. The time-average $\langle \mathbf{U} \rangle$ is obviously zero. We may define a Reynolds number $\mathcal{R}e = \langle \mathbf{U}^2 \rangle^{1/2} a/\nu$, and we suppose for simplicity that $\mathcal{R}e \ll 1$ and that $\omega_0 \ll \nu/Z_0^2$, so that inertia effects in the fluid are completely negligible; the methods of 'thin-film' lubrication theory (Batchelor 1967) are then applicable.

In these circumstances, the poloidal and toroidal components of the equations of motion are decoupled. We need here consider only the toroidal (φ-) component of velocity $v(s, z, t)$ within the layer. This is essentially a 'Couette-flow' situation, so that

$$v(s, z, t) = \omega_0 s \cos \sigma t \,(z/Z(t)) = \omega_0 s \cos \sigma t \,\frac{z}{Z_0(1 + \lambda \cos(\sigma t + \varphi_0))}\,. \tag{2.30}$$

The time average of this is not zero; in fact,

$$\langle v(s, z, t) \rangle = \omega_0 s \,(z/Z_0)\, \lambda^{-1} \left(1 - (1 - \lambda^2)^{-1/2}\right) \cos \varphi_0\,. \tag{2.31}$$

There is a corresponding mean torque $\langle G \rangle$ on the disc

$$\langle G \rangle = \tfrac{1}{2}\pi \mu \omega_0 \left(a^4/Z_0\right) \lambda^{-1} \left(1 - (1 - \lambda^2)^{-1/2}\right) \cos \varphi_0\,, \tag{2.32}$$

and an equal and opposite torque on the boundary $z = 0$.

If now this lower boundary is replaced by another disc of radius $b \geq a$ and large axial moment of inertia C (the fluid now filling the whole space), then this lower disc will rotate with a mean rate of rotation Ω determined by the equation

$$C\, \mathrm{d}\Omega/\mathrm{d}t = -\langle G \rangle - k\,\Omega\,, \tag{2.33}$$

where the term $-k\,\Omega$ represents extraneous frictional resistance. (The assumed large value of C ensures that periodic fluctuations in Ω are small and may be neglected.) Thus the boundary excitation (2.29) with zero average gives rise to a non-zero average effect. It is the pseudo-scalar correlation $\mathcal{X} = \langle Z(t)\omega(t) \rangle$ between the vertical displacement and the angular velocity, here a measure of chirality[7] analogous to mean helicity, that gives rise to this effect. If $\varphi_0 = 0$, this correlation is maximal, while if $\varphi_0 = \pi/2$ it is zero.

A similar effect occurs if a sphere S_a of radius a is placed inside a sphere S_b of radius $b > a$ and large moment of inertia C, the volume between these spheres being filled with viscous fluid. If S_a is caused to move so that its centre is at position $z = Z(t) = Z_0(1 + \lambda \cos(\sigma t + \phi_0))$ on the diameter $s = 0$ of S_b (with $0 < |Z_0 \lambda| < b - a$), and to rotate about this diameter with angular velocity $\omega(t) = \omega_0 \cos \sigma t$, then the sphere S_b will experience a mean torque, and will rotate at a non-zero mean angular velocity if free to do so. Again, it is the same pseudo-scalar correlation $\mathcal{X}$ that is responsible for this effect of chirality.

Note particularly the dependence of $\mathcal{X}$ ($\propto \cos \varphi_0$) on the phase φ_0 in the above

[7] The Greek symbol 'chi', $\mathcal{X}$, seems an appropriate notation here for this measure of **chi**rality.

analysis, and the corresponding need for a phase shift between the vertical and angular displacements of the upper disc (or inner sphere) in order to produce a non-zero mean effect. We shall find (§7.7) that a similar phase shift between velocity and magnetic fluctuations is required to generate the all-important α-effect in dynamo theory.

2.3 Magnetic field representations

2.3.1 Spherical polar coordinates

In a spherical geometry, the most natural coordinates to use are spherical polar coordinates (r, θ, φ) related to Cartesian coordinates (x, y, z) by

$$x = r\sin\theta\cos\varphi, \quad y = r\sin\theta\sin\varphi, \quad z = r\cos\theta. \tag{2.34}$$

Let us first recall some basic results concerning the use of this coordinate system.[8] Let $\psi(r, \theta, \varphi)$ be any scalar function of position. Then

$$\nabla^2\psi = \frac{1}{r^2}\frac{\partial}{\partial r}r^2\frac{\partial\psi}{\partial r} + \frac{1}{r^2}L^2\psi, \tag{2.35}$$

where

$$L^2\psi = \left(\frac{1}{\sin\theta}\frac{\partial}{\partial\theta}\sin\theta\frac{\partial}{\partial\theta} + \frac{1}{\sin^2\theta}\frac{\partial^2}{\partial\varphi^2}\right)\psi. \tag{2.36}$$

The vector identity

$$(\mathbf{x}\wedge\nabla)^2\psi \equiv r^2\nabla^2\psi - 2(\mathbf{x}\cdot\nabla)\psi - \mathbf{x}\cdot(\mathbf{x}\cdot\nabla)\nabla\psi \tag{2.37}$$

leads to the identification

$$L^2\psi = (\mathbf{x}\wedge\nabla)^2\psi. \tag{2.38}$$

L^2 is the negative square of the angular momentum operator of quantum mechanics. Its eigenvalues are $-n(n+1)$ $(n = 0, 1, 2, \ldots)$, and the corresponding eigenfunctions are spherical harmonics

$$Y_n^m(\theta, \varphi) = C_n^m\, P_n^m(\cos\theta)\,\mathrm{e}^{im\varphi}, \quad m \le n, \tag{2.39}$$

where $P_n^m(\cos\theta)$ are associated Legendre polynomials and the C_n^m are normalising constants;[9] i.e.

$$L^2Y_n^m = -n(n+1)Y_n^m. \tag{2.40}$$

The first few spherical harmonics as well as their representations are given in Table 2.1 and Figures 2.6 and 2.7.

[8] Some results concerning orthogonal curvilinear coordinates and some basic vector identities may be found in the Appendix on p. 482.

[9] The fully normalised spherical harmonics correspond to $C_n^m = [(2-\delta_{m0})(2n+1)(n-m)!/(n+m)!]^{1/2}$.

Table 2.1 *The First Spherical Harmonics for* $n = 0, 1, 2$

$Y_0^0 = 1$		
$Y_1^0 = \sqrt{3}\cos\theta$	$Y_1^1 = -\sqrt{3}\sin\theta\, e^{i\varphi}$	
$Y_2^0 = \frac{\sqrt{5}}{2}(3\cos^2\theta - 1)$	$Y_2^1 = -\sqrt{15}\cos\theta\sin\theta\, e^{i\varphi}$	$Y_2^2 = \frac{\sqrt{15}}{2}\sin^2\theta\, e^{2i\varphi}$

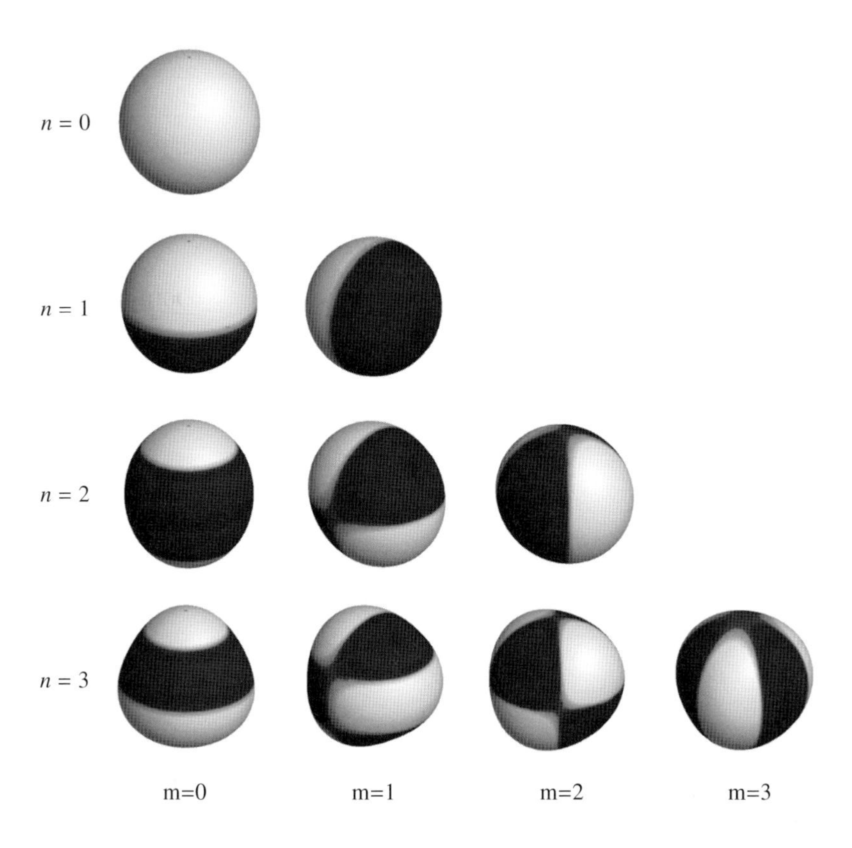

Figure 2.6 Spherical harmonics $\Re\mathrm{e}\,[Y_n^m(\theta,\varphi)]$ up to order and degree 3.

Now let $f(r,\theta,\varphi)$ be any smooth function having zero average over spheres $r =$ const., i.e.

$$\langle f \rangle \equiv \frac{1}{4\pi}\int_0^{2\pi}\int_0^{\pi} f(r,\theta,\varphi)\sin\theta\,\mathrm{d}\theta\,\mathrm{d}\varphi = 0\,. \tag{2.41}$$

We may expand f in surface harmonics S_n:

$$f(r,\theta,\varphi) = \sum_{n=1}^{\infty} f_n(r)S_n(\theta,\varphi) \quad \text{where} \quad S_n = \sum_{m=0}^{n} Y_n^m\,, \tag{2.42}$$

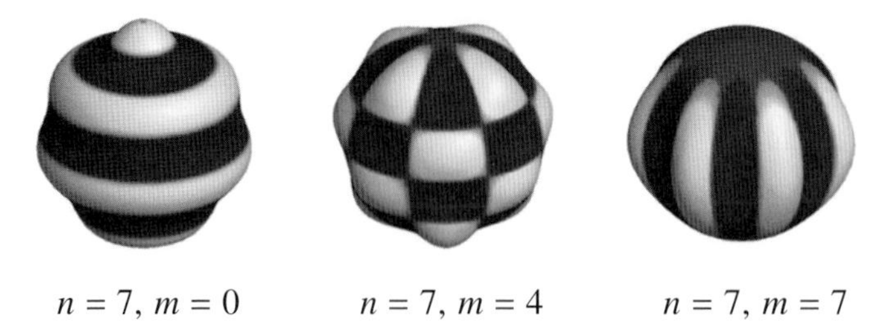

Figure 2.7 Examples of higher-degree harmonics: a zonal harmonic ($n = 7, m = 0$), a general case ($n = 7, m = 4$) and a sectorial harmonic ($n = 7, m = 7$).

the term with $n = 0$ being excluded by virtue of (2.41). The functions S_n satisfy the orthogonality relation

$$\langle S_n S_{n'} \rangle = 0, \quad (n \neq n'), \tag{2.43}$$

and so the coefficients $f_n(r)$ are given by

$$f_n(r) = \langle f\, S_n \rangle / \langle S_n^2 \rangle . \tag{2.44}$$

If now

$$L^2 \psi = f(r, \theta, \varphi), \tag{2.45}$$

then clearly the operator L^2 may be inverted to give

$$\psi = L^{-2} f = -\sum_{n=1}^{\infty} f_n(r)[n(n+1)]^{-1} S_n(\theta, \varphi), \tag{2.46}$$

the result also satisfying $\langle \psi \rangle = 0$.

Note that any function of the form $f = \mathbf{x} \cdot \nabla \wedge \mathbf{A}$, where $\mathbf{A}$ is an arbitrary smooth vector field, satisfies the condition (2.41); for

$$\int_{S(r)} \mathbf{x} \cdot \nabla \wedge \mathbf{A}\, \mathrm{d}S = r \int_{S(r)} \mathbf{n} \cdot \nabla \wedge \mathbf{A}\, \mathrm{d}S = r \int_{V(r)} \nabla \cdot (\nabla \wedge \mathbf{A})\, \mathrm{d}V = 0, \tag{2.47}$$

where $S(r)$ is the surface of the sphere of radius r and $V(r)$ its interior.

2.3.2 Toroidal/poloidal decomposition

A *toroidal magnetic field* $\mathbf{B}_T$ is any field of the form

$$\mathbf{B}_T = \nabla \wedge [\mathbf{x}\, T(\mathbf{x})] = -\mathbf{x} \wedge \nabla T, \tag{2.48}$$

where $T(\mathbf{x})$ is any scalar function of position. Note that addition of an arbitrary function of r to T has no effect on $\mathbf{B}_T$, so that without loss of generality we may suppose that $\langle T \rangle = 0$. Note further that $\mathbf{x} \cdot \mathbf{B}_T = 0$, so that the lines of force of $\mathbf{B}_T$ ('$\mathbf{B}_T$-lines') lie on the spherical surfaces $r =$ const.

A *poloidal magnetic field* $\mathbf{B}_P$ is any field of the form

$$\mathbf{B}_P = \nabla \wedge \nabla \wedge [\mathbf{x}\, P(\mathbf{x})] = -\nabla \wedge (\mathbf{x} \wedge \nabla P), \tag{2.49}$$

where $P(\mathbf{x})$ is any scalar function of position, which again may be assumed to satisfy $\langle P \rangle = 0$. $\mathbf{B}_P$ does in general have a non-zero radial component.

It is clear from these definitions that the curl of a toroidal field is poloidal. Moreover, reciprocally, the curl of a poloidal field is toroidal; for

$$\nabla \wedge \nabla \wedge \nabla \wedge (\mathbf{x}\, P) = -\nabla^2 \nabla \wedge (\mathbf{x}\, P) = -\nabla \wedge (\mathbf{x}\, \nabla^2 P)\,, \tag{2.50}$$

as can be trivially verified in Cartesian coordinates. Now suppose that

$$\mathbf{B} = \mathbf{B}_P + \mathbf{B}_T = \nabla \wedge \nabla \wedge (\mathbf{x}\, P) + \nabla \wedge (\mathbf{x}\, T)\,. \tag{2.51}$$

Then

$$\mathbf{x} \cdot \mathbf{B} = -(\mathbf{x} \wedge \nabla)^2 P, \quad \mathbf{x} \cdot (\nabla \wedge \mathbf{B}) = -(\mathbf{x} \wedge \nabla)^2 T\,, \tag{2.52}$$

so that, recalling (2.38), P and T may be obtained in the form

$$P = -L^{-2}[\mathbf{x} \cdot \mathbf{B}], \quad T = -L^{-2}[\mathbf{x} \cdot (\nabla \wedge \mathbf{B})]\,. \tag{2.53}$$

Conversely, given any solenoidal field $\mathbf{B}$, if we define P and T by (2.53), then (2.51) is satisfied, i.e. the decomposition of $\mathbf{B}$ into poloidal and toroidal ingredients is always possible.

If we 'uncurl' (2.51) we obtain the vector potential of $\mathbf{B}$ in the form

$$\mathbf{A} = \nabla \wedge (\mathbf{x}\, P) + \mathbf{x}\, T + \nabla U, \tag{2.54}$$

where U is a scalar 'function of integration'. Since $\nabla \cdot \mathbf{A} = 0$, U and T are related by

$$\nabla^2 U = -\nabla \cdot (\mathbf{x}\, T)\,. \tag{2.55}$$

By virtue of this condition, $\mathbf{x}\, T + \nabla U$ may itself be expressed as a poloidal field:

$$\mathbf{x}\, T + \nabla U = \nabla \wedge \nabla \wedge (\mathbf{x}\, S), \quad S = -L^{-2}(r^2 T + \mathbf{x} \cdot \nabla U). \tag{2.56}$$

The toroidal part of $\mathbf{A}$ is simply

$$\mathbf{A}_T = \nabla \wedge (\mathbf{x} P) = -\mathbf{x} \wedge \nabla P\,. \tag{2.57}$$

2.3.3 Axisymmetric fields

A field $\mathbf{B}$ is *axisymmetric* about a line Oz (the axis of symmetry) if it is invariant under rotations about Oz. In this situation, the defining scalars P and T of (2.51) are clearly independent of the azimuth angle φ, i.e. $T = T(r,\theta),\ P = P(r,\theta)$. The toroidal field $\mathbf{B}_T = -\mathbf{x} \wedge \nabla T$ then takes the form

$$\mathbf{B}_T = (0,\ 0,\ B_\varphi)\,, \quad B_\varphi = -\partial T/\partial\theta\,, \tag{2.58}$$

and so the $\mathbf{B}_T$-lines are circles about Oz, r = const., θ = const. Similarly,

$$\mathbf{A}_T = (0, 0,\ A_\varphi)\,, \quad A_\varphi = -\partial P/\partial\theta\,, \tag{2.59}$$

and, correspondingly, in spherical polars,

$$\mathbf{B}_P = \nabla \wedge \mathbf{A}_T = \left(-\frac{1}{r^2\sin\theta}\frac{\partial\chi}{\partial\theta},\ \frac{1}{r\sin\theta}\frac{\partial\chi}{\partial r},\ 0\right)\,, \tag{2.60}$$

where

$$\chi = -r\sin\theta\, A_\varphi = r\sin\theta\,\partial P/\partial\theta\,. \tag{2.61}$$

The scalar $\chi(r,\theta)$ is the analogue of the Stokes stream function $\psi(r,\theta)$ for incompressible axisymmetric velocity fields. The $\mathbf{B}_P$-lines are given by χ = const., and the differential

$$2\pi\, \mathrm{d}\chi = (B_\theta\, \mathrm{d}r - B_r r\, \mathrm{d}\theta)\, 2\pi r\, \sin\theta \tag{2.62}$$

represents the flux across the infinitesimal annulus obtained by rotating about Oz the line element joining (r,θ) and $(r + \mathrm{d}r,\ \theta + \mathrm{d}\theta)$. It is therefore appropriate to describe χ as the *flux-function* of the field $\mathbf{B}_P$.

When the context allows no room for ambiguity, we shall drop the suffix φ from B_φ and A_φ, and express $\mathbf{B}$ in the simple form

$$\mathbf{B} = B\,\mathbf{e}_\varphi + \nabla \wedge (A\,\mathbf{e}_\varphi)\,, \tag{2.63}$$

where $\mathbf{e}_\varphi$ is a unit vector in the φ-direction.[10]

2.3.4 Two-dimensional fields

Geometrical complications inherent in the spherical geometry frequently make it desirable to seek simpler representations. In particular, if we are concerned with processes in a spherical shell $r_1 < r < r_2$ with $r_2 - r_1 << r_1$, a local Cartesian representation $Oxyz$ is appropriate (Figure 2.8). Here Oz is now in the radial direction (i.e. the upward vertical direction in terrestrial and solar contexts), Ox is south and Oy east; hence (r,θ,φ) are replaced by (z, x, y).

[10] $\mathbf{e}_q$ will generally denote a unit vector in the direction of increasing q where q is any generalised coordinate.

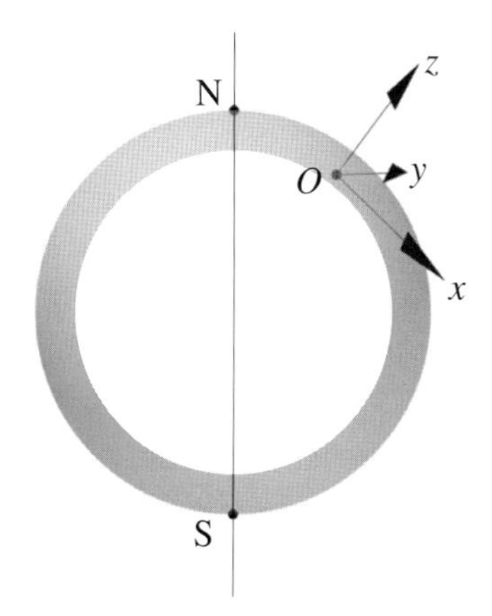

Figure 2.8 Local Cartesian coordinate system in a spherical shell geometry; Ox is directed south, Oy east and Oz vertically upwards.

The field decomposition analogous to (2.51) is then

$$\mathbf{B} = \mathbf{B}_P + \mathbf{B}_T = \nabla \wedge \nabla \wedge (\mathbf{e}_z P) + \nabla \wedge (\mathbf{e}_z T), \tag{2.64}$$

and P and T are given by

$$\nabla_2^2 P = -\mathbf{e}_z \cdot \mathbf{B}, \qquad \nabla_2^2 T = -\mathbf{e}_z \cdot (\nabla \wedge \mathbf{B}), \tag{2.65}$$

where ∇_2^2 is the two-dimensional Laplacian

$$\nabla_2^2 = (\mathbf{e}_z \wedge \nabla)^2 = \partial^2/\partial x^2 + \partial^2/\partial y^2. \tag{2.66}$$

If the field $\mathbf{B}$ is independent of the coordinate y (the analogue of axisymmetry), then $P = P(x, z)$, $T = T(x, z)$; the 'toroidal' field then becomes

$$\mathbf{B}_T = B\,\mathbf{e}_y, \quad B = -\partial T/\partial x, \tag{2.67}$$

and the 'poloidal' field becomes

$$\mathbf{B}_P = \nabla \wedge (A\,\mathbf{e}_y), \quad A = -\partial P/\partial x. \tag{2.68}$$

The $\mathbf{B}_P$-lines are now given by $A(x, z) = \text{const.}$, and A is the flux-function of the $\mathbf{B}_P$-field.

Quite apart from the spherical shell context, two-dimensional fields are of independent interest, and consideration of idealised two-dimensional situations can often provide valuable insights. It is then more natural to regard Oz as the direction of invariance of $\mathbf{B}$ and to express $\mathbf{B}$ in the form

$$\mathbf{B} = \nabla \wedge [A(x, y)\,\mathbf{e}_z] + B(x, y)\,\mathbf{e}_z. \tag{2.69}$$

2.4 Relations between electric current and magnetic field

2.4.1 *Ampère's law*

In a steady situation, the magnetic field $\mathbf{B}(\mathbf{x})$ is related to the electric current distribution $\mathbf{J}(\mathbf{x})$ by Ampère's law; in integral form this is

$$\oint_C \mathbf{B} \cdot d\mathbf{x} = \mu_0 \int_S \mathbf{J} \cdot \mathbf{n} \, dS \,, \tag{2.70}$$

where S is any open orientable surface spanning the closed curve C and μ_0 is a constant associated with the system of units used.[11] It follows from (2.70) that in any region where $\mathbf{B}$ and $\mathbf{J}$ are differentiable,

$$\nabla \wedge \mathbf{B} = \mu_0 \mathbf{J} \,, \quad \nabla \cdot \mathbf{J} = 0 \,, \tag{2.71}$$

and the corresponding jump conditions across surfaces of discontinuity are

$$[\![\mathbf{n} \wedge \mathbf{B}]\!] = \mu_0 \mathbf{J}_S \,, \quad [\![\mathbf{n} \cdot \mathbf{J}]\!] = 0, \tag{2.72}$$

where $\mathbf{J}_S$ ($\mathrm{A\,m^{-1}}$) represents surface current distribution. Surface currents (like concentrated vortex sheets) can survive only if dissipative processes do not lead to diffusive spreading, i.e. only in a perfect electrical conductor. In fluids or solids of finite conductivity, we may generally assume that $\mathbf{J}_S = 0$, and this together with (2.5) implies that all components of $\mathbf{B}$ are continuous:

$$[\![\mathbf{B}]\!] = 0 \,. \tag{2.73}$$

In an unsteady situation, (2.70) should generally be modified by the inclusion of Maxwell's displacement current; it is well known however that this effect is negligible in the treatment of phenomena whose time-scale is long compared with the time for electromagnetic waves to cross the region of interest. This condition is certainly satisfied in the planetary and stellar contexts, and we shall therefore neglect displacement current throughout; this has the effect of filtering electromagnetic waves from the system of governing equations. The resulting equations are entirely classical (i.e. non-relativistic) and are sometimes described as the 'pre-Maxwell equations'.

In terms of the vector potential $\mathbf{A}$ defined by (2.2), (2.71) becomes the (vector) Poisson equation

$$\nabla^2 \mathbf{A} = -\mu_0 \mathbf{J}. \tag{2.74}$$

Across discontinuity surfaces, $\mathbf{A}$ is in general continuous (since $\mathbf{B} = \nabla \wedge \mathbf{A}$ is in

[11] Permeability effect are generally unimportant in the topics to be considered and may be ignored from the outset. If $\mathbf{B}$ is measured in Tesla T (1 T= 10^4 Gauss) and $\mathbf{J}$ in $\mathrm{A\,m^{-2}}$, then $\mu_0 = 4\pi \times 10^{-7}\,\mathrm{T\,A^{-1}m}$.

general finite) and (2.73) implies further that the normal gradient of **A** must also be continuous; i.e. in general

$$[\![\mathbf{A}]\!] = 0\,, \qquad [\![(\mathbf{n}\cdot\nabla)\mathbf{A}]\!] = 0\,. \tag{2.75}$$

2.4.2 Multipole expansion of the magnetic field

Suppose now that S is a closed surface, with interior V and exterior $\hat{V}$, and suppose that $\mathbf{J}(\mathbf{x})$ is a current distribution entirely confined to V, i.e. $\mathbf{J} \equiv 0$ in $\hat{V}$; excluding the possibility of surface current on S, we must then also have $\mathbf{n}\cdot\mathbf{J} = 0$ on S. From (2.71), **B** is irrotational in $\hat{V}$, and so there exists a scalar potential $\Psi(\mathbf{x})$ such that, in $\hat{V}$,

$$\mathbf{B} = -\nabla\Psi\,, \qquad \nabla^2\Psi = 0\,. \tag{2.76}$$

Note that in general Ψ is not single-valued; it is, however, single-valued if $\hat{V}$ is simply connected, and we shall for the moment assume this to be the case. We may further suppose that $\Psi \to 0$ as $|\mathbf{x}| \to \infty$.

Relative to an origin O in V, the general solution of (2.76) vanishing at infinity may be expressed in the form

$$\Psi(\mathbf{x}) = \sum_{n=1}^{\infty} \Psi^{(n)}(\mathbf{x}), \quad \Psi^{(n)}(\mathbf{x}) = -\mu^{(n)}_{ij\ldots s}\left(\frac{1}{r}\right)_{,ij\ldots s}. \tag{2.77}$$

Here $\mu^{(n)}_{ij\ldots s}$ is the multipole moment tensor of rank n, $r = |\mathbf{x}|$, and a suffix i after the comma indicates differentiation with respect to x_i. The term with $n = 0$ is omitted by virtue of (2.1). The terms with $n = 1, 2$ are the dipole and quadrupole terms respectively; in vector notation

$$\Psi^{(1)}(\mathbf{x}) = -\boldsymbol{\mu}^{(1)}.\,\nabla(r^{-1}), \quad \Psi^{(2)}(\mathbf{x}) = -\boldsymbol{\mu}^{(2)} : \nabla\nabla(r^{-1})\,, \tag{2.78}$$

and similarly for higher terms.

The field **B** in $\hat{V}$ clearly has the expansion

$$\mathbf{B}(\mathbf{x}) = \sum_{n=1}^{\infty} \mathbf{B}^{(n)}(\mathbf{x}), \quad B^{(n)}_{\alpha}(\mathbf{x}) = \mu^{(n)}_{ij\ldots s}(r^{-1})_{ij\ldots s\,\alpha}\,, \tag{2.79}$$

and, since $\nabla^2(r^{-1}) = 0$ in $\hat{V}$, the expression for $\mathbf{B}^{(n)}$ may be expressed in the form $\mathbf{B}^{(n)} = \nabla \wedge \mathbf{A}^{(n)}$ where

$$\mathbf{A}^{(n)}_q = -\epsilon_{qmi}\,\mu^{(n)}_{mj\ldots s}(r^{-1})_{ij\ldots s}\,. \tag{2.80}$$

The first two terms of the expansion for the vector potential, corresponding to (2.79), thus have the form

$$\mathbf{A}^{(1)}(\mathbf{x}) = -\boldsymbol{\mu}^{(1)} \wedge \nabla(r^{-1}), \quad \mathbf{A}^{(2)}(\mathbf{x}) = -\boldsymbol{\mu}^{(2)}\{\wedge\,\cdot\}\nabla\nabla(r^{-1})\,. \tag{2.81}$$

The awkward, and slightly ambiguous, structure of the second term here makes it clear why the suffix notation of equation (2.80) is preferable.

The tensors $\boldsymbol{\mu}^{(n)}$ may be determined as linear functionals of $\mathbf{J}(\mathbf{x})$ as follows. The solution of (2.74) vanishing at infinity is

$$\mathbf{A}(\mathbf{x}) = \frac{\mu_0}{4\pi}\int_V |\mathbf{x}-\mathbf{x}'|^{-1}\mathbf{J}(\mathbf{x}')\,\mathrm{d}V'\,, \tag{2.82}$$

(and it may be readily verified using $\nabla\cdot\mathbf{J} = 0$, $\mathbf{n}\cdot\mathbf{J} = 0$ on S, that this satisfies $\nabla\cdot\mathbf{A} = 0$). The function $|\mathbf{x}-\mathbf{x}'|^{-1}$ has the Taylor expansion

$$|\mathbf{x}-\mathbf{x}'|^{-1} = \sum_n \frac{(-1)^n}{n!} x'_i x'_j \ldots x'_s (r^{-1})_{,ij\ldots s}\,. \tag{2.83}$$

Substitution in (2.82) leads immediately to $\mathbf{A} = \sum \mathbf{A}^{(n)}(\mathbf{x})$ where

$$A_q^{(n)}(\mathbf{x}) = c_{ij\ldots sq}(r^{-1})_{ij\ldots s}\,,\quad c_{ij\ldots sq} = \frac{\mu_0}{4\pi}\frac{(-1)^n}{n!}\int_V x_i x_j \ldots x_s J_q(\mathbf{x})\,\mathrm{d}V\,. \tag{2.84}$$

Comparison with (2.80) then gives the equivalent relations

$$c_{ij\ldots sq} = -\epsilon_{qmi}\mu^{(n)}_{mj\ldots s}\,,\qquad \mu^{(n)}_{mj\ldots s} = -\tfrac{1}{2}\epsilon_{qmi}c_{ij\ldots sq}\,. \tag{2.85}$$

In particular, we have for the terms $n = 1, 2$ the dipole and quadrupole moments[12]

$$\boldsymbol{\mu}^{(1)} = \frac{\mu_0}{8\pi}\int_V \mathbf{x}\wedge\mathbf{J}\,\mathrm{d}V\,,\quad \mu^{(2)}_{mj} = -\frac{\mu_0}{16\pi}\int_V x_j(\mathbf{x}\wedge\mathbf{J})_m\,\mathrm{d}V\,. \tag{2.86}$$

2.4.3 Axisymmetric fields

If $\mathbf{J}(\mathbf{x})$ is axisymmetric about the direction of the unit vector $\mathbf{e}_z = \mathbf{e}$, then these results can be simplified. Choosing spherical polar coordinates (r, θ, φ) based on the polar axis Oz, we have evidently

$$\boldsymbol{\mu}^{(1)} = \mu^{(1)}\mathbf{e}\,,\quad \mu^{(1)} = \frac{\mu_0}{4}\iint J_\varphi(r,\,\theta) r^3 \sin^2\theta\,\mathrm{d}r\,\mathrm{d}\theta\,. \tag{2.87}$$

Likewise, $\mu^{(2)}_{mj}$ must be axisymmetric about Oz; and since $\mu^{(2)}_{jj} = 0$ from (2.86*b*), it must therefore take the form

$$\mu^{(2)}_{mj} = \tfrac{3}{2}\mu^{(2)}\left(e_m e_j - \tfrac{1}{3}\delta_{mj}\right)\,. \tag{2.88}$$

Putting $m = 3$, $j = 3$ in (2.86) and (2.88) then gives $\mu^{(2)}$ in the form

$$\mu^{(2)} = -\frac{\mu_0}{8}\iint J_\varphi(r,\,\theta) r^4 \sin^2\theta\,\cos\theta\,\mathrm{d}r\,\mathrm{d}\theta\,. \tag{2.89}$$

[12] The dipole moment as defined by (2.86) corresponds to the magnetic moment. The 'current dipole moment' $\mathbf{m}$ is usually defined as $\mathbf{m} = 4\pi\boldsymbol{\mu}^{(1)}/\mu_0$.

In this axisymmetric situation, the expansion (2.77) clearly has the form

$$\Psi(\mathbf{x}) = -\sum_{n=1}^{\infty} \mu^{(n)} \frac{\partial^n}{\partial z^n}\left(\frac{1}{r}\right). \tag{2.90}$$

2.5 Force-free fields

We shall have frequent occasion to refer to magnetic fields for which $\mathbf{B}$ is everywhere parallel to $\mathbf{J} = \mu_0 \nabla \wedge \mathbf{B}$ and it will therefore be useful at this stage to gather together some properties of such fields[13] which are described as 'force-free' (Lust & Schlüter, 1954) since the associated Lorentz force $\mathbf{J} \wedge \mathbf{B}$ is of course identically zero. For any force-free field, there exists a scalar function of position $K(\mathbf{x})$ such that

$$\nabla \wedge \mathbf{B} = K\mathbf{B}, \quad \mathbf{B} \cdot \nabla K = 0, \tag{2.91}$$

the latter following from $\nabla \cdot \mathbf{B} = 0$. K is therefore constant on $\mathbf{B}$-lines, and if $\mathbf{B}$-lines cover surfaces then K must be constant on each such surface. A particularly simple situation is that in which K is constant everywhere; in this case, taking the curl of (2.91) immediately leads to the Helmholtz equation

$$\left(\nabla^2 + K^2\right) \mathbf{B} = 0. \tag{2.92}$$

Note however that this process cannot in general be reversed: a field $\mathbf{B}$ that satisfies (2.92) does not necessarily satisfy either of the equations $\nabla \wedge \mathbf{B} = \pm K\mathbf{B}$.

The simplest example of a force-free field with $K =$ const. is, in Cartesian coordinates,

$$\mathbf{B} = B_0(\sin Kz,\ \cos Kz,\ 0). \tag{2.93}$$

The property $\nabla \wedge \mathbf{B} = K\mathbf{B}$ is trivially verified. The $\mathbf{B}$-lines, as indicated in Figure 2.9a, lie in the x–y plane and their direction rotates with increasing z in a sense that is left-handed or right-handed according as K is positive or negative. The vector potential of $\mathbf{B}$ is simply $\mathbf{A} = K^{-1}\mathbf{B}$, so that its helicity density is uniform:

$$\mathbf{A} \cdot \mathbf{B} = K^{-1}\mathbf{B}^2 = K^{-1}B_0^2. \tag{2.94}$$

If we imagine the $\mathbf{B}$-lines closed by the dashed lines as indicated in the figure, then the resulting linkages are consistent with the discussion of §2.1.

A second example (Figure 2.9b) of a force-free field, with K again constant, is, in cylindrical polars (s, φ, z),

$$\mathbf{B} = B_0\left(0,\ J_1(Ks),\ J_0(Ks)\right), \tag{2.95}$$

[13] In general, a vector field $\mathbf{B}(\mathbf{x})$, with the property that $\nabla \wedge \mathbf{B}$ is everywhere parallel to $\mathbf{B}$, is known as a *Beltrami field*.

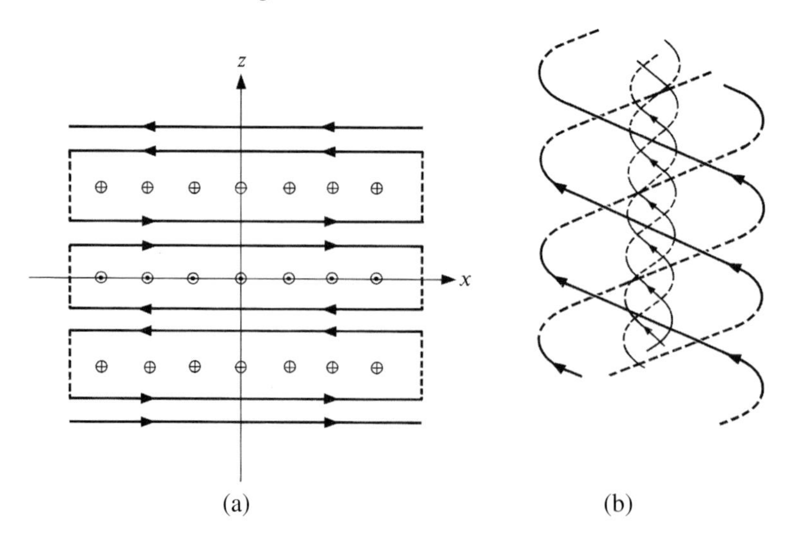

Figure 2.9 (a) **B**-lines of the field (2.93) with $K > 0$. ⊙ indicates a line in the positive y-direction (i.e. into the paper), and ⊕ a line in the negative y-direction; the **B**-lines rotate in a left-handed sense with increasing z; closing the **B**-lines by means of the dashed segments leads to linkages consistent with the positive helicity of the field. (b) Typical helical **B**-lines of the field (2.95); the linkages as illustrated are negative, and therefore correspond to a negative value of K in (2.95).

where J_n is the Bessel function of order n. Here the **B**-lines are helices on the cylinders s = const. Again $\mathbf{A} = K^{-1}\mathbf{B}$, and

$$\mathbf{A}\cdot\mathbf{B} = K^{-1}\mathbf{B}^2 = K^{-1}B_0^2\left[(J_1(Ks))^2 + (J_0(Ks))^2\right], \tag{2.96}$$

and again any simple closing of **B**-lines leads to linkages (which are negative, corresponding to a negative value of K, in Figure 2.9b).

In both these examples, the **J**-field extends to infinity. There are in fact *no* force-free fields, other than $\mathbf{B} \equiv 0$, for which **J** is confined (as in §2.4) to a finite volume V and **B** is everywhere differentiable and $\mathcal{O}(r^{-3})$ at infinity.[14] To prove this, let T_{ij} be the Maxwell stress tensor, given by

$$T_{ij} = \mu_0^{-1}\left(B_iB_j - \tfrac{1}{2}\mathbf{B}^2\delta_{ij}\right), \tag{2.97}$$

with the properties

$$(\mathbf{J}\wedge\mathbf{B})_i = T_{ij,j}, \quad T_{ii} = -(2\mu_0)^{-1}\mathbf{B}^2, \tag{2.98}$$

[14] This condition means of course that the only source for **B** is the current distribution **J**, and there are no further 'sources at infinity'. The leading term of the expansion (2.79) is clearly $\mathcal{O}(r^{-3})$. It is perhaps worth noting that the proof still goes through under the weaker condition $\mathbf{B} = \mathcal{O}\left(r^{-(3/2+\varepsilon)}\right)$, $(\varepsilon > 0)$, corresponding to finiteness of the magnetic energy $-\int T_{ii}\,\mathrm{d}V$.

and suppose that $\mathbf{J} \wedge \mathbf{B} \equiv 0$. Then

$$0 = \int x_i T_{ij,j}\,\mathrm{d}V = \int_{S_\infty} n_j x_i T_{ij}\,\mathrm{d}S - \int T_{ii}\,\mathrm{d}V\,, \tag{2.99}$$

the volume integrals being over all space. Now since $\mathbf{B} = \mathcal{O}(r^{-3})$ as $r \to \infty$, $T_{\mathbf{i}j} = \mathcal{O}(r^{-6})$, and so the integral over S_∞ vanishes. Hence the integral of $\mathbf{B}^2$ vanishes, and so $\mathbf{B} \equiv 0$. The proof fails if surfaces of discontinuity of $\mathbf{B}$ (and so of T_{ij}) are allowed, since then further surface integrals which do not in general vanish must be included in (2.99).

2.5.1 *Force-free fields in spherical geometry*

We can however construct solutions of (2.91) which are force-free in a finite region V and current-free in the exterior region $\hat{V}$ and which do *not* vanish at infinity. This can be done explicitly (Chandrasekhar 1956) in the important case when V is the sphere $r < R$, as follows. First let

$$K = \begin{cases} K_0 & (r < R) \\ 0 & (r > R) \end{cases} \tag{2.100}$$

where K_0 is constant. We have seen in §2.3 that, under the poloidal and toroidal decomposition

$$\mathbf{B} = \nabla \wedge (\nabla \wedge \mathbf{x} P) + \nabla \wedge \mathbf{x} T\,, \tag{2.101}$$

we have

$$\nabla \wedge \mathbf{B} = -\nabla \wedge \left(\mathbf{x}\,\nabla^2 P\right) + \nabla \wedge \nabla \wedge (\mathbf{x} T)\,, \tag{2.102}$$

and so (2.91) is satisfied provided

$$T = K P \quad \text{and} \quad \left(\nabla^2 + K^2\right) P = 0\,. \tag{2.103}$$

Here K is discontinuous across $r = R$; but continuity of $\mathbf{B}$ requires that T, P and $\partial P/\partial r$ must be continuous across $r = R$, or equivalently that

$$P = 0, \quad [\![\partial P/\partial r]\!] = 0 \quad \text{on} \quad r = R. \tag{2.104}$$

The simplest solution of (2.103), (2.104) is given in spherical polars (r, θ, φ) by

$$P(r, \theta) = \begin{cases} A\, r^{-\frac{1}{2}} J_{3/2}(K_0 r) \cos\theta & (r < R) \\ -B_0(r - R^3/r^2) \cos\theta & (r > R) \end{cases} \tag{2.105}$$

where

$$J_{3/2}(K_0 R) = 0 \quad \text{and} \quad 3B_0 = -A \frac{\mathrm{d}}{\mathrm{d}R}\left(R^{\frac{1}{2}} J_{3/2}(K_0 R)\right), \tag{2.106}$$

these conditions following from (2.104). The corresponding flux-function $\chi(r, \theta)$ is then given by (2.61); the $\mathbf{B}_P$-lines, given by $\chi = $ const., are shown in Figure 2.10a

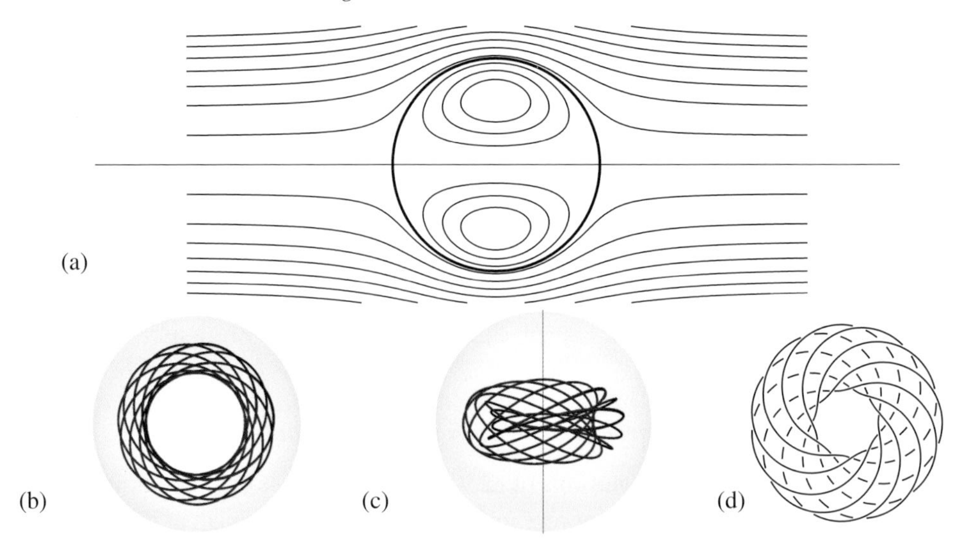

Figure 2.10 (a) The $\mathbf{B}_P$-lines (χ = const.) of the force-free field given by (2.105), with K_0R ($\approx$ 4.493) the smallest zero of $J_{3/2}(x)$. (b) A typical **B**-line (which follows a helical path on a toroidal surface); the axis of symmetry is here perpendicular to the paper. (c) A different projection of the same **B**-line. This particular **B**-line is very close to the torus knot $K_{8,11}$ shown in (d).

for the case where $K_0R \approx 4.493$ is the smallest positive zero of $J_{3/2}(x)$. For $r > R$, the **B**-lines are identical with the streamlines in irrotational flow past a sphere, and $\mathbf{B} \sim \mathbf{B}_0 = B_0\,\mathbf{e}_z$ as $r \to \infty$. For $r < R$, the surfaces χ =const. form a family of nested toroidal surfaces (or 'tori'), and the **B**-lines lie on these surfaces; one such **B**-line is shown in Figure 2.10b. This **B**-line nearly closes on itself after 8 circuits in the φ-direction and 11 circuits round the smaller section of the torus; i.e. it is close to the torus knot $K_{8,11}$.

The poloidal field has a zero where χ is extremal. At this point,

$$\frac{\mathrm{d}}{\mathrm{d}r}\left(r^{\frac{1}{2}} J_{3/2}(K_0 r)\right) = 0, \quad \text{i.e. } r = r_c \approx 0.548\,R. \tag{2.107}$$

The circle $r = r_c$, $\theta = \pi/2$ is the degenerate torus of the family (described as the 'magnetic axis' of the field). Each **B**-line is a helix and the pitch of the helices changes continuously as we move outwards across the family of tori from the magnetic axis to the limiting sphere $r = R$.

Note once again that $\mathbf{A} = K_0^{-1}\mathbf{B}$ for $r < R$, and therefore that

$$\int_{r<R} \mathbf{A}\cdot\mathbf{B}\,\mathrm{d}V = K_0^{-1}\int_{r<R} \mathbf{B}^2\,\mathrm{d}V \neq 0\,, \tag{2.108}$$

consistent with the fact that there is an indisputable degree of linkage in the **B**-lines within the sphere; e.g. each line of force winds round the magnetic axis which is a particular **B**-line of the field. The **B**-lines in $r < R$ in general cover the tori; but if the pitch p, defined as the increase $\Delta\varphi$ in the azimuth angle as a torus is circumscribed once by the **B**-line, is $2\pi m/n$ where m and n are co-prime integers, then the **B**-line is closed; moreover, if $m \geq 2$ and $n \geq 3$, the curve is knotted. The corresponding knot is known as the torus knot $K_{m,n}$; $K_{2,3}$ is just the trefoil knot of Figure 2.1b. It is an intriguing property of this **B**-field that if we take a subset of the **B**-lines consisting of one **B**-line on each toroidal surface, then an infinity of distinct torus knots are represented in this subset, since $p/2\pi$ passes through an infinity of rational numbers m/n in its continuous variation across the family of tori; and yet the closed **B**-lines are exceptional in that they constitute a subset of measure zero of the set of all **B**-lines inside the sphere![15]

More complicated force-free fields can be constructed either by choosing higher zeros of $J_{3/2}(K_0 R)$ (in which case there is more than one magnetic axis) or by replacing (2.105) by more general solutions of (2.103). The above example is however quite sufficient as a sort of prototype for spatial structures with which we shall later be concerned.

2.6 Lagrangian variables and magnetic field evolution

We must now consider magnetic field evolution in a moving fluid conductor. Let us specify the motion in terms of the displacement field $\mathbf{x}(\mathbf{a}, t)$, which represents the position at time t of the fluid particle that passes through the point $\mathbf{a}$ at a reference instant $t = 0$; in particular

$$\mathbf{x}(\mathbf{a}, 0) = \mathbf{a}\,, \quad (\partial x_i/\partial a_j)_{t=0} = \delta_{ij}\,. \tag{2.109}$$

Each particle is labelled by its initial position **a**. The mapping $\mathbf{x} = \mathbf{x}(\mathbf{a}, t)$ is clearly one-to-one for a real motion of a continuous fluid, and we can equally consider the inverse mapping $\mathbf{a} = \mathbf{a}(\mathbf{x}, t)$. The velocity of the particle **a** is

$$\mathbf{u}^L(\mathbf{a}, t) = (\partial \mathbf{x}/\partial t)_{\mathbf{a}} = \mathbf{u}(\mathbf{x}, t)\,; \tag{2.110}$$

$\mathbf{u}^{\mathrm{L}}(\mathbf{a}, t)$ is the Lagrangian representation, and $\mathbf{u}(\mathbf{x}, t)$ the more usual Eulerian representation. We shall use the superfix L in this way whenever fields are expressed as functions of $(\mathbf{a}, t)$, e.g.

$$\mathbf{B}^L(\mathbf{a}, t) = \mathbf{B}(\mathbf{x}(\mathbf{a}, t), t) \tag{2.111}$$

[15] See Bogoyavlenskij (2017) and Moffatt (2017) concerning this structure and correcting an erroneous statement in (Moffatt 1978a).

represents the magnetic field referred to Lagrangian variables. Defining the usual Lagrangian (or material) derivative by

$$\frac{\mathrm{D}}{\mathrm{D}t} \equiv \left(\frac{\partial}{\partial t}\right)_{\mathbf{a}} = \left(\frac{\partial}{\partial t}\right)_{\mathbf{x}} + \mathbf{u}\cdot\nabla, \tag{2.112}$$

it is clear that in particular

$$\frac{\mathrm{D}\mathbf{B}}{\mathrm{D}t} \equiv \frac{\partial \mathbf{B}}{\partial t} + (\mathbf{u}.\nabla)\mathbf{B} = \left(\frac{\partial \mathbf{B}^L}{\partial t}\right)_{\mathbf{a}}. \tag{2.113}$$

A material curve C_L is one consisting entirely of fluid particles, and which is therefore convected and distorted with the fluid motion. If p is a parameter on the curve at time $t = 0$, so that $\mathbf{a} = \mathbf{a}(p)$, then the parametric representation at time t is given by

$$\mathbf{x} = \mathbf{x}(\mathbf{a}(p), t)\,; \tag{2.114}$$

the curve C_L is closed if $\mathbf{a}(p)$ is a periodic function of p. A material surface S_L may be defined similarly and described in terms of two parameters.

An infinitesimal material line element may be described by the differential

$$\mathrm{d}x_i = E_{ij}(\mathbf{a}, t)\,\mathrm{d}a_j, \quad E_{ij}(\mathbf{a}, t) = \partial x_i/\partial a_j\,. \tag{2.115}$$

The symmetric and antisymmetric parts of E_{ij} describe respectively the distortion and rotation of the fluid element initially at $\mathbf{a}$. The material derivative of E_{ij} is

$$\mathrm{D}E_{ij}/\mathrm{D}t = \partial u_i^L/\partial a_j\,, \tag{2.116}$$

and so it follows that

$$\mathrm{D}\,\mathrm{d}\mathbf{x}/\mathrm{D}t = \mathrm{d}a_j\,\partial \mathbf{u}^L/\partial a_j = (\mathrm{d}\mathbf{x}\cdot\nabla)\mathbf{u}\,, \tag{2.117}$$

a result that is equally clear from elementary geometrical considerations.

2.6.1 Change of flux through a moving circuit

Suppose now that

$$\Phi(t) = \int_{S_L} \mathbf{B}\cdot\mathbf{n}\,\mathrm{d}S = \oint_{C_L} \mathbf{A}\cdot\mathrm{d}\mathbf{x}\,, \tag{2.118}$$

where C_L is a material curve spanned by S_L. In order to calculate $\mathrm{d}\Phi/\mathrm{d}t$ we should use Lagrangian variables:

$$\Phi(t) = \oint_{C_L} A_i^L(\mathbf{a},\ t)(\partial x_i/\partial a_j)(\mathrm{d}a_j/\mathrm{d}p)\,\mathrm{d}p\,. \tag{2.119}$$

We can then differentiate under the integral keeping $\mathbf{a}(p)$ constant. This gives, using (2.117) and standard manipulation,

$$\frac{\mathrm{d}\Phi}{\mathrm{d}t} = \oint_{C_L}\left(\frac{\mathrm{D}\mathbf{A}}{\mathrm{D}t}\cdot\mathrm{d}\mathbf{x} + \mathbf{A}\cdot(\mathrm{d}\mathbf{x}\cdot\nabla)\mathbf{u}\right) = \oint_{C_L}\left(\frac{\partial\mathbf{A}}{\partial t} - \mathbf{u}\wedge(\nabla\wedge\mathbf{A}) + \nabla(\mathbf{A}.\mathbf{u})\right)\cdot\mathrm{d}\mathbf{x}\,. \tag{2.120}$$

The term involving $\nabla(\mathbf{A}\cdot\mathbf{u})$ makes zero contribution to the integral since $\mathbf{A}\cdot\mathbf{u}$ is single-valued; and we have therefore

$$\frac{\mathrm{d}\Phi}{\mathrm{d}t} = \oint_{C_L}\left(\frac{\partial\mathbf{A}}{\partial t} - \mathbf{u}\wedge\mathbf{B}\right)\cdot\mathrm{d}\mathbf{x}\,. \tag{2.121}$$

2.6.2 Faraday's law of induction

In its most fundamental form, Faraday's law states that if $\Phi(t)$ is defined as above for any moving closed curve C_L, then

$$\frac{\mathrm{d}\Phi}{\mathrm{d}t} = -\oint_{C_L}(\mathbf{E} + \mathbf{u}\wedge\mathbf{B})\cdot\mathrm{d}\mathbf{x}\,, \tag{2.122}$$

where $\mathbf{E}(\mathbf{x}, t)$ (V m^{-1}) is the electric field relative to some fixed frame of reference. Comparison of (2.121) and (2.122) then shows that $\mathbf{E}$ differs from $-\partial\mathbf{A}/\partial t$ by at most the gradient of a single-valued scalar $\phi(\mathbf{x}, t)$:

$$\mathbf{E} + \partial\mathbf{A}/\partial t = -\nabla\phi\,. \tag{2.123}$$

The curl of this gives the familiar Maxwell equation

$$\partial\mathbf{B}/\partial t = -\nabla\wedge\mathbf{E}\,. \tag{2.124}$$

The corresponding jump condition across discontinuity surfaces is, from (2.122),

$$[\![\mathbf{n}\wedge(\mathbf{E} + \mathbf{u}\wedge\mathbf{B})]\!] = 0\,. \tag{2.125}$$

2.6.3 Galilean invariance of the pre-Maxwell equations

The following simple property of the three equations

$$\nabla\cdot\mathbf{B} = 0, \quad \nabla\wedge\mathbf{B} = \mu_0\mathbf{J}, \quad \partial\mathbf{B}/\partial t = -\nabla\wedge\mathbf{E} \tag{2.126}$$

is worth noting explicitly. Under the Galilean transformation

$$\mathbf{x}' = \mathbf{x} - \mathbf{V}t, \quad t' = t\,, \tag{2.127}$$

the equations transform to

$$\nabla'\cdot\mathbf{B}' = 0\,, \quad \nabla'\wedge\mathbf{B}' = \mu_0\mathbf{J}'\,, \quad \partial\mathbf{B}'/\partial t' = -\nabla'\wedge\mathbf{E}'\,, \tag{2.128}$$

where

$$\mathbf{B}' = \mathbf{B}\,, \quad \mathbf{J}' = \mathbf{J}\,, \quad \mathbf{E}' = \mathbf{E} + \mathbf{V} \wedge \mathbf{B}\,. \tag{2.129}$$

(These are the non-relativistic limiting forms of the more general Lorentz transformations of the full Maxwell equations.) It is important to note that **B** and **J** are invariant under Galilean transformation, but that **E** is not. For a fluid moving with velocity $\mathbf{u}(\mathbf{x}, t)$, the field

$$\mathbf{E}' = \mathbf{E} + \mathbf{u} \wedge \mathbf{B} \tag{2.130}$$

is the electric field as measured by an observer moving with the fluid, and the right-hand side of (2.122) can therefore be regarded as (minus) the *effective* electromotive force in the moving circuit.

2.6.4 Ohm's law in a moving conductor

We shall employ throughout the simplest form of Ohm's law which provides the relation between electric current and electric field. In an element of fluid moving with velocity **u**, the relation between the field $\mathbf{J}'$ and $\mathbf{E}'$ in a frame of reference moving with the element is the same as if the element were at rest (on the assumption that acceleration of the element is insufficient to affect molecular transport processes), and we shall take this relation to be $\mathbf{J}' = \sigma\mathbf{E}'$ where σ is the electric conductivity of the fluid (measured in A $\mathrm{V}^{-1}\mathrm{m}^{-1}$). Relative to the fixed reference frame, this relation becomes

$$\mathbf{J} = \sigma(\mathbf{E} + \mathbf{u} \wedge \mathbf{B})\,. \tag{2.131}$$

It must be emphasised that, unlike the equations (2.126) which are fundamental, (2.131) is a phenomenological relationship with a limited range of validity. Its justification, and determination of the value of σ in terms of the parameters describing the molecular structure of the fluid, are the subject of transport theory in statistical mechanics, and are outside the scope of this book.

If we now combine (2.74), (2.123) and (2.131), we obtain immediately

$$\partial\mathbf{A}/\partial t = \mathbf{u} \wedge (\nabla \wedge \mathbf{A}) - \nabla\phi + \eta\nabla^2\mathbf{A}\,, \tag{2.132}$$

where $\eta = (\mu_0\sigma)^{-1}$ is the *magnetic diffusivity* of the fluid. Clearly, like any other diffusivity, η has dimensions length^{2}time^{-1}; unless stated otherwise, we shall always assume that η is uniform and constant. The divergence of (2.132), using $\nabla \cdot \mathbf{A} = 0$, gives

$$\nabla^2\phi = \nabla \cdot (\mathbf{u} \wedge \mathbf{B})\,. \tag{2.133}$$

The curl of (2.132) gives the well-known *induction equation* of magnetohydrodynamics

$$\partial \mathbf{B}/\partial t = \nabla \wedge (\mathbf{u} \wedge \mathbf{B}) + \eta \nabla^2 \mathbf{B}\,. \tag{2.134}$$

It is clear that if $\mathbf{u}(\mathbf{x}, t)$ is prescribed then, subject to appropriate initial and boundary conditions, this fundamental equation determines the evolution of $\mathbf{B}(\mathbf{x}, t)$. We shall consider in detail some of the properties of (2.134) in the following chapter.

2.7 Kinematically possible velocity fields

The velocity field $\mathbf{u}(\mathbf{x}, t)$ is related to the density field in a moving fluid by the equation of mass conservation

$$\partial \rho/\partial t + \nabla \cdot (\rho \mathbf{u}) = 0\,. \tag{2.135}$$

We have also the associated boundary condition

$$\mathbf{u} \cdot \mathbf{n} = 0 \quad \text{on} \quad S_r\,, \tag{2.136}$$

where S_r is any fixed boundary that may be present. These are kinematic (as opposed to dynamic) constraints, and we describe the joint field $\{\mathbf{u}(\mathbf{x}, t), \rho(\mathbf{x}, t)\}$ as *kinematically possible* if (2.135) and (2.136) are satisfied. It is of course only a small subset of such fields that are also *dynamically possible* under, say, the Navier–Stokes equations with prescribed body forces; but many useful results can be obtained without reference to the dynamical equations, and these results are generally valid for any kinematically possible flows.

Equation (2.135) may be written in the equivalent Lagrangian form

$$\mathrm{D}\rho/\mathrm{D}t = -\rho \, \nabla \cdot \mathbf{u}\,. \tag{2.137}$$

We shall frequently be concerned with contexts where the fluid may be regarded as incompressible, i.e. for which $\mathrm{D}\rho/\mathrm{D}t = 0$. In this case, a kinematically possible flow $\mathbf{u}(\mathbf{x}, t)$ is simply one which satisfies

$$\nabla \cdot \mathbf{u} = 0\,, \quad \mathbf{u} \cdot \mathbf{n} = 0 \quad \text{on} \quad S_r\,. \tag{2.138}$$

Conservation of mass may equivalently be represented by the Lagrangian equation for mass differentials,

$$\rho(\mathbf{x}, t)\, \mathrm{d}^3\mathbf{x} = \rho(\mathbf{a}, 0)\, \mathrm{d}^3\mathbf{a}\,. \tag{2.139}$$

Now[16]

$$\epsilon_{lmn}\, \mathrm{d}^3\mathbf{x} = \epsilon_{ijk} \frac{\partial x_i}{\partial a_l} \frac{\partial x_j}{\partial a_m} \frac{\partial x_k}{\partial a_n}\, \mathrm{d}^3\mathbf{a}\,, \tag{2.140}$$

[16] This is equivalent to the statement $\mathrm{d}^3\mathbf{x} = J\,\mathrm{d}^3\mathbf{a}$; $J = \partial(x_1, x_2, x_3)/\partial(a_1, a_2, a_3)$ is the Jacobian of the transformation $\mathbf{x} = \mathbf{x}(\mathbf{a}, t)$.

so that (2.139) becomes

$$\rho(\mathbf{x},t)\,\epsilon_{ijk}\,\frac{\partial x_i}{\partial a_l}\,\frac{\partial x_j}{\partial a_m}\,\frac{\partial x_k}{\partial a_n} = \rho(\mathbf{a},0)\,\epsilon_{lmn}\,. \tag{2.141}$$

For incompressible flow, this of course becomes simply

$$\epsilon_{ijk}\,\frac{\partial x_i}{\partial a_l}\,\frac{\partial x_j}{\partial a_m}\,\frac{\partial x_k}{\partial a_n} = \epsilon_{lmn}\,. \tag{2.142}$$

2.8 Free decay modes

In the absence of fluid motion, a current field $\mathbf{J}(\mathbf{x},t)$ confined to a finite region V, and its associated magnetic field $\mathbf{B}(\mathbf{x},t)$, decays under the action of magnetic ('ohmic') diffusion. Consideration of this straightforward effect is a useful preliminary to the topics that will be considered in later chapters. Suppose then that $\mathbf{u}\equiv 0$ in V, so that from (2.134), $\mathbf{B}$ satisfies the diffusion equation

$$\partial\mathbf{B}/\partial t = \eta\nabla^2\mathbf{B} \quad \text{in } V\,. \tag{2.143}$$

Suppose further that the external region $\hat{V}$ is non-conducting so that

$$\mu_0\mathbf{J} = \nabla\wedge\mathbf{B} = 0 \;\text{ in }\; \hat{V}\,. \tag{2.144}$$

We have also the boundary conditions

$$[\![\mathbf{B}]\!] = 0 \;\text{ on } S, \quad |\mathbf{B}| = \mathcal{O}(r^{-3}) \;\text{ as } r\to\infty\,, \tag{2.145}$$

where S is the surface of V.

The natural decay modes for this problem are defined by

$$\mathbf{B}(\mathbf{x},t) = \mathbf{B}^{(\alpha)}(\mathbf{x})\,\mathrm{e}^{p_\alpha t}\,, \tag{2.146}$$

where $\mathbf{B}^{(\alpha)}(\mathbf{x})$ satisfies

$$\left.\begin{array}{ll} \left(\nabla^2 - p_\alpha/\eta\right)\mathbf{B}^{(\alpha)} = 0 \;\text{ in } V\,, & \nabla\wedge\mathbf{B}^{(\alpha)} = 0 \;\text{ in } \hat{V}\,, \\ [\![\mathbf{B}^{(\alpha)}]\!] = 0 \;\text{ on } S\,, & \left|\mathbf{B}^{(\alpha)}\right| \sim \mathcal{O}(r^{-3}) \;\text{ as } r\to\infty\,. \end{array}\right\} \tag{2.147}$$

The equations (2.147) constitute an eigenvalue problem, the eigenvalues being p_α and the corresponding eigenfunctions $\mathbf{B}^{(\alpha)}(\mathbf{x})$. These eigenfunctions form a complete set, from the general theory of elliptic partial differential equations, and an initial field $\mathbf{B}(\mathbf{x},0)$ corresponding to an arbitrary initial current distribution $\mathbf{J}(\mathbf{x},0)$ in V may be expanded as a sum of eigenfunctions:

$$\mathbf{B}(\mathbf{x},0) = \sum_\alpha a_\alpha\mathbf{B}^{(\alpha)}(\mathbf{x})\,. \tag{2.148}$$

For $t > 0$, the field is then given by

$$\mathbf{B}(\mathbf{x}, t) = \sum_{\alpha} a_{\alpha} \mathbf{B}^{(\alpha)}(\mathbf{x}) \exp p_{\alpha} t \,. \tag{2.149}$$

Standard manipulation of (2.147) shows that

$$p_{\alpha} = -\eta \int_{V\infty} \left(\nabla \wedge \mathbf{B}^{(\alpha)}\right)^2 \mathrm{d}V \Big/ \int_{V} \left(\mathbf{B}^{(\alpha)}\right)^2 \mathrm{d}V \,, \tag{2.150}$$

where $V_{\infty} = V \cup \hat{V}$. Hence all the p_{α} are real and negative, and they may be ordered so that

$$0 > p_{\alpha 1} \geqslant p_{\alpha 2} \geqslant p_{\alpha 3} \geqslant \dots \,. \tag{2.151}$$

When V is the sphere $r > R$, as in the consideration of the force-free modes of §2.5, the poloidal and toroidal decomposition is appropriate. Suppose then that

$$\mathbf{B} = \nabla \wedge \nabla \wedge (\mathbf{x}\, P(\mathbf{x}, t)) + \nabla \wedge (\mathbf{x}\, T(\mathbf{x}, t)) \,, \tag{2.152}$$

with associated current distribution $\mathbf{J}$, where

$$\mu_0 \mathbf{J} = -\nabla \wedge (\mathbf{x}\, \nabla^2 P) + \nabla \wedge \nabla \wedge (\mathbf{x}\, T) \,. \tag{2.153}$$

The equations (2.143)–(2.145) are then satisfied provided

$$\left.\begin{array}{r} \partial P/\partial t = \eta \nabla^2 P \,, \;\; \partial T/\partial t = \eta \nabla^2 T \,, \;\; \text{in } r < R \\ \nabla^2 P = 0 \,, \;\; T = 0 \,, \;\; \text{in } r > R \\ [\![P]\!] = [\![\partial P/\partial r]\!] = [\![T]\!] = 0 \;\; \text{on } r = R \end{array}\right\} . \tag{2.154}$$

2.8.1 Toroidal decay modes

The field T may always be expanded in surface harmonics,

$$T(r, \theta, \varphi, t) = \sum T^{(n)}(r, t)\, S_n(\theta, \varphi) \,, \tag{2.155}$$

where, from (2.154), $T^{(n)}(r, t)$ satisfies

$$\frac{\partial T^{(n)}}{\partial t} = \eta \left(\frac{1}{r^2} \frac{\partial}{\partial r} r^2 \frac{\partial T^{(n)}}{\partial r} - \frac{n(n+1)}{r^2} T^{(n)} \right) \;\text{ for } r < R \,, \quad T^{(n)} = 0 \;\text{ for } r \geqslant R \,. \tag{2.156}$$

Putting $T^{(n)}(r, t) = f^{(n)}(r) \exp p_{\alpha} t$, we obtain a modified form of Bessel's equation for $f^{(n)}(r)$, with solution (regular at $r = 0$)

$$f^{(n)}(r) \propto r^{-1/2} J_{n+\frac{1}{2}}(k_{\alpha} r), \quad k_{\alpha}^2 = -p_{\alpha}/\eta \,. \tag{2.157}$$

The boundary condition $f^{(n)}(R) = 0$ is then satisfied provided

$$J_{n+\frac{1}{2}}(k_{\alpha} R) = 0 \,. \tag{2.158}$$

Table 2.2 *Zeros x_{nq} of $J_{n+\frac{1}{2}}(x)$, correct to three decimal places*

$n\downarrow$ $q\rightarrow$	1	2	3	4	5	6
0	π	2π	3π	4π	5π	6π
1	4.493	7.725	10.904	14.007	17.221	20.371
2	5.763	9.095	12.323	15.517	18.689	21.854
3	6.988	10.417	13.698	16.924	20.122	23.304
4	8.813	11.705	15.040	18.301	21.525	24.728

Let $x_{nq}\,(q = 1, 2, \ldots)$ denote the zeros of $J_{n+\frac{1}{2}}(x)$ (see Table 2.2); then the decay rates p_α of toroidal modes are given by

$$p_\alpha = -\eta R^{-2} x_{nq}^2 \qquad (n = 1, 2, \ldots;\ q = 1, 2, \ldots)\,, \tag{2.159}$$

where α is now a symbol for the pair $\{n, q\}$. The general solution for T has the form (in $r < R$)

$$T(r, \theta, \varphi, t) = \sum_n S_n(\theta, \varphi) \sum_q A_{nq}\, r^{-1/2} J_{n+\frac{1}{2}}(k_{nq} r)\, \mathrm{e}^{p_\alpha t}\,. \tag{2.160}$$

2.8.2 Poloidal decay modes

We may similarly expand P in the form

$$P(r, \theta, \varphi, t) = \sum P^{(n)}(r, t) S_n(\theta, \varphi)\,. \tag{2.161}$$

Now, however, since $\nabla^2 P = 0$ for $r > R$, we have

$$P^{(n)}(r, t) = c_n(t) r^{-(n+1)}, \quad (r > R)\,. \tag{2.162}$$

Hence continuity of $P^{(n)}$ and $\partial P^{(n)}/\partial r$ on $r = R$ requires that

$$P^{(n)}(R, t) = c_n(t) R^{-(n+1)} \quad \text{and} \quad \partial P^{(n)}/\partial r\big|_{r=R} = -(n+1) c_n(t) R^{-(n+2)}\,, \tag{2.163}$$

or, eliminating $c_n(t)$,

$$\partial P^{(n)}/\partial r + (n+1) R^{-1} P^{(n)} = 0 \ \text{ on } \ r = R\,. \tag{2.164}$$

Putting $P^{(n)}(r, t) = g_n(r)\, \mathrm{e}^{p_\alpha t}$ for $r < R$, we now obtain

$$g_n(r) \propto r^{-1/2} J_{n+\frac{1}{2}}(k_\alpha r)\,, \quad k_\alpha^2 = -p_\alpha/\eta\,, \tag{2.165}$$

as for the toroidal modes, but now the condition (2.164) reduces to[17]

$$J_{n-\frac{1}{2}}(k_\alpha R) = 0\,, \tag{2.166}$$

[17] This reduction requires use of the recurrence relation $xJ'_\nu(x) + \nu J_\nu(x) = xJ_{\nu-1}(x)$.

which is to be contrasted with (2.158). The decay rates for the poloidal modes are therefore

$$p_\alpha = -\eta R^{-2} x^2_{(n-1)q} \quad (n = 1, 2, \ldots; q = 1, 2, \ldots), \tag{2.167}$$

and the general solution for P, analogous to (2.160), is

$$P(r, \theta, \varphi, t) = \begin{cases} \sum_n S_n(\theta, \varphi) \sum_q B_{nq}\, r^{-1/2} J_{n+\frac{1}{2}}\left(k_{(n-1)q} r\right) e^{p_{(n-1)q}t} & (r < R) \\ \sum_n c_n(t) S_n(\theta, \varphi)\, r^{-(n+1)} & (r > R) \end{cases} \tag{2.168}$$

where, by (2.164),

$$c_n(t) = \sum_q B_{nq}\, J_{n+\frac{1}{2}}\left[k_{(n-1)q} R\right] R^{n+1/2} \exp\left[p_{(n-1)q} t\right]. \tag{2.169}$$

2.8.3 Behaviour of the dipole moment

The slowest decaying mode is the poloidal mode with $n = 1$, $q = 1$ for which (2.167) gives $p_\alpha = -\eta R^{-2} x^2_{01}$. This is a mode with dipole structure for $r > R$; if we choose the axis $\theta = 0$ to be in the direction of the dipole moment vector $\boldsymbol{\mu}^{(1)}(t)$, then clearly the angular dependence in the associated contribution to the defining scalar P involves only the particular axisymmetric surface harmonic $S_1(\theta, \varphi) = \cos\theta$.

It is interesting to enquire what happens in the case of a magnetic field which is initially totally confined to the conducting region $r < R$ (i.e. $\mathbf{B}(\mathbf{x}, 0) \equiv 0$ for $r > R$). The dipole moment of this field (as well as all the multipole moment tensors) are then evidently zero since the magnetic potential Ψ given by (2.77) must be zero to all orders for $r > R$. It is sufficient to consider the case in which the angular dependence of $\mathbf{B}(\mathbf{x}, 0)$ is the same as that of a dipole; i.e. suppose that only the term with $n = 1$ is present in the above analysis for the poloidal field. The dipole moment is clearly related to the coefficient $c_1(t)$. In fact, for $r > R$, using $\nabla^2 P = 0$, we have

$$\mathbf{B}_P = \nabla \wedge \nabla \wedge (\mathbf{x} P) = -\nabla^2(\mathbf{x} P) + \nabla\nabla \cdot (\mathbf{x} P) = \nabla\Psi, \tag{2.170}$$

where

$$\Psi = -P - (\mathbf{x} \cdot \nabla)P = -P - r\,\partial P/\partial r, \tag{2.171}$$

and with $P = c_1(t) r^{-2} \cos\theta$ (from (2.162)), this gives $\Psi = c_1(t) r^{-2} \cos\theta$ also. Hence in fact the dipole moment is

$$\boldsymbol{\mu}^{(1)}(t) = c_1(t)\, \mathbf{e}_z, \tag{2.172}$$

and its variation with time is given by (2.169) with $n = 1$. Under the assumed conditions we must have

$$c_1(0) = \sum_q B_{1q}\, J_{3/2}(k_{0q} R)\, R^{3/2} = 0. \tag{2.173}$$

For $t > 0$, $|c_1(t)|$ will depart from zero, rising to a maximum value in a time of order $R^2\eta^{-1}$, and will then again decay to zero in a time of this same order of magnitude, the term corresponding to $q = 1$ ultimately dominating.

It is important to note from this example that diffusion can result in a temporary increase in the dipole moment as well as leading to its ultimate decay if no regenerative agent is present. It is tempting to think that a linear supposition of exponentially decaying functions must inevitably decrease with time; consideration of the simple function $e^{-t} - e^{-2t}$ will remove this temptation;[18] the function $c_1(t)$ in the above example exhibits similar behaviour.

This possibility of *diffusive increase of the dipole moment* is so important that it is desirable to give it an alternative, and perhaps more transparent, formulation. To this end, we must first obtain an alternative expression for $\boldsymbol{\mu}^{(1)}$, which from (2.86a) is given by

$$8\pi\boldsymbol{\mu}^{(1)} = \int_V \mathbf{x} \wedge (\nabla \wedge \mathbf{B})\, dV\,. \tag{2.174}$$

First we decompose $\mathbf{B}$ into its poloidal dipole ingredient[19] $\mathbf{B}_1$ and the rest, $\mathbf{B}'$ say, i.e. $\mathbf{B} = \mathbf{B}_1 + \mathbf{B}'$, where $\mathbf{B}_1 = \mathcal{O}(r^{-3}), \mathbf{B}' = \mathcal{O}(r^{-4})$ as $r \to \infty$. Since $\nabla \wedge \mathbf{B}' = 0$ for $r > R$, we can rewrite (2.174) in the form

$$8\pi\boldsymbol{\mu}^{(1)} = \int_V \mathbf{x} \wedge (\nabla \wedge \mathbf{B}_1)\, dV + \int_{V_\infty} \mathbf{x} \wedge (\nabla \wedge \mathbf{B}')\, dV\,, \tag{2.175}$$

where as usual V_∞ is the whole space. The second integral can be manipulated by the divergence theorem giving

$$\int_{V_\infty} \mathbf{x} \wedge (\nabla \wedge \mathbf{B}')\, dV = \int_{S_\infty} (\mathbf{x} \wedge (\mathbf{n} \wedge \mathbf{B}') + 2(\mathbf{n} \cdot \mathbf{B}')\mathbf{x})\, dS\,, \tag{2.176}$$

and since $\mathbf{B}' = O(r^{-4})$ at infinity this integral vanishes as expected. Similarly the first integral may be transformed using the divergence theorem and we obtain

$$8\pi\boldsymbol{\mu}^{(1)} = \int_S \mathbf{x} \wedge (\mathbf{n} \wedge \mathbf{B}_1)\, dS + 2\int_V \mathbf{B}_1\, dV\,, \tag{2.177}$$

where S is the surface $r = R$. Now $\mathbf{n} \wedge \mathbf{B}_1$ is continuous across $r = R$, and on $r = R+$, $\mathbf{B}_1 = \nabla(\boldsymbol{\mu}^{(1)} \cdot \nabla)r^{-1}$; the surface integral in (2.177) may then be readily calculated and it is in fact equal to $(8\pi/3)\boldsymbol{\mu}^{(1)}$; hence (2.177) becomes

$$\boldsymbol{\mu}^{(1)} = \frac{3}{8\pi}\int_V \mathbf{B}_1\, dV\,. \tag{2.178}$$

[18] *Transient instability* of flows such as plane Couette flow results from a similar superposition of damped (i.e. stable) normal modes (Landahl 1980; Moffatt 2010).

[19] By 'dipole ingredient' we shall mean the ingredient having the same angular dependence as a dipole field.

The rate of change of $\boldsymbol{\mu}^{(1)}$ is therefore given by

$$\frac{8\pi}{3}\frac{\mathrm{d}\boldsymbol{\mu}^{(1)}}{\mathrm{d}t} = \int_V \frac{\partial \mathbf{B}_1}{\partial t}\,\mathrm{d}V = \eta\int_V \nabla^2\mathbf{B}_1\,\mathrm{d}V = \eta\int_S (\mathbf{n}\cdot\nabla)\mathbf{B}_1\,\mathrm{d}S\ . \tag{2.179}$$

Hence change in $\boldsymbol{\mu}^{(1)}$ can be attributed directly to the diffusion of $\mathbf{B}_1$ across S due to the normal gradient $(\mathbf{n}\cdot\nabla)\,\mathbf{B}_1$ on $r = R-$.

In the case of a sphere, $\int_V \mathbf{B}_1\,\mathrm{d}V = \int_V \mathbf{B}\,\mathrm{d}V$, and (2.178) becomes

$$\boldsymbol{\mu}^{(1)} = \frac{3}{8\pi}\int_V \mathbf{B}\,\mathrm{d}V\,. \tag{2.180}$$

2.9 Fields exhibiting Lagrangian chaos

As mentioned in §2.1, the **B**-lines of a general three-dimensional magnetic field need not be confined to a family of surfaces; only fields having some degree of symmetry, like the axisymmetric field described in §2.5 (see Figure 2.10) have this property. A family of fields confined to a sphere whose field lines do not generally lie on surfaces, but which in fact exhibit the phenomenon of 'Lagrangian chaos', has been studied by Bajer & Moffatt (1990). This family consists of fields quadratic in the Cartesian coordinates $x_i = (x, y, z)$, given in suffix notation by

$$B_i = a_i + b_{ij}x_j + c_{ijk}x_jx_k\,, \tag{2.181}$$

where the coefficients a_i, b_{ij}, c_{ijk} are chosen in such a way that

$$\nabla\cdot\mathbf{B} = 0, \quad \text{and} \quad \mathbf{n}\cdot\mathbf{B} = 0 \ \text{on} \ |\mathbf{x}| = 1\,. \tag{2.182}$$

A particular family studied in some detail was

$$\mathbf{B} = \left(\alpha z - 8xy,\, 11x^2 + 3y^2 + z^2 + xz - 3,\, -\alpha x + 2yz - xy\right), \tag{2.183}$$

where α is a parameter that can be freely chosen. It can be readily verified that the conditions (2.182) are satisfied. Note that $\mathbf{n}\wedge\mathbf{B} \neq 0$ on $r = |\mathbf{x}| = 1$, so that, if $\mathbf{B} = 0$ for $r > 1$, the field (2.183) is necessarily associated with a surface current $\mathbf{J}_S$ on $r = 1$, as given by (2.72).

The **B**-lines of the field (2.183) are found by integrating the equations

$$\left.\begin{aligned}\mathrm{d}x/\mathrm{d}t &= B_x = \alpha z - 8xy\\ \mathrm{d}y/\mathrm{d}t &= B_y = 11x^2 + 3y^2 + z^2 + xz - 3\\ \mathrm{d}z/\mathrm{d}t &= B_z = -\alpha x + 2yz - xy\end{aligned}\right\} \tag{2.184}$$

with an initial condition $\mathbf{x} = (x_0, y_0, z_0)$ at $t = 0$, where t is simply a parameter on the chosen field line. Figure 2.11a shows a typical **B**-line for the choice $\alpha = 0.05$. Although the field (2.183) is perfectly smooth locally, its global structure is evidently quite complex. The field line looks as if it may be confined to a surface,

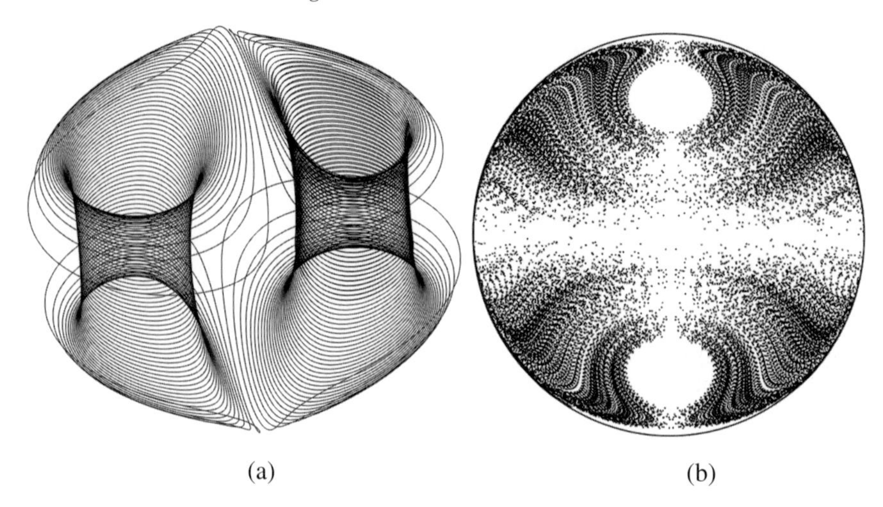

Figure 2.11 (a) Field line for the field (2.183), with $\alpha = 0.05$, for initial condition $(x_0, y_0, z_0) = (0.5, 0.5, 0.5)$ and for $0 < t < 200$. (b) Poincaré section for the same field: the diagram shows points where a single field line makes 40,000 successive crossings of a diametral plane. The fact that these points do not lie on a curve provides clear evidence that the field line is not confined to a surface. [From Bajer & Moffatt 1990.]

but in fact it is not so confined. To see this, it is necessary to integrate for a much longer time.

Field lines such as that illustrated by Figure 2.11a cross any diametral plane an arbitrarily large number of times if followed far enough. If each point of intersection on such a plane is marked with a point, the resulting diagram, known as a 'Poincaré section', provides an indication of the space-filling character (or degree of 'chaos') of the field. Figure 2.11b shows such a Poincaré section after 40,000 traverses of a diametral plane. The fact that the points are distributed over a large area of the plane is an indication of 'non-integrability' of the system (2.184), a phenomenon known by the term 'Lagrangian chaos'. In such a field, the field lines passing through two neighbouring points in the chaotic region diverge exponentially if followed far enough (in either direction!) away from these points. A magnetic tube of force will similarly diverge, so the concept of a tube of force has at best only local significance when the field exhibits this type of chaotic behaviour.

2.10 Knotted flux tubes

2.10.1 Twist surgery

As shown in §2.1, two untwisted, unknotted, but linked magnetic flux tubes contribute an amount $\pm\mathcal{N}\Phi_1\Phi_2$ to the helicity of the field, where $\mathcal{N}$ is the linking

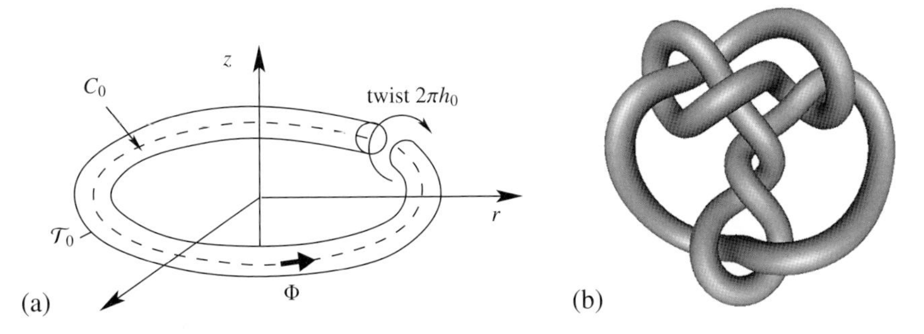

Figure 2.12 (a) Flux tube $\mathcal{T}_0$ with circular axis, carrying flux Φ. Helicity $\mathcal{H} = h_0\Phi^2$ is introduced by cutting the tube at a section φ =const., twisting a free end through an angle $2\pi h_0$, and reconnecting. (b) Flux tube in the form of an arbitrary knot; arbitrary internal twist may be added. [From Scharein: knotplot.]

number of the two tubes, and the sign is chosen according as the linkage is right- or left-handed. The situation concerning knotted tubes is more subtle (Moffatt & Ricca 1992).

Consider first a circular flux tube (Figure 2.12a), within which the field lines are unlinked circles parallel to the tube axis. Imagine that we cut the tube, give one of the free ends a right-handed twist through an angle 2π, and reconnect the tube. Each field line now encircles the tube axis once in its passage round the tube. We can think of this twisted tube as being built up in stages from the axis outwards by adding incremental flux $\delta\Phi_1$ to the flux Φ_1 already accumulated; the corresponding increment in helicity is $\delta\mathcal{H} = 2\Phi_1\delta\Phi_1$, and so the total helicity of the twisted tube (Berger & Field 1984) is

$$\mathcal{H} = \int_0^{\Phi} 2\Phi_1 \mathrm{d}\Phi_1 = \Phi^2 . \tag{2.185}$$

If we apply a twist $2\pi h$ instead of 2π, where h is any real number, it is obvious that the helicity of the twisted tube, being proportional to this twist, must now be

$$\mathcal{H} = h\,\Phi^2 . \tag{2.186}$$

If h is rational, i.e. $h = m/n$ with m, n co-prime integers, then each field line is a closed curve (a torus knot $K_{m,n}$). If h is irrational, then each field line covers a toroidal surface (cf. the discussion of §2.5.1). By this means, we may inject an arbitrary amount of 'internal helicity' into any flux tube, an operation that may be appropriately described as 'twist surgery' .

2.10.2 Helicity of a knotted flux tube

Consider now a magnetic field confined to a flux tube T whose axis C is knotted in the form of a knot K (Figure 2.12b).[20] It is clear that the helicity of the field is not determined solely by the knot topology, because this helicity may be augmented (or diminished) by an arbitrary amount by twist surgery, as described above. In order to make progress, we first need some elementary differential geometry.

Let s represent arc length on C measured from some point O, and let the point on C with parameter s be $\mathbf{x} = \mathbf{X}(s)$; then for each s, $\mathbf{X}(s+L) = \mathbf{X}(s)$ where L is the length of C. By analogy with the Gauss integral (2.17), we define the 'writhe' Wr of C by

$$Wr = \frac{1}{4\pi}\oint_C \oint_C \frac{(\mathbf{x}-\mathbf{x}')\cdot(\mathrm{d}\mathbf{x}\wedge\mathrm{d}\mathbf{x}')}{|\mathbf{x}-\mathbf{x}'|^3} . \tag{2.187}$$

Despite the apparent singularity at $\mathbf{x} = \mathbf{x}'$, this integral is convergent, because the numerator also has a triple zero at this point. Unlike the Gauss integral, Wr is not a topological invariant. It has to be augmented by a twist contribution which may be found as follows.

The unit tangent vector on C is $\mathbf{t} = \mathrm{d}\mathbf{X}/\mathrm{d}s$. The curvature $c(s)$ is defined by $c(s) = |\mathrm{d}\mathbf{t}/\mathrm{d}s| \geq 0$ and the unit principal normal $\mathbf{n}(s)$ is defined (at points of C where $c > 0$) by the equation

$$\mathrm{d}\mathbf{t}/\mathrm{d}s = c\,\mathbf{n} . \tag{2.188}$$

The plane defined by the unit vectors $\{\mathbf{t},\mathbf{n}\}$ is the 'osculating plane'; the circle on this plane with centre at $\mathbf{X}(s) + c^{-1}\mathbf{n}(s)$ and radius c^{-1} makes three-point contact with C at $\mathbf{X}(s)$, and is the 'circle of curvature' at $\mathbf{X}(s)$.

The unit binormal $\mathbf{b}(s)$ is defined by $\mathbf{b}(s) = \mathbf{t}(s)\wedge\mathbf{n}(s)$, so that $\{\mathbf{t},\mathbf{n},\mathbf{b}\}$ is a right-handed orthogonal triad of unit vectors (the 'Frenet triad') defined at each point of C where $c \neq 0$. Any point where $c = 0$ is an 'inflexion point' of the curve. In general, a closed curve C will not have any inflexion points. However if a curve is deformed continuously in time, then it may contain one or more inflexion points at particular instants; we say that it 'passes through an inflexional configuration' at such instants. As s increases along C, the Frenet triad rotates continuously provided $c \neq 0$; but if we pass through an inflexion point, the triad 'flips' through an angle π about the tangent vector $\mathbf{t}$, because the normal $\mathbf{n}$ points towards the centre of curvature which moves 'to infinity' on one side of C (on the osculating plane) and reappears 'from infinity' on the other side.

Obviously, $\mathbf{n}$ changes direction along C; if C is a plane curve, then $\mathrm{d}\mathbf{n}/\mathrm{d}s = -c\,\mathbf{t}$ (complementary to (2.188)). However, in general C twists out of the plane, and

[20] This figure is from Robert Scharein's 'knotplot' – see <http://knotplot.com> – which contains a mine of useful information.

there is then an additional term:

$$\mathrm{d}\mathbf{n}/\mathrm{d}s = -c\,\mathbf{t} + \tau\,\mathbf{b}\,, \tag{2.189}$$

where $\tau(s)$ is the 'torsion' of C, a measure of this twisting. Coupled with this is the equation

$$\mathrm{d}\mathbf{b}/\mathrm{d}s = -\tau\,\mathbf{n}\,. \tag{2.190}$$

Note that it follows from (2.189) that $\mathbf{n(s)} \wedge \mathbf{n'(s)} = c\,\mathbf{b} + \tau\,\mathbf{t}$, so that

$$\tau = \mathbf{n(s)} \wedge \mathbf{n'(s)} \cdot \mathbf{t(s)}\,; \tag{2.191}$$

this may be adopted as the definition of τ. The three equations (2.188)–(2.190) describing the rotation of the $\{\mathbf{t}, \mathbf{n}, \mathbf{b}\}$ triad as s increases round C are known as the 'Frenet–Serret equations'. They may be written in matrix notation

$$\frac{\mathrm{d}}{\mathrm{d}s}\begin{pmatrix} \mathbf{t} \\ \mathbf{n} \\ \mathbf{b} \end{pmatrix} = \begin{pmatrix} 0 & c & 0 \\ -c & 0 & \tau \\ 0 & -\tau & 0 \end{pmatrix}\begin{pmatrix} \mathbf{t} \\ \mathbf{n} \\ \mathbf{b} \end{pmatrix}. \tag{2.192}$$

Now consider a 'ribbon' whose edges are the curve C and a curve C_ε with parametric representation $\mathbf{x} = \mathbf{X}(s) + \varepsilon\,\mathbf{n(s)}$, where ε is the width of the ribbon, assumed small. The total twist of this ribbon in going round the curve is

$$\mathcal{T} = \frac{1}{2\pi}\oint_C \tau(s)\,\mathrm{d}s = \frac{1}{2\pi}\oint_C \mathbf{n(s)} \wedge \mathbf{n'(s)} \cdot \mathbf{t(s)}\,\mathrm{d}s\,. \tag{2.193}$$

Suppose now that we cut this ribbon, give one of the cut ends a right-handed rotation through an angle $2\pi\mathcal{N}$ (with $\mathcal{N}$ an integer), and reconnect. The total twist, denoted Tw, is now

$$Tw = \mathcal{T} + \mathcal{N}. \tag{2.194}$$

It is a well-known result of differential geometry (Călugăreanu 1959) that the sum of writhe and twist $Wr + Tw$ is invariant under continuous deformation of the ribbon, in fact it is just the linking number of C and C_ε in the limit $\varepsilon \to 0$. As shown by Moffatt & Ricca (1992), for a uniformly twisted flux tube, the coefficient h in (2.186) is precisely

$$h = Wr + Tw = Wr + \mathcal{T} + \mathcal{N}\,. \tag{2.195}$$

Thus the helicity of a knotted flux tube is invariant under continuous deformation of the tube. We shall show in the following chapter that this is a special case of a more general result concerning the invariance of magnetic helicity in a perfectly conducting fluid.

It is interesting to enquire what happens when the curve C (and with it the tube

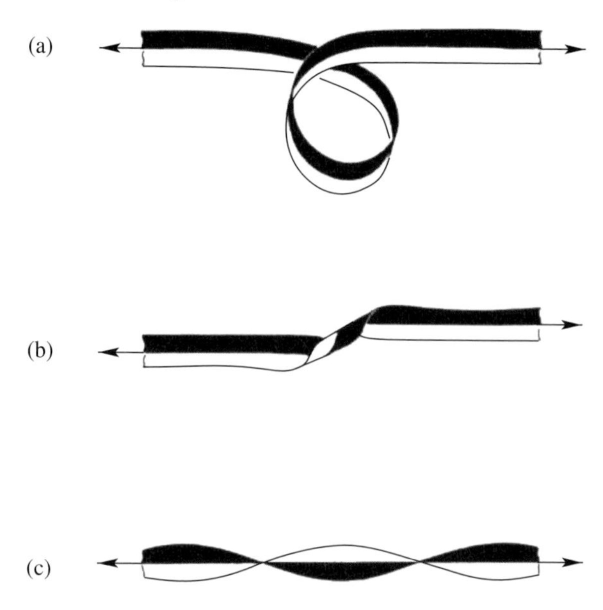

Figure 2.13 Conversion of writhe to twist by continuous distortion. In (a), the ribbon lies nearly in a plane, $Wr = 1$, $\mathcal{T} = 0$ and $\mathcal{N} = 0$ (imagine the ribbon closed by a large circle in the plane); in (c), $Wr = 0$, $\mathcal{T} = 0$ and $\mathcal{N} = 1$. Since $\mathcal{N}$ is an integer (zero or one) throughout the distortion, there must be an instant when $\mathcal{N}$ jumps discontinuously from zero to one, and when $\mathcal{T}$ makes a compensating jump from one to zero; this can only be achieved by passage through an inflexional configuration. [From Moffatt & Ricca 1992.]

T) is distorted through an inflexional configuration. Both Wr and Tw vary continuously through such distortion, but τ has an integrable singularity which is just such that $\mathcal{T}$ jumps by ± 1 in the passage through the inflexion point, this jump being compensated by an equal and opposite jump ∓ 1 in the winding number $\mathcal{N}$ of lines of force in the tube, relative to the Frenet frame on C. This behaviour is well illustrated by the diagram of Figure 2.13 (from Moffatt & Ricca 1992, where these results are proved), which shows a possible distortion of a portion of a ribbon, under which writhe is continuously converted to twist, the sum necessarily remaining constant.

3

Advection, Distortion and Diffusion

3.1 Alfvén's theorem and related results

In this section we shall consider certain basic properties of the equations derived in the previous chapter in the idealised limit of perfect conductivity, $\sigma \to \infty$, or equivalently $\eta \to 0$. First, from eqns. (2.122) and (2.131), we have that

$$d\Phi/dt = -\oint_{C_L} \sigma^{-1}\mathbf{J}\cdot d\mathbf{x}\,, \tag{3.1}$$

so that in the limit $\sigma \to \infty$, provided $\mathbf{J}$ remains finite on C_L, $\Phi =$ cst. This applies to every closed material curve C_L. In particular, consider a flux tube consisting of the aggregate of lines of force through a small closed curve. Since any and every curve embracing the flux tube conserves its flux as it moves with the fluid, it is a linguistic convenience to say that the flux tube itself moves with the fluid (or is *frozen* in the fluid) and that its flux is conserved. This is Alfvén's theorem (Alfvén 1942), analogous to Kelvin's circulation theorem in inviscid fluid dynamics.

A more formal derivation of the 'frozen-field' property can of course be devised. When $\eta = 0$, the right-hand side of (2.134) can be expanded, giving the equivalent equation

$$\mathrm{D}\mathbf{B}/\mathrm{D}t = (\mathbf{B}\cdot\nabla)\,\mathbf{u} - \mathbf{B}\,(\nabla\cdot\mathbf{u})\,. \tag{3.2}$$

If we combine this with (2.137), we obtain

$$\frac{\mathrm{D}}{\mathrm{D}t}\left(\frac{\mathbf{B}}{\rho}\right) = \frac{1}{\rho}\frac{\mathrm{D}\mathbf{B}}{\mathrm{D}t} - \frac{\mathbf{B}}{\rho^2}\frac{\mathrm{D}\rho}{\mathrm{D}t} = \left(\frac{\mathbf{B}}{\rho}\cdot\nabla\right)\mathbf{u}\,. \tag{3.3}$$

Hence $\mathbf{B}/\rho$ satisfies the same equation as that satisfied by the line element $d\mathbf{x}(\mathbf{a},t)$ (2.117) and the solution (c.f. (2.115)) is therefore

$$\frac{B_i(\mathbf{x},t)}{\rho(\mathbf{x},t)} = \frac{B_j(\mathbf{a},0)}{\rho(\mathbf{a},0)}\frac{\partial x_i}{\partial a_j}\,, \tag{3.4}$$

a result essentially due to Cauchy.[1] Suppose now that the line element $\mathrm{d}\mathbf{a}$ at time $t = 0$ is directed along a line of force of the field $\mathbf{B}(\mathbf{a}, 0)$, so that $\mathbf{B}(\mathbf{a}, 0) \wedge \mathrm{d}\mathbf{a} = 0$. Then for times $t > 0$, using (2.115), (3.4) and (2.141) we have

$$[\mathbf{B}(\mathbf{x}, t) \wedge \mathrm{d}\mathbf{x}(\mathbf{a}, t)]_i = \epsilon_{ijk} B_j(\mathbf{x}, t) \mathrm{d}x_k(\mathbf{a}, t) = \epsilon_{ijk} \frac{\rho(\mathbf{x}, t)}{\rho(\mathbf{a}, 0)} B_m(\mathbf{a}, 0) \frac{\partial x_j}{\partial a_n} \frac{\partial x_k}{\partial a_n} \mathrm{d}a_n$$
$$= \epsilon_{lmn} \frac{\partial a_l}{\partial x_i} B_m(\mathbf{a}, 0) \mathrm{d}a_n \,. \tag{3.5}$$

Hence

$$[\mathbf{B}(\mathbf{x}, t) \wedge \mathrm{d}\mathbf{x}(\mathbf{a}, t)]_i = [\mathbf{B}(\mathbf{a}, 0) \wedge \mathrm{d}\mathbf{a}]_l \partial a_l / \partial x_i = 0 \,, \tag{3.6}$$

so that the line element $\mathrm{d}\mathbf{x}$ is directed along a line of force $\mathbf{B}(\mathbf{x}, t)$. It follows that if a material curve C_L coincides with a $\mathbf{B}$-line at time $t = 0$, then when $\eta = 0$ it coincides with a $\mathbf{B}$-line also for all $t > 0$, and each $\mathbf{B}$-line may therefore be identified in this way with a material curve.

It is likewise clear from the above discussion that, when $\eta = 0$, any motion which stretches a line element $\mathrm{d}\mathbf{x}$ on a line of force will increase $\mathbf{B}/\rho$ proportionately. In an incompressible flow ($\mathrm{D}\rho/\mathrm{D}t = 0$) this means that stretching of $\mathbf{B}$-lines implies proportionate field intensification. This need not be the case in compressible flow; for example, in a uniform spherically symmetric expansion, with velocity field $\mathbf{u} = (\alpha r, 0, 0)$ ($\alpha > 0$) in spherical polar coordinates (r, θ, φ), any material line element increases in magnitude linearly with its distance r from the origin, while the density of a fluid element at position $r(t)$ decreases as r^{-3}; hence the field $\mathbf{B}$ following a fluid element *decreases* as r^{-2}. Conversely in a spherically symmetric contraction ($\alpha < 0$), the field following a fluid element *increases* as r^{-2}.

3.1.1 Conservation of magnetic helicity

The fact that $\mathbf{B}$-lines are frozen in the fluid implies that the topological structure of the field cannot change with time. One would therefore expect the integrals $\mathcal{H}_m$ defined by (2.13) to remain constant under any kinematically possible fluid motion, when $\eta = 0$. The following is a generalisation (Moffatt 1969) of a result anticipated by Woltjer (1958) .

We first obtain an expression for $\mathrm{D}(\rho^{-1}\mathbf{A} \cdot \mathbf{B})/\mathrm{D}t$. From (2.132) with $\eta = 0$ we have

$$\frac{\mathrm{D}A_i}{\mathrm{D}t} \equiv \frac{\partial A_i}{\partial t} + u_j \frac{\partial}{\partial x_j} A_i = u_j \frac{\partial}{\partial x_i} A_j - \frac{\partial \phi}{\partial x_i} \,. \tag{3.7}$$

[1] Cauchy obtained the equivalent result in the context of the vorticity equation for inviscid flow – see §3.2 below; for the history of this discovery, see Frisch & Villone (2014).

Combining this with (3.3), we have

$$\frac{\mathrm{D}}{\mathrm{D}t}\left(\frac{\mathbf{A}\cdot\mathbf{B}}{\rho}\right) = \mathbf{A}\cdot\frac{\mathrm{D}}{\mathrm{D}t}\left(\frac{\mathbf{B}}{\rho}\right) + \left(\frac{\mathbf{B}}{\rho}\right)\cdot\frac{\mathrm{D}\mathbf{A}}{\mathrm{D}t} = \left(\frac{\mathbf{B}}{\rho}\cdot\nabla\right)(\mathbf{A}\cdot\mathbf{u}-\phi)\,. \tag{3.8}$$

Now let S_m (interior V_m) be a material surface on which (permanently) $\mathbf{n}\cdot\mathbf{B} = 0$; then, since $\mathrm{D}(\rho\mathrm{d}V)/\mathrm{D}t = 0$,

$$\frac{\mathrm{d}\mathcal{H}_m}{\mathrm{d}t} = \int_{V_m}\frac{\mathrm{D}}{\mathrm{D}t}\left(\frac{\mathbf{A}\cdot\mathbf{B}}{\rho}\right)\rho\,\mathrm{d}V = \int_{V_m}(\mathbf{B}\cdot\nabla)(\mathbf{A}\cdot\mathbf{u}-\phi)\,\mathrm{d}V, \tag{3.9}$$

and hence

$$\frac{\mathrm{d}\mathcal{H}_m}{\mathrm{d}t} = \int_{S_m}(\mathbf{n}\cdot\mathbf{B})(\mathbf{A}\cdot\mathbf{u}-\phi)\,\mathrm{d}S = 0\,, \tag{3.10}$$

so that $\mathcal{H}_m$ is, as expected, constant.

If $\eta \neq 0$, this result is of course no longer true; for a localised field with total helicity $\mathcal{H}$, retention of the diffusion terms leads to the equation

$$\frac{\mathrm{d}\mathcal{H}}{\mathrm{d}t} = \eta\int_{V_\infty}(\mathbf{B}\cdot\nabla^2\mathbf{A} + \mathbf{A}\cdot\nabla^2\mathbf{B})\,\mathrm{d}V\,. \tag{3.11}$$

Now

$$\int\mathbf{A}\cdot\nabla^2\mathbf{B}\,\mathrm{d}V = \int\mathbf{B}\cdot\nabla^2\mathbf{A}\,\mathrm{d}V = -\int\mathbf{B}\cdot\nabla\wedge\mathbf{B}\,\mathrm{d}V\,, \tag{3.12}$$

the integrals being over all space. Hence (3.11) becomes

$$\frac{\mathrm{d}\mathcal{H}}{\mathrm{d}t} = -2\eta\int\mathbf{B}\cdot\nabla\wedge\mathbf{B}\,\mathrm{d}V\,. \tag{3.13}$$

The integral $\int\mathbf{B}\cdot\nabla\wedge\mathbf{B}\,\mathrm{d}V$ is the helicity of the field $\nabla\wedge\mathbf{B}$; it has been described by Hide (2002) as the 'superhelicity' of $\mathbf{B}$ (or sometimes 'current helicity').

Equation (3.13) describes the change with time of $\mathcal{H}$ under the action of diffusion. This implies a corresponding change in the topological structure of the field, and there is no way in which a particular line of force can be 'followed' unambiguously from one instant to the next. Attempts have sometimes been made to define an effective 'velocity of slip' $\mathbf{w}(\mathbf{x}, t)$ of field lines relative to fluid due to the action of diffusion; if such a concept were globally valid then field evolution when $\eta \neq 0$ would be equivalent to field evolution in a non-diffusive fluid with velocity field $\mathbf{u} + \mathbf{w}$; this would imply conservation of all knots and linkages in field lines which is inconsistent in general with (3.13); it must be concluded that the concept of a 'velocity of slip', although physically appealing, is also dangerously misleading when complicated field structures are considered.

3.2 The analogy with vorticity

The induction equation (2.134),

$$\partial \mathbf{B}/\partial t = \nabla \wedge (\mathbf{u} \wedge \mathbf{B}) + \eta \nabla^2 \mathbf{B} \quad (\nabla \cdot \mathbf{B} = 0), \tag{3.14}$$

bears a close formal resemblance to the equation for the vorticity $\omega = \nabla \wedge \mathbf{u}$ in the flow of a barotropic fluid (for which pressure depends only on density, $p = p(\rho)$) under conservative body forces, viz.

$$\partial \omega/\partial t = \nabla \wedge (\mathbf{u} \wedge \omega) + \nu \nabla^2 \omega \quad (\nabla \cdot \omega = 0)\,, \tag{3.15}$$

where ν is the kinematic viscosity. The analogy, first pointed out by Elsasser (1946) and exploited by Batchelor (1950) in the consideration of the action of turbulence on a weak random magnetic field, is a curious one, in that, since ω is related to $\mathbf{u}$ through $\omega = \nabla \wedge \mathbf{u}$, (3.15) is a non-linear equation for the evolution of ω, whereas (3.14) is undoubtedly linear in $\mathbf{B}$ when $\mathbf{u}(\mathbf{x}, t)$ is regarded as given. The fact that ω is restricted (through its additional relationship to $\mathbf{u}$) while $\mathbf{B}$ is not, means that the analogy has a sort of one-way character: general results obtained on the basis of (3.14) relating to the $\mathbf{B}$-field usually have a counterpart in the *more particular* context of (3.15). By contrast, results obtained on the basis of (3.15) may not have a counterpart in the *more general* context of (3.14).

We have already noted the parallel between Alfvén's theorem when $\eta = 0$ and Kelvin's circulation theorem when $\nu = 0$. We have also, when $\nu = 0$, results analogous to the further theorems of §3.1, viz.

(i) In the notation of §3.1,

$$\frac{\omega_i(\mathbf{x}, t)}{\rho(\mathbf{x}, t)} = \frac{\omega_j(\mathbf{a}, 0)}{\rho(\mathbf{a}, 0)} \frac{\partial x_i}{\partial a_j}\,. \tag{3.16}$$

This result, due to Cauchy, is sometimes described as the 'solution of the vorticity equation': this is perhaps a little misleading, since in the vorticity context $\partial x_i/\partial a_j$ is not known until $\mathbf{x}(\mathbf{a}, t)$ is known, and this can be determined only after $\omega(\mathbf{x}, t)$ is determined. Equation (3.16), far from providing a solution of (3.15) (with $\nu = 0$), is rather a reformulation of the equation. Contrast the situation in the magnetic context where $\partial x_i/\partial a_j$ and $B_i(\mathbf{x}, t)$ are truly independent (in so far as Lorentz forces are negligible) and where (3.4) provides a genuine solution of (3.2) or (3.3).

(ii) The integral

$$\mathcal{H}_m\{\omega\} = \int_{V_m} \mathbf{u} \cdot \omega \, \mathrm{d}V \tag{3.17}$$

is constant if $\nu = 0$ and $\omega \cdot \mathbf{n} = 0$ on the material surface S_m of V_m (Moffatt 1969). This integral admits interpretation in terms of linkages of vortex tubes (exactly as in §2.1), and conservation of $\mathcal{H}_m\{\omega\}$ is of course attributable to the fact that, when $\nu = 0$ and $p = p(\rho)$, vortex lines are frozen in the fluid. $\mathcal{H}_m\{\omega\}$ is the helicity of the velocity field within V_m; we shall use the term 'kinetic helicity' to distinguish it from the magnetic helicity $\mathcal{H}_m\{\mathbf{B}\}$ already introduced. Kinetic helicity is of profound importance in dynamo theory, as will become apparent in later chapters.

The relative importance of the two terms on the right of (3.14) is given by the well-known Reynolds number $\mathcal{R}e$ of conventional fluid mechanics; if u_0 is a typical scale for the velocity field $\mathbf{u}$, and ℓ_0 is a typical length-scale over which it varies, then

$$|\nabla \wedge (\mathbf{u} \wedge \omega)|/|\nu \nabla^2 \omega| = \mathcal{O}(\mathcal{R}e)\,, \quad \text{where} \quad \mathcal{R}e = u_0 \ell_0/\nu\,. \tag{3.18}$$

Similarly, if ℓ_0 is the scale of variation of $\mathbf{B}$ as well as of $\mathbf{u}$, then the ratio of the two terms on the right of (3.14) is

$$|\nabla \wedge (\mathbf{u} \wedge \mathbf{B})|/|\eta\, \nabla^2 \mathbf{B}| = \mathcal{O}(\mathcal{R}_{\mathrm{m}})\,, \quad \text{where} \quad \mathcal{R}_{\mathrm{m}} = u_0\, \ell_0/\eta = \mu_0 \sigma u_0 \ell_0\,. \tag{3.19}$$

$\mathcal{R}_{\mathrm{m}}$ is known as the *magnetic Reynolds number*, and it can be regarded as a dimensionless measure of the fluid conductivity in a given flow situation. If $\mathcal{R}_{\mathrm{m}} \gg 1$, then the diffusion term is relatively unimportant, and the frozen-field picture of §3.1 should be approximately valid. If $\mathcal{R}_{\mathrm{m}} \ll 1$, then diffusion dominates, and the ability of the flow to distort the field from whatever distribution it would have under the action of diffusion alone is severely limited.

These conclusions are of course of an extremely preliminary nature and will require modification in particular contexts. Two situations where the estimate (3.19) will be misleading may perhaps be anticipated. First, if the scale L of $\mathbf{B}$ is much greater than the scale ℓ_0 of $\mathbf{u}$, then

$$|\nabla \wedge (\mathbf{u} \wedge \mathbf{B})|/|\eta\, \nabla^2 \mathbf{B}| = \mathcal{O}(\mathcal{R}_{\mathrm{m}} L/\ell_0)\,, \quad \mathcal{R}_{\mathrm{m}} = u_0\, \ell_0/\eta\,. \tag{3.20}$$

Hence, even if $\mathcal{R}_{\mathrm{m}} \ll 1$, the induction term $\nabla \wedge (\mathbf{u} \wedge \mathbf{B})$ may nevertheless be of dominant importance if L/ℓ_0 is sufficiently large. Secondly, in any region of rapid variation of $\mathbf{B}$ (e.g. across a thin diffusing current sheet) the relevant scale δ of $\mathbf{B}$ may be *small* compared with ℓ_0; in this case we can have

$$|\nabla \wedge (\mathbf{u} \wedge \mathbf{B})|/|\eta\, \nabla^2 \mathbf{B}| = \mathcal{O}(\mathcal{R}_{\mathrm{m}} \delta/\ell_0) \ \text{ or } \ \mathcal{O}\left(\mathcal{R}_{\mathrm{m}} (\delta/\ell_0)^2\right)\,, \tag{3.21}$$

depending on the precise geometry of the situation. In such layers of rapid change, diffusion can be important even when $\mathcal{R}_{\mathrm{m}} \gg 1$. In any event, care is generally needed in the use that is made of estimates of the type (3.19), which should always be subject to retrospective verification.

3.3 The analogy with scalar transport

A further analogy that is sometimes illuminating (Batchelor 1952) is that between equation (3.14) for the 'transport' of the 'vector contaminant' $\mathbf{B}(\mathbf{x}, t)$, and the equation

$$\mathrm{D}\Theta/\mathrm{D}t \equiv \partial\Theta/\partial t + \mathbf{u}\cdot\nabla\Theta = \kappa\nabla^2\Theta\,, \tag{3.22}$$

which describes the transport of a scalar contaminant $\Theta(\mathbf{x}, t)$ (which may be, for example, temperature or dye concentration) subject to molecular diffusivity κ. The vector $\mathbf{G} = \nabla\Theta$ satisfies the equation

$$\partial\mathbf{G}/\partial\mathbf{t} = -\nabla(\mathbf{u}\cdot\mathbf{G}) + \kappa\nabla^2\mathbf{G} \quad (\nabla\wedge\mathbf{G} = \mathbf{0})\,, \tag{3.23}$$

which is the counterpart of (3.14) for an irrotational (rather than a solenoidal) vector field.

When $\kappa = 0$, the Lagrangian solution of (3.22) is simply

$$\Theta(\mathbf{x}, t) = \Theta(\mathbf{a}, 0)\,, \tag{3.24}$$

and surfaces of constant Θ are frozen in the field. The counterpart of the magnetic Reynolds number is the Péclet number

$$\mathcal{P}e = u_0\ell_0/\kappa\,, \tag{3.25}$$

and diffusion is dominant or negligible according as $\mathcal{P}e \ll$ or $\gg 1$.

3.4 Maintenance of a flux rope by uniform irrotational strain

A simple illustration of the combined effects of advection and diffusion is provided by the action on a magnetic field of the irrotational incompressible velocity field

$$\mathbf{u} = (\alpha x, \beta y, \gamma z), \quad \alpha + \beta + \gamma = 0\,. \tag{3.26}$$

The rate of strain tensor $\partial u_i/\partial x_j$ is uniform and its principal values are α, β, γ. We suppose further that α is positive and β and γ negative so that all fluid line elements tend to become aligned parallel to the x-axis. Likewise $\mathbf{B}$-lines tend to become aligned in the same way, so let us suppose that

$$\mathbf{B} = (B(y, z, t), 0, 0)\,. \tag{3.27}$$

Equation (3.14) then has only an x-component which becomes

$$\frac{\partial B}{\partial t} + \beta y\frac{\partial B}{\partial y} + \gamma z\frac{\partial B}{\partial z} = \alpha B + \eta\nabla^2 B\,, \tag{3.28}$$

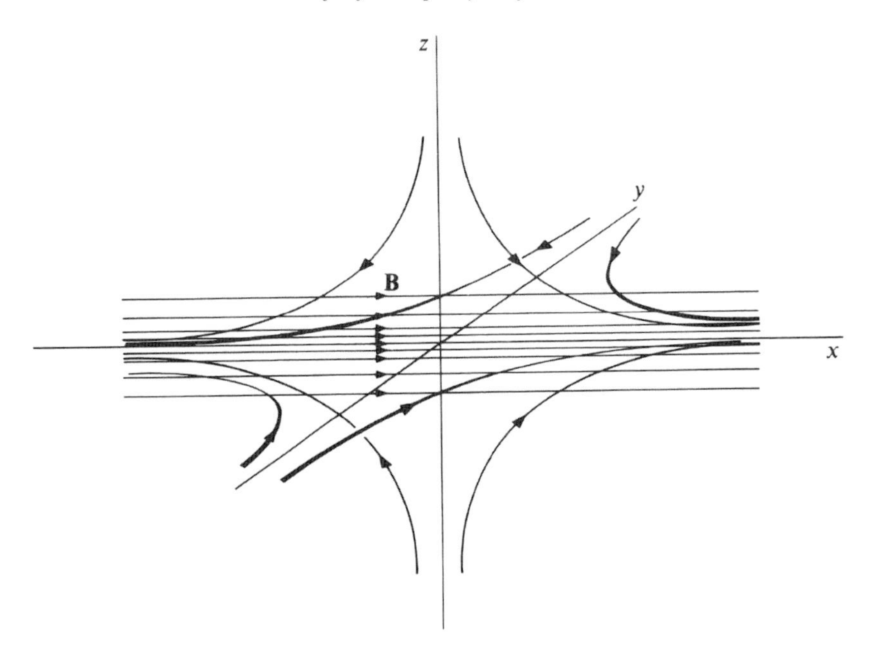

Figure 3.1 Flux rope maintained by the action of the uniform straining action (3.26).

an equation studied in various special cases by Clarke (1964, 1965). It may be easily verified that (3.28) admits the steady solution

$$B(y,z) = B_0 \exp\{-(|\beta|y^2 + |\gamma|z^2)/2\eta\}\,, \tag{3.29}$$

representing a flux rope of elliptical structure aligned along the x-axis (Figure 3.1). The total flux in the rope is

$$\Phi = \int_{-\infty}^{\infty}\int_{-\infty}^{\infty} B(y,z)\,dy\,dz = 2\pi B_0\eta(\beta\gamma)^{-1/2}\,. \tag{3.30}$$

Advection of the field towards the axis is exactly balanced by diffusion outwards. It is in fact not difficult to show by Fourier transform methods that (3.29) is the asymptotic steady solution of (3.28) for arbitrary initial conditions; the constant B_0 is related, as in (3.30), to the total initial flux of **B** across any plane $x =$ const., a quantity that is conserved during the subsequent stretching and diffusion process.

If the velocity field is axisymmetric about the x-axis, then $\beta = \gamma = -\frac{1}{2}\alpha$, and in this case (3.29) becomes

$$B(y,z) = B_0 \exp\left\{-\alpha(y^2 + z^2)/4\eta\right\}\,. \tag{3.31}$$

The flux rope has Gaussian structure, with characteristic radius $\delta = \mathcal{O}((\eta/\alpha)^{1/2})$.

3.5 A stretched flux tube with helicity

We may in a similar way consider, in cylindrical polar coordinates (s, φ, z), the action of a uniform strain field $\mathbf{U} = (-\frac{1}{2}\gamma s,\ 0,\ \gamma z)$ on an initially helical flux tube.

Helical vortex tube

It is illuminating to consider first the problem of a helical vortex tube for which

$$\mathbf{u} = \Big(0,\ u_\varphi(s,\ t),\ u_z(s,\ t)\Big), \quad \omega = \nabla \times \mathbf{u} = \Big(0,\ \omega_\varphi(s,\ t),\ \omega_z(s,\ t)\Big). \tag{3.32}$$

The z-component of the vorticity equation is then

$$\frac{\partial \omega_z}{\partial t} - \tfrac{1}{2}\gamma s \frac{\partial \omega_z}{\partial s} = \gamma \omega_z + \nu \frac{1}{s}\frac{\partial}{\partial s}\left(s \frac{\partial \omega_z}{\partial s}\right), \tag{3.33}$$

and this, by analogy with (3.31) has the steady solution

$$\omega_z(s) = \frac{\gamma \Gamma}{2\pi\nu} \exp\left[-\frac{\gamma s^2}{4\nu}\right] \quad \text{where } \Gamma = \int_0^\infty \omega_z(s)\, 2\pi s\, \mathrm{d}s. \tag{3.34}$$

Γ is the total flux of vorticity, i.e. the circulation of the vortex tube. The stretched vortex (3.34) is generally known as the 'Burgers vortex' after Burgers (1948) .

But now we envisage a superposed velocity $u_z(s,\ t)$ with similar Gaussian structure, parallel to the vortex. The z-component of the Navier–Stokes equation is

$$\frac{\partial u_z}{\partial t} - \tfrac{1}{2}\gamma s \frac{\partial u_z}{\partial s} + \gamma u_z = \nu \frac{1}{s}\frac{\partial}{\partial s}\left(s \frac{\partial u_z}{\partial s}\right), \tag{3.35}$$

and note here the difference between this and (3.33). However, if we define $\hat{u}_z(s,\ t) = \mathrm{e}^{2\gamma t} u_z(s,\ t)$, then it follows from (3.35) that

$$\frac{\partial \hat{u}_z}{\partial t} - \tfrac{1}{2}\gamma s \frac{\partial \hat{u}_z}{\partial s} = \gamma \hat{u}_z + \nu \frac{1}{s}\frac{\partial}{\partial s}\left(s \frac{\partial \hat{u}_z}{\partial s}\right), \tag{3.36}$$

which does now have the same structure as (3.33). It follows that the solution of (3.35) analogous to (3.34) is

$$u_z(s,\ t) = \frac{\gamma Q(t)}{2\pi\nu} \exp\left[-\frac{\gamma s^2}{4\nu}\right] \quad \text{where } Q(t) = \int_0^\infty u_z(s,\ t)\, 2\pi s\, \mathrm{d}s = Q_0\, \mathrm{e}^{-2\gamma t}. \tag{3.37}$$

Here, $Q_0 = Q(0)$ is the initial flow rate (or 'flux') along the vortex tube, and the flux $Q(t)$ at time t is exponentially decreasing as a result of the stretching process.

The helicity per unit length of the tube is

$$\mathcal{H}(t) = 2\int_0^\infty u_z \omega_z\, 2\pi s\, \mathrm{d}s = \frac{\Gamma Q_0}{\pi\nu} \mathrm{e}^{-2\gamma t}. \tag{3.38}$$

This exponential decrease is associated with similar decrease of ω_φ,

$$\omega_\varphi = -\frac{\partial u_z}{\partial s} = \frac{\gamma^2 Q_0}{4\pi\nu^2}\, s \exp\left[-\frac{\gamma s^2}{4\nu}\right] e^{-2\gamma t}. \tag{3.39}$$

The circular ω_φ-lines are advected towards the axis $s = 0$ by the straining flow, whereas the ω_z-lines are persistently stretched in the z-direction, this stretching being exactly compensated by viscous diffusion.

The analogous helical magnetic flux tube

For this situation, we simply replace ν by η, $\mathbf{u}$ by $\mathbf{A}$, and $\boldsymbol{\omega} = \nabla\times\mathbf{u}$ by $\mathbf{B} = \nabla\times\mathbf{A}$, where now

$$\mathbf{A} = (0,\, A_\varphi(s,\, t),\, A_z(s,\, t)), \quad \mathbf{B} = (0,\, B_\varphi(s,\, t),\, B_z(s,\, t)). \tag{3.40}$$

In this case, it is the magnetic helicity per unit length of tube that decreases exponentially:

$$\mathcal{H}_M(t) = 2\int_0^\infty A_z B_z\, 2\pi s\, \mathrm{d}s = \frac{\Phi\Psi_0}{\pi\eta} e^{-2\gamma t}\,, \tag{3.41}$$

where Φ is the constant magnetic flux in the z-direction, and Ψ_0 is the initial magnetic flux per unit length of tube round the tube axis (i.e. in the φ-direction).

More interest here attaches to the current distribution $\mathbf{j} = -\mu_0^{-1}\nabla\times\mathbf{B}$ in the tube, with components

$$j_\varphi = -\frac{1}{\mu_0}\frac{\partial B_z}{\partial s} = \frac{\gamma^2\Phi}{4\pi\mu_0\eta^2}\, s \exp\left[-\frac{\gamma s^2}{4\eta}\right], \tag{3.42}$$

and

$$j_z = \frac{1}{\mu_0 s}\frac{\partial\left(sB_\varphi\right)}{\partial s} = \frac{\gamma^2\Psi_0}{2\pi\mu_0\eta^2}\left(1 - \frac{\gamma s^2}{4\eta}\right)\exp\left[-\frac{\gamma s^2}{4\eta}\right] e^{-2\gamma t}. \tag{3.43}$$

Thus, j_z decreases exponentially as a result of the stretching process. Note that the integral of j_z over the tube cross section is zero (if it were not so, then B_φ would fall off like s^{-1} outside the tube, rather than exponentially).

3.6 An example of accelerated ohmic diffusion

Suppose now that the uniform strain of §3.4 is two-dimensional, i.e. that

$$\mathbf{u} = (\alpha x, -\alpha y, 0) \quad (\alpha > 0)\,, \tag{3.44}$$

and that at time $t = 0$

$$\mathbf{B} = (0, 0, B_0 \sin k_0 y)\,. \tag{3.45}$$

For $t > 0$, the **B**-lines (which are parallel to the z-axis) are swept in towards the plane $y = 0$. It is evident that both the wave number and the amplitude of the field must change with time.[2] We may seek a solution of (3.14) in the form

$$\mathbf{B} = (0, 0, B(t)\sin k(t)y)\,, \tag{3.46}$$

where $k(0) = k_0$, $B(0) = B_0$. Substitution in (3.14) leads to

$$(dB/dt)\sin ky + (dk/dt)\,yB\cos ky = \alpha k\,yB\cos ky - \eta k^2 B\sin ky\,, \tag{3.47}$$

and since this must hold for all y, it follows that

$$\mathrm{d}k/\mathrm{d}t = \alpha k\,, \quad \text{and} \quad \mathrm{d}B/\mathrm{d}t = -\eta k^2 B\,. \tag{3.48}$$

Hence

$$k(t) = k_0 e^{\alpha t}, \quad \text{and} \quad B(t) = B_0 \exp\left\{-\eta k_0^2(e^{2\alpha t} - 1)/2\alpha\right\}\,. \tag{3.49}$$

The wavelength of the non-uniformity of the **B**-field evidently decreases as $e^{-\alpha t}$ due to the advection of all variation towards the plane $y = 0$; in consequence the natural decay of the field is greatly accelerated. Note that in this case, with **B** oriented along the axis of zero strain rate, there is no tendency to stretch the **B**-lines ($(\mathbf{B}\cdot\nabla)\mathbf{u} = 0$), but merely a tendency to advect them ($(\mathbf{u}\cdot\nabla)\mathbf{B} \neq 0$).

3.7 Equation for vector potential and flux-function under particular symmetries

Suppose now that **u** is a solenoidal velocity field, and let

$$\mathbf{u} = \mathbf{u}_P + \mathbf{u}_T \quad \text{and} \quad \mathbf{B} = \mathbf{B}_P + \mathbf{B}_T \tag{3.50}$$

be the poloidal and toroidal decompositions of **u** and **B**. Suppose further that both **u** and **B** are either two-dimensional (i.e. independent of the Cartesian coordinate z) or axisymmetric (i.e. invariant under rotations about the axis of symmetry Oz). Then it is clear that $\mathbf{u}_T \wedge \mathbf{B}_T = 0$ and so

$$\mathbf{u} \wedge \mathbf{B} = (\mathbf{u}_P \wedge \mathbf{B}_T + \mathbf{u}_T \wedge \mathbf{B}_P) + (\mathbf{u}_P \wedge \mathbf{B}_P)\,, \tag{3.51}$$

the first bracketed term on the right being poloidal and the second toroidal. The poloidal ingredient of (3.14) is then

$$\partial \mathbf{B}_P/\partial t = \nabla \wedge (\mathbf{u}_P \wedge \mathbf{B}_P) + \eta\nabla^2 \mathbf{B}_P\,. \tag{3.52}$$

Writing $\mathbf{B}_P = \nabla \wedge \mathbf{A}_T$, we may 'uncurl' this equation obtaining what is in effect the toroidal ingredient of (2.132),

$$\partial \mathbf{A}_T/\partial t = \mathbf{u}_P \wedge (\nabla \wedge \mathbf{A}_T) + \eta\nabla^2 \mathbf{A}_T\,, \tag{3.53}$$

[2] This type of behaviour was recognised in the context of hydrodynamic stability by Kelvin; solutions of such problems for which both wave vector and amplitude are time-dependent may be described as 'Kelvin waves'.

there being no toroidal contribution from the term $-\nabla\phi$ of (2.132).[3] Similarly, the toroidal ingredient of (3.14) is

$$\partial\mathbf{B}_T/\partial t = \nabla \wedge (\mathbf{u}_P \wedge \mathbf{B}_T + \mathbf{u}_T \wedge \mathbf{B}_P) + \eta\nabla^2\mathbf{B}_T \,. \tag{3.54}$$

Equations (3.53) and (3.54) are quite convenient since both $\mathbf{A}_T$ and $\mathbf{B}_T$ have only one component each in both two-dimensional and axisymmetric situations. These situations however now require slightly different treatments.

3.7.1 Two-dimensional case

In this case $\mathbf{A}_T = A(x, y)\,\mathbf{e}_z$ and $\nabla \wedge \mathbf{A}_T = -\mathbf{e}_z \wedge \nabla A$; hence (3.53) becomes

$$\partial A/\partial t + \mathbf{u}_P \cdot \nabla A = \eta\,\nabla^2 A \,, \tag{3.55}$$

so that A behaves like a scalar quantity (cf. (3.22)). Similarly, with $\mathbf{B}_T = B(x, y)\,\mathbf{e}_z$ and $\mathbf{u}_T = u_z(x, y)\,\mathbf{e}_z$, (3.54) becomes

$$\partial B/\partial t + \mathbf{u}_P \cdot \nabla B = (\mathbf{B}_P \cdot \nabla)\,u_z + \eta\,\nabla^2 B \,. \tag{3.56}$$

Here B also behaves like a scalar, but with a ‘source term’ $(\mathbf{B}_P \cdot \nabla)\,u_z$ on the right-hand side. The interpretation of this term is simply that if u_z varies along a $\mathbf{B}_P$-line, then it will tend to shear the $\mathbf{B}_P$-line in the z-direction, i.e. to generate a toroidal field component.

3.7.2 Axisymmetric case

The differences here are purely associated with the curved geometry, and are in this sense trivial. First, with $\mathbf{A}_T = A(s, z)\,\mathbf{e}_\varphi$ in cylindrical polar coordinates (z, s, φ) (with $s = r\sin\theta$), we have

$$\nabla^2\mathbf{A}_T = \mathbf{e}_\varphi(\nabla^2 - s^{-2})A \,, \quad \mathbf{u}_P \wedge (\nabla \wedge \mathbf{A}_T) = -s^{-1}(\mathbf{u}_P \cdot \nabla)(sA)\,\mathbf{e}_\varphi \,, \tag{3.57}$$

so that (3.53) becomes

$$\partial A/\partial t + s^{-1}(\mathbf{u}_P \cdot \nabla)(sA) = \eta\,(\nabla^2 - s^{-2})A \,. \tag{3.58}$$

Similarly, with $\mathbf{B}_T = B(s, z)\,\mathbf{e}_\varphi$, $\mathbf{u}_T = u_\varphi(s, z)\,\mathbf{e}_\varphi$, we have

$$\left.\begin{array}{l} \nabla \wedge (\mathbf{u}_P \wedge \mathbf{B}_T) = -\mathbf{e}_\varphi s(\mathbf{u}_P \cdot \nabla)(s^{-1}B) \\ \nabla \wedge (\mathbf{u}_T \wedge \mathbf{B}_P) = \mathbf{e}_\varphi s(\mathbf{B}_P \cdot \nabla)(s^{-1}u_\varphi) \end{array}\right\} , \tag{3.59}$$

[3] This is because ϕ is independent of the azimuth angle φ in an axisymmetric situation. In the two-dimensional case, a uniform electric field E_z in the z-direction could be present, but this requires sources of field ‘at infinity’, and we disregard this possibility.

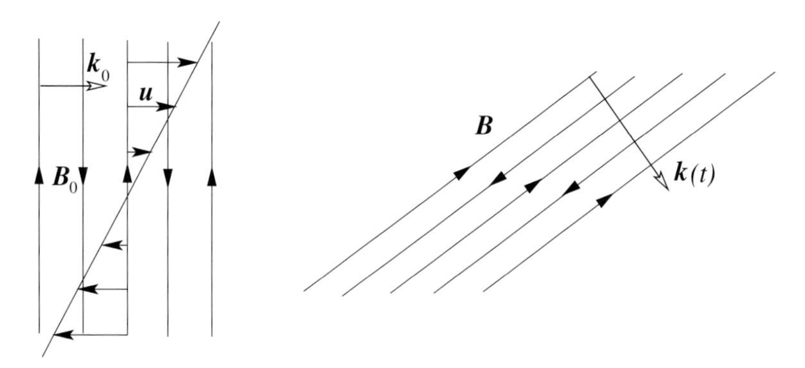

Figure 3.2 Shearing of the space-periodic magnetic field (3.65).

so that (3.54) becomes

$$\partial B/\partial t + s(\mathbf{u}_P \cdot \nabla)(s^{-1}B) = s(\mathbf{B}_P \cdot \nabla)(s^{-1}u_\varphi) + \eta\,(\nabla^2 - s^{-2})B\,. \tag{3.60}$$

Again there is a source term in the equation for B, but now it is variation of the angular velocity $\varpi(s,z) = s^{-1}u_\varphi(s,z)$ along a $\mathbf{B}_P$-line which gives rise, by field distortion, to the generation of toroidal field. This phenomenon will be considered in detail in §3.14 below.

Sometimes it is convenient to use the flux-function $\chi(s,z) = sA(s,z)$ (see (2.61)). From (3.58), the equation for χ is

$$\partial\chi/\partial t + (\mathbf{u}_P \cdot \nabla)\chi = \eta\,\mathcal{D}^2\chi\,, \tag{3.61}$$

where

$$\mathcal{D}^2\chi = s(\nabla^2 - s^{-2})(s^{-1}\chi) = (\nabla^2 - 2s^{-1}\partial/\partial s)\chi\,. \tag{3.62}$$

The operator $\mathcal{D}^2$, known as the Stokes operator, occurs frequently in problems with axial symmetry. In spherical polars (r,θ,φ), it takes the form

$$\mathcal{D}^2 = \frac{\partial^2}{\partial r^2} + \frac{\sin\theta}{r^2}\frac{\partial}{\partial\theta}\frac{1}{\sin\theta}\frac{\partial}{\partial\theta}\,. \tag{3.63}$$

Note, from (3.62), that

$$\mathcal{D}^2\chi = \nabla\cdot\mathbf{f} \;\text{ where }\; \mathbf{f} = \nabla\chi - 2s^{-1}\chi\,\mathbf{e}_s\,. \tag{3.64}$$

3.8 Shearing of a space-periodic magnetic field

We describe here a further example of accelerated diffusion (Moffatt & Kamkar, 1983) as a preliminary to the important topic of 'flux expulsion' to be considered in §3.11. We suppose that an initial field of the form

$$\mathbf{B}(\mathbf{x}, 0) = (0, B_0 \cos k_0 x, 0) \tag{3.65}$$

is subjected to a uniform shearing velocity $\mathbf{u} = (\alpha x, 0, 0)$. Here the relevant magnetic Reynolds number is

$$\mathcal{R}_m = \alpha k_0^{-2}/\eta . \tag{3.66}$$

The initial vector potential is $\mathbf{A} = (0, 0, A)$, with

$$A(x, y, 0) = -k_0^{-1} B_0 \,\Im\mathrm{m}\, \exp(ik_0 x) , \tag{3.67}$$

and, as in §3.6, we may expect both the wave vector and the amplitude to be time-dependent. Thus we look for a Kelvin-wave solution of (3.55) of the form

$$A(x, y, t) = -k_0^{-1} B_0 \,\Im\mathrm{m}\, \{a(t) \exp(i\,\mathbf{k}(t) \cdot \mathbf{x})\} , \tag{3.68}$$

where

$$a(0) = 1, \quad \mathbf{k}(0) = (k_0, 0, 0). \tag{3.69}$$

Substituting in (3.55), it is easily shown that

$$\mathbf{k}(t) = (k_0, -\alpha k_0 t, 0), \tag{3.70}$$

so that the wave-fronts are progressively tilted by the shear as shown in Figure 3.2. From (3.55), the amplitude $a(t)$ then satisfies the equation

$$da/dt = -\eta\, \mathbf{k}^2\, a = -\eta\, k_0^2\, (1 + \alpha^2 t^2)\, a , \tag{3.71}$$

and integration of this equation with the initial condition $a(0) = 1$ gives

$$a(t) = \exp\left\{-\eta\, k_0^2 \left(t + \tfrac{1}{3}\alpha^2 t^3\right)\right\} . \tag{3.72}$$

The magnetic field at time t is given by

$$\mathbf{B} = \nabla \wedge \mathbf{A} = B_0\, a(t)\, (-\alpha t,\ 1,\ 0) \cos k_0 (x - \alpha t\, y) . \tag{3.73}$$

Note that the component B_x grows linearly in time for so long as diffusion effects are negligible.

The situation is most interesting in the high-conductivity situation $\mathcal{R}_m \gg 1$. For $\alpha t \ll 1$, (3.72) shows the familiar decay $a \sim \exp\{-\eta\, k_0^2 t\}$ on the diffusive time-scale

$$t_d = (\eta\, k_0^2)^{-1} = \alpha^{-1} \mathcal{R}_m . \tag{3.74}$$

For $\alpha t \gg 1$ however, the accelerated decay law $a \sim \exp\{-\eta\, k_0^2 \alpha^2 t^3/3\}$ takes over, on a time-scale

$$t_a = (\alpha^2 \eta\, k_0^2)^{-1/3} = \alpha^{-1} {\mathcal{R}_m}^{1/3} . \tag{3.75}$$

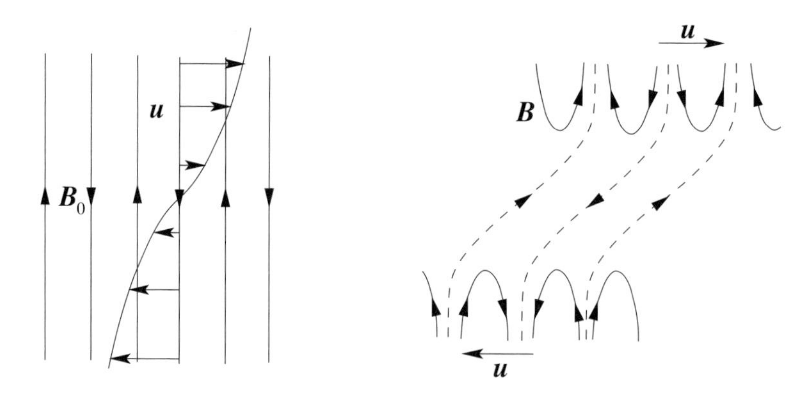

Figure 3.3 Flux expulsion by non-uniform shear.

This accelerated decay occurs because of the decreasing length-scale of the magnetic field under the influence of the shear. For $\mathcal{R}_m \gg 1$, the difference between (3.74) and (3.75) is obviously very significant; for example, if $\mathcal{R}_m = 10^6$, then the time-scale of decay is reduced by a factor $\mathcal{O}(10^4)$ by this shearing effect. When $t = \mathcal{O}(t_a)$, the x-component of magnetic field has order of magnitude

$$B_{x\,\max} = \mathcal{O}({\mathcal{R}_m}^{1/3} B_0); \tag{3.76}$$

thus a considerable intensification of field occurs when $\mathcal{R}_m \gg 1$, before diffusion sets in.

If the shear is non-uniform, e.g. if $\mathbf{u} = (\alpha\delta \tanh x/\delta, 0, 0)$, then the shearing effect obviously still occurs in the region $|x| \lesssim \delta$, with destruction of the magnetic field on the time-scale $\alpha^{-1}{\mathcal{R}_m}^{1/3}$, while in the regions $|x| \gtrsim \delta$, the field decays on the much longer diffusion time-scale $\alpha^{-1}\mathcal{R}_m$, always assuming $\mathcal{R}_m \gg 1$. For intermediate times in the range $\alpha^{-1}{\mathcal{R}_m}^{1/3} \ll t \ll \alpha^{-1}\mathcal{R}_m$, the field must then have the qualitative structure sketched in Figure 3.3, the flux being effectively expelled from the region where the shear is strong.

The behaviour is actually more complex than suggested by this sketch. Figure 3.4 shows the true development of the **B**-line pattern, under shearing by a velocity field $\mathbf{u} = (\alpha\delta \tanh y/\delta, 0, 0)$, with $\mathcal{R}_m = \alpha\delta^2/\eta = 10^3$ and $k\delta = 1$. Here $\mathcal{R}_m^{1/3} = 10$, and at time $\alpha t = 200$ the presence of the small closed field loops is an indication of the effect of weak diffusion; by time $\alpha t = 1000$, these closed loops have disappeared, and the field is effectively expelled from the inner region, much as anticipated in Figure 3.3.

We note that the analogy with vorticity described in §3.2 is applicable here. We may start with a vorticity field

$$\omega(\mathbf{x}, 0) = (0, \omega \cos k_0 x, 0), \tag{3.77}$$

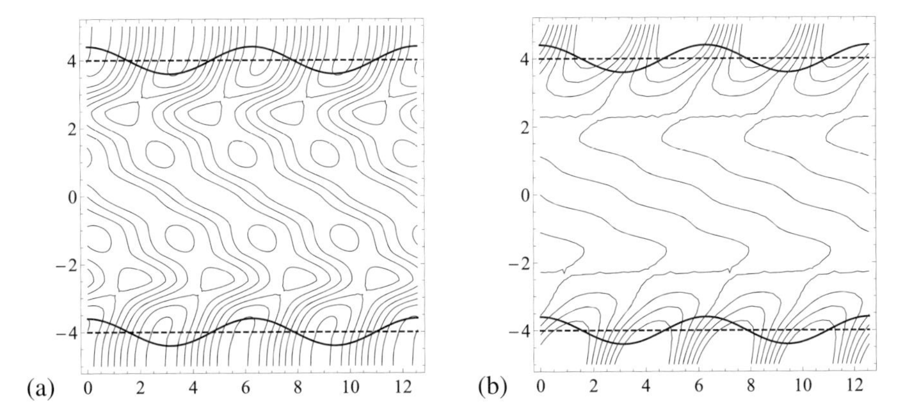

Figure 3.4 Shearing of the magnetic field (3.65) by the velocity field $\mathbf{u} = (\alpha\delta \tanh y/\delta, 0, 0)$ at $\mathcal{R}_m = 10^3$, $k\delta = 1$; the initial field profile is indicated by the sine-curves; (a) $\alpha t = 200$: field loops caused by magnetic reconnection are evident; (b) $\alpha t = 1000$: the field is completely expelled from the central region; only the 'singular' **B**-line, $A = 0$, remains visible in this region.

in a viscous fluid and subject this to the uniform shear $\mathbf{U} = (\alpha x, 0, 0)$. The 'perturbation velocity associated with the vorticity field (3.77) is

$$\mathbf{u} = (0, 0, -u_0 \sin k_0 x), \quad \text{with} \quad u_0 = \varpi_0/k_0\,, \tag{3.78}$$

(so that $\nabla \wedge (\mathbf{u} \wedge \boldsymbol{\omega}) = 0$, a condition that persists for $t > 0$). With Reynolds number $\mathcal{R}e = \alpha/\nu k_0^2$ assumed large, the field evolving from (3.77) decays on a time-scale identified by Rhines & Young (1982), $t_{RY} = \alpha^{-1}\mathcal{R}e^{1/3}$ (cf. (3.75)), an important result in the context of two-dimensional vorticity dynamics.

3.9 Oscillating shear flow

Consider a pure shear flow with components $\mathbf{u} = (Sy,\ 0,\ 0)$ (cf. §3.8). This has stream-function $\psi(x,\ y) = \frac{1}{2}Sy^2$ with $u_x = \partial\psi/\partial y = Sy$, $u_y = -\partial\psi/\partial x = 0$. Now, following Dormy & Gérard-Varet (2008), suppose that this shear flow oscillates about the axis Oz so that the direction of the flow is at an angle $\theta(t) = \alpha \sin\beta t$ to the x-axis. Let $x' = x\cos\theta + y\sin\theta$, $y' = -x\sin\theta + y\cos\theta$. Then the stream-function of the oscillating flow is $\psi = \frac{1}{2}Sy'^2 = \frac{1}{2}S(x\sin\theta - y\cos\theta)^2$, and so the velocity components in the fixed Oxyz frame of reference are now

$$\left.\begin{aligned} u_x = \partial\psi/\partial y \;\; &= -S\,[x\sin\theta(t) - y\cos\theta(t)]\cos\theta(t) \\ u_y = -\partial\psi/\partial x &= -S\,[x\sin\theta(t) - y\cos\theta(t)]\sin\theta(t) \end{aligned}\right\}. \tag{3.79}$$

This is a linear flow of the form $u_i = c_{ij}(t)x_j$, but the time dependence of the coefficients $c_{ij}(t)$ give it some unexpected properties.

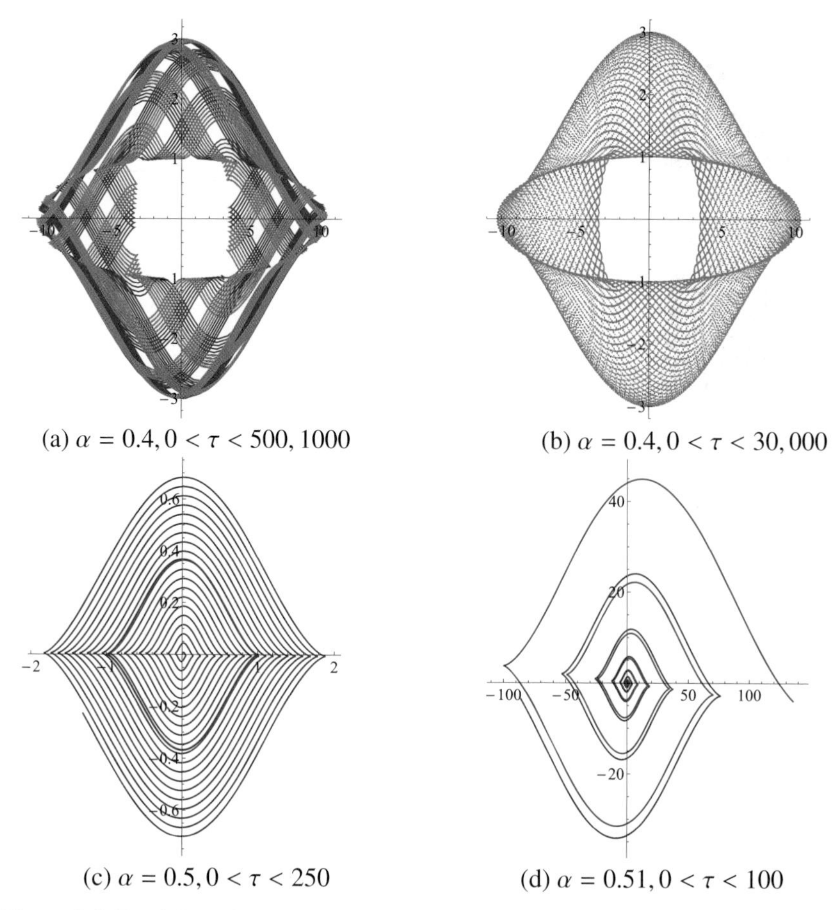

Figure 3.5 Particle trajectories in the velocity field (3.79) (with $S = \beta = 1$); (a) $\alpha = 0.4$; $\mathbf{x}(0) = (0, 1)$ (blue), $(10, 0)$ (red); both trajectories are bounded and quasi-periodic; (b) Poincaré section for the blue trajectory of (a); points are recorded at unit intervals for $0 < \tau < 30{,}000$; (c) $\alpha = 0.5$; the threshold of instability: $\mathbf{x}(0) = (1, 0)$ (red, periodic trajectory); $\mathbf{x}(0) = (0, 0.005)$ (blue, transient growth, linear in t); (d) $\alpha = 0.51$; $\mathbf{x}(0) = (0, 0.0001)$ (blue) and $\mathbf{x}(0) = (1, 0)$ (red); both trajectories are unstable, with exponentially growing amplitudes.

Figure 3.5 shows some sample particle trajectories for varying α (all with $S = \beta = 1$). When $\alpha = 0.4$ (panel (a)), two trajectories (red and blue) are shown starting from different initial positions. Both trajectories are evidently bounded in extent. Panel (b) shows a Poincaré section, constructed by placing points at times $t = 0, 1, 2, 3, \dots, 30{,}000$; the fact that these points lie on a curve (and are not 'space-filling') indicates that the trajectories are quasi-periodic, not chaotic.

Panel (c) shows trajectories for a critical value of α, $\alpha = 0.5$, at the onset of instability. The red curve is a periodic orbit starting from position $\mathbf{x}(0) = (1, 0)$; the blue curve starts from $\mathbf{x}(0) = (0, 0.005)$ and shows transient instability (linear

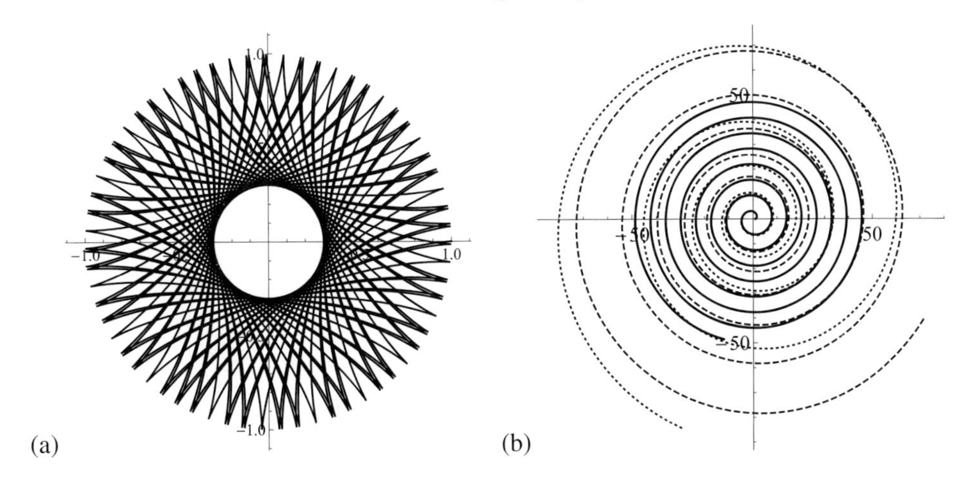

Figure 3.6 Particle trajectories when a pure shear $(Sy, 0, 0)$ rotates about the z-axis with steady angular velocity ω; (a) a quasi-periodic trajectory when $\sigma \equiv S/\omega = 10$, and (b) three unstable trajectories for $\sigma = -1$ (solid, $0 < t < 50$), -1.002 (dashed, $0 < t < 45$), and -1.008 (dotted, $0 < t < 31$); the case $\sigma = -1$ is critical, the instability being transient, with distance r from the origin increasing linearly; for $\sigma = -1.002$ and -1.008, the growth is exponential.

in time t). Panel (d) shows two unstable trajectories at the slightly increased value $\alpha = 0.51$; these trajectories exhibit exponential growth of distance from the origin (note the dramatic increase of scale on both x- and y-axes).

This is interesting in regard to what happens to a uniform magnetic field in the (x, y)-plane. The field remains uniform under the action of the uniform (although time-dependent) shear, and responds to it by instantaneous stretching of the component perpendicular to the instantaneous velocity. Since the magnetic field satisfies the same equation as a material line element, it shows exactly the same behaviour as any material line initially parallel to it. The field therefore varies periodically when $\alpha = 0.4$ and is unstable with oscillations of exponentially growing amplitude when $\alpha = 0.51$. The behaviour is actually quite complex over the full range of values of α, but enough has been said here to indicate a rich range of possible behaviour (see Dormy & Gérard-Varet 2008 for details).

3.9.1 The case of steady rotation of the shearing direction

The case of steady rotation of the shearing direction for which $\theta = \omega t$ with ω constant is equally interesting. In this case, the equations governing particle trajectories are

$$\left.\begin{aligned} \mathrm{d}x/\mathrm{d}t &= -S\,(x\sin\omega t - y\cos\omega t)\cos\omega t \\ \mathrm{d}y/\mathrm{d}t &= -S\,(x\sin\omega t - y\cos\omega t)\sin\omega t \end{aligned}\right\}. \tag{3.80}$$

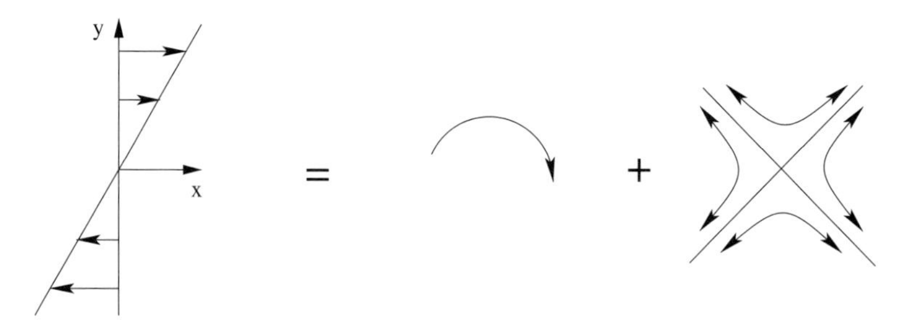

Figure 3.7 Decomposition of a shear flow $(S\,y, 0, 0)$ into rotation $(\frac{1}{2}S\,y, -\frac{1}{2}S\,x, 0)$ plus irrotational strain $(\frac{1}{2}S\,y, \frac{1}{2}S\,x, 0)$.

These equations can actually be solved explicitly (Dormy & Gérard-Varet 2008); here, we simply plot some resulting particle trajectories. Figure 3.6 shows (a) a quasi-periodic trajectory when $\sigma \equiv S/\omega = 10$, and (b) three unstable trajectories for $\sigma = -1$ (solid), -1.002 (dashed), and -1.008 (dotted). When $\sigma = -1$ (a critical case), the instability is transient with linear growth in time of the distance of the particle from the origin. For $\sigma < -1$, the instability is exponential, the growth rate increasing with increasing $|\sigma|$. This instability may be understood with reference to Figure 3.7. A steady plane shear with $S > 0$ as illustrated may be decomposed into a rigid-body rotation plus a uniform irrotational flow with hyperbolic streamlines. When this shearing flow rotates with a negative angular velocity ω (so that $S/\omega < 0$), there is a cooperative effect between S and ω that makes the hyperbolic ingredient of the flow dominate over the rotation, thus causing the instability. If $S/\omega > 0$, the rotation dominates, and the particle trajectories then remain bounded and oscillatory.

3.10 Field distortion by differential rotation

By *differential rotation*, we shall mean an incompressible velocity field axisymmetric about Oz, and with circular streamlines about this axis. Such a motion has the form (in cylindrical polars)[4]

$$\mathbf{u} = \varpi(s, z)\,\mathbf{e}_z \wedge \mathbf{x}\,. \tag{3.81}$$

If $\nabla\varpi = 0$, then we have rigid body rotation which clearly rotates a magnetic field without distortion. If $\nabla\varpi \neq 0$, lines of force are in general distorted in a way that depends both on the appropriate value of $\mathcal{R}_\mathrm{m}$ and on the orientation of the field

[4] Throughout this chapter, we shall use the symbol ϖ ('curly pi' or 'pomega') to denote the angular velocity of the fluid ring at (s, z). This should not be confused with vorticity: if $\varpi = \varpi(s) = u_\varphi/s$, then the vorticity is $\omega(s)\,\mathbf{e}_z$, where $\omega(s) = s^{-1}\partial(su_\varphi)/\partial s = 2\varpi + s\partial\varpi/\partial s$.

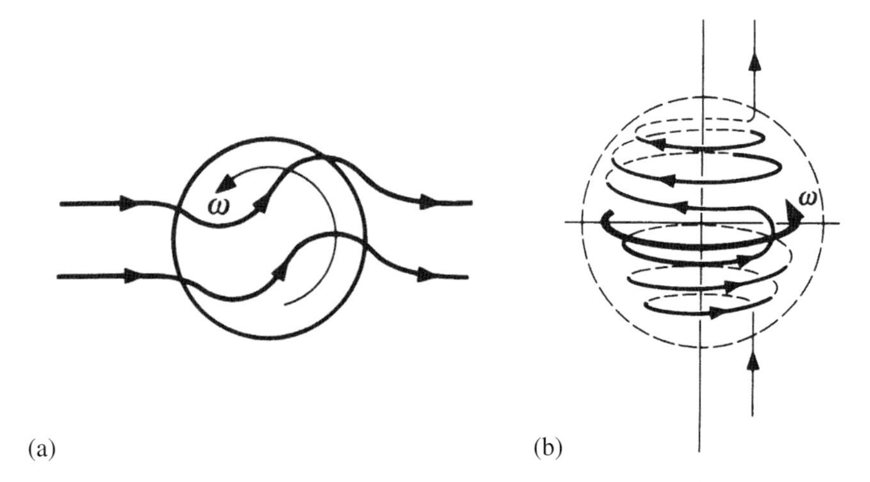

Figure 3.8 Qualitative action of differential rotation on an initially uniform magnetic field: (a) vorticity $\omega\mathbf{k}$ perpendicular to field; (b) vorticity $\omega\mathbf{k}$ parallel to field.

relative to the vector $\mathbf{e}_z$. The two main possibilities are illustrated in Figure 3.8. In (a), ϖ is a function of s alone, and the $\mathbf{B}$-field lies in the x-y plane perpendicular to $\mathbf{e}_z$; the effect of the motion, neglecting diffusion, is to wind the field into a tight double spiral in the x-y plane. In (b), $\varpi = \varpi(r)$, where $r^2 = s^2 + z^2$, and $\mathbf{B}$ is initially axisymmetric and poloidal; the effect of the rotation, neglecting diffusion, is to generate a toroidal field, the typical $\mathbf{B}$-line becoming helical in the region of differential rotation.

Both types of distortion are important in both solar and geomagnetic context, and have been widely studied. We discuss first the type (a) distortion (first studied in detail by E.N. Parker 1963), and the important related phenomenon of flux expulsion from regions of closed streamlines.

3.11 Effect of plane differential rotation on an initially uniform field: flux expulsion

Suppose then that $\varpi = \varpi(s)$ so that the velocity field given by (3.81) is independent of z, and suppose that at time $t = 0$ the field $\mathbf{B}(\mathbf{x}, 0)$ is uniform and equal to $\mathbf{B}_0$. We take the axis Ox in the direction of $\mathbf{B}_0$. For $t > 0, \mathbf{B} = -\mathbf{e}_z \wedge \nabla A$, where, from (3.55), A satisfies

$$\partial A/\partial t + \varpi(s)(\mathbf{x} \wedge \nabla A)_z = \eta\, \nabla^2 A\,. \qquad (3.82)$$

It is natural to use plane polar coordinates defined here by

$$x = s\cos\varphi\,, \quad y = s\sin\varphi\,, \tag{3.83}$$

in terms of which (3.82) becomes

$$\partial A/\partial t + \varpi(s)\partial A/\partial\varphi = \eta\nabla^2 A\,. \tag{3.84}$$

The initial condition $\mathbf{B}(\mathbf{x},0) = (B_0,0,0)$ is equivalent to

$$A(s,\varphi,0) = B_0 s\sin\varphi\,, \tag{3.85}$$

and the relevant solution of (3.84) clearly has the form

$$A(s,\varphi,t) = \Im\mathrm{m}\left[\,B_0\, f(s,t)\, e^{i\varphi}\right]\,, \tag{3.86}$$

where

$$\frac{\partial f}{\partial t} + i\varpi(s)f = \eta\left(\frac{1}{s}\frac{\partial}{\partial s}s\frac{\partial}{\partial s} - \frac{1}{s^2}\right)f\,, \tag{3.87}$$

and

$$f(s,0) = s\,. \tag{3.88}$$

3.11.1 The initial phase

When $t = 0$, the field $\mathbf{B}$ is uniform and there is no diffusion; it is therefore reasonable to anticipate that diffusion will be negligible during the earliest stages of distortion. With $\eta = 0$ the solution of (3.87) satisfying the initial condition (3.88) is $f(s,t) = s\,\mathrm{e}^{-i\varpi(s)t}$, so that from (3.85)

$$A(s,\varphi,t) = B_0 s\sin(\varphi - \varpi(s)t)\,. \tag{3.89}$$

This solution is of course just the Lagrangian solution $A(\mathbf{x},t) = A(\mathbf{a},0)$, since for the motion considered, the particle whose coordinates are (s,φ) at time t originated from position $(s,\varphi-\varpi(s)t)$ at time zero. The components of $\mathbf{B} = -\mathbf{e}_z\wedge\nabla A$ are now given by

$$\left.\begin{aligned} B_s &= s^{-1}\partial A/\partial\varphi = B_0\cos(\varphi - \varpi(s)t)\,,\\ B_\varphi &= -\partial A/\partial s = -B_0\sin(\varphi - \varpi(s)t) + B_0 s\varpi'(s)t\cos(\varphi - \varpi(s)t)\,. \end{aligned}\right\} \tag{3.90}$$

If $\varpi'(s) = 0$, i.e. if the motion is a rigid body rotation, then as expected the field is merely rotated with the fluid. If $\varpi'(s) \neq 0$, the φ-component of $\mathbf{B}$ increases linearly with time as a result of the stretching process, just as in the model problem of §3.8. Figure 3.9 shows the contours A = const. as given by (3.89) for the particular choice

$$\varpi(s) = \varpi_0\exp\left(-(s/s_0)^2\right) \tag{3.91}$$

at time $\varpi_0 t = 20$, by which time a fairly tight double spiral has indeed formed.

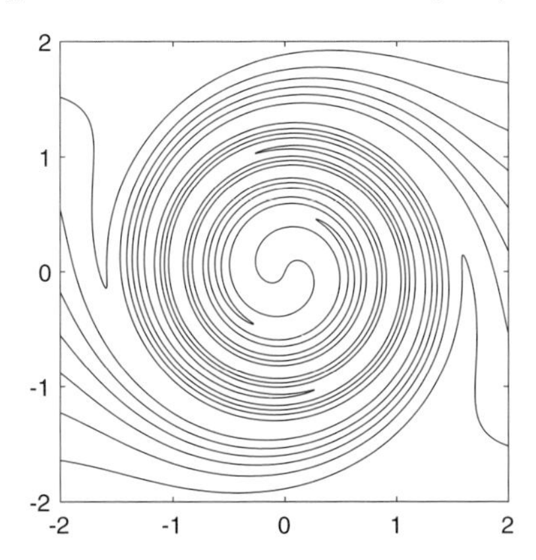

Figure 3.9 Initial phase of wind-up of an initially uniform magnetic field into a tight double spiral by the differential rotation (3.91); $s_0 = 1$ and $\varpi_0 t = 20$ (after three full rotations of inner region).

Note that in such a double spiral the field alternates rapidly in direction with increasing radius, again just as in §3.8. We may therefore expect that diffusion will in fact become operative on a time-scale $t_a = \varpi_0^{-1}\mathcal{R}_m^{1/3}$ with here $\mathcal{R}_m = \varpi_0 s_0^2/\eta$, where s_0 is the radial scale of the differential rotation. This is in agreement with the estimate first obtained by Parker (1963). Further evidence is provided by the results of numerical solution of (3.87) shown in Figure 3.13.[5]

3.11.2 The ultimate steady state

It is to be expected that when $t \to \infty$ the solution of (3.87) will settle down to a steady form $f_1(s)$ satisfying

$$\frac{i\varpi(s)}{\eta} f_1 = \left(\frac{1}{s}\frac{\mathrm{d}}{\mathrm{d}s} s \frac{\mathrm{d}}{\mathrm{d}s} - \frac{1}{s^2} \right) f_1 \, , \tag{3.92}$$

and that, provided $\varpi(s) \to 0$ as $s \to \infty$, the outer boundary condition should be that the field at infinity is undisturbed, i.e.

$$f_1(s) \sim s \text{ as } s \to \infty \, . \tag{3.93}$$

[5] The estimate $t_a = \varpi_0^{-1}\mathcal{R}_m^{1/2}$ given in Moffatt (1978a) was incorrect; however, as shown by Bajer et al. (2001), closed loops of field do survive very near the centre of rotation for a time of order $\varpi_0^{-1}\mathcal{R}_m^{1/2}$, due to the reduced gradient of angular velocity in this region.

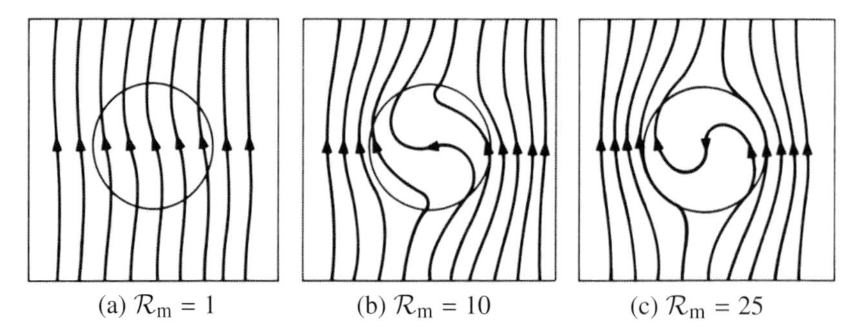

Figure 3.10 Ultimate steady state field distributions for three values of $\mathcal{R}_m$; when $\mathcal{R}_m$ is small, the field distortion is small, while when $\mathcal{R}_m$ is large, the field tends to be excluded from the rotating region. The sense of the rotation is anticlockwise.

The situation is adequately illustrated by the particular choice[6]

$$\frac{\varpi(s)}{\eta} = \begin{cases} k_0^2 & (s < s_0) \\ 0 & (s > s_0) \end{cases} \tag{3.94}$$

where k_0 is constant. The solution of the problem (3.92)–(3.94) is then straightforward:

$$f_1(s) = \begin{cases} s + C\,s^{-1} & (s > s_0) \\ D\,J_1(ps) & (s < s_0) \end{cases} \tag{3.95}$$

where $p = (1-i)k_0/\sqrt{2}$. The constants C and D are determined from the conditions that B_s and B_φ (and hence f_1 and f_1') should be continuous across $s = s_0$; these conditions yield

$$D = \frac{2}{pJ_0(ps_0)}, \qquad C = \frac{s_0(2J_1(ps_0) - ps_0J_0(ps_0))}{pJ_0(ps_0)}, \tag{3.96}$$

and this completes the formal determination of $f_1(s)$ from which A and hence B_s and B_φ may be determined. The **B**-lines are drawn in Figure 3.10 for $\mathcal{R}_m = 1, 10$ and 25; note the increasing degree of distortion as $\mathcal{R}_m$ increases.

The nature of the solution is of particular interest when $\mathcal{R}_m \gg 1$; in this situation $|ps_0| \gg 1$, and the asymptotic formulae

$$J_0(z) \sim (2/\pi z)^{1/2}\sin(z + \pi/4)\,, \quad J_1(z) \sim -(2/\pi z)^{1/2}\cos(z + \pi/4) \tag{3.97}$$

may be used[7] both in (3.96) with $z = ps_0$, and in (3.95) with $z = ps$. After some

[6] Note that for this discontinuous choice, there can be no initial phase of the type discussed above; diffusion must operate as soon as the motion commences to eliminate the incipient singularity in the magnetic field on $s = s_0$. Note also that it is only the variation with s of the ratio ϖ/η that affects the ultimate field distribution; in particular if $\varpi = 0$ for $s > s_0$, then η may be an arbitrary (strictly positive) function of s for $s > s_0$ without affecting the situation.

[7] There is a small neighbourhood of $s = 0$ where, strictly, the asymptotic formulae (3.97) may not be used, but it is evident from the nature of the result (3.98) that this is of no consequence.

simplification the resulting formula for A from (3.85) takes the form

$$A \sim \begin{cases} B_0\left(s - s_0^2/s\right)\sin\varphi + \frac{2B_0 s_0^2}{k_0 s}\sin(\varphi + \pi/4) & (s > s_0)\,, \\ \frac{2B_0}{k_0}\exp\left[-\frac{k_0(s_0-s)}{\surd 2}\right]\sin\left[\varphi + \frac{k_0(s_0-s)}{\surd 2} + \pi/4\right] & (s < s_0)\,. \end{cases} \tag{3.98}$$

In the limit $\mathcal{R}_m = \infty$ ($k_0 = \infty$), this solution degenerates to

$$A \sim \begin{cases} B_0\left(s - s_0^2/s\right)\sin\varphi & (s > s_0)\,, \\ 0 & (s < s_0)\,. \end{cases} \tag{3.99}$$

The lines of force $A =$ const. are then identical with the streamlines of an irrotational flow past a cylinder. In this limit of effectively infinite conductivity, the field is totally excluded from the rotating region $s < s_0$; the tangential component of field suffers a discontinuity across the surface $s = s_0$ which consequently supports a current sheet.

This form of field exclusion is related to the skin effect in conventional electromagnetism. Relative to axes rotating with angular velocity ϖ_0, the problem is that of a field rotating with angular velocity $-\varpi_0$ outside a cylindrical conductor. (As observed in footnote 6, the conductivity is irrelevant for $s > s_0$ in the steady state so that we may treat the medium as insulating in this region.) A rotating field may be decomposed into two perpendicular components oscillating out of phase, and at high frequencies these oscillating fields are excluded from the conductor. The same argument of course applies to the rotation of a conductor of any shape in a magnetic field, when the medium outside the conductor is insulating; at high rotation rate, the field is always excluded from the conductor when it has no component parallel to the rotation vector.

The additional terms in (3.98) describe the small perturbation of the limiting form (3.99) that results when the effects of finite conductivity in the rotating region are included. The field does evidently penetrate a small distance δ into this region, where

$$\delta = \mathcal{O}(k_0^{-1}) = \mathcal{O}\left(\mathcal{R}_m{}^{-1/2}\right) s_0\,. \tag{3.100}$$

The current distribution (confined to the region $s < s_0$) is now distributed through a layer of thickness $\mathcal{O}(\delta)$ in which the field falls to an effectively zero value. The behaviour is already evident in the field line pattern for $\mathcal{R}_m = 25$ in Figure 3.10c.

3.11.3 Flow distortion by the flow due to a line vortex

An important special case in which equation (3.92) may be solved exactly is that in which the flow is due to a point vortex of circulation Γ. In this case, the circulating

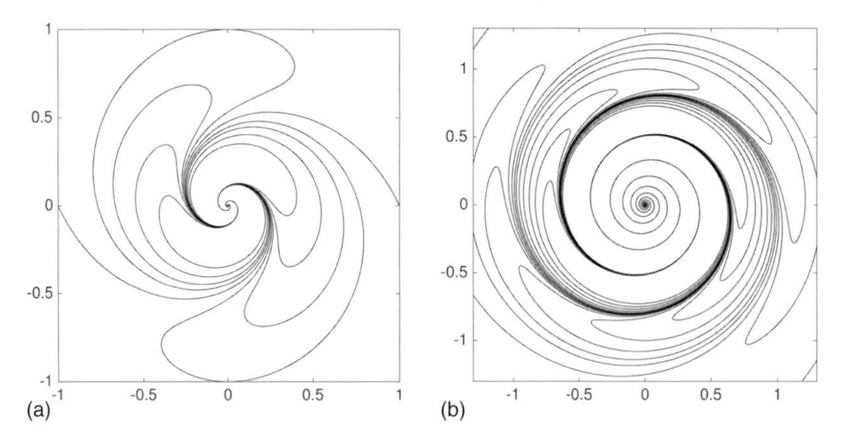

Figure 3.11 (a) Field distortion by a point vortex at $\mathcal{R}_m = \Gamma/2\pi\eta = 10$; (b) the same at $\mathcal{R}_m = 100$.

velocity is $u_\varphi = \Gamma/2\pi s$, obviously singular at the position of the vortex $s = 0$. For such a vortex, $\varpi(s) = \Gamma/2\pi s^2$, and (3.92) becomes

$$\mathrm{i}\mathcal{R}_m \frac{f_1}{s^2} = \left(\frac{1}{s}\frac{\mathrm{d}}{\mathrm{d}s}s\frac{\mathrm{d}}{\mathrm{d}s} - \frac{1}{s^2}\right) f_1 , \tag{3.101}$$

where $\mathcal{R}_m = \Gamma/2\pi\eta$. This equation, being homogeneous in s, admits solutions of the form $f_1 = s^{-\lambda}$, where the complex exponent λ satisfies

$$\mathrm{i}\mathcal{R}_m = \lambda^2 - 1 \quad \text{so} \quad \lambda = \pm\sqrt{1 + \mathrm{i}\mathcal{R}_m}\,. \tag{3.102}$$

For $\mathcal{R}_m \gg 1$, the relevant solution, singular at $s = 0$ and falling to zero as $s \to \infty$, is that for which $\lambda \sim (1 + \mathrm{i})(\mathcal{R}_m/2)^{1/2}$. More generally, for arbitrary $\mathcal{R}_m > 0$,

$$\lambda = (p + \mathrm{i}q)/\sqrt{2} \quad \text{where } p^2 = \sqrt{1+\mathcal{R}_m^2}+1,\ q^2 = \sqrt{1+\mathcal{R}_m^2}-1. \tag{3.103}$$

Figure 3.11 shows the resulting **B**-line pattern when (a) $\mathcal{R}_m = 10$, $\lambda \approx 2.35 + 2.13\,\mathrm{i}$, and (b) $\mathcal{R}_m = 100$, $\lambda \approx 7.11 + 7.04\,\mathrm{i}$. The structure is not resolved near $s = 0$, but in any case, in reality, a 'singular' line vortex would always be slightly diffused by viscosity in this neighbourhood. These patterns are self-similar, scaling as s^{-p} with distance from the origin. Comparison with the galactic structure, shown in Figure 5.11, is suggestive!

3.11.4 The intermediate phase

The full time-dependent problem described by (3.87) and (3.88) has been solved by R.L. Parker (1966) for the case of a rigid body rotation $\varpi = \varpi_0$ in $s < s_0$ and zero conductivity ($\eta = \infty$) in $s > s_0$. In this case, there are no currents for $s > s_0$

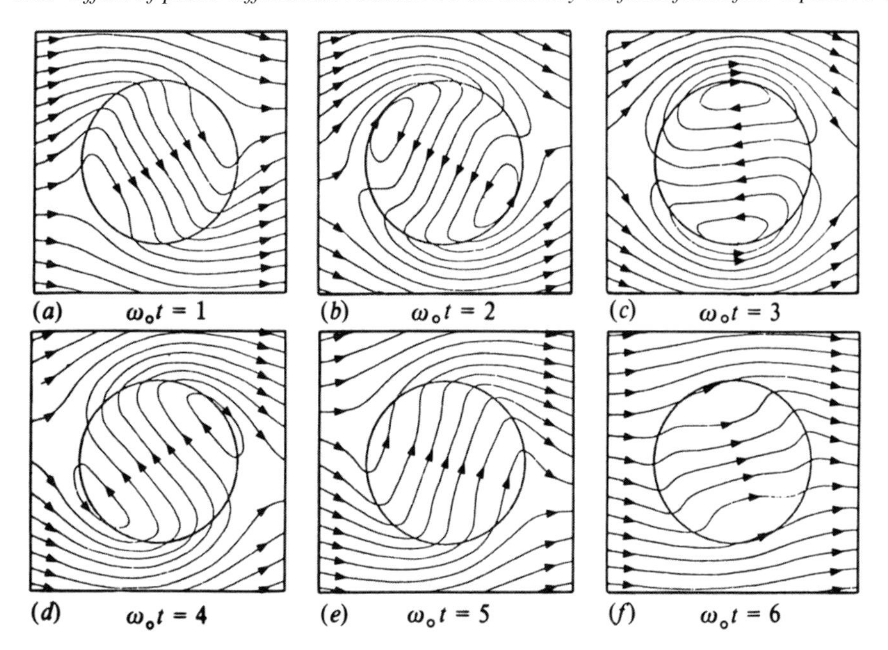

Figure 3.12 Development of lines of force A = const. due to rotation of cylinder with angular velocity ϖ_0; the sense of rotation is clockwise. The sequence (a)–(f) shows one almost complete rotation of the cylinder, with magnetic Reynolds number $\mathcal{R}_m = \varpi_0 a^2/\eta = 100$. [Republished with permission of The Royal Society, from R.L. Parker, 1966; permission conveyed through Copyright Clearance Center, Inc.]

so that $\nabla^2 A = 0$ in this region (for all t), and hence (cf. 3.95a)

$$f(s,t) = s + C(t)s^{-1}\,. \tag{3.104}$$

This function satisfies

$$f + s\,\partial f/\partial s = 2s\,, \tag{3.105}$$

and so continuity of f and $\partial f/\partial s$ across $s = s_0$ provides the boundary condition

$$f + s_0\partial f/\partial s = 2s_0 \quad \text{on} \quad s = s_0 \tag{3.106}$$

for the solution of (3.87). Setting $f = f_1(s) + g(s,t)$, the transient function $g(s,t)$ may be found as a sum of solutions separable in s and t. The result (obtained by Parker by use of the Laplace transform) is

$$g(s,t) = \sum_{n=1}^{\infty} \frac{4s_0 \exp(-i\varpi_0 t - (\varpi_0 t\sigma_n^2/\mathcal{R}_m))J_1(\sigma_n s/s_0)}{\sigma_n^2(i + (\sigma_n^2/\mathcal{R}_m))J_1(\sigma_n)} \tag{3.107}$$

where σ_n is the nth zero of $J_0(\sigma)$.

The lines of force $A =$ const. as computed by Parker for $\mathcal{R}_m = 100$ and for various values of $\varpi_0 t$ during the first revolution of the cylinder are reproduced in

Figure 3.12. According to the estimate of §3.11, diffusion effects may be expected when $\varpi_0 t \sim \mathcal{R}_m{}^{1/3} \sim 4.6$; this is not inconsistent with the behaviour observed.

Note the appearance of closed loops when $\varpi_0 t \approx 2$ and their subsequent disappearance when $\varpi_0 t \approx 5$; this process, of diffusive origin, is clearly responsible for the destruction of flux within the rotating region. The process is repeated in subsequent revolutions, flux being repeatedly expelled until the ultimate steady state is reached. Parker has in fact shown that, when $\mathcal{R}_m = 100$, closed loops appear and disappear during each of the first 15 revolutions of the cylinder but not subsequently. He has also shown that the number of revolutions during which the closed loop cycle occurs increases as $\mathcal{R}_m{}^{3/2}$ for large $\mathcal{R}_m$; in other words it takes a surprisingly long time for the field to settle down in detail to its ultimate form.

3.11.5 Flux expulsion with dynamic back-reaction

As flux expulsion progresses, the back-reaction on the flow due to the Lorentz force grows during the wind-up stage like t^2 and tends to restrict this process. Competing effects can then arise, as analysed by Gilbert et al. (2016). When the initially uniform magnetic field B_0 is very weak, flux expulsion occurs just as described above. When it is very strong however, flux expulsion does not occur; instead, the initial velocity distribution splits into two localised Alfvèn waves, which propagate along the magnetic field in both directions. In the ideal fluid limit ($\eta = \nu = 0$), these are exact solutions of the non-linear MHD equations (see §12.1). For intermediate field strengths in the range $\mathcal{R}_m^{-3/4} \ll B_0/[(\mu_0\rho)^{1/2}\varpi_0 s_0] \ll \mathcal{R}_m^{-1/3}$, the field is strong enough to suppress the differential rotation, but a core is created near $s = 0$ where the angular velocity is nearly uniform; within this core, field loops that are disconnected from the external 'applied' field are rotated without being significantly sheared, and so decay slowly on the natural ohmic time-scale $\mathcal{R}_m\varpi_0^{-1}$.

3.11.6 Flux expulsion by Gaussian angular velocity distribution

Figure 3.13 shows the result of numerical computation with the Gaussian angular velocity distribution (3.91), $\varpi(s) = \varpi_0 \exp\left(-(s/s_0)^2\right)$. The left-hand column is for $\mathcal{R}_m = 10^3$, the central column for $\mathcal{R}_m = 10^6$, and the right-hand column for $\eta = 0$, i.e. $\mathcal{R}_m = \infty$. For finite $\mathcal{R}_m$, the field evolution ultimately settles down to a steady state, as shown in Figure 3.14; in this state, the radius s_f of the 'flux-free zone' is determined in order-of-magnitude by equating the effects of shear and diffusion, a condition that yields

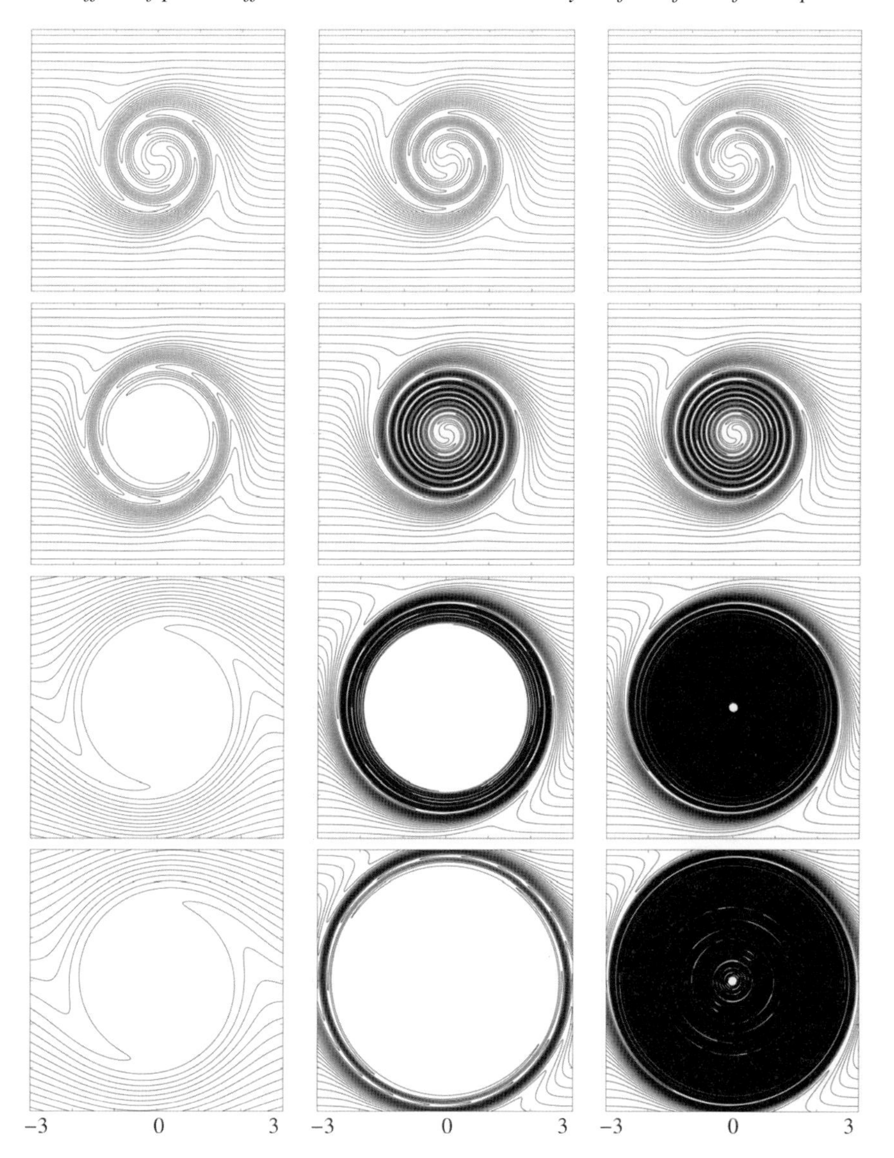

Figure 3.13 Field evolution with Gaussian angular velocity distribution, from left to right $\mathcal{R}_m = 10^3$, $\mathcal{R}_m = 10^6$, $\mathcal{R}_m \sim \infty$, at times $\varpi_0 t = 10, 31.6, 1000, 10{,}000$.

$$s_f/s_0 \sim (\log[\mathcal{R}_m/\log\mathcal{R}_m])^{1/2} \quad \text{as} \quad \mathcal{R}_m \to \infty, \tag{3.108}$$

giving $s_f/s_0 \sim 2.2$ for $\mathcal{R}_m = 10^3$, and $s_f/s_0 \sim 3.3$ for $\mathcal{R}_m = 10^6$.

When $\mathcal{R}_m = \infty$, flux expulsion of course does not occur. The field is wound up more and more tightly into the double spiral already evident at early times, and the radius $s_c(t)$ of the black region where the field is nearly circular grows

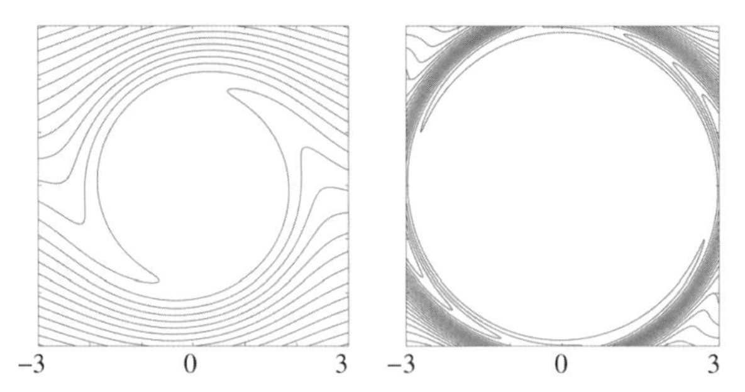

Figure 3.14 Steady state solution as $t \to \infty$, with Gaussian angular velocity distribution for $\mathcal{R}_{\mathrm{m}} = 10^3$ and $\mathcal{R}_{\mathrm{m}} = 10^6$.

monotonically, although extremely slowly:

$$s_c(t)/s_0 \sim (\log[\varpi_0 t/\log \varpi_0 t])^{1/2} \quad \text{as} \quad \varpi_0 t \to \infty. \tag{3.109}$$

When both $\mathcal{R}_{\mathrm{m}}$ and $\varpi_0 t$ are large, it is clear that the order in which the limiting processes $\lim_{\mathcal{R}_{\mathrm{m}}\to\infty}$ and $\lim_{\varpi_0 t\to\infty}$ are applied is of crucial importance!

3.12 Flux expulsion for general flows with closed streamlines

A variety of solutions of (3.55) have been computed by Weiss (1966)[8] for steady velocity fields representing either a single eddy or a regular array of eddies. The computed lines of force develop in much the same way as described for the particular flow of the previous section, closed loops forming and decaying in such a way as to gradually expel all magnetic flux from any region in which the streamlines are closed. The following argument (Proctor 1975), analogous to that given by Batchelor (1956) for vorticity, shows why the field must be zero in the final steady state in any region of closed streamlines in the limit of large $\mathcal{R}_{\mathrm{m}}$ (i.e. $\eta \to 0$).[9]

We consider a steady incompressible velocity field derivable from a stream function $\psi(x, y)$:

$$\mathbf{u} = (\partial\psi/\partial y, -\partial\psi/\partial x, 0)\,. \tag{3.110}$$

In the limit $\eta \to 0$ and under steady conditions, (3.55) becomes $\mathbf{u}\cdot\nabla A = 0$, and so A is constant on streamlines, or equivalently

$$A = A(\psi)\,. \tag{3.111}$$

If η were exactly zero, then any function $A(x, y)$ of the form (3.111) would remain

[8] These were 'state-of-the-art' computations at that time.

[9] A general tendency for two-dimensional turbulence to expel magnetic flux was noted by Zel'dovich (1957), who interpreted this behaviour in terms of a 'diamagnetic' analogy.

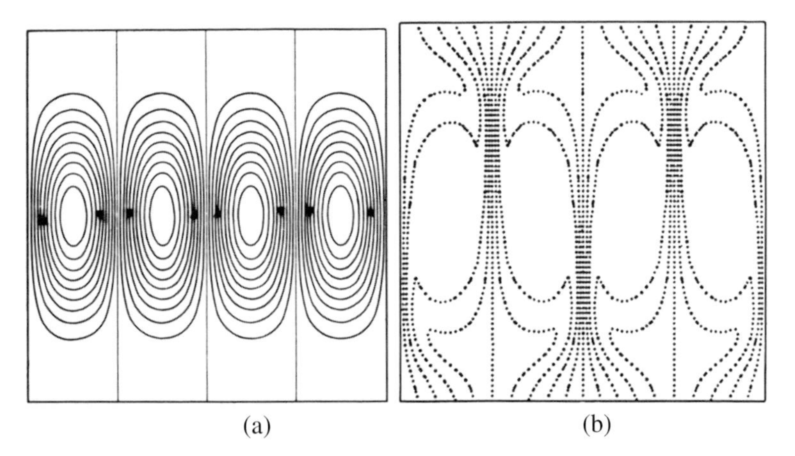

Figure 3.15 Concentration of flux into ropes by an advective layer ($\mathcal{R}_m = 10^3$): (*a*) streamlines ψ =cst, where ψ is given by (3.114); (b) lines of force of the resulting steady magnetic field. [Republished with permission of the Royal Society, from Weiss 1966; permission conveyed through Copyright Clearance Center, Inc.]

steady. However, the effect of non-zero η is to eliminate any variation in A across streamlines. To see this, we integrate the exact steady equation

$$\mathbf{u} \cdot \nabla A \equiv \nabla \cdot (\mathbf{u} A) = \eta \nabla^2 A \tag{3.112}$$

over the area inside any closed streamline C. Since $\mathbf{n}\cdot\mathbf{u} = 0$ on C, where $\mathbf{n}$ is normal to C, the left-hand side integrates to zero, while the right-hand side becomes (with s representing arc length)

$$\oint_C \eta \mathbf{n} \cdot \nabla A \mathrm{d}s = \eta A(\psi) \oint_C (\partial\psi/\partial n) ds = \eta K_C A(\psi) , \tag{3.113}$$

where K_C is the circulation round C. It follows that $A(\psi) = 0$; hence $A =$ const., and so $\mathbf{B} \equiv 0$ throughout the region of closed streamlines.

We have seen in §3.11 that the flux does in fact penetrate a distance δ into the region of closed streamlines, where $\delta = \mathcal{O}\left(\ell_0 \mathcal{R}_m{}^{-1/2}\right)$ and ℓ_0 is the scale of the region. Within this thin layer, the diffusion term in (3.112) is $\mathcal{O}(\eta A/\delta^2)$ and this is of the same order of magnitude as the advective term $\mathbf{u} \cdot \nabla A = \mathcal{O}(u_0 A/\ell_0)$.

The phenomenon of flux expulsion has interesting consequences when a horizontal band of eddies acts on a vertical magnetic field. Figure 3.15, reproduced from Weiss (1966), shows the steady state field structure when

$$\psi(x, y) = -(u_0/4\pi\ell_0)\left(1 - 4y^2/\ell_0^2\right)^4 \sin(4\pi x/\ell_0) , \tag{3.114}$$

and when $\mathcal{R}_m = u_0\ell_0/\eta = 10^3$. The field is concentrated into sheets of flux of thickness $\mathcal{O}(\mathcal{R}_m{}^{-1/2})$ along the vertical planes between neighbouring eddies. The field at the centre of these sheets has order of magnitude $\mathcal{R}_m{}^{1/2} B_0$ where B_0 is the uniform vertical field far from the eddies; this result follows from the fact that the total vertical magnetic flux must be independent of height. This behaviour is comparable with that described by the flux rope solution (3.29), particularly in the situation $\beta = 0$, when the 'rope' becomes a 'sheet'.

3.13 Expulsion of poloidal field by meridional circulation

We consider now the axisymmetric analogue of the result obtained in the preceding section. Let $\mathbf{u}$ be a steady poloidal axisymmetric velocity field with Stokes stream function $\psi(s,z)$ and let $\mathbf{B}$ be a poloidal axisymmetric field with flux-function $\chi(s,z,t)$. Then from (3.61), we have

$$\mathrm{D}\chi/\mathrm{D}t \equiv \partial\chi/\partial t + \mathbf{u}\cdot\nabla\chi = \eta\,\mathcal{D}^2\chi\,, \tag{3.115}$$

with the immediate consequence that when $\eta = 0$, in Lagrangian notation, $\chi(\mathbf{x},t) = \chi(\mathbf{a},0)$. In a region of closed streamlines in meridian planes, steady conditions are therefore possible in the limit $\mathcal{R}_{\mathrm{m}} = \infty$ only if

$$\chi = \chi\,(\psi(s,z))\ . \tag{3.116}$$

Again, as in the plane case, the effect of weak diffusion is to eliminate any variation of χ as a function of ψ. This may be seen as follows.

Using $\nabla\cdot\mathbf{u} = 0$ and the representation (3.64) for $\mathcal{D}^2\chi$, the exact steady equation for χ may be written

$$\nabla\cdot(\mathbf{u}\chi) = \eta\nabla\cdot(\nabla\chi - 2s^{-1}\chi\mathbf{e}_s)\,. \tag{3.117}$$

Let C be any closed streamline in the s-z (meridian) plane, and let S and $\mathcal{T}$ be the surface and interior of the torus described by rotation of C about Oz. Then $\mathbf{u}\cdot\mathbf{n} = 0$ on S, and integration of (3.117) throughout $\mathcal{T}$ leads to

$$\int_S \mathbf{n}\cdot\nabla\chi \mathrm{d}S = \int_S 2s^{-1}\chi\mathbf{e}_s\cdot\mathbf{n}\,\mathrm{d}S\ . \tag{3.118}$$

With $\chi = \chi(\psi)$, and noting that $\mathbf{e}_s\cdot\mathbf{n}\,\mathrm{d}s = \mathbf{e}_z\cdot\mathbf{t}\,\mathrm{d}s = \mathbf{e}_z\cdot\mathrm{d}\mathbf{x}$ on S, where $\mathbf{t}$ is a unit vector tangent to C, (3.118) gives

$$\chi'(\psi)\int_S \mathbf{n}\cdot\nabla\psi\mathrm{d}S = 4\pi\chi\oint_C \mathbf{e}_z\cdot\mathrm{d}\mathbf{x} = 4\pi\chi\oint_C \mathrm{d}z = 0\,, \tag{3.119}$$

so that $\chi'(\psi) = 0$ and hence $\mathbf{B} \equiv 0$ in the region of closed streamlines.

Poloidal magnetic flux is therefore expelled by persistent meridional circulation from regions of closed meridional streamlines in much the same manner as for the plane two-dimensional configuration of §3.12.

3.14 Generation of toroidal field by differential rotation

Consider now an axisymmetric situation in which the velocity field is purely toroidal, i.e.

$$\mathbf{u} = \mathbf{u}_T = s\,\varpi(s,z)\,\mathbf{e}_\varphi\,, \tag{3.120}$$

and a steady poloidal field $\mathbf{B}_P(s,z)$ is maintained by some unspecified mechanism. From (3.54), the toroidal field $\mathbf{B}_T$ then evolves according to the equation

$$\partial \mathbf{B}_T/\partial t = \nabla \wedge (\mathbf{u}_T \wedge \mathbf{B}_P - \eta \nabla \wedge \mathbf{B}_T)\,, \tag{3.121}$$

or equivalently, with $\mathbf{B}_T = B\,\mathbf{e}_\varphi$, from (3.60),

$$\partial B/\partial t = s\,(\mathbf{B}_P \cdot \nabla)\,\varpi + \eta\,(\nabla^2 - s^{-2})\,B\,. \tag{3.122}$$

Note first that if ϖ is constant on $\mathbf{B}_P$-lines so that $\mathbf{B}_P \cdot \nabla\varpi = 0$ and if $B = 0$ at time $t = 0$, then $B = 0$ for $t > 0$ also. This is the *law of isorotation*, one of the earliest results of magnetohydrodynamics (Ferraro 1937). For a perfectly conducting fluid, the result is of course self-evident: if ϖ is constant on $\mathbf{B}_P$-lines, then each $\mathbf{B}_P$-line is rotated without distortion about the axis Oz, and there is no tendency to generate toroidal field.

If $s\,(\mathbf{B}_P \cdot \nabla)\varpi \neq 0$, then undoubtedly a field $B(s,z,t)$ does develop from a zero initial condition according to (3.122). It is not immediately clear whether a net flux of $\mathbf{B}_T$ across the whole meridian plane can develop by this mechanism. Let S_m denote the meridian plane $(0 \leqslant s < \infty, -\infty < z < \infty)$ and let C_m denote its boundary consisting of the z-axis and a semi-circle at infinity. Then integration of (3.121) over S_m gives

$$\frac{\mathrm{d}}{\mathrm{d}t}\int_{S_m} \mathbf{B}_T \cdot \mathbf{n}\,\mathrm{d}S = \oint_{C_m} (\mathbf{u}_T \wedge \mathbf{B}_P - \eta \nabla \wedge \mathbf{B}_T) \cdot \mathrm{d}\mathbf{x}\,. \tag{3.123}$$

We shall assume that $\varpi(s,z)$ is finite on the axis $s = 0$ and, for the sake of simplicity, identically zero outside some sphere of finite radius R; then $\mathbf{u}_T \equiv 0$ on C_m; moreover, as will become apparent from the detailed solutions that follow, $\nabla \wedge \mathbf{B}_T = \mathcal{O}(r^{-3})$ as $r = (s^2 + z^2)^{1/2} \to \infty$, so that (3.123) becomes

$$\frac{\mathrm{d}}{\mathrm{d}t}\int_{S_m} \mathbf{B}_T \cdot \mathbf{n}\,\mathrm{d}S = -\int_{-\infty}^{\infty} \mu_0\,\eta\,(\mathbf{J}_P \cdot \mathbf{e}_z)_{s=0}\,\mathrm{d}z\,. \tag{3.124}$$

Hence a net toroidal flux *can* develop, but only as a result of diffusion, and only if the term $s(\mathbf{B}_P \cdot \nabla)\varpi$ in (3.122) is not antisymmetric about any plane $z =$ const. (in which case B would be similarly antisymmetric, and its total flux would vanish trivially).

Again, as in the discussion at the end of Chapter 2, the crucially important role of diffusion is evident. If $\eta = 0$, then, although toroidal field develops, the integrated toroidal flux across meridian planes remains equal to zero. Only if $\eta \neq 0$ is it possible for *net* toroidal flux to develop. Note incidentally that (3.121) and (3.123) are equally valid if η is an arbitrary function of s and z.

Let us now examine in detail the behaviour of solutions of (3.122), with initial condition $B = 0$ at $t = 0$. As in the discussion of §3.11, there is an initial phase when diffusion effects may be neglected and an ultimate steady state in which diffusion effects are all-important.

3.14.1 The initial phase

Putting $\eta = 0$, the solution of (3.122) is simply

$$B(s, z, t) = s(\mathbf{B}_P \cdot \nabla)\varpi(s, z)\, t \,. \tag{3.125}$$

Physically, it is as if the $\mathbf{B}_P$-lines are gripped by the fluid and 'cranked round the z-axis' in regions where ϖ is greatest. In the important special case when $\mathbf{B}_P = B_0\,\mathbf{e}_z$, B_0 being constant, (3.125) becomes

$$B(s, z, t) = sB_0(\partial\varpi/\partial z)t \,, \tag{3.126}$$

and it is evident that if ϖ is symmetric about the plane $z = 0$ then B is antisymmetric, and vice versa. In general if the $\mathbf{B}_P$-lines are symmetrical about the plane $z = 0$, then B exhibits the opposite symmetry from the product $B_z\,\varpi$.

The neglected diffusion term of (3.122) has order of magnitude $\eta\, B/\ell_0^2$, where ℓ_0 is the length-scale of the region over which ϖ varies appreciably. This becomes comparable with the retained term $s(\mathbf{B}_P \cdot \nabla)\varpi$ when $t = \mathcal{O}(\ell_0^2/\eta)$; if $\varpi = \mathcal{O}(\varpi_0)$ and $|\nabla\varpi| = \mathcal{O}(\varpi_0/\ell_0)$, then at this stage (contrast (3.76))

$$B = \mathcal{O}(\mathcal{R}_{\mathrm{m}})B_0 \quad \text{where} \quad \mathcal{R}_{\mathrm{m}} = \varpi_0\ell_0^2/\eta \,. \tag{3.127}$$

3.14.2 The ultimate steady state

When $t \gg \ell_0^2/\eta$, B may be expected to reach a steady state given by

$$(\nabla^2 - s^{-2})B = -\eta^{-1}\, s\,(\mathbf{B}_P \cdot \nabla)\varpi \,. \tag{3.128}$$

To solve this, note first that if $\mathbf{B}_T = \nabla \wedge (\mathbf{x}T)$ then, from equation (2.58), $B = -\partial T/\partial\theta$, and it follows easily that

$$(\nabla^2 - s^{-2})B = -\partial(\nabla^2 T)/\partial\theta \,. \tag{3.129}$$

Suppose that T has the expansion

$$T = \sum_n f_n(r)\, P_n(\cos\theta) \,. \tag{3.130}$$

Then

$$\nabla^2 T = \sum_n g_n(r)\, P_n(\cos\theta), \quad \text{where} \quad g_n(r) = r^{-2}[(r^2 f_n')' - n(n+1) f_n]. \quad (3.131)$$

If the right-hand side of (3.128) has the expansion

$$-\eta^{-1}\, s\, (\mathbf{B}_P \cdot \nabla)\, \varpi = -\sum_1^\infty g_n(r)\, \mathrm{d}P_n(\cos\theta)/\mathrm{d}\theta\,, \qquad (3.132)$$

then (3.130) gives the appropriate form of $T(r,\theta)$, and the corresponding $B(r,\theta)$ is given by

$$B(r,\theta) = -\sum_1^\infty f_n(r)\, \mathrm{d}P_n(\cos\theta)/\mathrm{d}\theta\,. \qquad (3.133)$$

If $(\mathbf{B}_P \cdot \nabla)\, \varpi$ is regular at $r = 0$, then clearly from (3.132), $g_n(0) = 0$ for each n. We shall moreover suppose that, for each n, $g_n(r) = \mathcal{O}(r^{-3})$ at most[10] as $r \to \infty$. The solution $f_n(r)$ of (3.131b) for which $f_n(0)$ is finite and $f_n(\infty) = 0$ is then

$$f_n(r) = \frac{-1}{2n+1}\left\{\frac{1}{r^{n+1}}\int_0^r x^{n+2} g_n(x)\, \mathrm{d}x + r^n \int_r^\infty \frac{g_n(x)}{x^{n-1}}\, \mathrm{d}x\right\}. \qquad (3.134)$$

A case of particular interest is that in which $\nabla \wedge \mathbf{B}_P = 0$, so that $\mathbf{B}_P = -\nabla\Psi$, with $\nabla^2\Psi = 0$. Then Ψ, being axisymmetric and finite at $r = 0$, has the expansion

$$\Psi = \sum_{m=1}^\infty \Psi_m\,, \quad \Psi_m = -A_m r^m P_m(\cos\theta)\,. \qquad (3.135)$$

Let us suppose further, to simplify matters, that ϖ is a function of $r = (s^2 + z^2)^{1/2}$ only. Then, using the recurrence relation for Legendre polynomials $(1+2m)P_m(\mu) = P'_{m+1}(\mu) - P'_{m-1}(\mu)$, we have

$$\begin{aligned} s(\mathbf{B}_P \cdot \nabla)\varpi &= -r\sin\theta \frac{\partial\Psi}{\partial r}\varpi'(r) = \sum_{m=1}^\infty m A_m r^m \varpi'(r) \sin\theta\, P_m(\cos\theta) \\ &= \sum_{m=1}^\infty \frac{m A_m}{2m+1} r^m \varpi'(r) \frac{\mathrm{d}}{\mathrm{d}\theta}(P_{m+1}(\cos\theta) - P_{m-1}(\cos\theta))\,. \end{aligned} \qquad (3.136)$$

Hence comparing with (3.132), we have for $n = 1, 2, \ldots$,

$$\eta\, g_n(r) = -\left\{\frac{(n-1)A_{n-1} r^{n-1}}{2n-1} - \frac{(n+1)A_{n+1} r^{n+1}}{2n+3}\right\}\varpi'(r)\,, \qquad (3.137)$$

[10] This assumption is of course much less restrictive than the assumption $g_n(r) \equiv 0$ for $r > R$, but it includes this possibility; the effect of the differential rotation is *localised* provided the $g_n(r)$ fall off sufficiently rapidly with r.

wherein we may take $A_0 = 0$. From (3.134), the corresponding expression for $f_n(r)$ (after integrating by parts and simplifying) is given by

$$\eta f_n = \frac{-(n-1)A_{n-1}}{(2n-1)r^{n+1}} \int_0^r x^{2n}\, \varpi(x)\, \mathrm{d}x + \frac{(n+1)A_{n+1}}{(2n+1)r^{n+1}} \int_0^r x^{2n+2} \varpi(x)\, \mathrm{d}x + \frac{2(n+1)A_{n+1}r^n}{(2n+3)(2n+1)} \int_r^\infty x\varpi(x)\, \mathrm{d}x\,. \tag{3.138}$$

In the particular case when $\mathbf{B}_P = B_0\mathbf{e}_z$, we have only the term $\Psi = \Psi_1$, with $A_1 = B_0$. Then from (3.133) and (3.138),

$$B(r,\theta) = -\eta^{-1} B_0 \sin\theta \cos\theta\, r^{-3} \int_0^r x^4 \varpi(x)\, \mathrm{d}x\,. \tag{3.139}$$

Note that the condition $\varpi = \mathcal{O}(x^{-6})$ as $x \to \infty$ is sufficient in this case to ensure that $B = \mathcal{O}(r^{-3})$ at infinity.

Similarly, if $\Psi = -A_2 r^2 P_2(\cos\theta)$, then B is given by

$$B(r,\theta) = \eta^{-1} A_2 \left\{ \frac{2}{3} r^{-2} \int_0^r x^4 \varpi(x) \mathrm{d}x + \frac{4}{15} r \int_r^\infty x\varpi(x)\, \mathrm{d}x \right\} \sin\theta + \frac{3}{5}\eta^{-1} A_2 \left\{ r^{-4} \int_0^r x^6 \varpi(x)\, \mathrm{d}x \right\} (1 - 5\cos^2\theta)\sin\theta\,. \tag{3.140}$$

The most important thing to notice about this rather complicated expression is its asymptotic behaviour as $r \to \infty$: if $\varpi(x) = \mathcal{O}(x^{-6})$ as $x \to \infty$, then

$$B(r,\theta) \sim \frac{2A_2 \sin\theta}{3\eta r^2} \int_0^\infty x^4 \varpi(x)\, \mathrm{d}x \quad \text{as} \quad r \to \infty\,. \tag{3.141}$$

This is to be contrasted with the behaviour (3.139) (which implies $B \propto r^{-3}$ as $r \to \infty$) in the former case. The expression (3.140) exhibits an *infinite* toroidal flux over the meridian plane S_m. As is clear from the introductory discussion in this section, this flux arises through the action of diffusion, which of course has an infinite time to operate before the steady field (3.140) can be established throughout all space.

The slower decrease of B with r given by (3.141) as compared with that given by (3.139) is attributable to the symmetry of the B-field about the plane $z = 0$ ($\theta = \pi/2$); the simple $\sin\theta$ dependence of (3.141) makes the associated B less vulnerable to the influence of diffusion than the more complicated $\sin\theta\cos\theta$ structure of (3.139) and the field therefore diffuses further from the region of differential rotation. This has the important consequence that, in general, if $\mathbf{B}_P$ is non-uniform, it is the gradient of $\mathbf{B}_P$ in the neighbourhood of the centre of rotation (rather than its local average value) that determines the toroidal field generated at a very great distance.

Finally note that in all cases, in the steady state,

$$\max |B| = \mathcal{O}(\mathcal{R}_{\mathrm{m}})|\mathbf{B}_P| \,. \tag{3.142}$$

This means that, unlike the situation considered in §3.11, the toroidal field here increases to values of order $\mathcal{R}_{\mathrm{m}} B_P$ and then levels off through the diffusion process without further change in order of magnitude, the whole process taking a time of order ℓ_0^2/η. There is no suggestion of any flux expulsion mechanism here; flux expulsion does not occur if the 'applied field' $\mathbf{B}_P$ is symmetric about the axis of rotation.

The analysis given above can be modified to cope with the situation when the poloidal field $\mathbf{B}_P$ has non-axisymmetric as well as axisymmetric ingredients (see Herzenberg & Lowes 1957 for the case of a rigid spherical rotator imbedded in a solid conductor). The analysis is complicated by the appearance of spherical Bessel functions in the inversion of the operator $\nabla^2 - s^{-2}$, but the result is not unexpected: the non-axisymmetric ingredients are expelled from the rotating region when $\mathcal{R}_{\mathrm{m}} \gg 1$, and the axisymmetric ingredient is distorted, without expulsion, in the manner described above.

3.15 Topological pumping of magnetic flux

A fundamental variant of the flux expulsion mechanism discussed in §3.11 was discovered by Drobyshevski & Yuferev (1974). This study was motivated by the observation that in steady thermal convection between horizontal planes, the lower plane being heated uniformly, the convection cell pattern generally exhibits what may be described as a topological asymmetry about the centre-plane: fluid generally rises at the centre of the convection cells and falls on the periphery, so that regions of rising fluid are separated from each other whereas regions of falling fluid are all connected. The reason for this type of behaviour must be sought in the non-linear dynamical stability characteristics of the problem: it certainly cannot be explained in terms of linear stability analysis since if $\mathbf{u}(\mathbf{x}, t)$ is any velocity field satisfying linearised stability equations about a state of rest, then $-\mathbf{u}(\mathbf{x}, t)$ is another solution. However, it would be inappropriate to digress in this manner here; in the spirit of the present kinematic approach, let us simply assume that a steady velocity field $\mathbf{u}(\mathbf{x})$ exhibiting the above kind of topological asymmetry is given, and we consider the consequences for an initial horizontal magnetic field $\mathbf{B}$ which is subject to diffusion and to convection by $\mathbf{u}(\mathbf{x})$.

Suppose that the fluid is contained between the two planes $z = 0, z_0$. Near the upper plane $z = z_0$ the rising fluid must diverge, so that the flow is everywhere directed towards the periphery of the convection cells. A horizontal $\mathbf{B}$-line near $z = z_0$ will then tend to be distorted by this motion so as to lie everywhere near

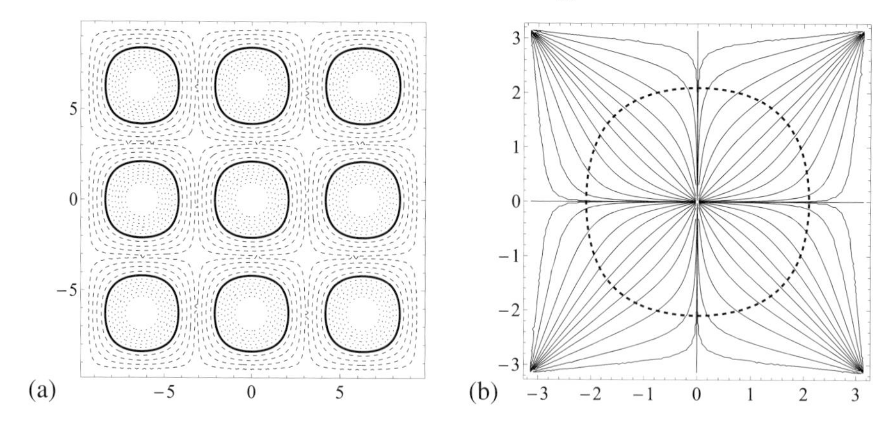

Figure 3.16 (a) Contours of vertical velocity $u_z = \cos x + \cos y + \cos x \cos y =$ const. in the region $\{-3\pi < x,\ y < 3\pi\}$, covering nine cells of the flow; $u_z > 0$ in the disconnected regions inside the solid contours, and $u_z < 0$ in the connected region outside these contours. (b) The family of curves (3.145) defining the cylindrical surfaces that contain the streamlines of the velocity field (3.143) within a single cell of the flow $\{-\pi < x, y < \pi\}$.

cell peripheries, where it can then be convected downwards. A horizontal **B**-line near $z = 0$, by contrast, cannot be distorted so as to lie everywhere in a region of rising fluid, since these regions are disconnected. Hence a **B**-line *cannot* be convected upwards (although loops of field can be lifted by each rising blob of fluid). It follows that, as far as the horizontal average $\mathbf{B}_0(z,t) = \langle \mathbf{B}(x,y,z,t) \rangle$ is concerned, there is a valve effect which permits downward transport but prohibits upward transport. This effect will be opposed by diffusion; but one would expect on the basis of this physical argument that an equilibrium distribution $\mathbf{B}_0(z)$ will develop, asymmetric about $z = \frac{1}{2}z_0$, and with greater flux in the lower half, the degree of asymmetry being related to the relevant magnetic Reynolds number.

A regular cell pattern over the horizontal plane can be characterised by cell boundaries that are either triangular, square or hexagonal. In normal Bénard convection, the hexagonal pattern is preferred (again for reasons associated with the non-linear dynamics of the system). A velocity field with square cell boundaries however allows simpler analysis, and the qualitative behaviour is undoubtedly the same whether hexagons or squares are chosen. Drobyshevski & Yuferev (1974) chose as their velocity field $\mathbf{u} = (u_x, u_y, u_z)$, where

$$\left.\begin{aligned} u_x &= -u_0 \sin\hat{x}(1 + \tfrac{1}{2}\cos\hat{y})\cos\hat{z} \\ u_y &= -u_0 \sin\hat{y}(1 + \tfrac{1}{2}\cos\hat{x})\cos\hat{z} \\ u_z &= u_0(\cos\hat{x} + \cos\hat{y} + \cos\hat{x}\cos\hat{y})\sin\hat{z} \end{aligned}\right\} . \tag{3.143}$$

Here $\hat{\mathbf{x}} = \pi\mathbf{x}/z_0$ (and we immediately drop the hats for simplicity of notation). The magnetic Reynolds number is $\mathcal{R}_m = u_0 z_0/\eta$. The velocity field (3.143) satisfies $\nabla\cdot\mathbf{u} = 0$, and $\mathbf{n}\cdot\mathbf{u} = 0$ on $z = 0,\ \pi$; has square cell boundaries at $x = (2n+1)\pi,\ y = (2m+1)\pi,\ (n,\ m = 0,\ \pm1,\ \pm2,\ldots)$; and has the basic topological asymmetry referred to above. Contours of vertical velocity $u_z = \cos x + \cos y + \cos x \cos y = \text{const.}$ (the same at each level z) are shown in Figure 3.16a. The fluid rises in the disconnected regions inside the solid contours, and falls in the connected region outside these contours. On any streamline of the flow, we have $\mathrm{d}x/u_x = \mathrm{d}y/u_y = \mathrm{d}z/u_z$, so that

$$\frac{(1+\frac{1}{2}\cos x)\,\mathrm{d}x}{\sin x\cos z} = \frac{(1+\frac{1}{2}\cos y)\,\mathrm{d}y}{\sin y\cos z} = \frac{-(1+\frac{1}{2}\cos x)(1+\frac{1}{2}\cos y)\,\mathrm{d}z}{(\cos x+\cos y+\cos x\cos y)\sin z}, \quad (3.144)$$

which provides a first integral

$$\mathcal{I}(x,y) \equiv \frac{\cos^2(x/2)\sin y}{\cos^2(y/2)\sin x} = \text{const.} \quad (3.145)$$

Some members of the family of curves $\mathcal{I}(x,y) = \text{const.}$ are shown in Figure 3.16b for a single cell of the flow $\{-\pi < x, y < \pi\}$. Any fluid particle follows a closed curve, rising on a surface $\mathcal{I}(x,y) = \text{const.}$ inside the dashed curve, and falling outside it.

Under steady conditions, the magnetic field $\mathbf{B}(\mathbf{x})$ must satisfy (3.14) (with $\partial\mathbf{B}/\partial t = 0$). Putting $\hat{\mathbf{u}} = \mathbf{u}/u_0$, and dropping the hat also on $\hat{\mathbf{u}}$, this becomes,

$$\nabla^2\mathbf{B} = -\varepsilon\nabla\wedge(\mathbf{u}\wedge\mathbf{B})\,,\quad \varepsilon = \mathcal{R}_m/\pi = z_0\,u_0/\pi\,\eta\,. \quad (3.146)$$

In order to solve this equation, we need boundary conditions on $\mathbf{B}$. Following Drobyshevski & Yuferev, we assume that the solid regions $z < 0$ and $z > \pi$ are perfect electrical conductors in which $\mathbf{E}$ and $\mathbf{B}$ vanish; then from (2.5),

$$B_z = 0 \ \text{ on } \ z = 0\,,\ \pi\,; \quad (3.147)$$

also, from (2.125) and (2.131), we have that $\mathbf{n}\wedge\mathbf{J} = 0$ on the boundaries, or equivalently, since $\mu_o\mathbf{J} = \nabla\wedge\mathbf{B}$,

$$\partial B_x/\partial z = \partial B_y/\partial z = 0 \ \text{ on } \ z = 0\,,\ \pi\,. \quad (3.148)$$

(There is a surface current on the boundaries under these conditions.) Finally we suppose that in the absence of fluid motion (or equivalently if $\mathcal{R}_m = 0$) the field $\mathbf{B}$ is uniform and in the x-direction, $(B_0, 0, 0)$; the flux $\Phi_0 = z_0 B_0$ is trapped between the perfectly conducting planes, and the problem is to determine the distribution of this flux according to (3.146) when $\mathcal{R}_m \neq 0$.

If $\varepsilon \ll 1$, the problem can be solved in power series

$$\mathbf{B}(\mathbf{x}) = \sum_{n=0}^{\infty} \varepsilon^n\,\mathbf{B}^{(n)}(\mathbf{x})\,, \quad (3.149)$$

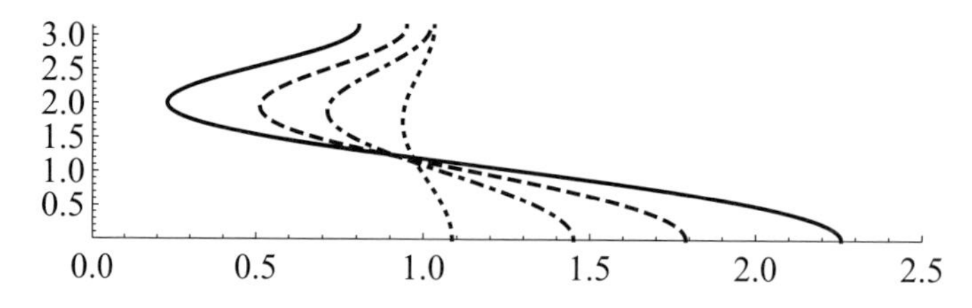

Figure 3.17 Mean field $\langle B \rangle (z)$, as given by (3.152), for $\mathcal{R}_m = \pi\varepsilon = 2, 4, 5$ and 6; as $\mathcal{R}_m$ increases, the downward pumping of magnetic flux becomes very prominent.

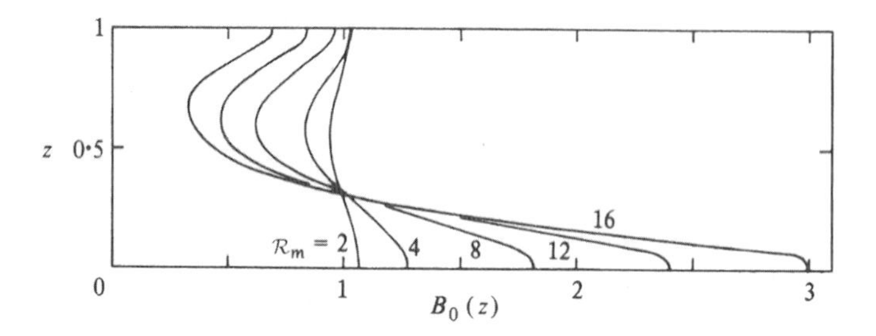

Figure 3.18 Mean field distribution for different values of $\mathcal{R}_m$ as computed by Drobyshevski & Yuferev (1974). Note (i) the almost symmetric, but weak, field expulsion effect when $\mathcal{R}_m = 2$; (ii) the strong concentration of field near the lower boundary when $\mathcal{R}_m = 16$.

where $\mathbf{B}^{(0)} = (B_0, 0, 0)$, and from (3.146),

$$\nabla^2 \mathbf{B}^{(n+1)} = -\nabla \wedge (\mathbf{u} \wedge \mathbf{B}^{(n)}) \quad (n = 0, 1, 2, \ldots) \,. \tag{3.150}$$

Since the right-hand side is always a space-periodic function, inversion of the operator ∇^2 simply requires repeated use of the identity

$$\begin{aligned} &\nabla^{-2} \cos(lx + p) \cos(my + q) \cos(nz + r) \\ &\quad = -(l^2 + m^2 + n^2)^{-1} \cos(lx + p) \cos(my + q) \cos(nz + r) \,. \end{aligned} \tag{3.151}$$

The horizontal averages $\mathbf{B}_0^{(n)}(z) = \langle \mathbf{B}^{(n)}(x, y, z) \rangle$ can then be derived. The procedure is somewhat tedious, and it is necessary to go as far as the term $\mathbf{B}_0^{(3)}(z)$ before the pumping effect appears. The result at that level of approximation is $\langle \mathbf{B} \rangle (z) = \langle B \rangle (z)\, \mathbf{e}_x$, where

$$\langle B \rangle (z) = B_0 \left(1 + \frac{7\varepsilon^2}{48} \cos 2z + \frac{\varepsilon^3}{240} (28 \cos z - 3 \cos 3z) + \mathcal{O}(\varepsilon^4) \right) . \tag{3.152}$$

This function is shown in Figure 3.17, which clearly shows the downward pumping of magnetic flux. In (3.152), the term of order ε^2 is *symmetric* about the centre plane $z = \pi/2$; at this level there is therefore symmetrical flux expulsion of the type

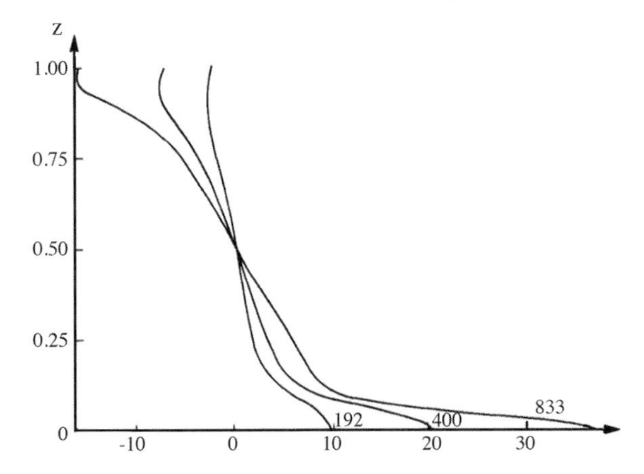

Figure 3.19 Mean field $\langle B\rangle(z)$ for $\mathcal{R}_m = 192, 400$ and 833. [From Arter 1985.]

obtained for two-dimensional flows by Weiss (1966) (but here of course the effect is weak when $\mathcal{R}_m$ is small). The term of order ε^3 is however antisymmetric about $z = \pi/2$, and shows as expected that there is a net downward transport of flux per unit width; in fact the difference in flux between the lower half and the upper half is, returning to dimensional variables,

$$\Delta\Phi = \left(\int_0^{z_0/2} - \int_{z_0/2}^{z_0}\right) B_0\left(\frac{\pi z}{z_0}\right) dz = \frac{B_0 z_0}{\pi}\left(0.24\,\varepsilon^3 + \mathcal{O}(\varepsilon^5)\right) . \tag{3.153}$$

It may be noted that if the expansion (3.152) is continued, only terms involving odd powers of ε can contribute to the asymmetric flux pumping effect since the even-power terms are invariant under the sign change $u_0 \to -u_0$.

The function $\langle B\rangle(z)$ was computed directly from (3.146) (by a computational method involving truncation of Fourier series) by Drobyshevski & Yuferev for 5 values of $\mathcal{R}_m$ up to $\mathcal{R}_m = 16$. The qualitative agreement of Figures 3.17 and 3.18 is striking; but in fact the truncated expansion (3.152) seriously overestimates the topological pumping effect when $\mathcal{R}_m \gtrsim 1$.

The computations have been carried to much higher $\mathcal{R}_m$ by Arter et al. (1982); see also Arter (1985). Figure 3.19 shows the mean field $\langle B\rangle(z)$ for $\mathcal{R}_m = 192, 400$ and 833. What is remarkable here is that the flux is negative for $1/2 < z < 1$, this being compensated by greater positive flux for $0 < z < 1/2$. Similar behaviour has been found by Galloway & Proctor (1983) for a hexagonal cell pattern exhibiting the same topological asymmetry between upward and downward flow. In both situations, a flux concentration develops on the lower boundary as $\mathcal{R}_m \to \infty$, in a boundary layer of thickness $\mathcal{O}(\mathcal{R}_m{}^{-1/2})$, but, unlike the situation when the cellular flow is two-dimensional, there is no evidence of significant flux expulsion from the interior region as $\mathcal{R}_m$ increases.

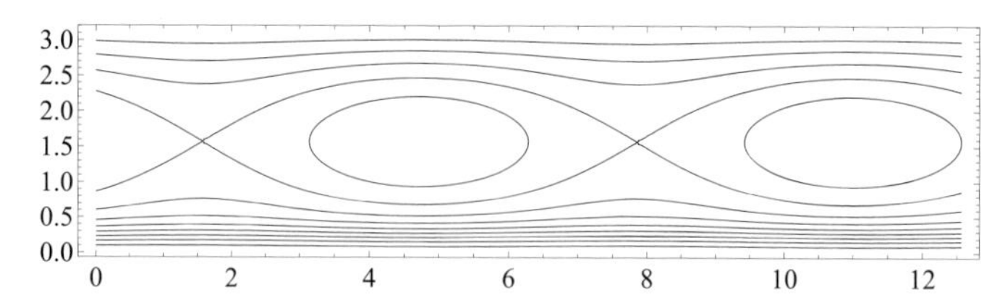

Figure 3.20 Presumed field structure of y-averaged **B**-field, when negative flux is present in the upper half.

Curiously, the low-$\mathcal{R}_{\mathrm{m}}$ formula (3.152) also gives negative flux in the upper-half region when $\mathcal{R}_{\mathrm{m}} \gtrsim 5$, but this truncation cannot be trusted for such large $\mathcal{R}_{\mathrm{m}}$. It is difficult to understand this phenomenon in physical terms; but presumably, when averaged over y, the field lines must have a structure like that sketched in Figure 3.20. Galloway & Proctor (1983) argue that diffusion is responsible for the appearance of X-type neutral points which are a necessary concomitant of the reversed field structure.

4

The Magnetic Field of the Earth and Planets

4.1 Planetary magnetic fields in general

Over the last half-century, we have witnessed extraordinary achievements in space travel and planetary exploration, which now make it possible to see the Earth's magnetic field in the proper context of planetary magnetic fields in general. Among these achievements, we may note in particular NASA's Voyager missions. Voyager 1, launched in 1977, flew by Jupiter (a 'close flyby') in March 1979 and by Saturn in November 1980. Voyager 2, also launched in 1977, flew by Jupiter in July 1979 and by Saturn in August 1981, and went on to a close flyby of Uranus (January 1986) and Neptune (August 1989); by 2013, both spacecraft were at the outer limit of the solar system, bound for interstellar space, and still sending signals back to Earth,[1] a stunning engineering achievement well-timed for the extended dawn of this third millennium.

More recently, NASA's Messenger spacecraft was launched in August 2004 for the purpose of scientific exploration of the innermost planet Mercury; following three preliminary flybys, it was dynamically tuned to enter into orbit of Mercury in March 2011, and has been sending scientific data back to Earth since then, again an outstanding engineering achievement.

Certain obviously relevant properties of the planets, for all of which, as a result of these and earlier space missions, magnetic field measurements are now available, are summarised in Table 4.1. Of these planets, Jupiter has the strongest mean surface field (as measured by the quantity μ/R^3), followed by the Earth. At the other extreme, the field of Venus, if it exists at all, is extremely weak and below the threshold of present detectors.

It is now widely agreed that the mechanism of generation and maintenance of the Earth's field is to be sought in the inductive motion, strongly influenced by Coriolis forces, in its liquid core. Seismological studies, coupled with knowledge

[1] This 'space-mail' communication is expected to continue until 2025.

Table 4.1 *Properties of the planets Mercury, Venus, Earth, Mars, Jupiter, Saturn, Uranus and Neptune*

Planet	Radius R km	Mean density kg m^{-3}	Rotation period days	Angular velocity Ω s^{-1}	Dipole moment μ T km^3	μ/R^3 G$(10^{-4}$T)
Mercury	2440	5400	59	1.23×10^{-6}	4.8×10^3	3.3×10^{-3}
Venus	6050	5200	243	-2.99×10^{-7}	$< 4 \times 10^3$	$< 1.8 \times 10^{-4}$
Earth	6371	5500	1.0	7.27×10^{-5}	7.72×10^6	0.299
Mars	3390	3900	1.026	7.09×10^{-5}	2.47×10^3	6.36×10^{-4}
Jupiter	71400	1240	0.41	1.77×10^{-4}	1.45×10^{11}	3.98
Saturn	60300	620	0.44	1.65×10^{-4}	4.6×10^9	0.21
Uranus	25900	1250	0.72	1.00×10^{-4}	3.9×10^8	0.23
Neptune	24800	1600	0.671	1.08×10^{-4}	3.9×10^8	0.23

The data for Mercury are derived from Ness et al. (1975), for Mars from Dolginov et al. (1973), for Jupiter from Warwick (1963) and Smith et al. (1974), for Saturn from Smith et al. (1980), for Uranus from Ness et al. (1986) and for Neptune from Connerney et al. (1991). The rotation of Venus is retrograde relative to its sense of rotation round the Sun. The rotations of the other planets are all prograde. The quantity μ/R^3 in the final column provides a measure of mean surface field strength.

of the density distribution within the Earth and the relative abundances of chemical compounds of which it is composed, lead to the conclusion that the gross structure of the Earth is as indicated in Figure 4.1 (see, for example, Jacobs 1975).

The radial profiles in Figure 4.1 are based on the Preliminary Reference Earth Model (PREM) (Dziewonski & Anderson 1981) which, though described as 'preliminary', nevertheless provides a reliable model of the spherically averaged properties of the Earth's interior. The increase of density ρ with depth is shown in the left part of Figure 4.1a, and the resulting profile of gravity g is shown in the right part; this is approximately linear in the core.

The PREM also provides radial profiles of seismic wave velocities, for both compression (V_p) and shear (V_s) waves. The resulting bulk modulus $K = \rho(V_p^2 - \frac{4}{3}V_s^2)$ is shown in the left part of Figure 4.1b; the inverse of this quantity provides a measure of compressibility. The approximate pressure profile within the Earth is part of the PREM model, and is shown in the right part of Figure 4.1b. Comparison with the bulk modulus indicates that compression of material by its own weight is not entirely negligible within the Earth, although compressibility resulting from dynamic effects can safely be ignored.

If we take $R_{\rm E} = 6371$ km as the mean radius of the Earth (there are of course slight departures from exact sphericity), then Figure 4.1 reveals distinct transitions at $r = R_{\rm C}$ (the Gutenberg discontinuity) and at $r = R_{\rm I}$ (the Bullen or Lehmann discontinuity), where

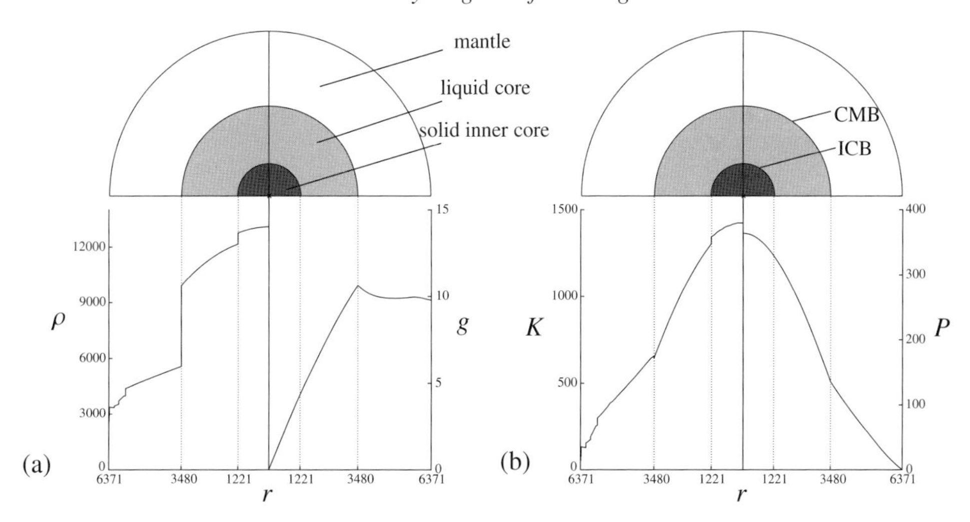

Figure 4.1 Interior structures of the Earth ($R_E = 6371$ km) from the centre to the surface: the solid inner core, iron/nickel alloy (black); the liquid outer core, iron/nickel and some lighter elements (grey); the 'solid' mantle (white). Radial profiles of (a) density (kg m^{-3}) and gravity (m s^{-2}), (b) bulk modulus (GPa) and pressure (GPa), as derived from the PREM model. [Dziewonski & Anderson 1981.]

$$R_C \approx 0.55 R_E, \quad R_I \approx 0.35 R_C. \tag{4.1}$$

The radius $r = R_C$ marks the core boundary; the outer core, $R_I < r < R_C$, consists of molten metal (shear waves cannot propagate through this region). We know from the analysis of iron meteorites and the observation of the Earth's moment of inertia, that the primary constituent of the Earth's core is iron, alloyed with some 5% nickel and some lighter elements (such as carbon, oxygen, silicon and sulphur).

The inner core $r < R_I$ is solid, most probably an alloy of iron and nickel, formed by slow crystallisation from the outer core. During this process, the lighter elements are expelled from the growing inner core, providing a compositional source of buoyancy in the outer core. The mantle $r > R_C$ is also solid, although subject to viscoplastic deformation on time-scales of the order of millions of years.

From the standpoint of dynamo theory in the terrestrial context, we are therefore faced with the problem of fluid flow in a rotating spherical shell, $R_I < r < R_C$, and the electric currents and magnetic fields that such flow may generate. The fluid may reasonably be regarded as incompressible (although density-stratified under its own weight), and in a purely kinematic approach any kinematically possible velocity fields, satisfying merely $\nabla \cdot \mathbf{u} = 0$, and $\mathbf{n} \cdot \mathbf{u} = 0$ on $r = R_I,\ R_C$, may be considered. At a subsequent stage it is of course essential to consider the nature of the forces (or of the sources of energy) that may be available to drive the motions. We defer to §4.5 further consideration of the physical state of the Earth's interior.

The inference that planetary dynamo action requires both a conducting fluid core and a 'sufficient' degree of rotation (i.e. a sufficient influence of Coriolis forces) is to some extent supported (in a most preliminary way) by the information contained in Table 4.1. Venus, with approximately the same radius and mean density as the Earth, has presumably a comparable structure (see Figure 4.2); it rotates very slowly, however, as compared with the Earth, and it is reasonable to suppose that this is why it exhibits no magnetic field.

Mars on the other hand rotates at approximately the same angular velocity as the Earth and has no global magnetic field. However, the Mars Global Surveyor (1996-2006) detected a crustal magnetic field on Mars more than 30 times stronger than that of the Earth, suggesting that Mars had a global field earlier in its evolution, possibly with polarity reversals, as suggested by the textured nature of the crustal field. The presence of a metallic core in Mars, dominated by iron-nickel-sulphur, has been established by martian meteorite geochemistry. Precise determination of the solar tidal deformation (Yoder et al. 2003) has provided constraints on structural models of the planet, requiring a core radius of 1520 to 1840 km, which may be compared with its mean planetary radius of 3390 km (see Figure 4.2). The same data also indicate that the present metallic core cannot be completely solid (Yoder et al. 2003). These data are however consistent with a range of possibilities between an entirely liquid core and a mainly solidified core with only a small outer liquid region. Stewart et al. (2007) point to absence of crystallisation in most, if not all, of the present martian core. If the core of Mars is largely liquid, a possible explanation for the absence of a global field (applicable also to Venus) could be that reduced heat flux from the interior, as indicated by the lack of plate tectonics, provides less energy to drive convection in the core (Nimmo & Stevenson 2000; Nimmo 2002).[2]

Mercury is the only inner (terrestrial) planet of the solar system other than Earth with an active internally generated magnetic field. Mariner 10 revealed that Mercury possesses a magnetic field sufficient to deflect the solar wind (Connerney & Ness 1988). Messenger observations (Anderson et al. 2008; Anderson et al. 2011) revealed that Mercury's field is dipolar, axial (the magnetic axis is tilted by less than 3° from the rotation axis) and weak (the surface field strength is about 1% that of the Earth). The large longitude libration of Mercury points to the existence of a molten core (Margot et al. 2007). Low-altitude measurements of crustal magnetisation suggest that Mercury's field is almost as old as the planet itself, i.e. ~ 3.8 billion years (Johnson et al. 2015). The planet rotates slowly, and its substantial dipole moment (relative to size) is therefore something of a surprise. The energy budget

[2] The InSight mission to Mars, which landed in November 2018, is expected to provide more detailed information on the structure of the planet.

to drive Mercury's dynamo remains an issue. Intermediate between the Earth and Venus in both its rotation rate and its mean surface field strength, Mercury provides a key test case for dynamo models (Christensen 2006). The BepiColombo mission (joint ESA and JAXA), launched in October 2018 and due to arrive at Mercury in December 2025, should provide crucial new information on Mercury's magnetic field.

The giant planets Jupiter and Saturn are of great interest in the dynamo context. Their low mean density indicates an internal constitution totally different from that of the four terrestrial planets. The hypothetical structure of Jupiter, based on total mass, moments of inertia, and general arguments concerning relative abundance of elements in the proto-planetary medium, is indicated in Figure 4.2. The vast bulk of the planet consists of liquid hydrogen with possibly a small admixture of helium; in the core region $r \lesssim 46{,}000$ km (excluding a very small central region where heavier elements may be concentrated), high pressure (of the order of 3×10^6 atmospheres and greater) causes dissociation of the hydrogen molecules into atoms, i.e. hydrogen is then in its liquid metallic phase with an electrical conductivity comparable with that of other liquid metals. The planet rotates at more than twice the angular velocity of the Earth, and Coriolis forces are undoubtedly important in its internal dynamics. It is reasonable to conclude that Jupiter's magnetic field, like that of the Earth, is attributable to dynamo action in its liquid conducting core region.

The situation for Saturn is very similar, although it is arguable that the outer part of the liquid metallic region may be stably stratified due to a negative gradient in the proportion of helium (Stevenson 1983). The field of Saturn is very nearly axisymmetric: the measurements made by Pioneer 11 and Voyager 1 and 2 are consistent with a purely axisymmetric field model. This has been confirmed by Cao et al. (2011) using measurements from the Cassini mission; Cao et al. (2017) find that the tilt ψ of Saturn's dipole is less than 0.06°! This fact, as we shall see, poses a challenge for dynamo theory. A possible explanation that had been proposed much earlier by Stevenson (1982) is that differential rotation in the conducting stable layers above Saturn's dynamo region acts to symmetrise the observable surface magnetic field.

As for the outer giant planets, Uranus and Neptune, the Voyager flybys revealed magnetic fields of internal origin for both planets; the big surprise in each case was the discovery of the large angle of tilt of their dipoles relative to the axis of spin, 59° for Uranus, 47° for Neptune. The internal structure of these planets is conjectured to consist of "a rocky core, a water-ice and ammonia layer, and a gaseous envelope of hydrogen and helium, comprising the outer 30% of the planet" (Russell & Luhmann 1997); as in Jupiter and Saturn, the water-ice and ammonia layer is presumably in a liquid metallic phase, and it is in this inner region of the planets that dynamo action must be located. This is consistent with the description

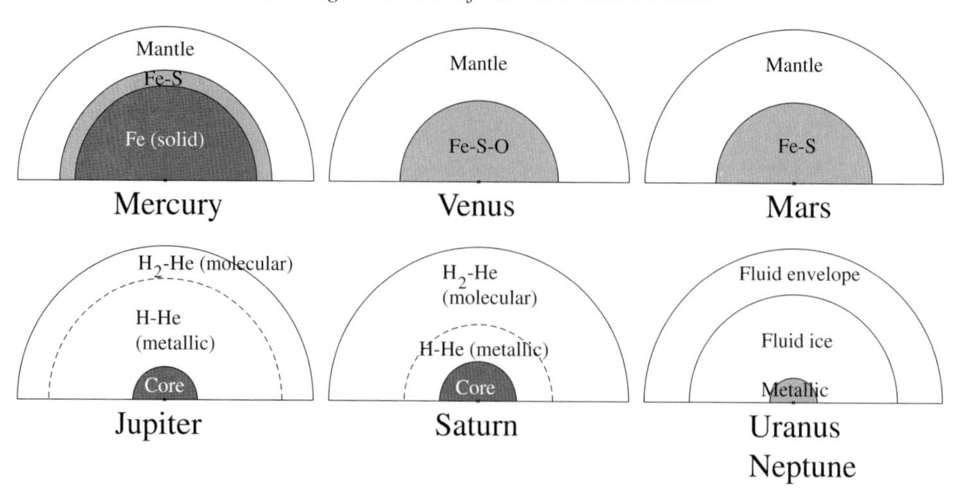

Figure 4.2 Internal structure of planets. [After Stevenson 1983.]

that had been advanced by Stevenson (1983) some years before the Voyager flybys (see Figure 4.2); but see also Nellis (2015).

4.2 Satellite magnetic fields

Interestingly, dynamo action is not restricted to planets; present or fossil magnetic fields of internal origin have also been detected in satellites. While the Moon does not currently have a magnetic field of internal origin, it has been known since the Apollo mission that lunar samples are magnetised. Recent palaeomagnetic studies of such samples (Weiss & Tikoo 2014; Buz et al. 2015) indicate that much of this magnetisation is the product of a core dynamo that existed on the Moon from at least 4.2 to 3.56 billion years ago. The samples indicate the presence of a rather intense field, up to $70\,\mu$T, before 3 billion years. This is in strong contrast with samples less then 2 billion years old, which indicate that the lunar magnetic field did not disappear completely, but persisted, albeit at least 10 times weaker. The lunar dynamo probably died away about 1 billion years ago. The available palaeomagnetic recordings are presented in Figure 4.3. They must be interpreted with caution because of the large error bars both on the age of the sample and on the associated palaeomagnetic intensity (often a factor of two), yet the evidence for a strong field with an intensity close to $50\,\mu$T before 3 billion years ago appears robust, as is the evidence from more recent samples of a field of much lower intensity.

The mechanisms powering the lunar dynamo are uncertain and much debated. They may include mechanical stirring (due to precession or large meteoroid

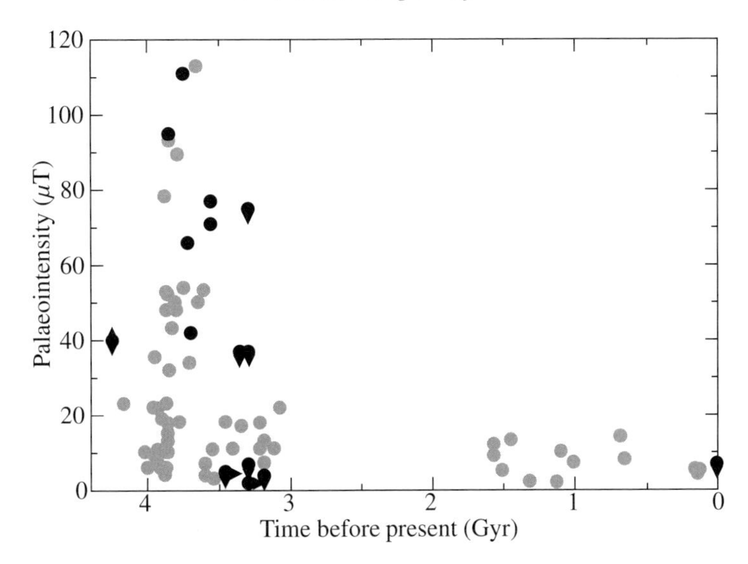

Figure 4.3 Palaeointensity of the lunar field as reconstructed from samples from the Apollo mission. The light symbols correspond to measurements performed at the time of the mission, whereas the black symbols correspond to recent analysis of the same samples (arrowheads are used to indicate upper limits either on palaeointensity or on age). [Data from Weiss & Tikoo 2014.]

impacts), or possibly compositional convection driven by the release of light elements associated with inner-core crystallisation. The evidence for both a strong-field and a weak-field branch on the Moon, if confirmed by further measurements, may be described by the same sort of strong and weak field force balances for the Earth that will be described in Chapter 12.

Jupiter's moons have also been investigated, thanks to data from the Galileo mission covering the period December 1995 to September 2003. Ganymede was found to have a magnetic field of internal origin, produced within its iron core (Kivelson et al. 1996). However, as Ganymede is close to Jupiter, it is also strongly influenced by Jupiter's magnetic field, and it is natural to question whether Ganymede is a genuine dynamo or is simply acting as an amplifier of Jupiter's magnetic field. Sarson et al. (1997, 1999) argued that dynamo action was more likely. Jones (2011) writes that 'Ganymede has an icy crust, a warm silicate mantle, and a metallic core. The radius is 2634 km, but the metallic core radius is believed (from gravity measurements) to be approximately 500 km. The dipole moment is 1.3×1013 Tm^3, about three times that of Mercury. At the surface the field is typically 700 nT, but at its core–mantle boundary this rises to 0.12 mT, weaker than the Earth's field (at its CMB) but much stronger than Mercury's field'. As for the Moon, identifying the energy source to power such a dynamo is a delicate issue.

Table 4.2 *First eight Gauss coefficients in the 1975 and 2015 IGRF-12 model of the Earth's surface magnetic field*

		1975		2015	
n	m	g_n^m	h_n^m	g_n^m	h_n^m
1	0	−30.100		−29.4420	
1	1	−2.013	5.675	−1.5010	4.7971
2	0	−1.902		−2.4451	
2	1	3.010	−2.067	3.0129	−2.8456
2	2	1.632	−0.068	1.6767	−0.6419

Note: In units of micro Tesla, μT.

The other moons of Jupiter (Callisto, Io and Europa) do not appear to have working dynamos; they do however exhibit fields induced by the field of Jupiter (Khurana et al. 1998; Kivelson et al. 2001). An inductive process of this kind differs fundamentally from self-excited dynamo action, and the polarity of the inducing field determines that of the induced field that is generated by this process (Sarson et al. 1999).

4.3 Spherical harmonic analysis of the Earth's field

The magnetic field at the surface of the Earth is due in part to currents in the interior and in part to currents in the outer conducting layers of the Earth's atmosphere. Measurements of all three field components on the surface provide a means of separating out these contributions, and it has been demonstrated by this process that by far the dominant contribution is of internal origin.

The magnetic potential due to the internal currents has the spherical polar expansion (cf. (2.77))

$$\Psi(r,\,\theta,\,\varphi) = R_{\rm E}\sum_{n=1}^{\infty}(R_{\rm E}/r)^{n+1}S_n(\theta,\,\varphi) = \sum_{n=1}^{\infty}\Psi^{(n)}\,, \tag{4.2}$$

where

$$S_n(\theta,\,\varphi) = \sum_{m=0}^{n}\left(g_n^m(t)\cos m\varphi + h_n^m(t)\sin m\varphi\right)\,P_n^m(\cos\theta),$$

and where $P_n^m(\cos\theta)$ is the associated Legendre polynomial with Schmidt normalisation (Winch et al. 2005)

$$\int_{-1}^{1} P_n^m(\cos\theta)\,P_{n'}^{m'}(\cos\theta)\,{\rm d}(\cos\theta) = \frac{2\,(2-\delta_{m0})}{2n+1}\,\delta_{nn'}\,\delta_{mm'}\,.$$

The coefficients $g_n^m(t),\ h_n^m(t)$ are the 'geomagnetic elements' in conventional notation. The axis of reference $\theta = 0$ is taken to be the geographical axis (i.e. the axis of rotation) of the Earth; the origin of longitude ($\varphi = 0$) is the meridian through Greenwich.

The potential Ψ is conveniently split into dipole and non-dipole ingredients:

$$\Psi = \Psi_{\mathrm{d}} + \Psi_{\mathrm{nd}}\,, \quad \Psi_{\mathrm{d}} = \Psi^{(1)}\,, \quad \Psi_{\mathrm{nd}} = \sum_{n=2}^{\infty} \Psi^{(n)}\,. \tag{4.3}$$

The dipole ingredient corresponds to a fictitious dipole $\boldsymbol{\mu}$ at the Earth's centre where, with Cartesian coordinates (x, y, z) related to (r, θ, φ) by (2.34),

$$\mu_x = R_{\mathrm{E}}^3 g_1^1, \quad \mu_y = R_{\mathrm{E}}^3 h_1^1, \quad \mu_z = R_{\mathrm{E}}^3 g_1^0. \tag{4.4}$$

This dipole is tilted at an angle $\psi(t)$ relative to the geographical axis Oz where

$$g_1^0 \tan\psi = \left[\left(g_1^1\right)^2 + \left(h_1^1\right)^2\right]^{1/2}. \tag{4.5}$$

The International Geomagnetic Reference Field (IGRF) provides a spherical harmonic description of the geomagnetic field going back to 1900 (we rely here on the IGRF-12, see Thébault et al. 2015). It is computed to order and degree 10 up to 1995, and up to 13 since 2000 thanks to satellite measurements. The first eight coefficients for 1975 and 2015 are listed in Table 4.2. The radial component of the geomagnetic field in 1975 (as presented in Moffatt 1978b) and in 2015 (the latest model currently available) are shown in Figure 4.4. Close examination of these reveals a tendency for the magnetic structures to drift westward and a strengthening of the so-called South Atlantic Anomaly. Both effects are more clearly evident in the radial fields at the core–mantle boundary (found by downward extrapolation) shown in Figure 4.5. A similar spherical harmonic decomposition of surface fields can be performed (though with much less resolution) for the other planets of our system. These fields are represented in Figure 4.6.

Using Table 4.2 for the year 2015, and equation (4.5) gives for the 'tilt angle' $\psi \approx 9.7^{\circ}$. The 'North Magnetic Pole' (NMP) is defined as that point in the northern hemisphere where the magnetic field points vertically downwards.[3] Since its first direct observation by Ross (1834), the NMP has been located in the Canadian Arctic and has been drifting north-north-west. After more than a century and a half of slow drift at less than 15 km/yr, the NMP suddenly accelerated around 1990. This phenomenon was first detected through local surveys (Newitt & Barton 1996; Newitt et al. 2002). According to direct observation of the NMP in 2007, its drift rate appears to have since stabilised at just over 50 km/yr (Newitt et al. 2009). Its location is affected by the non-dipole ingredients of the field, so provides only

[3] The north pole of a compass needle points north, so the magnetic pole of 'magnet Earth' in the northern hemisphere is actually its south pole; it is nevertheless customary to refer to it (paradoxically!) as the 'North Pole'.

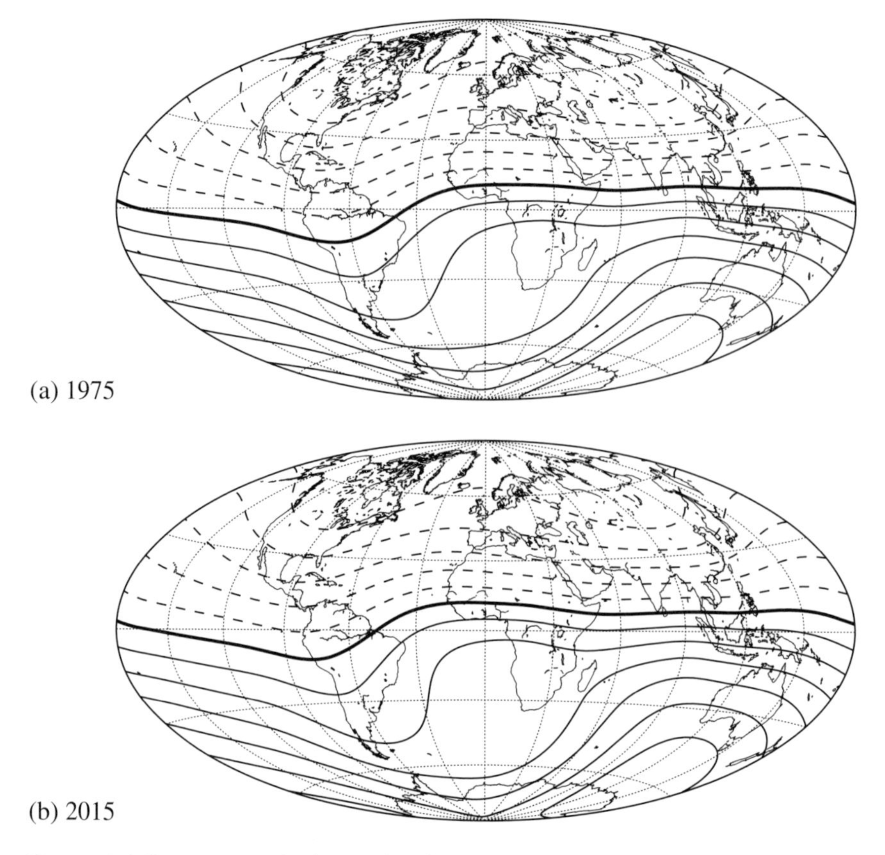

Figure 4.4 Isocontours (at intervals of $10\mu T$) of the normal component of the geomagnetic field at the surface of the Earth in (a) 1975 and (b) 2015.

an approximation for the direction of the dipole $\boldsymbol{\mu}$ (see Figure 4.7). Furthermore, the horizontal component of the field can be shown to be very weak over a wide elongated area near the North Pole (but not under the South Pole (SMP)) so that small changes in the field can yield huge changes in the pole's position (Mandea & Dormy 2003). Rapid displacement of the pole without great change in the field can be explained in this way. Ground measurements to locate the SMP in 2000 (Barton 2002) serve to underline the very different behaviour of the two magnetic poles.

The series (4.2) converges quite rapidly for $r = R_{\mathrm{E}}$. If the source currents were strictly confined to the core region $r < R_{\mathrm{C}}$, then the series would converge for $r \geq R_{\mathrm{C}}$. However, various effects, among them magnetisation of the Earth's crust, screen out measurements of the internal field. Also, the mantle is by no means a perfect insulator, although its conductivity is certainly much less than that of the core, and some current may leak from core to mantle. Moreover, errors in the

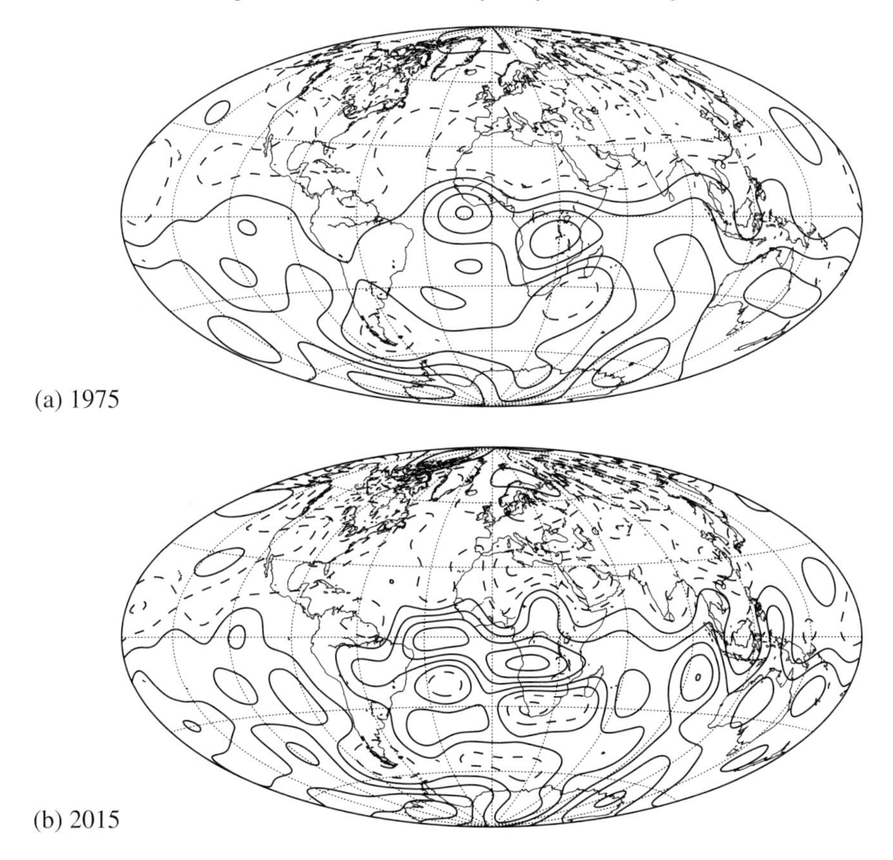

Figure 4.5 Isocontours (at intervals of 200 μT) of the normal component of the geomagnetic field at the core–mantle boundary in (a) 1975 and (b) 2015.

contributions from higher-degree harmonics are amplified more rapidly (like $(R_E/r)^n$) with depth.

For these reasons, one must exercise caution in using the expansion (4.2) near $r = R_C$. Nevertheless it is interesting to plot the mean-square contributions $\langle B_n^2 \rangle$, where $\mathrm{B}_n = -\nabla\Psi^{(n)}$ and the angular brackets represent averaging over a sphere $r =$ cst., for $n = 1, 2, \ldots, 8$ (Lowes, 1974) – see Figure 4.8. On $r = R_E$, the convergence is convincing, while on $r = R_C$, as expected, convergence is certainly not obvious; indeed it is evident that the harmonics up to $n = 13$ make roughly equal contributions to the mean-square field on $r = R_C$. Thus although the field at the Earth's surface is strongly dipolar in structure, the evidence is that this situation does not persist down to the neighbourhood of the core–mantle boundary (see also Figure 4.5). This difference between the field structure on $r = R_E$ and $r = R_C$ is even more marked in terms of the mean square of the time derivatives $\dot{B}_n^2$; in this

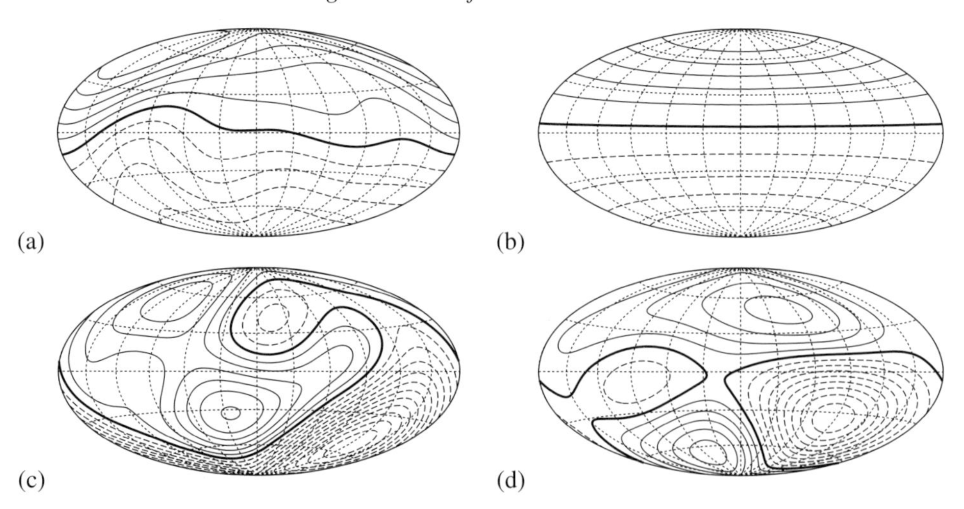

Figure 4.6 Models of planetary fields for (a) Jupiter (at intervals of 200μT), (b) Saturn, (c) Uranus and (d) Neptune (all three at intervals of 10 μT). These are based on low-order spherical harmonic expansions of the available data. [Data from Ness 2010.]

case, the contributions on $r = R_C$ actually show a distinct increase with increasing n (Figure 4.8b).

The non-dipole field is highlighted in Figure 4.5. Comparison of the maps of the radial field at the core–mantle boundary over the 40-year period 1975–2015 provides qualitative evidence of the evolution of the non-dipole field with time. Since $\langle B^2_{nd}\rangle^{1/2} \approx 0.02$ Gauss and $\langle \dot{B}^2_{nd}\rangle^{1/2} \approx 0.0005$ Gauss/year, the time-scale for this *secular variation* (discovered by Gellibrand 1635) at any fixed location is of the order of 40 years. The magnetic contours evolve on this time-scale, with centres of activity growing and decaying like isobars on a weather chart.

The secular variation is continuous, but marked by 'acceleration discontinuities', completed in a year or less, and separated by a few decades, often of worldwide extent (e.g. Courtillot et al. 1978). These events, called magnetic jerks or magnetic impulses, are still poorly understood, but may provide a signature of rapid events in the liquid core.

Measurements of the coefficients $g^m_n(t)$, $h^m_n(t)$ have been available only since the time of Gauss. The first magnetic observatories were established in 1832. More or less continuous observations of the direction of the field have been made, since the middle of the seventeenth century for the declination δ, and since the middle of the eighteenth century for the inclination[4] ϑ. Although these points are very

[4] In the notation of Figure 2.8, $\cos\delta = -B_z\left(B_x^2 + B_y^2 + B_z^2\right)^{-1/2}$ and $\cos\vartheta = B_y\left(B_x^2 + B_y^2\right)^{-1/2}$.

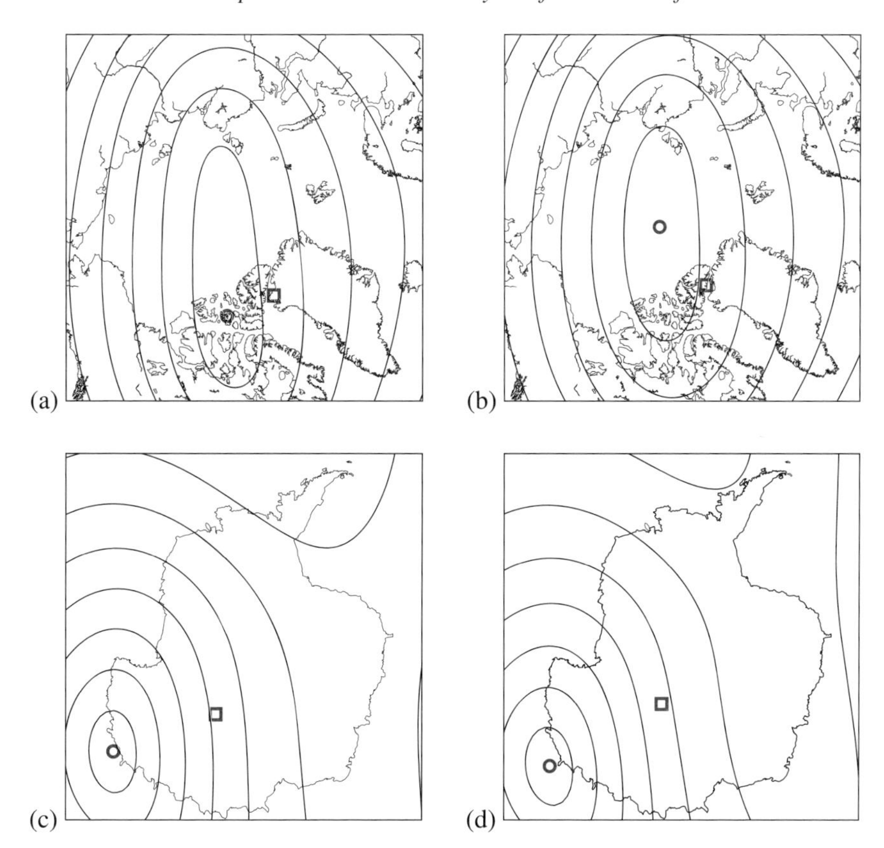

Figure 4.7 Positions of the north and south magnetic poles in 1975 (a, c) and 2015 (b, d). The dot indicates the dip poles, while the square indicates the dipole axis projection at the surface. Isocontours of the horizontal component of the magnetic field are also represented at $3\mu T$ intervals.

irregularly distributed over the Earth's surface, planetary models have been based on these data; the oldest is for the year 1590 (Jackson et al. 2000).

Over the secular time-scale, two characteristics of the geomagnetic field should be noted: the decrease in the axial dipole intensity (of the order of 5% per century) and the so-called westward drift of the non-dipole field. The dipole moment has decreased from 8.56×10^{22} A m^2 in 1835 (as measured by Gauss himself, see Chapman & Bartels 1940) to 7.72×10^{22} A m^2 in 2015 (IGRF-12). This decrease is illustrated in Figure 4.9 (relying on the 'gufm1' model of Jackson et al. 2000 for the period 1600–1990 and the International Geomagnetic Reference Field IGRF-12 of Thébault et al. 2015 going back to 1900). Because of this current decrease of the dipolar component of the field, the issue of a possible approaching reversal is sometimes raised. We shall come back to this in the next section.

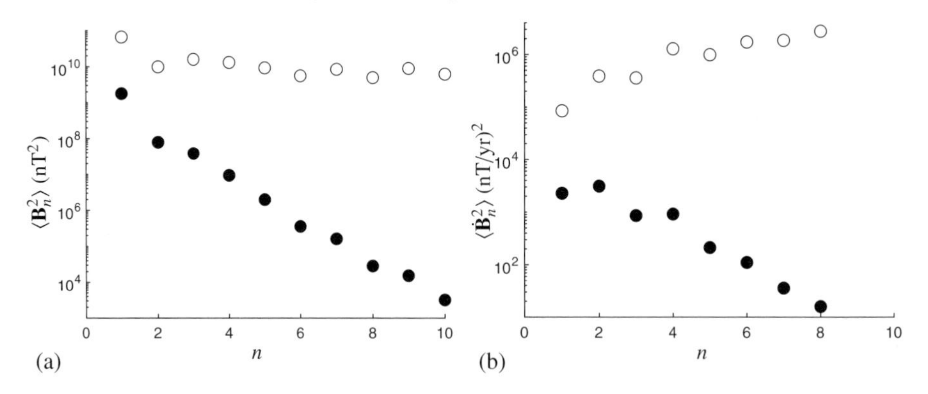

Figure 4.8 Contributions to (a) mean-square field $\langle \mathbf{B}_n^2 \rangle$ and (b) mean-square rate of change of field $\langle \dot{\mathbf{B}}_n^2 \rangle$ in 2015 on the surfaces $r = R_E$ (solid circles) and $r = R_C$ (open circles); $\mathbf{B}_n = -\nabla \Psi^{(n)}$, and $\langle \ldots \rangle$ indicates an average over a spherical surface; the unit of field strength is $1\text{nT}=10^{-9}\text{T}$. [Data from IGRF-12.]

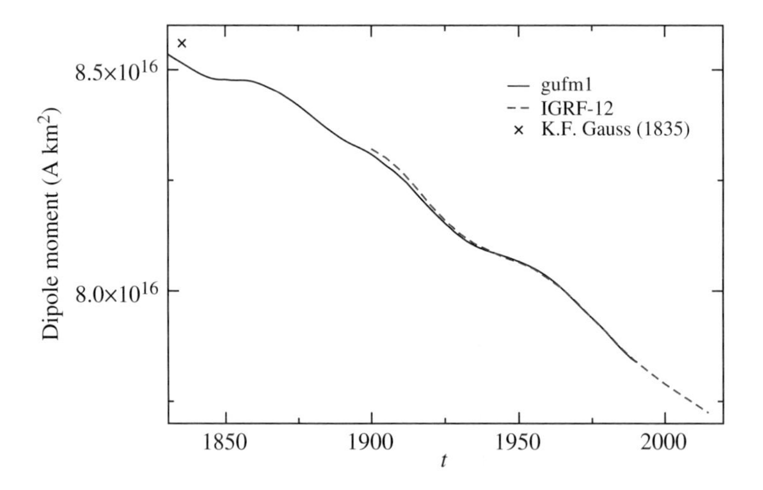

Figure 4.9 Dipole moment evolution with time from 1835 to 2015 interpolated from direct measurements. The dipole moment estimated from the historic measurement of Gauss in Göttingen in 1835 is also indicated.

The westward drift was discovered by Halley (1692) and quantified by Bullard et al. (1950), who evaluated its global rate as 0.18°/yr. Bloxham & Jackson (1992) have however emphasised the fact that, over the past three centuries, westward drift has been far from uniform; and Dormy et al. (2000) have stressed that, on much longer geological time-scales, westward drift is not a robust geomagnetic feature.

4.4 Variation of the dipole field over long time-scales

For evidence of field evolution in earlier eras, we have two important sources of information. First and most important, the science of rock magnetism (palaeomagnetism) provides estimates of **B** at geological epochs when rocks cooled below the Curie point (the temperature below which they retain their magnetisation); since some rocks are known on the basis of radiometric dating to be as old as 3.5×10^9 years, this provides information about the Earth's field from the earliest stages in its history, the estimated age of the Earth being of order 5×10^9 years. Studies of the magnetisation of sea-floor sediments provides similar information. The intensity of the ancient field can be determined using techniques developed particularly by Thellier (1938, 1981) (see also Thellier & Thellier 1959). The information from such sources is reliable only if the samples analysed can be accurately dated. Dating of ancient rocks is accurate only to within about 3%, so that, roughly speaking, the further back we go in geological time, the more uncertain the picture becomes.

Secondly, archaeomagnetism (i.e. the study of the remanent magnetism acquired by clay pots, kilns and other objects baked by man) gives information about the field intensity and orientation over the last 4000 years or so. Archeomagnetic data are not uniformly distributed over the surface of the Earth: there is however a reasonable density of observations in Europe, the Middle East and North America; Siberia and Asia are sparsely but fairly uniformly covered. In some regions, spatio-temporal studies of the geomagnetic field over several thousand years have been possible.

Certain broad conclusions are now widely accepted from such studies (Bullard 1968, Jacobs 1976, Dormy et al. 2000). First, from the archaeomagnetic investigations, it has been shown that the intensity fluctuates on a time-scale of order 10^4 years, and that it has been decreasing over the last few thousand years from a level about 50% greater than the present level. The direction of the dipole moment also changes ('dipole wobble'), though at a rate small compared with the westward drift of the non-dipole field; the indications are that the time average of the dipole moment $<\mu(t)>$ over periods of order 10^3–10^4 years is accurately in the north-south direction. The near alignment of the present magnetic axis and the geographic axis is an indication of the relevance of rotational constraints on the inductive fluid motions in the core; the fact that the long-time average of ψ is apparently zero, or very near zero, provides more emphatic evidence that these rotational constraints have a strong bearing on the magnetic-field-generation problem.[5]

More dramatically, the palaeomagnetic studies provide clear evidence of reversals in the polarity of the Earth's field that have occurred repeatedly over

[5] The situation for Uranus and Neptune is strikingly different, and still awaits plausible explanation. It may be that the dynamo process for these giant planets is more unstable than that of Earth, and that the secular time variation of $\mu(t)$ is much greater for reasons that are not as yet apparent.

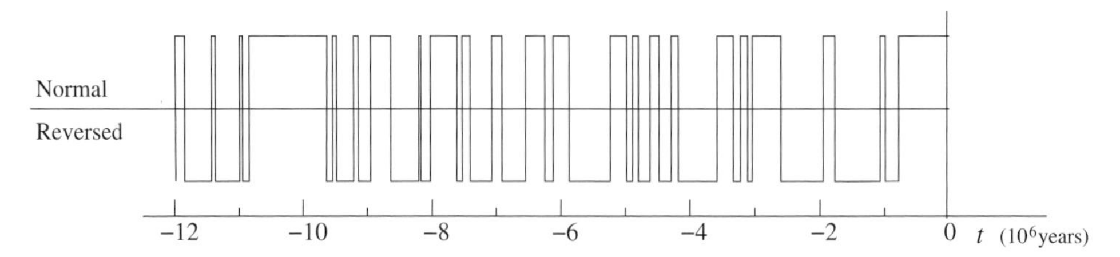

Figure 4.10 Record of reversals of the Earth's dipole polarity over the last 12×10^6 years. The figure indicates only the direction of the dipole vector, not its intensity.

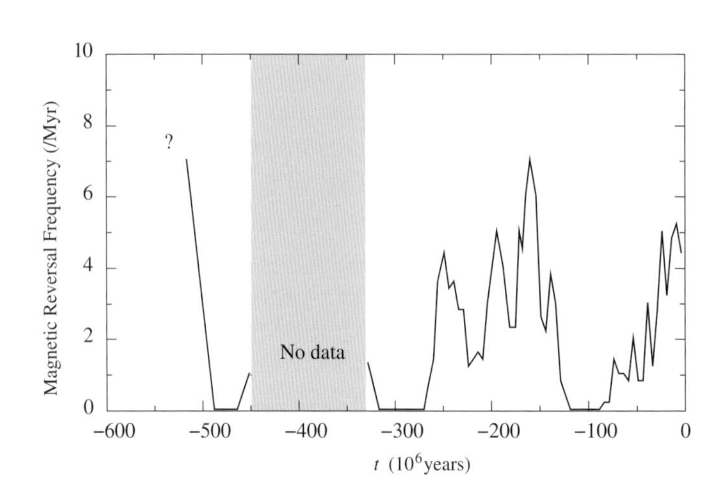

Figure 4.11 Variation in geomagnetic reversal frequency over $\sim 5 \times 10^8$ years as reconstructed from rock magnetisation. [After Pavlov & Gallet 2005.]

geological eras. Figure 4.10 shows the record of variations in polarity of the Earth's field over the past 12 million years. The periods between two reversals are known as *chrons*. The duration of chrons during the last 8×10^7 years, varies from 10^5 to 10^6 years. There have been periods as long as 60 million years with no reversals (*superchrons*). Reversals occur relatively quickly, the reversal time being of the order of 1000 years.

Figure 4.11 redrawn from Gallet et al. (2012), illustrates the changes in reversal frequency as recovered from marine magnetic anomalies over the past 5×10^8 yr (using sliding windows of 5×10^6 yr) and from paleaomagnetic data of earlier eras (the frequency estimates being computed over time intervals determined by the duration of the corresponding geological eras). The sequence of periods of chrons, marked by the reversals, does indeed include chrons of exceptional duration. An important issue is to assess whether such superchrons are extreme events

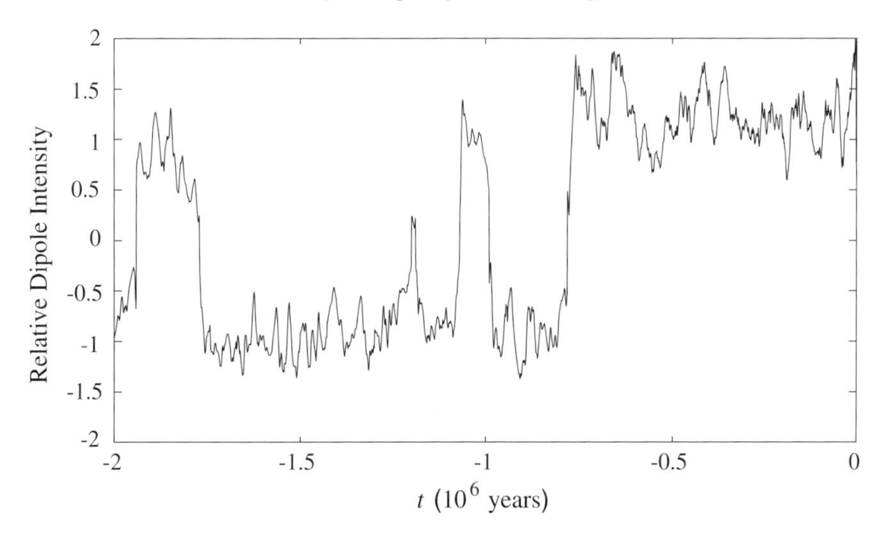

Figure 4.12 Dipole intensity variation from palaeomagnetic data for the past 2 million years; the present corresponds to $t = 0$. [Data from Valet et al. 2005.]

of a random distribution, occurring without change in the nature of the dynamo mechanism, or whether they require such a change; for example, it is possible to imagine scenarios in which the heat flow at the CMB varies, with time constants of the order of 10^7 years (Courtillot & Olson 2007) .

As illustrated in Figure 4.5, the dominance of the dipole field still prevails at the core–mantle boundary. This does not hold during reversals; but on average over a few thousand years, the magnetic field outside the core is that of an 'axial centred dipole' (located at the Earth's centre and aligned with the rotation axis). In contrast to reversals, which involve a full polarity change, *excursions* are deviations from the axial centred dipole that are considered to be large but do not lead to a change of polarity. Excursions and reversals share several common characteristics, which suggest that they can be treated as manifestations of similar processes within the core. The duration of excursions is difficult to quantify, but probably comparable with that of reversals (Bourne et al. 2012). In most cases, directional changes both for excursions and reversals take place in less than 5000 years. Insofar as excursions can be considered as 'failed reversals', they can provide valuable information on the mechanism of polarity reversals.

The intensity of the dipole field reconstructed from palaeomagnetic data over the last 2 million years (Valet et al. 2005) is shown in Figure 4.12, which reveals significant fluctuations. The average dipole strength over the last 8×10^5 years (i.e. since the last polarity reversal) is estimated at $7.5 \pm 1.7 \times 10^{22}$ A m^2. The present field, though decreasing, is therefore in fact quite large compared with the average. Figure 4.12 also reveals an asymmetric evolution in the neighbourhood of a reversal. The decay of the dipole intensity begins some 8×10^4 years in advance, while, after the reversal, its intensity is re-established quickly, and often at a higher absolute level than prevailed before the reversal.

4.5 Parameters and physical state of the lower mantle and core

Apart from chemical constitution, the fundamental thermodynamic variables in the interior of the Earth are density and temperature from which, in principle, other properties such as electrical conductivity may be deduced (although this requires bold extrapolation of curves based on available laboratory data).

It has been suggested more than once that the uppermost layer of the Earth's liquid core is stably stratified (Braginsky's 'hidden ocean of the core'), even though the fluid beneath it is in a state of vigorous convection (Braginsky 1993, 1999, 2006, Shearer & Roberts 1998, Lister & Buffett 1998, Buffett 2014). The fact that the outer core might be thermally stably stratified was recognised by Braginskii (1964c), who argued that convection might nevertheless be driven by the upward flotation of lighter elements (e.g. silicon) during the process of crystallisation of iron from the outer core onto the inner core. A similar process of density differentiation had been earlier suggested by Urey (1952): precipitation of iron from the mantle into the outer core and subsequent sedimentation of the iron towards the inner core (through the lighter core fluid mixture) could also lead to the generation of convection currents. In both processes, gravitational energy is released and a proportion of this is potentially available for conversion to magnetic energy by the dynamo mechanism.

Pozzo et al. (2012, 2014) estimate the electrical conductivity σ in the core to be $\sigma \simeq 1.5 \times 10^6\,\mathrm{S m^{-1}}$. With $\mu_0 = 4\pi \times 10^{-7}\,\mathrm{H\,m^{-1}}$, this gives a value for the magnetic diffusivity $\eta = (\mu_0\sigma)^{-1} \simeq 0.5\mathrm{m^2 s^{-1}}$. The value of the conductivity in the lower mantle can be estimated (see §4.7) to be $\sigma_\mathrm{m} \approx 10\,\mathrm{S m^{-1}}$, so that

$$\sigma/\sigma_\mathrm{m} = \eta_\mathrm{m}/\eta \sim 10^5 .$$

While the precise value of both σ and σ_m is uncertain, this ratio is large enough for it to be reasonable to treat the mantle as an insulator ($\sigma_\mathrm{m} = 0$) in a first approximation. Some phenomena however (for example, the electromagnetic coupling between core and mantle) depend crucially on the leakage of electric current from core to mantle, a process that is controlled by the weak mantle conductivity, which in such contexts must naturally be retained in the analysis.

The kinematic viscosity ν of the outer core was described (Roberts & Soward 1972) as the worst-determined quantity in the whole of geophysics. The best estimate so far is from de Wijs et al. (1998) who find $\nu = 1.4 \times 10^{-6}\,\mathrm{m^2\,s^{-1}}$. With this estimate, and the above estimate for η, we have for the dimensionless ratio ν/η (the 'magnetic Prandtl number') the estimate

$$\mathcal{P}_m = \nu/\eta \approx 10^{-6},$$

which may be compared with the value 1.5×10^{-7} for mercury under normal laboratory conditions. The small value of ν/η suggests that the dominant contribution to energy dissipation in the outer core will be ohmic rather than viscous, and that (except possibly in thin shear layers at the solid boundaries of the outer core or in its interior) viscous effects will be negligible in the governing dynamical equations.

4.6 The need for a dynamo theory for the Earth

If induction effects in the core were non-existent or negligible, the free decay time for any current distribution predominantly confined to the core would (from §2.8) be of order $t_d = R_c^2/\pi^2\eta$; with $R_c = 3400$ km and $\eta \simeq 0.5\,\mathrm{m^2s^{-1}}$, this gives $t_d \approx 2.2 \times 10^{12} s \approx 7 \times 10^4$ years. As mentioned in §4.4, studies of rock magnetism indicate that the field of the Earth has existed at roughly its present intensity (except possibly during rapid reversals) on a geological time-scale of order 10^9 years. It therefore clearly cannot be regarded as a relic of a field trapped during accretion of the Earth from interplanetary matter; such a field could not have survived throughout the long history of the Earth in the absence of any regenerative mechanism.

On the other hand, the fact that the field exhibits time variation on scales extremely short compared with geological time-scales of order 10^6 years and greater (as exemplified by the secular variation of the non-dipole field with characteristic time of order 40 years) indicates rather strongly that such variations are not attributable to general evolutionary properties of the Earth on these geological time-scales, but rather to relatively rapid processes most probably associated directly with core fluid motions.

The rate of the secular variation suggests (on the simplistic picture that magnetic perturbations are convected by the core fluid) that near the core–mantle boundary, the velocity of the core fluid relative to the mantle is of order $u_c \sim 10^{-4}\mathrm{ms^{-1}}$. A characteristic length-scale for magnetic perturbations associated with the secular variation is $\ell_c \approx 10^3$ km. A magnetic Reynolds number can be constructed on the basis of these figures:

$$\mathcal{R}_m = u_c\ell_c/\eta \approx 200\,. \tag{4.6}$$

This is by no means infinite, but is perhaps large enough to justify the frozen-field assumption for sufficiently rapid magnetic perturbations, at least in a first approximation.

4.7 The core–mantle boundary and interactions

The liquid core can interact with the mantle across the core–mantle boundary (CMB) in a number of ways: transfer of heat, mass, momentum, and electric current. These interactions significantly influence the dynamics and evolution of both the core and the mantle on a wide range of time-scales (from human to geological). Some evidence for rapid interactions is provided by observed variations of up to a few milliseconds in the 'length-of-day' (lod) on a decadal time-scale (Jault et al. 1988; Holme & Whaler 2001).

Based on the temperature jump across the thermal boundary layer at the base of the mantle (Buffett 2002), the heat flux across the CMB is estimated to be in the range 6 to 12 TW. This flux is a fundamental controlling parameter for the evolution and dynamics of the core. It is not expected to be homogeneous at the core–mantle boundary; indeed heterogeneities are to be expected due to slow convective patterns within the mantle and thermal plumes near its base. Such heterogeneities affect both core dynamics and magnetic polarity reversals (Glatzmaier et al. 1999; Courtillot & Olson 2007; Pétrélis et al. 2011). Indeed the long time-scale of reversal frequency modulation is comparable to the typical time-scale for mantle convection. It is possible therefore that the time-variation of reversal frequency (Figure 4.11) may actually be controlled by mantle dynamics, rather than being an intrinsic property of the core dynamics.

Chemical interactions between the mantle and the core are still controversial; they raise the issue of mass exchange between the core and mantle over geological time-scales (Brandon & Walker 2005), and are related to the possible presence of iron alloys at the base of the mantle.

Electromagnetic core–mantle interactions arise from the Lorentz force exerted by the dynamo-generated field on mantle currents, these being proportional to mantle conductivity. Studies of currents induced in the mantle by external fluctuations indicate a conductivity of the order of 1 Sm^{-1} at the top of the lower mantle (Schultz et al. 1993; Petersons & Constable 1996). High-pressure experiments indicate that it is of order 1 to 10 Sm^{-1} at greater depth. This is rather low, and a highly conducting layer needs to be assumed at the base of the mantle to allow currents to participate significantly in core–mantle coupling.

Direct mechanical coupling between core and mantle arises from flow over the topography (roughness) of the core–mantle boundary (Jault & Le Mouël 1989); this yields a torque, which transfers angular momentum between core and mantle (Moffatt & Dillon 1976, Mound & Buffet 2005; Mound & Buffett 2006). In this context, Gillet et al. (2010) have argued that torsional waves, triggered on the cylinder tangent to the inner core boundary, can propagate more rapidly than previously thought, and can cause the observed six-yearly variations of both magnetic field and length-of-day. This study has raised the possibility of a larger internal magnetic field than previously thought (3 mT). Another source of mechanical (gravitational) coupling arises from density anomalies (Jault & Le Mouël 1989).

4.8 Precession of the Earth's angular velocity

The angular velocity of the Earth's mantle $\mathbf{\Omega}(t)$ is not quite steady, but is subject to slow changes and weak perturbations in both magnitude and direction (Munk & MacDonald 1960). Chief among these in its relevance to core dynamics is the slow precession of $\mathbf{\Omega}$ about the normal to the plane of the Earth's rotation about the Sun caused by the net torque exerted by the Sun and the Moon on the Earth's equatorial bulge. The period of precession is known from astronomical observations to be about 25,800 years, the vector $\mathbf{\Omega}$ describing a cone of semi-angle 23.5° over this period. The precessional angular velocity $\mathbf{\Omega}_p$ has magnitude $7.71 \times 10^{-12}\,\mathrm{s}^{-1}$. Precession was advocated by Malkus (1963) as providing a plausible source of energy for core motions.

It is easy to see why precession must cause some departures from rigid body rotation in the liquid core region. Firstly, since the mean density of the core is substantially greater than the mean density of the mantle, the dynamic ellipticity $\varepsilon_{\mathrm{c}} = (C_{\mathrm{c}} - A_{\mathrm{c}})/C_{\mathrm{c}}$ of the core (where A_{c} and C_{c} are its equatorial and polar moments of inertia respectively) is less than the corresponding quantity ε_{m} for the mantle; in fact $\varepsilon_{\mathrm{c}} \approx \frac{3}{4}\varepsilon_{\mathrm{m}}$. If the core and the mantle were dynamically uncoupled, they would then precess at different angular velocities proportional to ε_{c} and ε_{m} respectively, i.e. the precessional angular velocity of the core would be $\frac{3}{4}\mathbf{\Omega}_p$, so that after about 10^5 years the angular velocities of core and mantle, though equal in magnitude, would be quite different in direction. This would imply large relative velocities between core and mantle of order $\Omega R_{\mathrm{C}} \approx 200\mathrm{m\,s}^{-1}$, for which there is no evidence whatsoever. The inference is that core and mantle are *not* dynamically uncoupled – indeed, as we saw in the previous section, there is no good reason to expect that they should be – but are quite strongly coupled through various mechanisms of angular-momentum transfer across the interface. This coupling must act in such a way as to equalise the precessional angular velocity of core and mantle, although in the absence of perfect coupling the mean angular velocity of the core may be expected to lag behind the angular velocity of the mantle in its precessional orbit. A difference in angular velocity requires a boundary layer at the core–mantle interface, a phenomenon studied in the purely viscous context by Stewartson & Roberts (1963), Toomre (1966) and Busse (1968), and in the hydromagnetic context by Rochester (1962) and Roberts (1972a).

A detailed analysis of the problem of coupling between mantle and core (Loper 1975) has indicated that the rate of supply of energy through the mechanism of precessional coupling is a factor of order 10^{-3} smaller than the estimated rate of ohmic dissipation in the core. Loper also made the point that a fraction of this energy supplied (necessarily 100% in a steady state model!) is dissipated, partly in the boundary layers through which angular momentum is transferred from mantle to core. The concept of a precessionally driven dynamo rests however on the existence of a turbulent flow in the core arising from instabilities of these boundary layers and the secondary flows to which they give rise (Malkus 1968), and it is by no means clear that Loper's arguments are applicable in this situation.

Theoretical analysis of flow in a precessing spheroid was initiated by Poincaré (1910), who showed that, if the fluid is regarded as inviscid, a flow of uniform vorticity is possible. This flow does not satisfy the 'no-slip' condition on the spheroidal boundary, and in a real viscous fluid an 'Ekman boundary layer' (Greenspan 1968) forms in order to accommodate this condition. As the rate of precession increases, this boundary layer becomes unstable, as predicted by Malkus (1968); this instability occurs near critical circles at latitudes $\pm 30^{\circ}$, from which shear layers germinate and grow on interior conical surfaces; these have been detected both experimentally (Noir et al. 2001a) and numerically (Hollerbach et al. 1995; Noir et al. 2001b). A different type of instability involving resonant interaction of inertial waves has been numerically identified by Lorenzani & Tilgner (2001). Both types of instability lead at sufficiently high precession rates to turbulence. An asymptotic approach to the analysis of emerging flow structures has been developed by Kida (2011, 2018).

Wu & Roberts (2009) have established numerically that dynamo action can occur as the result of flows driven by precession of the spheroidal core–mantle interface (see also Goto et al. 2011). Similar dynamo action has been explored by Lin et al. (2016), who focus particularly on generation by the large-scale cyclonic vortices identified by Lin et al. (2015). No doubt we may expect further exciting developments in this branch of dynamo theory in the years to come!

It would be inappropriate to enter at this stage into any of the detailed calculations undertaken by the above authors – this would require a disproportionate digression from the main theme of this book. The influence of precession is merely mentioned here as one element of the complicated dynamical background that will have to be fully understood before the dynamo problem in the terrestrial context can be regarded as solved. There is still disagreement about what the dominant source of energy for core motions may be, although convection due to buoyancy forces (of thermal or compositional origin as discussed in §4.5) still appears to be the principal candidate for serious consideration; and it will be this source that we focus on in later chapters of this book.

5
Astrophysical Magnetic Fields

5.1 The solar magnetic field

As for the Earth, it is now generally accepted among astrophysicists that the origin of the Sun's magnetic field, which is highly variable both in space and in time, must be sought in inductive motions; these are localised in its outer convection zone, which extends from the visible surface of the Sun ($r = R_\odot = 6.96\times10^5$ km) down to about $r = 0.7\,R_\odot$ (Figure 5.1). At the base of the convection zone, a relatively sharp shear layer known as the 'tachocline'[1] (Spiegel & Zahn 1992; Hughes et al. 2007) extends over about $0.03\,R_\odot$, and marks the boundary with the radiative interior.

The Sun is a good electrical conductor, being primarily composed of ionised hydrogen and helium. The natural decay time for the fundamental dipole mode is of order 4×10^9 years (Wrubel 1952), which is of the same order as the age of the solar system. If the solar magnetic field were steady on historic time-scales there would be no need to seek for a renewal mechanism; the field could simply be a fossil relic of a field frozen into the solar gas during the initial process of condensation from the galactic medium. Even local time-dependent phenomena, such as the evolution of sunspots (see §5.2 below), could be regarded as transient and localised events occurring in the presence of such a fossil field, and this was in fact the widely accepted view until the mid-1950s. The development of the solar magnetograph (Babcock & Babcock 1955) permitting direct measurement of the weak general poloidal field of the Sun, and the discovery of reversals of this field (see §5.4 below), first in the period 1957–8 and again in 1969–71 (in both cases either at or just after periods of maximum sunspot activity), have led to the view that the 22-year sunspot cycle is in fact a particular manifestation of a roughly periodic evolution of the Sun's general field. Such a periodic behaviour has no obvious explanation in a fossil theory. Curiously, whereas dynamo theory was originally conceived to explain the persistence of cosmic magnetic fields over very long time-scales, in the

[1] From the Greek *tacho*: speed and *klinein*: to slope.

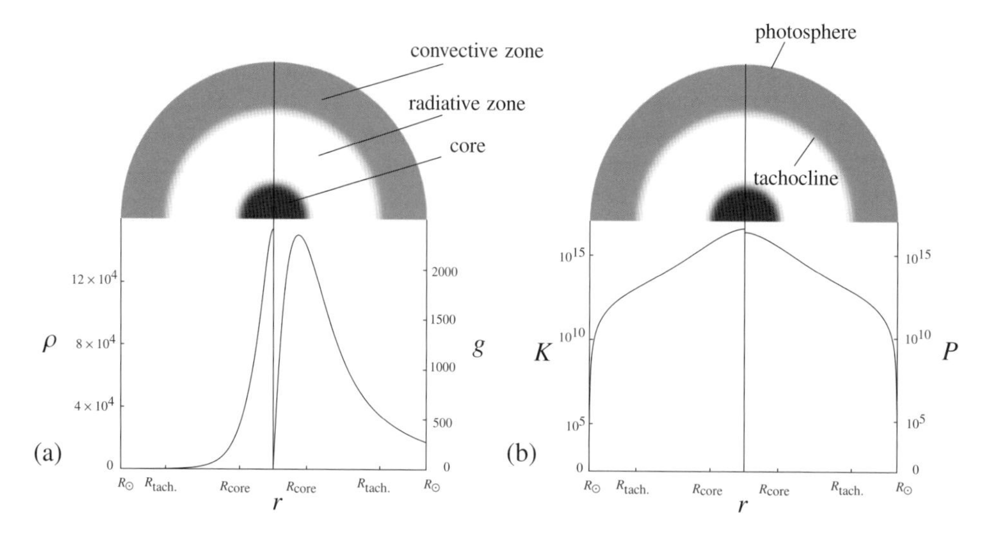

Figure 5.1 Interior structure of the Sun ($R_\odot = 6.957 \times 10^5$ km) from the centre to the surface: the core (black), radiative zone (white) and convection zone (grey). Radial profiles of (a) density ρ (kg m^{-3}) and gravity g (m s^{-2}), (b) bulk modulus K (Pa) and pressure P (Pa), as derived from model S of Christensen-Dalsgaard et al. (1996).

case of the solar field it is now invoked to explain the extremely rapid variations (relative to the cosmic time-scale) that at present attract no other equally plausible explanation.

5.2 Velocity field in the Sun

5.2.1 Surface observations

The Sun rotates about an axis inclined at an angle $7°15'$ to the normal of the Earth's orbital plane, with a period of approximately 27 days. The rotation period is however non-uniform, being greater (by about 4 days) near the poles than in a neighbourhood of the equatorial plane. The angular velocity $\varpi(\theta)$ on the surface $r = R_\odot$ as a function of heliographic co-latitude θ may be represented to good approximation by the formula

$$\varpi(\theta)/2\pi = 452 - 49\cos^2\theta - 84\cos^4\theta \text{ nHz}, \tag{5.1}$$

(Howard & Harvey 1970; Snodgrass 1984). This expression is based on 17 years of Doppler data from the Mount Wilson Observatory; a frequency of 450 nHz corresponds to a period of approximately 26 days. Small fluctuations $\varpi'(\theta, t)$, in the form of torsional oscillations superposed on (5.1), have also been detected.

The visible surface of the Sun is not uniformly bright, but exhibits a regular pattern of granulation, the scale of this pattern (i.e. the mean radius of the granules) being approximately 1000 km. The details of the pattern change on a time-scale of the order of a few minutes. An alternative angular velocity profile can be determined tracing magnetic features

$$\varpi(\theta)/2\pi = 462 - 74\cos^2\theta - 53\cos^4\theta \text{ nHz} . \tag{5.2}$$

This profile was obtained by cross-correlating Mount Wilson magnetograms over successive days (Snodgrass 1983). A possible explanation for the difference between (5.1) and (5.2) is that the Doppler techniques used to derive (5.1) measure the velocity at the surface, whereas tracers such as sunspots used to derive (5.2) are advected in part by a faster-rotating layer below the surface (Beck 2000; Howe 2009).

Velocity fluctuations (again detected by means of Doppler shift measurements) of the order of 1 km s^{-1} are associated with the granulation, and are the surface manifestation of convective turbulence which penetrates to a depth of the order of 1.5×10^5 km below the solar surface (Weiss 1976; Gough & Weiss 1976). Velocity fluctuations exist on larger scales (e.g. velocities of order 0.1 km s^{-1} are associated with 'supergranular' and 'giant cell' structures on scales of order 10^4–10^5 km) and likewise velocity fluctuations may be detected on all scales down to the limit of resolution (of order 10^2 km). If we adopt the scales

$$\ell_g \sim 10^3 \text{ km} , \quad u_g \sim 1 \text{ km s}^{-1} , \tag{5.3}$$

as being characteristic of the most energetic ingredients of the turbulent convection in the outermost layers of the Sun, then, with estimates of the kinematic viscosity and magnetic diffusivity given by[2]

$$\nu \sim 10^{-8} \text{ km}^2 \text{ s}^{-1}, \quad \eta \sim 10^{-1} \text{ km}^2 \text{ s}^{-1} , \tag{5.4}$$

we may construct a Reynolds number $\mathcal{R}e$ and magnetic Reynolds number $\mathcal{R}_{\mathrm{m}}$:

$$\mathcal{R}e = \frac{u_g \ell_g}{\nu} \sim 10^{11}, \quad \mathcal{R}_{\mathrm{m}} = \frac{u_g \ell_g}{\eta} \sim 10^4 . \tag{5.5}$$

If magnetic effects were dynamically negligible, then the traditional Kolmogorov picture of turbulence (see Chapter 15) would imply the existence of a continuous spectrum of velocity fluctuations on all scales down to the inner Kolmogorov scale $\ell_\nu \sim \mathcal{R}e^{-3/4} \ell_g \sim 1$ cm, a factor 10^7 below the limit of resolution! Lack of an adequate theory for the effects of small-scale turbulence has generally led to a

[2] These estimates are very depth-dependent; in particular, the magnetic Prandtl number $\mathcal{P}_{\mathrm{m}} = \nu/\eta = \mathcal{R}_{\mathrm{m}}/\mathcal{R}e$ increases rapidly with depth from $\mathcal{O}(10^{-7})$ at the photosphere to $\mathcal{O}(10^{-2})$ deep in the convection zone – see Figure 5.2.

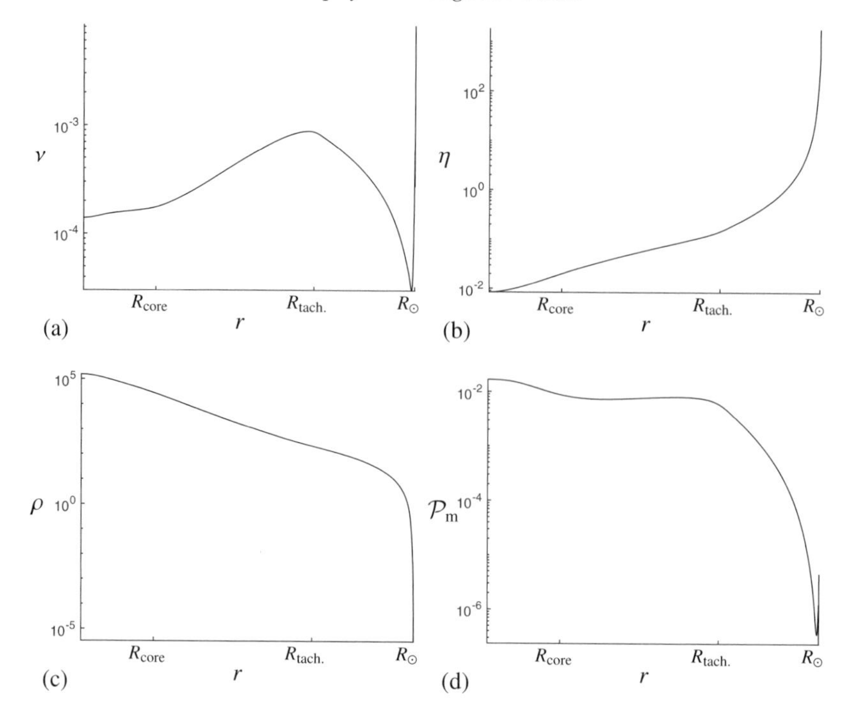

Figure 5.2 Interior structure of the Sun: (a), (b), (c), (d) kinematic viscosity ν, magnetic resistivity η, density ρ and magnetic Prandtl number $\mathcal{P}_{\rm m} \equiv \nu/\eta$ as functions of radius from the centre to the photosphere at $R = R_\odot$; the rapid decrease of ν with depth through the photosphere is due to the rapid increase of ρ; η decreases rapidly with depth through the photosphere, due mainly to the rapid increase of temperature with depth (derived from model S).

simple crude representation of the mean effects of the turbulence in terms of a turbulent (or 'eddy') viscosity ν_e and magnetic diffusivity η_e; if these are supposed to incorporate all the effects of turbulence on scales smaller than the granular scale $\ell_{\rm g}$, then they are given in order of magnitude by

$$\nu_e \sim \eta_e \sim u_{\rm g}\ell_{\rm g} \sim 10^3\,{\rm km^2 s^{-1}}\,. \tag{5.6}$$

The precise mechanism by which turbulence can lead to a 'cascade' of both kinetic and magnetic energies towards smaller and smaller length-scales and ultimately to the very small scales on which viscous and ohmic dissipation take place presents a difficult problem, some aspects of which will be considered in Chapter 15.

5.2.2 Helioseismology

Helioseismology is a technique, similar to seismology on the Earth, which is used to determine properties of the solar interior. It is based on the study of the many

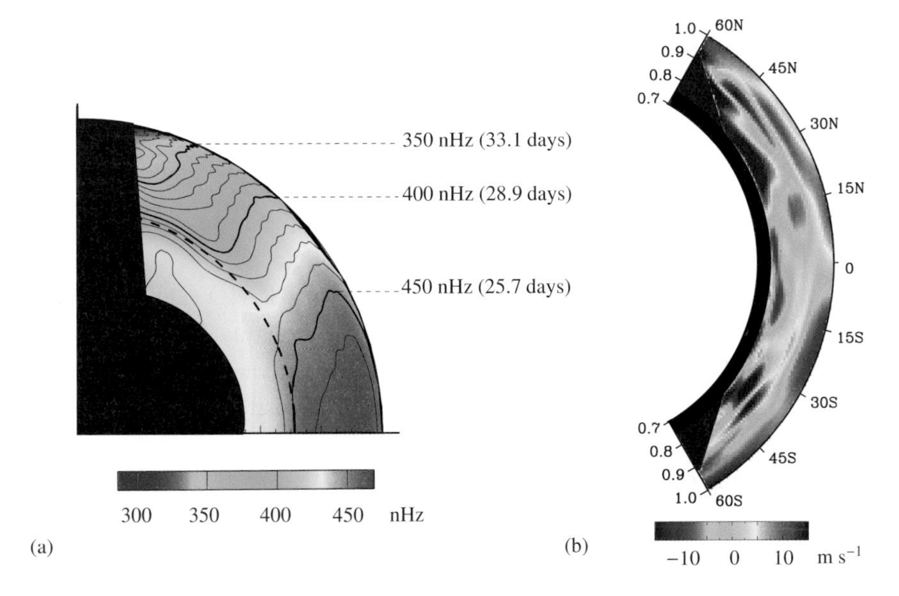

Figure 5.3 Mean flow in the Sun based on helioseismology: (a) contours of mean toroidal (zonal) angular velocity [adapted from Schou et al. 1998]; (b) cross section of the weaker meridional flow, with positive velocity (red) directed north. [Adapted from Zhao et al. 2013, ©AAS. Reproduced with permission.] In both cases the black area indicates regions where reconstruction of the flow is as yet unreliable.

modes of solar oscillations. Analysis of the characteristic frequencies of these modes provides constraints on the structure and dynamics of the solar interior. The radial distribution of density and sound speed, and constraints on internal differential rotation, have been determined in this way (Christensen-Dalsgaard 2002).

Figure 5.1 shows the radial structure of the Sun based on 'model S' (Christensen-Dalsgaard et al. 1996), a classic reference model for helioseismic inversion. In contrast to the Earth (Figure 4.1), most of the mass is concentrated in the radiative core of the Sun: the total mass in the convection zone, where dynamo action can occur, is small compared to that in the core region (Figure 5.1a). This results in a gravity profile in the convection zone varying as r^{-2}. Moreover, the bulk modulus K profile, estimated from $\rho(\partial P/\partial\rho)_{ad}$, is close to the pressure profile P (Figure 5.1b). This means that the Sun is strongly stratified against its own weight, and that the adiabatic gradient plays an important role. For these reasons, dynamo action in the Sun may be expected to differ greatly from that in the Earth.

Figure 5.2 shows the variation of kinematic viscosity $\nu = \mu/\rho$ (where μ is the dynamic viscosity), magnetic resistivity η, density ρ, and magnetic Prandtl number $\mathcal{P}_\mathrm{m} \equiv \nu/\eta$ as functions of radius from the centre to the photosphere at $R = R_\odot$. Here, $\eta \sim T^{-3/2}$ depends primarily on temperature and only very weakly on density (Spitzer1956), and this accounts for its rapid variation with depth through the surface layers of the Sun. Even more dramatic is the variation of $\nu \sim T^{3/2}/\rho$ (Maxwell 1867) with depth through the surface layers. The magnetic Prandtl number $\mathcal{P}_{\rm m}$ therefore also varies by several orders of magnitude (10^{-6} – 10^{-2}) with depth through the photosphere and convection zone. We may anticipate that dynamo behaviour will depend quite critically on the value of $\mathcal{P}_{\rm m}$, and that this strong variation may therefore have an important influence in determining the depth within the Sun where the dynamo process is most efficient.

Figure 5.3a (constructed from data collected by the SOHO satellite) shows the azimuthally averaged zonal flow (symmetric about the equator), and Figure 5.3b (based on data from the NASA Solar Dynamics Observatory (SDO)) shows the meridional circulation within the Sun. The abrupt transition region between the convection zone ($r > 0.7\,R_\odot$) and the uniformly rotating radiative interior ($r < 0.7\,R_\odot$) is the tachocline (Spiegel & Zahn 1992; Hughes et al. 2007) whose extent (too thin to be properly resolved by current seismic data) is estimated to be between 2% and 3% of the solar radius. The physical origin of this sharp layer is still a matter of debate.

5.3 Sunspots and the solar cycle

Sunspots are dark spots (typically of order 10^4 km in diameter) that appear and disappear on the surface of the Sun mainly within $\pm 35^{\rm o}$ of the equatorial plane. The number of spots visible at any time varies from day to day and from year to year, the most striking feature being the current periodicity (with approximate period 11 years) in the annual mean sunspot number. Figure 5.4 presents the record of sunspot numbers from 1610 to 1995; this is based on historical observations preserved in archives and libraries (Hoyt & Schatten 1998, and see Figure 1.2) supplemented by observatory data for the recent period. This figure also shows signs of a weak longer-term periodicity with period of order 100 years. The period 1650–1700, known in this context as the 'Maunder minimum' (Maunder 1904), was anomalous in that very few sunspots were recorded.

Sunspots commonly occur in pairs roughly aligned along a line of latitude $\theta =$ const., and they rotate with an angular velocity a little greater than that given by (5.1). The typical distance between spots in a pair (or more complicated spot group) is of order 10^5 km. The leading spot of a pair is slightly nearer the equatorial plane (in general) than the following spot; as pointed out by Hale et al. (1919), this tilt tends to increase with latitude (*Joy's law*).

Formation of sunspot pairs is a magnetohydrodynamic phenomenon, which may be understood in physical terms (Parker 1955a) as follows. Suppose that a weak poloidal magnetic field of, say, dipole symmetry is present and is maintained by some mechanism as yet unspecified. Any differential rotation that is present in the

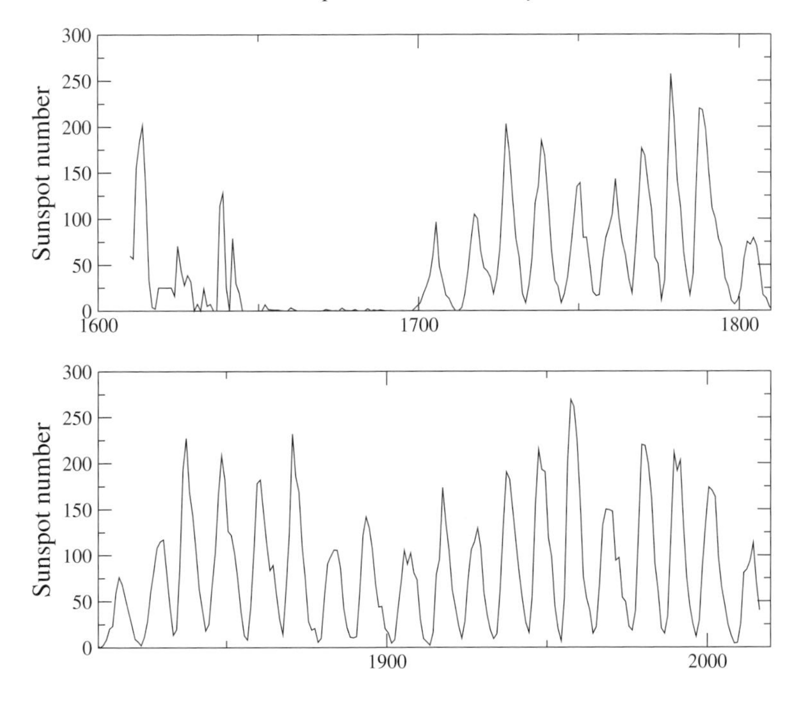

Figure 5.4 Annual mean sunspot number, 1610–2016. During the last 300 years, there has been a characteristic periodicity, with period approximately 11 years. The 'Maunder minimum', when few sunspots were recorded, ran from 1650 to 1700. Observations prior to 1650 were unsystematic, and there are large gaps in the data. [Based on data from Hoyt & Schatten 1998 and from WDC-SILSO, Royal Observatory of Belgium, Brussels.]

convective zone of the Sun (due to redistribution of angular momentum by thermal turbulence) will then tend to generate a toroidal field which is greater than the poloidal field by a large factor ($\mathcal{O}(\mathcal{R}_m)$ if the mechanism of §3.14 is of dominant importance). Consider then a tube of strong toroidal flux ($B_\varphi \sim 1$ T) immersed at some depth in the convective zone. If this tube is in dynamic equilibrium with its surroundings, then the total pressure $p + (2\mu_0)^{-1}B_\varphi^2$ inside the tube must equal the fluid pressure p_0 outside; hence the fluid pressure p is less inside the tube than outside, and so, if there is a simple monotonic relationship between pressure and density, the density inside must also be less than the density outside. The tube is then buoyant relative to its surroundings (like a toroidal bubble) and may be subject to instabilities if the decrease of density with height outside the flux tube is not too great. This phenomenon (described as 'magnetic buoyancy' by Parker 1955a) will be considered in some detail in §12.11. For the moment it is enough to say that instability can manifest itself in the form of large kinks in the toroidal tube of force, which may rise and break through the solar surface (Figure 5.5). Such

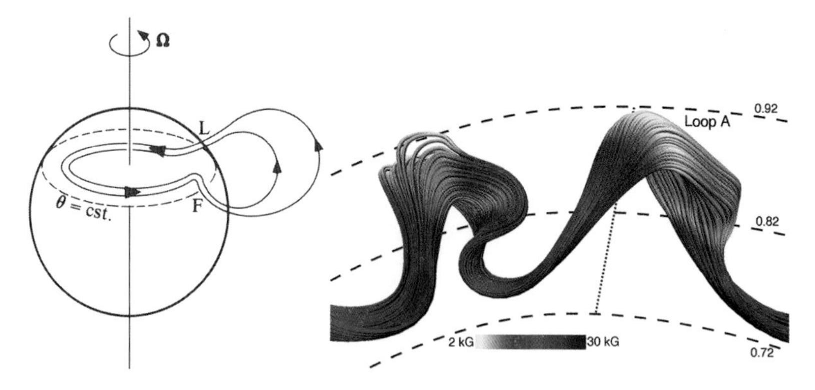

Figure 5.5 (a) Schematic of the formation of a sunspot pair by eruption of a subsurface toroidal tube of force. The vertical field is intensified at L and F, where the leading and following spots are formed. Expansion of the erupting gas leads to a deficit of the local vertical component of angular momentum relative to the rotating Sun, and so to a twist of the sunspot pair LF relative to the surface line of latitude θ = const. [After Parker 1955a.] (b) A pair of upwelling loops in an anelastic spherical-harmonic (ASH) simulation, as seen looking down along the rotation axis towards the equatorial plane; colour coding indicates field strength; dashed circles indicate radial position $R/R_{\odot}$. [From Nelson et al. 2011, AAS. Reproduced with permission.]

a rise leads to vertical stretching of the lines of force in the tube, as exemplified in the idealised solution of the induction equation discussed in §3.4. Also, since fluid in the neighbourhood of the rising kink may be expected to expand on rising (the total vertical extent of the phenomenon being large compared with the scale-height) conservation of angular momentum will make this fluid rotate in a left-handed sense (in the northern hemisphere) as it rises, thus providing a natural explanation for the preferred tilt of a spot pair relative to the line of latitude as mentioned above. The strong localised vertical fields thus created may be expected to suppress thermal turbulence and therefore to decrease the transport of heat to the solar surface – hence the darkening of a sunspot relative to its surroundings.

Magnetic fields were first detected in sunspots by Hale (1908); a sunspot field is typically of the order of 10^{-1} T, and may be as large as 0.4 T. The polarity of sunspot fields (i.e. whether the direction is radially outwards or inwards) is governed by *Hale's polarity laws*: (i) in any sunspot pair, the polarity is positive in one sunspot and negative in the other; (ii) all pairs in one hemisphere (with few exceptions) have the same polarity sense, indicating eruption from a subsurface toroidal field that is coherent over the hemisphere: and (iii) pairs of sunspots in opposite hemispheres generally have the opposite polarity, indicating that the toroidal field is generally antisymmetric about the equatorial plane. It is evident that these laws are consistent with the physical picture described above.

As mentioned earlier, sunspots appear only within about $\pm 35^{\circ}$ of the equatorial

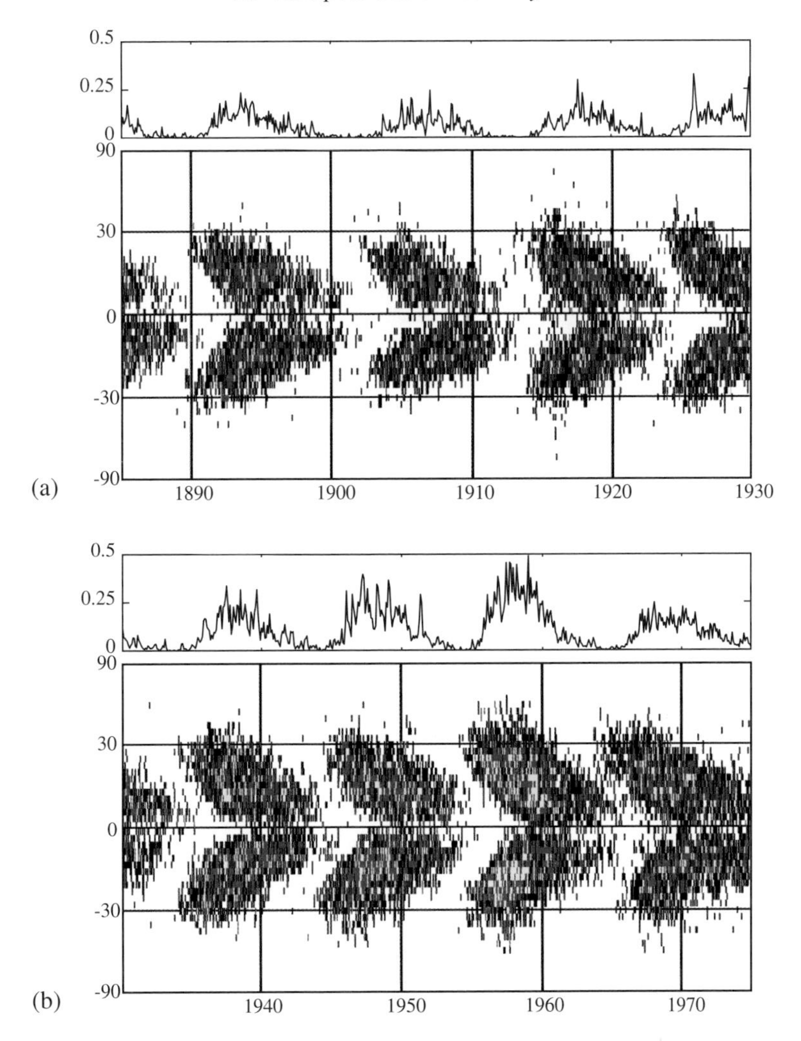

Figure 5.6 Butterfly diagram showing the incidence of sunspots as a function of latitude $\theta' = 90° - \theta$ and time t for the periods (a) 1885–1930 and (b) 1930–1975. In each case, the upper panel represents the average daily sunspot area (as a percentage of the visible hemisphere) as a function of time; the lower panel shows the sunspot areas as a function of latitude and time (the colour from black to red and yellow reflects the sunspot relative area in equal 1° latitude strips). The 11-year cycle and the migration of the 'active regions' towards the equatorial plane are clearly revealed. [Adapted from Hathaway 2015.]

plane $\theta = 90°$. In fact the distribution of spots in latitude varies periodically in time with the same period (~ 11 years) as that of the sunspot cycle. This behaviour is generally represented by means of the celebrated 'butterfly diagram' introduced by Maunder (1904). Following Maunder (1913), this diagram (Figures 5.6 and 5.7) is

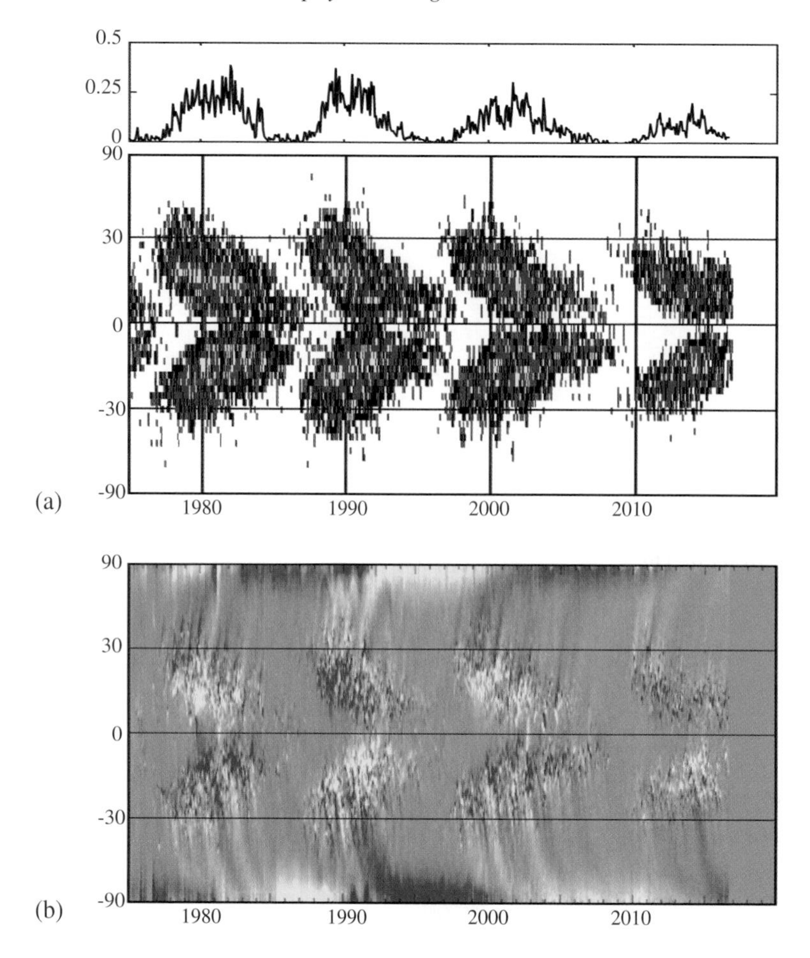

Figure 5.7 (a) Continuation of Figure 5.6 for the period 1975–2016: (b) a butterfly diagram constructed from the zonally averaged radial magnetic field. This panel illustrates Hale's polarity laws, Joy's law, polar field reversals and the transport of higher latitude magnetic field elements toward the poles. [Adapted from Hathaway 2015.]

constructed by inserting, for each rotation of the Sun, a set of vertical line segments corresponding to the bands of latitude in which sunspots are observed during the period of rotation. The 11-year periodicity is again clearly evident in such a diagram; also evident is a migration of the sunspot pattern towards the equatorial plane, reflecting a similar migration of the underlying toroidal field if the above qualitative description of sunspot formation is correct. The sense of polarity of sunspot pairs in either hemisphere is observed to change from one 11-year cycle to the next. It may be inferred that the underlying toroidal magnetic field is periodic in time with period approximately $2 \times 11 = 22$ years. If this toroidal field is

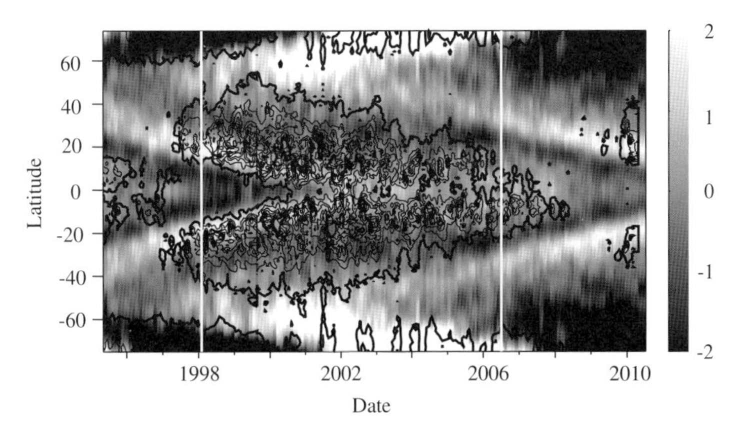

Figure 5.8 Time evolution of the zonal flow at $0.99\,R_\odot$. The perturbation in zonal flow $\delta\varpi/2\pi$ is represented in nHz in greyscale. The band-pattern of torsional oscillations is clearly visible. The superposed contours represent unsigned magnetic field strength. The bands appear to drift equatorwards in tandem with the active regions. [Adapted from Howe et al. 2011.]

generated from a poloidal field by steady differential rotation, then the poloidal field must also be time-periodic with the same period. The observational evidence for this behaviour is discussed in the following section.

The imprint of the solar cycle on the zonal flow at the surface of the Sun in the form of weak alternating bands known as torsional oscillations was identified by Howard & Labonte (1980). Figure 5.8 reveals the strong correlation between this oscillating zonal flow and the regions of magnetic activity (Howe et al. 2011, 2013).

5.4 The general poloidal magnetic field of the Sun

It is clear from Figure 5.7b (Hathaway 2015) that the field within 30° of the poles also follows a 22-year cycle, though out of phase with that of the butterfly diagram. This suggests that, although there are clearly strong random effects at work in the evolution of the solar magnetic field, there is nevertheless a significant coupling between the poloidal and toroidal ingredients of the field, and that these both exhibit (in a suitably averaged sense) time-periodic behaviour with period of order 22 years. At any rate, from the observations there is certainly sufficient motivation to study any dynamo models which involve coupling between toroidal and poloidal fields and which are capable of predicting such time-periodic behaviour. The coupling between mean poloidal field evolution and sunspot activity is rather strikingly revealed in Figure 5.7. During the half-cycle 1987–98, the polarity of the leading spots in the northern hemisphere was negative (indicating positive B_φ) (and similarly negative B_φ in the southern hemisphere). During this same period, B_r was negative in the northern hemisphere (in the sunspot zone) and positive in the southern hemisphere. The fact that B_r and B_φ are in this sense *out of phase* has important implications for possible dynamo mechanisms, as pointed out by Stix (1976). These implications will be discussed later in the context of particular dynamo models (see §9.9).

Let us now consider briefly the detailed spatial structure of the radial component of magnetic field as observed over the solar disc. Increasing refinement in spectroscopic detection techniques can now reveal the fine structure of this field down to scales of the order of 100 km and less. The remarkable fact is that, although the spatially averaged radial field is of the order of 10^{-4} T it is by no means uniformly spread over the solar surface, but appears to be concentrated in 'flux elements' with diameters of order 200 km and less, in which field strengths are typically of the order of 0.1–0.2 T (Stenflo 1976). Such flux elements naturally occupy a very small fraction ($\sim 0.1\%$) of the solar surface; the number of elements required to give the total observed flux is of order 10^5. The process of flux concentration can be understood in terms of the 'flux expulsion' mechanism discussed in §3.12; there are however difficulties in accepting this picture, in that flux expulsion as described in §3.12 occurs effectively when the flow pattern is steady over a time-scale large compared with the turn-over time of the constituent eddies, whereas observation of the granulation pattern suggests unsteadiness on a time-scale t_g somewhat *less* than the turn-over time given by $\ell_g/u_g \sim 10^3$s. The problem of the fine-scale structure of the solar field is a challenging one that is by no means as yet fully resolved.

Our main concern however in subsequent chapters will be with the evolution of the *mean* magnetic field (the mean being defined either spatially over scales large compared with ℓ_g or temporally over scales large compared with t_g). The difficulties associated with the fine-scale structure are to some extent concealed in this type of treatment; nevertheless the treatment is justified in that the first requirement in the solar (as in the terrestrial) context is to provide a theoretical framework for the treatment of the *gross* properties of the observed field; the treatment of the fine-structure must at this stage be regarded to some extent as a secondary problem, although in fact the two aspects are inextricably linked, and understanding of the dynamo mechanism will be complete only when the detailed fine-structure processes and their cumulative effects are fully understood.

5.5 Magnetic stars

The magnetic field of the Sun is detectable only by virtue of its exceptional proximity to the Earth as compared with other stars. Magnetic fields of distant stars are detectable only if strong enough to provide significant Zeeman splitting of their spectral lines; and this phenomenon can be observed only for stars whose spectral lines are sharply defined and not smeared out by other effects such as rapid rotation. The dramatic discovery Babcock (1947) of a magnetic field of the order of 0.05 T in the star 78 Virginis heralded a new era in the subject of cosmical electrodynamics. This star is one of the class of peculiar A-type (Ap) stars, which exhibit, via their spectroscopic properties, unusual chemical composition relative to the Sun. It is known (Preston 1967) that 78 Virginis is typical of Ap stars whose spectral lines are not too broadened by rotation to make Zeeman-splitting detection impracticable: virtually all such stars exhibit magnetic fields in the range 10^{-2} T – 3T; these fields are variable in time, and a proportion of these 'magnetic stars' show periodic behaviour with period typically of the order of several days.

The fields of these stars have an order of magnitude that is consistent with the simple and appealing conjecture that the stellar field is formed by compression of the general galactic field (which is of order 10^{-10} T) during the process of gravitational condensation of the star from the interstellar medium. For a sphere of gas of radius R, with mass M and trapped flux F, this gives simple relations for the mean density ρ and mean surface field strength B:

$$Br^2 \sim F, \quad \rho r^3 \sim M, \quad B \sim \left(F/M^{2/3}\right)\rho^{2/3}, \tag{5.7}$$

and hence

$$B_s/B_g = (\rho_s/\rho_g)^{2/3}, \tag{5.8}$$

where the suffixes s and g refer to the stellar and galactic fields respectively. A compression ratio of order 10^{15} would thus be sufficient to explain the order of magnitude of the observed fields ($B_s/B_g \sim 10^{10}$). In fact ρ_s/ρ_g is considerably greater than 10^{15}, and the problem is rather to explain the inferred loss of magnetic flux during the process of gravitational condensation and subsequent early stage of stellar evolution (Mestel 1967).

As in the case of the Sun, the natural (ohmic) decay time for the field of a magnetic star is of the same order as (or even greater than) the life-time of the star, and a possible view is that the field is simply a 'fossil' relic of the field created during the initial condensation of the star. If the dipole moment of the star is inclined to the rotation axis, the observed periodicities can be explained in a natural way in terms of a rotating dipole (the 'oblique rotator' model). Some of the observations (e.g. the irregular field variations observed in at least some Ap stars) cannot however be explained in this way, and it seems likely that dynamo processes, similar to those occurring in the Sun, may play an important part in determining field evolution in most cases. As pointed out by Preston (1967), a dynamo theory and an oblique rotator theory are not necessarily mutually exclusive possibilities; it is more likely that both types of process occur and interact in controlling the evolution of the observed fields.

Many stars are like the Sun in that they possess an outer convective envelope; such 'solar-type' stars exhibit 'starspots' (producing variations in luminosity) and a hot corona (detected at radio and X-ray wavelengths), signatures of their internal magnetic fields. In the Sun, as we shall see in Chapter 14, the tachocline is thought to play an important role in the dynamo process. In contrast, main-sequence stars with masses below 0.35 $M_\odot$ are fully convective and do not therefore have a tachocline. The dynamo process in such stars is therefore believed to differ significantly from that in the Sun, and may operate throughout the whole stellar interior.

Measurements of the surface magnetic fields of a number of M dwarfs (stars of mass between 0.08 $M_\odot$ and 0.7 $M_\odot$) are available from two approaches, described in detail by Morin et al. (2010). The first is based on spectroscopy: the average value of the magnetic field magnitude at the stellar surface is inferred from the Zeeman broadening of spectral lines (Saar 1988, Reiners & Basri 2006). The second uses time-series of circularly polarised spectra together with tomographic imaging techniques to produce spatially resolved maps of the large-scale vector magnetic field (up to spherical-harmonic degree in the range 6–30, depending on the rotational velocity) (Zeeman-Doppler Imaging (ZDI): Semel 1989; Donati & Brown 1997, Donati et al. 2006b). The first such observations of a fully convective M dwarf, V374Peg, (Donati et al. 2006a) have revealed that M dwarfs do possess magnetic fields characterised by a strong dipole component almost aligned with the axis of rotation. Such a magnetic field is much more like the field of a planet than that of the Sun or a solar-type star (Donati et al. 2003).

5.6 Magnetic interaction between stars and planets

Although it is usual in dynamo theory to consider planets and stars as isolated systems surrounded by vacuum, this is at best an approximation. In the case of the solar system, a flux of charged particles (mainly electrons and protons) known as the *solar wind* is emitted from the Sun, at speeds of some 4×10^5 m s^{-1}. The magnetic interactions between this solar wind and a planetary field occurs in the *magnetosphere*, a neighbourhood of the planet controlled by its own magnetic field. The boundary of the magnetosphere, determined by pressure balance between the pressure in the solar wind plasma and the magnetic pressure, is known as the *magnetopause*. The solar wind compresses magnetic field lines on the day-side and drags it away from the planet in a tail-like structure on the night-side. A standing 'bow shock' forms upstream of the magnetosphere in which the solar wind is compressed and heated. It is slowed down to the point where it flows around

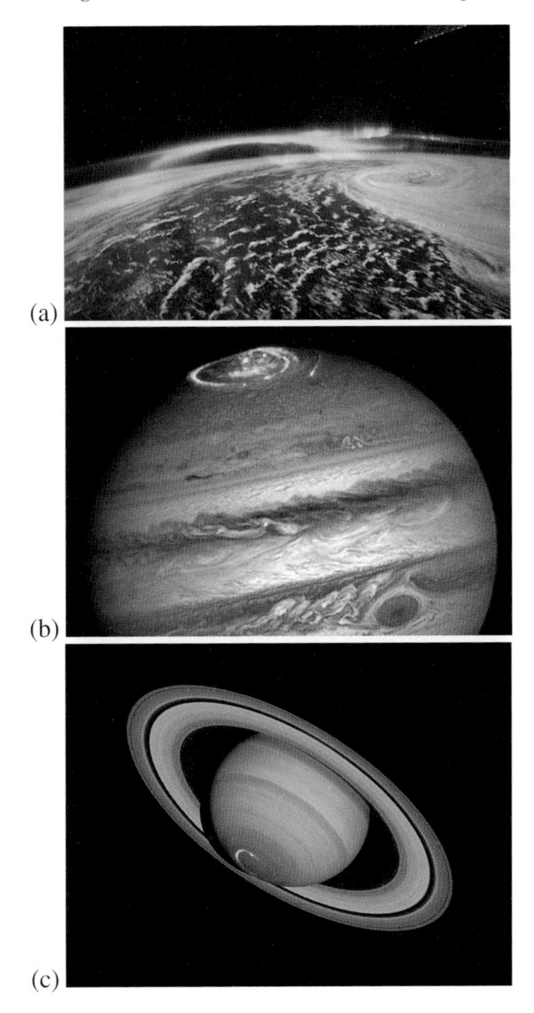

Figure 5.9 (a) True colour picture of the northern lights (aurora borealis) as seen from the International Space Station. (b) Composite pictures of visible light and ultraviolet images of auroræ on Jupiter and Saturn. [Credits: NASA, ESA, J. Clarke (Boston University), and Z. Levay (STScI).]

the magnetosphere in the so-called *magnetosheath*. The magnetosphere thus effectively shields the planet from the solar wind.

A fraction of the solar wind however diffuses across the magnetopause, so some magnetic field lines threading the plasma enter the planetary atmosphere. In this 'reconnection' process, known as the 'Dungey cycle' (Dungey 1961), particles can 'flow' along these field lines into the upper atmosphere of the polar regions. The impact of these energetic particles with the atoms of the atmosphere results in heating and light emission – the *aurora borealis* or *northern lights* phenomenon.

The wavelength of this light depends on the emitting atom: in the Earth's upper atmosphere, the green and red colours are emitted by oxygen and nitrogen atoms respectively (Figure 5.9a); in Jupiter and Saturn, the emission is mainly from hydrogen, and so ultra-violet (Figures 5.9b and c).

The trajectories of charged particles approaching the dipolar magnetic field of the Earth were originally calculated by Størmer (1937), who showed how the convoluted shapes of auroræ could be explained by the resulting spread of these trajectories. In the case of Jupiter, the auroræ reveal the *footprints* of the Galilean moons; that of Io in particular is very visible (see Figure 5.9b).

The idealised model of an isolated dynamo surrounded by vacuum is thus an approximation, necessary for mathematical tractability, but lacking some of the complexity of the real world.

5.7 Galactic magnetic fields

The presence of a magnetic field on the galactic scale and pervading the interstellar medium was inferred by Hiltner (1949). On such large scales, the relevant unit is the parsec, i.e. the distance from which the mean radius of the Earth's orbit subtends an angle of one second of arc (1 pc $\simeq 3.1 \times 10^{16}$ m $\simeq$ 3.2 light years).

Direct measurements of the galactic magnetic field are obviously not possible, but several indirect methods have been used to probe this field: measurements of Faraday rotation, synchrotron emission, starlight polarisation, Zeeman effect, ultra-high energy cosmic rays .... These measurements provide information on the strength and direction of the field; moreover they can also distinguish between the large-scale field and the small-scale turbulent fluctuations: the polarisation and Faraday rotation of radio synchrotron emission are used to trace the ordered field, whereas the unpolarised emission is used to measure the turbulent fluctuations (Zweibel & Heiles 1997).

Consider first what is known of the magnetic field of our galaxy, the Milky Way. The magnetic field component along the line of sight can be deduced from Faraday rotation data of compact extra-galactic polarised radio sources (see Figure 5.10a). The interpretation in terms of the large-scale magnetic field is complicated because we observe the system from within. The figure does however indicate an overall anti-symmetry with respect to both the galactic plane and the vertical axis: on the large scale, the Faraday depth tends to be more positive in the upper left and lower right quadrants and more negative in the other two. This feature is often argued to be the signature of a quadrupolar field (e.g. Oppermann et al. 2012). The Milky Way has also been recently mapped by the Planck satellite (Alves et al., 2014). Observation of the polarised thermal emission of galactic dust at sub-millimetre wavelengths yields a map of the orientation of the interstellar magnetic field

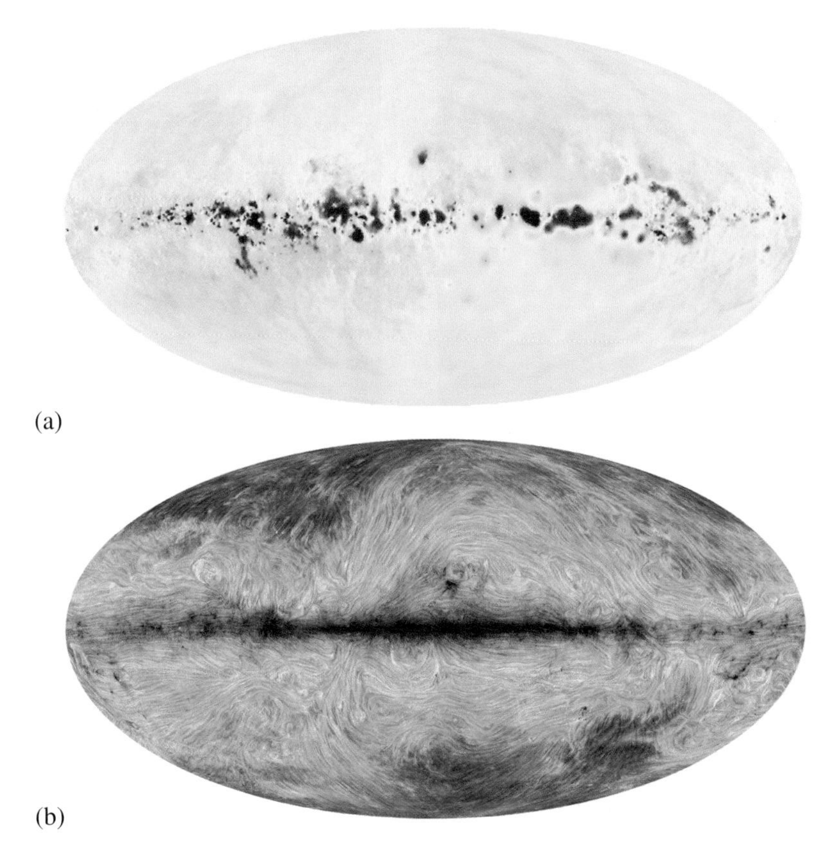

Figure 5.10 Two sky maps of the magnetic field of the Milky Way. In these representations, the plane of the galactic disc extends horizontally, and the centre of the galaxy lies at the centre of the maps. (a) The Faraday effect (Faraday depth) caused by the galactic magnetic field; red and blue respectively indicate regions in which the magnetic field points towards and away from the observer (from Oppermann et al. 2012, reproduced with permission ©ESO). (b) The direction of the field projected on the plane of the sky (i.e. perpendicular to the line of sight), represented as a texture; this is reconstructed from the polarisation of microwaves due to the preferred alignment of cosmic dust with the ambient magnetic field. The total dust emission intensity (here represented in shading) reveals the structure of interstellar clouds in the Milky Way. [ESA and the Planck Collaboration, Alves et al. 2014.]

averaged along the line of sight and projected on the plane of the sky (i.e. the component perpendicular to the line of sight). Figure 5.10 reveals the existence of a complex large-scale component of the magnetic field, which appears to be largely parallel to the plane of the Milky Way.

The Milky Way is believed to be a barred spiral galaxy. Its size is estimated to be ~ 20 kpc in radius and ~ 1 kpc in thickness, and the Sun is located approximately at mid-radius. The strength of the magnetic field in the neighbourhood of the Sun

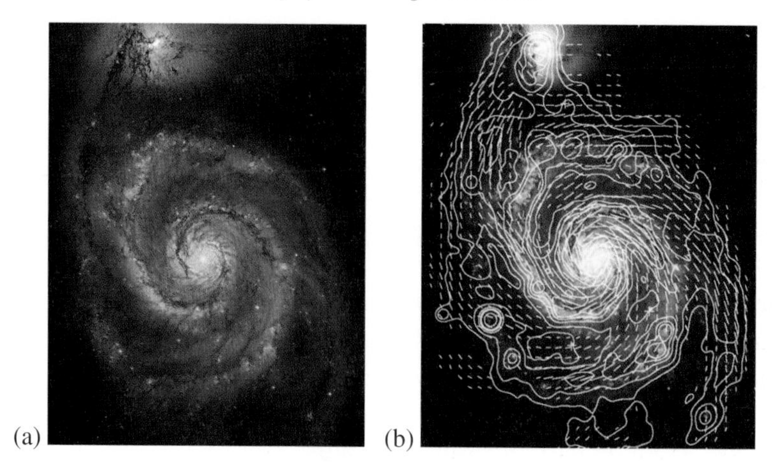

Figure 5.11 (a) The Whirlpool Galaxy (M51 or NGC 5194) as seen from the Hubble Space Telescope, characterised by its spiral arms and its companion galaxy (NGC 5195). (b) The magnetic field of M51 represented both by intensity contours and by the reconstructed magnetic field vectors (with length proportional to the polarised intensity). [After Fletcher et al. 2011 and by permission of Oxford University Press on behalf of the Royal Astronomical Society; image credit: NASA, ESA, S. Beckwith (STScI) and the Hubble Heritage Team (STScI/AURA).]

is about $2\,\mu$G for the large-scale field and about $6\,\mu$G for the small-scale fluctuating component. Within about 150 pc of the galactic center, Ferrière (2010) estimates the field strength to be about $10\,\mu$G and roughly poloidal in the diffuse medium, while in filaments and dense clouds, it is nearly aligned with the galactic plane and and as high as ~ 1 mG.

A reversed direction of the large-scale component of the field close to the Sun is sometimes interpreted as the signature of a bi-symmetric spiral pattern (i.e. an $m = 1$ dominant mode, in terms of azimuthal dependence). It is possible however that this observed reversed direction is associated with a 'super-bubble' (see Chapter 13).

A wide variety of galactic magnetic fields have now been surveyed (as reviewed by Beck 2012, 2016). The nearby spiral galaxy M51, also known as the 'Whirlpool Galaxy', is shown in Figure 5.11. This galaxy is about 10^4 kpc distant from the Milky Way and of similar size.

In more remote galaxies, field amplitudes of the order of a few μG up to 10μG have been measured. In the case of spiral galaxies, the strongest large-scale field is often observed in the regions between the spiral arms, its pitch angle $p = \tan^{-1}\left[\langle B_s\rangle/\langle B_\phi\rangle\right]$ ranging between 10° and 35°, matching that of the spiral arms.

There are many different types of galaxies with different geometries. Even restricting attention to spiral galaxies, no simple general picture emerges. Some galaxies such as IC342 exhibit an axisymmetric fields (i.e. an $m = 0$ azimuthal mode, often described as 'axisymmetric spirals' (ASS) in the galactic literature), whereas others, such as M81 or the Milky Way, are characterised by a non-axisymmetric $m = 1$ magnetic field (or BSS for bi-symmetric spirals) (Krause et al., 1989). The case of M51, pictured in Figure 5.11, appears even more problematic: the field in the disc of M51 contains both $m = 0$ and $m = 1$ components of comparable magnitudes, while the field in the surrounding halo rotates in a sense opposite to that in the disc (Berkhuijsen et al. 1997; see also Fletcher et al. 2011).

Some galaxies can be observed from different angles, some nearly face-on (as M51), and a few nearly edge-on (obviously the choice does not lie with the observer!). Edge-on galaxies are usually characterised by a field parallel to the disc near its plane, suggestive of quadrupole symmetry. A few studies indicate a vertical field component in the halo of the disc in the form of an X-shaped pattern (see again Beck 2012, 2016).

The relevance of dynamo theory to account for these observed large-scale fields is debatable. While for smaller compact objects, such as planets, dynamo action is required to counteract the effect of ohmic diffusion, on the galactic scale ohmic diffusion is of little relevance. One can for example estimate the ohmic decay time of a structure of size $\ell = 100$ pc. For a plasma of temperature $T = 10^4$ K, following Zeldovich et al. (1990), we take $\eta \simeq 10^9\, T^{-3/2}\, \mathrm{m^2 s^{-1}}$, giving

$$\tau_\eta = \ell^2/\eta \sim 10^{26} \text{ years}, \tag{5.9}$$

i.e. many orders of magnitude more than the age of the Universe.[3] Turbulent resistivity may greatly reduce this ohmic decay time. The key argument for the relevance of dynamo action however relies on the structure of the observed fields, which are not purely azimuthal as they would need to be for a fossil field sheared by differential rotation (§3.10).

Observations indicate that significant large-scale magnetic fields can exist in very early galaxies. By exploiting gravitational lensing, Mao et al. (2017) detected a coherent μG magnetic field in a galaxy beyond the 'local volume' (i.e. the volume occupied by our local cluster of galaxies), as seen 4.6 Gyr ago. It exhibits similar strength and geometry to local-volume galaxies. This is the most distant galaxy (of highest redshift) for which a large-scale magnetic field has been observed.

Note further that galaxies, like planets, are not isolated objects: they do interact and sometimes merge. This could be significant for the process of magnetic field amplification. Moreover, coherent magnetic fields of order μG have been reported

[3] The estimates $\eta \sim 1.2 \times 10^3\, \mathrm{m^2 s^{-1}}$, $\nu \sim 7.8 \times 10^{13}\, \mathrm{m^2 s^{-1}}$ (so $\eta/\nu \sim 1.5 \times 10^{-11}$) can be obtained from formulae given by Spitzer (1956); these were adopted by Moffatt (1963) in a study of disc-shaped magnetic structures in an incompressible fluid; the Mach number in hot HI1 gas clouds of the Milky Way was estimated at ~ 0.5, so the flow is subsonic, although not incompressible, and is likely to be turbulent at a rather modest Reynolds number $\sim 10^7$.

in clusters of galaxies on scales of some 400 kpc (Govoni & Feretti, 2004; Govoni et al., 2005).

5.8 Neutron stars

The simple relationship (5.8) suggests that, as massive stars evolve into yet more compressed states with decreasing radius $R(t)$, the associated surface fields should be intensified in proportion to R^{-2}, under the condition that the magnetic flux remains trapped during the collapse process.[4] Very strong magnetic fields (of order $10^3 - 10^4$ T) have been detected in white dwarfs ($R \sim 10^{-2} R_\odot$; see e.g. Landstreet & Angel 1974; Angel 1975). Fields of the order of 10^8 T and greater are to be expected under the extreme condensation conditions of neutron stars ($R \sim 10^{-5} R_\odot$; see e.g. Woltjer 1975). In fact, extremely high surface fields of order $10^{10} - 10^{11}$ T have been inferred from spin-down rates of certain neutron stars, called 'magnetars' (Reisenegger 2003), and from X-ray spectra (Ibrahim et al. 2003); these are the strongest magnetic fields as yet detected in the Universe.

Neutron stars are extremely compact objects ('10 million tons in a teaspoon'), with a solid crust extending down to about $0.9R$. Within this crust the velocity field is effectively zero. However the magnetic field can evolve as a result of the Hall effect, whereby the magnetic field is transported with a velocity

$$\mathbf{v}_H = -k\mu_0 \mathbf{j} = -k\nabla\times\mathbf{B}, \tag{5.10}$$

where k is a positive constant. Neglecting diffusion (the resistivity η being extremely small), the field then evolves according to the equation

$$\partial\mathbf{B}/\partial t = \nabla\times(\mathbf{v}_H\times\mathbf{B}) = -k\nabla\times[(\nabla\times\mathbf{B})\times\mathbf{B}]. \tag{5.11}$$

This equation obviously conserves magnetic helicity (because the field **B** is effectively frozen in flow of the current **j**). It also conserves magnetic energy:

$$\begin{aligned}\frac{\mathrm{d}}{\mathrm{d}t}\int \tfrac{1}{2}\mathbf{B}^2\,\mathrm{d}V &= \int \mathbf{B}\cdot\frac{\partial\mathbf{B}}{\partial t}\,\mathrm{d}V = \int \mathbf{B}\cdot\nabla\times(\mathbf{v}_H\times\mathbf{B})\,\mathrm{d}V \\ &= \int \nabla\times\mathbf{B}\cdot(\mathbf{v}_H\times\mathbf{B})\,\mathrm{d}V = -k^{-1}\int \mathbf{v}_H\cdot(\mathbf{v}_H\times\mathbf{B})\,\mathrm{d}V = 0.\end{aligned} \tag{5.12}$$

However, it does not conserve magnetic dipole moment: as will be proved in §6.3, the dipole moment can be increased by transport of surface poloidal field towards the poles; equally, it can be decreased by transport towards the equator. Either process can occur in the surface layers of a neutron star.

[4] In the case of neutron stars, some flux may be stripped away during the collapse process, according to the 'strip-tease hypothesis' of Reisenegger (2003).

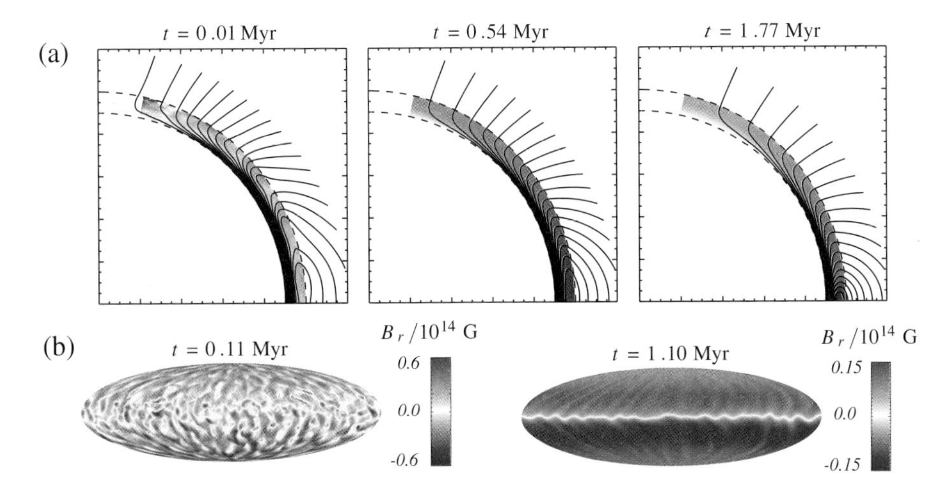

Figure 5.12 Magnetic field evolution in the solid crust of a neutron star. (a) Poloidal field lines, averaged over azimuth angle, showing transport towards equatorial plane; the shading indicates the corresponding electron angular velocity, i.e. $j_\varphi/n(r)\sin\theta$, where j_φ is toroidal current and $n(r)$ is the assumed electron number density. (b) Change in distribution of radial component B_r on the surface over a period of 1 Myr; note the change in colour scaling on the right; non-axisymmetric perturbations weaken over this period, as the flow relaxes to the Hall attractor. [Reprinted figure with permission from Wood & Hollerbach 2015. Copyright 2015 by the American Physical Society.]

A related property of (5.11) is worth noting: if $\mathbf{B}(\mathbf{x}, t)$ is a solution of this equation, then so is $-\mathbf{B}(\mathbf{x}, -t)$. This has the following consequence: suppose $\mathbf{B}_1(\mathbf{x}, t)$ is a solution starting from an initial condition $\mathbf{B}_1(\mathbf{x}, t) = \mathbf{B}_0(\mathbf{x})$ at time $t = -t_1$, and that this evolves to $\mathbf{B}_1(\mathbf{x}, 0)$ by time $t = 0$. If we now consider a field $\mathbf{B}_2(\mathbf{x}, t)$ with initial condition (at time $t = 0$) $\mathbf{B}_2(\mathbf{x}, 0) = -\mathbf{B}_1(\mathbf{x}, 0)$, then this field will evolve, as it were backwards, to the state $\mathbf{B}_2(\mathbf{x}, t_1) = \mathbf{B}_0(\mathbf{x})$ at time $t = +t_1$. Thus for every field for which the dipole moment systematically decreases, there is a complementary field for which the dipole moment systematically increases. Inclusion of weak diffusion may however skew the evolution one way or the other.

Steady (equilibrium) solutions of (5.11) must evidently satisfy a magnetostatic equation

$$\mathbf{j} \times \mathbf{B} = \nabla p, \tag{5.13}$$

for some effective pressure field $p(\mathbf{x})$; such a state has been described as a 'Hall attractor' by Gourgouliatos & Cumming (2014), who have explored axisymmetric evolution governed by (5.11) including the effects of weak diffusivity, and starting from an initial condition in which the poloidal field is in the fundamental (dipole) ohmic decay mode for a spherical shell. They showed that the asymptotic state includes both dipole and octupole ingredients, attained through the above transport mechanism. Non-axisymmetric perturbations of the initial state have been considered by Wood & Hollerbach (2015), who find preferential migration of small-scale features towards the equatorial plane; for the reasons given above, this must again be a consequence of weak diffusivity. Estimates of the magnetic stress generated in these simulations indicate that, when the neutron field exceeds 10^{10} T, the crust is subject to rupture ('starquakes'), a possible source of the gamma-ray flares and X-ray outbursts that have been observed (Thompson & Duncan 2001).

Part II

Foundations of Dynamo Theory

6

Laminar Dynamo Theory

6.1 Formal statement of the kinematic dynamo problem

As in previous chapters, V will denote a bounded region in $\mathbb{R}^3$, S its surface, and $\hat{V}$ the exterior region extending to infinity. Conducting fluid of uniform magnetic diffusivity η ($0 < \eta < \infty$) is confined to V, and the medium in $\hat{V}$ will be supposed insulating, so that the electric current distribution $\mathbf{J}(\mathbf{x}, t)$ is also confined to V. Let $\mathbf{u}(\mathbf{x}, t)$ be the fluid velocity, satisfying

$$\mathbf{u} \cdot \mathbf{n} = 0 \quad \text{on } S. \tag{6.1}$$

Steady velocity fields $\mathbf{u}(\mathbf{x})$ are of particular interest; we shall also be concerned in subsequent chapters with turbulent velocity fields having statistical properties that are time-independent. For the most part we shall limit attention to incompressible flows for which $\nabla \cdot \mathbf{u} = 0$; we shall also assume, unless explicitly stated otherwise, that $\mathbf{u}$ is differentiable in V, and that the total kinetic energy $E(t)$ of the motion is permanently bounded:

$$E(t) = \tfrac{1}{2} \int_V \rho\, \mathbf{u}^2 \mathrm{d}V \; < \infty\,. \tag{6.2}$$

Concerning the magnetic field $\mathbf{B}(\mathbf{x}, t)$, we assume that this is produced entirely by the current distribution $\mathbf{J}$, which is without artificial singularities; then $\mathbf{B}$ is also without singularities, and it satisfies the outer condition

$$\mathbf{B} = \mathcal{O}(r^{-3}) \quad \text{as } r = |\mathbf{x}| \to \infty\,. \tag{6.3}$$

The condition $\nabla \cdot \mathbf{B} = 0$ is of course always satisfied. The field $\mathbf{B}$ then evolves according to the equations

$$\left.\begin{aligned} \partial\mathbf{B}/\partial t = \nabla \wedge (\mathbf{u} \wedge \mathbf{B}) + \eta\nabla^2\mathbf{B} \quad & \text{in } V \\ \nabla \wedge \mathbf{B} = 0, \text{ so } \mathbf{B} = -\nabla\Psi \quad & \text{in } \hat{V} \\ [\![\mathbf{B}]\!] = 0 \quad & \text{across } S \end{aligned}\right\}, \tag{6.4}$$

and subject to an initial condition, compatible with (6.4), of the form

$$\mathbf{B}(\mathbf{x}, 0) = \mathbf{B}_0(\mathbf{x}). \tag{6.5}$$

The total magnetic energy $M(t)$ is given by

$$M(t) = (2\mu_0)^{-1} \int_{V_\infty} \mathbf{B}^2 \, \mathrm{d}V\,, \tag{6.6}$$

where V_∞ as usual represents the whole space. Under the assumed conditions, $M(t)$ is finite, and we suppose that $M(0) = M_0 > 0$. We know that if $\mathbf{u} \equiv 0$, then $M(t) \to 0$ as $t \to \infty$, the time-scale for this natural process of ohmic decay (§2.8) being $t_d = L^2/\eta$ where L is a scale characterising V. A natural definition of dynamo action is then the following: for given V and η, the velocity field $\mathbf{u}(\mathbf{x}, t)$ *acts as a dynamo* if $M(t) \nrightarrow 0$ as $t \to \infty$, i.e. if it successfully counteracts the erosive action of ohmic dissipation. Under these circumstances, $M(t)$ may tend to a constant (non-zero) value, or may fluctuate about such a value either regularly or irregularly, or it may increase without limit.[1]

A given velocity field $\mathbf{u}(\mathbf{x}, t)$ may act as a dynamo for some, but not all, initial field structures $\mathbf{B}_0(\mathbf{x})$ and for some, but not all, values of the parameter η. The field $\mathbf{u}(\mathbf{x}, t)$ may be described as *capable of dynamo action* if there exists an initial field structure $\mathbf{B}_0(\mathbf{x})$ and a finite value of η for which, under the evolution defined by (6.4), $M(t) \nrightarrow 0$ as $t \to \infty$. Under this definition, a velocity field is either capable of dynamo action or it is not, and a primary aim of dynamo theory must be to develop criteria by which a given velocity field may be 'tested' in this respect.

6.2 Rate-of-strain criterion

Since magnetic field intensification is associated with stretching of magnetic lines of force, it is physically clear that a necessary condition for successful dynamo action must be that in some sense (for given η) the rate of strain associated with $\mathbf{u}(\mathbf{x}, t)$ must be sufficiently intense. The precise condition ((6.14) below) was obtained by Backus (1958). Suppose for simplicity that V is the sphere $r < R$, and that $\mathbf{u}$ is steady and solenoidal and zero on $r = R$. The rate of change of magnetic energy is given by

$$\frac{\mathrm{d}M}{\mathrm{d}t} = -\frac{1}{\mu_0} \int_{V_\infty} \mathbf{B} \cdot (\nabla \wedge \mathbf{E}) \, \mathrm{d}V = -\frac{1}{\mu_0} \int_{V_\infty} \mathbf{E} \cdot (\nabla \wedge \mathbf{B}) \, \mathrm{d}V\,, \tag{6.7}$$

[1] Such 'unphysical' behaviour would imply a growing importance of the Lorentz force $\mathbf{J} \wedge \mathbf{B}$ whose effect on $\mathbf{u}(\mathbf{x}, t)$ would ultimately have to be taken into account.

since there is zero flux of the Poynting vector $\mathbf{E} \wedge \mathbf{B}$ out of the sphere at infinity. Hence since $\nabla \wedge \mathbf{B} = 0$ for $r > R$,

$$\mathrm{d}M/\mathrm{d}t = \mathcal{P} - \mathcal{J}\,, \tag{6.8}$$

where

$$\mu_0\mathcal{P} = \int_V (\mathbf{u} \wedge \mathbf{B}) \cdot (\nabla \wedge \mathbf{B})\,\mathrm{d}V = \int_V \mathbf{B} \cdot \nabla \wedge (\mathbf{u} \wedge \mathbf{B})\,\mathrm{d}V = \int_{\mathbf{v}} \mathbf{B} \cdot (\mathbf{B} \cdot \nabla)\,\mathbf{u}\,\mathrm{d}V\,, \tag{6.9}$$

and

$$\mu_0\mathcal{J} = \eta \int_V (\nabla \wedge \mathbf{B})^2\,\mathrm{d}V = \mu_0 \int_V \sigma^{-1}\mathbf{J}^2\,\mathrm{d}V\,. \tag{6.10}$$

Here, $\mathcal{P}$ represents the rate of production of magnetic energy by the velocity field $\mathbf{u}$, and $\mathcal{J}$ represents the rate of ohmic dissipation.

Bounds may be put on both these integrals as follows. First, let $e_{ij} = \frac{1}{2}(u_{i,\,j} + u_{j,i})$ be the rate-of-strain tensor,[2] with principal values $\left\{e^{(1)}, e^{(2)}, e^{(3)}\right\}$ and with $e_{ii} = e^{(1)} + e^{(2)} + e^{(3)} = \nabla \cdot \mathbf{u} = 0$. Then from (6.9),

$$\mu_0\mathcal{P} = \int_V B_i B_j e_{ij}\,\mathrm{d}V \leq e_m \int_V \mathbf{B}^2\,\mathrm{d}V = 2e_m\mu_0 M\,, \tag{6.11}$$

where $e_m = \max_{\mathbf{x}\in V}\left\{\left|e^{(1)}\right|, \left|e^{(2)}\right|, \left|e^{(3)}\right|\right\}$. Secondly, by standard methods of the calculus of variations, it may easily be seen that the quotient $\mathcal{J}/M$ is minimised when $\mathbf{B}$ has the simplest free-decay mode structure discussed in §2.8, and consequently, from (2.150) and using $x_{01} = \pi$, that

$$\mathcal{J} \geq 2\pi^2(\eta/R^2)M\,. \tag{6.12}$$

Hence from (6.8), (6.11) and (6.12),

$$\mathrm{d}M/\mathrm{d}t \leq 2(e_m - \pi^2\eta/R^2)M\,, \tag{6.13}$$

and so M certainly decays to zero (and the motion fails as a dynamo) if $e_m < \eta\pi^2/R^2$. Conversely a necessary (though by no means sufficient) condition for dynamo action is that

$$e_m R^2/\eta \geq \pi^2\,. \tag{6.14}$$

Equality is possible in (6.14) only if e_{ij} is everywhere uniform, $\mathbf{B}$ is everywhere aligned with the direction of maximum rate of strain and the structure of $\mathbf{B}$ is that of a free decay mode. In general these conditions cannot be simultaneously satisfied, and it is likely that the condition (6.14) grossly underestimates the order of magnitude of the rate-of-strain intensity that is really needed for successful dynamo action.

[2] A suffix after a comma will as usual denote space differentiation, e.g. $u_{i,\,j} = \partial u_i/\partial x_j$.

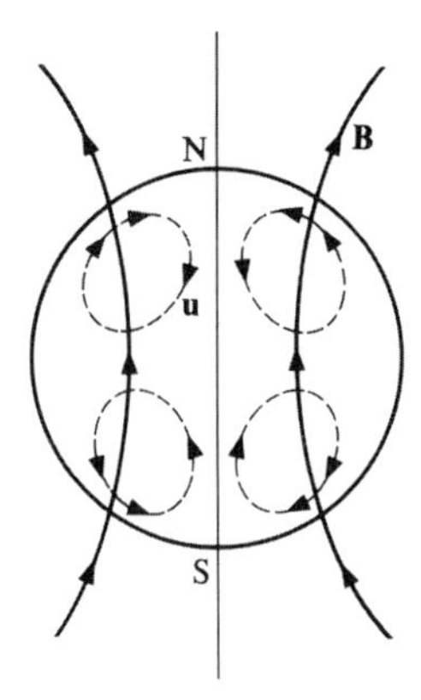

Figure 6.1 The motion illustrated tends to increase the dipole moment due to the sweeping of surface field towards the North and South poles (Bondi & Gold, 1950).

6.3 Rate of change of dipole moment

We have seen in §2.8 that, when V is a sphere, the dipole moment of the current distribution in V may be expressed in the form

$$\mu^{(1)}(t) = \frac{3}{8\pi} \int_V \mathbf{B}(\mathbf{x}, t)\, \mathrm{d}V\,. \tag{6.15}$$

Its rate of change is therefore given by

$$\frac{8\pi}{3} \frac{\mathrm{d}\boldsymbol{\mu}^{(1)}}{\mathrm{d}t} = -\int_V \nabla \wedge \mathbf{E}\, \mathrm{d}V = -\int_S \mathbf{n} \wedge \mathbf{E}\, \mathrm{d}S\,. \tag{6.16}$$

With $\mathbf{E} = -\mathbf{u} \wedge \mathbf{B} + \eta \nabla \wedge \mathbf{B}$, and $\mathbf{u} \cdot \mathbf{n} = 0$ on S, this becomes

$$\frac{8\pi}{3} \frac{\mathrm{d}\boldsymbol{\mu}^{(1)}}{\mathrm{d}t} = \int_S \mathbf{u}\,(\mathbf{n} \cdot \mathbf{B}) \mathrm{d}S - \eta \int_S \mathbf{n} \wedge (\nabla \wedge \mathbf{B}) \mathrm{d}S\,. \tag{6.17}$$

In the perfectly conducting limit ($\eta = 0$), the dipole moment $\left|\boldsymbol{\mu}^{(1)}\right|$ can therefore be increased by any motion $\mathbf{u}$ having the property that $\left(\mathbf{u} \cdot \boldsymbol{\mu}^{(1)}\right)(\mathbf{n} \cdot \mathbf{B}) \geq 0$ at all points of S, i.e. such as to sweep the flux lines towards the polar regions defined by the direction of $\boldsymbol{\mu}^{(1)}$ (Figure 6.1). As pointed out by Bondi & Gold (1950), possible increase of $\left|\boldsymbol{\mu}^{(1)}\right|$ by this mechanism is limited, and $\left|\boldsymbol{\mu}^{(1)}\right|$ in fact reaches a maximum finite value when the flux lines crossing S are entirely concentrated at the poles.[3]

[3] Bondi & Gold argued that this type of limitation in the growth of $\left|\boldsymbol{\mu}^{(1)}\right|$ when $\eta = 0$ need not apply if V has toroidal (rather than spherical) topology, and they quoted the homopolar disc dynamo as an example of a dynamo of toroidal topology which will function when $\eta = 0$. As already pointed out in Chapter 1, however, even this simple system requires $\eta > 0$ in the disc if the flux $\Phi(t)$ across it is to change with time; and the same *flux limitation* will certainly apply for general toroidal systems.

A small but non-zero diffusivity η may totally transform the situation since, as shown in §2.8, diffusion even on its own may temporarily increase $|\boldsymbol{\mu}^{(1)}|$ if $(\mathbf{n}\cdot\nabla)\mathbf{B}$ is appropriately distributed over S. A *sustained* increase in $|\boldsymbol{\mu}^{(1)}|$ may be envisaged if the role of the velocity $\mathbf{u}$ is to maintain a distribution of $(\mathbf{n}\cdot\nabla)\mathbf{B}$ over S that implies diffusive increase of $|\boldsymbol{\mu}^{(1)}|$. Thus, paradoxically, diffusion must play the primary role in increasing $|\boldsymbol{\mu}^{(1)}|$ if this increase is to be sustained, and induction, though impotent in this respect in isolation, plays a crucial subsidiary role in making (sustained) diffusive increase of $|\boldsymbol{\mu}^{(1)}|$ a possibility.

6.4 The impossibility of axisymmetric dynamo action

When both $\mathbf{u}$ and $\mathbf{B}$ (and the associated vectors $\mathbf{A}$, $\mathbf{E}$, $\mathbf{J}$) have a common axis of symmetry Oz, we have seen (§3.7) that the toroidal component of Ohm's law becomes

$$\frac{\partial \mathbf{A}_T}{\partial t} = \mathbf{u}_P \wedge \mathbf{B}_P - \eta \nabla \wedge \mathbf{B}_P \,, \tag{6.18}$$

the corresponding equation for the flux function being

$$\frac{\partial \chi}{\partial t} + \mathbf{u}_P \cdot \nabla \chi = \eta \mathcal{D}^2 \chi \,. \tag{6.19}$$

The absence of any 'source' term in this scalar equation is an indication that a steady state is not possible, a result established by Cowling (1934).

The following proof is a slight modification of that given by Braginskii (1964b). We shall allow $\eta\,(> 0)$ to be an axisymmetric function of position satisfying

$$\mathbf{u}_P \cdot \nabla \eta = 0 \,, \tag{6.20}$$

i.e. η is constant on streamlines of the $\mathbf{u}$-field. (Note that (6.18) and (6.19) remain valid when η is non-uniform.) It is natural to impose the condition (6.20), since variations in conductivity (and so in η) tend to be convected with the fluid, so that $\eta(\mathbf{x})$ can be steady only if (6.20) is satisfied. The condition (6.20) covers the situation when η is constant in V and is an arbitrary axisymmetric function of position in $\hat{V}$, and in particular includes the limiting case when the medium in $\hat{V}$ is insulating.

We now multiply (6.19) by $\eta^{-1}\chi$ and integrate throughout all space; using (6.20) and $\nabla \cdot \mathbf{u}_P = 0$, it follows that $\eta^{-1}\chi\, \mathbf{u}_P \cdot \nabla \chi = \frac{1}{2}\nabla \cdot (\mathbf{u}_P \eta^{-1} \chi^2)$, and so

$$\frac{\mathrm{d}}{\mathrm{d}t}\int_{V_\infty} \frac{1}{2\eta}\chi^2 \,\mathrm{d}V + \int_{S_\infty} \frac{1}{2\eta}\chi^2\, \mathbf{u}_p \cdot \mathbf{n}\,\mathrm{d}S = \int_{V_\infty} \chi \mathcal{D}^2 \chi \,\mathrm{d}V \,. \tag{6.21}$$

For a field that is at most dipole at infinity, $\chi = \mathcal{O}(r^{-1})$ as $r \to \infty$, and the surface integral vanishes provided $\eta^{-1}|\mathbf{u}_P| \to 0$ as $r \to \infty$, a condition that is of course trivially satisfied when $\mathbf{u}_P \equiv 0$ outside a finite volume V. Moreover, using (3.64) we have

$$\int_{V_\infty} \chi \mathcal{D}^2 \chi \, dV = \int_{V_\infty} \chi \nabla \cdot \mathbf{f} \, dV = -\int_{V_\infty} \mathbf{f} \cdot \nabla \chi \, dV = -\int_{V_\infty} (\nabla \chi)^2 \, dV \,, \qquad (6.22)$$

where we have used $\chi = \mathcal{O}(r^{-1})$, $|\mathbf{f}| = \mathcal{O}(r^{-2})$ to discard integrals over the surface at infinity; the final step also requires use of the identity $\chi^2 \nabla \cdot (\mathbf{e}_s/s) = 0$. Hence (6.21) becomes

$$\frac{d}{dt} \int_{V_\infty} \frac{1}{2\eta} \chi^2 \, dV = -\int_{V_\infty} (\nabla \chi)^2 \, dV = -\int_{V_\infty} s^2 \mathbf{B}_P^2 \, dV \,. \qquad (6.23)$$

It is clear from this equation that a steady state with $\chi \not\equiv 0$ is not possible and that ultimately

$$\int s^2 \mathbf{B}_P^2 \, dV \to 0 \,. \qquad (6.24)$$

Excluding the unphysical possibility that singularities in $\mathbf{B}_P$ develop as $t \to \infty$, we are driven to the conclusion that $\mathbf{B}_P \to 0$ everywhere.

Meridional circulation $\mathbf{u}_P$ cannot therefore prevent the decay of an axisymmetric poloidal field.[4] As is clear from the structure of (6.19), the role of $\mathbf{u}_P$ is to redistribute the poloidal flux but it cannot regenerate it. If the meridional circulation is strong, it seems likely that the process of redistribution will in general lead to greatly accelerated decay for the following reason. We have seen in §3.13 that when $\mathcal{R}_m \gg 1$ ($\mathcal{R}_m$ being based on the scale a and intensity u_0 of the meridional circulation) poloidal flux tends to be excluded from regions of closed streamlines of the $\mathbf{u}_P$-field. The time-scale of this flux expulsion is of order $\mathcal{R}_m^{1/3} a/u_0$, and this is much less than the natural decay time $\mathcal{R}_m \, a/u_0$ (when $\mathcal{R}_m \gg 1$). A modest amount of meridional circulation ($\mathcal{R}_m = \mathcal{O}(1)$ or less) may on the other hand lead to a modest delaying action in the decay process (Backus 1957).

6.4.1 Ultimate decay of the toroidal field

Under axisymmetric conditions, the toroidal component of the induction equation is

$$\frac{\partial B}{\partial t} + s(\mathbf{u}_P \cdot \nabla)\left(\frac{B}{s}\right) = s(\mathbf{B}_P \cdot \nabla)\varpi - \nabla \wedge [\eta(\nabla \wedge B \, \mathbf{e}_\varphi)] \cdot \mathbf{e}_\varphi \,, \qquad (6.25)$$

the notation being as in §3.10. We have seen that while $\mathbf{B}_P \neq 0$, the source term $s(\mathbf{B}_P \cdot \nabla)\varpi$ on the right of (6.25) can lead to the generation of a strong toroidal

[4] It must be emphasised that the result as proved here relates to a situation in which both $\mathbf{B}$ and $\mathbf{u}$ are axisymmetric with the same axis of symmetry. Steady maintenance of a non-axisymmetric field by an axisymmetric $\mathbf{u}$-field is not excluded; indeed an example of such a dynamo will be considered in §6.10.

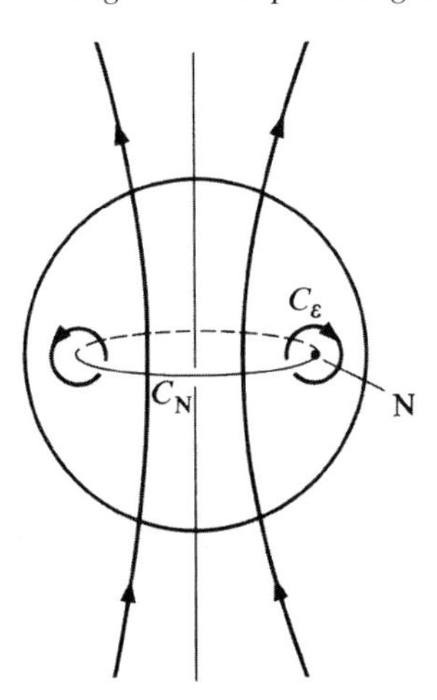

Figure 6.2 In the neighbourhood of the neutral point N of the $\mathbf{B}_P$-field, induction must fail to maintain the field against ohmic decay.

field $B\,\mathbf{e}_\varphi$. Ultimately however, $\mathbf{B}_P$ tends uniformly to zero, and from (6.25) we then deduce that

$$\frac{\mathrm{d}}{\mathrm{d}t}\int \tfrac{1}{2}(B/s)^2\,\mathrm{d}V = -\int \eta\,[\nabla(B/s)]^2\,\mathrm{d}V\,, \tag{6.26}$$

and hence, by arguments similar to those used above,

$$\mathbf{B}_T = B\,\mathbf{e}_\varphi \to 0 \text{ everywhere.} \tag{6.27}$$

6.5 Cowling's neutral point argument

The proof of the impossibility of axisymmetric dynamo action given in the previous section rests on the use of global properties of the field, as determined by the integrals appearing in (6.21). It is illuminating to supplement this type of proof with a purely local argument, as devised by Cowling (1934). The flux-function χ is zero at infinity and is zero (by symmetry) on the axis Oz. If $\mathbf{B}_P$ (and so χ) is not identically zero, there must exist at least one neutral point, N say, in the (s, z) plane where χ is maximal or minimal; at N, $\mathbf{B}_P$ vanishes and the $\mathbf{B}_P$-lines are closed in the neighbourhood of N, i.e. N is an O-type neutral point of the field $\mathbf{B}_P$.

Let C_ε be a closed $\mathbf{B}_P$-line near N of small length ε, and let S_ε be the surface in the (s, z) plane spanning C_ε (Figure 6.2). Suppose that the field $\mathbf{B}_P$ is steady. Then from (6.18), with $\partial\mathbf{A}_T/\partial t = 0$,

$$\int_{S_\varepsilon}(\mathbf{u}_P \wedge \mathbf{B}_P)\cdot\mathrm{d}\mathbf{S} = \oint_{C_\varepsilon}\eta\,\mathbf{B}_P\cdot\mathrm{d}\mathbf{x}\,. \tag{6.28}$$

Let B_ε be the average value of $|\mathbf{B}_P|$ on C_ε; the right-hand side of (6.28) is then $\eta_{\mathrm{N}}\,\varepsilon B_\varepsilon$ to leading order, where η_{N} is the value of η at N. Moreover, since $B_\varepsilon \to 0$ as $\varepsilon \to 0$, the mean value of $|\mathbf{B}_P|$ over S_ε is evidently less than B_ε when ε is sufficiently small; hence

$$\left|\int_{S_\varepsilon} (\mathbf{u}_P \wedge \mathbf{B}_P)\cdot \mathrm{d}S\right| \le U B_\varepsilon S_\varepsilon\,, \tag{6.29}$$

where U is the maximum value of $|\mathbf{u}|$ in V. Hence

$$\eta_{\mathrm{N}}\,\varepsilon \le U S_\varepsilon\,. \tag{6.30}$$

But, as $\varepsilon \to 0$, $S_\varepsilon = \mathcal{O}(\varepsilon^2)$ and this is clearly incompatible with (6.30) for any finite values of U/η_{N}. The implication is that the induction effect, represented by $\mathbf{u}_P \wedge \mathbf{B}_P$ in (6.18), cannot compensate for the diffusive action of the term $-\eta\,\nabla\wedge\mathbf{B}_P$ in the neighbourhood of the neutral point. Of course if there is more than one such O-type neutral point, then the same behaviour occurs in the neighbourhood of each.

The circle C_{N} obtained by rotating the point N about the axis of symmetry is a closed **B**-line (on which $\mathbf{B}_P = 0$, $\mathbf{B}_T \neq 0$) and the failure of dynamo action in the neighbourhood of N may equally be interpreted in terms of the inability of the induced electromotive force $\mathbf{u}\wedge\mathbf{B}$ to drive current along such a closed **B**-line. In general, if we integrate Ohm's law (2.131) along a closed **B**-line on the assumption that **E** is a steady electrostatic field (therefore making no contribution to the integral) we obtain trivially

$$\oint_{\mathbf{B}\text{−line}} \eta\,(\nabla\wedge\mathbf{B})\cdot \mathrm{d}\mathbf{x} = 0\,. \tag{6.31}$$

Hence if $\mathbf{B}\cdot(\nabla\wedge\mathbf{B})$ is non-zero over any portion of a closed **B**-line, then positive values must be compensated by negative values so that (6.31) can be satisfied.

More generally, let S_B (interior V_B) be any closed 'magnetic surface' on which $\mathbf{n}\cdot\mathbf{B} = 0$. Then, from $\eta\,\nabla\wedge\mathbf{B} = -\nabla\phi + \mathbf{u}\wedge\mathbf{B}$, we have

$$\int_{V_B} \eta\,(\mathbf{B}\cdot\nabla\wedge\mathbf{B})\,\mathrm{d}V = -\int_{V_B} \mathbf{B}\cdot\nabla\phi\,\mathrm{d}V = -\int_{S_B} (\mathbf{n}\cdot\mathbf{B})\,\phi\,\mathrm{d}S = 0\,. \tag{6.32}$$

This result reduces to (6.31) in the particular case when a toroidal surface S_B shrinks onto a closed curve C; for then $\mathbf{B}\,\mathrm{d}V \to \Phi\,\mathrm{d}\mathbf{x}$ where Φ is the total flux of **B** round the torus.

Similarly if S_u (interior V_u) is a closed surface on which $\mathbf{n}\cdot\mathbf{u} = 0$, and if $\nabla\cdot\mathbf{u} = 0$ in V_u, then by the same reasoning, under steady conditions,

$$\int_{V_u} \eta\,\mathbf{u}\cdot(\nabla\wedge\mathbf{B})\,\mathrm{d}V = 0\,. \tag{6.33}$$

The results (6.32) and (6.33) hold even if η is non-uniform.

6.6 Some comments on the situation $\mathbf{B} \cdot (\nabla \wedge \mathbf{B}) \equiv 0$

If $\mathbf{B}$ is an axisymmetric poloidal field, then clearly it satisfies

$$\mathbf{B} \cdot \nabla \wedge \mathbf{B} \equiv 0 \,. \tag{6.34}$$

Similarly an axisymmetric toroidal field satisfies (6.34). We know from §6.4 that in either case dynamo action is impossible. It seems likely that dynamo action is impossible for *any* field satisfying (6.34), but this has not been proved.

The condition (6.34) is well known as the necessary and sufficient condition for the existence of a family of surfaces everywhere orthogonal to $\mathbf{B}$, or equivalently for the existence of functions $\alpha(\mathbf{x})$ and $\beta(\mathbf{x})$ such that

$$\mathbf{B} = \beta(\mathbf{x})\nabla\alpha(\mathbf{x}), \quad \nabla \wedge \mathbf{B} = \nabla\beta \wedge \nabla\alpha \,, \tag{6.35}$$

the orthogonal surfaces then being α =const. In the external region $\hat{V}$ where $\nabla \wedge \mathbf{B} = 0$, we may clearly take $\beta = 1$.

It is easy to show that there is no smooth field of the form (6.35) with α and β single-valued, and satisfying the conditions $\mathbf{B} = \mathcal{O}(r^{-3})$ at infinity and $\mathbf{B}^2 > 0$ for all finite $\mathbf{x}$. For suppose there is such a field; in this case, $\beta > 0$ for all finite $\mathbf{x}$, and so $\beta^{-1}\mathbf{B}^2 > 0$ everywhere. But

$$\int_{V_\infty} \beta^{-1}\mathbf{B}^2 \,\mathrm{d}V = \int_{V_\infty} \mathbf{B} \cdot \nabla\alpha \,\mathrm{d}V = \int_{S_\infty} (\mathbf{n} \cdot \mathbf{B})\,\alpha \,\mathrm{d}S = 0 \,, \tag{6.36}$$

and we have a contradiction.

It follows that any field of the form (6.35) and at most dipole at infinity must vanish for at least one finite value of $\mathbf{x}$. The simplest topology (an arbitrary distortion of the axisymmetric poloidal case) is that in which every point of a closed curve C in V is an O-type neutral point of $\mathbf{B}$. If there is just one such curve, then all the surfaces α = const. intersect on C. In this situation, $\beta = 0$ on C and α is not single-valued, and the simple statement (6.36) is certainly not applicable.

Pichakhchi (1966) claimed to prove that dynamo action is impossible if $\mathbf{E} \equiv 0$, i.e. if $\eta \nabla \wedge \mathbf{B} = \mathbf{u} \wedge \mathbf{B}$ in V, (6.34) being an immediate consequence. His argument however rested on the unjustified (and generally incorrect) assertion that the surfaces α = const. do not intersect. The result claimed nevertheless seems plausible, and a correct proof would be of considerable interest.

6.7 The impossibility of dynamo action with purely toroidal motion

It is physically plausible that poloidal velocities are necessary to regenerate poloidal magnetic fields, and that a velocity field that is purely toroidal will therefore be incapable of sustained dynamo action. This result was first discovered by Elsasser (1946), and again by Bullard & Gellman (1954) as a by-product of their treatment of the induction equation by spherical-harmonic decomposition. A simpler direct proof was given by Backus (1958), whose method we follow here. We revert to the standard situation in which η is assumed uniform in V, and we restrict attention to incompressible velocity fields for which $\nabla \cdot \mathbf{u} = 0$. We assume for simplicity that V is the sphere $R \leqslant r$. From the induction equation in the form

$$\mathrm{D}\mathbf{B}/\mathrm{Dt} = \mathbf{B} \cdot \nabla \mathbf{u} + \eta \nabla^2 \mathbf{B}, \tag{6.37}$$

we may then immediately deduce an equation for the scalar $\mathbf{x} \cdot \mathbf{B}$, viz.

$$\frac{\mathrm{D}}{\mathrm{D}t}(\mathbf{x} \cdot \mathbf{B}) \equiv \mathbf{x} \cdot \frac{\mathrm{D}\mathbf{B}}{\mathrm{D}t} + \mathbf{u} \cdot \mathbf{B} = (\mathbf{B} \cdot \nabla)(\mathbf{x} \cdot \mathbf{u}) + \eta \nabla^2 (\mathbf{x} \cdot \mathbf{B}). \tag{6.38}$$

Hence if the motion is purely toroidal, so that $\mathbf{x} \cdot \mathbf{u} = 0$, the quantity $Q = \mathbf{x} \cdot \mathbf{B}$ satisfies the diffusion equation

$$\mathrm{d}Q/\mathrm{d}t = \eta \nabla^2 Q \quad \text{in } V. \tag{6.39}$$

Moreover $\nabla^2 Q = 0$ in $\hat{V}$, and Q and $\partial Q/\partial n$ are continuous across the surface S of V. Standard manipulation then gives the result

$$\frac{\mathrm{d}}{\mathrm{d}t} \int_V Q^2 \, \mathrm{d}V = -2\eta \int_{V_\infty} (\nabla Q)^2 \, \mathrm{d}V, \tag{6.40}$$

so that $Q \to 0$ everywhere.

Now, in the standard poloidal-toroidal decomposition of $\mathbf{B}$, we have $L^2 P = -\mathbf{x} \cdot \mathbf{B}$ (from (2.52)); and the equation $L^2 P = 0$ with outer boundary condition $P = \mathcal{O}(r^{-2})$ at infinity has only the trivial solution $P \equiv 0$. It follows that a steady state is possible only if $\mathbf{B}_P \equiv 0$. The equation for the toroidal field $\mathbf{B}_T$ then becomes

$$\frac{\partial \mathbf{B}_T}{\partial t} = \nabla \wedge (\mathbf{u}_T \wedge \mathbf{B}_T) + \eta \nabla^2 \mathbf{B}_T. \tag{6.41}$$

With $\mathbf{B_T} = -\mathbf{x} \wedge \nabla T$, and $\mathbf{x} \cdot \mathbf{u_T} = 0$, we have

$$\mathbf{u}_T \wedge \mathbf{B}_T = -\mathbf{x}(\mathbf{u}_T \cdot \nabla) T, \tag{6.42}$$

so that (6.41) becomes

$$-(\mathbf{x} \wedge \nabla)\frac{\partial T}{\partial t} = (\mathbf{x} \wedge \nabla)\mathbf{u}_T \cdot \nabla T - \eta(\mathbf{x} \wedge \nabla)\nabla^2 T, \tag{6.43}$$

using the commutativity of $\mathbf{x} \wedge \nabla$ and ∇^2. Hence

$$\frac{\mathrm{d}T}{\mathrm{d}t} \equiv \frac{\partial T}{\partial t} + \mathbf{u}_T \cdot \nabla T = \eta \nabla^2 T + f(r), \tag{6.44}$$

for some function $f(r)$. When we multiply by T and integrate (6.44) throughout V, the term involving $f(r)$ makes no contribution since T integrates to zero over each surface $r = \text{const}$. Moreover $T = 0$ on the surface S of V, and so we obtain

$$\frac{\mathrm{d}}{\mathrm{d}t} \int_V T^2 \, \mathrm{d}V = -2\eta \int_{\mathbf{V}} (\nabla T)^2 \, \mathrm{d}V. \tag{6.45}$$

Hence T (and so $\mathbf{B}_T$ also) ultimately decays to zero.

By further manipulations of (6.38), Busse (1975b) has succeeded in obtaining an estimate for the magnitude of the radial velocity field that would be required to prevent the diffusive decay of the radial magnetic field. Comparison of the two terms on the right of (6.38) gives the preliminary estimate

$$u_r \sim \eta B_r/|\mathbf{B}|R\,. \tag{6.46}$$

If the term involving $\mathbf{x}\cdot\mathbf{u}$ in (6.38) is retained, then Busse shows that (6.40) may be replaced by the inequality

$$\frac{\mathrm{d}}{\mathrm{d}t}\int_V Q^2\,\mathrm{d}V \le -2\left[\eta - \max(\mathbf{u}\cdot\mathbf{x})\left(\frac{M}{2M_P}\right)^{1/2}\right]\int_{V_\infty}(\nabla Q)^2\,\mathrm{d}V\,, \tag{6.47}$$

where M represents the total magnetic energy, and M_P the energy of the poloidal part of the magnetic field. Hence a necessary condition for amplification of $\int_V Q^2\,\mathrm{d}V$, is

$$\max(\mathbf{u}\cdot\mathbf{x}) > \eta\,(2M_P/M)^{1/2}\,, \tag{6.48}$$

which may be compared with (6.46). For a given magnetic field distribution, and therefore a given value of the ratio M_P/M, (6.48) clearly provides a necessary, although again by no means sufficient, condition that a velocity field must satisfy if it is to maintain the magnetic field steadily against ohmic decay.

The condition (6.48) is of particular interest and relevance in the terrestrial context in view of the arguments in favour of a stably stratified core (§4.5). Stable stratification implies inhibition of radial (convective) velocity fields, but may nevertheless permit wave motions (e.g. internal gravity waves modified by Coriolis forces) if perturbing force fields are present. The condition (6.48) indicates a minimum level for radial fluctuation velocities to maintain a given level of poloidal field energy.

It is instructive to interpret (6.48) in terms of magnetic Reynolds numbers $\mathcal{R}_{\mathrm{m}T}$ and $\mathcal{R}_{\mathrm{m}P}$ characterising the toroidal and poloidal motions. We have seen that when $\mathcal{R}_{\mathrm{m}T} \gg 1$ the toroidal magnetic energy builds up to $\mathcal{O}(\mathcal{R}_{\mathrm{m}T}^2)$ times the poloidal magnetic energy, so that $(M_P/M)^{1/2} = \mathcal{O}(\mathcal{R}_{\mathrm{m}T}^{-1})$; hence, with $\mathcal{R}_{\mathrm{m}P} = \max(\mathbf{u}\cdot\mathbf{x})/\eta$, (6.48) becomes simply

$$\mathcal{R}_{\mathrm{m}P}\mathcal{R}_{\mathrm{m}T} \gtrsim 1\,. \tag{6.49}$$

It is interesting to note that the result that a purely toroidal flow is incapable of dynamo action has no analogue in a cylindrical geometry, i.e. when the velocity is confined to cylindrical (rather than spherical) surfaces. The reason is that diffusion in a cylindrical geometry with coordinates (s,φ,z) introduces a coupling between the radial component B_s and the azimuthal component B_φ (see §6.12 below). There is however an analogous result in a Cartesian configuration; this is obtained in the following section.

6.8 The impossibility of dynamo action with plane two-dimensional motion

The analogue of purely toroidal motion in a Cartesian configuration is a motion $\mathbf{u}$ such that $\mathbf{k}\cdot\mathbf{u} = 0$ where $\mathbf{k}$ is a unit vector, say $(0, 0, 1)$. Under this condition the field component $\mathbf{B}\cdot\mathbf{k}$ satisfies

$$\frac{\mathrm{D}}{\mathrm{D}t}(\mathbf{B}\cdot\mathbf{k}) = \eta\,\nabla^2(\mathbf{B}\cdot\mathbf{k})\,, \tag{6.50}$$

and so $\mathbf{B}\cdot\mathbf{k}$ tends everywhere to zero in the absence of external sources.

Suppose then that $\mathbf{B} = \nabla\times(\mathbf{k}\,T(x, y, z)) = -\mathbf{k}\times\nabla T$, where we may suppose that $\int T(x, y, z)\,\mathrm{d}x\,\mathrm{d}y = 0$, the integral being over any plane z =const. Then, since $\mathbf{u}\times\mathbf{B} = -\mathbf{u}\times(\mathbf{k}\times\nabla T) = -(\mathbf{u}\cdot\nabla T)\mathbf{k}$, the induction equation becomes

$$(\mathbf{k}\times\nabla)\frac{\partial T}{\partial t} = -(\mathbf{k}\times\nabla)(\mathbf{u}\cdot\nabla T) + \eta(\mathbf{k}\times\nabla)\nabla^2 T\,, \tag{6.51}$$

which integrates to give

$$\frac{DT}{Dt} = \eta\nabla^2 T + f(z) \tag{6.52}$$

for some function $f(z)$. Assuming that $T = \mathcal{O}(r^{-2})$ at ∞, it follows that $\nabla T \to 0$ everywhere (cf. the argument following (6.44)), and so $\mathbf{B}\to 0$ everywhere.

Note that this result holds whether $\mathbf{B}$ is z-dependent or not, provided solely that $\mathbf{k}\cdot\mathbf{u} = 0$. The result was proved in a weaker form (assuming $\partial\mathbf{B}/\partial z = 0$) by Lortz (1968b). An analogous result was first discussed in the context of two-dimensional turbulence by Zel'dovich (1957).

6.9 Rotor dynamos

We now turn to some examples of kinematically possible motions in a homogeneous conductor which *do* give rise to steady dynamo action. In order to avoid the consequences of the foregoing anti-dynamo theorems, such motions must necessarily be quite complicated, and the associated analysis is correspondingly complex. Nevertheless it was a key challenge of the 1950s to find at least one explicit example of successful dynamo action, if only to be confident that there could be no all-embracing anti-dynamo theorem. The first such example, provided by Herzenberg (1958), was significant in that, for the first time, it provided unequivocal proof that steady motions $\mathbf{u}(\mathbf{x})$ do exist in a sphere of conducting fluid which can maintain a steady magnetic field $\mathbf{B}(\mathbf{x})$ against ohmic decay, and which give a non-zero dipole moment outside the sphere. Herzenberg's model consisted of two spherical rotors imbedded in a conducting sphere of fluid otherwise at rest; within each rotor the angular velocity was constant and the radius of the rotors was small compared with the distance between their centres which in turn was small compared with the radius of the conducting sphere. Herzenberg's analysis was subsequently elucidated by Gibson (1968a,b), Gibson & Roberts (1967) and Roberts (1971), and the following discussion is based largely on these papers.

Let S_α ($\alpha = 1, 2, \ldots, n$) denote the n spheres $|\mathbf{x} - \mathbf{x}_\alpha| = a$, and suppose that for each pair (α, β), $|\mathbf{x}_\alpha - \mathbf{x}_\beta| \gg a$, i.e. the spheres are all far apart relative to their radii. We define a velocity field

$$\mathbf{u}(\mathbf{x}) = \begin{cases} \boldsymbol{\omega}_\alpha \wedge (\mathbf{x} - \mathbf{x}_\alpha) & \text{when } |\mathbf{x} - \mathbf{x}_\alpha| < a \quad (\alpha = 1, 2, \ldots, n)\,, \\ 0 & \text{otherwise}\,, \end{cases} \tag{6.53}$$

where the $\boldsymbol{\omega}_\alpha$ are constants; i.e. the fluid inside each S_α rotates with uniform angular velocity $\boldsymbol{\omega}_\alpha$, and the fluid outside all n spheres is at rest. We enquire under what circumstances such a velocity field in a fluid of infinite extent and uniform conductivity can maintain a steady magnetic field $\mathbf{B}(\mathbf{x})$, at most $\mathcal{O}(r^{-3})$ at infinity. If such a field $\mathbf{B}(\mathbf{x})$ exists, then in the neighbourhood of the sphere S_α it may be decomposed into its poloidal and toroidal parts

$$\mathbf{B}_P^\alpha = \nabla \wedge \nabla \wedge (\mathbf{r}_\alpha P(\mathbf{r}_\alpha))\,, \quad \mathbf{B}_T^\alpha = \nabla \wedge (\mathbf{r}_\alpha T(\mathbf{r}_\alpha))\,, \tag{6.54}$$

where $\mathbf{r}_\alpha = \mathbf{x} - \mathbf{x}_\alpha$. Let $P^s(\mathbf{r}_\alpha)$ and $T^s(\mathbf{r}_\alpha)$ be the average of $P(\mathbf{r}_\alpha)$ and $T(\mathbf{r}_\alpha)$ over the azimuth angle about the rotation vector $\boldsymbol{\omega}_\alpha$, and let $P^a(\mathbf{r}_\alpha)$, $T^a(\mathbf{r}_\alpha)$ be defined by

$$P = P^s + P^a\,, \quad T = T^s + T^a\,, \tag{6.55}$$

(the superfixes s and a indicating symmetry and asymmetry about the axis of rotation). Let the corresponding decomposition of $\mathbf{B}_P^\alpha$ and $\mathbf{B}_T^\alpha$ be

$$\mathbf{B}_P^\alpha = \mathbf{B}_P^{\alpha s} + \mathbf{B}_P^{\alpha a}\,, \quad \mathbf{B}_T^\alpha = \mathbf{B}_T^{\alpha s} + \mathbf{B}_T^{\alpha a}\,. \tag{6.56}$$

We know from §6.4 that the toroidal motion of the sphere S_α has no direct regenerative effect on $\mathbf{B}_P^\alpha$, which is maintained entirely by the inductive effects of the other $n-1$ spheres. Hence $\mathbf{B}_P^\alpha$ can be regarded as an 'applied' field in the neighbourhood of S_α, and the rotation of S_α in the presence of this applied field determines the structure of the toroidal field $\mathbf{B}_T^\alpha$. The asymmetric part of the total field $\mathbf{B}^\alpha$ is merely excluded from the rotating region essentially by the process of §3.11. The symmetric part of $\mathbf{B}_P^\alpha$ on the other hand interacts with the differential rotation (here concentrated on the spherical surface $r = |\mathbf{x} - \mathbf{x}_\alpha| = a$) to provide $\mathbf{B}_T^{\alpha s}$, which, by the arguments of §3.14, is $\mathcal{O}(\mathcal{R}_{m\alpha})|\mathbf{B}_P^{\alpha s}|$, where $\mathcal{R}_{m\alpha} = \omega_\alpha a^2/\eta$. If $\mathcal{R}_{m\alpha} \gg 1$, as we shall suppose, then $\mathbf{B}_\mathbf{T}^{\alpha s}$ is the dominant part of the total field $\mathbf{B}^\alpha$ in a large neighbourhood of S_α.

Now $\mathbf{B}_P^{\alpha s}$ may be expanded about the point $\mathbf{x} = \mathbf{x}_\alpha$ in Taylor series:

$$\begin{aligned} B_{Pi}^{\alpha s}(\mathbf{x}) &= B_{Pi}^{\alpha s}(\mathbf{x}_\alpha) + r_{\alpha j} B_{Pi,j}^{\alpha s}(\mathbf{x}_\alpha) + \mathcal{O}(r_\alpha^2) \\ &= B_{Pi}^{\alpha s}(\mathbf{x}_\alpha) + \tfrac{1}{2} r_{\alpha j}(B_{Pi,j}^{\alpha s}(\mathbf{x}_\alpha) + B_{Pj,i}^{\alpha s}(\mathbf{x}_\alpha)) + \mathcal{O}(r_\alpha^2), \end{aligned} \quad (6.57)$$

since $\nabla \wedge \mathbf{B}_P^\alpha$ is the toroidal current which vanishes at $\mathbf{x}_\alpha$; this may be expressed in the equivalent form

$$\mathbf{B}_P^{\alpha s}(\mathbf{x}) = \nabla\Phi + \mathcal{O}(r_\alpha^2), \quad (6.58)$$

where

$$\Phi = A_1^\alpha r P_1(\cos\theta) + a^{-1} A_2^\alpha r^2 P_2(\cos\theta), \quad (6.59)$$

and (with the convention that repeated Greek suffices are *not* summed)

$$A_1^\alpha = \omega_\alpha^{-1}\, \boldsymbol{\omega}_\alpha \cdot \mathbf{B}_P^{\alpha s}(\mathbf{x}_\alpha), \quad A_2^\alpha = \tfrac{1}{2} a \omega_\alpha^{-2}\, \boldsymbol{\omega}_\alpha \cdot \nabla \mathbf{B}_P^{\alpha s}(\mathbf{x}_\alpha) \cdot \boldsymbol{\omega}_\alpha . \quad (6.60)$$

Hence from the results of §3.14, for $r_\alpha \gg a$,

$$\mathbf{B}_T^{\alpha s}(\mathbf{x}) = -\frac{1}{5} a^3 A_1^\alpha \mathcal{R}_{m\alpha} \frac{(\boldsymbol{\omega}_\alpha \cdot r_\alpha)(\boldsymbol{\omega}_\alpha \wedge \mathbf{r}_\alpha)}{\omega_\alpha^2 r_\alpha^5} + \frac{2}{15} a^2 A_2^\alpha \mathcal{R}_{m\alpha} \frac{\boldsymbol{\omega}_\alpha \wedge \mathbf{r}_\alpha}{\omega_\alpha r_\alpha^3} + \mathcal{O}(r_\alpha^{-4}). \quad (6.61)$$

The field in the neighbourhood of S_β is the sum of fields of the form (6.61) resulting from the presence of all the other spheres S_α ($\alpha \neq \beta$), i.e.

$$\mathbf{B}^\beta(\mathbf{x}) = \sum_{\alpha(\neq\beta)} \mathbf{B}_T^{\alpha s}(\mathbf{x}). \quad (6.62)$$

For self-consistency, we must have, for $\beta = 1, 2, \ldots, n$,

$$A_1^\beta = \omega_\beta^{-1} \boldsymbol{\omega}_\beta \cdot \mathbf{B}_P^{\beta s}(\mathbf{x}_\beta) = \omega_\beta^{-1} \boldsymbol{\omega}_\beta \cdot \sum_{\alpha(\neq\beta)} \mathbf{B}_T^{\alpha s}(\mathbf{x}_\beta), \quad (6.63)$$

and, similarly,

$$A_2^\beta = \tfrac{1}{2} a \omega_\beta^{-2}\, \boldsymbol{\omega}_\beta \cdot \sum_{\alpha(\neq\beta)} \nabla \mathbf{B}_T^{\alpha s}(\mathbf{x}_\beta) \cdot \boldsymbol{\omega}_\beta . \quad (6.64)$$

If the terms $\mathcal{O}(r_\alpha^{-4})$ in (6.61) are neglected, these provide $2n$ linear equations for A_1^α and A_2^α ($\alpha = 1, \ldots, n$), and the determinant of the coefficients must vanish for a non-trivial solution.

6.9.1 The 3-sphere dynamo

The procedure is most simply followed for the case of three spheres (Gibson 1968b) in the configuration of Figure 6.3. Let

$$\mathbf{x}_1 = (d, 0, 0), \quad \mathbf{x}_2 = (0, d, 0), \quad \mathbf{x}_3 = (0, 0, \mathrm{d}), \quad (6.65)$$

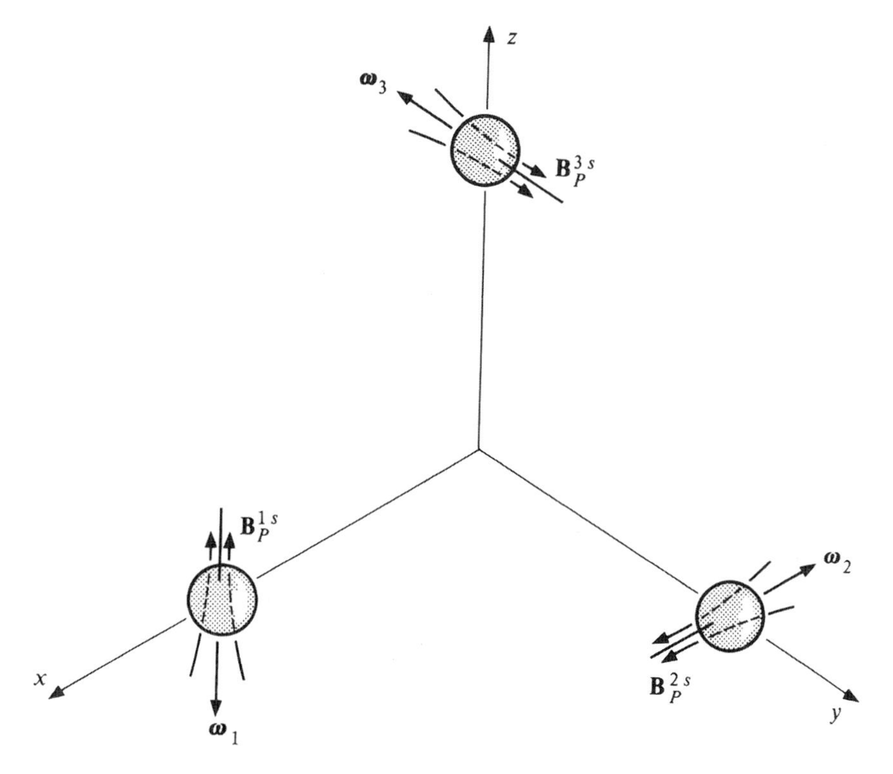

Figure 6.3 The three rotor dynamo of Gibson (1968b) for the particular configuration that is invariant under rotations of $2\pi/3$ and $4\pi/3$ about the direction $(1, 1, 1)$.

and let

$$\omega_1 = (0, 0, -\omega), \quad \omega_2 = (-\omega, 0, 0), \quad \omega_3 = (0, -\omega, 0), \tag{6.66}$$

where $\omega > 0$. The reason for this choice of sign will emerge below. This configuration has a three-fold symmetry, in that it is invariant under rotations of $2\pi/3$ and $4\pi/3$ about the direction $(1, 1, 1)$. Let us therefore look only for magnetic fields $\mathbf{B}(\mathbf{x})$ which exhibit this same degree of symmetry. In particular, in the above notation, we have

$$A_1^1 = A_1^2 = A_1^3 = A_1 \text{ say}, \quad A_2^1 = A_2^2 = A_2^3 = A_2 \text{ say}. \tag{6.67}$$

There are then only two conditions deriving from (6.63) and (6.54) relating A_1 and A_2.

In the neighbourhood of the sphere S_3

$$\mathbf{r}_1 = \mathbf{x}_3 - \mathbf{x}_1 \approx (-d,\ 0,\ d) \quad \text{and} \quad \mathbf{r}_2 = \mathbf{x}_3 - \mathbf{x}_2 \approx (0,\ -d,\ d), \tag{6.68}$$

and so, after some vector manipulation, eqns. (6.63) and (6.64), with $\beta = 3$, become

$$A_1\left(1 + \frac{\mathcal{R}_m}{10}\left(\frac{a}{R}\right)^3\right) = \frac{4}{15\sqrt{2}}\mathcal{R}_m\left(\frac{a}{R}\right)^3 A_2\frac{R}{a}, \tag{6.69}$$

and

$$A_2\left(1 - a^3\mathcal{R}_m/5R^3\right) = 0, \tag{6.70}$$

where $\mathcal{R}_m = \omega a^2/\eta$, and $R = \sqrt{2}d$ is the distance between the sphere centres. Hence (6.70) gives a critical magnetic Reynolds number

$$\mathcal{R}_{mc} = 5(R/a)^3, \tag{6.71}$$

and (6.69) then becomes

$$A_2 = 9A_1 a\big/\left[4\sqrt{2}R\right]. \tag{6.72}$$

The configuration of the $\mathbf{B}_P^{\alpha s}$-lines in the neighbourhood of each sphere are as indicated in Figure 6.3. (It would be difficult to portray the full three-dimensional field pattern.)

Certain points in the above calculation deserve particular comment. First note that since $a \ll R$, A_2 is an order of magnitude smaller than A_1. This means that the field $\mathbf{B}_P^{\alpha s}$ is approximately uniform in the neighbourhood of S_α. It would be quite wrong however to treat it as exactly uniform; this would correspond to putting $A_2 = 0$ in (6.69) and (6.70) with the erroneous conclusion that dynamo action will occur if $\mathcal{R}_m = -10(R/a)^3$ (requiring $\omega < 0$ in (6.61)). The small field gradient in the neighbourhood of each sphere is important because of the phenomenon noted in §3.14 and evident in the expression (6.61) that terms in $\mathbf{B}_T^{\alpha s}$ arising from the gradient of $\mathbf{B}_P^{\alpha s}$ fall off more slowly with distance than do terms arising from the magnitude of $\mathbf{B}_P^{\alpha s}$ itself in the sphere neighbourhood. This effect compensates for the smaller value of the coefficient A_2.

Secondly, note that the directions of $\boldsymbol{\omega}_1$, $\boldsymbol{\omega}_2$ and $\boldsymbol{\omega}_3$ in Figure 6.3 were chosen for maximum simplicity, but the same method would work if $\boldsymbol{\omega}_1$ were taken in any direction and $\boldsymbol{\omega}_2$ and $\boldsymbol{\omega}_3$ were obtained from $\boldsymbol{\omega}_1$ by rotations of $2\pi/3$ and $4\pi/3$ about (1, 1, 1). What matters is that the pseudo-scalars $[\boldsymbol{\omega}_1, \boldsymbol{\omega}_2, \boldsymbol{\omega}_3]$ and $[\mathbf{x}_1, \mathbf{x}_2, \mathbf{x}_3]$ should have opposite signs if the system is to act as a dynamo.

Thirdly, there is no real need for $\boldsymbol{\omega}$ to be uniform throughout each sphere. The results of §3.14 indicate that if $\boldsymbol{\omega} = \omega(r)\,\mathbf{k}$, then, as far as the induced field far from the sphere is concerned, the only quantity that matters is $\overline{\omega}$ where

$$\tfrac{1}{5}a^5\overline{\omega} = \int_0^a r^4\omega(r)\,\mathrm{d}r\,. \tag{6.73}$$

If ω varies with radius within each sphere, but $\overline{\omega}$ is the same for each sphere, then the above results apply with $\mathcal{R}_m = \overline{\omega}a^2/\eta$.

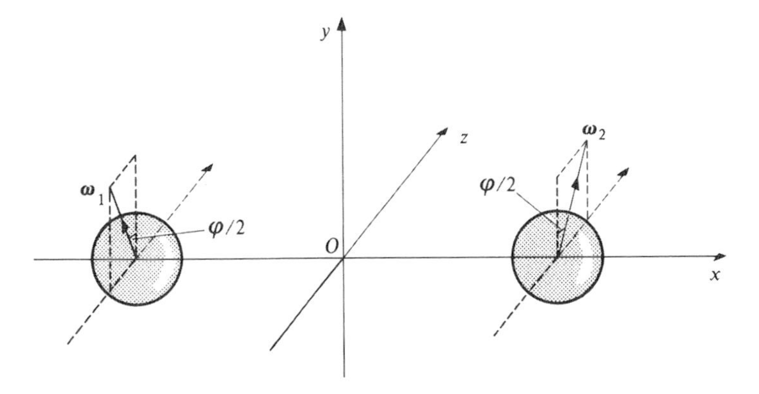

Figure 6.4 Two-sphere dynamo configuration defined by (6.74) and (6.75).

6.9.2 The 2-sphere dynamo

If there are only two spherical rotors, then taking origin at the mid-point of the line joining their centres we may take

$$\mathbf{x}_1 = (-d/2,\ 0, 0), \quad \mathbf{x}_2 = (d/2, 0, 0)\,. \tag{6.74}$$

Suppose (Figure 6.4) that

$$\boldsymbol{\omega}_1 = \omega(0, \cos\varphi/2, -\sin\varphi/2),\ \boldsymbol{\omega}_2 = \omega(0, \cos\varphi/2, \sin\varphi/2)\,. \tag{6.75}$$

The configuration is then invariant under a rotation of π about the axis Oy, and we may therefore look for a magnetic field exhibiting this same two-fold symmetry. Putting $A_1^1 = A_1^2 = A_1$ and $A_2^1 = A_2^2 = A_2$, the conditions (6.63) and (6.64) reduce to[5]

$$A_1 = -(2a^2/15d^2)A_2\mathcal{R}_m \sin\varphi \quad \text{and} \quad A_2 = (a^4/10d^4)A_1\mathcal{R}_m \sin\varphi\cos\varphi\,. \tag{6.76}$$

Hence

$$\left(\tfrac{1}{5}\mathcal{R}_m(a/d)^3 \sin\varphi\right)^2 (-\cos\varphi) = 3\,, \tag{6.77}$$

so that dynamo action is possible if $\pi/2 < \varphi < 3\pi/2$ (excluding $\varphi = \pi$), and then from (6.76)

$$A_2 = \tfrac{1}{2}A_1(a/d)(-3\cos\varphi)^{1/2}\,. \tag{6.78}$$

The dynamo action is most efficient i.e. the resulting value of $\mathcal{R}_m$ is least) when $\tan\varphi = \pm\sqrt{2}$, i.e. $\varphi \approx 54.7^{o}$ or 125.3^{o}; only the latter value is relevant, and for this, $\cos\varphi = -0.577$, and then, from (6.78),

$$A_2 \approx 0.66\,A_1(a/d)\,. \tag{6.79}$$

[5] The antisymmetric possibility $A_1^1 = -A_1^2$ and $A_2^1 = -A_2^2$ leads to the same value of $\mathcal{R}_m$ as given by (6.77), and a field structure which is transformed into its inverse ($\mathbf{B}(\mathbf{x}) \to -\mathbf{B}(\mathbf{x})$) under rotation through π about Oy.

The result (6.77) is a particular case of the result obtained by Herzenberg (1958) who studied the more general configuration for which.

$$\left.\begin{aligned}\boldsymbol{\omega}_1 &= \omega\,(\cos\theta_1,\ \sin\theta_1\cos\varphi/2,\ -\sin\theta_1\sin\varphi/2)\\ \boldsymbol{\omega}_2 &= \omega\,(\cos\theta_2,\ \sin\theta_2\cos\varphi/2,\ +\sin\theta_2\sin\varphi/2)\end{aligned}\right\}. \tag{6.80}$$

Unless $\theta_1 = \pm\theta_2$, this configuration does not exhibit two-fold symmetry about any axis, and there is therefore no *a priori* justification for putting $A_1^1 = A_1^2$ and $A_2^1 = A_2^2$ in general. There are therefore four equations linear in these quantities obtained from (6.63) and (6.64); vanishing of the determinant of the coefficients yields the condition

$$\lambda^2 \equiv \left\{\left[\tfrac{1}{5}\mathcal{R}_m(a/d)^3\sin\theta_1\sin\theta_2\sin\varphi\right]^2(\cos\theta_1\cos\theta_2 - \sin\theta_1\sin\theta_2\cos\varphi) - 3\right\}^2 = 0. \tag{6.81}$$

This reduces to (6.77) when $\theta_1 = \theta_2 = \pi/2$. However Herzenberg observed that terms neglected in the expansion scheme (those denoted $\mathcal{O}(r_\alpha^{-4})$ in (6.61)) could conceivably, if included, give a negative contribution, $-\varepsilon^2$ say, on the right of (6.81). We would then have $\lambda = \pm i\varepsilon$, and the resulting magnetic Reynolds number would be complex indicating that steady dynamo action is not in fact possible for the given velocity field.

Herzenberg resolved this difficulty by taking account of fields reflected from the distant spherical boundary of the conductor. He found that these modified the relationship (6.81) to the form

$$(\lambda - \lambda_1)(\lambda - \lambda_2) = 0\,, \tag{6.82}$$

where λ_1 and λ_2 are small numbers determined by these distant boundary effects. In general $\lambda_1 \neq \lambda_2$, and the possibilities $\lambda = \lambda_1$ and $\lambda = \lambda_2$ give two distinct values of $\mathcal{R}_m$ at which steady dynamo action is possible, with two corresponding (but different) field structures. When the conductor extends to infinity in all directions, the double root of (6.81) for $\mathcal{R}_m$ suggests that, even when the configuration has no two-fold symmetry about any axis, there is nevertheless a degeneracy of the eigenvalue problem, i.e. corresponding to the critical value $\mathcal{R}_m$ there are two linearly independent eigenfunctions $\mathbf{B}_1(\mathbf{x})$ and $\mathbf{B}_2(\mathbf{x})$; when the configuration has the two-fold symmetry of Figure 6.4, it seems altogether plausible that one of these should be symmetric and one anti-symmetric in the sense that $\mathbf{B}_1(\mathbf{x}) \to \mathbf{B}_1(\mathbf{x})$, $\mathbf{B}_2(\mathbf{x}) \to -\mathbf{B}_2(\mathbf{x})$ under a rotation of π about Oy. Gibson (1968a) has in fact shown that the degeneracy implicit in (6.81) is not removed by the retention of the term of order r_α^{-4} in (6.61); it seems likely that the degeneracy persists to all orders, at any rate for the symmetric case (6.75).

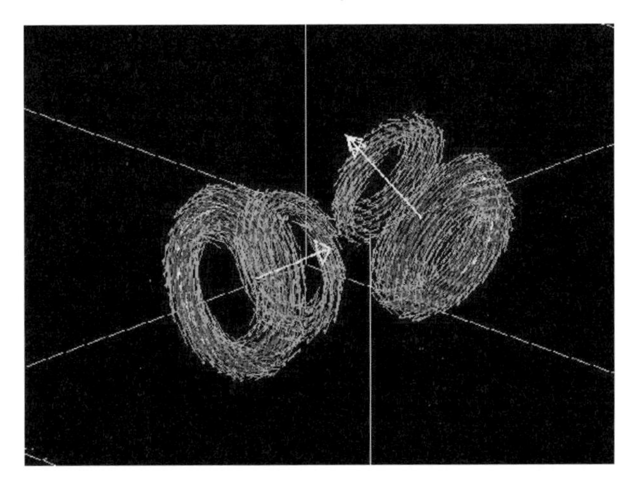

Figure 6.5 Magnetic field structure for the two-rotor Herzenberg dynamo; the directions of the angular velocities of the rotors is indicated by the arrows. [From Brandenburg et al. 1998; Brandenburg & Subramanian 2005.]

6.9.3 Numerical treatment of the Herzenberg configuration

A numerical study of the Herzenberg configuration, but without limitation on the value of the ratio a/d, has been carried out by Brandenburg et al. (1998), who placed the two spheres in a 'periodic box' of scale large compared with the sphere separation, enabling the use of periodic boundary conditions; with the sphere centres at $\mathbf{x}_{1,2} = (\pm d, 0, 0)$, they adopted angular velocities for the two rotors as in (6.75) but with a smoothed rotation profile of the form

$$\omega = \omega_0 \exp\left[-(|\mathbf{x} - \mathbf{x}_{1,2}|/a)^n\right], \tag{6.83}$$

with (arbitrarily) $n = 10$. Figure 6.5, in which the location and orientation of the rotors is indicated by the two arrows, shows the distribution of magnetic field vectors in regions where the field strength exceeds 80% of the maximum, confirming that it is indeed the toroidal field in the neighbourhood of each rotor that dominates. In broad outline, these computations confirmed the general validity of Herzenberg's analysis, each rotor indeed providing the poloidal field on which the other rotor acts to generate its locally toroidal field by the differential rotation mechanism.

An interesting feature of this numerical study, also confirmed by asymptotic analysis, was that oscillatory dynamo modes (i.e. modes with complex growth rate) exist when the angle φ is outside the range for which non-oscillatory modes (with real growth rate) exist. For good measure, the authors also discuss the possible relevance of the Herzenberg model to binary stars considered as two rotors immersed in a stationary conducting medium, as earlier proposed by Dolginov & Urpin (1979).

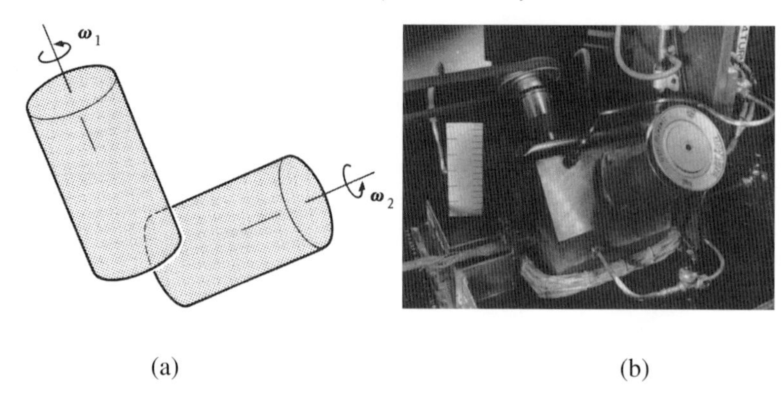

(a) (b)

Figure 6.6 (a) Rotating cylinder configuration of Lowes & Wilkinson (1963). (b) Photograph of the original apparatus, courtesy of F. J. Lowes.

6.9.4 The rotor dynamo of Lowes and Wilkinson

Lowes & Wilkinson (1963) constructed a laboratory dynamo consisting of two solid cylindrical rotors imbedded in a solid block of the same material, electrical contact between the rotors and the block being provided by a thin lubricating film of mercury. Figure 6.6a shows a typical orientation of the cylinders; the photograph in Figure 6.6b, included here for its historical interest, shows the original 1963 apparatus at the University of Newcastle-upon-Tyne. Lowes (2011) has provided an account of the genesis of this experiment. The principle on which the dynamo operates is essentially that of the Herzenberg model: the 'applied' poloidal field of rotor A is the induced toroidal field of rotor B, and reciprocally. The use of cylinders rather than spheres was dictated by experimental expediency; but the interaction between two cylinder ends is in fact stronger than that between two spheres since, roughly speaking, the toroidal field lines generated by the rotation of a cylinder in a nearly uniform axial field all have the same sense in the neighbourhood of the cylinder ends (Herzenberg & Lowes 1957) (contrast the case of the rotating sphere, where the toroidal field changes sign across the equatorial plane). A reasonably small value of the magnetic diffusivity $\eta = (\mu\sigma)^{-1}$ was achieved through the use of ferromagnetic material (large permeability μ) for the rotors and the block.

For the most favourable orientation of the cylinders, dynamo action was found to occur at a critical angular velocity of the rotors of about 400 rpm. corresponding to a critical magnetic Reynolds number $\mathcal{R}_m \approx 200$. As the angular velocities are increased to the critical value, dynamo action manifests itself as a sudden increase of the magnetic field measured outside the block. The currents in the block (and the corresponding field) increase until the retarding torque on the rotors associated with the Lorentz force distribution is just sufficient to prevent the angular velocities from increasing above the critical value. In the steady state (Lowes & Wilkinson

1963), the power supplied to the rotors equals the sum of the rates of ohmic and viscous dissipation in the lubricating films of mercury. In the later improved model of Lowes & Wilkinson (1968), the rate of viscous dissipation was much reduced, and the system exhibited oscillatory behaviour about the possible steady states, and reversals of polarity. This type of behaviour, which has a clear bearing on the question of reversals of the Earth's dipole field (§4.4), involves dynamical effects which will be considered in later chapters – see particularly Chapter 11.

6.10 Dynamo action associated with a pair of ring vortices

A further ingenious example of dynamo action associated with a pair of rotors was analysed by Gailitis (1970). The velocity field, as sketched in Figure 6.7, is axisymmetric, has circular streamlines, and is confined to the interior of two toroidal rings $\mathcal{J}_1$ and $\mathcal{J}_2$. We know from Cowling's theorem that such a velocity field cannot support a steady axisymmetric magnetic field vanishing at infinity; it can however, under certain circumstances, support a non-axisymmetric field proportional to $e^{im\varphi}$, where φ is the azimuth angle about the common axis of the toroids and m is an integer. The figure shows the field lines in the neighbourhood of each torus when $m = 1$ and indicates in a qualitative way how the rotation within each torus can generate a magnetic field which acts as the inducing field for the other torus.

The analysis of Gailitis (with slight changes of notation) proceeds as follows. Let (s, φ, z) be the usual cylindrical polar coordinates, and let (χ, φ, σ) be displaced polar coordinates (Roberts 1971) defined by

$$s = c - \sigma \cos\chi, \quad z = \sigma \sin\chi . \tag{6.84}$$

Let $\mathcal{T}_1$ be the torus $\sigma = a$ where $a \ll c$, and let $\mathcal{T}_2$ be the torus obtained by reflecting $\mathcal{T}_1$ in the plane $z = \frac{1}{2}d$ where $a \ll d$ also. Terms of order a/c and a/d are neglected throughout. Let the velocity field be

$$\mathbf{u} = \mathbf{u}_1(\mathbf{x}) + \mathbf{u}_2(\mathbf{x}) , \tag{6.85}$$

where

$$\mathbf{u}_1(\mathbf{x}) = \begin{cases} v_1(\sigma)\, \mathbf{e}_\chi & \text{inside } \mathcal{T}_1 , \\ 0 & \text{outside } \mathcal{T}_1 , \end{cases} \tag{6.86}$$

and $\mathbf{u}_2(\mathbf{x})$ is similar defined relative to $\mathcal{T}_2$. The total velocity field is zero except in the two toroids. Note that the assumption $a \ll c$ allows us to neglect the small variation of $v_1(\sigma)$ with χ that would otherwise arise from the incompressibility condition $\nabla \cdot \mathbf{u}_1 = 0$.

The steady induction equation is

$$\eta\nabla^2\mathbf{B} + \nabla \wedge (\mathbf{u} \wedge \mathbf{B}) = 0 , \tag{6.87a}$$

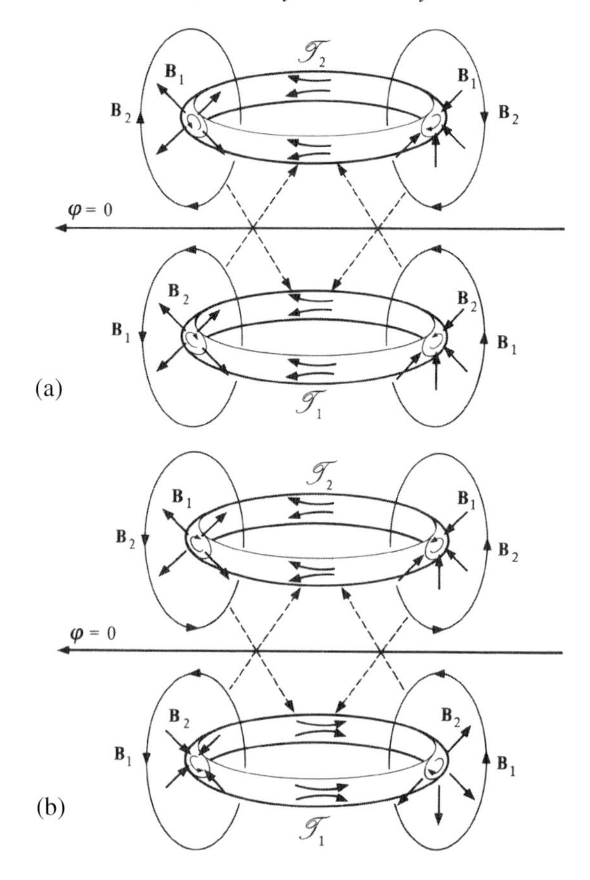

Figure 6.7 Field maintenance by a pair of ring vortices (after Gailitis 1970): (a) dipole configuration; (b) quadrupole configuration. In each case, rotation within each torus induces a field in the neighbourhood of the other torus which has the radial divergence indicated.

and this is formally satisfied by $\mathbf{B} = \mathbf{B}_1 + \mathbf{B}_2$, where

$$\eta\nabla^2\mathbf{B}_1 + \nabla \wedge (\mathbf{u}_1 \wedge \mathbf{B}_1) = -\nabla \wedge (\mathbf{u}_1 \wedge \mathbf{B}_2), \tag{6.87b}$$

$$\eta\nabla^2\mathbf{B}_2 + \nabla \wedge (\mathbf{u}_2 \wedge \mathbf{B}_2) = -\nabla \wedge (\mathbf{u}_2 \wedge \mathbf{B}_1). \tag{6.87c}$$

We regard $\mathbf{B}_1$ as the field *induced* by the motion $\mathbf{u}_1$ *and* as the *inducing field* for the motion $\mathbf{u}_2$; similarly for $\mathbf{B}_2$.

The essential idea behind the Gailitis analysis is very similar to that applying in the Herzenberg model. Let $\mathbf{B}_2^s$ be the symmetric ingredient of $\mathbf{B}_2$ obtained by averaging its (σ, χ, φ) components over the angle χ, and let

$$\mathbf{B}_2^a = \mathbf{B}_2 - \mathbf{B}_2^s \tag{6.88}$$

be the asymmetric ingredient: the components of $\mathbf{B}_2^a$ average to zero over χ. The corresponding solution of (6.87b) may be denoted by $\mathbf{B}_1^a + \mathbf{B}_1^s$. We know from the analysis of §3.13 that the effect of the closed streamline motion $\mathbf{u}_1$ is simply to expel the asymmetric field from the rotating region when the magnetic Reynolds number $\mathcal{R}_m$, based on a and an appropriate average of $v(\sigma)$ (see (6.93) below), is large, i.e.

$$\mathbf{B}_1^a + \mathbf{B}_2^a \approx 0 \quad \text{in } \mathcal{T}_1 \,. \tag{6.89}$$

On the other hand, by the differential rotation mechanism of §3.10, the motion $\mathbf{u}_1$ generates from $\mathbf{B}_2^s$ a field $\mathbf{B}_1^s$ satisfying

$$\left|\mathbf{B}_1^s\right| = \mathcal{O}(\mathcal{R}_m)\left|\mathbf{B}_2^s\right| . \tag{6.90}$$

We anticipate that $\mathcal{R}_m \gg 1$, and (6.90) then provides the dominant contribution to $\mathbf{B}_1$, so that in calculating $\mathbf{B}_1$ from (6.87b) we may regard $\mathbf{B}_2$ as symmetric with respect to χ in the neighbourhood of $\mathcal{T}_1$.

Suppose then that in $\mathcal{T}_1$

$$B_{2\varphi} = B_0 e^{im\varphi} . \tag{6.91}$$

The condition $\nabla \cdot \mathbf{B}_2 = 0$ then implies that in $\mathcal{T}_1$

$$B_{2\sigma} = -(i\sigma m/2c) B_0 e^{im\varphi} , \tag{6.92}$$

and so, from (6.86),

$$\mathbf{u}_1 \wedge \mathbf{B}_2 = B_0 v_1 e^{im\varphi} (\mathbf{e}_\sigma + (im\sigma/2c)\,\mathbf{e}_\varphi) \quad \text{in } \mathcal{T}_1 , \tag{6.93}$$

and so

$$\mathbf{g} \equiv \nabla \wedge (\mathbf{u}_1 \wedge \mathbf{B}_2) = -\frac{imB_0 e^{im\varphi}}{2c} \sigma^2 \frac{d}{d\sigma}\left(\frac{v}{\sigma}\right) \mathbf{e}_\chi \quad \text{in } \mathcal{T}_1 . \tag{6.94}$$

As expected, the motion $\mathbf{u}_1$ generates a field in the χ-direction as a result of differential rotation within $\mathcal{T}_1$. It follows that $\mathbf{u}_1 \wedge \mathbf{B}_1 = 0$, and so the solution of (6.87b) (Poisson's equation) is

$$\mathbf{B}_1(\mathbf{x}') = \frac{1}{4\pi\eta} \int_{\mathcal{T}_1} \frac{\mathbf{g}(\mathbf{x})}{|\mathbf{x} - \mathbf{x}'|}\, dV . \tag{6.95}$$

We now wish to evaluate the φ-component of this integral on the curved axis of $\mathcal{T}_2$ to see whether $\mathbf{B}_1$ can act as the inducing field for $\mathcal{T}_2$. Let $\mathbf{x}_0$ be a point on the axis of $\mathcal{T}_1$ with cylindrical polar coordinates $(0, c, \varphi)$, $\mathbf{x}_0'$ a point on the axis of $\mathcal{T}_2$ with coordinates (d, c, φ'), and let $\psi = \varphi' - \varphi$. Then

$$\left|\mathbf{x}_0 - \mathbf{x}_0'\right| = [d^2 + 2c^2(1 - \cos\psi)]^{1/2} , \tag{6.96}$$

and, for $\mathbf{x} \in \mathcal{T}_1$,

$$\left|\mathbf{x} - \mathbf{x}_0'\right|^{-1} = \left|\mathbf{x}_0 - \mathbf{x}_0'\right|^{-1} \left(1 - \frac{\sigma\, \mathbf{e}_\sigma \cdot (\mathbf{x}_0 - \mathbf{x}_0')}{|\mathbf{x}_0 - \mathbf{x}_0'|^3} + \mathcal{O}(\sigma^3)\right). \tag{6.97}$$

Moreover

$$\mathbf{e}_{\varphi'} \cdot \mathbf{e}_\chi = -\sin\chi \sin\psi\,. \tag{6.98}$$

When (6.97) and (6.98) are substituted in the equation

$$\mathbf{B}_1(\mathbf{x}_0') \cdot \mathbf{e}_{\varphi'} = \frac{1}{4\pi\eta} \int_{\mathcal{T}_1} \frac{\mathbf{g}(\mathbf{x}) \cdot \mathbf{e}_{\varphi'}}{|\mathbf{x} - \mathbf{x}_0'|} c\sigma \,\mathrm{d}\sigma \,\mathrm{d}\chi \,\mathrm{d}\varphi\,, \tag{6.99}$$

only a term proportional to $\sin^2\chi$ gives a non-zero contribution when integrated over χ. Using (6.94) the result simplifies to

$$\mathbf{B}_1(\mathbf{x}_0') \cdot \mathbf{e}_{\varphi'} = B_0' \mathrm{e}^{im\varphi'}\,, \tag{6.100}$$

where

$$B_0' = B_0 \left(\frac{a}{c}\right)^2 \frac{V_1 a}{\eta} F_m\left(\frac{d}{c}\right), \quad V_1 a^3 = \int_0^a \sigma^2 v_1(\sigma)\,\mathrm{d}\sigma\,, \tag{6.101}$$

and

$$F_m(q) = \tfrac{1}{2} mq \int_0^{2\pi} \frac{\sin\psi \sin m\psi \,\mathrm{d}\psi}{(q^2 + 2 - 2\cos\psi)^{3/2}} \quad (> 0 \text{ for } q > 0)\,. \tag{6.102}$$

Similarly, by analysing the inductive effect of the motion in $\mathcal{T}_2$, we obtain

$$B_0 = -B_0' \left(\frac{a}{c}\right)^2 \frac{V_2 a}{\eta} F_m\left(\frac{d}{c}\right), \tag{6.103}$$

where V_2 is defined like V_1 but in relation to the motion in $\mathcal{T}_2$. The results (6.101) and (6.103) are compatible only if V_1 and V_2 have opposite signs, so that the net circulations (weighted according to (6.101b)) must be opposite in the two toroids. Defining

$$\mathcal{R}_\mathrm{m} = +(-V_1 V_2)^{1/2} a/\eta\,, \tag{6.104}$$

we then have the condition for steady dynamo action in the form

$$\mathcal{R}_\mathrm{m} = \left(\frac{c}{a}\right)^2 T_m\left(\frac{d}{c}\right), \tag{6.105}$$

where $T_m = F_m^{-1}$. When $m = 1$ the possibility $V_1 = -V_2 < 0$ corresponds to the field configuration of Figure 6.7a for which the field has a steady dipole moment perpendicular to Oz and in the plane $\varphi = 0$, and the possibility $V_1 = -V_2 > 0$ corresponds to the field configuration of Figure 6.7b for which the dipole moment is evidently zero and the far field is that of a quadrupole.

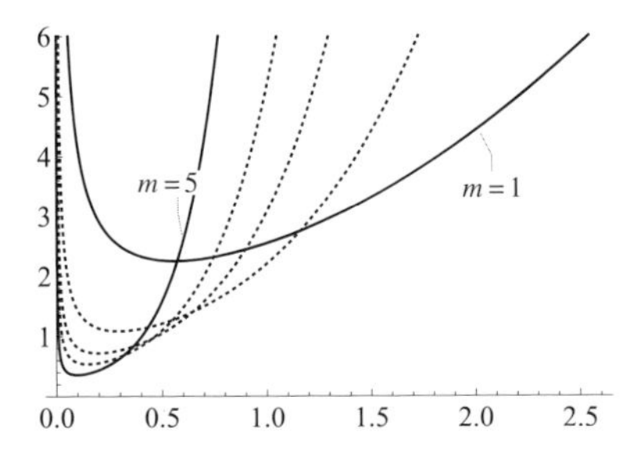

Figure 6.8 The functions $T_m(q) = (F_m(q))^{-1}$ for $m = 1, 2, \ldots, 5$. [Adapted from Gailitis 1970.]

The functions $T_m(q)$ ($m = 1, 2, \ldots, 5$) are shown in Figure 6.8. It is evident that, for $d/c > 1.16$, $T_m(z_\mathbf{o}/c)$ is least (corresponding to the most easily excited magnetic mode) for $m = 1$. As d/c is decreased, the value of m corresponding to the most easily excited mode increases; this is physically plausible in that as the rings approach each other, each becomes more sensitive to the detailed field structure within the other. The analysis of course breaks down if d is decreased to values of order a.

The above analysis is only approximate in that terms of order a/c and a/d are neglected throughout. Strictly a formal perturbation procedure in terms of these small parameters is required, and a rigorous proof of dynamo action would require strict upper bounds to be put on the neglected terms of the Gailitis analysis. There seems little doubt however that such a procedure (paralleling the procedure followed by Herzenberg 1958) would confirm the validity of the 'zero-order' analysis presented above. It might be thought that the Gailitis dynamo is nearer to physical reality than the Herzenberg dynamo, in that vortex rings are a well-known dynamically realisable phenomenon in nearly inviscid fluids, whereas spherical rotors are not. However it must be recognised that the velocity field $\mathbf{u}_1(\mathbf{x})$ given by (6.86) is unlike that of a real ring vortex in that the net flux of vorticity around $\mathcal{T}_1$ (including a possible surface contribution) is zero; if it were non-zero (as in a real vortex ring) then the vortex would necessarily be accompanied by an irrotational flow outside $\mathcal{T}_1$. Two real vortices oriented as in Figure 6.7 would as a result of this irrotational flow either separate and contract (case (a)) or approach each other and expand (case (b)). There can be no question of maintenance of a *steady* magnetic field by an unsteady motion of either kind.

6.11 Dynamo action with purely meridional circulation

Motivated by the above theory of Gailitis (1970), more general dynamo action associated with purely meridional circulation has been investigated by Moss (2006), who considered flows contained in a spherical annulus $r_0 < r < 1$, with Stokes

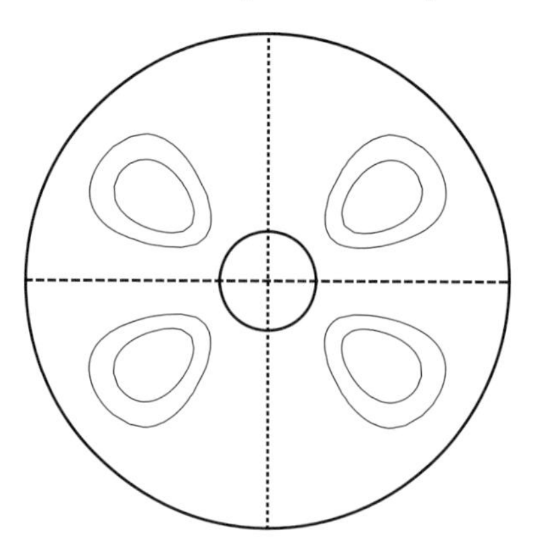

Figure 6.9 Streamlines ψ = const. in a meridian plane for the flow (6.106) with $m = 6, n = 3$; the flow is confined to the thick spherical shell $0.2 < r < 1$; the axis of symmetry is vertical, and the flow is radially outwards or inwards near this axis accordingly as $A >$ or < 0. An equatorial dipole field is preferentially excited in the former case, and an equatorial quadrupole in the latter. [Adapted from Moss 2006.]

streamfunction $\psi(r, \theta)$ of the form

$$\psi(r, \theta) = A(r - r_0)^2(r - 1)^2 \sin^m\theta \cos^n\theta, \tag{6.106}$$

with corresponding velocity components

$$u_r = \frac{1}{r^2 \sin\theta} \frac{\partial\psi}{\partial\theta}, \quad u_\theta = -\frac{1}{r \sin\theta} \frac{\partial\psi}{\partial r}. \tag{6.107}$$

Here m and n are positive integers, and A is a constant proportional to the magnetic Reynolds number. The flow is symmetric or antisymmetric about the equatorial plane according as n is odd or even; and with n odd, the flow is radially outwards or inwards near the axis of symmetry according as $A >$ or < 0.

The streamlines of one such flow (for $m = 6$, $n = 3$) are shown in Figure 6.9 (cf. Figure 6.7); here $r_0 = 0.2$, the axis of symmetry is dotted, and the equatorial plane is dashed. The boundary conditions are that the field $\mathbf{B}(\mathbf{x}, t)$ should match smoothly to a potential field (satisfying $\nabla \wedge \mathbf{B} = 0$) in the external region $r > 1$, and that, with the usual poloidal/toroidal decomposition, $\mathbf{B}_T = 0$ and $\mathbf{n} \cdot \mathbf{B}_P = 0$ at the inner boundary $r = r_0$. The induction equation was discretised in r and θ, and integrated forward in time starting from a seed field, until the fastest growing mode emerged, growing exponentially without change of structure. The results were in qualitative agreement with those of Gailitis (1970), and cover a much wider family of flows. Both equatorial dipole and quadrupole modes emerged from the computations, depending simply on the sign of A, again in conformity with the results of

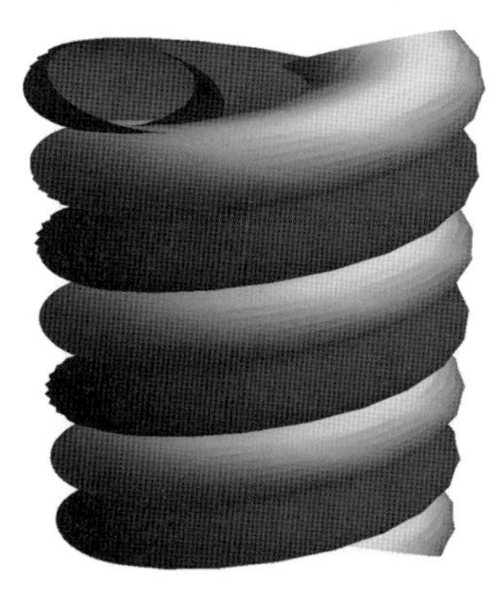

Figure 6.10 Schematic helical structure of magnetic flux tubes in the Ponomarenko dynamo; the field direction alternates between up and down from one tube to the next and is sinusoidal in both φ and z directions.

Gailitis. As a bonus, Moss found new oscillatory dynamo modes when n is even, the flow being then antisymmetric about the equatorial plane.

6.12 The Ponomarenko dynamo

Let us now consider a steady velocity field in a fluid of infinite extent, of the form

$$\mathbf{u} = (0, s\varpi(s), w(s)) \tag{6.108}$$

in cylindrical polar coordinates (s, φ, z). The streamlines of such a flow are helices on cylinders $s = \text{const.}$, with pitch $2\pi w(s)/\varpi(s)$. This type of helical flow was first considered by Lortz (1968a) and by Ponomarenko (1973); any dynamo driven by such a flow is now generally known as a Ponomarenko dynamo (see for example Gilbert 1988, 2003).

Consider first the situation in which $\varpi(s)$ and $w(s)$ are smooth functions of the radial coordinate s. The differential rotation and axial shear of the flow have a tendency to draw out any magnetic field into a similar helical pattern; it is therefore natural to consider magnetic structures of the form

$$\mathbf{b}(s, \varphi, z, t) = [b_s(s, t)\,\mathbf{e}_s + b_\varphi(s, t)\,\mathbf{e}_\varphi + b_z(s, t)\,\mathbf{e}_z]\,\mathrm{e}^{\mathrm{i}(kz+m\varphi)} \tag{6.109}$$

(real part understood), where k is a positive constant and m a non-zero integer. A pictorial representation of such a structure (with $m = 1$) is shown in Figure 6.10.

The induction equation may be written in the expanded form

$$\frac{\partial \mathbf{b}}{\partial t} + \mathbf{u} \cdot \nabla \mathbf{b} = \mathbf{b} \cdot \nabla \mathbf{u} + \eta \nabla^2 \mathbf{b}\,. \tag{6.110}$$

In substituting (6.109), we must be careful to take account of the variation of the unit vectors $\mathbf{e}_s$ and $\mathbf{e}_\varphi$ with φ; for example

$$\frac{\partial}{\partial\varphi}(b_s\,\mathbf{e}_s) = \mathrm{i}\,m\,b_s\,\mathbf{e}_s + b_s\,\mathbf{e}_\varphi\,, \qquad \frac{\partial}{\partial\varphi}(b_\varphi\,\mathbf{e}_\varphi) = \mathrm{i}\,m\,b_\varphi\,\mathbf{e}_\varphi - b_\varphi\,\mathbf{e}_s\,. \tag{6.111}$$

Using such results repeatedly, we obtain from (6.110) the component equations

$$\frac{\partial b_s}{\partial t} + \mathrm{i}\,[k\,w(s)+m\,\varpi(s)]b_s = \eta\left(\nabla^2 b_s - \frac{1}{s^2}b_s - \frac{2\mathrm{i}mb_\varphi}{s^2}\right), \tag{6.112}$$

$$\frac{\partial b_\varphi}{\partial t} + \mathrm{i}\,[k\,w(s)+m\,\varpi(s)]b_\varphi = \eta\left(\nabla^2 b_\varphi - \frac{1}{s^2}b_\varphi + \frac{2\mathrm{i}mb_s}{s^2}\right) + s\varpi'(s)b_s\,, \tag{6.113}$$

$$\frac{\partial b_z}{\partial t} + \mathrm{i}\,[k\,w(s)+m\,\varpi(s)]b_z = \eta\nabla^2 b_z + w'(s)b_s\,, \tag{6.114}$$

where here

$$\nabla^2 = \frac{\partial^2}{\partial s^2} + \frac{1}{s}\frac{\partial}{\partial s} - \frac{m^2}{s^2} - k^2\,. \tag{6.115}$$

Note the diffusive interaction between the components b_s and b_φ represented by the terms $-2\mathrm{i}\,\eta\,mb_\varphi/s^2$ and $+2\mathrm{i}\,\eta\,mb_s/s^2$; note also the effects of differential rotation and axial shear acting on the radial field, represented by the terms $s\varpi'(s)\,b_s$ and $w'(s)\,b_s$ respectively.

The condition $\nabla\cdot\mathbf{b} = 0$ (which is satisfied for all $t > 0$ if satisfied at $t = 0$) gives

$$\mathrm{i}kb_z = -\frac{1}{s}\frac{\partial}{\partial s}(sb_s) - \frac{\mathrm{i}m}{s}b_\varphi\,, \tag{6.116}$$

an equation that can be used to determine b_z from b_s and b_φ; this means that we can focus on just equations (6.112) and (6.113).

The case of concentrated shear

Ponomarenko (1973) analysed the particular case for which

$$\left.\begin{aligned} \varpi(s) &= \Omega\,, \quad w(s) = U, \quad \text{for } s < a \\ \varpi(s) &= 0, \quad\; w(s) = 0, \quad \text{for } s > a \end{aligned}\right\}, \tag{6.117}$$

where Ω and U are constants; this is a rigid-body motion for $s < a$ in an otherwise stationary conductor of uniform diffusivity η. We may use a as the unit of length, so that in effect $a = 1$. Appropriate dimensionless numbers are then the 'swirl' parameter (related to helicity density) $h = U/\Omega$ and the swirl magnetic Reynolds number $\mathcal{R}_\mathrm{s} = \Omega a^2/\eta$ (with $a = 1$). The differential rotation and axial shear are concentrated in a helical vortex sheet on $s = 1$, and if dynamo action occurs, it is attributable to processes in the neighbourhood of this sheet. Although somewhat artificial (as in the previous Herzenberg and Gailitis models), this special choice permits analytical progress, and is illuminating as a prototype of helical dynamo action. It therefore merits detailed description; here we follow Gilbert (2003).

The magnetic field is continuous across $s = 1$, but the field gradients are discontinuous. Integration through the sheet provides the jump conditions

$$[\![\partial b_\varphi/\partial s]\!]_-^+ = \Omega b_s \quad \text{and} \quad [\![\partial b_z/\partial s]\!]_-^+ = U b_s \quad \text{at } s = 1. \tag{6.118}$$

From (6.116), we also have

$$[\![\partial b_s/\partial s]\!]_-^+ = 0. \tag{6.119}$$

The situation is simplified if we assume that

$$kU + m\Omega = 0, \tag{6.120}$$

i.e. that the wave-crests of the field $\mathbf{b}$ are perfectly aligned with the streamlines of the flow (equivalently, $(\mathbf{u}\cdot\nabla)\mathbf{b} = 0$). We look for normal modes for which $\mathbf{b}(s,t) \sim \hat{\mathbf{b}}(s)\mathrm{e}^{\lambda t}$. Then, both inside and outside the cylinder, (6.112) and (6.113) may be rearranged to give

$$\left.\begin{aligned} p^2\hat{b}_s &= \left(\Delta_m - s^{-2}\right)\hat{b}_s - 2\mathrm{i}ms^{-2}\hat{b}_\varphi \\ p^2\hat{b}_\varphi &= \left(\Delta_m - s^{-2}\right)\hat{b}_\varphi + 2\mathrm{i}ms^{-2}\hat{b}_s \end{aligned}\right\}, \tag{6.121}$$

where $\Delta_m = \partial^2/\partial s^2 + s^{-1}\partial/\partial s - m^2 s^{-2}$, $p^2 = k^2 + \lambda/\eta$, and we require that $\mathrm{Re}\, p \geq 0$. Defining $b_\pm = \hat{b}_s \pm \hat{b}_\varphi$, these transform to the beautifully compact form

$$\Delta_{m\pm 1} b_\pm = p^2 b_\pm, \tag{6.122}$$

and the relevant jump conditions across $s = 1$ transform to

$$[\![b_\pm]\!]_-^+ = 0, \quad [\![\partial b_\pm/\partial s]\!]_-^+ = \pm\mathrm{i}\Omega a\,[b_+ + b_-]\,. \tag{6.123}$$

The solution of (6.122) satisfying the first of (6.123), and the conditions that $b_\pm$ be both finite at $s = 0$ and asymptotically zero as $s \to \infty$ is

$$b_\pm = A_\pm \frac{I_{m\pm 1}(sp)}{I_{m\pm 1}(p)} \quad (s < 1), \quad \text{and } b_\pm = A_\pm \frac{K_{m\pm 1}(sp)}{K_{m\pm 1}(p)} \quad (s > 1)\,, \tag{6.124}$$

where $I_m(z)$ and $K_m(z)$ are modified Bessel functions in standard notation. Application of the second of (6.123) now gives two linear equations for the coefficients $A_\pm$. The determinant compatibility condition reduces to

$$2\eta = \mathrm{i}\Omega\left[I_{m-1}(p)K_{m-1}(p) - I_{m+1}(p)K_{m+1}(p)\right]. \tag{6.125}$$

In principle, this equation determines $p = p_r + \mathrm{i}\,p_i$ in terms of $\mathcal{R}_\mathrm{s} = \Omega/\eta$ and m, and hence $\lambda/\eta = p^2 - k^2 = p^2 - m^2/h^2$ in terms of $\mathcal{R}_\mathrm{s}$, m and h.

In general, $\lambda = \lambda_r + \mathrm{i}\,\lambda_i$, and dynamo action occurs if $\lambda_r > 0$. The threshold for dynamo action is therefore determined by the condition $\lambda_r = 0$, or equivalently

$$p_r^2 - p_i^2 = m^2/h^2. \tag{6.126}$$

Thus, for example, when $m = 1$ and $\mathcal{R}_s = 17.0$, computation gives $p = 1.511 + 1.268\,\mathrm{i}$, giving $h = 1.22$ and $\lambda_i/\eta = 3.8319$; this last determines the speed of propagation of the 'dynamo wave' that can propagate in this critical condition. We could of course use a magnetic Reynolds based on the maximum speed $\mathcal{R}_m = \mathcal{R}_s(1 + h^2)^{1/2}$; then when $\mathcal{R}_s = 17.0$ and $h = 1.3$, $\mathcal{R}_m = 26.8$; this is in fact close to the minimum value of $\mathcal{R}_m$ for dynamo action under the condition (6.120).

If k/m is unconstrained however, a lower critical magnetic Reynolds number is found (Gailitis & Freiberg 1976). Figure 6.11a shows the computed neutral curve $\lambda_r = 0$ (hatched) in the plane of the variables $\mathcal{R}_m$ and $-k$ (for $m = 1$ and $h = U/\Omega = 1.3$). Dynamo action occurs in the regime to the right of the hatched curve. The vertical tangent is at $\mathcal{R}_m = \mathcal{R}_m^* = 17.7$, indicating the minimum critical Reynolds number; the point of tangency is at $-k \approx 0.4$. The dashed curves are contours $\lambda_i =$ const. For all unstable modes, $\lambda_i \neq 0$, and they are therefore oscillatory in character: these modes propagate as helical waves of exponentially growing amplitude.

Figure 6.11b shows a modification of the Ponomarenko dynamo, in which a return flow is envisaged in an outer annulus, the total axial flux through inner cylinder and outer annulus being zero (here $h = 1$). There is now the possibility that $p_r = p_i = 0$ but with $k = k_r + \mathrm{i}\,k_i$ now complex. The envelope of the family of curves $p_r = 0$, $k_i =$ const. is shown hatched. The vertical tangent is at $\mathcal{R}_m = \mathcal{R}_m^{**} = 15.74$, and this is the minimum magnetic Reynolds number for 'absolute' instability, i.e. for modes of instability that do not propagate, but rather remain within a cylinder of finite length as they amplify, as relevant for the experimental realisation of this type of dynamo (Gailitis 1990).

Interest also attaches to the behaviour when $\mathcal{R}_s \to \infty$. Anticipating that $|p|$ also becomes large in this limit, we may use the asymptotic formula (Abramowitz & Stegun 1970)[6]

$$I_m(z)K_m(z) \sim \frac{1}{2z}\left[1 - \frac{1}{2}\frac{4m^2-1}{(2z)^2} + \frac{3}{8}\frac{(4m^2-1)(4m^2-9)}{(2z)^4} - \cdots\right] \quad (|\arg z| < \tfrac{1}{2}\pi), \tag{6.127}$$

to evaluate both terms on the right of (6.125). With some simplification, this yields $p \sim (\mathrm{i}m\Omega/2\eta)^{1/3}$, and hence at leading order, provided $|m| = \mathcal{O}(1)$,

$$\lambda = \lambda_r + \mathrm{i}\lambda_i = \eta(p^2 - k^2) \sim 2^{-5/3}\left(\eta\, m^2\, \Omega^2\right)^{1/3}(1 + \sqrt{3}\,\mathrm{i}). \tag{6.128}$$

This is therefore a 'slow' dynamo, in the sense that the growth rate $\lambda_r \to 0$ as $\eta \to 0$. However, as recognised by Gilbert (1988), λ_r as given by (6.128) increases with

[6] Formula 9.7.5.

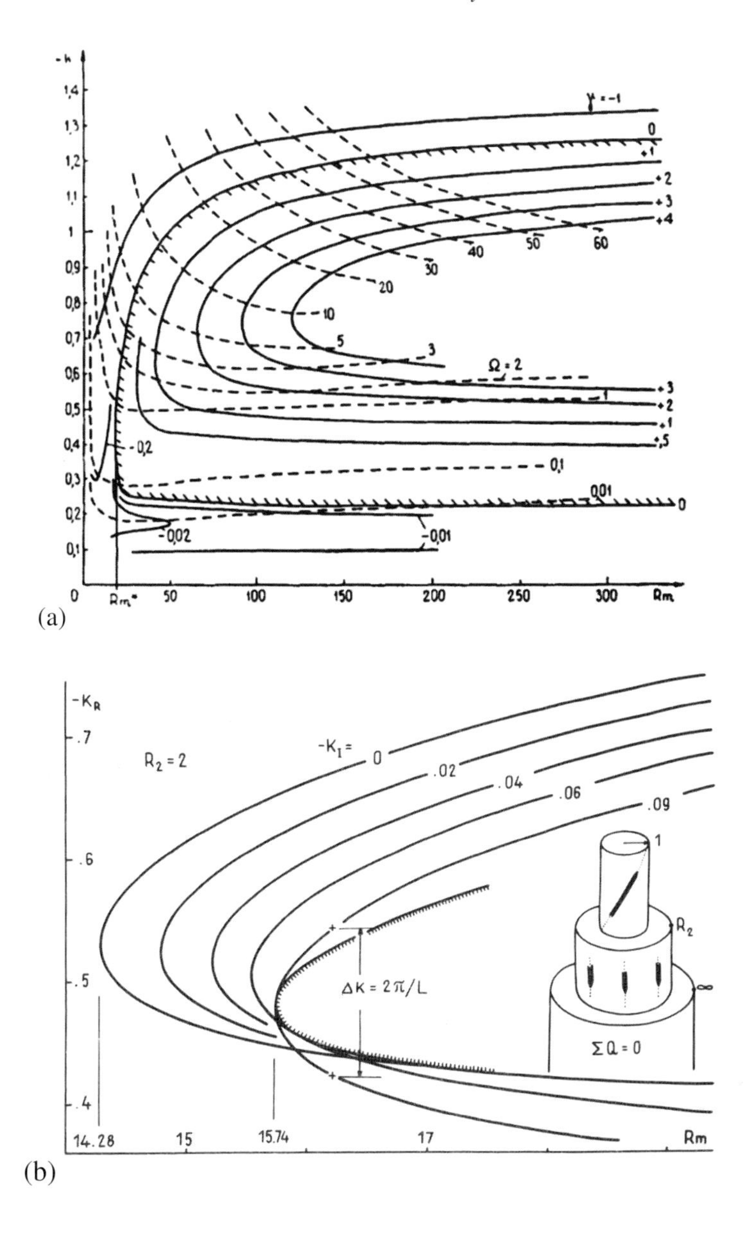

Figure 6.11 (a) Family of curves λ_r = const. (solid) and λ_i = const. (dashed) in $\{-k, \mathcal{R}_{\mathrm{m}}\}$-plane; $m = 1, h = 1.3$. The hatched curve is $\lambda_r = 0$, and oscillatory dynamo instability ($\lambda_i \neq 0$) occurs to the right of this curve; the vertical tangent is at the minimum critical $\mathcal{R}_{\mathrm{m}} = \mathcal{R}_{\mathrm{m}}^{*} = 17.7$, and the point of tangency is at $-k \approx 0.4$. (b) Family of curves $p_r = 0, k_i$ =const. in the $\mathcal{R}_{\mathrm{m}}, k_r$-plane, for a 'return-flow' helical dynamo as shown (with $h = 1$). The vertical tangent to the leftmost curve $k_i = 0$ is now at $\mathcal{R}_{\mathrm{m}}^{*} = 14.28$, and oscillatory dynamo instability occurs to the right of this curve; the hatched curve is the envelope of the family k_i =const., and 'absolute' dynamo instability ($\lambda_r > 0, \lambda_i = 0, k_r \neq 0, k_i \neq 0$) occurs to the right of *this* curve, where $\mathcal{R}_{\mathrm{m}} > \mathcal{R}_{\mathrm{m}}^{**} = 15.74$. [From Gailitis & Freiberg 1976; Gailitis 1990.]

$|m|$, i.e. with decreasing θ-scale of the **b**-field, and closer investigation is needed if $|m|$ is allowed to increase in tandem with $\mathcal{R}_s$. We shall return to this 'fast-dynamo' subtlety in §10.5.

At the next order of approximation for small η, (6.128) gives

$$\lambda \sim 2^{-5/3}\left(\eta m^2 \Omega^2\right)^{1/3}(1+\sqrt{3}\,\mathrm{i}) - \eta\left(m^2/h^2 + 8m^2 - 2\right), \tag{6.129}$$

so the critical condition $\lambda_r = 0$ requires that $h \sim \eta^{1/3}$; in terms of $\mathcal{R}_s$, the exact asymptotic relation for criticality (always supposing $|m| = \mathcal{O}(1)$) is

$$h \sim 2^{5/6}\left(m^2/\mathcal{R}_s\right)^{1/3}. \tag{6.130}$$

In this limit, the pitch $2\pi h$ of the helical streamlines is small: they are tilted only slightly from planes z = const., much as illustrated in Figure 6.10.

6.13 The Riga dynamo experiment

The Riga dynamo experiment (Gailitis et al. 2000, 2001) was developed over a number of years on the basis of the above theoretical considerations, and provided the first demonstration of dynamo action in a conducting fluid – here liquid sodium at a temperature of order 200°C, with $\eta \approx 0.1\mathrm{m}^2/\mathrm{s}$. The apparatus is shown in Figure 6.12. The propellor could be driven at speed Ω_P up to about 2100 rpm, generating a downward swirling flow rate[7] of order 15m/s. The flow returned up through a surrounding cylindrical annulus (denoted '4') in the figure), and an outer annulus (5) contained liquid sodium at rest. A magnetic Reynolds number $\mathcal{R}_m \sim 30$ was achieved, somewhat above the critical for absolute dynamo instability. Figure 6.13 shows the nature of the resulting dynamo action; the inset shows the initial instability; the field amplitude grows until the back-reaction of the Lorentz force establishes a saturated state when the radial field component is of order 1mT. The saturation level increases as Ω_P is further increased, the power input being predominantly balanced by the rate of Joule dissipation in the fluid.

6.14 The Bullard–Gellman formalism

Suppose now that the domain of conducting fluid V is the sphere $r < R$, and that $\mathbf{u}(\mathbf{x})$ is a given steady solenoidal velocity field, satisfying $\mathbf{u}\cdot\mathbf{n} = 0$ on $r = R$. It is convenient to use R as the unit of length, $u_m = \max|\mathbf{u}|$ as the unit of velocity,[8] and R^2/η as the unit of time, and to define $\mathcal{R}_m = u_m R/\eta$.

[7] The flow was presumably turbulent, but this appears to have had little effect on the dynamo action.
[8] Different authors adopt different conventions here and care is needed in making detailed comparisons.

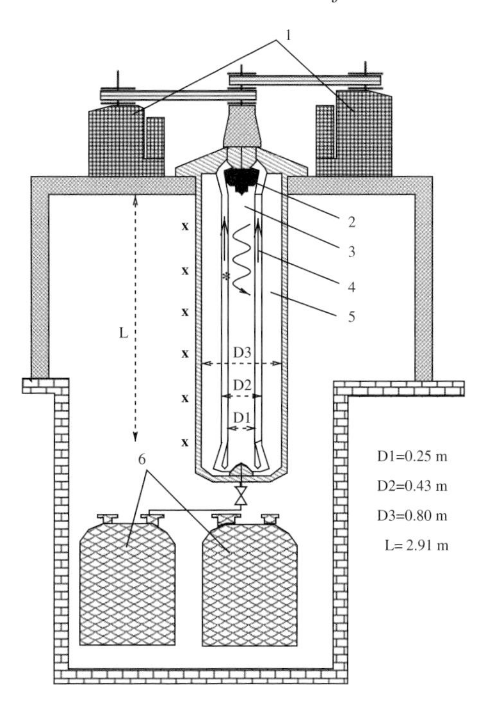

Figure 6.12 Schematic of the Riga dynamo facility: two 55 kW motors (1) drive the propellor (2), which propels a helical flow of liquid sodium down the inner cylinder (3) and a return flow up the surrounding annulus (4); more liquid sodium is at rest in an outer annulus (5) and is stored in tanks (6); an induction coil (a'flux-gate sensor') is placed at position $*$ and six 'Hall sensors' are placed at the positions marked x. The dimensions D1–D3 and L are as indicated. [From Gailitis et al. 2001.]

The problem (6.4) then takes the dimensionless form

$$\left.\begin{aligned} \partial\mathbf{B}/\partial t = \mathcal{R}_{\mathrm{m}}\nabla\wedge(\mathbf{u}\wedge\mathbf{B}) + \nabla^2\mathbf{B} \qquad &\text{for } r<1\,, \\ \nabla\wedge\mathbf{B}=0 \quad \text{for } r>1, \quad [\![\mathbf{B}]\!]=0 \quad &\text{across } r=1\,. \end{aligned}\right\} \tag{6.131}$$

As in §2.8, the problem admits solutions of the form

$$\mathbf{B}(\mathbf{x},t) = \mathbf{B}(\mathbf{x})\,\mathrm{e}^{pt}\,, \tag{6.132}$$

where

$$\left.\begin{aligned} (p-\nabla^2)\mathbf{B} = \mathcal{R}_{\mathrm{m}}\nabla\wedge(\mathbf{u}\wedge\mathbf{B}) \qquad &\text{for } r<1\,, \\ \nabla\wedge\mathbf{B}=0 \quad \text{for } r>1\,, \quad [\![\mathbf{B}]\!]=0 \quad &\text{across } r=1\,. \end{aligned}\right\} \tag{6.133}$$

This may be regarded as an eigenvalue problem for the parameter p, the eigenvalues $p_1,\ p_2,\ \ldots$, being functions of $\mathcal{R}_{\mathrm{m}}$ (as well as depending of course on the structural properties of the **u**-field). When $\mathcal{R}_{\mathrm{m}} = 0$, the eigenvalues p_α are given by the free-decay-mode theory of §2.8, and they are all real and negative. As $\mathcal{R}_{\mathrm{m}}$

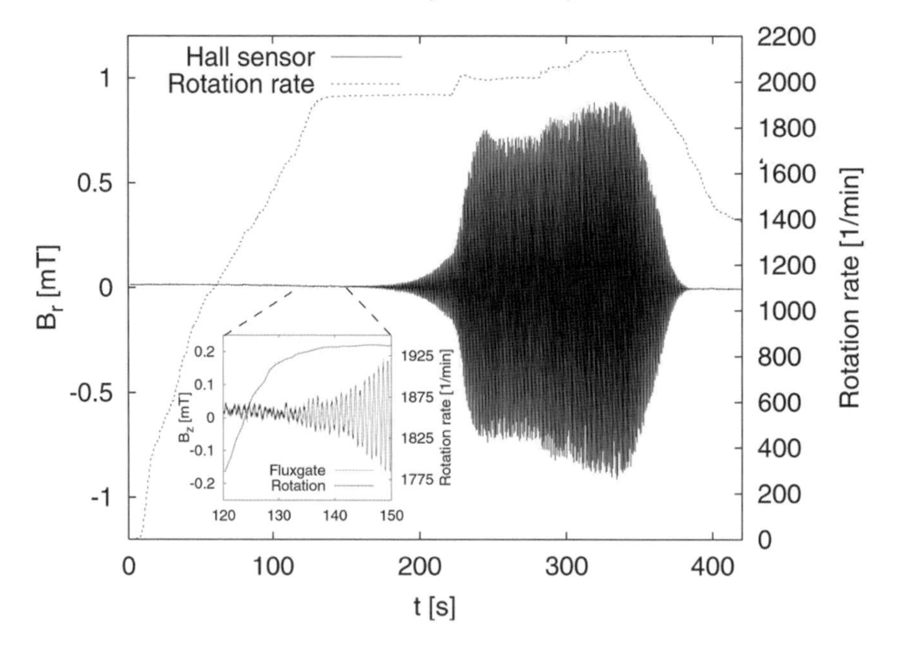

Figure 6.13 A typical measurement of the component B_r of a rotating magnetic field generated by dynamo action in the Riga facility. The propellor rotation rate $\Omega_P(t)$ is shown by the faint dotted curve; the inset shows the threshold of instability for the component B_z at $t \sim 132$ as detected by an induction coil placed in the annulus (4). [After Gailitis et al. 2001.]

increases from zero (for a given structure $\mathbf{u}(\mathbf{x})$), each may be expected to vary continuously and may become complex. If $\mathcal{R}e\, p_\alpha$ becomes positive for some finite value of $\mathcal{R}_\mathrm{m}$, then the corresponding field structure $\mathbf{B}^{(\alpha)}(\mathbf{x})$ is associated with an exponential growth factor in (6.132), and dynamo action occurs. If all the $\mathcal{R}e\, p_\alpha$ remain negative (as would happen if for example $\mathbf{u}$ were purely toroidal) then dynamo action does not occur for any value of $\mathcal{R}_\mathrm{m}$.

The natural procedure for solving (6.133) is a direct extension of that adopted in §2.8 (when $\mathcal{R}_\mathrm{m} = 0$). Let P and T be the defining scalars of $\mathbf{B}_P$ and $\mathbf{B}_T$, and let

$$P(r,\, \theta,\, \varphi) = \sum_{n,m} \left(P_n^{mc}(r) \cos m\varphi + P_n^{ms}(r) \sin m\varphi\right) P_n^m(\cos\theta)\,, \tag{6.134}$$

and similarly for T. Then, as in §2.8, the conditions in $r \geq 1$ may be replaced by boundary conditions

$$T_n^m(r) = 0,\; \mathrm{d}P_n^m/\mathrm{d}r + (n+1)P_n^m = 0 \quad \text{on } r = 1\,, \tag{6.135}$$

where $T_n^m(r)$ denotes either $T_n^{mc}(r)$ or $T_n^{ms}(r)$, and similarly for $P_n^m(r)$. The equations for $T_n^m(r),\ P_n^m(r)$ in $r < 1$ take the form

$$\left(p - \frac{1}{r^2}\frac{\mathrm{d}}{\mathrm{d}r}r^2\frac{\mathrm{d}}{\mathrm{d}r} + \frac{n(n+1)}{r^2}\right)\left\{\begin{array}{c} T_n^{ms/c} \\ P_n^{ms/c} \end{array}\right\} = \mathcal{R}_\mathrm{m}\left\{\begin{array}{c} I_n^{ms/c} \\ J_n^{ms/c} \end{array}\right\}, \tag{6.136}$$

where $I_n^m(r),\ J_n^m(r)$ are terms that arise through the interaction of **u** and **B**. The determination of these interaction terms requires detailed prescription of **u**, and is in general an intricate matter. It is clear however that, in view of the linearity of the induction equation (in **B**), the terms $I_n^m,\ J_n^m$ can each (for given **u**) be expressed as a sum of terms linear in $T_{n'}^{m'},\ P_{n'}^{m'}$, over a range of values of m', n' dependent on the particular choice of **u**. In general we thus obtain an infinite set of coupled linear second-order ordinary differential equations for the functions $T_n^m,\ P_n^m$.

From a purely analytical point of view, there is little more that can be done, and recourse must be had to numerical methods to make further progress. From a numerical point of view also, the problem is quite formidable. The method usually adopted is to truncate the system (6.136) by ignoring all harmonics having $n > N$, where N is some fixed integer (m is of course limited by $0 \le m \le n$). The radial derivatives are then replaced by finite differences,[9] the range $0 < r < 1$, being divided into, say, M segments. Determination of p is then reduced to a numerical search for the roots of the discriminant of the resulting set of linear algebraic equations. Interest centres on the value p_1 having largest real part. This value depends on N and M, and the method can be deemed successful only when N and M are sufficiently large that further increase in N and/or M induces negligible change in p_1. Actually, as demonstrated by Gubbins (1973), the eigenfunction $\mathbf{B}^{(1)}(\mathbf{x})$ is much more sensitive than the eigenvalue p_1 to changes in N and M, and a more convincing criterion for convergence of the procedure is provided by requiring that this eigenfunction show negligible variation with increasing N and M.

The velocity field $\mathbf{u}(\mathbf{x})$ can, like **B**, be expressed as the sum of poloidal and toroidal parts, each of which may be expanded in surface harmonics. The motion selected for detailed study by Bullard & Gellman (1954) was of the form

$$\mathbf{u} = \varepsilon\nabla\wedge(\mathbf{x}Q_T(r)P_1(\cos\theta)) + \nabla\wedge\nabla\wedge\left(\mathbf{x}Q_P(r)P_2^2(\cos\theta)\cos 2\varphi\right), \tag{6.137}$$

where Q_P and Q_T had simple forms, e.g.

$$Q_P \propto r^3(1-r)^2, \quad Q_T \propto r^2(1-r). \tag{6.138}$$

We may use the notation $\mathbf{u} = \left\{\varepsilon\mathbf{T}_1 + \mathbf{P}_2^{2c}\right\}$ as a convenient abbreviation for (6.137). Interaction of **u** and **B** for this choice of motion is depicted diagrammatically in

[9] There are other possibilities here, e.g. functions of r may be expanded as a series of spherical Bessel functions (the free decay modes) with truncation after m terms as suggested by Elsasser (1946). This procedure has been followed by Pekeris et al. (1975).

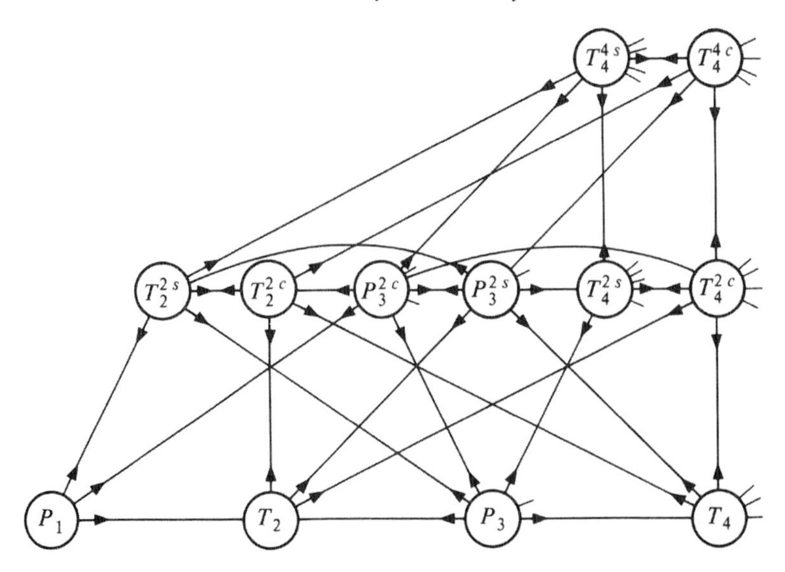

Figure 6.14 Diagrammatic representation of the interaction of harmonics of velocity and magnetic fields when the velocity field consists of a $\mathbf{T}_1$ ingredient and a $\mathbf{P}_2^{2c}$ ingredient. Each circle indicates an excited magnetic mode; coupling along the rows is provided by the $\mathbf{T}_1$-motion and coupling between the rows by the $\mathbf{P}_2^{2c}$-motion. [Republished with permission of the Royal Society, from Bullard & Gellman 1954; permission conveyed through Copyright Clearance Center, Inc.]

Figure 6.14, in which the small circles represent excited magnetic modes. The $\mathbf{T}_1$-motion introduces the coupling along the rows: for example interaction of a $\mathbf{T}_1$ motion with a $\mathbf{P}_1$-field generates a $\mathbf{T}_2$-field (this is just the process of generation of toroidal field by differential rotation analysed in §3.14). Similarly the $\mathbf{P}_2^{2c}$-motion introduces the coupling between the rows, interaction of a $\mathbf{P}_2^{2c}$ motion with a $\mathbf{P}_1$-field generating a $\left\{\mathbf{T}_2^{2s} + \mathbf{P}_3^{2c}\right\}$ field. The figure is 'truncated' for $n \geq 5$; even at this low level of truncation, the complexity of the interactions is impressive! The particular shape of the interaction diagram is entirely determined by the choice of $\mathbf{u}$: to each possible choice of $\mathbf{u}$ there corresponds one and only one such diagram.

It was shown by Gibson & Roberts (1969) that the Bullard & Gellman velocity field (6.137) could not in fact sustain a dynamo: the procedure outlined above failed to converge as M was increased. A similar convergence failure has been demonstrated by Gubbins (1973) in respect of the more complex motion $\left\{\mathbf{T}_1 + \mathbf{P}_2^{2c} + \mathbf{P}_2^{2s}\right\}$ proposed as a dynamo model by Lilley (1970). Positive results have however been obtained by Gubbins (1973) who considered axisymmetric velocity fields of the form $\varepsilon\mathbf{P}_n + \mathbf{T}_n$ with $n = 6, 4, 2$ and $\varepsilon = \frac{1}{30}, \frac{1}{30}, \frac{1}{10}$ respectively. Such a motion cannot maintain a field axisymmetric about the same axis (by Cowling's theorem), but may conceivably maintain a non-axisymmetric

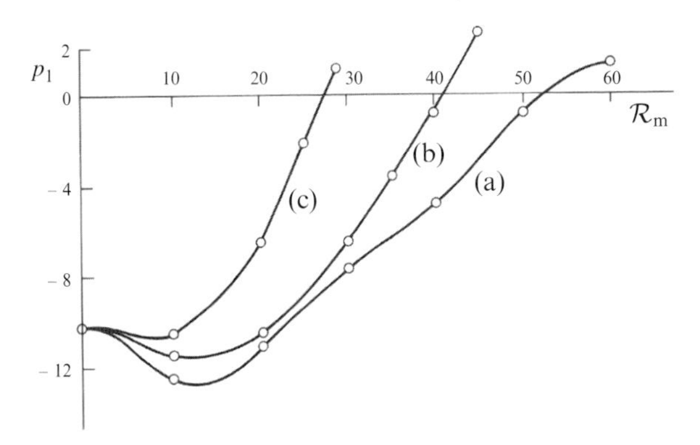

Figure 6.15 Dependence of growth rate p_1 on $\mathcal{R}_m$ for a motion of the form $\mathbf{T}_n + \varepsilon \mathbf{P}_n$: (a) $n = 2, \varepsilon = \frac{1}{10}$; (b) $n = 4, \varepsilon = \frac{1}{30}$; (c) $n = 6, \varepsilon = \frac{1}{30}$ (replotted from Gubbins 1973). It may be shown (Gubbins, private communication) that $\mathrm{d}p_1/\mathrm{d}\mathcal{R}_m = 0$ at $\mathcal{R}_m = 0$.

field (cf. the Gailitis dynamo discussed in §6.9). The dependence of p_1 on $\mathcal{R}_m$ as obtained numerically by Gubbins for the three cases is shown in Figure 6.15; in each case p_1 remains real as $\mathcal{R}_m$ increases and changes sign at a critical value $\mathcal{R}_{mc}$ of $\mathcal{R}_m$ ($\mathcal{R}_{mc} = 27,\ 41,\ 53$ in the three cases). It is noteworthy that p_1 first decreases slightly in all three cases as $\mathcal{R}_m$ increases from zero (indicating accelerated decay) before increasing to zero; the reason for this behaviour is by no means clear.

Positive results have also been obtained by Pekeris et al. (1975), who studied kinematic dynamo action associated with velocity fields satisfying the 'maximal helicity' condition $\nabla \wedge \mathbf{u} = k\mathbf{u}$ in the sphere $r < 1$. Such motions are interesting in that they can be made to satisfy the equations of inviscid incompressible fluid mechanics (the Euler equations) and the condition $\mathbf{u} \cdot \mathbf{n} = 0$ on $r = 1$ (Moffatt 1969; Pekeris 1972). The defining scalars for both poloidal and toroidal ingredients of the velocity field are then both proportional to $r^{1/2} J_{n+1/2}(kr) S_n(\theta, \varphi)$ where k satisfies $J_{n+1/2}(k) = 0$ (see Table 2.2 on p. 50). Pekeris et al. studied particularly the case $n = 2$, with $S_n(\theta, \varphi) \propto \sin^2\theta \cos 2\varphi$ and found steady dynamo action for each $k = x_{2q}$ $(q = 1, 2, \ldots, 20)$ in the notation of Table 2.2. The corresponding critical magnetic Reynolds number $\mathcal{R}_m$ decreased with increasing q (i.e. with increasing radial structure in the velocity field) from 99.2 (when $q = 1$) to 26.9 when $q = 6$, and 26.4 when $q = 20$. The values of N and M used in the numerical calculations were $N = 10, M = 100$.

The results of Pekeris et al. have been independently confirmed by Kumar & Roberts (1975), who also studied the numerical convergence of eigenvalues and eigenfunctions for a range of motions of the form $\left\{\mathbf{T}_1 + \varepsilon_1 \mathbf{P}_2 + \varepsilon_2 \mathbf{P}_2^{2c} + \varepsilon_3 \mathbf{P}_2^{2s}\right\}$. The helicity density $\mathbf{u} \cdot \nabla \wedge \mathbf{u}$ of such motions is antisymmetric about the equatorial

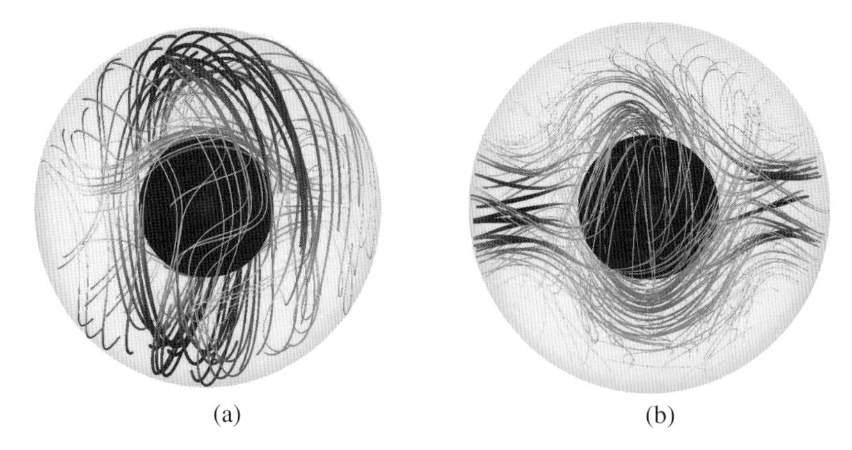

Figure 6.16 Magnetic field structures for a particular axisymmetric reversible flow in a spherical shell having the same dynamo growth rate under complementary boundary conditions (a) $\mathbf{n} \cdot \mathbf{B} = 0$ and (b) $\mathbf{n} \times \mathbf{B} = 0$; the field strength is indicated by the shading of **B**-lines. [Reprinted figure with permission from Favier & Proctor 2013. Copyright 2013 by the American Physical Society.]

plane (unlike the motions studied by Pekeris et al.) and in this respect are more relevant in the geophysical context[10] – see the later discussion in §9.3.3.

A systematic search for both steady and unsteady solutions of the induction equation in a sphere was carried out by Dudley & James (1989), who found that quite simple flows involving axisymmetric rolls with both poloidal and toroidal ingredients (and so non-zero helicity density) are capable of dynamo action, generating a non-axisymmetric field, provided $\mathcal{R}_m$ is sufficiently large and the poloidal to toroidal ratio lies in an appropriate range for each flow considered. This was a definitive treatment using the Bullard & Gellman approach.

An interesting result in this context has been obtained by Favier & Proctor (2013), who proved that the growth rate (eigenvalue) for dynamo action for a given flow $\mathbf{u}(\mathbf{x})$ in any domain $\mathcal{D}$ with a perfectly conducting boundary (on which $\mathbf{n} \cdot \mathbf{B} = 0$) is the same as that for the reversed flow $-\mathbf{u}(\mathbf{x})$ under the 'infinite permeability' boundary condition $\mathbf{n} \times \mathbf{B} = 0$. Some flows in a sphere or spherical shell, described as 'reversible', have the property that $\mathbf{u}(\mathbf{x})$ can be transformed to $-\mathbf{u}(\mathbf{x})$ under rotation; for any such flow, the growth rates are the same for the two different boundary conditions. The corresponding eigenfunctions are however very different, as of course they must be, given the different boundary condition. Figure 6.16 shows the eigenfunctions for a particular axisymmetric reversible flow (one of the flows considered by Dudley & James 1989), illustrating this striking difference.

[10] In this context, see also Bullard & Gubbins (1973) who have computed eigenvalues and eigenfunctions for further velocity fields having an axisymmetric structure similar to that driven by thermal convection in a rotating sphere (Weir 1976); the fields maintained by such motions are of course non-axisymmetric.

The Bullard & Gellman procedure has been developed and refined over the years, and has been extended to cover the full MHD dynamo problem incorporating the Lorentz force in the Navier–Stokes equation in a convecting, rotating fluid (see for example Hollerbach 2000). For the moment however, we continue to restrict attention to the purely kinematic problem governed by the induction equation with prescribed velocity field. In this spirit, we now describe one more famous early example through which dynamo action by a laminar, albeit unsteady, flow was rigorously established, the so-called stasis dynamo of Backus (1958).

6.15 The stasis dynamo

As noted above, it is difficult in general to justify truncation of the spherical-harmonic expansion of $\mathbf{B}$, and erroneous conclusions can result from such truncation. There are two situations however in which truncation (of one form or another) *can* be rigorously justified. The first of these is the rotor dynamo situation of Herzenberg (1958) considered in §6.9, in which the radius of each rotor is small compared with the distance between rotors: in this situation the spatial attenuation of higher harmonics of the field induced by each rotor permits the imposition of a rigorous upper bound on the influence of these higher harmonics and hence permits rigorous justification of the process described (without due respect for rigour!) in §6.9.

The second situation in which 'unwanted' higher harmonics may be dropped without violation of mathematical rigour was conceived by Backus (1958), and invokes temporal rather than spatial attenuation. We know from the theory of free decay modes that higher harmonics (i.e. those corresponding to higher values of n and q in the notation of §2.8) decay faster than lower harmonics when $\mathbf{u} \equiv 0$, and that the fundamental harmonic (i.e. that corresponding to the lowest available values of n and q) will survive the longest and will ultimately dominate during a period of free decay (when $\mathbf{u} \equiv 0$).

Suppose then that at some initial instant we start with a poloidal field $\mathbf{B}_{P11}$ where the suffix $_{11}$ indicates that only the fundamental ('dipole') harmonic $n = q = 1$ is present. Suppose that we subject this field to the influence of the following time-dependent velocity field (devised and justifiable in terms of mathematical expediency rather than physical plausibility): (i) a short period of intense differential rotation $\mathbf{u}_T$, thus generating a strong toroidal field $\mathbf{B}_T$ by the mechanism analysed in §3.14; (ii) a period of 'stasis' ($\mathbf{u} \equiv 0$) so that all but the fundamental harmonic $\mathbf{B}_{T11}$ of $\mathbf{B}_T$ decay to a negligible level; (iii) a short period of intense non-axisymmetric poloidal motion $\mathbf{u}_P$ generating a poloidal field $\mathbf{B}_P^*$ from $\mathbf{B}_T$ through the mechanism described (at least in part) by (6.38); (iv) a second period of stasis to allow all but the fundamental harmonic $\mathbf{B}_{P11}^*$ of $\mathbf{B}_P^*$ to decay to a negligible level; (v) a

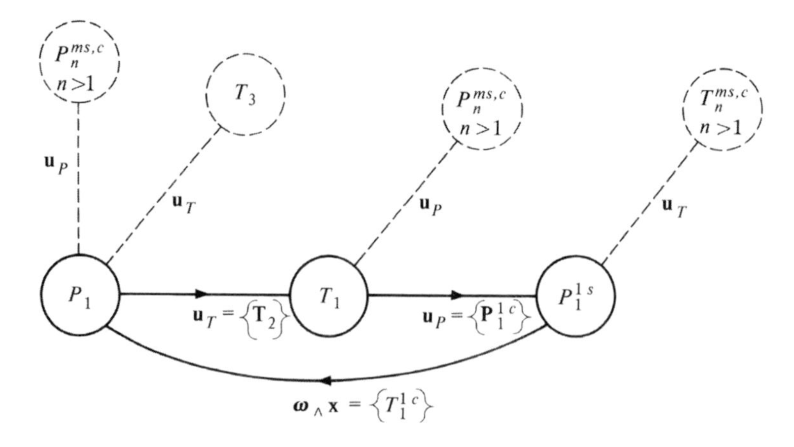

Figure 6.17 Interaction diagram for the stasis dynamo of Backus (1958). The motion considered was of the form $\mathbf{u} = \{\mathbf{T}_2 + \mathbf{P}_1^{1c} + \mathbf{T}_1^{1c}\}$ in the notation of §6.14, with periods of stasis between separate phases of application of the three ingredients. The dotted lines and circles indicate excitations which are evanescent due to their relatively rapid decay during the periods of stasis. The field ingredients represented by $\mathbf{P}_1$, $\mathbf{T}_1$ and $\mathbf{P}_1^{1s}$ then provide the dominant contributions in the closed dynamo cycle.

rapid rigid body rotation $\omega \wedge \mathbf{x}$ to bring $\mathbf{B}^*_{P11}$ (plus whatever remains of $\mathbf{B}_{P11}$) into alignment with the original direction of $\mathbf{B}_{P11}$. If the fields $\mathbf{u}_T$, $\mathbf{u}_P$ and $\omega \wedge \mathbf{x}$, and the durations $t_1, t_2, \ldots, t_5$ of the phases (i), ..., (v) are suitably chosen, then a net field amplification (with arbitrarily little change of structure) can be guaranteed. The interaction diagram corresponding to the particular velocity fields $\mathbf{u}_T$, $\mathbf{u}_P$ and $\omega \wedge \mathbf{x}$ chosen by Backus to give substance to this skeleton procedure is indicated in Figure 6.17, in which the evanescent (and impotent) higher harmonics suppressed (relatively) by the periods of stasis are indicated by dotted lines.

The Backus dynamo, like the Herzenberg dynamo, makes no claims to dynamic plausibility. Although artificial from a dynamical viewpoint, the enduring interest of dynamo models incorporating spatial or temporal decay (of which Herzenberg and Backus almost simultaneously devised the respective prototypes) resides in the fact that by either technique the existence of velocity fields capable of dynamo action (as defined in §6.1) can be rigorously demonstrated. In this respect, the Herzenberg and Backus dynamos have provided corner-stones (i.e. reliable positive results) that have acted both as a test and a basis for subsequent developments in which mathematical rigour has necessarily given way to the more pressing demands of physical plausibility.

7

Mean-Field Electrodynamics

7.1 Turbulence and random waves

We have so far treated the velocity field $\mathbf{U}(\mathbf{x}, t)$ as a known function of position and time.[1] In this chapter we consider the situation of greater relevance in both solar and terrestrial contexts when $\mathbf{U}(\mathbf{x}, t)$ includes a random ingredient whose statistical (i.e. average) properties are assumed known, but whose detailed (unaveraged) properties are too complicated for either analytical description or observational determination. Such a velocity field generates random perturbations of electric current and magnetic field, and our aim is to determine the evolution of the statistical properties of the magnetic field (and in particular of its local mean value) in terms of the 'given' statistical properties of the $\mathbf{U}$-field.

The random velocity field may be a field of turbulence as normally understood, or it may consist of a random superposition of interacting wave motions. The distinction can be most easily appreciated for the case of a thermally stratified fluid. If the stratification is unstable (i.e. if the fluid is strongly heated from below) then thermal turbulence will ensue, the net upward transport of heat being then predominantly due to turbulent convection. If the stratification is stable (i.e. if the temperature either increases with height, or decreases at a rate less than the adiabatic rate) then turbulence will not occur, but the medium may support internal gravity waves which will be present to a greater or lesser extent, in proportion to any random influences that may be present, distributed either throughout the fluid or on its boundaries. For example, if the outer core of the Earth is stably stratified as maintained by Higgins & Kennedy 1971 (even just in the outer layers, as argued by Braginsky 1993 – see §4.5), then random inertial waves may be excited either by sedimentation of iron-rich material released from the mantle across the core–mantle interface or by flotation of light compounds (rich in silicon or sulphur) released by chemical separation at the interface between inner core and outer core, or possibly

[1] From now on, we shall use $\mathbf{U}$ to represent the total velocity field, reserving $\mathbf{u}$ for its random ingredient.

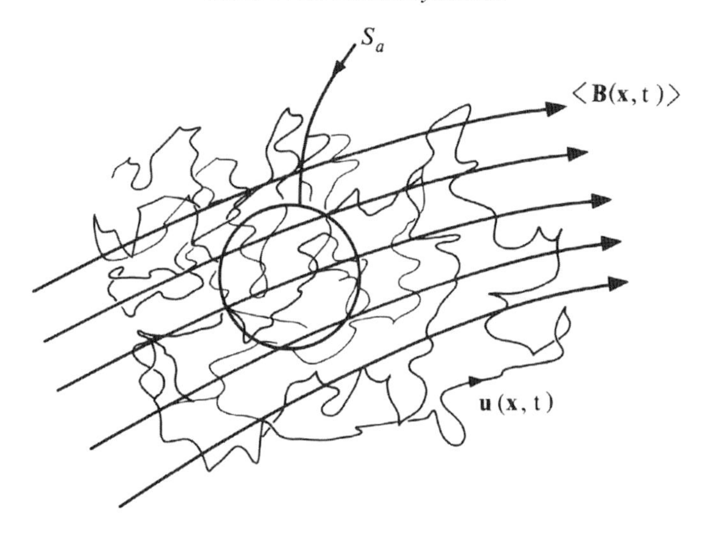

Figure 7.1 Schematic of the random velocity field $\mathbf{u}(\mathbf{x}, t)$ varying on the small length-scale l_0 and the mean magnetic field varying on the large-scale L. The mean is defined as an average over the sphere S_a of radius a where $l_0 \ll a \ll L$.

by shear-induced instability in the boundary layers and shear layers formed as a result of the slow precession of the Earth's angular velocity vector. Such effects may generate radial perturbation velocities whose amplitude is limited by the stabilising buoyancy forces; the flow fields will then have the character of a field of forced weakly interacting internal waves, i.e. a field of 'weak turbulence', rather than of strongly non-linear turbulence of the 'conventional' type.

We shall throughout this chapter suppose that the random ingredient of the motion is characterised by a length-scale l_0 which is small compared with the 'global' scale L of variation of mean quantities (Figure 7.1). L will in general be of the same order of magnitude as the linear dimension of the region occupied by the conducting fluid, e.g. $L = \mathcal{O}(R)$ when the fluid is confined to a sphere of radius R. In the case of turbulence, l_0 may be loosely defined as the scale of the 'energy-containing eddies' (see e.g. Batchelor 1953). Likewise, in the case of random waves, l_0 may be identified in order of magnitude with the wavelength of the constituent waves of maximum energy. On any intermediate scale a satisfying the double inequality

$$l_0 \ll a \ll L, \tag{7.1}$$

the global variables (e.g. mean velocity and mean magnetic field) may be supposed nearly uniform; here the 'mean', which we shall denote by angular brackets, may reasonably be defined as an average over a sphere of intermediate radius a: i.e. for any $\psi(\mathbf{x}, \mathbf{t})$, we define

$$\langle\psi(\mathbf{x}, t)\rangle_a = \frac{3}{4\pi a^3}\int_{|\boldsymbol{\xi}|<a} \psi(\mathbf{x}+\boldsymbol{\xi}, t)\,\mathrm{d}^3\boldsymbol{\xi}\,, \tag{7.2}$$

with the expectation that this average is insensitive to the precise value of a provided merely that (7.1) is satisfied. The statistical (i.e. mean) properties of the **U**-field are weakly varying on the scale a; the methods of the theory of homogeneous turbulence (Batchelor 1953) may therefore be employed in calculating effects on these intermediate scales.

We could equally use time-scales rather than spatial scales in defining mean quantities. If T is the time-scale of variation of global fields and t_0 is the time-scale characteristic of the fluctuating part of the **U**-field, then we shall require as a matter of consistency that T be large compared with t_0. If analysis reveals that in any situation T and t_0 are comparable, then the general approach (as applied to that situation) must be regarded as suspect. When $T \gg t_0$, then for any intermediate time-scale τ satisfying

$$t_0 \ll \tau \ll T\,, \tag{7.3}$$

we can define

$$\langle\psi(\mathbf{x}, t)\rangle_\tau = \frac{1}{2\tau}\int_{-\tau}^{\tau} \psi(\mathbf{x}, t+\tau')\mathrm{d}\tau'\,, \tag{7.4}$$

with again a reasonable expectation that this is insensitive to the value of τ provided (7.3) is satisfied. We shall use the notation $\langle\psi(\mathbf{x}, t)\rangle$ without suffix a or τ to denote either average (7.2) or (7.4), which from a purely mathematical point of view may both be identified with an 'ensemble' average (in the asymptotic limits $l_0/L \to 0$, $t_0/T \to 0$, respectively).

Having thus defined a mean, the velocity and magnetic fields may be separated into mean and fluctuating parts:

$$\mathbf{u}(\mathbf{x}, t) = \mathbf{U}_0(\mathbf{x}, t) + \mathbf{u}(\mathbf{x}, t), \quad \langle\mathbf{u}\rangle = 0\,, \tag{7.5}$$

$$\mathbf{B}(\mathbf{x}, t) = \mathbf{B}_0(\mathbf{x}, t) + \mathbf{b}(\mathbf{x}, t), \quad \langle\mathbf{b}\rangle = 0\,. \tag{7.6}$$

Likewise the induction equation (3.14) may be separated into *its* mean and fluctuating parts:

$$\partial\mathbf{B}_0/\partial t = \nabla\wedge(\mathbf{U}_0\wedge\mathbf{B}_0) + \nabla\wedge\boldsymbol{\mathcal{E}} + \eta\nabla^2\mathbf{B}_0\,, \tag{7.7}$$

$$\partial\mathbf{b}/\partial t = \nabla\wedge(\mathbf{U}_0\wedge\mathbf{b}) + \nabla\wedge(\mathbf{u}\wedge\mathbf{B}_0) + \nabla\wedge\boldsymbol{\mathcal{F}} + \eta\nabla^2\mathbf{b}\,, \tag{7.8}$$

where

$$\boldsymbol{\mathcal{E}} = \langle\mathbf{u}\wedge\mathbf{b}\rangle, \quad \boldsymbol{\mathcal{F}} = \mathbf{u}\wedge\mathbf{b} - \langle\mathbf{u}\wedge\mathbf{b}\rangle\,. \tag{7.9}$$

Note that in (7.7) there now appears a term $\nabla \wedge \mathcal{E}$ associated with a product of random fluctuations. The mean electromotive force $\mathcal{E}$ is a quantity of central importance in the theory: the aim must clearly be to find a way to express $\mathcal{E}$ in terms of the mean fields $\mathbf{U}_0$ and $\mathbf{B}_0$ so that, for given $\mathbf{U}_0$, (7.7) may be integrated.

The idea of averaging equations involving random fluctuations is of course well known in the context of conventional turbulence theory: averaging of the Navier–Stokes equations likewise leads to the appearance of the important quadratic mean $-\langle u_i u_j \rangle$ (the Reynolds stress tensor) which is the counterpart of the $\langle \mathbf{u} \wedge \mathbf{b} \rangle$ of the present context. There is no satisfactory theory of turbulence that succeeds in expressing $\langle u_i u_j \rangle$ in terms of the mean field $\mathbf{U}_0$. By contrast, there *is* now a satisfactory body of theory for the determination of $\mathcal{E}$. The reason for this (comparative) degree of success can be attributed to the linearity (in $\mathbf{B}$) of the induction equation. There is no counterpart of this linearity in the dynamics of turbulence.

The two-scale approach in the context of the induction equation was first introduced by Steenbeck et al. (1966), and many ideas of the present chapter can be traced either to this pioneering paper, or to the series of papers by the same authors that followed (for a comprehensive treatment, see Krause & Rädler 1980). The enduring influence of these papers has been marked by publication of a special issue of the *Journal of Plasma Physics* devoted entirely to developments in the theory of mean-field electrodynamics (Tobias et al. 2018).

7.2 The linear relation between $\mathcal{E}$ and $\mathbf{B}_0$

The term $\nabla \wedge (\mathbf{u} \wedge \mathbf{B}_0)$ in (7.8) acts as a source term generating the fluctuating field $\mathbf{b}$. If we suppose that $\mathbf{b} = 0$ at some initial instant $t = 0$, then the linearity of (7.8) guarantees that the fields $\mathbf{b}$ and $\mathbf{B}_0$ are linearly related. It follows that the fields $\mathcal{E} = \langle \mathbf{u} \wedge \mathbf{b} \rangle$ and $\mathbf{B}_0$ are likewise linearly related, and since the spatial scale of $\mathbf{B}_0$ is by assumption large (compared with scales involved in the detailed solution of (7.8)), we may reasonably anticipate that this relationship may be developed as a rapidly convergent series of the form[2]

$$\mathcal{E}_i = \alpha_{ij} B_{0j} + \beta_{ijk} \frac{\partial B_{0j}}{\partial x_k} + \gamma_{ijkl} \frac{\partial^2 B_{0j}}{\partial x_k \partial x_l} + \cdots , \tag{7.10}$$

where the coefficients α_{ij}, β_{ijk}, ..., are pseudo-tensors (‘pseudo’ because $\mathcal{E}$ is a polar vector whereas $\mathbf{B}_0$ is an axial vector). It is clear that, since the solution $\mathbf{b}(\mathbf{x}, t)$ of (7.8) depends on $\mathbf{U}_0$, $\mathbf{u}$ and η, the pseudo-tensors α_{ij}, β_{ijk}, ..., may be expected to depend on, and indeed are totally determined by (i) the mean field $\mathbf{U}_0$, (ii) the statistical properties of the random field $\mathbf{u}(\mathbf{x}, t)$, and (iii) the value of the parameter

[2] Terms involving time derivatives $\partial B_{0j}/\partial t$, $\partial^2 B_{0j}/\partial t^2$, ..., may also appear in the expression for $\mathcal{E}_i$. Such terms may however always be replaced by terms involving only space derivatives by means of (7.7).

η. These pseudo-tensors will in general vary on the 'macroscale' L, but in the conceptual limit $L/\ell_0 \to \infty$, when $\mathbf{U}_0$ becomes uniform and $\mathbf{u}$ becomes statistically strictly homogeneous, the pseudo-tensors $\alpha_{ij}, \beta_{ijk}, \ldots$, which may themselves be regarded as statistical properties of the $\mathbf{u}$-field, also become strictly uniform.

If $\mathbf{U}_0$ is uniform, then it is natural to take axes moving with velocity $\mathbf{U}_0$, and to redefine $\mathbf{u}(\mathbf{x}, t)$ as the random velocity in this frame of reference. With this convention, (7.7) becomes

$$\frac{\partial B_{0i}}{\partial t} = \epsilon_{ijk}\frac{\partial}{\partial x_j}\left(\alpha_{kl}B_{0l} + \beta_{klm}\frac{\partial B_{0l}}{\partial x_n} + \cdots\right) + \eta\nabla^2 B_{0i}\,. \tag{7.11}$$

It is immediately evident that if $\mathbf{B}_0$ is also uniform (and if $\mathbf{u}$ is statistically homogeneous so that $\alpha_{kl}, \beta_{klm}, \ldots$ are uniform) then $\partial\mathbf{B}_0/\partial t = 0$, i.e. infinite length-scale for $\mathbf{B}_0$ implies infinite time-scale. We may therefore in general anticipate that as $L/\ell_0 \to \infty$, $T/t_0 \to \infty$ also in the notation of §7.1, and the notions of spatial and temporal means are compatible.

When $\mathbf{B}_0$ is weakly non-uniform, it is evident from (7.11) that the first term on the right (incorporating α_{kl}) is likely to be of dominant importance when $\alpha_{kl} \neq 0$, since it involves the lowest derivative of $\mathbf{B}_0$; the second term (incorporating β_{klm}) is also of potential importance, and cannot in general be discarded, since it involves second spatial derivatives of $\mathbf{B}_0$ and so has status comparable with the natural diffusion term $\eta\nabla^2\mathbf{B}_0$. Subsequent terms indicated by $\ldots$ in (7.11) should however be negligible provided the scale of $\mathbf{B}_0$ is sufficiently large for the series (7.10) to be rapidly convergent. We shall in the following two sections consider some general properties of the α- and β-terms of (7.11) and we shall go on in subsequent sections of this chapter to evaluate α_{ij} and β_{ijk} explicitly in certain limiting situations.

7.3 The α-effect

Let us now focus attention on the leading term of the series (7.10), viz.

$$\mathcal{E}_i^{(0)} = \alpha_{ij}B_{0j}\,. \tag{7.12}$$

The pseudo-tensor α_{ij} (which is uniform insofar as the $\mathbf{u}$-field is statistically homogeneous) may be decomposed into symmetric and antisymmetric parts:

$$\alpha_{ij} = \alpha_{ij}^{(s)} - \epsilon_{ijk}\,a_k \quad \text{where} \quad a_k = -\tfrac{1}{2}\epsilon_{ijk}\,\alpha_{ij}\,, \tag{7.13}$$

and correspondingly, from (7.12),

$$\mathcal{E}_i^{(0)} = \alpha_{ij}^{(s)}B_{0j} + (\mathbf{a} \wedge \mathbf{B}_0)_i\,. \tag{7.14}$$

It is clear that the effect of the antisymmetric part is merely to provide an additional ingredient **a** (evidently a polar vector) to the *effective mean velocity* that acts upon the mean magnetic field: if $\mathbf{U}_0$ is the actual mean velocity, then $\mathbf{U}_0+\mathbf{a}$ is the *effective mean velocity* (as far as the field $\mathbf{B}_0$ is concerned).

The nature of the symmetric part $\alpha_{ij}^{(s)}$ is most simply understood in the important special situation when the **u**-field is (statistically) isotropic[3] as well as homogeneous. In this situation, by definition, all statistical properties of the **u**-field are invariant under rotations (as well as translations) of the frame of reference, and in particular α_{ij} must be isotropic, i.e.

$$\alpha_{ij} = \alpha\,\delta_{ij}\,, \quad \alpha = \tfrac{1}{3}\alpha_{ii}^{s}\,, \tag{7.15}$$

and of course in this situation $\mathbf{a} = 0$.

The parameter α is a *pseudo-scalar* (like the mean helicity $\mathcal{H} \equiv \langle \mathbf{u} \cdot \boldsymbol{\omega} \rangle$) and it must therefore change sign under any transformation from a right-handed to a left-handed frame of reference ('parity transformations'). Since α is a statistical property of the **u**-field, it can be non-zero only if the **u**-field itself is not statistically invariant under such a transformation. The simplest such transformation is reflection in the origin $\mathbf{x}' = -\mathbf{x}$; we shall say that the **u**-field is *reflectionally symmetric* if all its statistical properties are invariant under parity transformations, and in particular under reflection in the origin.[4] Otherwise the **u**-field *lacks reflectional symmetry*. Only in this latter case can α be non-zero.

Combination of (7.12) and (7.15) gives the very simple result

$$\boldsymbol{\mathcal{E}}^{(0)} = \alpha\,\mathbf{B}_0\,, \tag{7.16}$$

and, from Ohm's law (2.131), we have a corresponding contribution to the mean *current* density

$$\mathbf{J}^{(0)} = \sigma\boldsymbol{\mathcal{E}}^{(0)} = \sigma\alpha\,\mathbf{B}_0\,. \tag{7.17}$$

This possible appearance of a current parallel to the local mean field $\mathbf{B}_0$ is in striking contrast to the conventional situation in which the induced current $\sigma\,\mathbf{U} \wedge \mathbf{B}$ is perpendicular to the field **B**. It may appear paradoxical that two fields (**B** and $\mathbf{U}\wedge\mathbf{B}$) that are everywhere perpendicular may nevertheless have mean parts that are not perpendicular (and that may even be parallel); and to demonstrate beyond doubt

[3] We shall use the word 'isotropic' in the weak sense to indicate 'invariant under rotations but not necessarily under reflections' of the frame of reference; in simple terms this means that there is 'no preferred direction' in the statistics of the turbulence.

[4] Alternatively, parity transformation of the kind $x' = -x,\ y' = y,\ z' = z$ representing mirror reflection in the plane $x = 0$ could be adopted as the basis of a definition of 'mirror symmetry' (a term in frequent use in many published papers). Care is however needed: the mirror transformation can be regarded as a superposition of reflection in the origin followed by a rotation; a **u**-field that is reflectionally symmetric but anisotropic will then not in general be mirror symmetric since its statistical properties will be invariant under the reflection in the origin but not under the subsequent rotation.

that this is in fact a real possibility it is necessary to obtain an explicit expression for the parameter α and to show that it can indeed be non-zero (see §7.8). The appearance of an electromotive force of the form (7.16) was described by Steenbeck & Krause (1966) as the 'α-effect', a terminology that, although arbitrary,[5] is now well established. It is this effect that is at the heart of all modern dynamo theory. The reason essentially is that it provides an obvious means whereby the 'dynamo cycle' $\mathbf{B}_P \leftrightarrows \mathbf{B}_T$ may be completed. We have seen that toroidal field $\mathbf{B}_T$ may very easily be generated from poloidal field $\mathbf{B}_P$ by the process of differential rotation. If we think now in terms of mean fields, then (7.16) indicates that the α-effect will generate toroidal current (and hence poloidal field) from the toroidal field. It is this latter step $\mathbf{B}_T \rightarrow \mathbf{B}_P$ that is so hard to describe in laminar dynamo theory; in the turbulent (or random wave) context it is brought to the same simple level as the much more elementary process $\mathbf{B}_P \rightarrow \mathbf{B}_T$. Cowling's anti-dynamo theorem no longer applies to *mean* fields, and analysis of axisymmetric mean-field evolution is then both possible and promising (see Chapter 9).

As noted above, α can be non-zero only if the $\mathbf{u}$ field lacks reflectional symmetry, and in this situation the mean helicity $\langle \mathbf{u} \cdot \boldsymbol{\omega} \rangle$ will in general be non-zero also. To understand the physical nature of the α-effect, consider the situation depicted in Figure 7.2 (as conceived essentially by Parker 1955b). Following Parker, we define a 'cyclonic event' as a velocity field $\mathbf{u}(\mathbf{x}, t)$ that is localised in space and time and for which the integrated helicity $\mathcal{H} = \int\int (\mathbf{u} \cdot \nabla \wedge \mathbf{u})\, \mathrm{d}V\, \mathrm{d}t$ is non-zero. For definiteness suppose that (in a right-handed frame of reference) $\mathcal{H} > 0$. Such an event tends to distort a line of force of an initial field $\mathbf{B}_0$ in the manner indicated in Figure 7.2, the process of distortion being resisted more or less by diffusion. The normal $\mathbf{n}$ to the field loop generated has a component parallel to $\mathbf{B}_0$, with $\mathbf{n} \cdot \mathbf{B}_0$ less than or greater than zero depending on the net angle of twist of the loop, the former being certainly more likely if diffusion is strong or if the events are very short-lived.

Suppose now that cyclonic events all with $\mathcal{H} > 0$ are randomly distributed in space and time (a possible idealisation of a turbulent velocity field with positive mean helicity). Each field loop generated can be associated with an elemental perturbation current in the direction $\mathbf{n}$, and the spatial mean of these elemental currents[6] will have the form $\mathbf{J}^{(0)} = \sigma\alpha\mathbf{B}_0$, where (if the case $\mathbf{n} \cdot \mathbf{B}_0 < 0$ dominates) the coefficient α will be negative. We shall in fact find below that, in the diffusion dominated situation, $\alpha \langle \mathbf{u} \cdot \boldsymbol{\omega} \rangle < 0$ consistent with this picture.

As noted above, non-zero mean helicity is not absolutely necessary to ensure a non-zero α-effect. In fact a wide variety of flows exist for which $\mathbf{u} \cdot \boldsymbol{\omega} \equiv 0$, and yet

[5] The effect was in fact first isolated by Parker (1955b) who introduced, on the basis of physical arguments, a parameter Γ which may be identified (almost) with the α of Steenbeck & Krause (1966).

[6] This interpretation, although appealing and frequently encountered, should be treated with caution, because any localised current $\mathbf{j}_{loc}(\mathbf{x})$ satisfying $\nabla \cdot \mathbf{j}_{loc} = 0$ must also satisfy $\int \mathbf{j}_{loc}(\mathbf{x})\mathrm{d}V = 0$, so that positive elemental current in any small volume is necessarily compensated by negative current elsewhere.

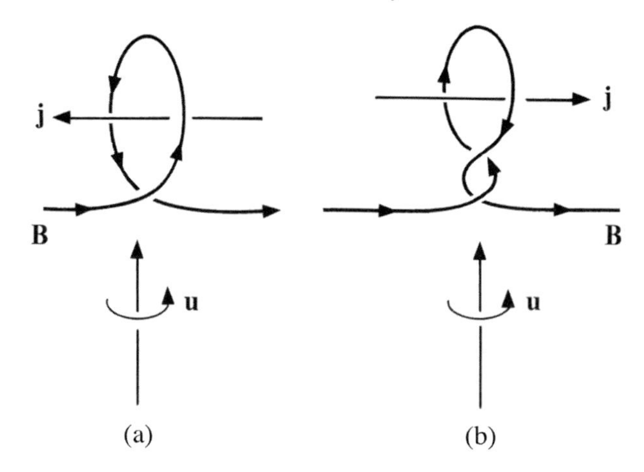

Figure 7.2 Field distortion by a localised helical disturbance (a 'cyclonic event' in the terminology of Parker 1970). In (a) the loop is twisted through an angle $\pi/2$ and the associated current is anti-parallel to **B**; in (b) the twist is $3\pi/2$, and the associated current is parallel to **B**.

$\alpha \neq 0$ (Rasskazov et al. 2018). These authors have concluded that "kinetic helicity and helicity spectrum are not the quantities controlling the dynamo properties of a flow regardless of whether scale separation is present or not". As we have emphasised, it is lack of reflectional symmetry that is necessary to provide a non-zero α, and to this extent we would agree with this conclusion. We would however argue that flows that lack reflectional symmetry and yet have zero helicity are rather unusual, requiring quite skilful construction; and moreover, that under Navier–Stokes evolution in a rotating fluid in convective motion, a flow will inevitably develop non-zero helicity, even if this is zero at some initial instant. The flows devised by Rasskazov et al. (2018), while interesting in a purely kinematic context, would therefore seem to be of little relevance, as recognised by the authors themselves, in realistic geophysical or astrophysical situations.

If the turbulence is not isotropic, then the simple relationship (7.15) of course does not hold. The symmetric pseudo-tensor $\alpha_{ij}^{(s)}$ may however be referred to its principal axes:

$$\alpha_{ij}^{(s)} = \begin{pmatrix} \alpha^{(1)} & \cdot & \cdot \\ \cdot & \alpha^{(2)} & \cdot \\ \cdot & \cdot & \alpha^{(3)} \end{pmatrix}, \tag{7.18}$$

and the corresponding contribution to $\boldsymbol{\mathcal{E}}^{(0)}$ is, from (7.14),

$$\boldsymbol{\mathcal{E}}^{(0s)} = \left(\alpha^{(1)} B_{01},\ \alpha^{(2)} B_{02},\ \alpha^{(3)} B_{03}\right). \tag{7.19}$$

Again under the reflection $\mathbf{x}' = -\mathbf{x}$, the pseudo-scalars $\alpha^{(1)}$, $\alpha^{(2)}$ and $\alpha^{(3)}$ must change sign; and so in general $\alpha_{ij}^{(s)}$ can be non-zero only if the **u**-field lacks reflectional symmetry.[7]

7.4 Effects associated with the coefficient β_{ijk}

Consider now the second term of the series (7.10), viz.

$$\mathcal{E}_i^{(1)} = \beta_{ijk}\,\partial B_{0j}/\partial x_k\,. \tag{7.20}$$

In the simplest situation, in which the **u**-field is isotropic, β_{ijk} is also isotropic, and so

$$\beta_{ijk} = \beta\,\epsilon_{ijk}\,, \tag{7.21}$$

where β is a pure scalar. Equation (7.20) then becomes

$$\boldsymbol{\mathcal{E}}^{(1)} = -\beta\,\nabla\wedge\mathbf{B}_0 = -\beta\,\mu_0\mathbf{J}_0\,, \tag{7.22}$$

where $\mathbf{J}_0$ is the mean current. Hence also (if β is uniform)

$$\nabla\wedge\boldsymbol{\mathcal{E}}^{(1)} = \beta\,\nabla^2\mathbf{B}_0\,, \tag{7.23}$$

and it is evident from (7.7) that the net effect of the emf $\boldsymbol{\mathcal{E}}^{(1)}$ is simply to alter the value of the effective magnetic diffusivity, which becomes $\eta + \beta$ rather than simply η. We shall find that in nearly all circumstances in which β can be calculated explicitly, it is positive, consistent with the simple physical notion of an 'eddy diffusivity': one would expect random mixing (due to the **u**-field) to enhance (rather than detract from) the process of molecular diffusion that gives rise to a positive value of η. However there is no general proof that β must inevitably and in all circumstances be positive, and there are some indications (see §7.11) that it may in fact in some extreme circumstances be negative; if $\beta < -\eta$, there would of course be dramatic consequences as far as solutions of (7.11) are concerned.[8]

If the **u**-field is not isotropic, then departures from the simple relationship (7.21) are to be expected. Suppose for example that the **u**-field is (statistically) invariant under rotations about an axis defined by the unit (polar) vector **e**, but not under general rotations, i.e. **e** defines a 'preferred direction'. Turbulence with this property is described as axisymmetric about the direction of **e**. (The situation is of

[7] Note that here, if $\mathbf{B}_0$ is weakly varying, then $\nabla\cdot\boldsymbol{\mathcal{E}}^{(0s)} \neq 0$, so the mean current $\mathbf{J}$ cannot be simply $\sigma\,\boldsymbol{\mathcal{E}}^{(0s)}$; it must include a potential contribution $-\sigma\nabla\Phi^{(0s)}$ to ensure that $\nabla\cdot\mathbf{J} = 0$.

[8] Families of flows for which $\beta + \eta < 0$ have been constructed by Rasskazov et al. (2018); however, for such flows, (7.11) together with (7.23) is 'ill-posed' in the sense of Hadamard (1902), showing sensitivity to initial conditions with explosive instabilities at the smallest scales; in these circumstances, the two-scale approach adopted at the outset would appear to be untenable.

course of potential importance in the context of turbulence that is strongly influenced by Coriolis forces in a system rotating with angular velocity $\mathbf{\Omega}$: in this situation $\mathbf{e} = \pm\mathbf{\Omega}/\Omega$.) The pseudo-tensor β_{ijk}, which is then also axisymmetric about the direction of $\mathbf{e}$, may be expressed in the form

$$\begin{aligned}\beta_{ijk} = {} & \beta_0\epsilon_{ijk} + \beta_1\epsilon_{mjk}e_n e_i + \beta_2\epsilon_{imk}e_n e_j + \beta_3\epsilon_{ijm}e_n e_k \\ & + \tilde{\beta}_0 e_i e_j e_k + \tilde{\beta}_1 e_i\delta_{jk} + \tilde{\beta}_2 e_j\delta_{ki} + \tilde{\beta}_3(e_k\delta_{ij} - e_j\delta_{ki}) ,\end{aligned} \tag{7.24}$$

where $\beta_0, \ldots, \beta_3$ are pure scalars and $\tilde{\beta}_0, \ldots, \tilde{\beta}_3$ are pseudo-scalars, which can be non-zero only if the $\mathbf{u}$-field lacks reflectional symmetry. The corresponding expression for $\boldsymbol{\mathcal{E}}^{(1)}$ from (7.20) is

$$\begin{aligned}\boldsymbol{\mathcal{E}}^{(1)} = {} & -\beta_0\nabla\wedge\mathbf{B}_0 + \beta_1\,\mathbf{e}\,\nabla\cdot(\mathbf{e}\wedge\mathbf{B}_0) + \beta_2\,(\mathbf{e}\wedge\nabla)(\mathbf{e}\cdot\mathbf{B}_0) - \beta_3(\mathbf{e}\cdot\nabla)(\mathbf{e}\wedge\mathbf{B}_0) \\ & + \tilde{\beta}_0\,\mathbf{e}\,(\mathbf{e}\cdot\nabla)(\mathbf{e}\cdot\mathbf{B}_0) + \tilde{\beta}_2\,\nabla(\mathbf{e}\cdot\mathbf{B}_0) + \tilde{\beta}_3\,\mathbf{e}\wedge(\nabla\wedge\mathbf{B}_0) .\end{aligned} \tag{7.25}$$

The complexity of this type of expression as compared with the simple isotropic relationship (7.22) is noteworthy (and it is not hard to see that if the assumption of axisymmetry is relaxed and *two* preferred directions $\mathbf{e}^{(1)}$ and $\mathbf{e}^{(2)}$ are introduced, the relevant expression for $\boldsymbol{\mathcal{E}}^{(1)}$ becomes still more complex with a dramatic increase in the number of scalar and pseudo-scalar coefficients).

It seems likely that the terms of (7.25) involving β_1, β_2 and β_3 admit interpretation in terms of contributions to a non-isotropic (eddy) diffusivity for the mean magnetic field. These terms do not however appear to have been given detailed study, and it may be that more interesting effects may be concealed in their structure. As for the terms involving the pseudo-scalars $\tilde{\beta}_0$, $\tilde{\beta}_2$ and $\tilde{\beta}_3$, that involving $\tilde{\beta}_3$ has been singled out for detailed examination by Rädler (1969a). This term indicates the possible generation of a mean emf perpendicular to the mean current $\mathbf{J}_0 = \mu_0^{-1}\nabla\wedge\mathbf{B}_0$, and is of particular significance again in the context of the closure of the dynamo cycle: through this 'Rädler-effect', a toroidal emf (and so a toroidal current) may be directly generated from a poloidal current, or equivalently a poloidal field may be generated from a toroidal field. In conjunction with the complementary effect of differential rotation, a closed dynamo cycle may be envisaged, and has indeed been demonstrated by Rädler (1969b).

It must be emphasised however that $\tilde{\beta}_3$, being a pseudo-scalar, can be non-zero only if the $\mathbf{u}$-field lacks reflectional symmetry,[9] and in this situation the pseudo-tensor α_{ij} will in general be non-zero also. Since the dominant term of the series (7.10) involves α_{ij}, it seems almost inevitable that whenever the Rädler-effect is operative, it will be dominated by the α-effect.

[9] This conclusion is at variance with that of Rädler (1969a) who expressed the argument throughout in terms of an axial vector $\mathbf{\Omega}$ rather than a polar vector $\mathbf{e}$, a procedure that (from a purely kinematic point of view) is hard to justify.

7.5 First-order smoothing

We now turn to the detailed solution of (7.8) and the subsequent derivation of $\mathcal{E} = \langle \mathbf{u} \wedge \mathbf{b} \rangle$. We suppose for the remainder of this chapter that $\mathbf{U}_0 = 0$, and that the $\mathbf{u}$-field is statistically homogeneous. Effects associated with non-zero $\mathbf{U}_0$ (in particular with strong differential rotation in a spherical geometry) will be deferred to Chapter 8. The difficulty in solving (7.8) in general arises from the term $\nabla \wedge \mathcal{F}$ involving the interaction of the fluctuating fields $\mathbf{u}$ and $\mathbf{b}$, and it is natural first to consider circumstances in which this awkward term may be neglected. There are two distinct circumstances when this neglect (*the first-order smoothing approximation*)[10] would appear to be justified. The order of magnitude of the terms in (7.8) (with $\mathbf{U}_0 = 0$) is indicated in (7.26):

$$\begin{array}{ccccccc} \partial \mathbf{b}/\partial t & = & \nabla \wedge (\mathbf{u} \wedge \mathbf{B}_0) & + & \nabla \wedge \mathcal{F} & + & \eta \nabla^2 \mathbf{b}\,. \\ \mathcal{O}(b_0/t_0) & & \mathcal{O}(B_0 u_0/\ell_0) & & \mathcal{O}(u_0 b_0/\ell_0) & & \mathcal{O}(\eta b_0/\ell_0^2) \end{array} \tag{7.26}$$

Here, as usual, ℓ_0 and t_0 are length- and time-scales characteristic of the $\mathbf{u}$-field, and u_0 and b_0 are, say, the root-mean-square values of $\mathbf{u}$ and $\mathbf{b}$:

$$u_0 = \langle \mathbf{u}^2 \rangle^{1/2}, \quad b_0 = \langle \mathbf{b}^2 \rangle^{1/2}\,. \tag{7.27}$$

Here we must distinguish between two situations:

$$\text{conventional turbulence:} \qquad u_0 t_0/\ell_0 = \mathcal{O}(1)\,, \tag{7.28}$$

$$\text{random waves or `\textit{weak turbulence}':} \qquad u_0 t_0/\ell_0 \ll 1\,. \tag{7.29}$$

If (7.29) is satisfied, then it is immediately clear from (7.26) that $|\nabla \wedge \mathcal{F}| \ll |\partial \mathbf{b}/\partial t|$, and that a good first approximation is given by

$$\partial \mathbf{b}/\partial t = \nabla \wedge (\mathbf{u} \wedge \mathbf{B}_0) + \eta \nabla^2 \mathbf{b}\,, \tag{7.30}$$

an equation first studied (with $\mathbf{B}_0$ uniform and $\mathbf{u}$ random) by Liepmann (1952).

If on the other hand (7.28) is satisfied, then $|\partial \mathbf{b}/\partial t|$ and $|\nabla \wedge \mathcal{F}|$ are of the same order of magnitude, and both are negligible compared with $|\eta \nabla^2 \mathbf{b}|$ provided

$$\mathcal{R}_m = u_0 \ell_0/\eta = \mathcal{O}(\ell_0^2/\eta t_0) \ll 1\,. \tag{7.31}$$

Under this assumption of small (turbulent) magnetic Reynolds number, a legitimate first approximation to (7.26) is

$$0 = \nabla \wedge (\mathbf{u} \wedge \mathbf{B}_0) + \eta \nabla^2 \mathbf{b}\,. \tag{7.32}$$

Although the physical situations described by (7.30) and (7.32) are rather different, both equations say essentially the same thing: the fluctuating field $\mathbf{b}(\mathbf{x}, t)$ is generated by the interaction of $\mathbf{u}$ with the local mean field $\mathbf{B}_0$. In (7.32) this process

[10] Some authors (e.g. Kraichnan 1976b) instead use the term 'quasilinear approximation'.

is instantaneous because of the dominant influence of diffusion, whereas in (7.30) $\mathbf{b}(\mathbf{x}, t)$ can evidently depend on the previous history of $\mathbf{u}(\mathbf{x}, t)$ (i.e. on $\mathbf{u}(\mathbf{x}, t')$ for all $t' \leq t$). It may be anticipated that solutions of (7.30) will approximate to solutions of (7.32) when (7.31) is satisfied. We may therefore focus attention on the more general equation (7.30), bearing in mind that, in application to the turbulent (as opposed to random wave) situation, the study is relevant only if the additional condition (7.31) is satisfied.

7.6 Spectrum tensor of a stationary random vector field

Before considering the consequences of (7.30), we must digress briefly to recall certain basic properties of a random velocity field $\mathbf{u}(\mathbf{x}, t)$ that is statistically homogeneous in $\mathbf{x}$ and stationary in t. We may define (in the sense of generalised functions – see for example Lighthill 1959) the Fourier transform (with $d\mathbf{x} \equiv d^3\mathbf{x}$)

$$\tilde{\mathbf{u}}(\mathbf{k}, \omega) = \frac{1}{(2\pi)^4} \iint \mathbf{u}(\mathbf{x}, t) e^{-i(\mathbf{k}\cdot\mathbf{x}-\omega t)}\, d\mathbf{x}\, dt\,, \tag{7.33}$$

which satisfies the inverse relation

$$\mathbf{u}(\mathbf{x}, t) = \iint \tilde{\mathbf{u}}(\mathbf{k}, \omega) e^{i(\mathbf{k}\cdot\mathbf{x}-\omega t)}\, d\mathbf{k}\, d\omega\,. \tag{7.34}$$

Since $\mathbf{u}$ is real, we have for all $\mathbf{k}, \omega$,

$$\tilde{\mathbf{u}}(-\mathbf{k},\ -\omega) = \tilde{\mathbf{u}}^*(\mathbf{k}, \omega)\,, \tag{7.35}$$

where the star denotes a complex conjugate. Moreover, if $\mathbf{u}$ satisfies $\nabla \cdot \mathbf{u} = 0$, then

$$\mathbf{k} \cdot \tilde{\mathbf{u}}(\mathbf{k}, \omega) = 0 \quad \text{for all } \mathbf{k}\,. \tag{7.36}$$

Now consider the mean quantity

$$\langle \tilde{u}_i^*(\mathbf{k}, \omega) \tilde{u}_j(\mathbf{k}', \omega') \rangle = \frac{1}{(2\pi)^8} \iiiint \langle u_i(\mathbf{x}, t) u_j(\mathbf{x}', t') \rangle e^{i(\mathbf{k}\cdot\mathbf{x}-\mathbf{k}'\cdot\mathbf{x}'-\omega t+\omega' t')}\, d\mathbf{x}\, d\mathbf{x}'\, dt\, dt'. \tag{7.37}$$

If $\mathbf{x}' = \mathbf{x} + \mathbf{r}$ and $t' = t + \tau$, then, under the assumption of homogeneity and stationarity,

$$\langle u_i(\mathbf{x}, t) u_j(\mathbf{x}', t') \rangle = R_{ij}(\mathbf{r}, \tau)\,, \tag{7.38}$$

the correlation tensor of the field $\mathbf{u}$. Using the basic property of the δ-function

$$\iint e^{i(\mathbf{k}-\mathbf{k}')\cdot\mathbf{x}} e^{-i(\omega-\omega')t}\, d\mathbf{x}\, dt = (2\pi)^4 \delta(\mathbf{k} - \mathbf{k}')\, \delta(\omega - \omega')\,, \tag{7.39}$$

(7.37) then takes the form

$$\langle \tilde{u}_i^*(\mathbf{k}, \omega) \tilde{u}_j(\mathbf{k}', \omega') \rangle = \Phi_{ij}(\mathbf{k}, \omega)\, \delta(\mathbf{k} - \mathbf{k}')\, \delta(\omega - \omega')\,, \tag{7.40}$$

where

$$\Phi_{ij}(\mathbf{k},\omega) = \frac{1}{(2\pi)^4}\iint R_{ij}(\mathbf{r},\tau)\mathrm{e}^{-\mathrm{i}(\mathbf{k}\cdot\mathbf{r}-\omega\tau)}\,\mathrm{d}\mathbf{r}\,\mathrm{d}\tau. \tag{7.41}$$

The relation inverse to (7.41) is

$$R_{ij}(\mathbf{r},\ \tau) = \iint \Phi_{ij}(\mathbf{k},\ \omega)\mathrm{e}^{\mathrm{i}(\mathbf{k}\cdot\mathbf{r}-\omega\tau)}\,\mathrm{d}\mathbf{k}\,\mathrm{d}\omega. \tag{7.42}$$

The tensor $\Phi_{ij}(\mathbf{k},\omega)$ is the *spectrum tensor* of the field $\mathbf{u}(\mathbf{x},t)$, and it plays a fundamental role in the subsequent theory. From (7.35), it satisfies the condition of Hermitian symmetry

$$\Phi_{ij}(\mathbf{k},\omega) = \Phi_{ji}(-\mathbf{k},-\omega) = \Phi^{*}_{ji}(\mathbf{k},\omega)\,, \tag{7.43}$$

while, from (7.36), if $\nabla\cdot\mathbf{u} = 0$ then, for all $\mathbf{k}$,

$$k_j\Phi_{ij}(\mathbf{k},\omega) = 0, \quad k_i\Phi_{ij}(\mathbf{k},\omega) = 0\,. \tag{7.44}$$

The energy spectrum function $E(k,\omega)$ is defined by

$$E(k,\omega) \equiv \tfrac{1}{2}\int_{S_k} \Phi_{ii}(\mathbf{k},\omega)\,\mathrm{d}S\ , \tag{7.45}$$

where the integration is over the surface of the sphere S_k of radius k in $\mathbf{k}$-space. Note that

$$\tfrac{1}{2}\langle\mathbf{u}^2\rangle = \tfrac{1}{2}R_{ii}(0,0) = \tfrac{1}{2}\iint \Phi_{ii}(\mathbf{k},\omega)\mathrm{d}\mathbf{k}\,\mathrm{d}\omega = \iint E(k,\omega)\mathrm{d}k\,\mathrm{d}\omega\,, \tag{7.46}$$

where the integral over $k = |\mathbf{k}|$ naturally runs from 0 to ∞. Hence $\rho E(k,\omega)\mathrm{d}k\,\mathrm{d}\omega$ is the contribution to the kinetic energy density from the wave number range $(k, k + \mathrm{d}k)$ and frequency range $(\omega,\ \omega + \mathrm{d}\omega)$. Note that the scalar quantity $\Phi_{ii}(\mathbf{k},\omega)$ is non-negative for all $\mathbf{k},\omega$. (If it could be negative, integration of (7.40) over an infinitesimal neighbourhood of $(\mathbf{k},\omega)$ would give a contradiction.) Hence also

$$E(k,\omega) \geq 0 \quad \text{for all}\ \ k,\omega\,. \tag{7.47}$$

The vorticity field $\boldsymbol{\omega} = \nabla\wedge\mathbf{u}$ evidently has Fourier transform $\tilde{\omega} = i\,k\wedge\overline{\mathbf{u}}$, and spectrum tensor

$$\Omega_{ij}(\mathbf{k},\omega) = \epsilon_{imn}\epsilon_{jpq}k_n k_p\Phi_{nq}(\mathbf{k},\omega)\,. \tag{7.48}$$

In particular, using (7.44),

$$\Omega_{ii}(\mathbf{k},\omega) = k^2\Phi_{ii}(\mathbf{k},\omega)\,, \tag{7.49}$$

and as an immediate consequence

$$\tfrac{1}{2}\langle\boldsymbol{\omega}^2\rangle = \iint k^2 E(k,\omega)\mathrm{d}k\,\mathrm{d}\omega\,. \tag{7.50}$$

By analogy with the definition of $E(k,\omega)$, we define the *helicity spectrum function* $\mathcal{H}(k,\omega)$ by

$$\mathcal{H}(k,\omega) = \mathrm{i}\int_{S_k} \epsilon_{ikl}k_k\Phi_{il}(\mathbf{k},\omega)\,\mathrm{d}S\,, \tag{7.51}$$

so that, with $\boldsymbol{\omega} = \nabla\wedge\mathbf{u}$,

$$\langle\mathbf{u}\cdot\boldsymbol{\omega}\rangle = \mathrm{i}\epsilon_{ikl}\iint k_k\Phi_{il}(\mathbf{k},\omega)\,\mathrm{d}\mathbf{k}\,\mathrm{d}\omega = \iint \mathcal{H}(k,\omega)\,\mathrm{d}k\,\mathrm{d}\omega\,. \tag{7.52}$$

The function $\mathcal{H}(k,\omega)$ is real (by virtue of (7.43)) and is a pseudo-scalar, and so vanishes if the **u**-field is reflectionally symmetric. We have seen however in the previous sections that lack of reflectional symmetry is likely to be of crucial importance in the dynamo context, and it is important therefore to consider situations in which $\mathcal{H}(k,\omega)$ may be non-zero. The mean helicity $\langle\mathbf{u}\cdot\boldsymbol{\omega}\rangle$ is the simplest (although by no means the only) measure of the lack of reflectional symmetry of a random **u**-field.

Unlike $E(k,\omega)$, $\mathcal{H}(k,\omega)$ can be positive or negative. It is however limited in magnitude; in fact from the Schwarz inequality in the form

$$\left|\int_{S_k}\langle\tilde{\mathbf{u}}\cdot\tilde{\boldsymbol{\omega}}^* + \tilde{\mathbf{u}}^*\cdot\tilde{\boldsymbol{\omega}}\rangle\,\mathrm{d}S\right|^2 \le 4\int_{S_k}\langle|\tilde{\mathbf{u}}|^2\rangle\,\mathrm{d}S\int_{S_k}\langle|\tilde{\boldsymbol{\omega}}|^2\rangle\,\mathrm{d}S\,, \tag{7.53}$$

together with (7.46), (7.49) and (7.52), we may deduce that[11]

$$|\mathcal{H}(k,\omega)| \le 2kE(k,\omega) \qquad \text{for all}\ \ k,\omega\,. \tag{7.54}$$

If the **u**-field is statistically isotropic,[12] as well as homogeneous, then the functions $E(k,\omega)$ and $\mathcal{H}(k,\omega)$ are sufficient to completely specify $\Phi_{ij}(\mathbf{k},\omega)$. In fact the most general isotropic form for $\Phi_{ij}(\mathbf{k},\omega)$ consistent with (7.44), (7.45) and (7.51) is

$$\Phi_{ij}(\mathbf{k},\omega) = \frac{E(k,\omega)}{4\pi k^4}(k^2\delta_{ij} - k_ik_j) + \frac{\mathrm{i}\mathcal{H}(k,\omega)}{8\pi k^4}\epsilon_{ijk}k_k\,. \tag{7.55}$$

The assumption of isotropy can lead to dramatic simplifications in the mathematical analysis. Since however turbulence (or a random wave field) that lacks reflectional symmetry can arise in a natural way only in a rotating system in which there is necessarily a preferred direction (the direction of the rotation vector $\boldsymbol{\Omega}$), it is perhaps unrealistic to place too much emphasis on the isotropic situation.

There are however 'unnatural' ways of generating isotropic non-reflectionally

[11] The results (7.47) and (7.54) are particular consequences of the fact that $X_iX_j^*\Phi_{ij}(\mathbf{k},\omega) \ge 0$ for arbitrary complex vectors **X** (Cramer's theorem); in the isotropic case, when $\Phi_{ij}(\mathbf{k},\omega)$ is given by (7.55), choosing **X** real gives (7.47), and choosing $\mathbf{X} = \mathbf{p} + \mathrm{i}\mathbf{q}$, where **p** and **q** are unit orthogonal vectors both orthogonal to **k**, gives (7.54).

[12] See footnote 3.

symmetric turbulence, and it may be useful to describe one such 'thought experiment' if only to fix ideas. Suppose that the fluid is contained in a large sphere whose surface S is perforated by a large number of small holes placed randomly. Suppose that a small right-handed screw propellor is freely mounted at the centre of each hole, and suppose that fluid is injected at high velocity through a random subset of the holes, an equal mass flux then emerging from the complement of this subset. In a neighbourhood of the centre of the sphere the turbulent velocity field that results from the interaction of the incoming swirling jets will be approximately homogeneous and isotropic (there is clearly no preferred direction at the centre.) The turbulence nevertheless certainly lacks reflectional symmetry since each fluid particle entering the sphere follows a right-handed helical path at the start of its trajectory, so that (presumably) $\langle \mathbf{u} \cdot \boldsymbol{\omega} \rangle$ will be positive throughout the sphere. Note that the net angular momentum generated will be zero since the torques exerted on the fluid by propellors at opposite ends of a diameter will tend to cancel, and the cancellation will be complete when the injection statistics are uniform over the surface of the sphere.[13]

If the **u**-field is not isotropic, but is nevertheless statistically axisymmetric about the direction of a unit vector **e**, then the most general form for $\Phi_{ij}(\mathbf{k}, \omega)$ compatible with (7.43) and (7.44) is (with $\mathbf{k} \cdot \mathbf{e} = k\mu = k \cos \theta$)

$$\begin{aligned}\Phi_{ij}(\mathbf{k}, \omega) = \varphi_1(k^2\delta_{ij} - k_ik_i) + \varphi_2(ke_ie_j + k\mu^2\delta_{ij} - \mu k_ie_j - \mu k_je_i) \\ + \mathrm{i}\tilde{\varphi}_3\epsilon_{ijk}k_k + \mathrm{i}\tilde{\varphi}_4\epsilon_{ijk}e_k + \tilde{\varphi}_5(\mathbf{k} \wedge \mathbf{e})_ik_j + \tilde{\varphi}_6(\mathbf{k} \wedge \mathbf{e})_ie_j \\ + \tilde{\varphi}_5^*(\mathbf{k} \wedge \mathbf{e})_jk_i + \tilde{\varphi}_6^*(\mathbf{k} \wedge \mathbf{e})_je_i\,,\end{aligned} \tag{7.56}$$

where

$$\mathrm{i}\tilde{\varphi}_4 + k^2\tilde{\varphi}_5 + \mu k\tilde{\varphi}_6 = 0\,. \tag{7.57}$$

Here, as usual, the star denotes a complex conjugate and the tilde denotes a pseudo-scalar. The functions $\varphi_1, \ldots, \tilde{\varphi}_6$ are functions of k, μ and ω; $\varphi_1, \varphi_2, \tilde{\varphi}_3$ and $\tilde{\varphi}_4$ are real, while $\tilde{\varphi}_5$ and $\tilde{\varphi}_6$ are complex. The energy and helicity spectrum functions defined by (7.45) and (7.51) are related to these functions by

$$E(k, \omega) = \pi \int_0^1 (2k^4\varphi_1 + k^3(1 + \mu^2)\varphi_2)\mathrm{d}\mu\,, \tag{7.58}$$

$$\mathcal{H}(k, \omega) = 4\pi \int_0^1 (k^4\tilde{\varphi}_3 + k^3\mu\,\tilde{\varphi}_4 + k^4(\mu^2 - 1)\,\Im\mathrm{m}\,\tilde{\varphi}_6)\,\mathrm{d}\mu\,. \tag{7.59}$$

It may of course happen that the **u**-field exhibits in its statistical properties more than one preferred direction. For example if both Coriolis forces and buoyancy

[13] We note in anticipation that the counter-rotating propellors of the VKS-experiment as depicted in Figure 9.21 approximate this scenario, although in a statistically axisymmetric geometry.

forces act on the fluid, the rotation vector $\mathbf{\Omega}$ and the gravity vector $\mathbf{g}$ provide independent preferred directions which will influence the statistics of any turbulence present. The general formulæ for Φ_{ij}, E and $\mathcal{H}$ corresponding to (7.56)–(7.59) can be readily obtained, and again (as is to be expected) only that part of Φ_{ij} involving pure scalar functions contributes to E, while only the part involving pseudo-scalar functions contributes to $\mathcal{H}$. We shall meet such situations in later chapters.

7.7 Determination of α_{ij} for a helical wave motion

Since the expansion (7.10) is valid for *any* field distribution $\mathbf{B}_0$ of sufficiently large length-scale, in calculating α_{ij} we may suppose that $\mathbf{B}_0$ is uniform, and therefore time-independent. Restricting attention to incompressible motions for which $\nabla\cdot\mathbf{u} = 0$, (7.30) then becomes

$$\partial\mathbf{b}/\partial t - \eta\nabla^2\mathbf{b} = (\mathbf{B}_0\cdot\nabla)\mathbf{u}\,. \tag{7.60}$$

Before treating the general random field $\mathbf{u}(\mathbf{x}, t)$, it is illuminating to consider first the effect of a single 'helical wave' given by

$$\mathbf{u}(\mathbf{x}, t) = u_0(\sin(kz - \omega t), \cos(kz - \omega t), 0) = \Re\mathrm{e}\,\mathbf{u}_0\,\mathrm{e}^{\mathrm{i}(\mathbf{k}\cdot\mathbf{x}-\omega t)}\,, \tag{7.61}$$

where

$$\mathbf{u}_0 = u_0(-\mathrm{i}, 1, 0), \quad \mathbf{k} = (0, 0, k), \tag{7.62}$$

and where for definiteness we suppose $k > 0,\ \omega > 0$. Note that for this motion

$$\nabla\wedge\mathbf{u} = k\mathbf{u}\,, \quad \mathbf{u}\cdot(\nabla\wedge\mathbf{u}) = ku_0^2 \quad \text{and} \quad \mathrm{i}\mathbf{u}_0\wedge\mathbf{u}_0^* = 2u_0^2(0, 0, 1). \tag{7.63}$$

The helicity density is evidently uniform and positive. With this choice of $\mathbf{u}$, the corresponding periodic solution of (7.60) has the form

$$\mathbf{b}(\mathbf{x}, t) = \Re\mathrm{e}\,\mathbf{b}_0\,\mathrm{e}^{\mathrm{i}(\mathbf{k}\cdot\mathbf{x}-\omega t)}\,, \tag{7.64}$$

where

$$\mathbf{b}_0 = \frac{\mathrm{i}\mathbf{B}_0\cdot\mathbf{k}}{-\mathrm{i}\omega + \eta k^2}\mathbf{u}_0\,. \tag{7.65}$$

Hence

$$\mathbf{b} = \frac{\mathbf{B}_0\cdot\mathbf{k}}{\omega^2 + \eta^2 k^4}(-\omega\mathbf{u} + \eta k^2\mathbf{v})\,, \tag{7.66}$$

where

$$\mathbf{v} = u_0(\cos(kz - \omega t),\ -\sin(kz - \omega t),\ 0) = \Re\mathrm{e}\,\mathrm{i}\,\mathbf{u}_0\,\mathrm{e}^{\mathrm{i}(\mathbf{k}\cdot\mathbf{x}-\omega t)}\,, \tag{7.67}$$

and we can immediately obtain

$$\boldsymbol{\mathcal{E}} = \langle\mathbf{u}\wedge\mathbf{b}\rangle = \frac{\eta(\mathbf{B}_0\cdot\mathbf{k})k^2}{\omega^2 + \eta^2 k^4}\langle\mathbf{u}\wedge\mathbf{v}\rangle = -\frac{\eta u_0^2(\mathbf{B}_0\cdot\mathbf{k})k^2}{\omega^2 + \eta^2 k^4}(0, 0, 1)\,. \tag{7.68}$$

Hence in this case $\mathcal{E}_i = \alpha_{ij}B_{0j}$, where

$$\alpha_{ij} = \alpha^{(3)}\delta_{i3}\delta_{j3}, \qquad \alpha^{(3)} = -\frac{\eta u_0^2 k^3}{\omega^2 + \eta^2 k^4}. \tag{7.69}$$

The α-effect is clearly non-isotropic because the velocity field (7.61) exhibits the preferred direction $(0, 0, 1)$. More significantly, note that $\alpha_{ij} \to 0$ as $\eta \to 0$, i.e. *some diffusion is essential to generate an α-effect*. The role of diffusion is evidently (from (7.66)) to shift the phase of $\mathbf{b}$ relative to that of $\mathbf{u}$, a process that is crucial in producing a non-zero value of $\boldsymbol{\mathcal{E}}$.

Note further that $\mathbf{u} \wedge \mathbf{b}$ is in fact uniform in the above situation, so that $\mathcal{F} = \mathbf{u} \wedge \mathbf{b} - \langle \mathbf{u} \wedge \mathbf{b} \rangle \equiv 0$. This means that the first-order smoothing approximation (in which the $\mathcal{F}$-term in (7.26) is ignored) is exact when the wave field contains only *one* Fourier component like (7.61). If more than one Fourier component is present, then $\mathcal{F}$ is no longer zero. We can however give slightly more precision to the condition for the validity of first-order smoothing. Suppose that the wave spectrum (discrete or continuous) is sharply peaked around a wave number k_0 and a frequency ω_0, and that $u_0 = \langle \mathbf{u}^2 \rangle^{1/2}$. Then from (7.29) and (7.31), the effects of the $\mathcal{F}$-term in (7.26) should be negligible provided that either

$$u_0/\eta k_0 \ll 1 \quad \text{or} \quad u_0 k_0/\omega_0 \ll 1. \tag{7.70}$$

Conversely, if $\eta \ll u_0/k_0$, then first-order smoothing must be regarded as a dubious approximation for all pairs $(\mathbf{k}, \omega)$ strongly represented in the wave spectrum for which $|\omega| \lesssim u_0 k$.

Note finally that the solution (7.64) does not of course satisfy an initial condition $\mathbf{b}(\mathbf{x}, 0) = 0$. If we insist on this condition, we must simply add to (7.64) the transient term

$$\mathbf{b}_1 = -\Re\mathrm{e}\, \mathbf{b}_0\, \mathrm{e}^{\mathrm{i}\mathbf{k}\cdot\mathbf{x}}\, \mathrm{e}^{-\eta k^2 t}. \tag{7.71}$$

This makes an additional contribution to $\boldsymbol{\mathcal{E}}$ which however decays to zero in a time of order $(\eta k^2)^{-1}$. The limit $\eta \to 0$ again poses problems: the influence of initial conditions is 'forgotten' only for $t \gtrsim (\eta k^2)^{-1}$, and the result obtained for $\boldsymbol{\mathcal{E}}$ will depend on the ordering of the limiting processes $\eta \to 0$ and $t \to \infty$ (cf. the problem discussed in §3.11). If we first let $t \to \infty$ (with $\eta > 0$) so that the transient effect disappears, then we obtain the result (7.68). Alternatively, if we first let $\eta \to 0$, then we obtain (with $\mathbf{b}(\mathbf{x}, 0) = 0$)

$$\mathbf{b}(\mathbf{x}, t) = -\omega^{-1} u_0 (\mathbf{B}_0 \cdot \mathbf{k})(\sin(kz - \omega t) - \sin\, kz\,, cos(kz - \omega t) - \cos kz\,,\ 0)\,, \tag{7.72}$$

and so

$$\boldsymbol{\mathcal{E}} = \langle \mathbf{u} \wedge \mathbf{b} \rangle = -\omega^{-1} u_0^2\, (\mathbf{B}_0 \cdot \mathbf{k}) \sin \omega t\, (0, 0, 1)\,, \tag{7.73}$$

and this does not settle down to any steady value as $t \to \infty$.

7.8 Determination of α_{ij} for a random u-field under first-order smoothing

Suppose now that $\mathbf{u}$ is a stationary random function of $\mathbf{x}$ and t with Fourier transform (7.33). The Fourier transform of (7.60) is

$$(-\mathrm{i}\omega + \eta k^2)\tilde{\mathbf{b}} = \mathrm{i}(\mathbf{B}_0 \cdot \mathbf{k})\tilde{\mathbf{u}}\,, \tag{7.74}$$

and we may immediately calculate

$$\boldsymbol{\mathcal{E}} = \langle \mathbf{u} \wedge \mathbf{b} \rangle = \iiiint \frac{\langle \tilde{\mathbf{u}}^*(\mathbf{k},\omega) \wedge \tilde{\mathbf{u}}(\mathbf{k}',\omega')\rangle\, \mathrm{i}\mathbf{B}_0 \cdot \mathbf{k}}{-\mathrm{i}\omega + \eta k^2} \mathrm{e}^{\mathrm{i}(\mathbf{k}-\mathbf{k}')\cdot\mathbf{x} - \mathrm{i}(\omega-\omega')t}\, \mathrm{d}\mathbf{k}\, \mathrm{d}\mathbf{k}'\, \mathrm{d}\omega\, \mathrm{d}\omega'\,. \tag{7.75}$$

Using (7.40) and noting that $\mathrm{i}\epsilon_{ikl}\Phi_{kl}(\mathbf{k},\omega)$ is real by virtue of (7.43), and that the imaginary part of (7.75) must vanish (since $\boldsymbol{\mathcal{E}}$ is real), we obtain $\mathcal{E}_i = \alpha_{ij}B_{0j}$ where

$$\alpha_{ij} = \mathrm{i}\,\eta\, \epsilon_{ikl} \iint (\omega^2 + \eta^2 k^4)^{-1} k^2 k_j \Phi_{kl}(\mathbf{k},\omega)\, \mathrm{d}\mathbf{k}\, \mathrm{d}\omega\,, \tag{7.76}$$

essentially a superposition of contributions like (7.69). Note that if we define $\alpha = \frac{1}{3}\alpha_{ii}$ (consistent with $\alpha_{ij} = \alpha\delta_{ij}$ in the isotropic situation) then from (7.51) and (7.76) we have

$$\alpha = -\tfrac{1}{3}\eta \iint \frac{k^2 \mathcal{H}(k,\omega)}{\omega^2 + \eta^2 k^4} \mathrm{d}k\, \mathrm{d}\omega\,, \tag{7.77}$$

a result that holds irrespective of whether the **u**-field is isotropic or not. It is here that the relation between α and helicity is at its most transparent: α is simply a weighted integral of the helicity spectrum function. As remarked earlier, $\mathcal{H}(k,\omega)$ can take positive or negative values; if however $\mathcal{H}(k,\omega)$ is non-negative for all (k,ω) (and not identically zero) so that $\langle \mathbf{u} \cdot \boldsymbol{\omega} \rangle > 0$, then evidently, from (7.77), $\alpha < 0$; likewise if $\mathcal{H}(k,\omega) \le 0$ for all (k,ω) (but not identically zero), then $\alpha > 0$.

In the case of turbulence, first-order smoothing is valid only if $\eta k^2 \gg |\omega|$ for all $(\mathbf{k},\omega)$ for which a significant contribution is made to the integral (7.76). Hence in this situation the factor $(\omega^2 + \eta^2 k^4)^{-1}$ may be replaced by $\eta^{-2}k^{-4}$, giving

$$\alpha_{ij} \approx \mathrm{i}\eta^{-1} \epsilon_{ikl} \int k^{-2} k_j \Phi_{kl}(\mathbf{k})\, \mathrm{d}\mathbf{k}\,, \tag{7.78}$$

where

$$\Phi_{kl}(\mathbf{k}) = \int \Phi_{kl}(\mathbf{k},\omega)\, \mathrm{d}\omega = \frac{1}{(2\pi)^3} \int R_{kl}(\mathbf{r},\ 0)\, \mathrm{e}^{-\mathrm{i}\mathbf{k}\cdot\mathbf{r}}\, \mathrm{d}\mathbf{r}\,. \tag{7.79}$$

Correspondingly (7.77) becomes

$$\alpha \approx -\frac{1}{3\eta} \int k^{-2} \mathcal{H}(k)\, \mathrm{d}k, \quad \mathcal{H}(k) = \int \mathcal{H}(k,\omega)\, \mathrm{d}\omega\,. \tag{7.80}$$

$\Phi_{kl}(\mathbf{k})$ is the conventional zero-time-delay spectrum tensor of homogeneous turbulence (see e.g. Batchelor 1953). The results (7.78) and (7.80) may be most simply obtained directly from (7.32) (Moffatt 1970).

In the random wave situation, the full expression (7.76) must be retained. Note again the property that if there are no 'zero frequency' waves in the wave spectrum, or more precisely if

$$\Phi_{kl}(\mathbf{k}, \omega) = \mathcal{O}(\omega^2) \quad \text{as } \omega \to 0\,, \tag{7.81}$$

then

$$\alpha_{ij} = \mathcal{O}(\eta) \quad \text{as } \eta \to 0\,. \tag{7.82}$$

If on the other hand $\Phi_{kl}(\mathbf{k},\ 0) \neq 0$ then, since

$$\int_{-\infty}^{\infty} \frac{\mathrm{d}\omega}{\omega^2 + \eta^2 k^4} = \frac{\pi}{\eta k^2}\,, \tag{7.83}$$

we obtain formally from (7.76)

$$\alpha_{ij} \sim \pi\,\mathrm{i}\,\epsilon_{ikl} \int k_j \Phi_{kl}(\mathbf{k}, 0)\,\mathrm{d}\mathbf{k} \quad \text{as } \eta \to 0\,. \tag{7.84}$$

Here however we must bear in mind the limitations of the first-order smoothing approximation. As indicated in the remark following (7.70), this approximation is suspect when η is small and $|\omega| \lesssim u_0 k$; since the asymptotic expression (7.84) is determined entirely by the spectral density at $\omega = 0$, the limiting procedure that yields (7.84) is in fact incompatible with the first-order smoothing approximation. The limiting result (7.84) is therefore of dubious validity.

If the field is isotropic, then from (7.55)

$$\mathrm{i}\epsilon_{ikl}\Phi_{kl}(\mathbf{k}, \omega) = -(4\pi k^4)^{-1} k_i\,\mathcal{H}(k, \omega)\,, \tag{7.85}$$

and so (7.76) becomes simply $\alpha_{ij} = \alpha\delta_{ij}$ where α is given by (7.77).

Under the weaker symmetry condition that the **u**-field is statistically axisymmetric, (7.56) gives

$$\mathrm{i}\epsilon_{ikl}\Phi_{kl}(\mathbf{k},\ \omega) = -2[k_i\,\tilde{\varphi}_3 + e_i\,\tilde{\varphi}_4 + (k^2 e_i - k\,\mu\,k_i)\,\Im\mathrm{m}\,\tilde{\varphi}_5 + (k\,\mu\,e_i - k_i)\,\Im\mathrm{m}\,\tilde{\varphi}_6]\,, \tag{7.86}$$

and (7.76) then reduces to the axisymmetric form

$$\alpha_{ij} = \alpha\delta_{ij} + \alpha_1(\delta_{ij} - 3e_i e_j)\,, \tag{7.87}$$

where $\alpha = \frac{1}{3}\alpha_{ii}$ is still given by (7.77), and

$$\alpha_1 = \tfrac{1}{2}\alpha + \eta \iiint \frac{k^3\mu}{\omega^2 + \eta^2 k^4}\left(k\,\mu\,\tilde{\varphi}_3 + \tilde{\varphi}_4 + k^2(1-\mu^2)\,\Im\mathrm{m}\,\tilde{\varphi}_5\right)\mathrm{d}k\,\mathrm{d}\mu\,\mathrm{d}\omega\,. \tag{7.88}$$

It is relevant to note here that the α-effect was detected in the laboratory in an

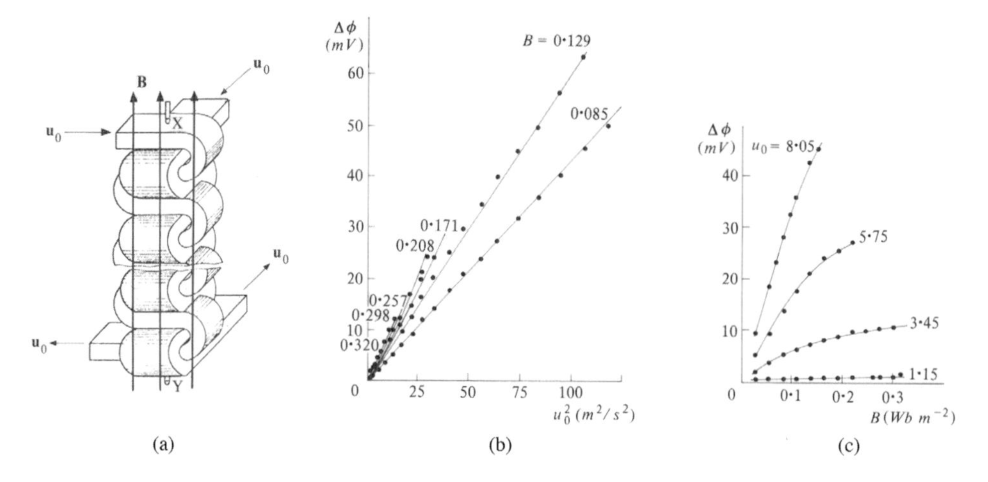

Figure 7.3 Experimental verification of the α-effect. (a) Duct configuration. (b) potential difference $\Delta\phi$ measured between the electrodes X and Y as a function of u_0^2 for various values of the applied field $\mathbf{B}$. (c) $\Delta\phi$ as a function of B for various values of u_0. [Steenbeck et al. 1967.]

early ingenious experiment of Steenbeck et al. (1967). A velocity field having a deliberately contrived negative mean helicity was generated in liquid sodium by driving the liquid through two linked copper ducts (Figure 7.3a). The effective magnetic Reynolds number was low, and the result (7.80) is therefore relevant. The streamline linkage was left-handed, and, assuming maximum helicity, $\langle \mathbf{u} \cdot \omega \rangle \approx -u_0^2/\ell_0$; from (7.80) (or (7.69) with $\omega = 0,\ k = \ell_0^{-1}$), an order-of-magnitude estimate for α is given by

$$\alpha \sim \ell_0 u_0^2/\eta\,, \tag{7.89}$$

where u_0 is the mean velocity through either duct. The total potential drop between electrodes at the points X and Y is then

$$\Delta\phi \sim (n\,\ell_0 u_0^2/\eta)B_0\,, \tag{7.90}$$

where n is the number of duct sections between X and Y ($n = 28$ in the experiment) and B_0 is the field applied (by external coils) parallel to the 'axis' XY. The measured values of $\Delta\phi$ ranged from zero up to 60 millivolts as u_0 and B_0 were varied. Figures 7.3b and 7.3c show the measured variation of $\Delta\phi$ with u_0^2 and with B_0. The linear relation between $\Delta\phi$ and u_0^2 is strikingly verified by these measurements. On the other hand the linear relation between $\Delta\phi$ and B_0 is evidently valid only when B_0 is weak ($\lesssim 0.1\text{Wb m}^{-2}$); reasons for the non-linear dependence of $\Delta\phi$ on B_0 when B_0 is strong must no doubt be sought in dynamical modifications of the (turbulent) velocity distribution in the ducts due to non-negligible Lorentz forces.

7.9 Determination of β_{ijk} under first-order smoothing

To determine β_{ijk}, we suppose now that the field $\mathbf{B}_0(\mathbf{x})$ in the expansion (7.10) is of the form

$$B_{0j}(\mathbf{x}) = x_k \left(\partial B_{0j}/\partial x_k\right), \tag{7.91}$$

where the field gradient $\partial B_{0j}/\partial x_k$ is uniform. Equation (7.30) then becomes

$$\left(\frac{\partial}{\partial t} - \eta \nabla^2\right) b_i = x_k \frac{\partial B_{0j}}{\partial x_k}\frac{\partial u_i}{\partial x_j} - u_j \frac{\partial B_{0i}}{\partial x_j}, \tag{7.92}$$

with Fourier transform

$$\left(-\mathrm{i}\omega + \eta k^2\right)\tilde{b}_i = -\frac{\partial}{\partial k_k}\left(k_j \tilde{u}_i\right)\frac{\partial B_{0j}}{\partial x_k} - \tilde{u}_j \frac{\partial B_{0i}}{\partial x_j}. \tag{7.93}$$

Construction of $\langle \mathbf{u}\wedge\mathbf{b}\rangle_i$ now leads to an expression of the form $\beta_{ijk}\partial B_{0j}/\partial x_k$, where (after some manipulation)

$$\beta_{ijk} = \Re\mathrm{e}\, \epsilon_{iml} \iint \frac{\mathrm{i}\omega + \eta k^2}{\omega^2 + \eta^2 k^4}\left\{\frac{\partial}{\partial k_k} k_j \Phi_{lm}(\mathbf{k},\omega) + \Phi_{lk}(\mathbf{k},\omega)\delta_{mj}\right\} \mathrm{d}\mathbf{k}\, \mathrm{d}\omega. \tag{7.94}$$

Note the appearance in this expression of the gradient in $\mathbf{k}$-space of the spectrum tensor. Again in the turbulence situation (7.94) must be replaced by

$$\beta_{ijk} \approx \Re\mathrm{e}\, \epsilon_{iml} \int \eta^{-1} k^{-2}\left\{\frac{\partial}{\partial k_k} k_j \Phi_{lm}(\mathbf{k}) + \Phi_{lk}(\mathbf{k})\delta_{mj}\right\} \mathrm{d}\mathbf{k}. \tag{7.95}$$

Now $\epsilon_{iml}\Phi_{lm}(\mathbf{k})$ is pure imaginary (by virtue of the Hermitian symmetry of Φ_{lm}), and so (7.95) reduces to

$$\beta_{ijk} \approx \Re\mathrm{e}\, \epsilon_{ijl} \int \eta^{-1} k^{-2} \Phi_{lk}(\mathbf{k}) \mathrm{d}\mathbf{k}. \tag{7.96}$$

In the case of an isotropic $\mathbf{u}$-field, with $\Phi_{ij}(\mathbf{k},\omega)$ given by (7.55), it is again only the second term under the integral (7.94) that makes a non-zero contribution, and we find $\beta_{ijk} = \beta\epsilon_{ijk}$ where

$$\beta = \tfrac{1}{6}\epsilon_{ijk}\beta_{ijk} = \tfrac{2}{3}\eta \iint \frac{k^2 E(k,\omega)}{\omega^2 + \eta^2 k^4} \mathrm{d}k\, \mathrm{d}\omega, \tag{7.97}$$

the corresponding expression in the turbulence limit being

$$\beta \approx \tfrac{2}{3}\eta^{-1}\int_0^\infty k^{-2} E(k) \mathrm{d}k. \tag{7.98}$$

Similarly, for the case of rotational invariance about a direction $\mathbf{e}$, substitution of (7.56) in (7.95) leads to explicit expressions for the coefficients in the axisymmetric form (7.24) of β_{ijk}. It is tedious to calculate these coefficients and the expressions will not be given here; it is enough to note that the scalar coefficients $\beta_0, \ldots, \beta_3$

emerge as linear functionals of $\varphi_1(k,\mu,\omega)$ and $\varphi_2(k,\mu,\omega)$ while the pseudo-scalar coefficients $\tilde{\beta}_0,\ldots,\tilde{\beta}_3$ emerge (like α and α_1 in (7.77) and (7.88)) as linear functionals of $\tilde{\varphi}_3,\ \ldots,\ \tilde{\varphi}_6$; and that, as commented earlier, in any circumstance in which $\tilde{\beta}_0,\ldots,\tilde{\beta}_3$ are non-zero, α and α_1 are generally non-zero also.

7.10 Lagrangian approach to the weak diffusion limit

For a turbulent velocity field $\mathbf{u}(\mathbf{x},t)$ with $u_0t_0/\ell_0 = \mathcal{O}(1)$, and in the weak diffusion limit $\mathcal{R}_\mathrm{m} = u_0\ell_0/\eta \gg 1$, the first-order smoothing approach described in the previous sections is certainly not applicable. Even in the random-wave situation ($u_0t_0/\ell_0 \ll 1$), we have seen that first-order smoothing may break down in the limit $\eta \to 0$ if the wave spectral density at $\omega = 0$ is non-zero. An alternative approach that retains the influence of the interaction term $\nabla \wedge \mathcal{F}$ in (7.8) is therefore desirable. The following approach (Parker 1971a, Moffatt 1974) is analogous to the traditional treatment of turbulent diffusion of a passive scalar field in the limit of vanishing molecular diffusivity (Taylor 1921).

The starting point is the Lagrangian solution of the induction equation, which is exact in the limit $\eta = 0$, viz.

$$B_i(\mathbf{x},t) = B_j(\mathbf{a},0)\partial x_i/\partial a_j\,, \tag{7.99}$$

in the Lagrangian notation of §2.6. Hence we have immediately

$$\mathcal{E}_i(\mathbf{x},t) = \langle \mathbf{u}\wedge\mathbf{b}\rangle_i = \langle \mathbf{u}\wedge\mathbf{B}\rangle_i = \epsilon_{ijk}\langle u_j^L(\mathbf{a},t)B_l(a,0)\partial x_k/\partial a_l\rangle\,. \tag{7.100}$$

7.10.1 Evaluation of α_{ij}

As in §7.8, we may most simply obtain an expression for α_{ij} on the assumption that $\mathbf{B}_0$ is uniform (and therefore constant). If $\mathbf{b}(\mathbf{x},0) = 0$, then $\mathbf{B}(\mathbf{a},0) = \mathbf{B}_0$, and (7.100) then has the expected form $\mathcal{E}_i = \alpha_{il}B_{0l}$, but now with α_{il} a function of t:

$$\alpha_{il}(t) = \epsilon_{ijk}\langle u_j^L(\mathbf{a},t)\partial x_k(\mathbf{a},t)/\partial a_l\rangle\,. \tag{7.101}$$

Now the displacement of a fluid particle is simply the time integral of its Lagrangian velocity, i.e.

$$\mathbf{x}(\mathbf{a},t) - \mathbf{a} = \int_0^t \mathbf{u}^L(\mathbf{a},\tau)\mathrm{d}\tau\,. \tag{7.102}$$

Hence, since $\langle \mathbf{u}^L(\mathbf{a},t)\rangle = 0$, (7.101) becomes

$$\alpha_{il}(t) = \epsilon_{ijk}\int_0^t \left\langle u_j^L(\mathbf{a},t)\partial u_k^L(\mathbf{a},\tau)/\partial a_l\right\rangle \mathrm{d}\tau\,. \tag{7.103}$$

The time-dependence here is of course associated with the imposition of the initial condition $\mathbf{b}(\mathbf{x}, 0) = 0$ which trivially implies that $\alpha_{il} = 0$ when $t = 0$. When $t \gg t_1$, where t_1 is a typical turbulence correlation time, one would expect the influence of initial conditions to be 'forgotten'; equivalently one would expect $\alpha_{il}(t)$ to settle down asymptotically to a constant value given by

$$\alpha_{il} = \epsilon_{ijk} \int_0^\infty \left\langle u_j^L(\mathbf{a}, t) \partial u_k^L(\mathbf{a}, \tau)/\partial a_l \right\rangle \mathrm{d}\tau \,. \tag{7.104}$$

There is however some doubt concerning the convergence of the integral (7.103) as $t \to \infty$ for a general stationary random field of turbulence, and the delicate question of whether the influence of initial conditions is ever forgotten (when $\eta = 0$) is still to some extent unanswered.[14]

Some light may be thrown on this question through comparison of (7.103) with the expression for the diffusion tensor of a passive scalar field, viz. (Taylor 1921)

$$D_{ij}(t) = \int_0^t \left\langle u_i^L(\mathbf{a}, t) u_j^L(\mathbf{a}, \tau) \right\rangle \mathrm{d}\tau \,. \tag{7.105}$$

For a statistically stationary field of turbulence, the integrand here (the Lagrangian correlation tensor) is a function of the time difference $t - \tau$ only:

$$\left\langle u_i^L(\mathbf{a}, t) u_j^L(\mathbf{a}, \tau) \right\rangle = R_{ij}^{(L)}(t - \tau) \,, \tag{7.106}$$

and for $t \gg t_1$, provided simply that

$$t^{1+\mu} R_{ij}^{(L)}(t) \to 0 \quad \text{as } t \to \infty \quad \text{for some } \mu > 0 \,, \tag{7.107}$$

(7.105) gives

$$D_{ij} \sim \int_0^\infty R_{ij}^{(L)}(t) \mathrm{d}t \,. \tag{7.108}$$

The condition (7.107) is a very mild requirement on the statistics of the turbulence.

There is however a crucial difference between (7.105) and (7.103) in that the latter contains the novel type of derivative

$$\partial u_k^L / \partial a_l = (\partial u_k / \partial x_n)(\partial x_n / \partial a_l) \,. \tag{7.109}$$

Although $\partial u_k / \partial x_n$ is statistically stationary in time, $\partial x_n / \partial a_i$ in general is not, since any two initially adjacent particles $(\mathbf{a}, \ \mathbf{a} + \delta\mathbf{a})$ tend to wander further and further apart – in fact $|\delta\mathbf{x}|/|\delta\mathbf{a}| \sim t^{1/2}$ as $t \to \infty$; it follows that $\partial u_k^L / \partial a_i$ is not statistically stationary in time in general, and so the integrand in (7.103) depends on t and τ independently and not merely on the difference $t - \tau$.

[14] Computer simulations were carried out by Kraichnan (1976a) with the aim of evaluating the integrals (7.115) and (7.116) below for the case of a statistically isotropic field with Gaussian statistics. The results show that the integrals for $\alpha(t)$ and $\beta(t)$ do in general converge as $t \to \infty$ to values of order u_0 and $u_0 \ell_0$ respectively; whether this is true for non-Gaussian statistics is not yet clear.

As in the discussion following (7.84), it is apparently the 'zero frequency' ingredients of the velocity field that are responsible for the possible divergence of (7.103). In any time-periodic motion (with zero mean), any two particles that are initially adjacent do not drift apart but remain permanently adjacent; it is spectral contributions in the neighbourhood of $\omega = 0$ which are responsible for the relative dispersion of particles in turbulent flow, and it is these same contributions that make it hard to justify the step between (7.103) and (7.104).

7.10.2 Evaluation of β_{ijk}

Suppose now that the mean field gradient $\partial B_{0i}/\partial x_l$ is uniform at time $t = 0$. From (7.11), we have

$$\frac{\partial}{\partial t}\frac{\partial B_{0i}}{\partial x_p} = \epsilon_{ijk}\frac{\partial}{\partial x_j}\left(\alpha_{kl}\frac{\partial B_{0l}}{\partial x_p} + \beta_{klm}\frac{\partial^2 B_{0l}}{\partial x_m \partial x_p} + \cdots\right) + \eta\nabla^2\frac{\partial B_{0i}}{\partial x_p}\,, \tag{7.110}$$

so that (with α_{kl}, β_{klm} uniform in space), $\partial B_{0i}/\partial x_p$ then remains uniform for $t > 0$. We may therefore integrate (7.11) to give

$$B_{0i}(\mathbf{x}, t) = B_{0i}(\mathbf{x}, 0) + \epsilon_{ijk}\frac{\partial B_{0l}}{\partial x_j}\int_0^t \alpha_{kl}(\tau)\mathrm{d}\tau\,. \tag{7.111}$$

From (7.100), we now obtain

$$\mathcal{E}_i(\mathbf{x}, t) = \epsilon_{ijk}\left\langle u_j^L(\mathbf{a}, t)\partial x_k/\partial a_l[B_{0l}(\mathbf{x}, 0) - (\mathbf{x} - \mathbf{a})_n\partial B_{0l}/\partial x_n]\right\rangle\,, \tag{7.112}$$

and, using (7.111), this is of the form

$$\mathcal{E}_i(\mathbf{x}, t) = \alpha_{il}(t)B_{0l}(\mathbf{x},\ t) + \beta_{ilm}(t)\partial B_{0l}/\partial x_n\,, \tag{7.113}$$

where $\alpha_{il}(t)$ is as given by (7.103), and

$$\begin{aligned}\beta_{ilm}(t) &= -\epsilon_{ijl}D_{jm}(t) - \epsilon_{ijk}\int_0^t\int_0^t\left\langle u_j^L(\mathbf{a}, t)\frac{\partial u_k^L(\mathbf{a}, \tau)}{\partial a_l}u_m^L(\mathbf{a}, \tau')\right\rangle \mathrm{d}\tau\,\mathrm{d}\tau' \\ &\quad + \epsilon_{jkm}\int_0^t \alpha_{ij}(t)\alpha_{kl}(\tau)\,d\tau,\end{aligned} \tag{7.114}$$

where $D_{jm}(t)$ is given by (7.105). This expression now involves the double integral of a triple Lagrangian correlation (whose convergence as $t \to \infty$ is open to the same doubts as expressed for the integral (7.103)). Moreover in the case of turbulence that lacks reflectional symmetry for which $\alpha_{ij}(t) \neq 0$, if $\alpha_{ij}(t)$ tends to a non-zero constant value as $t \to \infty$, then the final term of (7.114) is certainly unbounded as $t \to \infty$.[15] This result is of course due to the total neglect of molecular diffusivity

[15] The fact that the computer simulations of Kraichnan (1976a) give a finite value for $\beta(t) = \frac{1}{6}\epsilon_{ijk}\beta_{ijk}(t)$ as $t \to \infty$ (see footnote 14) implies that this divergence must be compensated by simultaneous divergence of the

effects; it is possible that inclusion of weak diffusion effects (i.e. small but non-zero η) will guarantee the convergence of $\alpha_{ij}(t)$ and $\beta_{ijk}(t)$ to constant values as $t \to \infty$. An alternative approach to the zero diffusivity limit will be described in §7.12 below.

7.10.3 *The isotropic situation*

From (7.103)

$$\alpha(t) = \tfrac{1}{3}\alpha_{ii}(t) = -\tfrac{1}{3}\int_0^t \left\langle \mathbf{u}^L(\mathbf{a},t)\cdot\nabla_{\mathbf{a}}\wedge\mathbf{u}^L(\mathbf{a},\tau)\right\rangle \mathrm{d}\tau\,, \tag{7.115}$$

and if $\mathbf{u}(\mathbf{x},t)$ is isotropic then $\alpha_{ij} = \alpha\delta_{ij}$. The operator $\nabla_{\mathbf{a}}$ indicates differentiation with respect to $\mathbf{a}$. The integrand here contains a type of Lagrangian helicity correlation; note again the appearance of the minus sign in (7.115).

Similarly in the isotropic case, $\beta_{ijk} = \beta(t)\epsilon_{ijk}$ where, from (7.114),

$$\beta(t) = \tfrac{1}{6}\epsilon_{ijk}\beta_{ijk} = \tfrac{1}{3}\int_0^t \left\langle \mathbf{u}^L(t)\cdot\mathbf{u}^L(\tau)\right\rangle \mathrm{d}\tau + \int_0^t \alpha(t)\alpha(\tau)\mathrm{d}\tau$$
$$+\tfrac{1}{6}\int_0^t\!\!\int_0^t \left\langle \mathbf{u}^L(t)\cdot\mathbf{u}^L(\tau')\nabla_{\mathbf{a}}\cdot\mathbf{u}^L(\tau) - \mathbf{u}^L(t)\cdot\nabla_{\mathbf{a}}\mathbf{u}^L(\tau)\cdot\mathbf{u}^L(\tau')\right\rangle \mathrm{d}\tau\,\mathrm{d}\tau'\,, \tag{7.116}$$

where the dependence of $\mathbf{u}^L$ on $\mathbf{a}$ is understood throughout. The first term here is the effective turbulent diffusivity for a scalar field, and the second and third terms describe effects that are exclusively associated with the vector character of $\mathbf{B}$. The structure of the second term, involving the product of values of α at different instants of time, suggests that *fluctuations* in helicity may have an important effect on the effective magnetic diffusivity. This suggestion, advanced by Kraichnan (1976b), will be examined further in the following section.

7.11 Effect of helicity fluctuations on effective turbulent diffusivity

When $\alpha_{kl} = \alpha\delta_{kl}$ and $\beta_{klm} = \beta\epsilon_{klm}$, (7.11) takes the form

$$\partial\mathbf{B}_0/\partial t = \nabla\wedge(\alpha\mathbf{B}_0) + \eta_1\nabla^2\mathbf{B}_0\,, \tag{7.117}$$

where $\eta_1 = \eta + \beta$, and β is assumed uniform. Let us now (following Kraichnan 1976b) consider the effect of spatial and temporal *fluctuations* in α on scales ℓ_α, t_α

second term of (7.114) involving the double integral. This fortuitous occurrence can hardly be of general validity, given the very different structure of the two terms, and may well be associated with the particular form of Gaussian statistics adopted by Kraichnan in the numerical specification of the velocity field. It may be noted that if the second and third terms on the right of (7.116) *exactly* compensate each other, then $\beta(t) = \frac{1}{3}D_{ii}(t)$, i.e. the magnetic turbulent diffusivity is equal to the scalar turbulent diffusivity. This was claimed as an exact result by Parker (1971b); however, the results of Kraichnan (1976a,b) indicate that although $\beta(t)$ and $D(t) = \frac{1}{3}D_{ii}(t)$ may be of the same order of magnitude as $t \to \infty$, they are *not* in general identically equal.

satisfying

$$\ell_0 \ll \ell_\alpha \ll L,\ t_0 \ll t_\alpha \ll T\,. \tag{7.118}$$

In order to handle such a situation, we need to define a double averaging process over scales a_1 and a_2 satisfying

$$\ell_0 \ll a_1 \ll \ell_\alpha \ll a_2 \ll L\,. \tag{7.119}$$

Preliminary averaging over the scale a_1 yields (7.117) as described in the foregoing sections. Now we treat $\alpha(\mathbf{x}, t)$ as a random function and examine the effect of averaging (7.117) over the scale a_2. (The process may also be interpreted in terms of an 'ensemble of ensembles': in each sub-ensemble α is constant, but it varies randomly from one sub-ensemble to another.) We shall use the notation $\langle\!\langle \cdots \rangle\!\rangle$ to denote averaging over the scale a_2 of quantities already averaged over the scale a_1. We shall suppose further that the **u**-field is globally reflectionally symmetric so that in particular $\langle\!\langle \alpha \rangle\!\rangle = 0$.

Spatial fluctuations in α will presumably occur in the presence of corresponding fluctuations in background helicity $\langle \mathbf{u} \cdot \boldsymbol{\omega} \rangle$. It is easy to conceive of a kinematically possible random velocity field exhibiting such fluctuations. A pair of vortex rings linked as in Figure 2.1a has an associated positive helicity; reversing the sign of one of the arrows gives a similar 'flow element' with negative helicity. We can imagine such elements distributed at random in space in such a way as to give a velocity field that is homogeneous and isotropic, and reflectionally symmetric if elements of opposite parity occur with equal probability. Clustering of right-handed and left-handed elements will however give spatial fluctuations in helicity on the scale of the clusters.

From a dynamic point of view, there may seem little justification for consideration of somewhat arbitrary models of this kind. The reason for doing so is the following. When α is constant, (7.117) has solutions that grow exponentially when the length-scale is sufficiently large (see §9.2 for details). In turbulence that is reflectionally symmetric, α is zero, and there then seems no possibility of growth of $\mathbf{B}_0$ according to (7.117). We have however encountered grave difficulties in calculating β (and so η_1) in any circumstances which are not covered by the simple first-order smoothing approximation, and it is difficult to exclude the possibility that the effective diffusivity may even be negative in some circumstances. The investigation of Kraichnan (1976b) was motivated by a desire to shed light on this question.

Let us then (following the same procedure as applied to the induction equation in §7.1) split (7.117) into mean and fluctuating parts. Defining

$$\mathbf{B}_0 = \langle\!\langle \mathbf{B} \rangle\!\rangle + \mathbf{b}_1(\mathbf{x}, t) \quad \text{with } \langle\!\langle \mathbf{b}_1 \rangle\!\rangle = 0\,, \tag{7.120}$$

we obtain

$$\partial\langle\langle \mathbf{B} \rangle\rangle/\partial t = \nabla \wedge \langle\langle \alpha \mathbf{b}_1 \rangle\rangle + \eta_1 \nabla^2 \langle\langle \mathbf{B} \rangle\rangle \,, \tag{7.121}$$

and

$$\partial \mathbf{b}_1/\partial t = -\langle\langle \mathbf{B} \rangle\rangle \wedge \nabla\alpha + \alpha\nabla \wedge \langle\langle \mathbf{B} \rangle\rangle + \nabla \wedge \mathcal{F}_1 + \eta_1 \nabla^2 \mathbf{b}_1 , \tag{7.122}$$

where

$$\mathcal{F}_1 = \alpha \mathbf{b}_1 - \langle\langle \alpha \mathbf{b}_1 \rangle\rangle \,. \tag{7.123}$$

Let us now apply the first-order smoothing method to (7.122). The term $\nabla \wedge \mathcal{F}_1$ is negligible provided either

$$\alpha_0 t_\alpha/\ell_\alpha \ll 1 \quad \text{or} \quad \alpha_0 \ell_\alpha/\eta_1 \ll 1 \,, \tag{7.124}$$

where $\alpha_0^2 = \langle\langle \alpha^2 \rangle\rangle$, the mean square of the fluctuation field $\alpha(\mathbf{x}, t)$. The Fourier transform of (7.122) (treating $\langle\langle \mathbf{B} \rangle\rangle$ and $\nabla \wedge \langle\langle \mathbf{B} \rangle\rangle$ as uniform) is then

$$(-\mathrm{i}\omega + \eta_1 k^2)\tilde{\mathbf{b}}_1 = -\mathrm{i}\langle\langle \mathbf{B} \rangle\rangle \wedge \mathbf{k}\,\tilde{\alpha} + \tilde{\alpha}\nabla \wedge \langle\langle \mathbf{B} \rangle\rangle \,, \tag{7.125}$$

from which we may readily obtain $\langle\langle \alpha \mathbf{b}_1 \rangle\rangle$ in the form

$$\langle\langle \alpha \mathbf{b}_1 \rangle\rangle = -\langle\langle \mathbf{B} \rangle\rangle \wedge \mathbf{Y} + X\,\nabla \wedge \langle\langle \mathbf{B} \rangle\rangle \tag{7.126}$$

where

$$X = \iint \frac{\eta_1 k^2 \Phi_\alpha(\mathbf{k}, \omega)}{\omega^2 + \eta_1^2 k^4} \mathrm{d}\mathbf{k}\, \mathrm{d}\omega \,, \quad \mathbf{Y} = -\iint \frac{\omega \mathbf{k} \Phi_\alpha(\mathbf{k}, \omega)}{\omega^2 + \eta_1^2 k^4} \mathrm{d}\mathbf{k}\, \mathrm{d}\omega \,, \tag{7.127}$$

where $\Phi_\alpha(\mathbf{k}, \omega)$ is the spectrum function of the field $\alpha(\mathbf{x}, t)$. Substitution of (7.127) in (7.121) now gives

$$\frac{\partial}{\partial t}\langle\langle \mathbf{B} \rangle\rangle = \nabla \wedge (\mathbf{Y} \wedge \langle\langle \mathbf{B} \rangle\rangle) + (\eta_1 - X)\nabla^2 \langle\langle \mathbf{B} \rangle\rangle \,. \tag{7.128}$$

The term involving $\mathbf{Y}$ here is not of great interest: it implies a uniform effective convection velocity $\mathbf{Y}$ of the field $\langle\langle \mathbf{B} \rangle\rangle$ relative to the fluid. If the α-field is statistically isotropic, then of course $\mathbf{Y} = 0$, since there is then no preferred direction.[16]

The term involving X is of greater potential interest. It is evident from (7.127) that $X > 0$, so that the helicity fluctuations do in fact make a negative contribution to the new effective diffusivity $\eta_2 = \eta_1 - X$. Let us estimate X when ℓ_α is large enough for the following inequalities to be satisfied:

$$\varepsilon_1 = \alpha_0 t_\alpha/\ell_\alpha \ll 1, \quad \varepsilon_2 = \eta_1 t_\alpha/\ell_\alpha^2 \ll 1 \,. \tag{7.129}$$

[16] Kraichnan (1976b) obtains only the X-term in (7.126) by (in effect) restricting attention to α-fields which, though possibly anisotropic, do have a particular statistical property that makes the integral (7.127b) vanish.

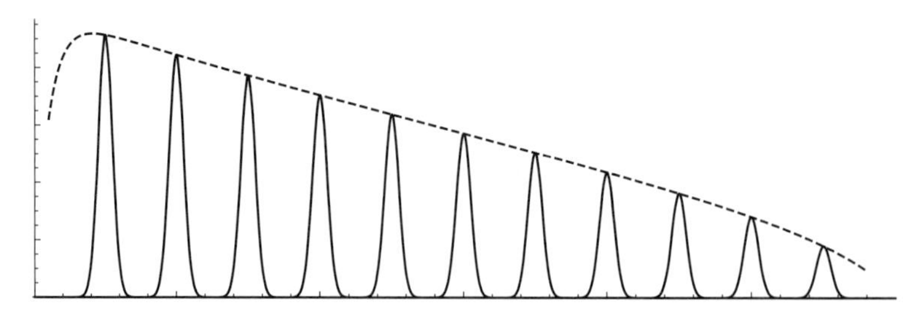

Figure 7.4 Sketch indicating the replacement of a continuous spectrum $E(k)$ (dashed) by a discrete spectrum centred on wave numbers $k_1, k_2, \ldots, k_N$ with $k_1 \gg k_2 \gg \cdots \gg k_N$ (the scale is logarithmic).

The first justifies the use of first-order smoothing. The second allows asymptotic evaluation of (7.127a) (cf. the process leading to (7.84)) in the form

$$X \sim \pi \int \Phi_\alpha(\mathbf{k}, 0)\mathrm{d}k = \mathcal{O}(\alpha_0^2 t_\alpha) \,. \tag{7.130}$$

Hence

$$X/\eta_1 = \mathcal{O}(\alpha_0^2 t_\alpha/\eta_1) = \mathcal{O}(\varepsilon_1^2/\varepsilon_2) \,, \tag{7.131}$$

and so apparently X *can* become of the same order as η_1 or greater provided $\varepsilon_1^2 \gtrsim \varepsilon_2$, or equivalently provided t_α (as well as ℓ_α) is sufficiently large.

The above argument, resting as it does on order-of-magnitude estimates, cannot be regarded as conclusive, but it is certainly suggestive, and the double ensemble technique merits further close study. As indicated in footnote 8, a negative diffusivity $\eta_2 = \eta_1 - X$ in (7.128) yields an ill-posed problem, with sensitive dependence on initial conditions; the two-scale analysis is no longer valid in such circumstances.

7.12 Renormalisation approach to the zero-diffusivity limit

The idea of repeated averaging over successively larger scales underlies the 're-normalisation' technique that has been developed with great success in the context of phase transitions (see, for example, Kadanoff 2000). In our present context, the spontaneous appearance of a large-scale magnetic field in a highly conducting medium in turbulent motion may be regarded as a type of phase transition, occurring when the parameter $\mathcal{R}_m$ increases through a critical value. This approach was first outlined in §11 of Moffatt (1983).

For simplicity, we assume that the turbulence is homogeneous and isotropic (but not reflectionally symmetric). For the purpose of successive averaging, we imagine the energy spectrum $E(k)$ of the turbulence to be discretised in wave-number space as indicated schematically in Figure 7.4; here, the energy level can be adjusted to the correct value by simply raising the level of the 'spikes' by an appropriate factor (while maintaining the ratios). We imagine the helicity spectrum $\mathcal{H}(k)$ (which must satisfy the realisability condition $|\mathcal{H}(k)| \le 2kE(k)$) to be similarly discretised. We shall use the notation $\langle \cdots \rangle_n$ $(n = 1, 2, \ldots)$ to denote averaging over the scale $l_n = k_n^{-1}$. Then $\mathbf{U}_n \equiv \langle \mathbf{u} \rangle_n$ is by definition uniform over any fluid volume of scale l_n; within any such volume, the fluctuation $\mathbf{u}_n \equiv \mathbf{u} - \mathbf{U}_n$ satisfies $\langle \mathbf{u}_n \rangle_n = 0$, and we assume that the random field $\mathbf{u}_n$ is also statistically homogeneous and isotropic. A strong assumption of this kind is needed to justify the steps that follow.

We may suppose that k_1 is chosen large enough for the magnetic Reynolds number $\mathcal{R}_{\mathrm{m}1} = u_1 l_1/\eta$ based on the scale l_1 and the characteristic velocity u_1 at that scale to satisfy[17] $\mathcal{R}_{\mathrm{m}1} \lesssim 1$. In these circumstance, we may use first-order smoothing to obtain an equation for the magnetic field $\mathbf{B}_1 = \langle \mathbf{B} \rangle_1$ in the form

$$\frac{\partial \mathbf{B}_1}{\partial t} = \nabla \wedge (\mathbf{U}_1 \wedge \mathbf{B}_1) + \alpha_1 \nabla \wedge \mathbf{B}_1 + (\eta + \beta_1)\nabla^2 \mathbf{B}_1 \,, \tag{7.132}$$

where, from (7.80) and (7.98)

$$\alpha_1 = -\frac{1}{3\eta}\int_{\Delta k_1} k^{-2}\mathcal{H}(k)\,\mathrm{d}k, \quad \beta_1 = \frac{2}{3\eta}\int_{\Delta k_1} k^{-2}E(k)\,\mathrm{d}k \,. \tag{7.133}$$

Now, treating α_1 and β_1 as constants, we take a 'second bite' of the energy spectrum $E(k_2)\delta k_2$, and apply first-order smoothing to (7.132). This gives increments $\delta\alpha_1$ and $\delta\beta_1$ to α and β, given (with $n = 1$) by

$$\begin{aligned}\delta\alpha_n \equiv \alpha_{n+1} - \alpha_n &= -\tfrac{1}{3}\int_{\Delta k_{n+1}} \frac{2\alpha_n E(k) + (\eta + \beta_n)\mathcal{H}(k)}{k^2(\eta+\beta_n)^2 - \alpha_n^2}\,\mathrm{d}k\,,\\ \delta\beta_n \equiv \beta_{n+1} - \beta_n &= +\tfrac{1}{3}\int_{\Delta k_{n+1}} \frac{2(\eta+\beta_n)E(k) + \alpha_n k^{-2}\mathcal{H}(k)}{k^2(\eta+\beta_n)^2 - \alpha_n^2}\,\mathrm{d}k\,,\end{aligned} \tag{7.134}$$

and a new mean-field equation now averaged over the scale $l_2 = k_2^{-1}$,

$$\frac{\partial \mathbf{B}_2}{\partial t} = \nabla \wedge (\mathbf{U}_2 \wedge \mathbf{B}_2) + \alpha_2 \nabla \wedge \mathbf{B}_2 + (\eta + \beta_2)\nabla^2 \mathbf{B}_2 \,. \tag{7.135}$$

Noting the similarity between (7.132) and (7.135), it is evident that we are now into a 'renormalisation' process that can be iterated for $n = 2, 3, 4, \ldots$ until the whole spectrum is covered. In this way, we arrive at differential equations for $\alpha(k)$ and $\beta(k)$ produced by turbulence at all wave numbers greater than k, i.e. at all scales less than k^{-1}; going to the limit in (7.134), and noting that k decreases as n

[17] Strictly we should require that $\mathcal{R}_{\mathrm{m}1} \ll 1$; however first-order smoothing is generally found to be a reasonable approximation when $\mathcal{R}_{\mathrm{m}}$ is $\mathcal{O}(1)$, even up to $\mathcal{R}_{\mathrm{m}} \approx 10$.

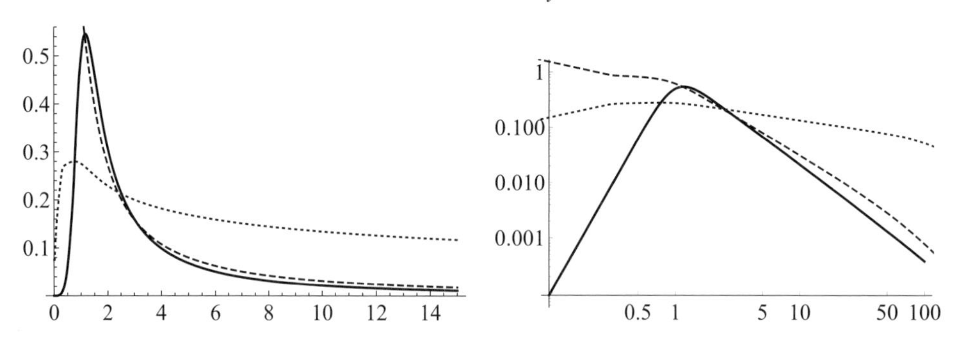

Figure 7.5 The result of integrating (7.136) with $E(k) = k^4(1 + k^{17/3})^{-1}$ and $\mathcal{H}(k) = kE(k)$: (a) $E(k)$ (solid), $-\alpha(k)$ (dotted), $\beta(k)$ (dashed); (b) the same on logarithmic scale for $0 < k < 10^3$.

increases, these equations are

$$\frac{\mathrm{d}\alpha}{\mathrm{d}k} = \frac{2\alpha E(k) + (\eta+\beta)\mathcal{H}(k)}{3\left[k^2(\eta+\beta)^2 - \alpha^2\right]}, \quad \frac{\mathrm{d}\beta}{\mathrm{d}k} = -\frac{2(\eta+\beta)E(k) + \alpha k^{-2}\mathcal{H}(k)}{3\left[k^2(\eta+\beta)^2 - \alpha^2\right]}. \tag{7.136}$$

For given $E(k)$ and $\mathcal{H}(k)$, these equations may be integrated inwards from $k = \infty$ where $\alpha(\infty) = \beta(\infty) = 0$.

By way of example, the result of this integration is shown in Figure 7.5 (natural scaling on the left, log-log plot on the right) for the non-dimensionalised choice $\eta = 10^{-3}$ (equivalent to $\mathcal{R}_{\mathrm{m}} = 10^3$) and with

$$E(k) = k^4\left(1 + k^{17/3}\right)^{-1}, \quad \mathcal{H}(k) = kE(k), \tag{7.137}$$

thus trivially satisfying $|\mathcal{H}(k)| \leq 2kE(k)$; this $E(k)$, shown solid in the figure, is chosen to have Kolmogorov form $\sim k^{-5/3}$ for $k \gg 1$. The dotted curve shows $-\alpha(k)$ (note that, as expected, $\alpha(k) < 0$, when $\mathcal{H}(k) > 0$). The dashed curve shows $\beta(k)$ which follows $E(k)$ quite closely. The important conclusion is that both $|\alpha|$ and β increase monotonically as k decreases all the way through the Kolmogorov regime, and are both $\mathcal{O}(1)$ when $k = \mathcal{O}(1)$. In dimensional terms, this means that $|\alpha| \sim u_0$ and $\beta \sim u_0 l_0$, where u_0 is the rms turbulent velocity and l_0 is the scale of the energy-containing eddies of the turbulence The same conclusion holds no matter how small η may be, provided only that $\eta > 0$: molecular diffusion is essential just to get the process started.

For the consistency of the above approach, it is important that the 'local turbulent magnetic Reynolds number' $\mathcal{R}_{\mathrm{m\,loc}}(k) = k^{-1}(kE(k))^{1/2}/(\eta + \beta(k))$ should remain $\lesssim \mathcal{O}(1)$ as k decreases, since otherwise the first-order smoothing approximation on which (7.136) is based would be illegitimate. $\mathcal{R}_{\mathrm{m\,loc}}(k)$ is shown plotted in Figure 7.6; this increases from about 0.1 at $k = 10^3$ as k decreases, but $\beta(k)$ increases in such a way as to keep $\mathcal{R}_{\mathrm{m\,loc}}(k)$ under control: it levels off at a value

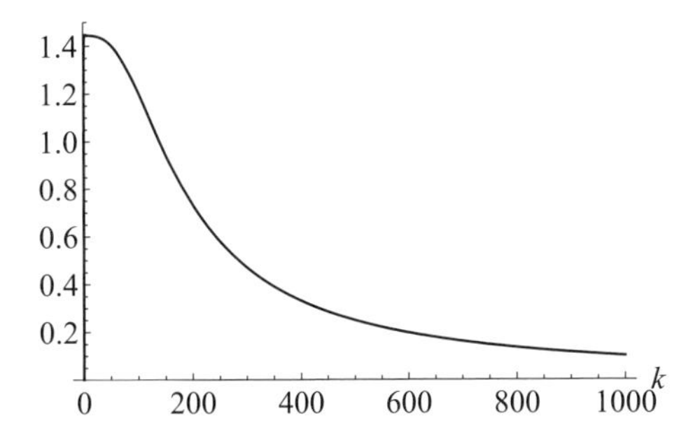

Figure 7.6 The local turbulent magnetic Reynolds number $\mathcal{R}_{\mathrm{m\,loc}}(k)$.

just less than 1.45 when $k = \mathcal{O}(1)$. First-order smoothing therefore appears to be quite reasonable, at least as a starting point for this type of investigation.

8

Nearly Axisymmetric Dynamos

8.1 Introduction

The general arguments of §§7.1 and 7.2 indicate that when a velocity field consists of a steady mean part $\mathbf{U}_0(\mathbf{x})$ together with a fluctuating part $\mathbf{u}(\mathbf{x}, t)$, the mean magnetic field evolves according to the equation

$$\partial \mathbf{B}_0/\partial t = \nabla \wedge (\mathbf{U}_0 \wedge \mathbf{B}_0) + \nabla \wedge \mathcal{E} + \eta \nabla^2 \mathbf{B}_0 \,, \tag{8.1}$$

where

$$\mathcal{E}_i = \alpha_{ij} B_{0j} + \beta_{ijk} \partial B_{0j}/\partial x_k + \cdots \,. \tag{8.2}$$

In §§7.5-7.11, we obtained explicit expressions for α_{ij} and β_{ijk} as quadratic functionals of the **u**-field, on the assumptions that $\mathbf{u}_0 = 0$ (or cst.) and that the statistical properties of the **u**-field are uniform in space and time. The approach was a two-scale approach, involving the definition of a mean as an average over the smaller length- (or time-) scale.

A very similar approach was developed by Braginskii (1964b,c) in an investigation of the effects of weak departures from axisymmetry in a spherical dynamo system. The idea motivating the study was that, although Cowling's theorem eliminates the possibility of axisymmetric dynamo action, if diffusive effects are weak (i.e. η small) then weak departures from axisymmetry in the velocity field (and consequently in the magnetic field) may provide a regenerative electromotive force of the kind required to compensate ohmic decay.

In this situation it is natural to define the mean fields in terms of averages over the azimuth angle φ. For any scalar $\psi(s, \varphi, z)$, we therefore define its *azimuthal mean* by

$$\psi_0(s, z) = \langle \psi(s, \varphi, z) \rangle_{\mathrm{az}} = \frac{1}{2\pi} \int_0^{2\pi} \psi(s, \varphi, z) \mathrm{d}\varphi \,, \tag{8.3}$$

and for any vector $\mathbf{f}(s, \varphi, z)$ we define similarly

$$\mathbf{f}_0(s, z) = \langle \mathbf{f} \rangle_{\mathrm{az}} = \langle f_s \rangle \, \mathbf{e}_s + \langle f_\varphi \rangle \, \mathbf{e}_\varphi + \langle f_z \rangle \, \mathbf{e}_z \,. \tag{8.4}$$

We shall also use the notation

$$\mathbf{f}_M = \mathbf{f} - (\mathbf{f}\cdot\mathbf{e}_\varphi)\,\mathbf{e}_\varphi = f_s\,\mathbf{e}_s + f_z\,\mathbf{e}_z \tag{8.5}$$

for the meridional projection of $\mathbf{f}$. Of course if $\mathbf{f}$ is axisymmetric, then $\mathbf{f}_0 = \mathbf{f}$; if in addition $\mathbf{f}$ is solenoidal, then $\mathbf{f}_M = \mathbf{f}_P$, the poloidal ingredient of $\mathbf{f}$.

Let us now consider a velocity field expressible in the form

$$\mathbf{U}_0(s,z) + \varepsilon\,\mathbf{u}'(s,\varphi,z,t)\,, \tag{8.6}$$

where $\varepsilon \ll 1$ and $\langle\mathbf{u}'\rangle_{\mathrm{az}} = 0$. Any magnetic field convected and distorted by this velocity field must exhibit at least the same degree of asymmetry about the axis Oz, and it is consistent to restrict attention to magnetic fields expressible in the form

$$\mathbf{B}_0(s,z) + \varepsilon\,\mathbf{b}'(s,\varphi,z,t),\ \langle\mathbf{b}'\rangle_{\mathrm{az}} = 0\,. \tag{8.7}$$

(It is implicit in the notation of (8.6) and (8.7) that $\mathbf{u}'$ and $\mathbf{b}'$ are $\mathcal{O}(1)$ as $\varepsilon \to 0$.) The azimuthal average of the induction equation is then (8.1) with

$$\boldsymbol{\mathcal{E}}(s,z) = \varepsilon^2\langle\mathbf{u}'\wedge\mathbf{b}'\rangle_{\mathrm{az}}\,. \tag{8.8}$$

If we separate (8.1) into its toroidal and poloidal parts, with the notation

$$\mathbf{B}_0 = B(s,z)\,\mathbf{e}_\varphi + \mathbf{B}_P(s,z),\quad \mathbf{U}_0 = U(s,z)\,\mathbf{e}_\varphi + \mathbf{U}_P(s,z)\,, \tag{8.9}$$

we have for the toroidal part (cf. (3.60))

$$\frac{\partial B}{\partial t} + s\,(\mathbf{U}_P\cdot\nabla)\,\frac{B}{s} = s(\mathbf{B}_P\cdot\nabla)\frac{U}{s} + (\nabla\wedge\boldsymbol{\mathcal{E}})_\varphi + \eta\left(\nabla^2 - \frac{1}{s^2}\right)B\,. \tag{8.10}$$

Also, as in §3.7, writing $\mathbf{B}_P = \nabla\wedge(A(s,z)\,\mathbf{e}_\varphi)$, the poloidal part may be 'uncurled' to give

$$\frac{\partial A}{\partial t} + \frac{1}{s}(\mathbf{U}_P\cdot\nabla)(sA) = \mathcal{E}_\varphi + \eta\left(\nabla^2 - \frac{1}{s^2}\right)A\,. \tag{8.11}$$

We have seen in §3.14 that the term $s\,(\mathbf{B}_P\cdot\nabla)(U/s)$ in (8.10) can act as an adequate source term for the production of toroidal field by the process of differential rotation. We now envisage a situation in which the term $\mathcal{E}_\varphi$ in (8.11) acts as the complementary source term for the production of poloidal field $\mathbf{B}_P$ via its vector potential $A\,\mathbf{e}_\varphi$. If the rate of production of toroidal field is to be adequate to compensate ohmic dissipation, then a simple comparison of the terms on the right of (8.11) suggests that η must be no greater than $\mathcal{O}(\varepsilon^2)$, and we shall assume this to be the case in the following sections; i.e. we put

$$\eta = \zeta_0\,\varepsilon^2\,, \tag{8.12}$$

and assume that $\zeta_0 = \mathcal{O}(1)$ in the limit $\varepsilon \to 0$. Equivalently if U_0 is a typical order of magnitude of the toroidal velocity $U(s,z)$ and L an overall length-scale, then

$$\mathcal{R}_\mathrm{m}^{-1} = \eta/U_0 L = \mathcal{O}(\varepsilon^2)\,, \tag{8.13}$$

or equivalently, $\varepsilon = \mathcal{O}(\mathcal{R}_\mathrm{m}^{-1/2})$.

The Braginskii approach is further based on the assumption that the dominant ingredient of the mean velocity $\mathbf{U}_0$ is the toroidal ingredient $U\,\mathbf{e}_\varphi$, and more specifically that

$$\mathbf{U}_P = \varepsilon^2 \mathbf{u}_P, \quad \mathbf{u}_P = \mathcal{O}(1)\,. \tag{8.14}$$

This means that a magnetic Reynolds number based on say $|\mathbf{U}_P|_{\max}$ will be $\mathcal{O}(1)$; this poloidal velocity will then redistribute toroidal and poloidal field, but will *not* be sufficiently intense to expel poloidal field from regions of closed $\mathbf{U}_P$-lines.

Since differential rotation tends to generate a toroidal field that is a factor $\mathcal{O}(\mathcal{R}_\mathrm{m})$ larger than the poloidal field, it is natural to introduce the further scaling (in parallel with (8.14))

$$\mathbf{B}_P = \varepsilon^2 \mathbf{b}_P, \quad \mathbf{b}_P = \mathcal{O}(1)\,, \tag{8.15}$$

the implication then being that $|\mathbf{b}_P|$ and B are of the same order of magnitude, as $\varepsilon \to 0$.

With the total velocity field now of the form

$$U\mathbf{e}_\varphi + \varepsilon\,\mathbf{u}' + \varepsilon^2 \mathbf{u}_P\,, \tag{8.16}$$

it is clear that any fluid particle follows a nearly circular path about the axis Oz; in these circumstances, there is good reason to anticipate that $\boldsymbol{\mathcal{E}} = \varepsilon^2 \langle \mathbf{u}' \wedge \mathbf{b}' \rangle_{\mathrm{az}}$ will be determined by *local* values of $B(s,z)$, $U(s,z)$ and azimuthal average properties of $\mathbf{u}'$. The situation is closely analogous to the two-scale approach of Chapter 7, the smaller scale l_0 now being the rms departure of a fluid particle from a circular path in its trajectory round the axis Oz. A local expansion of the form (8.2) is therefore to be expected, but now with $\boldsymbol{\mathcal{E}}(s,x)$, $\mathbf{B}_0(s,z)$ axisymmetric by definition. The feature that most distinguishes the Braginskii approach from the previous two-scale approach is the presence of the dominant azimuthal flow $U(s,z)\,\mathbf{e}_\varphi$; and a possible influence of local shear, given by $s\nabla(U/s)$, on $\boldsymbol{\mathcal{E}}$ cannot be ruled out. It turns out however, as described in the following sections, that the influence of the flow $U\mathbf{e}_\varphi$ on $\boldsymbol{\mathcal{E}}$ is simply accommodated through replacement of the fields $\mathbf{u}_P$ and $\mathbf{b}_P$ by 'effective fields' $\mathbf{u}_{eP}$ and $\mathbf{b}_{eP}$, the structure of equations (8.10) and (8.11) remaining otherwise unchanged. When this effect has been accounted for, the residual mean electromotive force (which is wholly diffusive in origin) is, with very minor modification, identifiable with that given by the first-order smoothing theory of §7.8. The theory that follows, although initiated by Braginskii (1964b,c) before the mean-field electrodynamics of Steenbeck et al. (1966), can best be regarded as a branch of mean-field electrodynamics that takes account of spatial inhomogeneity of mean velocity and of mean properties of the fluctuating ingredient of the velocity field. The close relationship between the two approaches was emphasised by Soward (1972), whose line of argument we follow in subsequent sections.

8.2 Lagrangian transformation of the induction equation when $\eta = 0$

Soward's (1972) approach is based on a simple property of invariance of the induction equation in the limit of zero diffusion. For reasons that emerge, it is useful to modify the notation slightly: let $\tilde{\mathbf{B}}(\tilde{\mathbf{x}}, t)$, $\tilde{\mathbf{U}}(\tilde{\mathbf{x}}, t)$ represent field and velocity at $(\tilde{\mathbf{x}}, t)$; when $\eta = 0$, we have

$$\partial \tilde{\mathbf{B}}/\partial t = \tilde{\nabla} \wedge (\tilde{\mathbf{U}} \wedge \tilde{\mathbf{B}}) . \tag{8.17}$$

From §2.6, we know that (when $\tilde{\nabla} \cdot \tilde{\mathbf{U}} = 0$) the Lagrangian solution of this equation is

$$\tilde{B}_i(\tilde{\mathbf{x}}, t) = \tilde{B}_j(\mathbf{a}, 0)\, \partial \tilde{x}_i / \partial a_j , \tag{8.18}$$

where $\tilde{\mathbf{x}}(\mathbf{a}, t)$ is the position of a fluid particle satisfying $\tilde{\mathbf{x}}(\mathbf{a}, 0) = \mathbf{a}$. Equivalently, the Lagrangian form of (8.17) is

$$\frac{\mathrm{D}}{\mathrm{D}t}\left(\tilde{B}_i(\tilde{\mathbf{x}}, t)\frac{\partial a_j}{\partial \tilde{x}_i}\right) = 0 , \tag{8.19}$$

where $\mathrm{D}/\mathrm{D}t$ represents differentiation keeping $\mathbf{a}$ constant.

Consider now an 'incompressible' change of variable

$$\mathbf{x} = \mathbf{x}(\tilde{\mathbf{x}}(\mathbf{a}, t), t) = \mathbf{X}(\mathbf{a}, t) \text{, say,} \tag{8.20}$$

the determinant of the transformation $\|\partial x_i/\partial \tilde{x}_j\|$ being equal to unity, a condition that may be expressed in the form

$$\epsilon_{ijk}\frac{\partial x_i}{\partial \tilde{x}_p}\frac{\partial x_j}{\partial \tilde{x}_q}\frac{\partial x_k}{\partial \tilde{x}_r} = \epsilon_{pqr} . \tag{8.21}$$

Equation (8.19) immediately transforms to

$$\frac{\mathrm{D}}{\mathrm{D}t}\left(B_k(\mathbf{x}, t)\frac{\partial a_j}{\partial x_k}\right) = 0 \quad \text{where} \quad B_k(\mathbf{x}, t) = \tilde{B}_i(\tilde{\mathbf{x}}, t)\partial x_k/\partial \tilde{x}_i ; \tag{8.22}$$

and, reversing the process that led from (8.17) to (8.19), we see that the Eulerian equivalent of (8.22) is

$$\partial \mathbf{B}/\partial t = \nabla \wedge (\mathbf{U} \wedge \mathbf{B}) \quad \text{where} \quad U_k(\mathbf{x}, t) = \left(\frac{\partial X_k}{\partial t}\right)_{\mathbf{a}} = \frac{\partial x_k}{\partial t} + \tilde{U}_i \frac{\partial x_k}{\partial \tilde{x}_i} . \tag{8.23}$$

Equation (8.17) is therefore invariant under the transformation defined by (8.20), (8.22) and (8.23). $\mathbf{B}(\mathbf{x},t)$ is (physically) the field that would result from $\tilde{\mathbf{B}}(\tilde{\mathbf{x}},t)$ under an instantaneous frozen-field distortion of the medium defined by (8.20); $\mathbf{U}(\mathbf{x},t)$ is the incompressible velocity field associated with the (hypothetical) Lagrangian displacement $\mathbf{X}(\mathbf{a},t)$.

In view of the discussion of §8.1, it is of particular interest to consider the effect of a mapping $\mathbf{x} \leftrightarrow \tilde{\mathbf{x}}$ which is nearly the identity. Such a mapping may be considered as the result of a steady Eulerian velocity field $\mathbf{v}(\mathbf{x})$ applied over some short time interval $\varepsilon\tau_0$ say. Then the particle path $\tilde{\mathbf{x}}(\mathbf{x},\tau)$ is given by

$$d\tilde{\mathbf{x}}/d\tau = \mathbf{v}(\tilde{\mathbf{x}}), \quad \tilde{\mathbf{x}}(\mathbf{x},0) = \mathbf{x}\,, \tag{8.24}$$

and, at time $\tau = \varepsilon\tau_0$, the net displacement $\tilde{\mathbf{x}}$ is given by

$$\tilde{\mathbf{x}} = \mathbf{x} + \varepsilon\boldsymbol{\zeta}(\mathbf{x}) + \tfrac{1}{2}\varepsilon^2(\boldsymbol{\zeta}\cdot\nabla)\boldsymbol{\zeta} + \tfrac{1}{3!}\varepsilon^3(\boldsymbol{\zeta}\cdot\nabla)^2\boldsymbol{\zeta} + \cdots\,, \tag{8.25}$$

where $\boldsymbol{\zeta}(\mathbf{x}) = \tau_0\mathbf{v}(\mathbf{x})$. If $\mathbf{v}(\mathbf{x})$ is incompressible, then $\nabla\cdot\boldsymbol{\zeta} = 0$, and, under this condition, the form of displacement given by (8.25) must automatically satisfy (8.21) to all orders in ε.

When $\mathbf{x}$ and $\tilde{\mathbf{x}}$ are related by (8.25), the instantaneous relation between $\mathbf{B}(\mathbf{x})$ and $\tilde{\mathbf{B}}(\mathbf{x})$ can be obtained as follows. Let $\tilde{\mathbf{B}}_\tau(\mathbf{x})$ be defined by

$$\partial\tilde{\mathbf{B}}_\tau/\partial\tau = \nabla\wedge(\mathbf{v}(\mathbf{x})\wedge\tilde{\mathbf{B}}_\tau)\,, \quad \tilde{\mathbf{B}}_0(\mathbf{x}) = \mathbf{B}(\mathbf{x})\,, \tag{8.26}$$

so that evidently $\tilde{\mathbf{B}}(\mathbf{x}) = \tilde{\mathbf{B}}_{\varepsilon\tau_0}(\mathbf{x})$. We may write (8.26) in the equivalent integral form

$$\tilde{\mathbf{B}}_\tau(\mathbf{x}) = \mathbf{B}(\mathbf{x}) + \int_0^\tau \nabla\wedge\left(\mathbf{v}(\mathbf{x})\wedge\tilde{\mathbf{B}}_{\tau'}(\mathbf{x})\right)d\tau'\,, \tag{8.27}$$

and solve iteratively to obtain

$$\tilde{\mathbf{B}}_\tau(\mathbf{x}) = \mathbf{B}(\mathbf{x}) + \tau\nabla\wedge(\mathbf{v}\wedge\mathbf{B}) + \tfrac{1}{2}\tau^2\nabla\wedge[\mathbf{v}\wedge\nabla\wedge(\mathbf{v}\wedge\mathbf{B})] + \cdots\,. \tag{8.28}$$

Putting $\tau = \varepsilon\tau_0$ and $\mathbf{v} = \boldsymbol{\zeta}/\tau_0$, we obtain

$$\tilde{\mathbf{B}}(\mathbf{x}) = \mathbf{B}(\mathbf{x}) + \varepsilon\nabla\wedge(\eta\wedge\mathbf{B}) + \tfrac{1}{2}\varepsilon^2\nabla\wedge[\eta\wedge\nabla\wedge(\eta\wedge\mathbf{B})] + \cdots\,. \tag{8.29}$$

This is the Eulerian equivalent of the Lagrangian statement

$$\tilde{B}_i(\tilde{\mathbf{x}}) = B_k(\mathbf{x})\left(\delta_{ik} + \varepsilon\frac{\partial\zeta_i}{\partial x_k} + \tfrac{1}{2}\varepsilon^2\frac{\partial}{\partial x_k}(\boldsymbol{\zeta}\cdot\nabla)\zeta_i + \cdots\right). \tag{8.30}$$

Since these relationships are instantaneous, they remain valid when $\boldsymbol{\zeta}$, $\tilde{\mathbf{B}}$ and $\mathbf{B}$ depend explicitly on t as well as on $\mathbf{x}$.

By virtue of (8.23), the corresponding relationship between $\tilde{\mathbf{U}}(\mathbf{x})$ and $\mathbf{U}(\mathbf{x})$ is, in compact notation,

$$\tilde{\mathbf{U}}(\mathbf{x}) = \sum_{n=0}^{\infty} \frac{\varepsilon^n}{n!}(\nabla \wedge \boldsymbol{\zeta} \wedge)^n \left(\mathbf{U}(\mathbf{x}) - \frac{\partial \mathbf{x}}{\partial t}\right). \tag{8.31}$$

Here, as in (8.23), $\partial \mathbf{x}/\partial t$ is to be evaluated 'keeping $\tilde{\mathbf{x}}$ constant'; hence, from (8.25),

$$\frac{\partial x_i}{\partial t} = -\varepsilon \frac{\partial \zeta_i}{\partial t} + \tfrac{1}{2}\varepsilon^2 \left(\frac{\partial \zeta_j}{\partial t}\frac{\partial \zeta_i}{\partial x_j} - \zeta_j \frac{\partial^2 \zeta_i}{\partial x_j \partial t}\right) + \mathcal{O}(\varepsilon^3). \tag{8.32}$$

8.3 Effective variables in a Cartesian geometry

To simplify the discussion, let us for the moment use Cartesian coordinates (x, y, z) $(\equiv (x_1, x_2, x_3))$ instead of cylindrical polars (s, φ, z). Following the discussion of §8.1 we consider velocity and magnetic fields of the form

$$\left.\begin{aligned} \tilde{\mathbf{u}}(\mathbf{x}, t) &= U(x, z)\, \mathbf{e}_y + \varepsilon\, \mathbf{u}'(\mathbf{x}, t) + \varepsilon^2\, \mathbf{u}_P(x, z) \\ \tilde{\mathbf{B}}(\mathbf{x}, t) &= B(x, z)\, \mathbf{e}_y + \varepsilon\, \mathbf{b}'(\mathbf{x}, t) + \varepsilon^2\, \mathbf{b}_P(x, z) \end{aligned}\right\}. \tag{8.33}$$

Here the fields $U\mathbf{e}_y$, $B\mathbf{e}_y$ are the dominant toroidal fields, $\mathbf{u}_P$ and $\mathbf{b}_P$ are poloidal (so that $\mathbf{e}_y \cdot \mathbf{b}_P = \mathbf{e}_y \cdot \mathbf{u}_P = 0$), and $\langle \mathbf{u}' \rangle = \langle \mathbf{b}' \rangle = 0$, the brackets $\langle \cdots \rangle$ now indicating an average over the coordinate y. The essence of Soward's (1972) approach is now to 'accommodate' the $\mathcal{O}(\varepsilon)$ terms in (8.33) through choice of $\boldsymbol{\zeta}(\mathbf{x}, t)$, in such a way that the transformed fields $\mathbf{U}$, $\mathbf{B}$ take the form

$$\left.\begin{aligned} \mathbf{U}(\mathbf{x}, t) &= U(x, z)\left(1 + \mathcal{O}(\varepsilon^2)\right) \mathbf{e}_y + \varepsilon^2\, \mathbf{u}_{eP}(x, z) + \varepsilon^2\, \mathbf{u}''(\mathbf{x}, t) \\ \mathbf{B}(\mathbf{x}, t) &= B(x, z)\left(1 + \mathcal{O}(\varepsilon^2)\right) \mathbf{e}_y + \varepsilon^2\, \mathbf{b}_{eP}(x, z) + \varepsilon^2\, \mathbf{b}''(\mathbf{x}, t) \end{aligned}\right\}, \tag{8.34}$$

where $\langle \mathbf{u}'' \rangle = \langle \mathbf{b}'' \rangle = 0$, and where the effective fields (or 'effective variables' as Braginskii 1964b,c called them) $\mathbf{u}_{eP}$, $\mathbf{b}_{eP}$ are to be determined. Substitution of (8.33) and (8.34) in (8.29) and equating terms of order ε gives immediately

$$\mathbf{b}'(\mathbf{x}, t) = \nabla \wedge (\boldsymbol{\zeta}(\mathbf{x}, t) \wedge B(x, z)\, \mathbf{e}_y) = B\frac{\partial \boldsymbol{\zeta}}{\partial y} - (\boldsymbol{\zeta} \cdot \nabla)B\, \mathbf{e}_y\,. \tag{8.35}$$

Similarly

$$\mathbf{u}'(\mathbf{x}, t) = \left(\frac{\partial}{\partial t} + U\frac{\partial}{\partial y}\right)\boldsymbol{\zeta} - (\boldsymbol{\zeta} \cdot \nabla)U\, \mathbf{e}_y\,, \tag{8.36}$$

an equation which in principle serves to determine $\boldsymbol{\zeta}$ if $\mathbf{u}'$ and U are given. It is in fact simpler now to regard $\boldsymbol{\zeta}(\mathbf{x}, t)$ as given, and the fluctuating part of the velocity field as then given by (8.36).

Similarly at the $\mathcal{O}(\varepsilon^2)$ level, the mean poloidal ingredient of (8.36) is given by

$$\mathbf{b}_P(x, z) = \mathbf{b}_{eP}(x, z) + \tfrac{1}{2}\nabla \wedge \left\{\langle \boldsymbol{\zeta} \wedge \partial \boldsymbol{\zeta}/\partial y\rangle_y\ B\, \mathbf{e}_y\right\}. \tag{8.37}$$

Alternatively, defining vector potentials $a\,\mathbf{e}_y$ and $a_e\,\mathbf{e}_y$ by

$$\mathbf{b}_P = \nabla \wedge (a\,\mathbf{e}_y), \quad \mathbf{b}_{eP} = \nabla \wedge (a_e\,\mathbf{e}_y)\,, \tag{8.38}$$

the relation between a and a_e is evidently

$$a_e = a + \varpi B \quad \text{where} \quad \varpi = -\tfrac{1}{2}\langle \zeta \wedge \partial\zeta/\partial y\rangle_{\mathrm{y}}\,. \tag{8.39}$$

The quantity ϖ is a pseudo-scalar (having the dimensions of a length). It is therefore non-zero only if the statistical (i.e. y-averaged) properties of the function $\zeta(\mathbf{x}, t)$ lack reflectional symmetry.

The relationship between $\mathbf{u}_{eP}$ and $\mathbf{u}_P$ is analogous to (8.37), but with the difference again that $B\,\partial/\partial y$ is replaced by $\partial/\partial t + U\,\partial/\partial y$, i.e.

$$\mathbf{u}_{eP} = \mathbf{u}_P - \tfrac{1}{2}\nabla \wedge \left\{\left\langle \zeta \wedge \left(\frac{\partial}{\partial t} + U\frac{\partial}{\partial y}\right)\zeta\right\rangle_{\mathrm{y}} \mathbf{e}_y\right\}. \tag{8.40}$$

If the displacement function ζ depends only on $\mathbf{x}$ (as is appropriate if the perturbation field is steady) then the analogy between (8.37) and (8.40) becomes precise.

The fact that the perturbation fields in (8.34) are relegated to $\mathcal{O}(\varepsilon^2)$ means that when the magnetic diffusivity $\eta = 0$ the relevant equations for the evolution of $\mathbf{b}(\mathbf{x}, t)$ at leading order are precisely the two-dimensional equations derived in §3.7, but now expressed in terms of the effective poloidal fields, viz.

$$\partial B/\partial t + \mathbf{U}_{eP}\cdot\nabla B = \mathbf{B}_{eP}\cdot\nabla U, \quad \partial A_e/\partial t + \mathbf{U}_{eP}\cdot\nabla A_e = 0, \tag{8.41}$$

where $\mathbf{U}_{eP} = \varepsilon^2\mathbf{u}_{eP}$, $\mathbf{B}_{eP} = \varepsilon^2\,\mathbf{b}_{eP}$, $A_e = \varepsilon^2 a_e$. This holds because of the basic invariance of the induction equation (8.17) under the frozen-field transformation (8.20). It is evident from (8.41) that the distinction between $\mathbf{B}_P$ and $\mathbf{B}_{eP}$ will be significant only if $\nabla U \neq 0$, i.e. only if the mean flow exhibits shear. Similarly it is evident from (8.40) that $\mathbf{u}_{eP}$ differs from $\mathbf{u}_P$ only when either $\nabla U \neq 0$ or ζ is statistically inhomogeneous in the (x, z) plane. It is therefore inhomogeneity in the y-averaged (or equivalently azimuthally averaged) flow properties that leads to the natural appearance of effective variables; this aspect of Braginskii's theory has no perceptible counterpart in the mean-field electrodynamics described in Chapter 7.

8.4 Lagrangian transformation including weak diffusion effects

Suppose now that we subject the full induction equation in the form

$$\left(\partial\tilde{\mathbf{B}}/\partial t - \tilde{\nabla}\wedge(\tilde{\mathbf{U}}\wedge\tilde{\mathbf{B}})\right)_i = -\eta\left(\tilde{\nabla}\wedge(\tilde{\nabla}\wedge\tilde{\mathbf{B}})\right)_i \tag{8.42}$$

to the transformation (8.20), (8.22) and (8.23). Multiplying by $\partial x_k/\partial\tilde{x}_i$, the left-hand side transforms by the result of §8.2 to

$$(\partial\mathbf{B}/\partial t - \nabla\wedge(\mathbf{U}\wedge\mathbf{B}))_k\,. \tag{8.43}$$

The right-hand side transforms to

$$-\eta \frac{\partial x_k}{\partial \tilde{x}_i} \epsilon_{ijm} \frac{\partial x_p}{\partial \tilde{x}_j} \frac{\partial}{\partial x_p} (\tilde{\nabla} \wedge \tilde{\mathbf{B}})_m = -\eta \, \epsilon_{kpq} \frac{\partial \tilde{x}_m}{\partial x_q} \frac{\partial}{\partial x_p} (\tilde{\nabla} \wedge \tilde{\mathbf{B}})_m \, , \tag{8.44}$$

using (8.21). Now

$$(\tilde{\nabla} \wedge \tilde{\mathbf{B}})_m = \epsilon_{mrs} \frac{\partial \tilde{B}_s}{\partial \tilde{x}_r} = \epsilon_{mrs} \frac{\partial}{\partial x_j} \left(B_j \frac{\partial \tilde{x}_s}{\partial x_i} \right) \frac{\partial x_i}{\partial \tilde{x}_r} \, . \tag{8.45}$$

Hence, using the identities

$$\epsilon_{kpq} \frac{\partial^2 \tilde{x}_m}{\partial x_p \partial x_q} = 0, \qquad \frac{\partial}{\partial x_j} \left(\frac{\partial \tilde{x}_s}{\partial x_i} \frac{\partial x_i}{\partial \tilde{x}_r} \right) = \frac{\partial}{\partial x_j} \delta_{rs} = 0 \, , \tag{8.46}$$

the expression (8.44) reduces to the form

$$(\nabla \wedge \mathcal{E}')_k + \eta (\nabla^2 \mathbf{B})_k \, , \tag{8.47}$$

where

$$\mathcal{E}'_i(\mathbf{x}, t) = \alpha'_{ij} B_j + \beta'_{ijk} \partial B_j / \partial x_k \, , \tag{8.48}$$

and where

$$\alpha'_{ij} = \eta \, \epsilon_{ikl} \frac{\partial x_k}{\partial \tilde{x}_p} \frac{\partial}{\partial x_j} \left(\frac{\partial x_l}{\partial \tilde{x}_p} \right), \quad \beta'_{ijk} = \eta \, \epsilon_{jip} \left(\frac{\partial x_p}{\partial \tilde{x}_r} \frac{\partial x_k}{\partial \tilde{x}_r} - \delta_{pk} \right), \tag{8.49}$$

and (8.42) becomes

$$\partial \mathbf{B} / \partial t - \nabla \wedge (\mathbf{U} \wedge \mathbf{b}) = \nabla \wedge \mathcal{E}' + \eta \nabla^2 \mathbf{B} \, . \tag{8.50}$$

The structure of (8.50) and the expression (8.48) for $\mathcal{E}'$ are now strongly reminiscent of equations for mean fields encountered in Chapter 7. Note however that in the present context the term $\nabla \wedge \mathcal{E}'$ is *wholly diffusive in origin*. As in the case of the random wave situation of §7.8, α'_{ij} and β'_{ijk} are $\mathcal{O}(\eta)$ as $\eta \to 0$. We have noted previously that although diffusion is responsible for the natural tendency of the field to decay, it is also of crucial importance in making field regeneration a possibility. In Braginskii's theory it is diffusion that, as in §7.7, shifts the phase of field perturbations relative to velocity perturbations leading to the appearance of a mean toroidal electromotive force (proportional to η) which is sufficient to provide closure of the dynamo cycle.

8.5 Dynamo equations for nearly rectilinear flow

As in §8.3, we again specialise the argument to the nearly two-dimensional situation. The y-average of $\mathcal{E}'$ is then

$$\mathcal{E}'_i(x, z) = \alpha_{ij} B_j + \beta_{ijk} \partial B_j / \partial x_k \, , \tag{8.51}$$

where, to leading order in ε,

$$\alpha_{ij} = \langle\alpha'_{ij}\rangle = \eta\varepsilon^2\epsilon_{ikl}\langle\zeta_{k,m}\zeta_{l,mj}\rangle \quad\text{and}\quad \beta_{ijk} = \langle\beta'_{ijk}\rangle = \eta\varepsilon^2\epsilon_{ijp}\langle\zeta_{p,r}\zeta_{k,r}\rangle\,. \tag{8.52}$$

With $\eta = \mathcal{O}(\varepsilon^2)$, both of these pseudo-tensors are $\mathcal{O}(\varepsilon^4)$. With $\mathbf{B}$ given by (8.34), we have, at leading order,

$$\mathcal{E}'_y \sim \alpha_{22}B = \mathcal{O}(\varepsilon^4)\,. \tag{8.53}$$

Note that the β-term in (8.51) makes no contribution at the $\mathcal{O}(\varepsilon^4)$ level. The counterpart of (8.41) incorporating diffusion effects now therefore takes the form (with $\alpha_{22} = \alpha$)

$$\partial A_e/\partial t + \mathbf{U}_{eP}\cdot\nabla A_e = \alpha B + \eta\,\nabla^2 A_e\,. \tag{8.54}$$

Note that all three terms contributing to $\partial A_e/\partial t$ are $\mathcal{O}(\varepsilon^4)$.

Similarly $(\nabla\wedge\boldsymbol{\mathcal{E}})_y = \mathcal{O}(\varepsilon^4)$ in general, and this is negligible compared with the natural diffusive term $\eta\,\nabla^2 B$ in the mean y-component of (8.50). The counterpart of (8.41) is therefore simply

$$\partial B/\partial t + \mathbf{U}_{eP}\cdot\nabla B = \mathbf{B}_{eP}\cdot\nabla U + \eta\,\nabla^2 B\,. \tag{8.55}$$

In this equation, all three contributions to $\partial B/\partial t$ are $\mathcal{O}(\varepsilon^2)$.

Apart from the appearance of effective variables, equations (8.54) and (8.55) are just what would be obtained on the basis of the mean-field electrodynamics of Chapter 7, provided that in the 'toroidal' equation (8.55) the production of toroidal field by the term $(\mathbf{B}_{eP}\cdot\nabla)\,U$ dominates over possible production by the α-effect. By contrast, the α-effect term αB in (8.54) is the only source term for the field A_e and hence for the poloidal field $\mathbf{B}_{eP}$, and it is therefore of central importance.

The expression in (8.52) for α_{ij} is closely related to the expression (7.76) obtained on the basis of first-order smoothing theory. To see this, let us suppose for the moment that all y-averaged properties of the velocity field are independent of x and z (or so weakly dependent that they may be treated as locally uniform). In this situation, from (8.36),

$$\mathbf{u}' = \left(\frac{\partial}{\partial t} + U\frac{\partial}{\partial y}\right)\zeta\,, \tag{8.56}$$

and the corresponding Fourier transforms in the notation of §7.8 are related by

$$\tilde{\mathbf{u}}' = -\mathrm{i}(\omega - Uk_2)\tilde{\zeta}\,. \tag{8.57}$$

The Fourier transform of $\zeta_{k,m}$ is $\mathrm{i}k_m\tilde{\zeta}_k$ and of $\zeta_{l,mj}$ is $-k_mk_j\tilde{\zeta}_l$, and so the expression for α_{ij} in (8.52) may be translated into spectral terms as

$$\alpha_{ij} = \mathrm{i}\eta\epsilon_{ikl}\iint\frac{k^2k_j\Phi_{kl}(\mathbf{k},\omega)\,\mathrm{d}\mathbf{k}\,\mathrm{d}\omega}{(\omega - Uk_2)^2}\,, \tag{8.58}$$

where $\Phi_{kl}(\mathrm{k}, \omega)$ is the spectrum tensor of the perturbation velocity field $\varepsilon \mathbf{u}'$. This expression agrees with (7.76) in the weak diffusion limit $\eta k^2 \ll \omega$ (for all relevant $(\mathbf{k}, \omega)$) and under the natural replacement of ω^2 by $(\omega - Uk_2)^2$ to take account of mean convection in the y-direction with velocity U. Note that in this important domain of overlap of Braginskii's theory and mean-field electrodynamics, the former theory undoubtedly gives the result $\alpha_{ij} \to 0$ as $\eta \to 0$. We have already noted that this result holds (see (7.82)) when the velocity spectrum contains no zero frequency ingredients (or, in the notation of the present section, ingredients with wave speed ω/k_2 equal to the mean velocity component U). If the velocity spectrum *does* contain such ingredients, then the displacement $\boldsymbol{\zeta}(\mathbf{x}, t)$ will not remain bounded for all t, and the Braginskii approach, based implicitly on the assumption $|\boldsymbol{\zeta}| = \mathcal{O}(1)$, is no longer valid.

In general, of course, if U does depend significantly on x and z (as it must do if the production term $(\mathbf{B}_P \cdot \nabla)U$ is to be important) then mean quantities such as $\langle \zeta_{k,m} \zeta_{l,mj} \rangle$ will also vary with x and z because of the dynamical interaction of mean and perturbation fields; hence in general α_{ij} as given by (8.52) is also a function of x and z.

8.6 Corresponding results for nearly axisymmetric flows

Let us now return to the notation of §8.1, with cylindrical polar coordinates (s, φ, z). The relation between $\mathbf{b}'$ and $\boldsymbol{\zeta}$ is now (cf. (8.35))

$$\mathbf{b}' = \nabla \wedge (\boldsymbol{\zeta} \wedge \mathbf{e}_\varphi B) . \tag{8.59}$$

The meridional projection of this equation (see (8.5)) is

$$\mathbf{b}'_M = \frac{B}{s} \frac{\partial_1}{\partial \varphi} \boldsymbol{\zeta}_M , \tag{8.60}$$

where the operator $\partial_1/\partial\varphi$ is defined for any vector $\mathbf{f}$ by

$$\frac{\partial_1 \mathbf{f}}{\partial \varphi} = \mathbf{e}_s \frac{\partial f_s}{\partial \varphi} + \mathbf{e}_\varphi \frac{\partial f_\varphi}{\partial \varphi} + \mathbf{e}_z \frac{\partial f_z}{\partial \varphi} = \frac{\partial \mathbf{f}}{\partial \varphi} - \mathbf{e}_z \wedge \mathbf{f} . \tag{8.61}$$

Similarly the relation between $\mathbf{u}'_M$ and $\boldsymbol{\zeta}_M$ is

$$\mathbf{u}'_M = \left(\frac{\partial}{\partial t} + \frac{U}{s} \frac{\partial_1}{\partial \varphi} \right) \boldsymbol{\zeta}_M , \tag{8.62}$$

a relation which in principle determines $\boldsymbol{\zeta}_M$ if $\mathbf{u}'_M$ and U are given.

Effective variables are now given by formulae closely analogous to (8.37) and (8.40), viz.

$$\mathbf{b}_{eP} = \mathbf{b}_P + \nabla \wedge (\varpi B \mathbf{e}_\varphi), \quad \mathbf{u}_{eP} = \mathbf{u}_P + \nabla \wedge (\varpi_1 U \mathbf{e}_\varphi) , \tag{8.63}$$

where now

$$\varpi = -\frac{1}{2}\left\langle \zeta \wedge \frac{1}{s}\frac{\partial_1 \zeta}{\partial \varphi}\right\rangle_{\text{az}}, \qquad \varpi_1 = -\frac{1}{2U}\left\langle \zeta \wedge \left(\frac{\partial}{\partial t} + \frac{U}{s}\frac{\partial_1}{\partial \varphi}\right)\zeta\right\rangle_{\text{az}}. \tag{8.64}$$

The evolution equations for $B(s, z, t)$ and $A_e(s, z, t)$ analogous to (8.55) and (8.54) are

$$\frac{\partial B}{\partial t} + s(\mathbf{U}_{eP} \cdot \nabla)\frac{B}{s} = s(\mathbf{B}_{eP} \cdot \nabla)\frac{U}{s} + \eta\left(\nabla^2 - \frac{1}{s^2}\right)B, \tag{8.65}$$

and

$$\frac{\partial A_e}{\partial t} + \frac{1}{s}(\mathbf{U}_{eP} \cdot \nabla)sA_e = \alpha B + \eta\left(\nabla^2 - \frac{1}{s^2}\right)A_e, \tag{8.66}$$

and the expression for α analogous to (8.37) (Soward 1972) is given by

$$\frac{\alpha}{2\eta\varepsilon^2} = \frac{1}{s}\left\langle \zeta_{z,s}\zeta_{s,s\varphi} + s^{-2}\zeta_{z,\varphi}\zeta_{s,\varphi\varphi} + \zeta_{z,z}\zeta_{s,z\varphi}\right\rangle + \frac{1}{s^2}\left\langle s^{-1}\zeta_s\zeta_{z,\varphi} + \zeta_{z,z}\zeta_{z,\varphi}\right\rangle. \tag{8.67}$$

Here the first group of terms is precisely analogous to the group appearing in (8.52); the second group arises from the curvature of the metric in cylindrical polar coordinates, and can be derived only by first converting the expression $\alpha_{ij}B_j$ in (8.51) to this coordinate system.

The equations (8.65) and (8.66) describe the evolution of the fields $\varepsilon^{-2}A_e$ and B correctly to order ε^2. The equations were obtained in this form by Braginskii (1964b,c) by a direct expansion procedure which made no appeal to the Lagrangian invariance property. The labour involved in this procedure is very considerable, and the pseudo-Lagrangian approach of Soward (1972) can now be seen as providing an important and illuminating simplification. The Braginskii expansion was continued to the next level ($\mathcal{O}(\varepsilon^3)$) by Tough (1967) (see also Tough & Gibson 1969) who showed that the structure of the equations is unaltered, but that ϖ, ϖ_1 and α as given by (8.64) and (8.51) need small corrections; these corrections can be obtained by including the ε^3 terms in the expansions (8.29) and (8.31). Similarly the correction to α ($= \alpha_{22}$) can be obtained by retaining terms up to $\mathcal{O}(\varepsilon^3)$ in (8.49); it is evident that this procedure leads to an expression of the form

$$\alpha/\eta\varepsilon^2 = \Gamma_0 + \varepsilon\Gamma_1, \tag{8.68}$$

where Γ_0 is the right-hand side of (8.67) and Γ_1 involves the mean of an expression cubic in ζ.[1]

If terms of order ε^4 are retained (Soward 1972) then the structure of (8.65) and (8.66) *is* modified, firstly through the effects of components of α_{ij} other than α_{22} and secondly through the β_{ijk} terms of (8.51).

[1] An analogous correction to the first-order smoothing result (7.76) can be obtained by systematic expansion in powers of the amplitude of the velocity fluctuations; this correction likewise involves a cubic functional of the velocity field (a weighted integral in **k**-space of the Fourier transform of a triple velocity correlation).

8.7 A limitation of the pseudo-Lagrangian approach

The arguments of Soward (1972) are expressed in general terms that do not for the most part invoke the limitation of small displacement functions. It is assumed that the magnetic Reynolds number $\mathcal{R}_{\mathrm{m}}$ is large, and velocity fields[2] $\tilde{\mathbf{U}}(\tilde{\mathbf{x}})$ are considered having the property that there exists a continuous (1-1) mapping $\tilde{\mathbf{x}} \to \mathbf{x}$ such that the related velocity field $\mathbf{U}(\mathbf{x})$ given by (8.23) has the form

$$\mathbf{U}(\mathbf{x}) = U(s,z)\mathbf{e}_{\varphi} + \mathcal{R}_{\mathrm{m}}^{-1}\mathbf{u}_M(s,z) + \mathcal{R}_{\mathrm{m}}^{-1}\mathbf{u}''(\mathbf{x})\,, \tag{8.69}$$

where $\langle \mathbf{u}'' \rangle_{\mathrm{az}} = 0$. The wide choice of mappings implies a correspondingly wide family of velocity fields $\tilde{\mathbf{U}}(\tilde{\mathbf{x}})$ that can be subjected to this treatment. Even so, such velocity fields are rather special for the following reason.

Under the instantaneous transformation

$$U_k(\mathbf{x}) = \tilde{U}_i(\tilde{\mathbf{x}})\partial x_k/\partial \tilde{x}_i\,, \tag{8.70}$$

it may readily be ascertained (with the help of (8.21)) that

$$(\mathrm{d}\mathbf{x} \wedge \mathbf{U}(\mathbf{x}))_i = (\mathrm{d}\tilde{\mathbf{x}} \wedge \tilde{\mathbf{U}}(\tilde{\mathbf{x}}))_j \partial \tilde{x}_j/\partial x_i\,, \tag{8.71}$$

and hence it follows that the streamlines of the $\tilde{\mathbf{U}}$-field map onto streamlines of the $\mathbf{U}$-field. Let $\mathbf{A}$ and $\tilde{\mathbf{A}}$ be vector potentials of $\mathbf{U}$ and $\tilde{\mathbf{U}}$, and let

$$\mathcal{H}_U = \int \mathbf{U}\cdot\mathbf{A}\,\mathrm{d}V = \int \tilde{\mathbf{U}}\cdot\tilde{\mathbf{A}}\mathrm{d}V\,, \tag{8.72}$$

the integral being throughout a volume V on whose (fixed) surface S it is assumed that $\mathbf{n}\cdot\mathbf{U} = \mathbf{n}\cdot\tilde{\mathbf{U}} = 0$. Equality of the two integrals in (8.72) follows from the interpretation of $\mathcal{H}_U$ as a topological measure for the $\mathbf{U}$ or $\tilde{\mathbf{U}}$-field (cf. the discussion of §2.1), a quantity that is evidently invariant under continuous mappings $\mathbf{x} \to \tilde{\mathbf{x}}$ which conserve the identity of streamlines. Now the vector potential of $U(s,z)\mathbf{e}_{\varphi}$ is evidently poloidal, and so, from (8.69),

$$\int \mathbf{U}\cdot\mathbf{A}\,\mathrm{d}V = \mathcal{O}(\mathcal{R}_{\mathrm{m}}^{-1})\,. \tag{8.73}$$

Hence $\int \tilde{\mathbf{U}}\cdot\tilde{\mathbf{A}}\,\mathrm{d}V = \mathcal{O}(\mathcal{R}_{\mathrm{m}}^{-1})$ also. Conversely, if we consider a velocity field $\tilde{\mathbf{U}}(\mathbf{x})$ for which $\int \tilde{\mathbf{U}}\cdot\tilde{\mathbf{A}}\mathrm{d}V = \mathcal{O}(1)$ as $\mathcal{R}_{\mathrm{m}} \to \infty$, then there exists no continuous transformation $\mathbf{x} = \mathbf{x}(\tilde{\mathbf{x}})$ for which the related velocity $\mathbf{U}(\mathbf{x})$ is given by (8.69).

In simpler terms, and rather loosely speaking, in the limit $\mathcal{R}_{\mathrm{m}} \to \infty$, the velocity field (8.69) exhibits a vanishingly small degree of streamline linkage, and only velocity fields exhibiting the same small degree of linkage are amenable to Soward's

[2] For simplicity the discussion is here restricted to steady velocity fields, although Soward's general theory covers unsteady fields also.

Lagrangian analysis. If

$$\tilde{\mathbf{U}}(\mathbf{x}) = U(s,z)\,\mathbf{e}_\varphi + \varepsilon\,\mathbf{u}'(\mathbf{x}) + \varepsilon^2\mathbf{u}_P(s,z)\,, \tag{8.74}$$

where $\langle\mathbf{u}'\rangle_{\mathrm{az}} = 0$ and $\varepsilon = \mathcal{O}(\mathcal{R}_{\mathrm{m}}^{-1/2})$, as effectively assumed in the foregoing sections, then

$$\int \tilde{\mathbf{U}}\cdot\tilde{\mathbf{A}}\mathrm{d}V = \mathcal{O}(\varepsilon^2) = \mathcal{O}(\mathcal{R}_{\mathrm{m}}^{-1})\,, \tag{8.75}$$

and the existence of the transformation function $\mathbf{x}(\tilde{\mathbf{x}})$ is not therefore ruled out by this argument.

8.8 Matching conditions and the external field

In the context of the Earth's magnetic field, let us assume that the core–mantle boundary S is the sphere $r = R_C$, and that the medium in the external region $\hat{V}$ $(r > R_C)$ is insulating (effects of weak mantle conductivity can be incorporated in an improved theory). The field $\hat{\mathbf{B}}(\mathbf{x},t)$ in $\hat{V}$ is then a potential field, which matches continuously to all orders in ε to the total field $\tilde{\mathbf{B}}(\tilde{\mathbf{x}},t)$ in the interior V. At leading order, the interior field in Braginskii's theory is the purely toroidal field $B(s,z)\,\mathbf{e}_\varphi$, and since the external field is purely poloidal (cf. §6.14), B must satisfy

$$B(s,z) = 0 \quad \text{on } r^2 \equiv s^2 + z^2 = R_C^2\,. \tag{8.76}$$

The mean field in the exterior region, $\hat{\mathbf{B}}_P(s,z)$ say, is given by

$$\hat{\mathbf{B}}_P(s,z) = \nabla\Psi(s,z), \quad \nabla^2\Psi = 0\,, \tag{8.77}$$

and satisfies the boundary condition

$$\hat{\mathbf{B}}_P = \varepsilon^2\mathbf{b}_P(s,z) = \varepsilon^2\nabla\wedge(a\,\mathbf{e}_\varphi) \quad \text{on } r = R_C\,. \tag{8.78}$$

Now the displacement function $\boldsymbol{\zeta}(\mathbf{x},t)$ satisfies $\mathbf{n}\cdot\boldsymbol{\zeta} = 0$ on S (and $\mathbf{n}\wedge\boldsymbol{\zeta} = 0$ also if viscous effects are taken into consideration) and so the pseudo-scalar ϖ defined by (8.64) vanishes on S. Also

$$(\mathbf{n}\cdot\nabla)\varpi B = \varpi(\mathbf{n}\cdot\nabla)B + B(\mathbf{n}\cdot\nabla)\varpi = 0 \quad \text{on } S, \tag{8.79}$$

and it follows that

$$a_e = a \quad \text{and} \quad (\mathbf{n}\cdot\nabla)a_e = (\mathbf{n}\cdot\nabla)a \quad \text{on } S, \tag{8.80}$$

where $a_e = a + \varpi B$, and so (8.78) becomes

$$\hat{\mathbf{B}}_P = \varepsilon^2\nabla\wedge(a_e\,\mathbf{e}_\varphi) \quad \text{on } r = R_C\,. \tag{8.81}$$

Hence the external field matches continuously to the internal *effective* field, and is $\mathcal{O}(\varepsilon^2)$ relative to the internal toroidal field.

The fluctuating ingredient of the external field, $\hat{\mathbf{B}}'$ say, is also of interest, since it is this ingredient that provides the observed secular variations and the slow drift of the dipole moment $\boldsymbol{\mu}^{(1)}(t)$ relative to the rotation axis. The fact that the magnetic and rotation axes are nearly coincident (see §4.4) provides evidence that Coriolis forces arising from the rotation are of dominant importance in controlling the structure of core motions; the fact that they are not exactly coincident provides evidence that systematic deviations from exact axisymmetry may be an essential ingredient in the Earth's dynamo process. The internal fluctuating field is $\mathcal{O}(\varepsilon)$ and this can penetrate to the external region only through the influence of diffusion; with $\eta = \mathcal{O}(\varepsilon^2)$, this gives a contribution to $\hat{\mathbf{B}}'$ of order ε.[3] There is a second contribution also $\mathcal{O}(\varepsilon^3)$ due to distortion by $\varepsilon\,\mathbf{u}'$ of the mean poloidal field $\varepsilon^2\mathbf{b}_P$. Braginskii (1964b) has shown that if the velocity field $\varepsilon\,\mathbf{u}'$ is steady (or at any rate steady over the time-scale R_C/U_0 characterising the mean toroidal flow), then the relevant boundary condition for the determination of $\hat{\mathbf{B}}'$ is, from the superposition of these two effects[3],

$$\hat{B}'_r = 2\varepsilon\frac{\eta}{U}\frac{\partial B}{\partial r}\frac{\partial \zeta_r}{\partial r} + \varepsilon^3\left(b_r\frac{\partial \zeta_r}{\partial r} - \frac{\zeta_\theta}{r}\frac{\partial b_r}{\partial \theta}\right) \quad \text{on } r = R_C\,, \tag{8.82}$$

where b_r is the radial component of $\mathbf{b}_P$. If $\boldsymbol{\zeta}(\mathbf{x}, t)$ is unsteady admitting decomposition into 'azimuthal waves' proportional to $\mathrm{e}^{\mathrm{i}m(\varphi-\omega t)}$, then each such Fourier component makes a contribution to $\hat{B}'_r$ of the form (8.82) with U replaced by $U - \omega s$, the toroidal mean velocity relative to the frame of reference which rotates at the angular phase velocity ω of the component considered (Braginskii 1964c).

The boundary condition (8.82) is sufficient to determine the exterior fluctuating field $\hat{\mathbf{B}}' = \nabla\Psi'$ uniquely (of course under the additional condition $\Psi' = \mathcal{O}(r^{-2})$ at infinity). The linearity of the relation between $\hat{\mathbf{B}}'$ and $\boldsymbol{\zeta}$ implies that each Fourier component $\sim \mathrm{e}^{\mathrm{i}m\varphi}$ in $\boldsymbol{\zeta}$ generates a corresponding Fourier component in $\hat{\mathbf{B}}'$. In particular, if $m = 1$, a contribution to Ψ' of the form

$$A_1\varepsilon^3 r^{-2} P_1^1(\cos\theta)\cos(\varphi - \omega t) \tag{8.83}$$

is generated, representing a dipole whose moment rotates in the equatorial plane with angular velocity ω. In conjunction with the axisymmetric dipole whose potential is of the form

$$A_0\varepsilon^2 r^{-2} P_1(\cos\theta)\,, \tag{8.84}$$

we have here the beginnings of a plausible explanation (in terms of core motions) for the tilt of the net vector dipole moment of the Earth, and the manner in which this drifts relative to the rotation axis. The observed angle of tilt ($\sim 10^\mathrm{o}$) is of course

[3] If the fluid is viscous then $U = 0$ on $r = R_c$ and an apparent singularity appears in the first term on the right of (8.82). However $\partial\zeta_r/\partial r = 0$ on $r = R_c$ also under the no-slip condition. The appropriate modification of (8.82) requires close examination of the viscous boundary layer on $r = R_C$.

not infinitesimal, and corresponds to a value of the ratio $A_1\varepsilon^3/A_0\varepsilon^2 = A_1\varepsilon/A_0$ of order 0.2 (see the figures of Table (4.2), i.e. $\mathcal{R}_m \sim \varepsilon^{-2} \sim 25$. This value is perhaps (as recognised by Braginskii) uncomfortably low for the applicability of 'large-$\mathcal{R}_m$' expansions, although these perhaps provide a useful first step in the right direction.

8.9 Related developments

As evident particularly in §8.4, the above treatment has a mixed Lagrangian/Eulerian character, in that a Lagrangian transformation, appropriate to the limit $\eta \to 0$, is used to generate 'effective' variables, and the resulting averaged equations (for $\eta > 0$) are then obtained in Eulerian form. Through this process, the α-effect appears in the averaged diffusion terms, as revealed by equation (8.50).

This type of hybrid Euler–Lagrange technique has been elaborated and extended in recent years – see particularly Roberts & Soward (2009), Soward & Roberts (2014), and references therein. The technique introduced by eqns. (8.24)–(8.32) above is described as 'Lie-dragging' (Schutz 1980) and can be generalised to describe the possible transformations of tensor fields that are transported by a flow in the diffusionless limit.

Similar ideas were introduced by Andrews & McIntyre (1978) in a treatment of the interaction of waves with a mean flow in a meteorological context; here, the non-linearity of the Navier–Stokes equation is a complicating feature, but the idea of a mean Lagrangian flow transporting a field (a wave field here rather than a magnetic field) is the fundamental idea underlying their treatment, which has had great impact in geophysical fluid dynamics (see, for example, Salmon 1988).

9

Solution of the Mean-Field Equations

9.1 Dynamo models of α^2- and $\alpha\omega$-type

In the previous two chapters it has been shown that in a wide variety of circumstances the mean field, which will now be denoted $\mathbf{B}(\mathbf{x}, t)$ (the mean being over time, or a Cartesian coordinate, or the azimuth angle as appropriate), satisfies an equation of the form

$$\partial \mathbf{B}/\partial t = \nabla \wedge (\mathbf{U} \wedge \mathbf{B}) + \nabla \wedge \boldsymbol{\mathcal{E}} + \eta \nabla^2 \mathbf{B}\,, \tag{9.1}$$

where

$$\mathcal{E}_i = \alpha_{ij} B_j + \beta_{ijk}\, \partial B_j / \partial x_k\,, \tag{9.2}$$

and where α_{ij} and β_{ijk} are determined by the mean velocity $\mathbf{U}(\mathbf{x}, t)$, the statistical properties of the fluctuating field $\mathbf{u}(\mathbf{x}, t)$, and the parameter η. In the mean-field approach of Chapter 7, a useful idealisation is provided by the assumption of isotropy, under which

$$\alpha_{ij} = \alpha\, \delta_{ij}\,, \quad \beta_{ijk} = \beta\, \epsilon_{ijk}\,, \quad \boldsymbol{\mathcal{E}} = \alpha\, \mathbf{B} - \beta \nabla \wedge \mathbf{B}\,. \tag{9.3}$$

It must be recognised however that these expressions are unlikely to be realistic if the background turbulence (or random wave field) is severely anisotropic, as will clearly be the case if Coriolis forces are of dominant importance in controlling the statistical properties of the $\mathbf{u}$-field.

In Braginskii's model as presented in Chapter 8, strong anisotropy is 'built in' at the outset through the assumption that the dominant ingredient of the velocity field is a strong toroidal flow $U(s, z)\, \mathbf{e}_\varphi$. In Braginskii's theory , a 'local' expression for $\boldsymbol{\mathcal{E}}$ of the form (9.2) is obtained, not because the scale of the velocity fluctuations is small (for example, fluctuations proportional to $e^{i\varphi}$ have a wave-length of the same order as the global scale L), but rather because the departure of a fluid particle from its φ-averaged position is small compared with L. A substantial part of the $\boldsymbol{\mathcal{E}}$ that is generated is 'absorbed' through the use of effective variables; the part of $\boldsymbol{\mathcal{E}}$ that cannot be thus absorbed is given by

$$\boldsymbol{\mathcal{E}} = \mathcal{E}_\varphi \mathbf{e}_\varphi, \quad \mathcal{E}_\varphi = \alpha B_\varphi\,, \tag{9.4}$$

in cylindrical polars coordinates (s, φ, z). The *effective* expressions for α_{ij} and β_{ijk} in Braginskii's theory at leading order in the small parameter $\eta^{1/2}$ are thus of the form

$$\alpha_{ij} = \begin{pmatrix} 0 & 0 & 0 \\ 0 & \alpha & 0 \\ 0 & 0 & 0 \end{pmatrix}, \quad \beta_{ijk} = 0\,, \tag{9.5}$$

relative to local Cartesian coordinates (x, y, z) in the directions of increasing (s, φ, z) respectively.

9.1.1 Axisymmetric systems

When $\mathbf{U}$, $\mathbf{B}$ and $\boldsymbol{\mathcal{E}}$ are axisymmetric, we have seen that (9.1) may be replaced by two scalar equations. Writing

$$\mathbf{U} = s\omega(s,z)\mathbf{e}_\varphi + \mathbf{U}_P\,, \quad \mathbf{B} = B(s,z)\mathbf{e}_\varphi + \mathbf{B}_P\,, \quad \boldsymbol{\mathcal{E}} = \mathcal{E}_\varphi \mathbf{e}_\varphi + \boldsymbol{\mathcal{E}}_P\,, \tag{9.6}$$

and with $\mathbf{B}_P = \nabla \wedge (A(s,z)\,\mathbf{e}_\varphi)$, these equations are

$$\left.\begin{aligned} \partial B/\partial t + s(\mathbf{U}_P \cdot \nabla)(s^{-1}B) &= s(\mathbf{B}_P \cdot \nabla)\omega + (\nabla \wedge \boldsymbol{\mathcal{E}}_P)_\varphi + \eta\left(\nabla^2 - s^{-2}\right)B \\ \partial A/\partial t + s^{-1}(\mathbf{U}_P \cdot \nabla)(sA) &= \mathcal{E}_\varphi + \eta\left(\nabla^2 - s^{-2}\right)A \end{aligned}\right\}, \tag{9.7}$$

and when $\boldsymbol{\mathcal{E}}$ is given by (9.3c), they become

$$\left.\begin{aligned} \partial B/\partial t + s(\mathbf{U}_P \cdot \nabla)(s^{-1}B) &= s(\mathbf{B}_P \cdot \nabla)\omega + (\nabla \wedge \alpha\mathbf{B}_P)_\varphi + \eta_e\left(\nabla^2 - s^{-2}\right)B \\ \partial A/\partial t + s^{-1}(\mathbf{U}_P \cdot \nabla)(sA) &= \alpha B + \eta_e\left(\nabla^2 - s^{-2}\right)A \end{aligned}\right\}, \tag{9.8}$$

where $\eta_e = \eta + \beta$ is the 'effective diffusivity' acting on the mean field; in (9.8) it is assumed for simplicity that β, and so η_e, are uniform.

There are two source terms involving $\mathbf{B}_P$ on the right of (9.8), and the type of dynamo depends crucially on which of these dominates. The ratio of these terms is in order of magnitude

$$\frac{|s(\mathbf{B}_P \cdot \nabla)\omega|}{|\nabla \wedge (\alpha \mathbf{B}_P)|} = \mathcal{O}\left(\frac{L^2\omega_0'}{\alpha_0}\right), \tag{9.9}$$

where α_0 is a typical value of α, and ω_0' a typical value of $|\nabla\omega|$. If $|\alpha_0| \gg |L^2\omega_0'|$, then the differential rotation term in (9.8) is negligible and we have simply

$$\partial B/\partial t + s(\mathbf{U}_P \cdot \nabla)(s^{-1}B) = [\nabla \wedge (\alpha \mathbf{B}_P)] \cdot \mathbf{e}_\varphi + \eta_e\left(\nabla^2 - s^{-2}\right)B\,. \tag{9.10}$$

The α-effect here acts both as the source of poloidal field (via the term αB in (9.8)) and as the source of toroidal field (via the term $[\nabla\wedge(\alpha\mathbf{B}_P)]\cdot\mathbf{e}_\varphi$ in (9.10)). Dynamos that depend on this reciprocal process are described as 'α^2-dynamos'.

If on the other hand $|\alpha_0| \ll |L^2\omega_0'|$, then the differential rotation term in (9.8) dominates, so that

$$\partial B/\partial t + s(\mathbf{U}_P\cdot\nabla)(s^{-1}B) = s(\mathbf{B}_P\cdot\nabla)\omega + \eta_e\left(\nabla^2 - s^{-2}\right)B\,. \tag{9.11}$$

Now, toroidal field is generated by differential rotation, and poloidal field is generated by the α-effect; dynamos that function in this way are described as '$\alpha\omega$-dynamos'. It will be noticed that, if the distinction between actual and effective variables is ignored, then (the second equation of) (9.8) and (9.11) are precisely the equations (8.65) and (8.66) obtained by Braginskii. The assumption $|\alpha_0| \ll |L^2\omega_0'|$ is implicit in Braginskii's analysis; in fact from (8.58), $|\alpha_0| \sim \eta u_0^2/LU_0^2 \sim \mathcal{R}_m^{-2}U_0$ so that, with $L^2|\omega_0'| \sim U_0$, in Braginskii's model with $\mathcal{R}_m \gg 1$,

$$\left|\alpha_0/L^2\omega_0'\right| \sim \mathcal{R}_m^{-2} \ll 1\,. \tag{9.12}$$

9.2 Free modes of the α^2-dynamo

To exhibit dynamo action in its simplest form, suppose that the fluid domain V is of infinite extent, that $\mathbf{U} \equiv 0$, and that $\boldsymbol{\mathcal{E}}$ is given by (9.3c) with α and β uniform and constant. Equation (9.1) then becomes

$$\partial\mathbf{B}/\partial t = \alpha\nabla\wedge\mathbf{B} + \eta_e\nabla^2\mathbf{B}\,. \tag{9.13}$$

Now let $\hat{\mathbf{B}}(\mathbf{x})$ be any field satisfying the 'force-free' condition

$$\nabla\wedge\hat{\mathbf{B}}(\mathbf{x}) = K\,\hat{\mathbf{B}}(\mathbf{x})\,, \tag{9.14}$$

where K is constant. Examples of such fields have been given in §2.5. For such a field,

$$\nabla^2\hat{\mathbf{B}} = -\nabla\wedge\nabla\wedge\hat{\mathbf{B}} = -K^2\mathbf{B}\,, \tag{9.15}$$

and it is evident from (9.13) that if $\mathbf{B}(\mathbf{x},0) = \hat{\mathbf{B}}(\mathbf{x})$, then

$$\mathbf{B}(\mathbf{x},t) = \hat{\mathbf{B}}(\mathbf{x})e^{pt} \tag{9.16}$$

where

$$p = \alpha K - \eta_e K^2\,. \tag{9.17}$$

Hence the field grows exponentially in strength (its force-free structure being preserved) provided

$$\alpha K > \eta_e K^2\,, \tag{9.18}$$

i.e. provided the initial scale of variation of the field $L = |K|^{-1}$ is sufficiently large.

To be specific, let us suppose that $\alpha > 0$, so that the growth condition is simply

$$0 < K < K_c = \alpha/\eta_e \,. \tag{9.19}$$

The maximum growth rate p_m occurs for $K = \frac{1}{2}K_c$, and is

$$p_m = \alpha^2/4\eta_e \,. \tag{9.20}$$

For self-consistency of the two-scale approach of Chapter 7, we require that $K_c\ell$ ($= \ell/L$) should be small. In the case of turbulence with $\mathcal{R}_{\rm m} \ll 1$, this condition is certainly satisfied; for in this case, from (7.89) and (7.98),

$$\alpha \sim \ell u_0^2/\eta, \quad \beta \sim (\ell u_0)^2/\eta \;\; (\ll \eta)\,, \tag{9.21}$$

and so

$$K_c\ell \sim (\ell u_0/\eta)^2 = \mathcal{R}_{\rm m}^{\;2} \ll 1 \,. \tag{9.22}$$

Similarly in the case of a random wave field with no zero-frequency ingredients, and with spectral peak at frequency $\omega_0 (= t_0^{-1})$ and wave number $k_0 (= \ell^{-1})$ satisfying

$$\omega_0 \gg \eta k_0^2 \,, \tag{9.23}$$

the expressions (7.77) and (7.97) lead to the estimates

$$\alpha \sim \eta t_0^2 u_0^2/\ell^3, \quad \beta \sim \eta(u_0 t_0/\ell)^2 \ll \eta \,, \tag{9.24}$$

and so in this case (under the condition (7.29))

$$K_c\ell \sim (u_0 t_0/\ell)^2 \ll 1 \,. \tag{9.25}$$

However there is a potential inconsistency when there *are* zero frequency ingredients in the wave spectrum if (as is by no means certain) α and β do tend to finite limits as $\eta \to 0$. If this is the case, then on dimensional grounds,

$$\alpha \sim u_0, \quad \beta \sim u_0\ell \;\; (\gg \eta)\,, \tag{9.26}$$

and so

$$K_c\ell = \alpha\ell/\eta_e = \mathcal{O}(1)\,. \tag{9.27}$$

In this situation, the medium would in fact be most unstable to magnetic modes whose length-scale is of the same order as the scale of the background **u**-field; this conclusion is incompatible with the two-scale approach leading to equation (9.13), and indicates again that conclusions that lean heavily on the estimates (9.26) must be treated with caution.

9.2.1 Weakly helical situation

Of course the estimate $\alpha \sim u_0$ is valid only if the **u**-field is *strongly helical* in the sense that

$$|\langle \mathbf{u} \cdot \omega \rangle| \sim \ell^{-1} \langle \mathbf{u}^2 \rangle . \tag{9.28}$$

In a *weakly helical* situation, with

$$|\langle \mathbf{u} \cdot \omega \rangle| = \varepsilon \ell^{-1} \langle \mathbf{u}^2 \rangle, \quad \varepsilon \ll 1 , \tag{9.29}$$

the estimate for α must be modified, while that for β remains unchanged:

$$\alpha \sim \varepsilon u_0, \quad \beta \sim u_0 \ell . \tag{9.30}$$

Now we have

$$K_c \ell = \mathcal{O}(\varepsilon) \ll 1 , \tag{9.31}$$

and the conclusions are once again compatible with the underlying assumptions.

9.2.2 Influence of higher-order contributions to $\mathcal{E}$

At this point, it may be useful to examine briefly the influence of subsequent terms in the expansion (7.10) of $\mathcal{E}$ in terms of derivatives of **B**. In the isotropic situation this expansion can only take the form

$$\mathcal{E} = \alpha \mathbf{B} - \beta \nabla \wedge \mathbf{B} + \gamma \nabla \wedge (\nabla \wedge \mathbf{B}) - \cdots , \tag{9.32}$$

where γ (like α) is a pseudo-scalar, and (9.13) is replaced by

$$\partial \mathbf{B}/\partial t = \alpha \nabla \wedge \mathbf{B} - \eta_e \nabla \wedge (\nabla \wedge \mathbf{B}) + \gamma \nabla \wedge \nabla \wedge (\nabla \wedge \mathbf{B}) - \cdots . \tag{9.33}$$

The eigenfunctions of this equation are still the force-free modes and (9.17) is replaced by

$$p = \alpha K - \eta_e K^2 + \gamma K^3 - \cdots . \tag{9.34}$$

If $\gamma > 0$, the last term can be destabilising if

$$K > \eta_e / \gamma ; \tag{9.35}$$

however, since η_e and γ are both determined by the statistical properties of the **u**-field, it is to be expected that η_e/γ is of order l^{-1} (at least) on dimensional grounds. The condition (9.35) is then incompatible with the condition $Kl \ll 1$, and the conclusion is that dynamo instabilities associated with the third (and subsequent) terms of (9.32) are unlikely to arise within the framework of a double-length-scale theory.

9.3 Free modes when α_{ij} is anisotropic

Suppose now that α_{ij} is no longer isotropic, but that it is still uniform and symmetric with principal values $\alpha^{(1)}$, $\alpha^{(2)}$, $\alpha^{(3)}$. We restrict attention here to the situation in which the β_{ijk} contribution to $\boldsymbol{\mathcal{E}}$ is negligible *cf.* (9.21b) or (9.24b). Then, with $\mathbf{U} = 0$, (9.1) becomes

$$\frac{\partial B_i}{\partial t} = \epsilon_{ijk}\alpha_{km}\frac{\partial B_m}{\partial x_j} + \eta\nabla^2 B_i\,. \tag{9.36}$$

This equation admits plane wave solutions of the form

$$\mathbf{B} = \hat{\mathbf{B}}\,\mathrm{e}^{pt}\,\mathrm{e}^{\mathrm{i}\mathbf{K}\cdot\mathbf{x}}\,, \quad \mathbf{K}\cdot\hat{\mathbf{B}} = 0\,, \tag{9.37}$$

and substitution in (9.36) gives

$$(p + \eta K^2)\hat{B}_i = \mathrm{i}\epsilon_{ijk}\alpha_{km}K_j\hat{B}_m\,. \tag{9.38}$$

If we refer to the principal axes of α_{km}, the first component of (9.38) becomes

$$(p + \eta K^2)\hat{B}_1 = \mathrm{i}\left(\alpha^{(3)}K_2\hat{B}_3 - \alpha^{(2)}K_3\hat{B}_2\right), \tag{9.39}$$

and the two other components are given by cyclic permutation of suffixes. For a non-trivial solution $(\hat{B}_1, \hat{B}_2, \hat{B}_3)$, the determinant of the coefficients must vanish. This gives a cubic equation for p with roots

$$p_0 = -\eta K^2\,, \quad p_1 = -\eta K^2 + Q\,, \quad p_2 = -\eta K^2 - Q\,, \tag{9.40}$$

where

$$Q^2 = \alpha^{(2)}\alpha^{(3)}K_1^2 + \alpha^{(3)}\alpha^{(1)}K_2^2 + \alpha^{(1)}\alpha^{(2)}K_3^2\,. \tag{9.41}$$

We are here only interested in the possibility of exponential growth of $\mathbf{B}$, for which $\Re\mathrm{e}\, p > 0$. The condition for this to occur is evidently

$$Q^2 > \eta^2 K^4\,. \tag{9.42}$$

The surface $Q^2 = \eta^2 K^4$ is sketched in Figure 9.1 in the axisymmetric situation $\alpha^{(1)} = \alpha^{(2)}$ (when it is a surface of revolution about the K_3-axis) for three values of the parameter $\mu = \alpha^{(1)}/\alpha^{(3)}$. Wave amplification now depends, as might be expected, on the direction as well as the magnitude of the wave vector $\mathbf{K}$. When $\mu > 0$, amplification occurs for all directions provided $|\mathbf{K}|$ is sufficiently small (as in the isotropic case). When $\mu < 0$ however, amplification can only occur for wavevectors within the cone $Q^2 > 0$ (and then only for sufficiently small $|\mathbf{K}|$).

The possibility $\alpha^{(1)}\alpha^{(3)} < 0$ is perhaps a little pathological in the context of turbulence as normally conceived; it would presumably arise in a situation in which $\langle u_1\omega_1\rangle$ and $\langle u_3\omega_3\rangle$ have opposite signs where $\mathbf{u}$ and $\boldsymbol{\omega}$ are as usual the random velocity and vorticity distributions. Although it is possible to conceive of artificial methods of generating such turbulence (cf. the discussion of §7.6), it is difficult to see how such a situation could arise without artificial helicity injection.

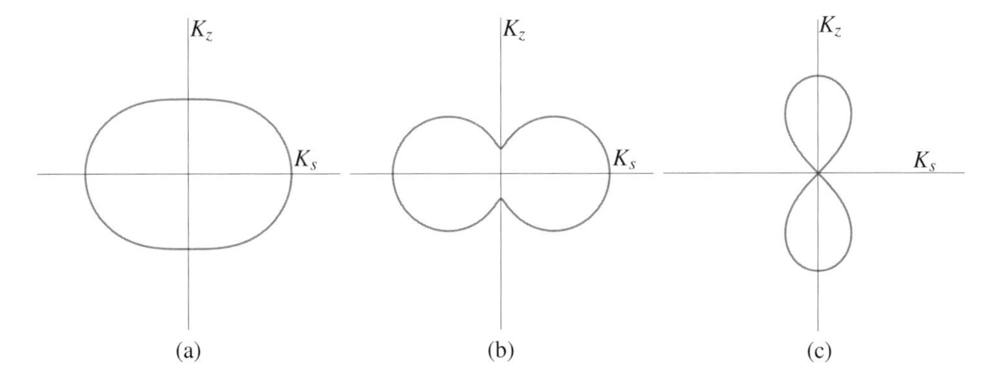

Figure 9.1 Representation of the surface $Q^2 = \eta^2 K^4$ in the (K_s, K_z) plane, with $K_s = (K_1^2 + K_2^2)^{1/2}$ and $K_z = K_3$. Q^2 is given by (9.41) and $\alpha^{(1)} = \alpha^{(2)}$. Field amplification occurs if the vector $\mathbf{K}$ is inside this surface. The plots (a), (b) and (c) correspond respectively to $\alpha^{(1)}/\alpha^{(3)} = 0.5$, 0.05 and -1.

9.3.1 Space-periodic velocity fields

Anisotropic field amplification as discussed in this section has been encountered in the closely related context of dynamo action due to velocity fields that are steady and strictly space-periodic (Childress 1970, G. O. Roberts 1970, 1972a); the methods of mean-field electrodynamics of course apply equally to this rather special situation, with the difference that the velocity spectrum tensor (which may be defined via the operation of spatial averaging) is discrete rather than continuous. In general this spectrum tensor (and similarly α_{ij}) will be anisotropic due to preferred directions that may be apparent in the velocity field. For example if

$$\mathbf{u} = u_0(\sin kz,\ \cos kz,\ 0)\,, \tag{9.43}$$

then Oz is clearly a preferred direction; and we have seen in §7.7 that in this situation

$$\alpha_{ij} = \alpha\delta_{i3}\delta_{j3} \quad \text{where } \alpha = -u_0^2/\eta k\,. \tag{9.44}$$

If however we choose a space-periodic velocity field that exhibits cubic symmetry (invariance under the group of rotations of the cube), e.g. the symmetric ABC-flow

$$\mathbf{u} = u_0(\sin kz + \cos ky,\ \sin kx + \cos kz,\ \sin ky + \cos kx) \tag{9.45}$$

(Childress 1970), then

$$\langle u_1\omega_1\rangle = \langle u_2\omega_2\rangle = \langle u_3\omega_3\rangle = ku_0^2\,, \tag{9.46}$$

and, as may be easily shown by the method of §7.7, provided $u_0/k\eta \ll 1$,

$$\alpha_{ij} = \alpha\delta_{ij} \quad \text{where } \alpha = -u_0^2/\eta k\,. \tag{9.47}$$

Hence although the velocity field (9.45) exhibits *three* preferred directions, the pseudo-tensor α_{ij} is nevertheless isotropic.[1] More generally, the velocity field

$$\mathbf{u} = u_0(\sin k_3 z + \cos k_2 y,\ \sin k_1 x + \cos k_3 z,\ \sin k_2 y + \cos k_1 x)\,, \tag{9.48}$$

for which

$$\frac{\langle u_1\omega_1\rangle}{k_2 + k_3} = \frac{\langle u_2\omega_2\rangle}{k_3 + k_1} = \frac{\langle u_3\omega_3\rangle}{k_1 + k_2} = \tfrac{1}{2}u_0^2\,, \tag{9.49}$$

yields the non-isotropic form

$$\alpha_{ij} = \alpha^{(1)}\delta_{i1}\delta_{j1} + \alpha^{(2)}\delta_{i2}\delta_{j2} + \alpha^{(3)}\delta_{i3}\delta_{j3}\,, \tag{9.50}$$

where

$$\alpha^{(1)}k_1 = \alpha^{(2)}k_2 = \alpha^{(3)}k_3 = -u_0^2/\eta\,. \tag{9.51}$$

Here k_1, k_2 and k_3 may be positive or negative, and all possible sign combinations of $\alpha^{(1)}$, $\alpha^{(2)}$ and $\alpha^{(3)}$ can therefore in principle occur.

9.3.2 The α^2-dynamo in a spherical geometry

Suppose now that α is uniform and constant within the sphere $r < R$, the external region $r > R$ being insulating. With $\mathbf{U} = 0$, equation (9.13) is still satisfied in the sphere and $\mathbf{B}$ must match to a potential field across $r = R$. This problem was first considered by Krause & Steenbeck (1967); details may be found in Chapter 14 of Krause & Rädler (1980).

We consider first the possibility of steady-state solutions of (9.13) where the current $\mathbf{J}(\mathbf{x}) = \mu_0^{-1}\nabla\wedge\mathbf{B}$ is confined to the sphere $r < R$; this current evidently satisfies[2]

$$\nabla\wedge\mathbf{J} = K\mathbf{J}\,, \quad \text{where } K = \alpha/\eta_e, \tag{9.52}$$

together with the boundary condition $\mathbf{n}\cdot\mathbf{J} = 0$ on $r = R$.[3] We seek to determine the smallest value of the magnetic Reynolds number

$$\mathcal{R}_\alpha = |\alpha|R/\eta_e, \tag{9.53}$$

for which steady dynamo action is possible.

Field structures satisfying (9.52) may be easily determined by the procedure of §2.5 (applied now to $\mathbf{J}$ rather than to $\mathbf{B}$). Letting

$$\mathbf{J} = \nabla\wedge\nabla\wedge(\mathbf{x}S) + K\nabla\wedge(\mathbf{x}S)\,, \tag{9.54}$$

[1] The situation may be compared with other familiar situations – e.g. cubic symmetry of a mass distribution is sufficient to ensure isotropy of its inertia tensor.

[2] It was assumed by Moffatt (1978a) that equation (9.52) is valid also for unsteady modes, but this is incorrect.

[3] Note that the corresponding fields are *not* force-free (see (9.59) below; we know from §2.5 that force-free fields continuous everywhere and $\mathcal{O}(r^{-3})$ at infinity do not exist. The non-existence theorem does not apply here to the field $\mathbf{J}$, for which $\mathbf{J}\wedge(\nabla\wedge\mathbf{J}) \equiv 0$, because this field has a tangential discontinuity on $r = R$.

(9.52) is satisfied provided

$$\left(\nabla^2 + K^2\right) S = 0, \quad S = 0 \quad \text{on } r = R. \tag{9.55}$$

Solutions have the form

$$S = A_n r^{-1/2} J_{n+\frac{1}{2}}(|K|r)\, S_n(\theta, \varphi), \quad (n = 1, 2, \ldots), \tag{9.56}$$

where possible values of K are determined by

$$J_{n+\frac{1}{2}}(|K|R) = 0\,. \tag{9.57}$$

Note that if α is negative, then K must be chosen negative also in (9.52) to give a growing current mode.

The roots x_{nq} of $J_{n+\frac{1}{2}}(x) = 0$ are shown in Table 2.2 (on p. 50). As we saw there, modes with $n = 1$ have dipole symmetry, those with $n = 2$ have quadrupole symmetry, and so on; while increasing q gives increasing radial structure. The smallest root (for $n \geqslant 1$) is $x_{11} = 4.493$; the minimum critical $\mathcal{R}_\alpha$ is therefore

$$\mathcal{R}_{\alpha,\min} = 4.493. \tag{9.58}$$

The field $\mathbf{B}$ corresponding to the current (9.54) may be easily determined. By uncurling (9.52) we obtain, for $r<R$,

$$\nabla \wedge \mathbf{B} = K\mathbf{B} + K\nabla\psi, \quad \text{i.e. } \nabla \wedge \mathbf{B}_P = K\mathbf{B}_T, \quad \nabla \wedge \mathbf{B}_T = K\mathbf{B}_P + K\nabla\psi, \tag{9.59}$$

for some function ψ satisfying

$$\nabla^2\psi = 0\,. \tag{9.60}$$

The toroidal part of $\mathbf{B}$ (cf. (2.48)) is simply

$$\mathbf{B}_T = \nabla \wedge (\mathbf{x}S) = -\mathbf{x} \wedge \nabla S\,, \tag{9.61}$$

and, from (9.59), the poloidal part is given by

$$\mathbf{B}_P = K^{-1}\nabla \wedge \mathbf{B}_T - \nabla\psi\,. \tag{9.62}$$

For $r > R$, $\mathbf{B}_P = -\nabla\hat{\psi}$, say, where $\nabla^2\hat{\psi} = 0$. The harmonic functions ψ and $\hat{\psi}$ must be chosen so that $\mathbf{B}_P$ is continuous across $r = R$.

When S is given by (9.56), we must have evidently

$$\psi = C_n(r/R)^n S_n(\theta, \varphi)\,, \quad \hat{\psi} = D_n(R/r)^{n+1} S_n(\theta, \varphi)\,, \tag{9.63}$$

the constants C_n and D_n being chosen so that $\mathbf{n} \cdot \mathbf{B}_P$ and $\mathbf{n} \wedge \mathbf{B}_P$ are continuous across $r = R$; these conditions give

$$C_n = D_n = -\frac{A_n}{2n+1}|K|R\frac{\mathrm{d}}{\mathrm{d}R}\left(R^{-1/2}J_{n+\frac{1}{2}}(|K|R)\right), \tag{9.64}$$

and this completes the determination of the field structures. As anticipated above, if

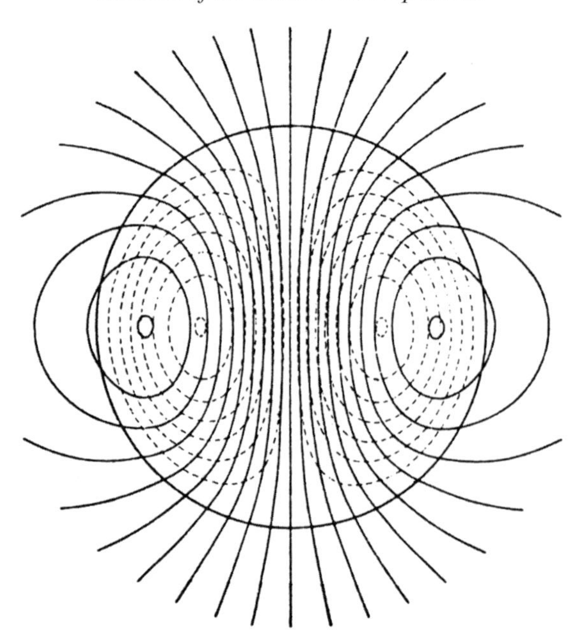

Figure 9.2 Lines of force of the poloidal field (solid) and lines of constant toroidal field (dotted) for the α^2-dynamo with α = const. in $r < R$ and with $n = q = 1$. [Krause & Steenbeck, 1967.]

$n = 1$ we have a dipole field outside the sphere, if $n = 2$ we have a quadrupole, and so on. The field structure for the dipole mode of simplest radial structure ($q = 1$) is shown in Figure 9.2.

If $\mathcal{R}_\alpha > \mathcal{R}_{\alpha,\mathrm{min}}$, then exponentially growing modes are to be expected. Assuming $\mathbf{B} = \mathbf{B}_0(\mathbf{x})\mathrm{e}^{pt}$, and with $c \equiv \mathcal{R}_\alpha$, it is found that, for modes of dipole structure ($n = 1$), possible values of $p = p_1$ are given implicitly by the equation

$$p_1 \sin \sqrt{c^2 - 4p_1} = \sqrt{c^2 - 4p_1}\left(\sin^2 \tfrac{1}{2}c - \sin^2 \tfrac{1}{2}\sqrt{c^2 - 4p_1}\right). \tag{9.65}$$

This equation determines the family of curves, symmetric about $c = 0$, shown in Figure 9.3a. Their asymptotic behaviour is given by

$$\sin \sqrt{c^2 - 4p_1} \to 0, \quad \text{i.e. } p_{1q} \sim \tfrac{1}{4}\left(c^2 - q^2\pi^2\right) \quad (q = 1, 2, 3, \ldots), \tag{9.66}$$

as shown by the dashed curves in Figure 9.3b. For fixed α and η_e, the growth rates apparently grow without limit as the sphere radius R (and so $c = \mathcal{R}_\alpha$) increases.

The number of unstable modes for large $\mathcal{R}_\alpha$ is of order $\mathcal{R}_\alpha/\pi$, and for any $\mathcal{R}_\alpha$, the growth rate p_{1q} is maximal for $q = 1$, and is a decreasing function of q.

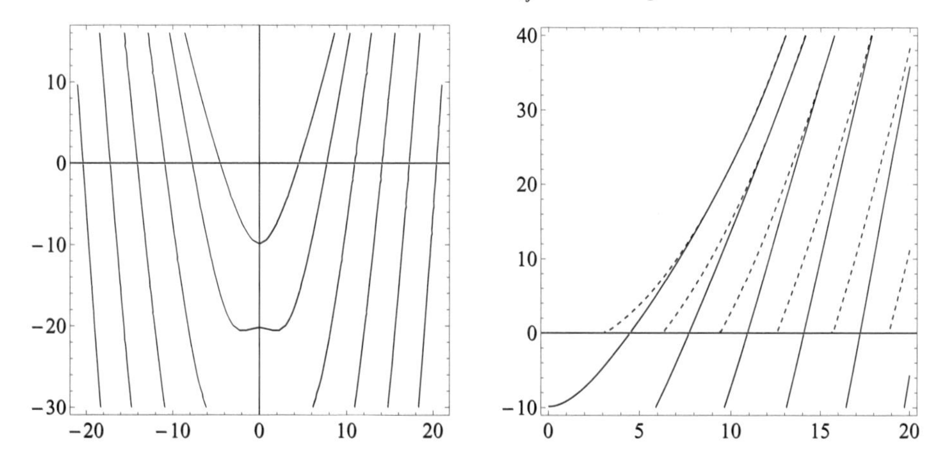

Figure 9.3 Growth rates p_{1q} of modes of dipole symmetry as a function of $\mathcal{R}_\alpha$. (a) Curves for $q = 1, 2, 3, \ldots, 6$. (b) Asymptotic behaviour as indicated by the dashed curves.

9.3.3 The α^2-dynamo with antisymmetric α

If attention is focussed on the possibility of *steady* dynamo action in a sphere due to an α-effect and still with zero mean velocity, then we are faced with the eigenvalue problem

$$\left.\begin{array}{ll} \nabla\wedge(\alpha\mathbf{B}) + \eta_e\nabla^2\mathbf{B} = 0 \ \ (r < R), & \nabla\wedge\mathbf{B} = 0 \ \ (r > R) \\ [\![\mathbf{B}]\!] = 0 \ \ \text{across } r = R\,, & \mathbf{B} = \mathcal{O}(r^{-3}) \ \ \text{at } \infty\,, \end{array}\right\}, \tag{9.67}$$

and the analysis of the preceding section in effect provides the solution of this problem in the particular situation when α is uniform. However, in a rotating body, in which the value of α is controlled in some way by Coriolis forces, a more realistic theory must allow for variation of α on scales large compared with the underlying scale l of the turbulence. In particular, it is to be expected that $\alpha(\mathbf{x})$ will have the same antisymmetry about the equatorial plane as the vertical component $\Omega\cos\theta$ of the rotation vector $\mathbf{\Omega}$ where θ is the colatitude.

A possible mechanism in the terrestrial context that makes the process physically explicit (Parker 1955b) is the following: suppose that a typical 'event' in the northern hemisphere consists of the rising of a blob of fluid (as a result of a density defect relative to its surroundings); fluid must be entrained from the sides and conservation of angular momentum (in the body of fluid which rotates as a whole) implies that the blob acquires positive helicity. Now suppose that these events occur at random throughout the body of fluid, the upward motion of the blobs being compensated by a downward flow between blobs (note the need for a topological asymmetry between upward and downward flow – cf. the discussion of §3.15); then a mean helicity distribution is generated, positive in the northern hemisphere, and (by the same argument) negative in the southern hemisphere. In the strong diffusion (or weak random wave) limit (§7.5), the corresponding value of α would be generally negative in the northern hemisphere and positive in the southern hemisphere; a simple assumption incorporating this antisymmetry is

$$\alpha(\mathbf{x}) = -\alpha_0\,\hat{\alpha}(\hat{r})\,\cos\theta\,, \tag{9.68}$$

where $\hat{r} = r/R$, and $\hat{\alpha}$ is dimensionless.[4]

Unfortunately, under any assumption of the form (9.68), the eigenvalue problem (9.67) is no longer amenable to simple analysis and recourse must be had to numerical methods. The problem was studied by Steenbeck & Krause (1966, 1969b) and P. H. Roberts (1972b) using series expansions for **B** and truncating after a few terms; Roberts found that six terms were sufficient to give eigenvalues to within 0.1% accuracy.

As in §9.3.2, the α-effect is here responsible for generating toroidal from poloidal field, *and* poloidal from toroidal field; the problem is essentially the determination of eigenvalues of the dimensionless parameter $R_\alpha = |\alpha_0|R/\eta_e$. The eigenvalues corresponding to fields of dipole and quadrupole symmetries about the equatorial plane as obtained by P. H. Roberts (1972b) for various choices of the function $\hat{\alpha}(\hat{r})$ are shown in Table 9.1. The most striking feature here is that the eigenvalues for fields of dipole and quadrupole symmetry are almost indistinguishable. This has been interpreted by Steenbeck & Krause (1969b) with reference to the field structures that emerge from the computed eigenfunctions. Figure 9.4 shows the poloidal field lines and the isotors (i.e. curves of constant toroidal fields) for the case $\hat{\alpha}(\hat{r}) = \hat{r}(3\hat{r} - 2)$. The toroidal current (giving rise to poloidal field) is concentrated in high latitudes ($|\theta|$, $|\pi-\theta| \lesssim \pi/10$), and the mutual inductance (or coupling) between these toroidal current loops is very small. Consequently the toroidal current in one hemisphere can be reversed without greatly affecting conditions in the other. This operation transforms a field of dipole symmetry into one of quadrupole symmetry and vice versa.

Note that the only O-type neutral points of the poloidal fields depicted in Figure 9.4 are situated in regions where $\alpha \neq 0$, and the field in the neighbourhood of these points can therefore be maintained by the α-effect In case (*a*), there is also a neutral point on the equatorial plane $\theta = \pi/2$ where $\alpha = 0$; this is however an X-type neutral point (i.e. a saddle point of the flux-function $\chi(s, z)$, and Cowling's neutral point argument (§6.5) does not therefore apply (Weiss 1971).

[4] It may be argued that whereas a rising blob will converge near the bottom of a convection layer (thus acquiring positive helicity in the northern hemisphere) it will diverge near the top of this layer, thus acquiring negative helicity (again in the northern hemisphere). There is therefore good reason to consider models in which $\hat{\alpha}(\hat{r})$ in (9.68) changes sign for an intermediate value of $\hat{r}$, as has been done by Yoshimura (1975).

Table 9.1 *Eigenvalues of $R_\alpha = |\alpha_0|R/\eta_e$, obtained by P. H. Roberts (1972b) for the problem (9.67) with $\alpha(\mathbf{x}) = -\alpha_0\,\hat{\alpha}(\hat{r})\cos\theta$.*

$\hat{\alpha}(\hat{r})$	R_α dipole symmetry	R_α quadrupole symmetry
1	7.64	7.81
$\hat{r}\,(3\hat{r}-2)$	24.95	24.93
$7.37\,\hat{r}^2\,(1-\hat{r})(5\hat{r}-3)$	14.10	14.10
$45.56\,\hat{r}^8\,(1-\hat{r}^2)^2$	13.04	13.11

Note: The numerical factors in $\hat{\alpha}(\hat{r})$ are chosen so that $\hat{\alpha}_{\max} = 1$. The figures given correspond to truncation of the spherical harmonic expansion of $\mathbf{B}$ at the level $n = 5$ for the first three case, and $n = 4$ for the last case, and to radial discretisation of the governing differential equations into 30 segments.

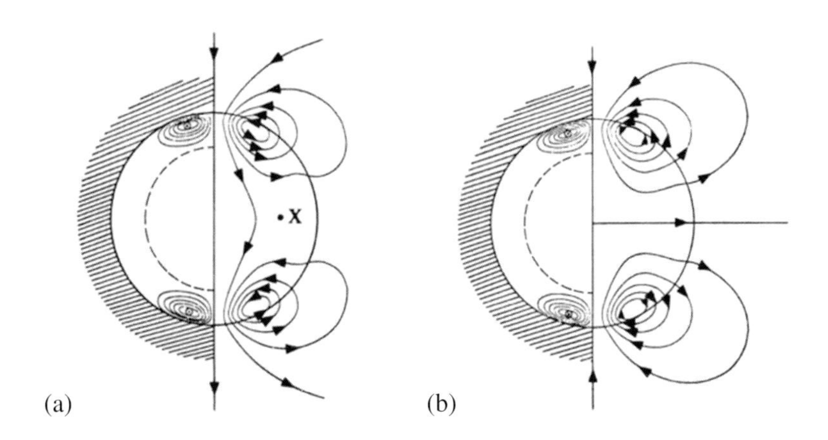

Figure 9.4 Fields excited by α^2-dynamo action when $\alpha = \alpha_0 R^{-2} r(3r - 2R)\cos\theta$ (Steenbeck & Krause 1969b): (*a*) dipole mode; (*b*) quadrupole mode. The poloidal field lines are shown on the right of each figure and the isotors on the left. Note that in case (a) the neutral point of $\mathbf{B}_P$ on the equatorial plane is X-type (rather than O-type), so that although $\alpha = 0$ at this point, Cowling's antidynamo theorem does not apply (Weiss 1971).

Steenbeck & Krause (1969b) also consider choices of $\hat{\alpha}(\hat{r})$ vanishing for values of $\hat{r}$ less than some value $\hat{r}_I$ between 0 and 1 in order to estimate the effect of the solid inner body of the Earth (for which $\hat{r}_I \approx 0.19$). The poloidal field lines and isotors were plotted for $\hat{r}_I = 0.5$ and were reproduced by Roberts & Stix (1971); the change in the eigenvalues was no more than might be expected from the reduced volume of fluid in which the α-effect is operative.

9.4 Free modes of the $\alpha\omega$-dynamo

The $\alpha\omega$-dynamo, as discussed in §9.1, is described by equations (9.8) and (9.11). Let us first consider the Cartesian analogue of these equations;[5] with mean velocity $U\mathbf{e}_y + \mathbf{U}_P$ and mean magnetic field $B\mathbf{e}_y + \mathbf{B}_P$, and with $\mathbf{B}_P = \nabla \wedge (A\mathbf{e}_y)$, these equations are

$$\left.\begin{aligned} \partial A/\partial t + \mathbf{U}_P \cdot \nabla A &= \alpha B + \eta \nabla^2 A \\ \partial B/\partial t + \mathbf{U}_P \cdot \nabla B &= \mathbf{B}_P \cdot \nabla U + \eta \nabla^2 B \end{aligned}\right\} . \tag{9.69}$$

These admit local solutions (Parker, 1955b) of the form

$$(A, B) = (\hat{A}, \hat{B}) \exp(pt + \mathrm{i}\mathbf{K} \cdot \mathbf{x}), \quad \mathbf{K} = (K_x, 0, K_z) , \tag{9.70}$$

over regions of limited extent in which $\mathbf{U}_P$, α and ∇U may all be treated as uniform. Substitution gives

$$\tilde{p}\hat{A} = \alpha \hat{B}, \quad \tilde{p}\hat{B} = -\mathrm{i}(\mathbf{K} \wedge \nabla U)_y \hat{A} , \tag{9.71}$$

where

$$\tilde{p} = p + \eta K^2 + \mathrm{i}\mathbf{U}_P \cdot \mathbf{K} .$$

Eliminating $\hat{A}$, $\hat{B}$ we obtain the dispersion relation

$$\tilde{p}^2 = 2\mathrm{i}\gamma \quad \text{where } \gamma = -\tfrac{1}{2}\alpha(\mathbf{K} \wedge \nabla U)_y . \tag{9.72}$$

The character of the solutions is largely determined by the sign of γ.

(i) *Case* $\gamma > 0$. In this case, $\tilde{p} = \pm(1 + \mathrm{i})\gamma^{1/2}$, and so

$$p = -\eta K^2 \pm \gamma^{1/2} + \mathrm{i}\left(\pm\gamma^{1/2} - \mathbf{U}_P \cdot \mathbf{K}\right) . \tag{9.73}$$

The solutions (9.70) do not decay if $\Re\mathrm{e}\, p \geqslant 0$, and this is satisfied by (9.73) (with the upper choice of sign) when $\gamma \geqslant \eta^2 K^4$, i.e. when

$$-\alpha(\mathbf{K} \wedge \nabla U)_y \geqslant 2\eta^2 K^4 . \tag{9.74}$$

The phase factor for this wave of growing (or at least non-decaying) amplitude is

$$\exp\left\{\mathrm{i}\mathbf{K} \cdot \mathbf{x} + \mathrm{i}\left(\gamma^{1/2} - \mathbf{U}_P \cdot \mathbf{K}\right) t\right\} , \tag{9.75}$$

and it propagates in the direction $\pm\mathbf{K}$ according as

$$\gamma^{1/2} - \mathbf{U}_P \cdot \mathbf{K} < \text{ or } > 0 . \tag{9.76}$$

When $\mathbf{U}_P = 0$, the field necessarily has an oscillatory character (contrast the steady free modes of the α^2-dynamo discussed in §9.2) and the wave propagates in the direction of the vector $-\mathbf{K}$.

[5] Here, and subsequently, η will be understood as including turbulent diffusivity effects when these are present.

(ii) *Case* $\gamma<0$. In this case, $\tilde{p} = \pm(1-\mathrm{i})|\gamma|^{1/2}$, and so

$$p = -\eta K^2 \pm |\gamma|^{1/2} + \mathrm{i}\left(\mp|\gamma|^{1/2} - \mathbf{U}_P \cdot \mathbf{K}\right), \tag{9.77}$$

and the growing field, when $\mathbf{U}_P = 0$, propagates in the $+\mathbf{K}$ direction.

As recognised by Parker (1955b), these results are strongly suggestive in the solar context in explaining the migration of sunspots towards the equatorial plane (as described in §5.3). In the outer convective layer of the Sun and in the northern hemisphere, let Oxyz again be locally Cartesian coordinates with Ox south, Oy east and Oz vertically upwards (Figure 9.5; cf. Figure 2.8) and suppose that vertical shear dominates in the toroidal flow so that

$$\gamma = \tfrac{1}{2}K_x\alpha\,\partial U/\partial z\,. \tag{9.78}$$

If magnetic disturbances can be represented in terms of simple migratory waves of the above kind, then the migration is towards or away from the equatorial plane according as $\alpha\,\partial U/\partial z <$ or >0. (The signs are reversed in the southern hemisphere.) The sign of the product $\alpha\,\partial U/\partial z$ (or of $\alpha\,\partial\omega/\partial r$ when we return to the spherical geometry) is of crucial importance. If sunspots are formed by distortion due to buoyant upwelling of any underlying toroidal field (see §12.1), then equatorial migration of sunspots reflects equatorial migration of the toroidal field pattern. This picture of equatorial migration as a result of the '$\alpha\omega$-effect' – i.e. the joint action of an α-effect and differential rotation (of appropriate sign) – is confirmed by numerical solutions which take due account of the spherical geometry and of spatial variation of α and $\nabla\omega$ (see §9.9 below).

If $\mathbf{U}_P \neq 0$ in (9.73) or (9.77), the phase velocity of the 'dynamo-wave' modes (9.70) is modified. If $|\gamma|^{1/2} = \mathbf{U}_P\cdot\mathbf{K}$, then the wave (9.73) (with upper sign) is stationary, while if $|\gamma|^{1/2} = -\mathbf{U}_P\cdot\mathbf{K}$, the wave (9.77) (with upper sign) is stationary. The inference is that an appropriate poloidal mean velocity may transform a situation in which an oscillating field $\hat{\mathbf{B}}(\mathbf{x})\mathrm{e}^{\mathrm{i}\omega t}$ is maintained by the $\alpha\omega$-effect into one in which the field $\mathbf{B}(\mathbf{x})$ may be maintained as a steady dynamo. The importance of meridional circulation $\mathbf{U}_P$ in determining whether the preferred mode of magnetic excitation has a steady or an oscillating character was recognised by Braginskii (1964c).

As in the case of the free modes discussed in §9.2, it is desirable for consistency that the length-scale of the most unstable mode should be large compared with the scale ℓ of the background random motions. Writing $(\mathbf{K}\wedge\nabla U)_y = KG$, where G is a representative measure of the mean shear rate, the critical wave number given by (9.74) has magnitude $K_c = (|\alpha G|/2\eta^2)^{1/3}$, and the maximum growth rate (i.e. maximum value of $\mathfrak{Re}\,p$ as given by (9.73)) occurs for $K = 2^{-4/3}K_c$. The scale of

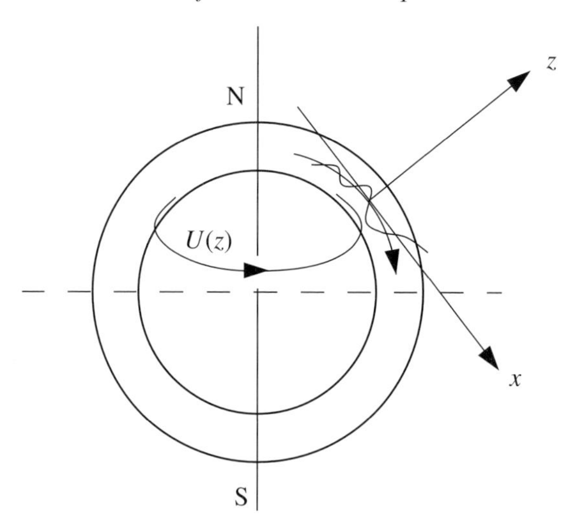

Figure 9.5 Parker's (1955b) dynamo wave, which propagates with increasing amplitude towards the equatorial plane when $\alpha\, \partial U/\partial z < 0$.

the most unstable modes is therefore given (in order of magnitude) by

$$L \sim K_c^{-1} \sim \left(\eta^2/|\alpha G|\right)^{1/3}, \tag{9.79}$$

and the consistency condition $L \gg \ell$ become

$$\eta^2/\ell^3|\alpha G| \gg 1\,. \tag{9.80}$$

If, in the weak diffusion limit, we adopt the estimates

$$|\alpha| \sim u_0, \quad \eta \sim u_0\ell\,, \tag{9.81}$$

then (9.80) become simply

$$u_0/\ell \gg |G|\,, \tag{9.82}$$

so that the treatment is consistent if the 'random shear' (of order u_0/ℓ) is large compared with the mean shear, a condition that is likely to be satisfied in the normal turbulence context. Note however that the $\alpha\omega$-model is appropriate only if $|\alpha| \ll L|G|$ (from the discussion of §9.1) so that, with the estimate $|\alpha| \sim u_0$, we in fact require the double inequality

$$|G| \ll u_0/\ell \ll |G|L/\ell\,. \tag{9.83}$$

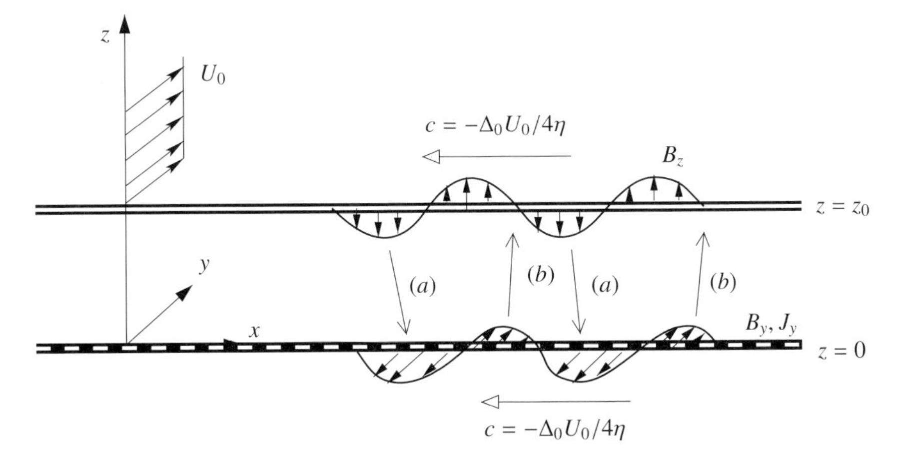

Figure 9.6 Oscillatory dynamo action with α-effect concentrated on $z = 0$ and shear concentrated on $z = z_0 > 0$. Shearing of the field B_z generates B_y which gives rise to J_y by the α-effect; this current is the source of a poloidal field $(B_x, 0, B_z)$. The complete field pattern propagates in the x-direction, and the wave amplitude grows exponentially if the wavelength is sufficiently large.

9.5 Concentrated generation and shear

In order to shed some further light on the structure of equations (9.69), consider now the situation $\mathbf{U}_P = 0$, and $\alpha = \alpha(z),\ U = U(z)$ where

$$\alpha(z) = \Delta_0\,\delta(z), \quad \mathrm{d}U/\mathrm{d}z = U_0\,\delta(z - z_0)\,, \tag{9.84}$$

so that the α-effect and the shear effect are concentrated in two parallel layers distance z_0 (> 0) apart, as illustrated in Figure 9.6. This is in fact not an unreasonable model of the type of dynamo action that can occur near the tachocline, at the base of the convection zone (see §9.10 below) (in which case z should be regarded as a vertical coordinate). It is of course an idealisation, but with the great merit that it permits simple mathematical analysis in a situation in which α and ∇U are non-uniform. More realistic distributions of α and U in terrestrial and solar contexts generally require recourse to the computer. It is useful to have some firm analytical results if only to provide points of comparison with computational studies. In the same spirit of idealisation, let us for the moment ignore the influence of fluid boundaries (as in §9.4), or equivalently assume that the fluid fills all space.

Under the assumptions (9.84), equations (9.69) become

$$\partial A/\partial t = \eta\nabla^2 A, \quad \partial B/\partial t = \eta\nabla^2 B, \quad (z \neq 0,\ z_0)\,. \tag{9.85}$$

The singularities (9.84) evidently induce discontinuities in $\partial A/\partial z$ across $z = 0$ and in $\partial B/\partial z$ across $z = z_0$, but A and B remain continuous across both layers.

Integration of (9.69) across the layers gives the jump conditions

$$\left.\begin{array}{lll} [\![\partial B/\partial z]\!] = 0, & \eta[\![\partial A/\partial z]\!] = -\Delta_0 B\,, & \text{on } z = 0 \\ [\![\partial A/\partial z]\!] = 0, & \eta[\![\partial B/\partial z]\!] = -U_0 \partial A/\partial x\,, & \text{on } z = z_0 \end{array}\right\}. \tag{9.86}$$

We suppose further that

$$A, B \to 0 \quad \text{as } |z| \to \infty\,. \tag{9.87}$$

We look for solutions of the form

$$(A, B) = \left(\hat{A}(z), \hat{B}(z)\right) \mathrm{e}^{pt+\mathrm{i}Kx}\,, \tag{9.88}$$

where K is real, but may be positive or negative. From (9.85), $\hat{A}$ and $\hat{B}$ can depend on z only through the factors $\mathrm{e}^{\pm q\hat{z}}$, where $\hat{z} = z/z_0$ and

$$\eta q^2 = z_0^2\left(p + \eta K^2\right), \quad \text{i.e.} \quad p/\eta = q^2 - K^2\,, \tag{9.89}$$

and where we may suppose that $\Re\mathrm{e}\, q > 0$. The conditions of continuity and the conditions at infinity then imply that

$$\hat{A}(\hat{z}) = \begin{cases} A_1 \mathrm{e}^{q\hat{z}} & (\hat{z} < 0), \\ A_1 \mathrm{e}^{-q\hat{z}} & (\hat{z} > 0), \end{cases} \qquad \hat{B}(\hat{z}) = \begin{cases} B_1 \mathrm{e}^{q(\hat{z}-1)} & (\hat{z} < 1), \\ B_1 \mathrm{e}^{-q(\hat{z}-1)} & (\hat{z} > 1). \end{cases} \tag{9.90}$$

The conditions (9.86) then give

$$2\eta(q/z_0)A_1 = \Delta_0 B_1 \mathrm{e}^{-q}, \quad 2\eta(q/z_0)B_1 = \mathrm{i}K\, U_0 A_1 \mathrm{e}^{-q}\,, \tag{9.91}$$

and elimination of $A_1 : B_1$ gives

$$q^2 \mathrm{e}^{2q} = \mathrm{i}\, X\,, \tag{9.92}$$

where $X = K\Delta_0 U_0\, z_0^2/4\eta^2$. This determines $q(X) = q_r + \mathrm{i}q_i$ as a function of X alone, and then $p(X, K)$ is determined by (9.89). Note that $p = p_r + \mathrm{i}p_i$ is now given by

$$p_r = \eta\left(q_r^2 - q_i^2 - K^2\right), \quad p_i = 2\eta\, q_r q_i\,. \tag{9.93}$$

Dynamo action ($p_r > 0$) can occur only if $q_r^2 > q_i^2$; and if this condition is satisfied, then dynamo action *will* occur provided

$$K^2 < Q \equiv q_r^2 - q_i^2\,. \tag{9.94}$$

Figure 9.7 shows a plot in the plane of the variables $\{q_r, q_i\}$, where the diagonal $q_r = q_i$ is dotted, with $q_r > q_i$ to the right of this diagonal. The curve $\Re\mathrm{e}\left[q^2\mathrm{e}^{2q}\right] = 0$ (as required by (9.92)) is dashed, and contours $\Im\mathrm{m}\left[q^2\mathrm{e}^{2q}\right] \equiv X = \text{const.}$ are shown solid, for representative values $X = 0.1, 1, 4$ and 16. Thus, for example, when $X = 1$, $q_r \approx 0.522$, $q_i \approx 0.284$, $Q \approx 0.192$, and dynamo action therefore occurs provided $K^2 < 0.192$. For small X, we have, asymptotically,

$$q_r \sim (X/2)^{1/2}, \quad q_i \sim q_r(1 - 2q_r), \tag{9.95}$$

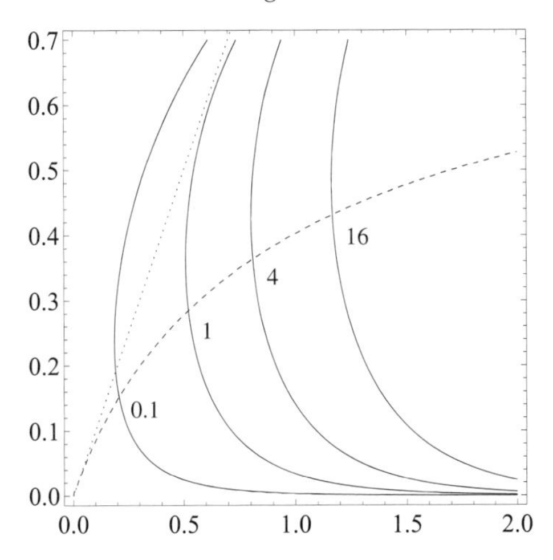

Figure 9.7 Contours X = const. (solid curves) in the $\{q_r, q_i\}$ plane ($q = q_r + \mathrm{i}q_i$), with contour levels 0.1, 1, 4, 16, as indicated. The diagonal $q_r = q_i$ is dotted. The dashed curve is $\Re\mathrm{e}\left[q^2 \mathrm{e}^{2q}\right] = 0$, and dynamo action occurs at points $\{q_r(X), q_i(X)\}$ where the X-contours intersect this curve, under the condition $(Kz_0)^2 < q_r(X)^2 - q_i(X)^2$.

and dynamo action occurs for

$$K^2 < q_r^2 - q_i^2 \sim \sqrt{2}\, X^{3/2}. \tag{9.96}$$

As in §9.4, we have here solutions of the dynamo equations (9.69) that represent migrating dynamo waves, with phase speed given by $c = -p_i/K = -2\eta q_r(X)\, q_i(X)/Kz_0^2$. For small X, this simplifies to

$$c \sim -\Delta_0 U_0/4\eta. \tag{9.97}$$

The physical nature of the dynamo process is as illustrated in Figure 9.6: shear distortion of the field component B_z at the layer $z = z_0$ generates the 'toroidal' field $B\,\mathbf{e}_y$ which diffuses (with a phase lag) to the neighbourhood of the layer $z = 0$; here the α-effect generates a toroidal current $J\,\mathbf{e}_y$ whose associated poloidal field diffuses (again with phase lag) back to the layer $z = z_0$, the net effect being regenerative as a wave of growing amplitude with phase speed c. Of course a uniform superposed fluid velocity $-c$ in the x–direction (analogous to meridional flow) would make the field pattern stationary.

9.5.1 Symmetric $U(z)$ and antisymmetric $\alpha(z)$

A case of greater interest in terrestrial and astrophysical contexts is that in which $U(z)$ is symmetric and $\alpha(z)$ is antisymmetric about the equatorial plane $z = 0$. With $\mathbf{U}_P = 0$ and $U = U(z)$, $\alpha = \alpha(z)$, equations (9.69) admit solutions of the form

(9.88) where

$$\left.\begin{aligned}(p+\eta K^2)\hat{A} &= \alpha(z)\hat{B} + \eta\hat{A}'(z)\\ (p+\eta K^2)\hat{B} &= \mathrm{i}KG(z)\hat{A} + \eta\hat{B}'(z)\end{aligned}\right\}, \tag{9.98}$$

with $G = dU/dz$. If we suppose that

$$\alpha(z) = -\alpha(-z), \quad G(z) = -G(-z), \tag{9.99}$$

then the equations (9.98) are satisfied by solutions of dipole symmetry

$$\hat{A}(z) = \hat{A}(-z), \quad \hat{B}(z) = -\hat{B}(-z), \tag{9.100}$$

or solutions of quadrupole symmetry

$$\hat{A}(z) = -\hat{A}(-z), \quad \hat{B}(z) = \hat{B}(-z). \tag{9.101}$$

The linearity of the equations (9.69) of course permits superposition of solutions of the different symmetries, each with its appropriate value of p and K.

An example similar to that treated above is provided, with $\hat{z} = z/z_0$, by the choice

$$\left.\begin{aligned}\alpha(\hat{z}) &= \Delta_0\delta(\hat{z}-\lambda) - \Delta_0\delta(\hat{z}+\lambda)\\ G(\hat{z}) &= U_0\delta(\hat{z}-(1+\lambda)) - U_0\delta(\hat{z}+(1+\lambda))\end{aligned}\right\}. \tag{9.102}$$

Here, we have placed the α-effect at $\hat{z} = \pm\lambda$, and the shear effect at $\hat{z} = \pm(1+\lambda)$, both antisymmetric about $\hat{z} = 0$. Solutions of dipole symmetry then have the form (for $\hat{z} > 0$)

$$\hat{A}(\hat{z}) = \begin{cases} A_1\dfrac{\cosh q\hat{z}}{\cosh q\lambda} & (\hat{z}<\lambda),\\ A_1\exp\left[-q(\hat{z}-\lambda)\right] & (\hat{z}>\lambda),\end{cases} \quad \hat{B}(\hat{z}) = \begin{cases} B_1\dfrac{\sinh q\hat{z}}{\sinh q(1+\lambda)} & (\hat{z}<\lambda+1),\\ B_1\exp\left[-q(\hat{z}-1-\lambda\right] & (\hat{z}>\lambda+1).\end{cases} \tag{9.103}$$

The conditions (9.86), applied now across the discontinuities at $\hat{z} = \lambda$ and $\hat{z} = 1+\lambda$ respectively, give two linear relations between A_1 and B_1, from which, after some simplification, we obtain the dispersion relation

$$\frac{q^2 e^{2q}}{1+e^{-4q}} = \mathrm{i}X. \tag{9.104}$$

This reduces to (9.92) for $\lambda \gg 1$, i.e. when the two 'double layers' are widely separated; in this limit, as might be expected, the behaviour settles down to that studied in §9.5.

This *same* dispersion relation may be derived for the quadrupole modes

$$\hat{A}(\hat{z}) = \begin{cases} A_1\dfrac{\sinh q\hat{z}}{\sinh q\lambda} & (\hat{z}<\lambda),\\ A_1\exp\left[-q(\hat{z}-\lambda)\right] & (\hat{z}>\lambda),\end{cases} \quad \hat{B}(\hat{z}) = \begin{cases} B_1\dfrac{\cosh q\hat{z}}{\cosh q(1+\lambda)} & (\hat{z}<\lambda+1),\\ B_1\exp\left[-q(\hat{z}-1-\lambda\right] & (\hat{z}>\lambda+1).\end{cases} \tag{9.105}$$

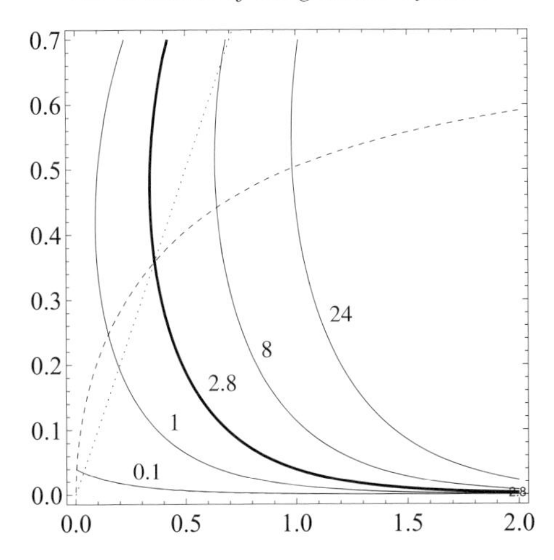

Figure 9.8 Contours $X = 0.1, 1, 2.8, 8, 24$, for $\lambda = 0.1$, otherwise as in Figure 9.7. The critical contour $X = 2.8$ is thick; this intersects the dashed curve $\Im\mathrm{m}[X] = 0$ on the diagonal $q_r = q_i$. Thus for $\lambda = 0.1$, dynamo action does not occur (i.e. all magnetic modes are damped) for $X \lesssim 2.8$.

indicating that for each unstable dipole mode there is a corresponding unstable quadrupole mode with the same (complex) growth rate.[6]

When λ decreases through values $\mathcal{O}(1)$, the behaviour is modified, as indicated in Figure 9.8, which shows (for $\lambda = 0.1$) the new contours $X = \text{const.}$, and their intersections with the dashed curve $\Im\mathrm{m}[X] = 0$ (cf. Figure 9.7). Note that now, for $X \lesssim 2.8$, the points $q_r(X), q_i(X)$ lie above the diagonal $q_r = q_i$, so dynamo action does not occur. For general $\lambda > 0$, there is similarly a minimum value $X_{min}(\lambda)$ below which dynamo action does not occur; and obviously $X_{\min}(\lambda) \to 0$ as $\lambda \to \infty$.

9.6 A model of the galactic dynamo

It has been suggested by Parker (1971a, 1971c)[7] that the galactic magnetic field may be maintained by a combination of toroidal shear (due to differential rotation in the galactic disc) and cyclonic turbulence. Cyclonic turbulence in the galaxy may be generated by the explosion of supernovae and resulting 'superbubbles' originating at or near the galactic plane of symmetry; as discussed by Ferrière (1992a), these superbubbles are efficient in generating radial (i.e. locally poloidal) field from azimuthal (i.e. toroidal) field, implying an effective distribution of α throughout the galactic disc. Although models based on the $\alpha\omega$-mechanism are by no means

[6] A discussion of more general circumstances which permit this type of correspondence between modes of dipole and quadrupole symmetries was given by Proctor (1977b).

[7] This was suggested independently by Vainshtein & Ruzmaikin (1971) (translated from Russian in Vainshtein & Ruzmaikin 1972), who considered a linear variation $\alpha(z) = \alpha_0 z$ in the range $(-z_0, z_0)$, cf. (9.107).

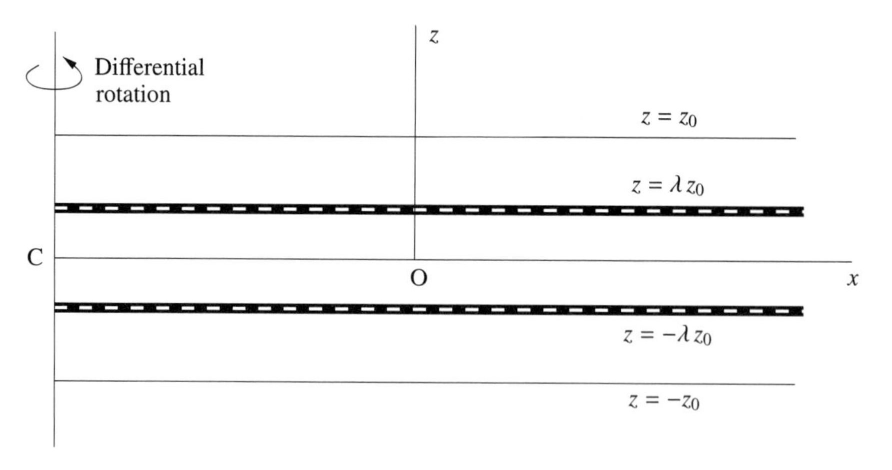

Figure 9.9 Idealised model for study of the galactic dynamo mechanism; C is the galactic centre, and O is a point on the plane of symmetry of the galactic disc. Ox is the radial extension of CO, and Oz is normal to the galactic disc. The disc boundary is represented by the planes $z = \pm z_0$; the α-effect is supposed concentrated in neighbourhoods of the planes $z = \pm\lambda z_0$ $(\lambda < 1)$. Differential rotation of the galaxy provides a mean velocity field, which in the neighbourhood of O has the form $(0, U(x), 0)$, with $\mathrm{d}U/\mathrm{d}x = G$ (const.).

universally accepted (see for example Piddington 1972a, 1972b), the analysis nevertheless provides useful insight into the structure of the dynamo equations (9.69), and is interesting in that it provides an example of dynamo action in which both steady and oscillatory dynamo modes can be excited. The model described below is more idealised than that of Parker, but permits a simpler analysis, while retaining the same essential physical features.

The origin O of a local Cartesian coordinate system is taken (Figure 9.9) at a point of the plane of symmetry of the galactic disc, with Oz normal to the disc, Ox radially outwards in the plane of the disc, and Oy in the azimuth direction. The toroidal velocity is taken to be a function of x only, with $G = \mathrm{d}U/\mathrm{d}x$ locally uniform, and equations (9.69) with $\mathbf{U}_P = 0$ then become

$$\frac{\partial A}{\partial t} = \alpha B + \eta\nabla^2 A\,, \qquad \frac{\partial B}{\partial t} = -G\frac{\partial A}{\partial z} + \eta\nabla^2 B\,. \tag{9.106}$$

For the same reasons as discussed in §9.3.3 in the solar and terrestrial contexts, it is appropriate to restrict attention to the situation in which $\alpha(z)$ is a (prescribed) *odd* function of z. Parker chose a step-function,

$$\alpha(z) = \begin{cases} +\alpha_0 & (0 < z < z_0) \\ -\alpha_0 & (-z_0 < z < 0)\,, \end{cases} \tag{9.107}$$

and he matched the resulting solutions of (9.106) to vacuum fields in the 'extra-galactic' region $|z| > z_0$, in order to obtain an equation for the (generally complex) growth rate p for field modes. Although the equations (9.106) appear simple enough, the dispersion relation turns out to be very complicated indeed and elaborate asymptotic procedures are required in the process of solution.

In order to simplify the notation, we shall set $z_0 = 1$, (i.e. we adopt z_0 as the unit of length) so that the disc boundaries are now at $z = \pm 1$. We further simplify the problem by supposing that, instead of (9.107), $\alpha(z)$ is given (as in §9.5.1) by

$$\alpha(z) = \alpha_0 \delta(z - \lambda) - \alpha_0 \delta(z + \lambda), \tag{9.108}$$

where $0 < \lambda < 1$, i.e. the α-effect is supposed concentrated in two layers $z = \pm\lambda$, the 'mean value' of $\alpha(z)$ throughout either half of the galactic disc being $\pm\alpha_0$ (as for (9.107)). Of course the choice (9.108) is an idealisation (as is (9.107) also) but in retaining the essential antisymmetry of $\alpha(z)$, it may be expected that the qualitative behaviour of the equations (9.106), should not be greatly affected.

As in the previous sections, we now investigate solutions of (9.106) of the form

$$A = \hat{A}(z) e^{pt + iKx}, \quad B = (G/\eta)\, \hat{B}(z) e^{pt + iKx}, \tag{9.109}$$

the factor G/η being introduced simply for convenience. Here again, the wave-number K may be positive or negative, and $p = p_r + ip_i$ is complex if the corresponding mode is oscillatory. For $|z| < 1$, $|z| \neq \lambda$, eqns. (9.106) become

$$\frac{d^2\hat{A}}{dz^2} - q^2\hat{A} = 0, \quad \frac{d^2\hat{B}}{dz^2} - q^2\hat{B} = \frac{d\hat{A}}{dz}, \tag{9.110}$$

where, just as in (9.89),

$$q^2 = p/\eta + K^2. \tag{9.111}$$

Using (9.108), we also have the jump conditions

$$[\![d\hat{A}/dz]\!] = -X\hat{B}, \quad [\![d\hat{B}/dz]\!] = 0 \quad \text{across } z = \lambda, \tag{9.112}$$

where

$$X = \alpha_0 G/\eta^2, \tag{9.113}$$

(and similarly across $z = -\lambda$). The dimensionless number X (Parker's 'dynamo number')[8] may be positive or negative; it provides a measure of the joint influence of the α-effect and the shear G in the layer, and we are chiefly concerned to find p as a function of X and in particular to determine whether p_r can be positive for any prescribed value of X.

Now from (9.111) with $q = q_r + iq_i$, we have

$$p_r + ip_i = \eta\left(q_r^2 - q_i^2 - K^2 + 2iq_rq_i\right), \tag{9.114}$$

so that we have dynamo action ($p_r > 0$) if $q_r^2 > q_i^2$ and $|K|$ is sufficiently small. This

[8] If the scale z_0 is retained, then $X = \alpha_0 G z_0^3/\eta^2$.

allows us to focus on the situation $|K| \ll 1$ in applying the remaining boundary conditions on $z = \pm 1$. In the vacuum region $|z| > 1$, the toroidal field B must vanish, and A must be a harmonic function, i.e. $\hat{A}(z) \propto e^{-|K|z}$; continuity of $\hat{A}$, $\hat{B}$ and $d\hat{A}/dz$ across $z = 1$ then gives

$$\hat{B} = 0, \quad d\hat{A}/dz \pm |K|\hat{A} = 0 \quad \text{on } z = \pm 1 . \tag{9.115}$$

For $|K| \ll 1$, these conditions become

$$\hat{B}(z) = 0, \quad d\hat{A}/dz = 0 \quad \text{on } z = \pm 1 , \tag{9.116}$$

and these are the conditions that we use in what follows.

It is clear that with $\alpha(z) = -\alpha(-z)$, the equations (9.106) admit solutions of either dipole or quadrupole symmetry (cf. (9.100) and (9.101)). These solutions satisfy the symmetry conditions

$$\hat{B} = d\hat{A}/dz = 0 \quad \text{on } z = 0 \quad \text{for 'dipole modes'}, \tag{9.117}$$

and

$$\hat{A} = d\hat{B}/dz = 0 \quad \text{on } z = 0 \quad \text{for 'quadrupole modes'}. \tag{9.118}$$

In imposing these conditions, we may restrict attention to the region $z \geqslant 0$.

9.6.1 Dipole modes

The solution of eqns. (9.110) satisfying (9.116) and (9.117), and with A and B continuous across $z = \lambda$, can be readily obtained in the form

$$\left.\begin{aligned} \hat{A}(z) &= A_1 \frac{\cosh qz}{\cosh q\lambda} \\ \hat{B}(z) &= \tfrac{1}{2} z\hat{A}(z) + \left(B_1 - \tfrac{1}{2}A_1\lambda\right)\frac{\sinh qz}{\sinh q\lambda} \end{aligned}\right\} \quad (z < \lambda) , \tag{9.119a}$$

$$\left.\begin{aligned} \hat{A}(z) &= A_1 \frac{\cosh q(z-1)}{\cosh q(\lambda - 1)} \\ \hat{B}(z) &= \tfrac{1}{2}(z-1)\hat{A}(z) + \left(B_1 - \tfrac{1}{2}A_1(\lambda - 1)\right)\frac{\sinh q(z-1)}{\sinh q(\lambda-1)} \end{aligned}\right\} \quad (\lambda < z < 1) , \tag{9.119b}$$

where A_1 and B_1 are constants, and where we may limit attention to the possibility $q_r \geqslant 0$. Application of the jump conditions (9.112) and elimination of A_1 and B_1 gives, after some manipulation,

$$X = \frac{4q \sinh^2 q}{2\lambda \sinh q \cosh q(2\lambda - 1) - \sinh 2q\lambda} = X_{\text{dip}}(q, \lambda), \ \text{say}, \tag{9.120}$$

from which $q(X, \lambda)$, and hence p from (9.111), may be determined.

The function $X_{\text{dip}}(q, \lambda)$ is antisymmetric about $\lambda = \frac{1}{2}$. Limiting forms of this

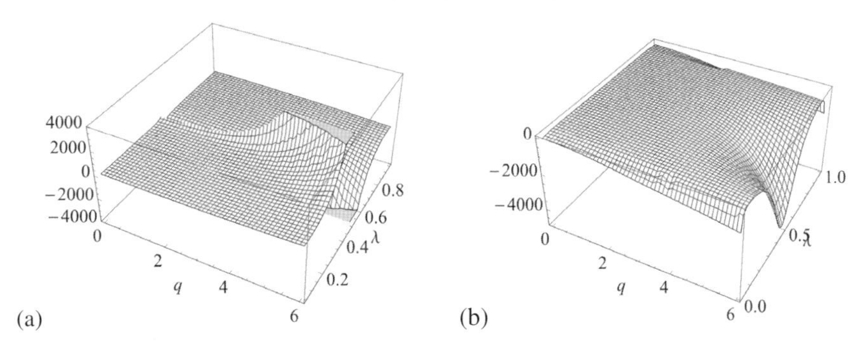

Figure 9.10 (a) Perspective plot of the function $X_{\mathrm{dip}}(q, \lambda)$; note the anti-symmetry about $\lambda = \frac{1}{2}$; $X_{\mathrm{dip}}(q, \lambda)$ is actually singular at $\lambda = 0$ and $\lambda = 1$ as well as at $\lambda = \frac{1}{2}$, although not apparent on the scale of this plot. (b) Similar plot for $X_{\mathrm{quad}}(q, \lambda)$, symmetric about $\lambda = \frac{1}{2}$; here the singular behaviour at $\lambda = 0$ and $\lambda = 1$ is quite evident.

function at $q = 0$ and (for real q) as $\lambda q \to \infty$ are

$$X_{\mathrm{dip}}(0, \lambda) = \frac{3}{\lambda(1-\lambda)(1-2\lambda)}\,, \tag{9.121}$$

and

$$X_{\mathrm{dip}}(q, \lambda) \sim (2/\lambda) q\, \mathrm{e}^{2\lambda q} \quad \text{as } \lambda q \to \infty \ (0 < \lambda < \tfrac{1}{2}). \tag{9.122}$$

For any λ in the range $0 < \lambda < \frac{1}{2}$, $X_{\mathrm{dip}}(q, \lambda)$ is an increasing function of q, minimal at $q = 0$. Figure 9.10a shows a 3D perspective plot of this function for real q in the range $0 < q < 6$ and for $0 < \lambda < 1$. For any value of λ in this range, and any $X > X_{\mathrm{dip}}(0, \lambda)$, one may read off from a plot of this kind a corresponding value of $q(X, \lambda) > 0$ at which non-oscillatory dynamo action of dipole symmetry occurs for all wave numbers K satisfying $K^2 < q^2$, i.e. for large enough field wavelengths in the x-direction; the maximum growth rate is then $p = \eta q^2$ corresponding to the limit $|K| \to 0$.

9.6.2 Quadrupole modes

Turning now to the quadrupole modes, the solution of (9.110) satisfying (9.116) and (9.118) has the form

$$\left.\begin{aligned} \hat{A}(z) &= A_2 \frac{\sinh qz}{\sinh q\lambda} \\ \hat{B}(z) &= \tfrac{1}{2} z \hat{A}(z) + \left(B_2 - \tfrac{1}{2} A_2 \lambda\right) \frac{\cosh qz}{\cosh q\lambda} \end{aligned}\right\} \quad (z < \lambda)\,, \tag{9.123a}$$

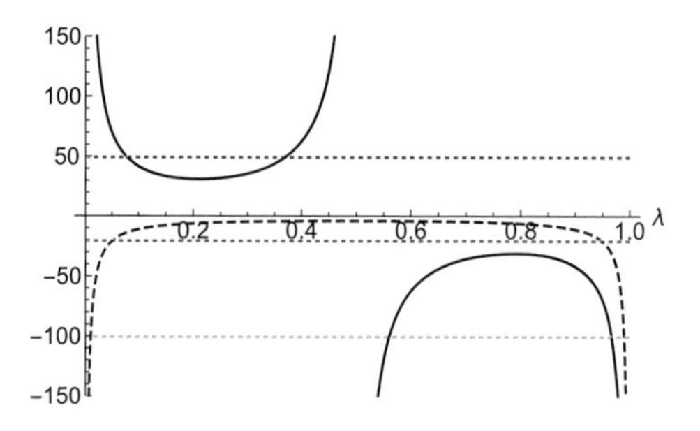

Figure 9.11 The functions $X_{\text{dip}}(0, \lambda)$ (solid curves) and $X_{\text{quad}}(0, \lambda)$ (dashed curve) given by (9.121) and (9.125); for any given value of X (represented by the dotted lines at $X = 50, -20$ and -100), dynamo action is possible for the range of λ on the intercept(s) between the intersections of these curves.

$$\left.\begin{aligned} \hat{A}(z) &= A_2 \frac{\cosh q(z-1)}{\cosh q(\lambda - 1)} \\ \hat{B}(z) &= \tfrac{1}{2}(z-1)\hat{A}(z) + \left(B_2 - \tfrac{1}{2}A_2(\lambda - 1)\right)\frac{\sinh q(z-1)}{\sinh q(\lambda-1)} \end{aligned}\right\} \quad (\lambda < z < 1) \qquad (9.123b)$$

where again A_2, B_2 are constants, and the relation between X and $\{q, \lambda\}$ corresponding to (9.120) takes the form

$$X = X_{\text{quad}}(q, \lambda) = \frac{4q\cosh^2 q}{2\lambda \cosh q \sinh q(2\lambda - 1) - \sinh 2q\lambda}. \qquad (9.124)$$

In this case, the function $X_{\text{quad}}(q, \lambda)$ is negative and symmetric about $\lambda = \frac{1}{2}$; the asymptotics here are

$$X_{\text{quad}}(0, \lambda) \sim -\frac{1}{\lambda(1-\lambda)}, \qquad (9.125)$$

and, for $0 < \lambda < \frac{1}{2}$,

$$X_{\text{quad}}(q, \lambda) \sim -(2/\lambda) q\, e^{2\lambda q} \quad \text{as } \lambda q \to \infty. \qquad (9.126)$$

Figure 9.10b shows the function $X_{\text{quad}}(q, \lambda)$ over the range $0<\lambda<1$, $0<q<6$, from which it is evident that for any λ in this range, and for any $X_{\text{quad}} < X_{\text{quad}}(0, \lambda)$, there exists a real positive q giving non-oscillatory dynamo action with quadrupole symmetry for $K^2 < q^2$.

Figure 9.11 shows the limiting functions $X_{\text{dip}}(0, \lambda)$ (solid) and $X_{\text{quad}}(0, \lambda)$ (dashed) given by (9.121) and (9.125). This indicates, for each X, the range of λ in which non-oscillatory dynamo action of dipole and/or quadrupole symmetry is possible. For example, for $X = 50$ (dotted blue line), dynamo action with dipole symmetry is possible for $0.077 < \lambda < 0.371$; while for $X = -20$ (dotted red line), dynamo action with quadrupole symmetry is possible for $0.053 < \lambda < 0.947$. For $X < -32$, dynamo modes of both symmetries are possible: for example, for $X = -100$ (green

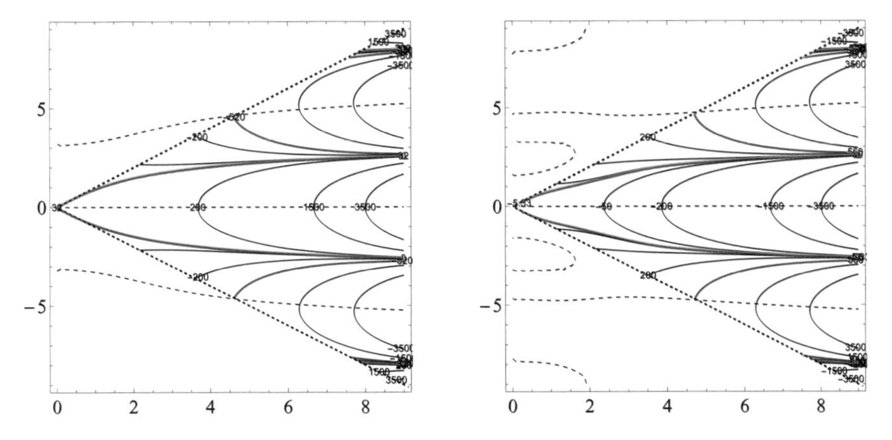

Figure 9.12 Oscillatory dynamo modes for the particular choice $\lambda = 0.25$, indicated by intersection of solid curves $\Re e[X] = \text{const.}$ and the dashed curves on which $\Im m[X] = 0$, in the dynamo region $q_r > |q_i|$ of the $\{q_r, q_i\}$ plane; both diagrams are symmetric about the q_r-axis. (a) Dipole modes: the blue contour is $X = 3((\lambda(1-\lambda)(1-2\lambda))^{-1}) = 32$, and contours for $X > 32$ intersect the q_r-axis at points corresponding to non-oscillatory modes; the red contour, which intersects the dashed curve on the diagonal $q_r = q_i$, is $X = -520$, and oscillatory dipole modes appear for $X < -520$. (b) Quadrupole modes: the blue contour is $X = -(\lambda(1-\lambda))^{-1} \approx -5.33$, and in this case contours for $X < -5.33$ intersect the q_r-axis at points corresponding to non-oscillatory modes; the red contour is $X = 560$, and oscillatory quadrupole modes appear for $X > 560$.

dotted line), there are dipole dynamo modes for $0.561 < \lambda < 0.966$, and quadrupole dynamo modes for $0.010 < \lambda < 0.990$.

9.6.3 Oscillatory dipole and quadrupole modes

For oscillatory modes, with p (and therefore q) complex, we may adopt the procedure of §§9.5 and 9.5.1 to describe them. Thus, for the dipole modes, we compute the contours of $\Re e[X_{\text{dip}}(q, \lambda)]$ for any fixed λ, in the plane of the variables $\{q_r, q_i\}$ and determine the points to the right of the dotted diagonal $q_r = q_i$ where these contours intersect the dashed contour $\Im m[X_{\text{dip}}(q, \lambda)] = 0$. Similarly for $\Re e[X_{\text{quad}}(q, \lambda)]$. Figure 9.12 shows the results for the case $\lambda = 1/4$, in the region $0 < q_r < 9$, $0 < q_i < q_r$. These contour plots, on which some of the contour levels are indicated, must be interpreted as described in the figure caption.

If these plots are extended to higher values of q_r and $|q_i|$, then additional dynamo branches appear for higher values of $|X|$ and the growth rates of all dynamo modes increase with increasing $|X|$. All in all, the situation is one of considerable complexity, and may serve as a warning of similar difficulties to be encountered in interpreting the results of more elaborate numerical models.

9.6.4 Oblate-spheroidal galactic model

A model of galactic dynamo action in which the galaxy was represented by a very flattened oblate spheroid was investigated numerically by Stix (1975, 1978), and by White (1978) with improved numerical methods. Soward (1978) provided an asymptotic analysis based on the slow decrease of galactic thickness with radius, and confirmed White's numerical results, which showed that steady dynamo modes can be much more easily excited than oscillatory modes, and that the quadrupole mode has a lower critical value of the dynamo number $|X|$ than the dipole mode and is therefore more easily excited. These results are consistent with the results of the model analysed above (as revealed by Figure 9.12 for the particular choice $\lambda = 0.25$ and by Figure 9.11 for all $\lambda \in (0, 1)$).

9.7 Generation of poloidal fields by the α-effect

In a spherical geometry, and still restricting attention to the situation $\mathbf{U}_P = 0$, the equations governing $\alpha\omega$-dynamo models are (from (9.8) and (9.11))

$$\left.\begin{aligned} \partial A/\partial t &= \alpha B + \eta\left(\nabla^2 - s^{-2}\right)A \\ \partial B/\partial t &= s\,(\mathbf{B}_P\cdot\nabla)\,\omega + \eta\left(\nabla^2 - s^{-2}\right)B \end{aligned}\right\}, \tag{9.127}$$

(where any eddy diffusivity effect is still assumed incorporated in the value of the parameter η). We have in §3.14 considered the process of generation of toroidal field $B\,\mathbf{e}_\varphi$ when the poloidal field $\mathbf{B}_P$ is supposed given. It is natural now to consider the complementary process of generation of poloidal field $\mathbf{B}_P = \nabla \wedge (A\,\mathbf{e}_\varphi)$ by the α-effect when the toroidal field $B\,\mathbf{e}_\varphi$ is supposed given. The similar structure of both equations (9.127) makes this a straightforward task, in the light of the results of §3.14.

Suppose that $B(r,\theta)\,\mathbf{e}_\varphi$ is a given steady toroidal field such that B/s is everywhere bounded (a natural condition to impose if the poloidal current associated with $B\,\mathbf{e}_\varphi$ is finite). Suppose further that $\alpha(r,\theta)$ is given as a function of r and θ and that $A = 0$ at time $t = 0$. Then for $t \ll L^2/\eta$, where L is the characteristic scale of the source term αB in (9.127), we have simply

$$A(r,\theta,t) = \alpha(r,\theta)\,B(r,\theta)\,t\,. \tag{9.128}$$

When t is of order L^2/η and greater, diffusion of the poloidal field thus generated becomes important, and for $t \gg L^2/\eta$, a steady state is approached in which

$$\left(\nabla^2 - s^{-2}\right)A = -\eta^{-1}\alpha B \tag{9.129}$$

within the conducting region. Let us as usual suppose that this is the region $r < R$,

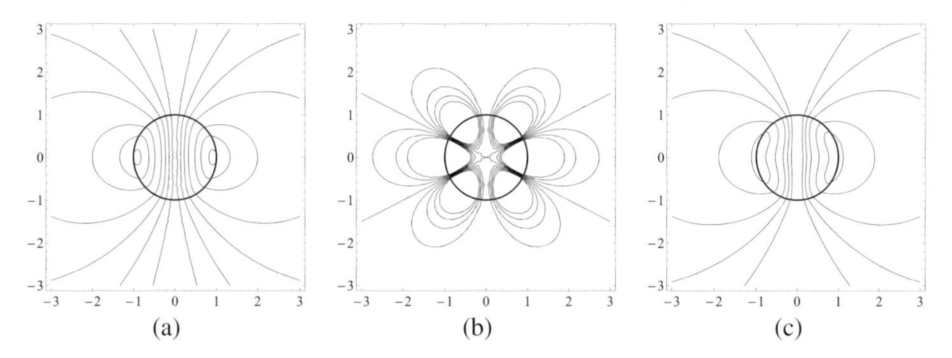

Figure 9.13 Field lines of (a) the dipole ingredient; (b) the sextupole ingredient in which the limitations of the plotting routing near $r = 0$ are evident; and (c) the total field, as given by equation (9.140).

and that the region $r > R$ is insulating so that

$$\left(\nabla^2 - s^{-2}\right) A = 0, \quad (r > R). \tag{9.130}$$

As in §3.14, it is apparent that if $\alpha B/\eta$ has the expansion

$$\alpha B/\eta = \sum_1^\infty F_n(r)\mathrm{d}P_n(\cos\theta)/\mathrm{d}\theta\,, \tag{9.131}$$

with $F_n(r) \equiv 0$ for $r > R$, then the solution of (9.129), (9.130) is

$$A(r,\theta) = -\sum_1^\infty G_n(r)\mathrm{d}P_n(\cos\theta)/\mathrm{d}\theta\,, \tag{9.132}$$

where

$$r^{-2}(r^2 G_n')' - r^{-2}n(n+1)G_n = F_n(r)\,. \tag{9.133}$$

Continuity of $\mathbf{B}_P$ across $r = R$ requires that

$$[G_n] = [G_n'] = 0 \quad \text{on} \quad r = R\,. \tag{9.134}$$

The solution (cf. (3.134)) is given by

$$G_n(r) = -\frac{1}{2n+1}\left\{\frac{1}{r^{n+1}}\int_0^r x^{n+2}F_n(x)\mathrm{d}x + r^n\int_r^\infty \frac{F_n(x)}{x^{n-1}}\mathrm{d}x\right\}. \tag{9.135}$$

Suppose for example that, for $r < R$,

$$B(r,\theta) = B_0(r/R)\cos\theta\sin\theta, \quad \alpha(r,\ \theta) = \alpha_0\cos\theta\,. \tag{9.136}$$

Then

$$\frac{\alpha B}{\eta} = \frac{\alpha_0 B_0 r}{\eta R}\cos^2\theta\sin\theta = Cr\frac{\mathrm{d}}{\mathrm{d}\theta}\left(2P_3(\cos\theta) + 3P_1(\cos\theta)\right), \tag{9.137}$$

where $C = -\alpha_0 B_0 R/15\eta$. Hence, for $r<R$,

$$F_1(r) = 3Cr, \quad F_3(r) = 2Cr, \tag{9.138}$$

and $F_n(r) \equiv 0\,(n \neq 1 \text{ or } 3)$. With $\hat{r} = r/R$, (9.135) now gives

$$G_1(\hat{r}) = \begin{cases} -(C\hat{r}/2)(1 - 3\hat{r}^2/5) & (\hat{r} < 1) \\ -C/5\hat{r}^2 & (\hat{r} > 1) \end{cases} \tag{9.139a}$$

$$G_3(\hat{r}) = \begin{cases} -2C\hat{r}^3\,(1 + 7\log(1/\hat{r}))\,/49 & (\hat{r} < 1) \\ -(2C/49)\hat{r}^{-4} & (\hat{r} > 1) \end{cases} \tag{9.139b}$$

and, from (9.132), we then have for $\hat{r} > 1$

$$A(\hat{r}, \theta) = CR\left[\frac{1}{5\,\hat{r}^2}\sin\theta + \frac{2}{49\,\hat{r}^4}(5\cos^2\theta - 1)\sin\theta\right], \tag{9.140}$$

and similarly for $\hat{r} < 1$. The terms on the right represent the fields of a dipole ($\mathcal{O}(r^{-2})$) and a sextupole ($\mathcal{O}(r^{-4})$) as $r \to \infty$.

The field lines of these dipole and sextupole ingredients are shown in Figures 9.13a,b. Figure 9.13c shows the field lines of the resultant total field, given by the contours $A(r, \theta) r \sin\theta = \text{const}$; this total field is dominated by the dipole ingredient and is only weakly perturbed by the sextupole (a consequence of the small factor 2/49 in (9.140)).

It is clear from this example not only that the α-effect can generate poloidal field directly from toroidal field in the interior region $r < R$, but also that this poloidal field will inevitably diffuse to the external region $r > R$. In combination with the effect of differential rotation in the interior, we have therefore the essential ingredients of a 'dynamo cycle'

$$\mathbf{B}_T \xrightarrow{\alpha\text{-effect}} \mathbf{B}_P \xrightarrow{\text{differential rotation}} \mathbf{B}_T. \tag{9.141}$$

This is the essence of the '$\alpha\omega$-dynamo', to which we now turn.

9.8 The $\alpha\omega$-dynamo with periods of stasis

In order to demonstrate that the α-effect in conjunction with differential rotation can indeed act as a dynamo, it is useful to adopt the artifice of 'stasis' as described in §6.15. Suppose that we start with a purely poloidal field $\mathbf{B}_{P0} = \nabla \wedge (A_0 \mathbf{e}_\varphi)$, with

$$A_0 = C_0 r^{-1/2} J_{3/2}(kr)\sin\theta \quad (r<R), \tag{9.142}$$

where kR is the first zero ($= \pi$) of $J_{1/2}(kR)$; i.e. we start with the fundamental dipole mode of lowest natural decay rate (§2.8). The field for $r > R$ is of course harmonic and matches smoothly with the interior field. Let us now subject this field to a short period t_1 of intense differential rotation $\omega(r)$. Diffusion is negligible if this process is sufficiently rapid, and from equation (3.125) a toroidal field is generated in $r<R$ and is given at $t = t_1$ by

$$\mathbf{B}_T = B\mathbf{e}_\varphi = r\sin\theta B_r\omega'(r)t_1\mathbf{e}_\varphi = 2C_0 r^{-1/2}J_{3/2}(kr)\omega'(r)\,t_1\sin\theta\cos\theta\,\mathbf{e}_\varphi. \quad (9.143)$$

We now stop the differential rotation and 'switch on' an intense α-effect with $\alpha = \alpha_0(r)\cos\theta$, which is maintained for a second short interval t_2 during which again diffusion is negligible. From (9.128), at time $t = t_1 + t_2$, an additional poloidal field $\mathbf{B}_P = \nabla\wedge(A\mathbf{e}_\varphi)$ has been generated, where

$$A = 5f(r)\sin\theta\cos^2\theta = -f(r)\frac{\mathrm{d}}{\mathrm{d}\theta}\left(P_1(\cos\theta) + \tfrac{2}{3}P_3(\cos\theta)\right), \quad (9.144a)$$

with

$$f(r) = \frac{2C_0}{5r^{1/2}}J_{3/2}(kr)\omega'(r)\alpha_0(r)t_1t_2\,. \quad (9.144b)$$

Finally we allow a period t_3 of stasis (i.e. pure ohmic decay) so that only the slowest decaying ingredient of (9.144a) will survive after a long time. The function $f(r)$ may be expanded as an infinite sum of radial functions

$$f(r) = \sum_{q=1}^{\infty} C_{nq}f_{nq}(r)\,, \quad (9.145)$$

with $n = 1$ or 3 as appropriate (n and q being the same labels as used in §2.8). The slowest decaying mode is that for which $n = q = 1$, i.e. precisely the mode (9.142) with which we started. By waiting long enough, we can ensure that the contamination by higher modes is negligible; and by increasing $|\omega'(r)\alpha_0(r)|$ (keeping other things constant), we can further ensure that at time $t = t_1 + t_2 + t_3$ the field that survives is more intense than the initial field given by (9.142). The process may then be repeated indefinitely to give sustained dynamo action.

This type of dynamo may be oscillatory or non-oscillatory, according as C_{11}/C_0 is less than or greater than zero. This depends on the precise radial dependence of the product $\omega'(r)\alpha_0(r)$. Reversing the sign of this product will clearly convert a non-oscillatory dynamo to an oscillatory dynamo, or vice versa.

9.9 Numerical investigations of the $\alpha\omega$-dynamo

A number of numerical studies have been made of equations (9.8) and (9.11) in a spherical geometry and with a variety of prescribed axisymmetric forms for the functions $\alpha(\mathbf{x})$, $\omega(\mathbf{x})$ and $\mathbf{U}_P(\mathbf{x})$. These studies are for the most part aimed at constructing plausible models for possible dynamo processes within either the liquid core of the Earth or the convective envelope of the Sun. A few general conclusions have emerged from these studies. In describing these, we may limit attention to the situation in which $\alpha(\mathbf{x})$ and $U_z(\mathbf{x})$ are odd functions of the coordinate z normal to the equatorial plane, and $\omega(\mathbf{x})$ and $U_s(\mathbf{x})$ are even functions of z. In this case equations (9.8) and (9.11) admit solutions of dipole symmetry (A even, B odd in z) or quadrupole symmetry (A odd, B even). In all cases of course the field must be matched to an irrotational field across the spherical boundary $r = R$.

The behaviour of the system is characterised (as in §9.6) by a dimensionless dynamo number (cf. 9.113),

$$X = \alpha_0 \omega_0' R^4 / \eta^2 , \tag{9.146}$$

where α_0 and ω_0' are typical values of α and $\partial\omega/\partial r$. To be specific, following the convention adopted by P. H. Roberts (1972b), let α_0 be the value of α at the point in the northern hemisphere where $|\alpha|$ has its maximum value (so that α_0 may be positive or negative) and let ω_0' be the value of $\partial\omega/\partial r$ where $|\partial\omega/\partial r|$ is maximal. X may then likewise be positive or negative.

Equations (9.8) and (9.11) admit solutions proportional to e^{pt}, and (as for the laminar dynamo theories discussed in §6.14) the problem that presents itself is essentially the determination of possible values of $p = p_r + ip_i$ as functions of X (and any other dimensionless parameter that may appear in the specification of the velocity field). Interest centres on that value of X for which p_r first becomes positive as $|X|$ increases continuously from zero. The corresponding field structure may have dipole or quadrupole symmetry – and we then say that the dipole (or quadrupole) mode is the *preferred* mode of excitation. Moreover if $p_i = 0$ when $p_r = 0$, this preferred mode is non-oscillatory, while if $p_i \neq 0$ when $p_r = 0$ it is oscillatory.

P. H. Roberts (1972b) studied a number of models, both with and without meridional circulation $\mathbf{U}_P$, for various smooth choices of $\alpha(\mathbf{x})$ and $\omega(\mathbf{x})$ (satisfying the symmetry conditions specified above) with the following conclusions:

(i) When $X > 0$, and $\mathbf{U}_P = 0$, the mode that is preferred is quadrupole and oscillatory; critical values of X ranging between 76^2 and 212^2 were obtained depending on the particular distribution of α and ω adopted. Introduction of a small amount of meridional circulation can however yield a preferred mode that is dipole and non-oscillatory, the critical value of X being reduced by a factor of order $\frac{1}{2}$ or less in the process. This rather dramatic effect of meridional velocity was first recognised by Braginskii (1964c) and presumably admits interpretation in terms of the convective effect of $\mathbf{U}_P$ on dynamo waves in a sense contrary to their natural phase velocity (cf. the discussion at the end of §9.4). Unfortunately however it appears (Roberts 1972b) that the *sense* of the meridional circulation (i.e. whether from poles to equator or equator to poles on the surface $r = R$) that will lead to a reduction in the critical value of X depends on the model (i.e. on the particular choice of $\alpha(\mathbf{x})$

and $\omega(\mathbf{x})$), and a simple physical interpretation of the effect of $\mathbf{U}_P$ therefore seems unlikely.

(ii) When $X < 0$, the conclusions are reversed; i.e. when $\mathbf{U}_P = 0$, the preferred mode is dipole and oscillatory, critical values of X ranging between -74^2 and -206^2 for the models studied; and introduction of suitable meridional velocity $\mathbf{U}_P$ (the sense being again model-dependent) substantially reduces the critical value of $|X|$, the preferred mode becoming quadrupole and non-oscillatory.

The oscillatory character of the modes when $\mathbf{U}_P = 0$ was confirmed by Jepps (1975) who carried out a direct numerical integration of equations (9.8) and (9.11) again with specified simple forms for $\alpha(\mathbf{x})$ and $\omega(\mathbf{x})$ with $|X|$ marginally greater than its critical value X_c as determined by the eigenvalue approach. For initial condition at time $t = 0$, Jepps assumed the field to be purely poloidal and in the fundamental decay mode for a sphere (§2.8); in the subsequent evolution, the field rapidly settled down to a time-periodic behaviour (modulated by the slow amplification expected as a result of the supercritical choice of X).

In the case $X < 0$, and when $\mathbf{U}_P = 0$, it is a characteristic feature of the periodic solutions that both poloidal and toroidal field ingredients appear to originate in the polar regions and then to amplify during a process of propagation towards the equatorial plane; there, diffusion eliminates toroidal field of opposite signs from the two hemispheres. This type of behaviour can be seen clearly in Figure 9.14 (from Roberts 1972b) which shows a half-cycle of both poloidal and toroidal fields for a model in which

$$\alpha = \frac{729}{16}\alpha_0 \hat{r}^8 (1-\hat{r}^2)^2 \cos\theta, \quad \omega = -\frac{19683}{40960}\omega_0'(1-\hat{r}^2)^5, \tag{9.147}$$

where $\hat{r} = r/R$, the coefficients being chosen so that the maximum values of $|\alpha|$ and $|\partial\omega/\partial r|$ are respectively $|\alpha_0|$ and $|\omega_0'|$. This apparent propagation from poles to equator is of course consistent with the behaviour of the plane-wave solutions discussed in §9.4: in that case also, the necessary condition for propagation towards the equatorial plane was $\alpha\, \partial\omega/\partial r < 0$.

On the assumption that sunspots form by a process of eruption when the subsurface toroidal field exceeds some critical value (see §12.11), a number of authors have, on the basis of time-periodic solutions of the dynamo equations such as that depicted in Figure 9.14, constructed butterfly diagrams (§5.3) defined in this context as the family of curves

$$B(r,\theta,t) = k\, B_{\max}(r) \tag{9.148}$$

in the plane of the variables θ and t, for a fixed representative value of r; here $B_{\max}$ is simply the maximum value of $B(r,\theta,t)$ for $0 \leqslant \theta \leqslant \pi$, $0 \leqslant t \leqslant 2\pi/p_i$, and k is a constant between 0 and 1. With the expectation that sunspots may be expected to form in any region where $|B| > k|B_{\max}|$ (for some k), these diagrams

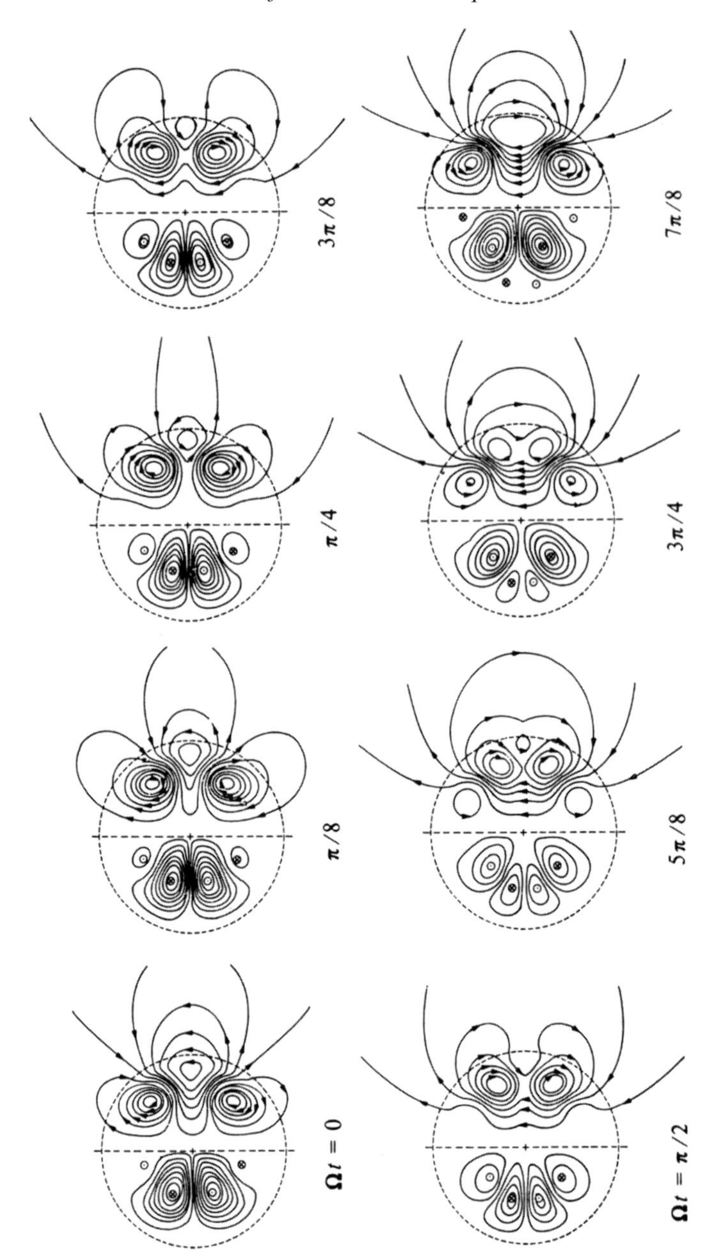

Figure 9.14 Evolution of the marginal dipole oscillation of the $\alpha\omega$-dynamo defined by (9.147), with $X = -206.1$, $\Omega = \eta^{-1}R^2 p_i = 47.44$. Meridian sections are shown, the symmetry axis being dotted. The $\mathbf{B}_P$ lines (χ = const.) are shown on the right of this axis, and the lines of constant toroidal field (B_φ = const.) on the left, at equal intervals of χ and B_φ respectively; B_φ is positive in regions marked $\odot$, and negative in regions marked $\otimes$. The progression of the pattern from poles to equator, as the half-cycle proceeds, is apparent. [Republished with permission of The Royal Society, from P. H. Roberts 1972b; permission conveyed through Copyright Clearance Center, Inc.]

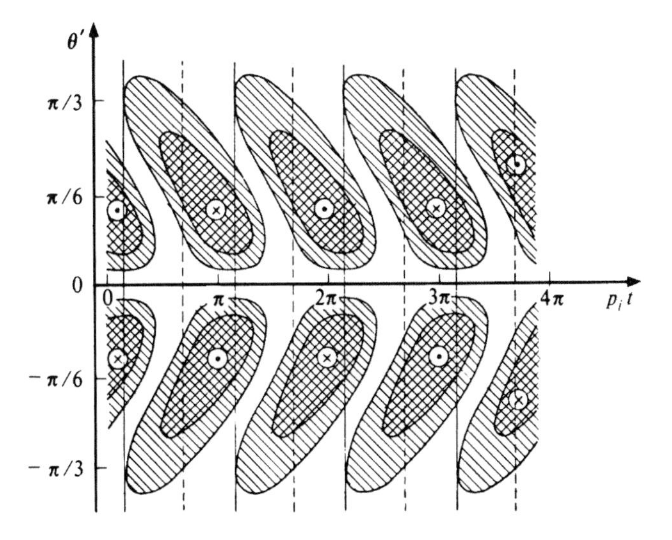

Figure 9.15 Butterfly diagram corresponding to the $\alpha\omega$-dynamo given by (9.149); the area $|B_\varphi| > \frac{1}{3}B_{\max}$ is hatched, and the area $|B_\varphi| > \frac{2}{3}B_{\max}$ is cross-hatched. The unbroken lines mark the phase at which the polarity at the poles changes; the dashed lines mark the phase at which the sign of the dipole moment changes. The phase $p_i t = n\pi$ ($n = 0, \pm 1, \pm 2, \dots$) corresponds to the maximum value of the toroidal field. [From Steenbeck and Krause 1969a.]

are directly comparable with Maunder's butterfly diagram (Figure 9.14) depicting observed occurrences of sunspots (again in the θt plane). Figure 9.15 is a diagram obtained by Steenbeck and Krause (1969a), with[9]

$$\left.\begin{aligned} 2\,\alpha &= \alpha_0\,(1 + \mathrm{erf}\,[40(\hat{r} - 0.9)/3)])\cos\theta \\ 2\,\omega &= \omega_0'\,(1 - \mathrm{erf}\,[40(\hat{r} - 0.7)/3)]) \end{aligned}\right\}. \tag{9.149}$$

The *qualitative* resemblance with Figure 5.6 is impressive. Quantitative comparison of course requires that the period $T = 2\pi/p_i$ of the theoretical solution be comparable with the period of the sunspot cycle, i.e. about 22 years. The various computed solutions (e.g. Roberts 1972b) give

$$p_i \approx 100\eta/R^2\,, \qquad T \approx 2\pi R^2/100\eta\,, \tag{9.150}$$

and with $R \approx 7 \times 10^5$ km, and a turbulent diffusivity[10] $\eta \approx u_0 l \approx 10^2\mathrm{km}^2\mathrm{s}^{-1}$, we get an estimated period $T \approx 10$ yr, which is of the right order of magnitude.

The condition $\alpha_0\omega_0' < 0$ is necessary to give propagation of field patterns towards the equatorial plane (and hence butterfly diagrams with the right qualitative

[9] These distributions of α and ω were chosen to provide a model in which the two types of inductive activity are separated in space; the choice is clearly arbitrary.

[10] If instead we choose $\eta \simeq 10^3\mathrm{km}^2\mathrm{s}^{-1}$ as suggested by the granulation scales (see Chapter 4), then we get $T \simeq 1$ yr, an order of magnitude smaller than the observed period.

characteristics). This condition would appear to leave open the two possibilities

$$\alpha_0 > 0,\ \omega_0' < 0 \quad \text{or} \quad \alpha_0 < 0,\ \omega_0' > 0\,. \tag{9.151}$$

(Recollect here that α_0 represents the extreme value of $\alpha(\mathbf{x})$ in the *northem* hemisphere.) It was however pointed out by Stix (1976) that the linear relation between the fields A and B (and hence between $\mathbf{B}_P$ and B) involves α and $\partial\omega/\partial r$ separately and not merely via their product $\alpha\,\partial\omega/\partial r$ (cf. equation (9.71)) and that the observed phase relation between the radial component of the Sun's general field and the toroidal component (as revealed by the sunspot pattern) is in fact incompatible with the second possibility in (9.151). Hence if the Sun does act as an $\alpha\omega$-dynamo, then the indications are that α is predominantly positive in the northern hemisphere (and negative in the southern), and that $\omega(r,\theta)$ increases with increasing depth.

There are independent arguments (Steenbeck et al. 1966) for the conclusion that $\alpha > 0$ in the northern hemisphere, based on simple dynamical considerations. We have already commented (§9.3.3) on one physical mechanism (viscous entrainment) that may generate positive helicity and so *negative* α in the northern hemisphere when blobs of hot fluid rise due to buoyancy forces. In the convective envelope of the Sun, there is however a second mechanism which has the contrary effect: compressibility. A blob of fluid rising through several scale heights will expand and so will tend to rotate in a sense opposite to the mean solar rotation (to conserve its absolute angular momentum); this leads to generation of negative helicity, and so positive α in the northern hemisphere, and it seems at least plausible that this is the dominant effect in the solar context. The detailed dynamical calculations of §12.11 below, in which the value of α associated with buoyancy instabilities is calculated, suggests that the true picture may be rather more complicated than suggested by these simple arguments.

Likewise, increase of ω with increasing depth is what one would naïvely expect as a result of conservation of angular momentum; in conjunction with meridional circulation, this might be expected to lead to a situation in which $\omega \propto r^{-2}$ in the convection zone. The same argument would suggest however that on the Sun's surface the rotation rate should be greater in the polar regions than in the equatorial zone, whereas the observed situation is quite the opposite. The dynamical theory of differential rotation of the Sun is a very large subject in its own right (see e.g. Durney 1976), and lies outside the scope of this book. Clearly however dynamo models can serve to eliminate those distributions $\omega(r,\theta)$ that are totally incompatible with observed solar magnetic activity. For example, Stix (1976) has commented that models in which $\omega = \omega(s)$ with $s = r\sin\theta$ are constrained by the observed distribution of ω on the solar surface $r = R$, and therefore satisfy $\partial\omega/\partial r > 0$; since, as mentioned above, this is incompatible with the observed phase relation between A and B on $r = R$, one may reasonably conclude that ω is *not* constant on

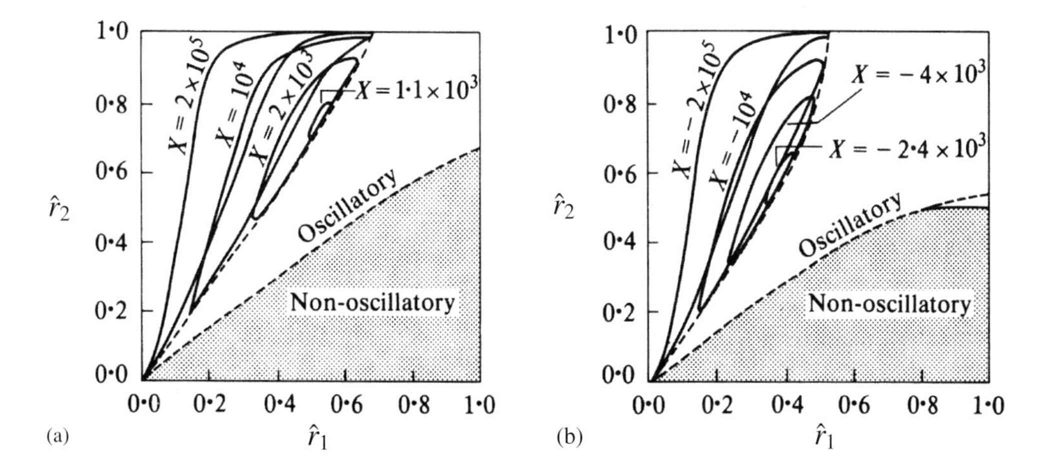

Figure 9.16 Regions of the $\hat{r}_1 - \hat{r}_2$ space for which oscillatory or non-oscillatory modes are preferred for the $\alpha\omega$-dynamo given by (9.152); both (a) (dipole symmetry) and (b) (quadrupole symmetry) are symmetric about the line $\hat{r}_1 = \hat{r}_2$. Variation of the critical dynamo number X for different values of $\hat{r}_1$ and $\hat{r}_2$ is indicated by the solid curves. [From Deinzer et al. 1974.]

cylindrical surfaces $s = s_0$ through the convection zone. This is indeed confirmed by the constraints on $\omega(r, \theta)$ determined by helioseismology – see §9.10 below.

The models discussed so far (particularly those of Roberts 1972b) have the property that, when $\mathbf{U}_P = 0$, the preferred modes are oscillatory. Examples are known however for which this is not the case. Deinzer et al. (1974) have studied the effect of concentrated layers of inductive activity, i.e.

$$\alpha = \alpha_0 \delta(\hat{r} - \hat{r}_2) \cos\theta\,, \quad \partial\omega/\partial r = \omega_0' \delta(\hat{r} - \hat{r}_1), \tag{9.152}$$

where $\hat{r} = r/R$, and $0 < \hat{r}_1, \hat{r}_2 < 1$. It appears that when $\hat{r}_1$ and $\hat{r}_2$ are sufficiently separated, the preferred modes are non-oscillatory. Figure 9.16 indicates the character of the preferred mode for the possible values of $\hat{r}_1$ and $\hat{r}_2$. Here again (as in Herzenberg's two-sphere dynamo discussed in §6.9) spatial separation of the regions of inductive activity appears to encourage the possibility of non-oscillatory dynamos through the filtering out of 'unwanted' harmonics of the main field. This effect would be most easily analysed for the distribution (9.152) of α and ω in the extreme case $\hat{r}_1 \ll \hat{r}_2 \ll 1$ in which only the leading field harmonics generated at either layer have any significant influence at the other layer.

A related situation was investigated by Levy (1972), viz.

$$\alpha = \alpha_0\, \delta(\hat{r}-\hat{r}_2)(\delta(\cos\theta-\cos\alpha)-\delta(\cos\theta+\cos\alpha))\,, \quad \partial\omega/\partial r = \omega_0'\, \delta(\hat{r}-\hat{r}_1)\,. \tag{9.153}$$

Here the α-effect is supposed concentrated in two 'rings of cyclonic activity' at

$\hat{r} = \hat{r}_2$, $\theta = \alpha, \pi - \alpha$. The same situation was further analysed by Stix (1973). The conclusion of these studies was that when $X > 0$, the preferred mode is non-oscillatory and dipole, while if $X < 0$, the preferred mode is non-oscillatory and quadrupole if $|\pi/2 - \alpha| \leqslant 54°$, otherwise oscillatory and dipole. It is perhaps not surprising that, as in the case of the δ-function model analysed in §9.6, the preferred mode can depend in quite a sensitive manner on just where the α-effect is concentrated. The true distribution of $\alpha(\mathbf{x})$ (and the related distribution of helicity density) is largely controlled by the *dynamics* of the background random motions, a topic treated in later chapters.

Finally, it is worth noting that the transition between oscillatory $\alpha\omega$-dynamo behaviour and non-oscillatory α^2-dynamo behaviour has been examined numerically by Roberts & Stix (1972). It is to be expected that if $|\alpha_0/\omega_0' R^2|$ is increased from small values, the $\alpha\omega$-behaviour will give way to the α^2-behaviour when this parameter reaches a value of order unity. For the particular model considered, Roberts & Stix found that this transition in behaviour in fact occurred when $|\alpha_0/\omega_0' R^2| \approx 0.1$. This treatment was based on an isotropic α-effect; it is perhaps worth remembering that in the presence of strong differential rotation, the assumption of isotropy in the background turbulence (or random wave motion) is really untenable; a severely non-isotropic α-effect of the form

$$\boldsymbol{\mathcal{E}} = \alpha \mathbf{B}_T + \alpha' \mathbf{B}_P, \qquad |\alpha'| \ll |\alpha|, \tag{9.154}$$

seems more plausible, and is moreover indicated by the approach of Braginskii as elaborated in Chapter 8.

Although results of the models discussed above, and in particular their ability to reproduce butterfly diagrams with the right qualitative structure, are suggestive in the solar context, nevertheless they must as yet be viewed with caution. To get the right time-scale a turbulent diffusivity of order $100\,\mathrm{km^2 s^{-1}}$ is necessary; as noted in Chapter 4, turbulence of the required intensity corresponds to a magnetic Reynolds number of order 10^4, and the turbulent magnetic fluctuations may then be expected to be an order of magnitude stronger than the mean field itself. Although the surface field of the Sun does exhibit intermittency on observable scales down to ~ 100 km, the associated observed fluctuations do not have the entirely random character that the theory strictly requires, and there is no direct evidence for field or velocity fluctuations on scales smaller than ~ 100 km. The mean field is heavily disguised by its fine-structure, and the analysis of Altschuler et al. (1974) suggests that in fact equatorial dipole and quadrupole ingredients are at least as strong, if not stronger, than the axial ingredients which appear most naturally in the theory of the $\alpha\omega$-dynamo, insofar as it can explain the solar cycle. As emphasised by Cowling (1975b), although mean-field electrodynamics is alluring in its relative simplicity,

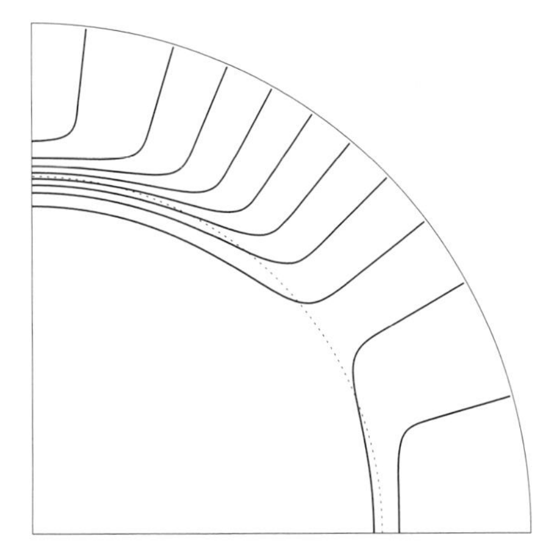

Figure 9.17 Contours of constant angular velocity based on the formula (9.155); cf. the actual angular velocity distribution of Figure 5.3a inferred from helioseismology. [From Charbonneau 2010.]

certain inconsistencies need to be resolved before it can be definitely accepted as providing the correct basic description of solar field generation.

9.10 More realistic modelling of the solar dynamo

Despite its limitations, the mean-field approach continues to provide the main approach to describing solar dynamo action. The situation has been thoroughly reviewed by Charbonneau (2010); see also Tobias & Weiss (2007). Helioseismology, as described in §5.2, has provided constraints on the distribution of toroidal velocity (i.e. of differential rotation) in the solar interior, a vital ingredient of the $\alpha\omega$ dynamo. The complex variation of toroidal velocity shown in Figure 5.3 reveals that the angular velocity $\omega(r, \theta)$ is almost independent of r throughout most of the convection zone $r_c \sim 0.7R_\odot < r < R_\odot$, but changes rapidly to the uniform angular velocity ω_c of the core through the tachocline, the rapid-shear layer of thickness $\delta \sim 0.05R_\odot$ at the base of the convection zone. This angular velocity distribution can be approximated by a formula

$$\omega(r, \theta) = \omega_c + \tfrac{1}{2}(\omega_s(\theta) - \omega_c)\,(1 - \mathrm{erf}\,[(r - r_c)/\delta])\ , \qquad (9.155)$$

where $\omega_s(\theta)$ is the surface angular velocity, as given by (5.1) or (5.2). Contours $\omega(r, \theta)$ =const. based on this formula are shown in Figure 9.17, in which the strong shear within the tachocline is very evident. The formula does not take account of the weak shear within the surface layer of the Sun, which is believed not to contribute significantly to the dynamo process.

The tachocline is the source of powerful instabilities associated both with the

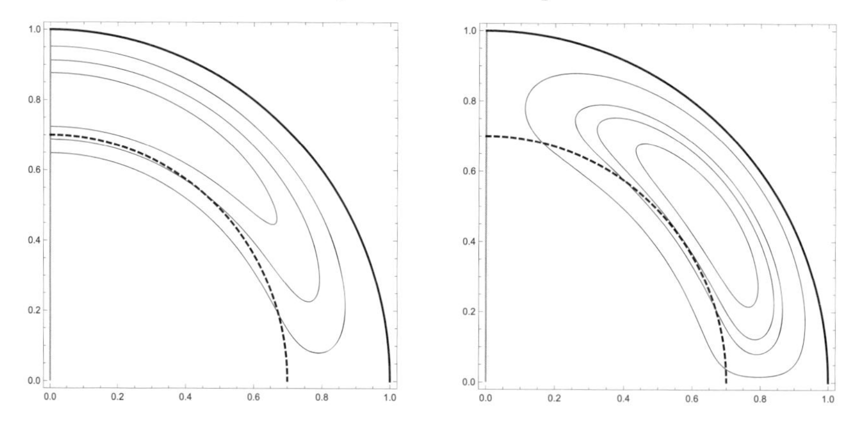

Figure 9.18 Contours α = const. for the choices (a) of equation (9.156) and (b) of equation (9.157). The choice (b), centred at $\cos\theta = 1/\sqrt{3}$, $\theta \approx 54^{\circ}$, is perhaps the more plausible.

strong shearing and also with magnetic buoyancy (to be considered in Chapter 11) in conjunction with thermal instability of a more traditional type; it is these instabilities that initiate the fully developed thermal turbulence of the overlying convective zone. There are two effects associated with such turbulence; first, the effective diffusivity $\eta(r)$ increases through the tachocline from its laminar core value to a much greater turbulent diffusivity η_e, as anticipated in Chapter 7, in the convective zone. Second, this thermal turbulence, which is influenced both by coriolis forces arising from the solar rotation and by the dynamo-generated magnetic field, must produce an α-effect, though no convincing derivation of the spatial variation of $\alpha(r,\theta)$ is as yet available; the choice of this function can best be described as a matter of inspired guesswork! Two possibilities of the type considered by Charbonneau (2010) are shown in Figure 9.18; these correspond to the choices, both antisymmetric about the equatorial plane,

$$(a) \quad \alpha(r,\theta) = \alpha_0 \exp\left[-(r-r_\alpha)^2/\delta^2\right]\cos\theta\,, \tag{9.156}$$

and

$$(b) \quad \alpha(r,\theta) = \alpha_0 \exp\left[-(r-r_\alpha)^2/\delta^2\right]\cos\theta\sin^2\theta\,, \tag{9.157}$$

where, in each case for the sake of illustration we choose $r_\alpha = 0.8R_\odot$ and $\delta = 0.1R_\odot$. The reason for localising the α-effect at some modest height above the tachocline is that (much as in §9.5) it is here that it interacts most effectively with the shear of the tachocline to generate an oscillatory dynamo through the $\alpha\omega$ mechanism. The choice (b) places the maximum of α at $r = r_\alpha$ and $\cos\theta = 1/\sqrt{3}$, i.e. $\theta \approx 54^{\circ}$, where both coriolis and magnetic effects may be expected to be strong. The parameters α_0, η_e and $r_\alpha/R_\odot$ may be adjusted to provide an oscillatory dynamo

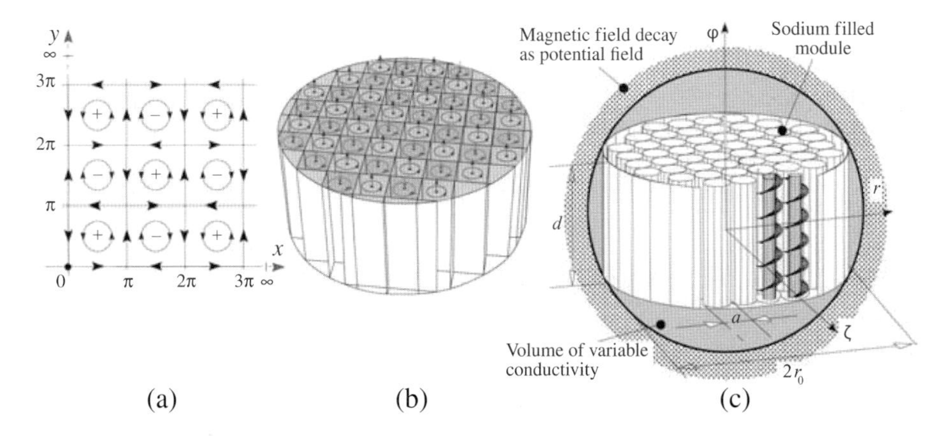

Figure 9.19 The Karlsruhe dynamo facility: (a) diagram of the flow, as conceived by G. O. Roberts (1970, 1972); (b) checkerboard array of tubes, all with right-handed helical veins (Busse 1992); (c) schematic of the actual facility. [From Müller et al. (2004).]

that mimics the 22-year periodicity of the solar dynamo, and the field migration towards the equator revealed by butterfly diagrams. Of course, such models can be criticised precisely because there are too many adjustable parameters.

The situation gets worse as more effects are incorporated, each involving one or more additional ill-determined parameters. Such effects are associated with meridional flow, non-isotropy of the α-tensor, field transport associated with an anti-symmetric contribution to the α-tensor, non-isotropic contributions to the β-tensor, topological pumping, as considered in §3.15, magnetic buoyancy (to be considered in §12.11), and, perhaps most importantly, the back-reaction of the Lorentz force on the turbulence which will presumably tend to 'quench', i.e. suppress, both α and β. Some or all of these effects have been included in models developed over recent decades; among them we may cite in particular Rüdiger & Brandenburg (1995) who provide an extensive treatment particularly of the effects of anisotropy. These authors attribute the α-effect to interaction of rotation $\mathbf{\Omega}$ and gradient of turbulent intensity. While it is true that there is a strong gradient of turbulent intensity through the tachocline, a gradient of mean density $\nabla\rho$ in the direction of the local gravitational acceleration $\mathbf{g}$, combining with rotation through the pseudo-scalar $\mathbf{g}\cdot\mathbf{\Omega}$, is presumably sufficient to produce the required non-isotropic α-effect. We shall find this to be the case in later chapters.

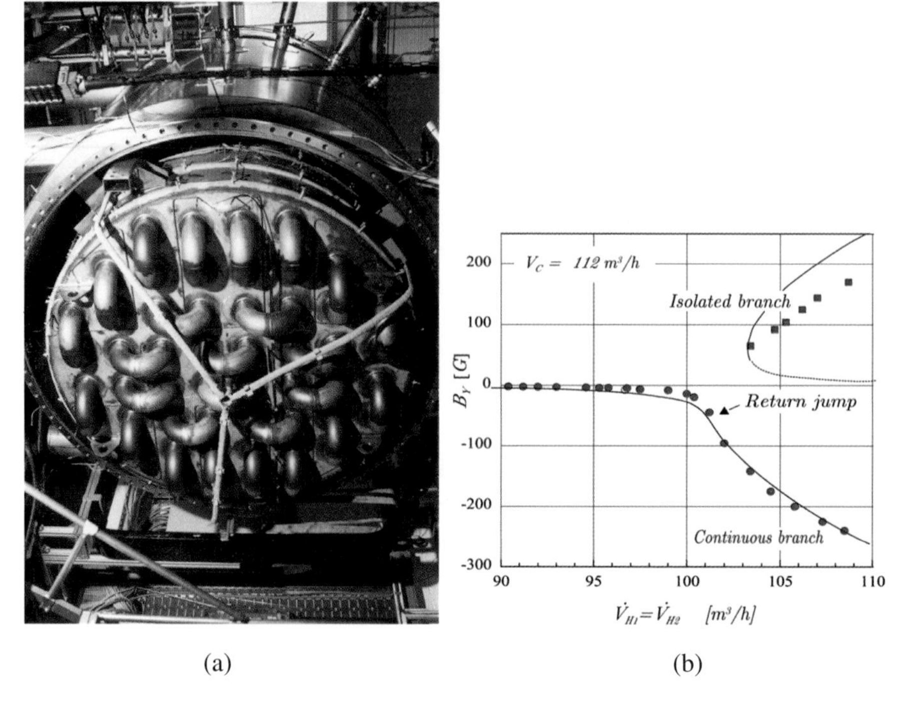

(a) (b)

Figure 9.20 (a) Photograph from above of the actual dynamo assembly at Karslruhe. (b) Measurements of the horizontal field component B_y for increasing volume flow rate through the tube system; the seed field is that of the ambient Earth's magnetic field (~0.3 Gauss); the dynamo generated field increases to values as high as 250 Gauss. [From Müller et al. 2008.]

9.11 The Karlsruhe experiment as an α^2-dynamo

As we have seen, α^2-dynamo action can in principle occur in any turbulent flow of non-zero mean helicity, or equally in any space-periodic flow, even when the magnetic Reynolds number based on the scale of periodicity is small. A family of such flows was conceived by G. O. Roberts (1970, 1972). These flows are illustrated in Figure 9.19a; they are steady space-periodic flows of the form $\mathbf{u} = (u(x, y), v(x, y), w(x, y))$, with mean helicity $\mathcal{H} = 2\langle \omega(x, y) w(x, y) \rangle$ where $\omega = \partial v/\partial y - \partial u/\partial x$.

Such a flow has been realised experimentally by the arrangement shown in Figures 9.19b and 9.19c (Müller et al. 2004) in which "the rectangular vortex-containing channels of the G. O. Roberts flow were replaced by vortex generators constructed as coaxial annular channels with helical baffles in the annulus to generate the helical flow [flow rate $\dot{V}_H$] and a central pipe through which the central flow [flow rate $\dot{V}_C$] could be varied independently". These vortex generators were arranged in checkerboard array, and liquid sodium was pumped through the system,

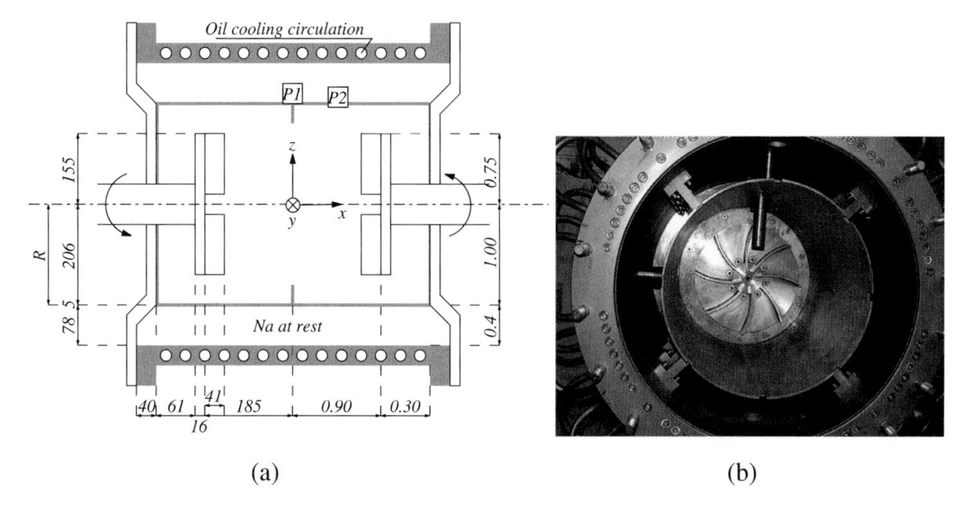

Figure 9.21 The von Karman Sodium (VKS) experiment: (a) sketch of the facility [reprinted figure with permission from Monchaux et al. 2007, copyright 2007 by the American Physical Society]; (b) photograph of the apparatus, from the side.

effectively up and down the tubes alternately; the helical baffles, all right-handed, ensure that the mean helicity is positive. Figure 9.20a is a photograph of the experimental assembly from above, showing tube connexions at the ends.

The field generated is almost entirely in the transverse (horizontal) plane; Figure 9.20b shows some measurements of the component B_y of the field for increasing values of the flow rate through the tubes. There are two branches of instability, one continuous, one isolated; the solid curves result from a low-order model dynamical system (Tilgner & Busse 2002), of a type to be considered in §13.3. The seed field in this case is just the ambient magnetic field of the Earth ($\sim$ 0.3 Gauss), whereas the generated field is $\sim$ 250 Gauss for the largest flow rates achieved. The field saturates at a level controlled by the back-reaction of the Lorentz force, a sure indication that this is a dynamo-generated field rather than merely a field resulting from amplification of the seed field (and therefore proportional to it).

The magnetic Reynolds number based on the diameter of each constituent tube is $\mathcal{O}(1)$ or less, but the Reynolds number is large and the flow through the tubes is turbulent at the threshold of dynamo instability. The theory of §9.2 is therefore broadly relevant.

9.12 The VKS experiment as an $\alpha\omega$-dynamo

The realisation of dynamo action in a liquid metal under laboratory conditions mimicking those of the Earth's liquid core has presented a major experimental challenge throughout the last half-century. The experiment that has come nearest to

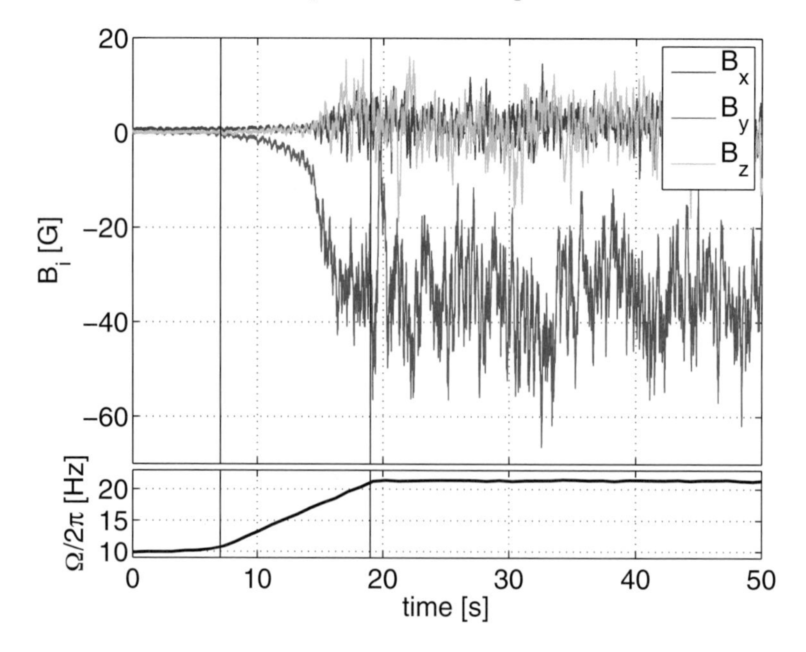

Figure 9.22 Dynamo action in the VKS experiment; the red curve shows the instantaneous toroidal component of **B** as measured at the cylinder boundary; this turbulent signal indicates the onset of dynamo action when the rotation frequency of both propellors is ramped up to the maximum of 26 Hz; the mean value of the weaker x and z components also increases, although this increase is swamped by the turbulent fluctuations. [Reprinted figure with permission from Monchaux et al. 2007. Copyright 2007 by the American Physical Society.]

achieving this objective is the 'von Karman Sodium' (or simply VKS) experiment at the French CEA site in Cadarache (Monchaux et al. 2007, 2009). A turbulent helical flow is generated by two counter-rotating propellors immersed in liquid sodium in a cylindrical copper container, as shown in Figure 9.21. Fluid is thrown radially outwards by the propellors and replaced by fluid sucked in in the axial direction – a mechanism first analysed by von Karman (1921) – hence the name of the experiment. The combination of rotational and axial flow is helical in character, with the same sign of helicity throughout the space between the propellors. The flow has very high Reynolds number, of the order of 10^7, and is fully turbulent; this turbulence presumably inherits helicity as well as energy from the mean flow. As the magnetic Reynolds number $\mathcal{R}_{\rm m}$ based on the rotational speed and span of the propellors is increased, a critical value $\mathcal{R}_{mc} \sim 32$ is reached above which a magnetic field grows spontaneously on the scale of the container from an initially low level, as shown in Figure 9.22. The field grows until the quadratic Lorentz force is strong enough to react back upon the turbulence, presumably suppressing the α-effect. The growing mean field is axisymmetric about the common axis of the propellors, and can be of either polarity, consistent with the invariance of the

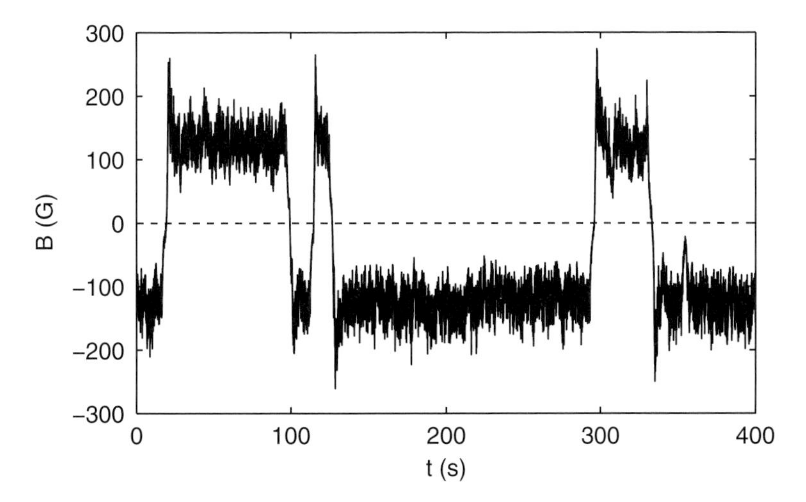

Figure 9.23 Field reversals in the VKS experiment, when the propellors are driven at angular velocities 16 Hz and −22 Hz; this dominant azimuthal component reverses direction at random times, the duration of a reversal being much less than the interval between reversals. (The weaker poloidal field components, not shown here, undergo synchronous reversals.) The vertical scale is in Gauss (1 Gauss = 10^{-4}T). There is no 'applied' field other than the ambient Earth's field, ~ 0.3 Gauss. [After Berhanu et al. 2007.]

MHD equations under the substitution $\mathbf{B} \to -\mathbf{B}$. The dynamo is almost certainly of $\alpha\omega$-type, requiring as it does both differential rotation driven by the propellors and turbulence that is undoubtedly chiral in character.

9.12.1 Field reversals in the VKS experiment

When the propellors are counter-rotated, but at different speeds, field reversals are realised, just as for the geomagnetic field (Berhanu et al. 2007). Figure 9.23 shows the time-dependence of the three field components, as measured by Hall probes placed at positions P1 and P2 in the sodium blanket (Figure 9.21). The behaviour bears a striking resemblance to the behaviour of the Earth's magnetic field over geological time-scales (Figures 4.10 and 4.12), and we have to conclude that similar mechanisms are at work despite the extreme difference in length- and time-scales in the two situations.

There is however one feature of the VKS experiment that is not yet well understood. Dynamo action has been achieved only when the propellors are made from soft iron, but not when they are made from copper. When dynamo action occurs, the iron itself becomes magnetised, and this evidently helps the dynamo process. It seems likely that the same type of dynamo will function with propellors of copper

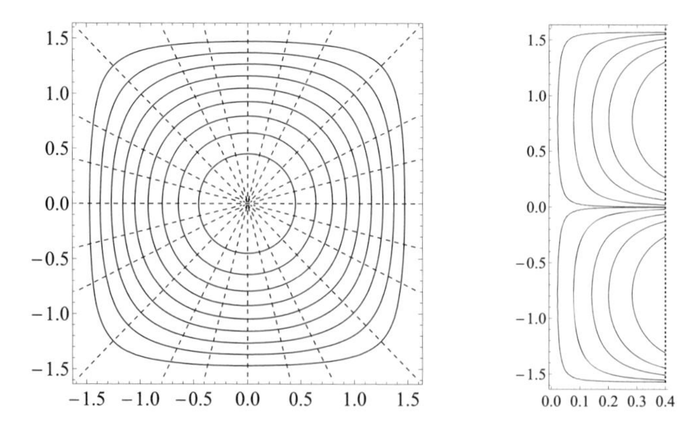

Figure 9.24 (a) Streamlines in any section z = const. of the initial Taylor–Green flow (9.158); the outward radial flow $\mathbf{u}_1$ (given by (9.161)) at $z = \pm\pi/2$ is indicated by the dashed lines. (b) Streamlines of the flow $\mathbf{u}_1$ in the $\{r, z\}$-plane near the axis $r = 0$; note the diverging flow at the boundaries $z = \pm\pi/2$ and the strong stretching on the axis for $|z| < \pi/4$.

(whose conductivity is comparable with that of liquid sodium) but at a significantly higher value of $\mathcal{R}_m$ than has yet been attained.

9.13 Dynamo action associated with the Taylor–Green vortex

The Taylor–Green vortex is a flow that evolves under either the Euler equations or the Navier–Stokes equations from an initial condition in which each velocity component is a periodic function of the cartesian coordinates $\{x, y, z\}$. This type of flow was introduced by Taylor & Green (1937) in an investigation of the growth of vorticity in a flow exhibiting some of the features of turbulence, and has since been intensively investigated (see, for example, Brachet et al. 1983) as a candidate for a finite-time singularity of the Euler equations.

We focus here on the particular choice of initial condition

$$\mathbf{u}_0(\mathbf{x}) = (\cos x \sin y \sin z, -\sin x \cos y \sin z, 0)\,, \tag{9.158}$$

satisfying $\nabla \cdot \mathbf{u}_0 = 0$, and we restrict attention to the box with boundaries $x = \pm\pi/2$, $y = \pm\pi/2, z = \pm\pi/2$ which must be considered impermeable and slip-free to be compatible with the assumed periodicity. The streamlines of this flow in any plane z = const. are given by $\cos x \cos y$ = const. as shown in Figure 9.24a. The fluid at $z = -\pi/2$ rotates in one sense, and at $z = +\pi/2$ in the opposite sense, so to this extent there is some similarity with the 'counter-rotating' situation of the VKS experiment considered in the previous section. In the perfectly conducting limit,

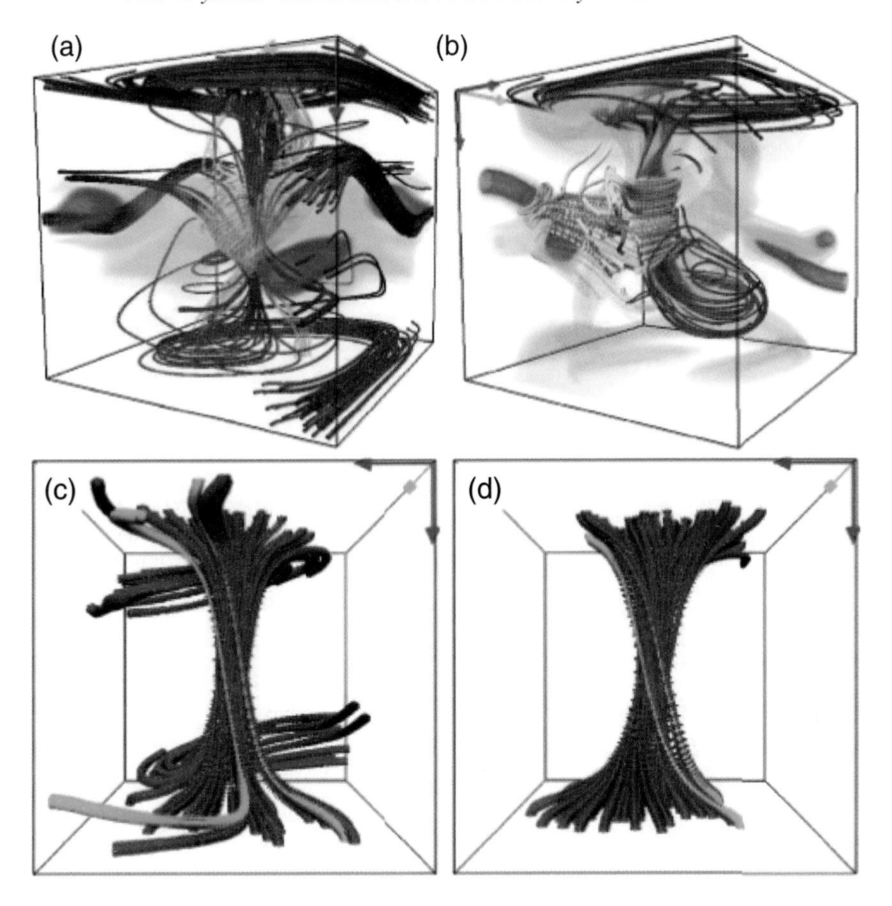

Figure 9.25 Magnetic fields generated by dynamo action in the Taylor–Green vortex; in all cases the top and bottom walls are insulating. (a) All side walls conducting, $\mathcal{R}_m = 30$. (b) One pair of opposite side walls insulating, the other pair conducting, $\mathcal{R}_m = 80$. (c) All walls insulating; magnetic mode during linear growth stage, $\mathcal{R}e = 20, \mathcal{R}_m = 80$. (d) The same when the field is saturated. [Reprinted figure with permission from Krstulovic et al. 2011. Copyright 2011 by the American Physical Society.]

any **B**-line initially parallel to the z-axis would be distorted into a helix by the flow (9.158) if this flow were maintained.

At both boundaries $z = \pm\pi/2$, the centrifugal force drives fluid particles radially outwards from the z-axis, with resulting suction towards both boundaries from the interior. If the development of the flow is represented by a time series[11] $\mathbf{u}(\mathbf{x}, t) = \sum t^n \mathbf{u}_n(\mathbf{x})$, then this effect appears already at the level $n = 1$. To see this, consider

[11] The convergence of this series is open to question. Taylor & Green (1937) calculated the $\mathbf{u}_n$ explicitly up to $n = 5$; Brachet et al. (1983) numerically up to $n = 80$.

the flow near the axis $x = y = 0$; in this neighbourhood,

$$\mathbf{u}_0 \sim (y \sin z, -x \sin z, 0) \quad \text{and so} \quad (\mathbf{u}_0 \cdot \nabla)\mathbf{u}_0 \sim -(x, y, 0) \sin^2 z. \qquad (9.159)$$

Under Euler (inviscid) evolution, $\mathbf{u}_1$ is given by

$$\mathbf{u}_1(\mathbf{x}) = -(\mathbf{u}_0 \cdot \nabla)\mathbf{u}_0 - \nabla p_0 , \qquad (9.160)$$

where p_0 must chosen to ensure that $\nabla \cdot \mathbf{u}_1 = 0$. The appropriate choice is $p_0 = \frac{1}{4}\cos 2z + \frac{1}{4}(x^2 + y^2)$, and this leads to the locally axisymmetric flow

$$\mathbf{u}_1(\mathbf{x}) = \tfrac{1}{2}\left(-x \cos 2z, -y \cos 2z, \sin 2z\right) , \qquad (9.161)$$

which is indeed radially outwards at $z = \pm\pi/2$ with a compensating inward flow near $z = 0$ (Figure 9.24b).

As time t increases, higher Fourier modes are excited, symmetries associated with the initial flow (9.158) are broken, and the flow in effect becomes turbulent. It is in this situation that dynamo action may occur (Nore et al. 1997, 2001). As explained by Krstulovic et al. (2011), the magnetic boundary conditions compatible with the assumed periodicity are that opposite boundaries are either perfect conductors ($\mathbf{B} \cdot \mathbf{n} = 0$) or ferromagnetic insulators ($\mathbf{B} \wedge \mathbf{n} = 0$). The critical magnetic Reynolds number $\mathcal{R}_{m,\mathrm{crit}}$ for the onset of dynamo action depends significantly on the boundary conditions imposed. Figure 9.25 shows the remarkable dynamo-generated magnetic field structures obtained in various situations; magnetic field lines are shown in red. In (a) and (b), a relatively strong field concentration may be seen centred on the z-axis; the top and bottom boundaries are insulating, so the field can't cross them. In (a), all the side walls are conducting, and $\mathcal{R}_\mathrm{m} = 30$; in (b), one pair of opposite side walls are conducting, the other pair insulating, and $\mathcal{R}_\mathrm{m} = 80$.

Even more striking is the pronounced tube structure when all the boundaries are insulating: Figure 9.25c shows the field for this case during the linear growth phase, and Figure 9.25d after saturation. This field structure is just what one might expect from the action of the flow ingredients $\mathbf{u}_0(\mathbf{x})$ and $\mathbf{u}_1(\mathbf{x})$ alone: a concentration of field near the axis with stretching in the central region and spreading out near both boundaries caused by $\mathbf{u}_1(\mathbf{x})$, together with twisting of the resulting flux tube caused by $\mathbf{u}_0(\mathbf{x})$.[12] The idea of a dynamo in which the magnetic field consists mainly of concentrated flux tubes is an attractive one, which we shall meet again in the following chapter.

[12] Note that in adjacent boxes in the x- or y-directions, the field direction in these tubes must alternate in direction, since the integrated magnetic flux across the xy-plane must vanish.

10

The Fast Dynamo

10.1 The stretch-twist-fold mechanism

The concept of the 'fast dynamo' was introduced by Vainshtein & Zel'dovich (1972), with particular reference to the following 'stretch-twist-fold' (or briefly 'STF') mechanism, operating in a perfectly conducting fluid ($\eta = 0$). Imagine a circular magnetic flux tube carrying flux Φ, and suppose this is stretched to double its length, then twisted through an angle π to the form of a figure-of-eight and folded back upon itself, just as one might do with an elastic band. The process is illustrated in Figure 10.1; here the central curve along the tube has parametric representation

$$\mathbf{x}(s,t) = (t\cos 2s - (1-t)\cos s,\ t\sin 2s - (1-t)\sin s,\ -2t(1-t)\sin s)\ , \quad (10.1)$$

where s ($0 \le s < 2\pi$) is a parameter on the curve, and the dimensionless time parameter t runs from zero to 1. The magnetic field $\mathbf{B}$ is obviously doubled in intensity through this process, so the magnetic energy is increased four-fold. If this process is repeated again and again, each step in the iteration taking the same time t_{STF}, say, then we have effectively exponential growth of the magnetic field on this time-scale. Figure 10.2 shows schematically the deformation of the tube centreline C after 2, 3 and 4 iterations; the structure becomes increasingly fine-scale (like $2^{-n/2}$ by virtue of flux conservation) as the number n of iterations increases. The behaviour here is similar to what occurs under the effect of turbulence that stretches magnetic flux tubes at a mean rate σ say, in the perfectly conducting limit $\eta = 0$. In the STF scenario however, the iterated deformation is contrived in an optimal way to provide coherent amplification of the total magnetic flux around the torus.

10.1.1 Writhe and twist generated by the STF cycle

However, we should note a small change in structure during even the first stage of the STF iteration: as discussed in §2.10.2 the writhe of the tube centreline changes by ± 1 according as the twist is right- or left-handed, and, since 'writhe plus twist' is conserved, there must be a compensating twist of ∓ 1 generated at this first stage by way of compensation. This twist means that a transverse field gradient must appear

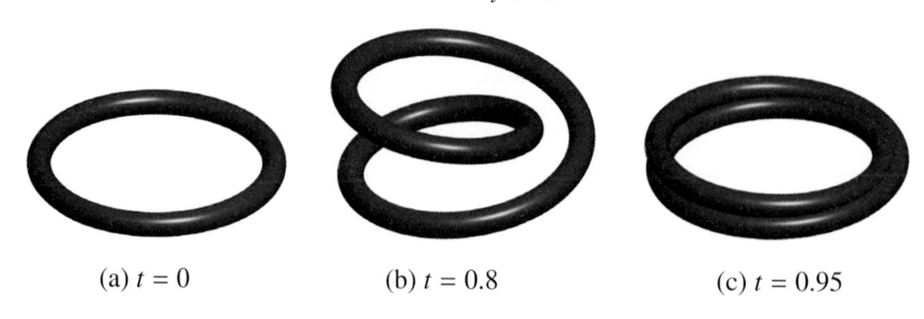

(a) $t = 0$ (b) $t = 0.8$ (c) $t = 0.95$

Figure 10.1 Hypothetical stretch-twist fold deformation of a magnetic flux tube; the centreline of the tube has parametric equation (10.1). (a) The initial circular flux tube. (b) The tube at time $t = 0.8$. (c) The nearly flattened state at time $t = 0.95$.

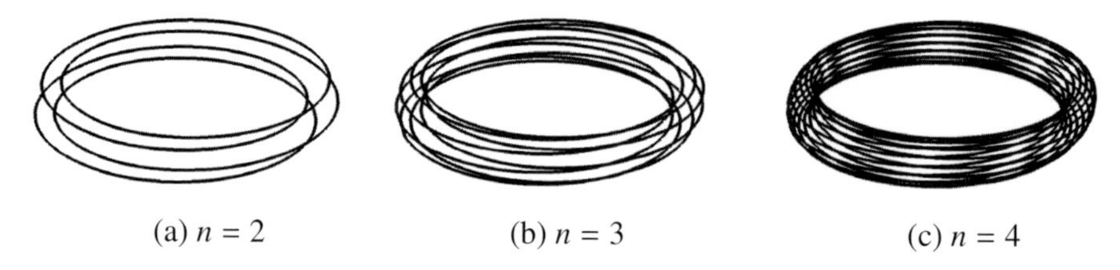

(a) $n = 2$ (b) $n = 3$ (c) $n = 4$

Figure 10.2 Form of the tube centreline after $n = 2$, 3 and 4 iterations; the transverse structure on the torus becomes increasingly fine scale as the number n of iterations increases.

where the two loops of the tube come together, and this gradient must be associated with a current distribution, possibly a localised sheet, in this neighbourhood. But things get more complicated with the second and subsequent iterations of the STF cycle. Let us suppose for definiteness that the imposed twist is right-handed at each stage. At the second iteration ($n = 2$) we cause two strands to cross two strands, as illustrated schematically in Figure 10.2a thus introducing a writhe of +4 and so internal twist of -4; at the third iteration ($n = 3$), we introduce a writhe of $4 \times 4 = 16$, so an internal twist of -16; and so on. At the nth iteration, a writhe $Wr_n = 2^{2(n-1)}$ is introduced; and each writhe (viewed as a strand crossing) must involve the appearance of a local elemental current distribution.

Of course if $\eta \neq 0$, these currents can decay, but with $\eta = 0$, they cannot, and with each iteration, the scale of variation of the transverse field structure is reduced by a factor of $\sqrt{2}$. This provides a clue to the somewhat pathological structure of the magnetic field in the fast dynamo limit $\eta \to 0$, as will emerge below. Incidentally, the curve deformation described by (10.1) involves passage through an inflexional configuration at time $t = 0.8$; this may be seen from the plot of the curvature as a function of s and t as shown in Figure 10.3: the curvature drops to zero at $s = 0$ at the instant $t = 0.8$, indicating passage through an inflexion point at this instant when integrated torsion is transmogrified to internal twist.

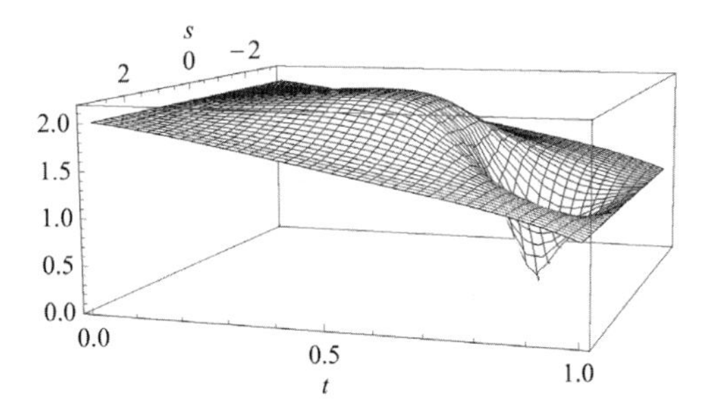

Figure 10.3 Curvature $c(s,t)$ for the curve defined by (10.1); the curvature dips to zero at $s = 0$, $t = 0.8$, indicating passage through an inflexion point at this instant, with associated creation of internal twist in the tube as described in §2.10.2.

This STF scenario may be thought to be somewhat artificial. After all, the whole purpose of dynamo theory as expounded so far is to explain how a velocity field, whether laminar or turbulent, can overcome the erosive effect of finite diffusivity $\eta > 0$. If $\eta = 0$, there is no problem! However, we should be careful to distinguish the situations $\eta \to 0$ and $\eta = 0$. In the former situation, resistive effects will always intervene no matter how small η may be, after a sufficient number of iterations. In fact, after n iterations, the transverse scale of the field will be reduced to $\mathcal{O}(2^{-n/2})$, so resistive diffusion may be expected to eliminate the elemental currents when $n \gtrsim \log \mathcal{R}_m / \log 2$. The axial field component (along the length of the tube) will however survive, the total flux around the torus after n iterations being $2^n\Phi$, indicating a genuine dynamo process provided $\eta \neq 0$.

10.1.2 Existence of a velocity field in $\mathbb{R}^3$ that generates the STF cycle

It is of course one thing to specify the deformation of a closed curve C by a parametric representation such as (10.1), but quite another to define a velocity field $\mathbf{u}(\mathbf{x},t)$ throughout the fluid which will have the effect of deforming C in precisely this way. For given deformation, such a velocity field can however be constructed, at least in principle, as the following argument demonstrates. We consider a tube $\mathcal{T}$ of small cross section having C as its axis, as illustrated in Figure 10.1. The velocity $\mathbf{U}(\mathbf{x},t)$ is then supposed given on the surface $\partial\mathcal{T}$ of $\mathcal{T}$. At each instant $t = t_1$, say, it is known that a unique quasi-static Stokes flow $\mathbf{u}(\mathbf{x},t_1)$ exists satisfying the boundary conditions

$$\mathbf{u}(\mathbf{x},t_1) = \mathbf{U}(\mathbf{x},t_1) \text{ on } \partial\mathcal{T}, \quad \mathbf{u}(\mathbf{x},t_1) \to 0 \text{ as } \mathbf{x} \to \infty. \tag{10.2}$$

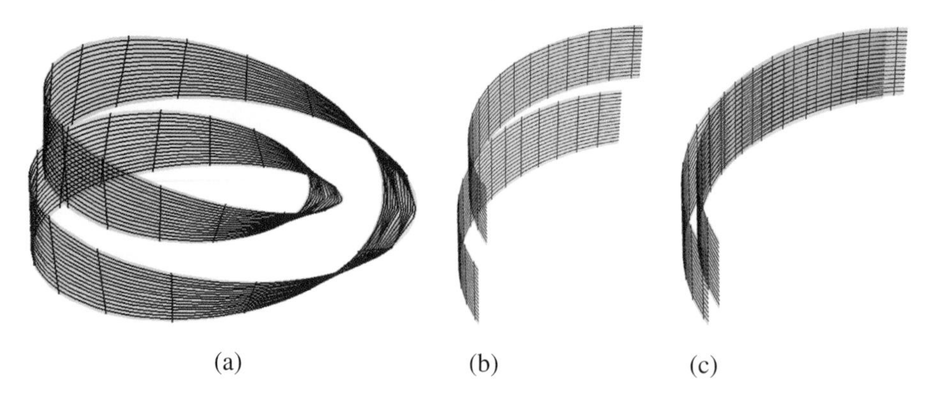

Figure 10.4 Stretch-twist-fold deformation of a ribbon bounded by any two neighbouring **B**-lines in the flux tube. (a) Near the end of the first STF sequence ($t = 0.9$), the writhe $Wr \sim +1$ is evident from the crossing, purple over green, on the left, and the compensating twist $Tw \sim -1$ appears distributed around the ribbon. (b) Situation near the crossing point. (c) The crossing is eliminated by diffusion and merging of the two strands, while the twist is unaffected.

Moreover, it is known that, among all kinematically possible flows satisfying (10.2), this Stokes flow has the least rate of dissipation of energy by viscosity (see, for example, Batchelor 1967). The velocity field defined at each instant in this way is therefore the field that dissipates the least energy (by viscosity) over the whole period of deformation, and which is therefore in some sense 'favoured'. Beware however of the singularity of $\mathbf{u}(\mathbf{x}, t)$ for the particular process (10.1) if it is continued all the way to the limit $t = 1$.

10.1.3 Tube reconnection and helicity cascade

The following discussion parallels to some extent the discussion in §5 of Gilbert (2002). The STF process is illustrated again in Figure 10.4 which shows a ribbon bounded by any two neighbouring **B**-lines in the flux tube. In this process, the most obvious effect of weak diffusivity must be to induce reconnection of field lines in the region where one strand crosses another; since the field is nearly in the same direction in the two strands, this slight topological transition destroys writhe but leaves twist unaffected. (For explicit analysis of such a diffusive writhe-destroying process, see Kimura & Moffatt 2014.) If we allow this reconnection at the end of the first STF-cycle, the result is again a single circular flux tube but now the field in the tube has a remanent twist -1, associated with an axial component $A_z(s, \varphi)$ of vector potential[1] and an axial current distribution $j_z(s, \varphi) = -\nabla^2 A_z$ flowing round the tube, where (s, φ, z) are local cylindrical coordinates. The integral of j_z over the tube cross section vanishes (since the field **B** is zero outside the tube). When

[1] As usual, we adopt the Coulomb gauge $\nabla \cdot \mathbf{A} = 0$ for the vector potential **A**.

we iterate the STF-cycle at stage $n = 2$, again allowing the obvious reconnection at the end of the stage, the $(n = 1)$-twist is reduced in scale by a factor $\sqrt{2}$, and a new $(n=1)$-twist is 'injected' on the cross-sectional scale δ_0 of the reconnected tube. With further iterations, each n-twist on length-scale $2^{-n/2}\delta_0$ is converted to an $(n+1)$-twist on scale $2^{-(n+1)/2}\delta_0$ and a new $(n=1)$-twist is injected at the scale δ_0. Ultimately when $n \sim \log \mathcal{R}_m/2 \log 2$, the currents on the scale $2^{-n/2}\delta_0$ must decay through the effect of the non-zero diffusivity η.

Although speculative, we see here the possibility of a very idealised cascade process: in a well-established state (i.e. after many iterations), twist Tw (and therefore the associated magnetic helicity $\mathcal{H}_M = Tw\,\Phi^2$), injected on the input scale δ_0, cascades through scales $2^{-n}\delta_0$, $(n = 1, 2, 3, \dots)$, and is ultimately dissipated on scales of order $\left(\mathcal{R}_m^{-1/2}\right)\delta_0$ and smaller. At the same time, magnetic energy is amplified by the dynamo process on the scale δ_0 and does not cascade to smaller scales.

10.2 Fast and slow dynamos

In view of the above discussion, it is legitimate to pose the following question. Let us focus for simplicity on the case of velocity fields $\mathbf{u}(\mathbf{x})$ that are steady, and either space-periodic or contained within a finite domain, the fluid having finite resistivity $\eta > 0$. Let u_0 and ℓ_0 be characteristic velocity and length-scales of the flow and $\mathcal{R}_m = u_0\ell_0/\eta$ the usual magnetic Reynolds number; time is then measured in units of ℓ_0/u_0. Suppose that for a given $\mathbf{u}(\mathbf{x})$ we have dynamo action with $\mathbf{B}(\mathbf{x}, t) = \hat{\mathbf{B}}(\mathbf{x}, \mathcal{R}_m) \exp[\sigma(\mathcal{R}_m)t]$ for at least some range of $\mathcal{R}_m > \mathcal{R}_{mc}$. Both the eigenmode $\hat{\mathbf{B}}(\mathbf{x}, \mathcal{R}_m)$ and the growth rate $\sigma(\mathcal{R}_m)$ may be expected to depend on $\mathcal{R}_m$ (and there is no guarantee that $\sigma(\mathcal{R}_m)$ must remain positive for *all* $\mathcal{R}_m > \mathcal{R}_{mc}$). The fast dynamo question however is this: how does $\sigma(\mathcal{R}_m)$ behave in the limit $\mathcal{R}_m \to \infty$?

In most of the situations that we have so far encountered, the growth rate $\sigma \to 0$ as $\mathcal{R}_m \to \infty$. In this case, following Vainshtein & Zel'dovich (1972), the dynamo is described as 'slow'. If, on the contrary, $\sigma \to$ const. > 0 as $\mathcal{R}_m \to \infty$, the dynamo is described as 'fast'. In this respect, it is hypothesised that the iterated STF process can act as a fast dynamo. But a secondary question in the general case is then 'what is the structure of the eigenmode $\hat{\mathbf{B}}(\mathbf{x}, \mathcal{R}_m)$ in the limit $\mathcal{R}_m \to \infty$?' Does this eigenmode necessarily have a pathological (non-smooth) structure?

It is perhaps necessary to emphasise that the above fast-dynamo question makes sense only in the context of kinematic dynamo theory, in which the velocity field is prescribed and either steady or possibly time-periodic as for the STF flow (in which case a mean growth-rate $\langle\sigma\rangle(t)$ can still be defined). This necessarily kinematic fast-dynamo problem has been comprehensively covered by Childress & Gilbert (1995), and only a very selective treatment will be given in the following sections.

10.3 Non-existence of smooth fast dynamos

Consider again the situation when $\eta = 0$, and let us restrict attention for the moment to fields $\mathbf{B}(\mathbf{x})$ that are smooth functions of position $\mathbf{x}$. Then we know that for any such localised magnetic field in fluid filling all space, the magnetic helicity $\mathcal{H}_M$ is invariant. If $\mathcal{H}_M \neq 0$, this invariance is clearly incompatible with the existence of a field of the form $\mathbf{B}(\mathbf{x}, t) = \hat{\mathbf{B}}(\mathbf{x}) \exp[\sigma t]$ with $\Re e\, \sigma \neq 0$, for then $\mathcal{H}_M \sim \exp[2 \Re e\, \sigma\, t]$; thus fast dynamo action is impossible under these circumstances; the possibility of a non-smooth eigenmode is not however excluded by this argument.

The following simple argument (Moffatt & Proctor 1985) shows that the same conclusion holds even if $\mathcal{H}_M = 0$. We suppose that the fluid is contained in a finite domain $\mathcal{D}$, and that the velocity field $\mathbf{u}(\mathbf{x})$ is steady, and satisfies $\mathbf{n} \cdot \mathbf{u} = 0$ on its surface $\partial\mathcal{D}$. Substituting $\mathbf{B}(\mathbf{x}, t) = \hat{\mathbf{B}}(\mathbf{x}) \exp[\sigma t]$ in the induction equation (2.134) (with $\eta = 0$) now gives

$$\sigma \hat{\mathbf{B}} = \nabla \times (\mathbf{u} \times \hat{\mathbf{B}}) . \tag{10.3}$$

This may be 'uncurled', with $\hat{\mathbf{B}} = \nabla \times \hat{\mathbf{A}}$, to give

$$\sigma \hat{\mathbf{A}} = \mathbf{u} \times \hat{\mathbf{B}} - \nabla \hat{\varphi} , \tag{10.4}$$

for some scalar field $\hat{\varphi}$. Here, we may choose the gauge of $\hat{\mathbf{A}}$ such that $\hat{\varphi} = 0$. It then follows immediately that

$$\hat{\mathbf{A}} \cdot \hat{\mathbf{B}} = \hat{\mathbf{A}} \cdot \nabla \times \hat{\mathbf{A}} = 0. \tag{10.5}$$

According to Frobenius' theorem (see, for example, Boothby 1986), this is the necessary and sufficient condition for the existence of (complex) scalar fields $f(\mathbf{x})$ and $g(\mathbf{x})$ such that

$$\hat{\mathbf{A}} = f \nabla g, \quad \text{so} \quad \hat{\mathbf{B}} = \nabla f \times \nabla g. \tag{10.6}$$

Substituting this back in (10.4) (with $\hat{\varphi} = 0$) gives

$$\sigma f \nabla g = \mathbf{u} \times (\nabla f \times \nabla g) = \nabla f (\mathbf{u} \cdot \nabla g) - \nabla g (\mathbf{u} \cdot \nabla f) . \tag{10.7}$$

Now crossing this equation with ∇f gives

$$(\sigma f + \mathbf{u} \cdot \nabla f)\, \hat{\mathbf{B}} = 0. \tag{10.8}$$

It follows that, at every point of $\mathcal{D}$, either $\sigma f + \mathbf{u} \cdot \nabla f = 0$ or $\hat{\mathbf{B}} = 0$.

Now let $\mathcal{D}_1$ be any fixed subdomain of $\mathcal{D}$ in which $\hat{\mathbf{B}} \neq 0$, with $\mathbf{n} \cdot \hat{\mathbf{B}} = 0$ on $\partial\mathcal{D}_1$ ($\mathcal{D}_1$ could be $\mathcal{D}$ itself if $\hat{\mathbf{B}} \neq 0$ throughout $\mathcal{D}$). Then $\mathbf{B}(\mathbf{x}, t) = \hat{\mathbf{B}}(\mathbf{x}) \exp[\sigma t]$ satisfies these conditions also, and so we must have $\mathbf{n} \cdot \mathbf{u} = 0$ on $\partial\mathcal{D}_1$, since otherwise the conditions on $\partial\mathcal{D}_1$ could not persist. It follows therefore that

$$\sigma f + \mathbf{u} \cdot \nabla f \equiv 0 \quad \text{in } \mathcal{D}_1 , \tag{10.9}$$

so that

$$(\sigma + \sigma^*) \int_{\mathcal{D}_1} |f|^2 \, \mathrm{d}V = - \int_{\partial \mathcal{D}_1} (\mathbf{u} \cdot \mathbf{n}) |f|^2 \, \mathrm{d}S = 0 \,, \tag{10.10}$$

and hence (since $f \not\equiv 0$ in $\mathcal{D}_1$), $\mathfrak{Re}\,\sigma \equiv \frac{1}{2}(\sigma + \sigma^*) = 0$. We may conclude that fast dynamos with smooth $\mathbf{B}$ (requiring $\mathfrak{Re}\,\sigma > 0$) are impossible when $\eta = 0$.

If however we allow for non-differentiable $\hat{\mathbf{B}}$ in the limit $\eta \to 0$, then there is no justification for the neglect of diffusion in (2.134). Indeed we may expect that eigenmodes $\hat{\mathbf{B}}(\mathbf{x})$ may exist with scale of variation $O\left(\mathcal{R}_{\mathrm{m}}^{-1/2}\right)$ as $\mathcal{R}_{\mathrm{m}} \to \infty$ nearly everywhere in the fluid domain, as exemplified by the STF dynamo.

10.4 The homopolar disc dynamo revisited

Consider again the disc dynamo introduced in Chapter 1. The simplistic description given there led to the equation

$$L \, \mathrm{d}I/\mathrm{d}t + R\,I = M\Omega\,I \,, \tag{10.11}$$

where, as will be recalled, R and L are the resistance and self-inductance of the current circuit and M is the mutual inductance between the wire loop and the rim of the disc. As the resistivity of the disc and the wire tend to zero (equivalently, in the limit $R \to 0$), it would appear that we still have a dynamo with growth rate $M\Omega/L$. Is this then a fast dynamo?

Certainly not! It shows rather an inadequacy of the model; for, when the conductivity of the disc tends to infinity, Alfvén's theorem tells us that the magnetic flux through the Lagrangian closed curve that here consists of the circular rim of the disc remains constant. The simple model of Chapter 1 fails to recognise that magnetic field cannot cross the rim of the disc when its conductivity is infinite; instead, a 'skin current' flowing round the rim of the disc is established, and this current has to be taken into account in an improved model. It is perhaps salutary to recognise that even the simplest arguments in the dynamo context are liable to lead to utterly false conclusions!

An improved model was proposed by Moffatt (1979), taking this skin current into account. In this model, illustrated in Figure 10.5, radial insulating strips are introduced in the disc thus permitting azimuthal current flow only in a narrow region near the rim of the disc. If $I(t)$ is still the current flowing in the wire loop (self-inductance L, resistance R) and $J(t)$ is the current flowing round the rim (self-inductance L', resistance R'), then the magnetic fluxes through these two loops are respectively

$$\Phi_1 = LI + MJ, \quad \Phi_2 = MI + L'J, \tag{10.12}$$

where M is still the mutual inductance as before; this satisfies $M^2 < LL'$, this

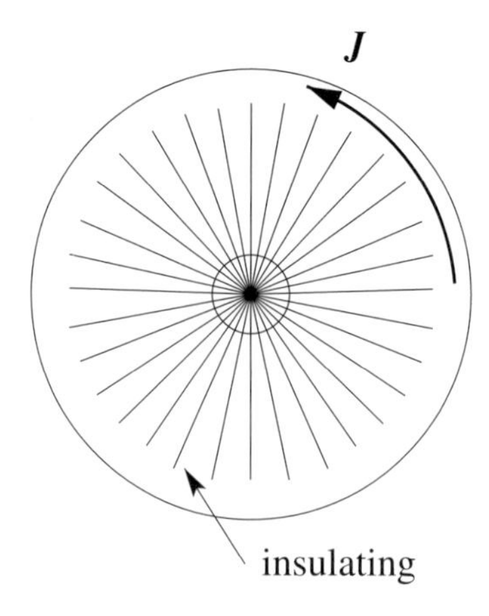

Figure 10.5 Disc segmented by insulating strips, so that azimuthal current J is confined to the narrow region near the rim.

inequality resulting from the positive-definiteness of the magnetic energy of the current system. These fluxes change according to Faraday's law of induction:

$$\frac{\mathrm{d}\Phi_1}{\mathrm{d}t} = \Omega\Phi_2 - RI, \qquad \frac{\mathrm{d}\Phi_2}{\mathrm{d}t} = -R'J = -\frac{R'}{L'}(\Phi_2 - MI), \tag{10.13}$$

equations that reduce to (10.11) when $R' = \infty$ (so $J = 0$). The second equation here indicates that when $R' \neq 0$, there is a phase lag between the current I and the flux Φ_2: this flux has to diffuse across the rim of the disc into the region where the inductive effect represented by the term $\Omega\Phi_2$ occurs. This is in fact a general requirement of dynamo theory: in a time-dependent situation, flux always needs to diffuse into any region where induction occurs, and this requires non-zero diffusivity $\eta > 0$.

Equations (10.12) and (10.13) admit exponential solutions $\{I, J, \Phi_1, \Phi_2\} \propto \exp pt$, where the growth rate p satisfies

$$\left(LL' - M^2\right)p^2 + (RL' + R'L)\,p - R'(M\Omega - R) = 0\,, \tag{10.14}$$

with roots

$$p_{1,2} = \frac{-(RL' + R'L) \pm \sqrt{(RL' + R'L)^2 + 4R'(M\Omega - R)(LL' - M^2)}}{2(LL' - M^2)}\,. \tag{10.15}$$

If $M\Omega > R$, the root p_1 is real and positive, thus still indicating dynamo action. However now

$$p_1 \sim R'(M\Omega - R)/RL' \quad \text{as } R' \to 0, \tag{10.16}$$

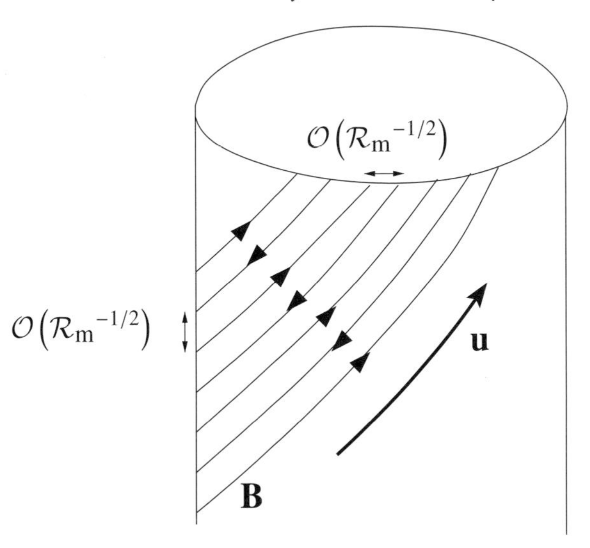

Figure 10.6 Sketch of the Ponomarenko dynamo configuration for large $\mathcal{R}_m$. [From Gilbert 1988.]

showing that this is *not* a fast dynamo; it is slow (R' being proportional to the disc diffusivity η). The apparent conflict with Alfvén's theorem is thus resolved.

Actually in the limit $\eta \to 0$, the insulating strips introduced as an artifact in the above model are not required, because the azimuthal current in the disc must in any case, because of the skin effect, be confined to an $O\left(\eta^{1/2}\right)$ neighbourhood of the rim during the phase of exponential growth. Of course, the exponential growth cannot continue indefinitely: eventually the Lorentz force in the disc provides a resisting torque, which, in conjunction with the driving torque G ultimately controls the rotation rate $\Omega(t)$. The dynamics of the segmented-disc system is then governed by a third-order non-linear dynamical system whose properties have been explored by Moffatt (1979); it was found that chaotic behaviour occurs in certain regimes of the parameter space. Details concerning this type of behaviour are deferred to §11.1.

10.5 The Ponomarenko dynamo in the limit $\eta \to 0$

Here we refer back to the Ponomarenko dynamo considered in §6.12, where we derived the equation

$$2\eta = \mathrm{i}\Omega\left[I_{m-1}(p)K_{m-1}(p) - I_{m+1}(p)K_{m+1}(p)\right] \tag{10.17}$$

that determines p and hence the dynamo growth rate $\lambda/\eta = p^2 - k^2$ (Gilbert 1988). For fixed $|m|$ and $m\Omega + kU = 0$ (so $h = U/\Omega = -m/k$), the asymptotic behaviour as $\eta \to 0$ was given by (6.128),

$$\lambda = \lambda_r + \mathrm{i}\lambda_i \sim 2^{-5/3}\left(\eta m^2 \Omega^2\right)^{1/3}(1 + \sqrt{3}\,\mathrm{i})\,. \tag{10.18}$$

As recognised by Gilbert, this increases with $|m|$, indicating that, as $\eta \to 0$, λ may remain $O(1)$ if $m = O(\Omega^{-1}\eta^{-1/2})$. With $p = mz$, Gilbert used asymptotic results for $I_m(mz)$ and $K_m(mz)$ for large m and fixed z to obtain the result

$$\lambda_r/\Omega \sim 2^{-5/3}\left(\eta m^2/\Omega\right)^{1/3} - \left(\eta m^2 \Omega\right)\left(1 + h^{-2}\right). \tag{10.19}$$

This is maximal when $\eta m^2/\Omega = h^3/12\sqrt{6}(1 + h^2)^{3/2}$, the maximum then being

$$\lambda_{r\,\max} = \frac{h\Omega}{6\sqrt{6}\,(1 + h^2)^{1/2}} = \frac{(\Omega^2 + U^2)^{1/2}}{6\sqrt{6}}\,\frac{h}{1 + h^2}\,. \tag{10.20}$$

This is indeed independent of η in the limit $\eta \to 0$, a symptom of fast dynamo action. However, consistent with the discussion of §10.3, the scale of variation of the eigenmode $\hat{\mathbf{B}}$ is $O(\eta^{1/2})$ as $\eta \to 0$, not only in the radial direction across the velocity discontinuity on $r = 1$, but also in the φ- and z-directions. Diffusion remains important for such modes no matter how small η may be. We note that, for fixed $(\Omega^2 + U^2)^{1/2}$, $\lambda_{r\,\max}$ is maximal when $h = 1$, and then

$$\lambda_{r\,\max|\,\max} = (\Omega^2 + U^2)^{1/2}/12\sqrt{6}\,. \tag{10.21}$$

This is the precisely the situation as indicated by the sketch of Figure 10.6, in which the **u**-lines and **B**-lines are inclined at an angle close to $\pi/4$ to the direction of the cylinder axis.

10.6 Fast dynamo with smooth space-periodic flows

Much attention has been devoted to the possibility of fast dynamo action associated with the space-periodic 'ABC-flow' (so-named after **A**rnold, **B**eltrami and **C**hildress):

$$\mathbf{u} = (C\sin z + B\cos y,\; A\sin x + C\cos z,\; B\sin y + A\cos x)\,. \tag{10.22}$$

For this flow, $\omega = \mathbf{u}$, so it is a Beltrami flow of maximal helicity $\mathcal{H} = \langle \mathbf{u}\cdot\omega\rangle = A^2 + B^2 + C^2$. The flow was first studied by Arnold (1965a) and Hénon (1966), and subsequently in greater detail by Dombre et al. (1986) who identified the heteroclinic orbits connecting stagnation points and the regions of chaos where the separation of initially neighbouring fluid particles increases exponentially; it is this aspect of 'material line-stretching' in conjuction with the helicity of the flow, that is expected to be conducive to dynamo action, through associated field-line stretching (Finn & Ott 1988; also Klapper & Young 1995 who prove that some degree of

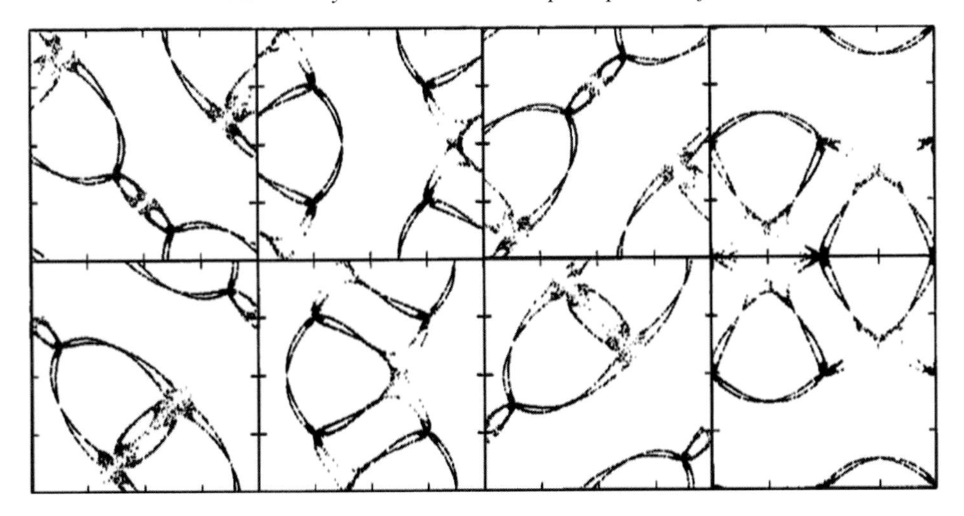

Figure 10.7 Poincaré sections by planes x = const. for the ABC-flow with $A = B = C = 1$. [From Dombre et al. 1986.]

chaos is necessary for fast dynamo action). The α-effect associated with this flow was identified by Childress (1970), leading to mean-field growth on a scale large compared with the periodicity 2π of the flow. At high $\mathcal{R}_m$ however, fields with the same 2π-periodicity of the flow can grow, and attention is focussed on such fields in the fast-dynamo context.

10.6.1 The symmetric case $A = B = C = 1$.

The case $A = B = C = 1$ has attracted particular interest, in view of the special symmetries of this flow. These are the symmetries of the group O_{24} of orientation-preserving transformations of a cube, which have been exploited very effectively by Jones & Gilbert (2014) to follow the growth rates of fields of different symmetries as functions of $\mathcal{R}_m$ for $\mathcal{R}_m \gg 1$. The aim is to determine asymptotic behaviour as $\mathcal{R}_m \to \infty$; this requires state-of-the-art computing and a huge investment of computing time. Recent work on the problem (Bouya & Dormy 2013, 2015) has pushed $\mathcal{R}_m$ up to 5×10^5 (for which more that 10^6 CPU hours of computing time were required), but even at this high value, an asymptotic regime has not yet been established.

Figure 10.7 (from Dombre et al. 1986) shows some typical Poincaré sections (see §2.9 for an indication of how such diagrams are constructed) for the case $A = B = C = 1$, showing a network of narrow bands, within which particle paths (and so streamlines for this steady flow) are chaotic. The fraction of the periodicity cube within which the flow is chaotic is evidently quite small. Between these bands are

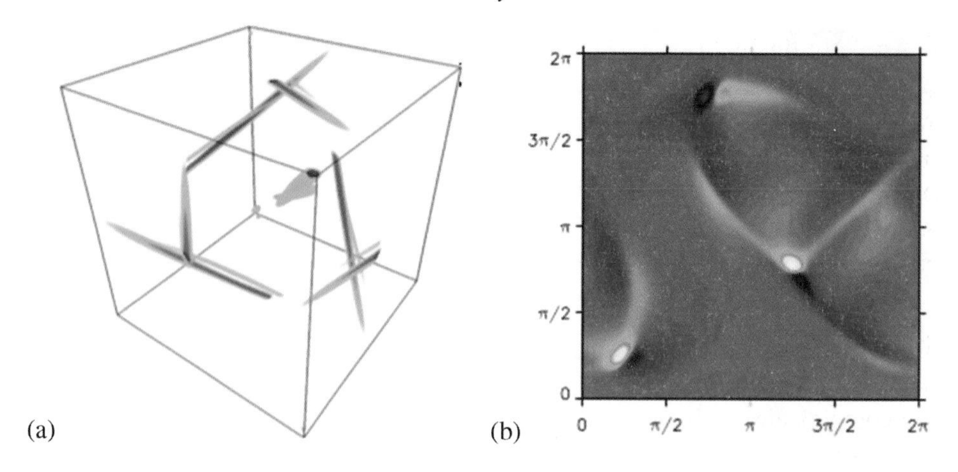

Figure 10.8 (a) Magnetic field strength (|**B**|), showing pronounced cigar-like structure for the ABC-flow ($A = B = C = 1$). (b) Field component b_z on the plane $z = \pi/4$ for the most rapidly growing mode at $\mathcal{R}_m = 250$. [From Jones & Gilbert 2014.]

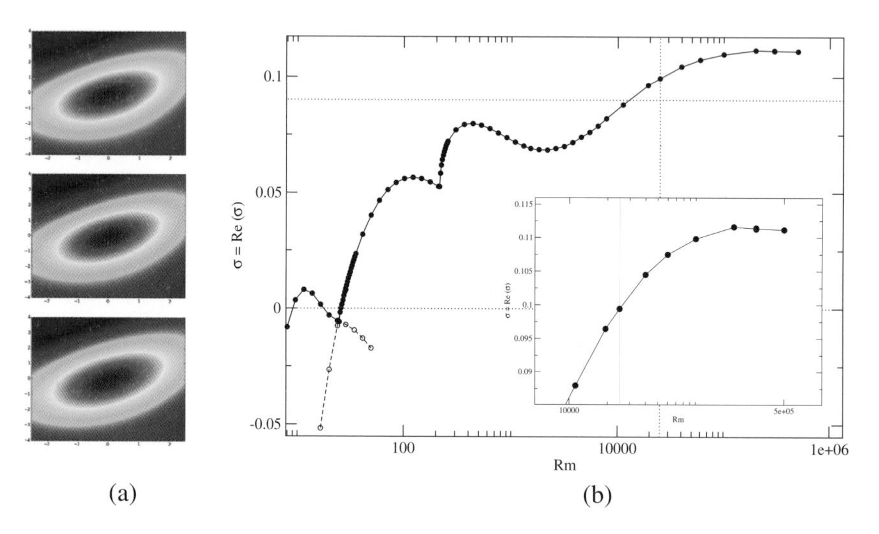

Figure 10.9 (a) Cross section of cigar-structures, stretched by a factor $\mathcal{R}_m^{-1/2}$, at $\mathcal{R}_m = 4\times10^4$, 2×10^5 and 5×10^5; no change of structure is evident. (b) Dynamo growth rate $\sigma(= \lambda_r)$ as a function of $\mathcal{R}_m$ up to $\mathcal{R}_m = 5\times10^5$. [From Bouya & Dormy 2015.]

major vortices which tend to expel magnetic flux by the flux-expulsion mechanism described in §3.11, leading to concentration of the dynamo-generated field near the heteroclinic orbits of the flow.

Figure 10.8 (from Jones & Gilbert 2014) shows the magnetic field structure that emerges for this flow at the relatively modest magnetic Reynolds number

$\mathcal{R}_m = 250$; this is for the mode of the particular symmetry that gives maximum growth rate, at least at this value of $\mathcal{R}_m$. What is most striking here is the concentration of magnetic field in tubes having a 'cigar-like' structure of relatively small cross section. The internal structure of these cigars has been computed by Bouya & Dormy (2015) (Figure 10.9a), who have shown that this structure scales as $\mathcal{R}_m^{-1/2}$ (with no apparent change of shape) in the range of $\mathcal{R}_m$ from 4×10^4 to 5×10^5. Figure 10.9b shows the growth rate (here σ) of the fastest growing dynamo mode as a function of $\mathcal{R}_m$ up to 5×10^5. The small decrease above $\sim 4 \times 10^5$ (which shows up more clearly in the inset panel) is an indication that the asymptotic stage has not yet been reached for this flow. The kink at $\mathcal{R}_m \approx 215$ is an indication that two complex eigenvalues (both having the same symmetry) coalesce, the dynamo being non-oscillatory for larger $\mathcal{R}_m$ (Bouya & Dormy 2013). [The low $\mathcal{R}_m$ mode of dynamo action evident in Figure 10.9b in the range $\mathcal{R}_m \sim 9-18$ was discovered by Arnold & Korkina 1983.]

The horizontal dotted line at $\sigma = 0.09$ in Figure 10.9b corresponds to a theoretical 'topological entropy' bound for the growth rate in the $\mathcal{R}_m = \infty$ limit (Childress & Gilbert 1995), a further indication that the asymptotic state has not yet been realised. It is remarkable that the growth rate exceeds this bound (essentially a bound on line-stretching) when $\mathcal{R}_m \gtrsim 1.5 \times 10^4$, raising the interesting question of just how it is that weak diffusion can contrive to increase, rather than decrease, this theoretically maximal growth rate.

The extent of chaos for the ABC-flow when $A = B = C$ is very limited, as evident in Figure 10.7. This extent is far greater as shown by exploration of the $\{A, B, C\}$ space (Galloway & O'Brian 1993; see also Alexakis 2011). For the case $A : B : C = 5 : 2 : 2$; this flow has no stagnation points, and the magnetic eigenfunctions were found to be sheet-like rather than filamentary. The extent of chaos can also be increased by the imposition of time-periodicity in the flow (Galloway & Proctor 1992, Otani 1993, Brummell et al. 2001); a particular case is described in the following section.

10.6.2 The Galloway–Proctor fast dynamo

Galloway & Proctor (1992) (in a paper reviewed among others by Galloway 2012) provided numerical evidence of fast dynamo action in certain flows that are periodic in both space and time; in particular in their 'circularly polarised' (CP) flow defined by the velocity field

$$\mathbf{u}(\mathbf{x}, t) = A\left[\sin(z + \sin\omega t) + \cos(y + \cos\omega t),\ \cos(z + \sin\omega t),\ \sin(y + \cos\omega t)\right], \tag{10.23}$$

with normalisation $A = \sqrt{3/2}$, $\omega = 1$. The 'horizontal' part of this field is

$$\mathbf{u}_H(\mathbf{x}, t) = A\left[0,\ \cos(z + \sin\omega t),\ \sin(y + \cos\omega t)\right], \tag{10.24}$$

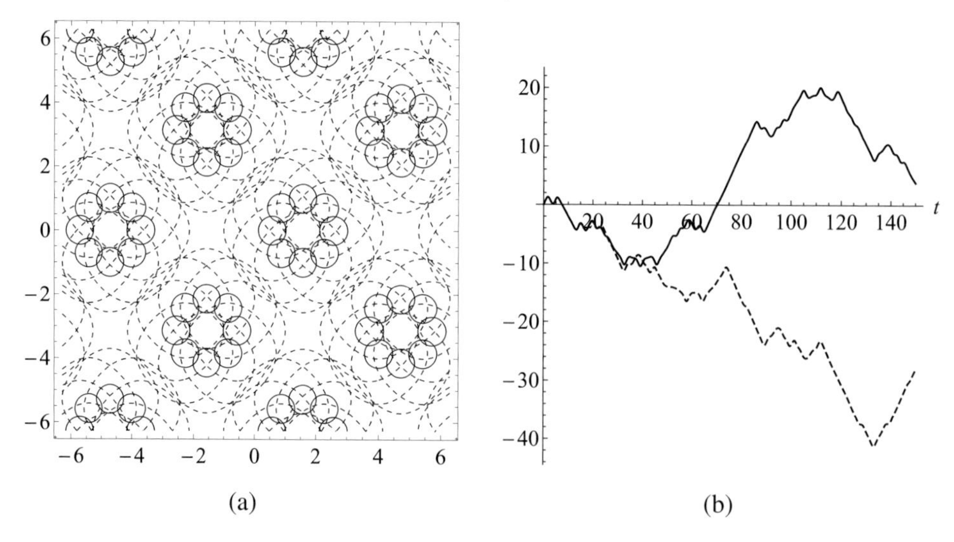

Figure 10.10 (a) Instantaneous streamlines $\psi_H(y, z, t) = \pm A$ (solid) and $\pm 1.9A$ (dashed) at times $t = n\pi/4$, $(n = 0, 1, 2, \ldots, 7)$. (b) Resulting component $z(t)$ of chaotic particle paths, starting from $(0, 0, 0)$ (solid) and $(0, 0, 0.01)$ (dashed).

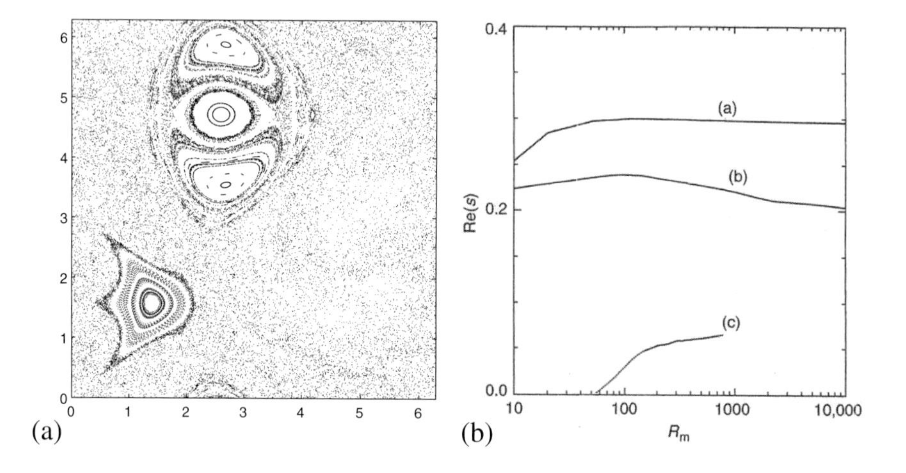

Figure 10.11 (a) Poincaré section on the plane $x = 0$; points are plotted at equal time intervals $2\pi/\omega$, and y and z values are plotted modulo 2π. (b) Resulting dynamo growth rate; the curve a corresponds to the flow (10.23); the curves b and c correspond to two other flows considered by Galloway & Proctor (1992). [From Galloway 2012.]

with stream-function $\psi_H(y, z, t) = A[\sin(z + \sin \omega t) + \cos(y + \cos \omega t)]$. The streamlines thus lie on cylindrical surfaces $\psi_H(y, z, t) = \text{const}$. This family of curves rotates in the yz-plane with angular velocity ω, as indicated in Figure 10.10a (hence the term 'circularly polarised'). The solid curves are $\psi_H(y, z, t) = \pm A$ for times

$t = n\pi/8$ ($n = 0, 1, 2, \ldots, 7$), and a particle starting on one of these curves can migrate from one cell to another, with associated chaotic particle path. The dashed curves are $\psi_H(y, z, t) = \pm 1.9A$ for the same time sequence; a particle starting on one of these remains within a single cell, and follows a regular closed orbit. Figure 10.10b shows the component $z(t)$ of particle paths starting from two neighbouring points $(0, 0, 0)$ (solid curves) and $(0, 0, 0.01)$ (dashed) in the chaotic region. The divergence of these curves is an indication of the exponential stretching of material line elements. The chaotic character of the flow is confirmed by the Poincaré section shown in Figure 10.11a, in which the (y, z) coordinates of a particle are plotted at time intervals $2\pi/\omega$; these coordinates are reduced modulo 2π, recognising the 2π-periodicity of the flow in the y- and z-directions. The chaotic region covers most of the domain, but with islands of regularity where the dots fall on closed curves.

Since the flow (10.23) does not depend on x, the induction equation admits solutions of the form $\mathbf{B} = \Re\mathrm{e}[\hat{\mathbf{B}}(y, z, t) \exp ikx]$, and the 2D problem for $\hat{\mathbf{B}}$ can be solved numerically. The long-term (complex) growth rate S (averaged over each short period $2\pi/\omega$) depends on $\mathcal{R}_m$ and on the wave number k, and $\Re\mathrm{e}\, S$ can be maximised over k; the maximum was found to occur at $k \approx 0.57$, varying very little with $\mathcal{R}_m$. This maximised growth rate is shown (curve a) in Figure 10.11b (the curves b and c correspond to two other modes considered by Galloway & Proctor 1992). The curve a levels off at ~ 0.3 for $\mathcal{R}_m \gtrsim 100$, again a symptom of fast dynamo action. The corresponding field structure (Figure 10.12) shows that the field $\hat{\mathbf{B}}(y, z)$ is concentrated in sheets; the width of these sheets is believed to scale like $\mathcal{R}_m^{-1/2}$, in conformity with the introductory discussion of §10.3.

10.7 Large-scale or small-scale fast dynamo?

The manner of transition from large- to small-scale dynamo action for space-periodic dynamos as $\mathcal{R}_m$ increases has been investigated by Shumaylova et al. (2017). For small $\mathcal{R}_m$, it was found that the mean-field approach giving an α^2-type dynamo (Childress 1970) gives a reasonably accurate value for the growth rate of dynamo modes on scales large compared with the scale of periodicity of the velocity field (the 'box-scale'). As $\mathcal{R}_m$ increases, the scale of the mode of maximum growth rate decreases, as found in §9.2, and small-scale dynamo modes are excited as indicated in Figure 10.9b; the field that is excited starting from random initial conditions will therefore be dominated in terms of magnetic energy by the small-scale modes. This is not to say however that the mean-field modes cease to exist; indeed averaging the excited magnetic field over the box-scale, will always filter out the small-scale modes leaving the weaker large-scale 'mean-field' ingredients.

The problem becomes more acute when the field grows to the point at which the

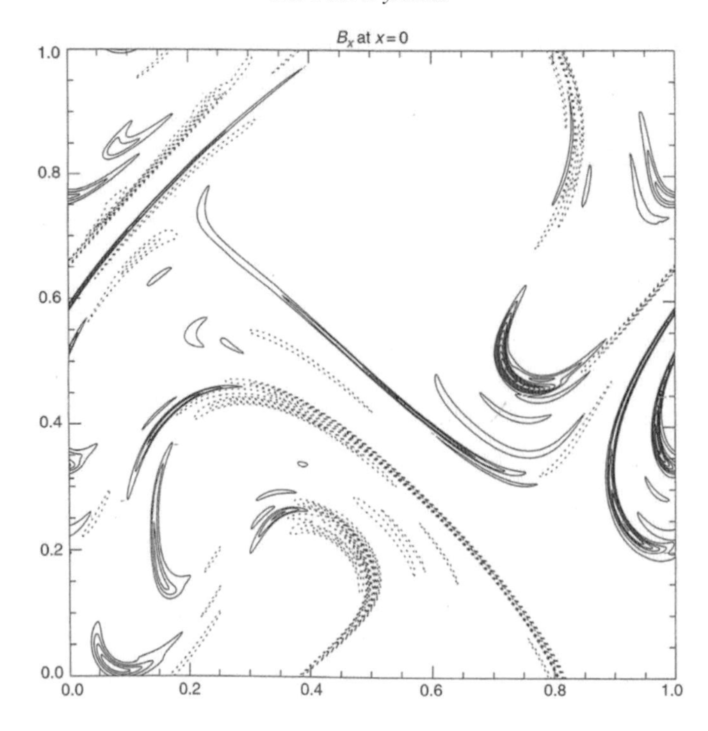

Figure 10.12 Structure of fastest growing mode of the Galloway–Proctor dynamo, represented by contours of B_x on the plane $x = 0$; $\omega = 1, k = 0.57, \mathcal{R}_m = 2000$. [From Galloway 2012.]

back-reaction of the Lorentz force has to be taken into account (i.e. the dynamic, as opposed to the purely kinematic regime; this regime is usually approached through imposition of a body force $\mathbf{f}(\mathbf{x})$ of ABC-structure in the Navier–Stokes equation, as will be described in Chapter 12). At this dynamic stage, the velocity will be influenced on the box-scale by the strong small-scale field, and this in turn will affect the parameter α of the α-effect, which is determined by this velocity field. The resulting effects, still a matter of investigation, are likely to be subtle, to say the least. The matter is important because it is incumbent on dynamo theory to explain how large-scale fields in astrophysical bodies in which $\mathcal{R}_m$ is exceedingly large, can nevertheless survive despite the dominance of small-scale fields.

10.8 Non-filamentary fast dynamo

There is also the related vexing problem that the fast-dynamo models described so far all result in fields that are exceedingly sparse in the fluid domain, being concentrated in sheets or filaments of transverse scale $O(\mathcal{R}_m^{-1/2})$ as $\mathcal{R}_m \to \infty$, in conformity with the theorem of §10.3. Although of great interest from a purely theoretical point of view, the relevance of such dynamos in astrophysical contexts is questionable. One escape is to recognise that small-scale turbulence is almost invariably present in such contexts, yielding an 'effective resistivity' η_e such that the effective magnetic Reynolds number $\mathcal{R}_{me} = (\eta/\eta_e)\mathcal{R}_m$ always remains $O(1)$. This however merely replaces one problem by another – that of understanding better the details of the turbulent process that is responsible for generating the effective diffusivity η_e; and we have already seen in Chapter 7 that this is an exceedingly challenging problem.

A new type of fast dynamo has been found in a numerical investigation of dynamo action in a *compressible* fluid taking account of the back-reaction of the Lorentz force on the driven motion (Archontis 2000, Archontis et al. 2003, 2007). This type of dynamo has been further investigated in an incompressible fluid (to which we restrict attention here) by Cameron & Galloway (2006). The dynamo is based on an initial velocity field

$$\mathbf{u}(\mathbf{x}) = (\sin z,\ \sin x,\ \sin y), \quad \omega(\mathbf{x}) = (\cos y,\ \cos z,\ \cos x), \tag{10.25}$$

which, unlike the ABC-flows, is not a Beltrami flow; this flow is in fact reflectionally symmetric ($\mathbf{u}(-\mathbf{x}) = -\mathbf{u}(\mathbf{x})$), and its mean helicity is zero! Early evidence of fast dynamo action for this flow was given by Galloway & Proctor (1992) (actually curve c of Figure 10.11b). The situation in the dynamic regime appears to be very different. Here, a force field

$$\mathbf{f}(\mathbf{x}) = \nu(\sin z,\ \sin x,\ \sin y), \tag{10.26}$$

is required to maintain a stationary state. This would drive the velocity field (10.25) if the Reynolds number $\mathcal{R}e$ were small. At high $\mathcal{R}e$ however, the flow structure evolves into something quite different; nevertheless, dynamo action still occurs, as shown in Figure 10.13a. Here, the lower (red) curve shows the rapid initial growth of magnetic energy attaining a level $\sim 2/3$ of the kinetic energy (green curve). These energies continue to grow however, on a slow diffusive time-scale, and the magnetic and velocity field structures continue to evolve as indicated by the upper (black) curve of cross-helicity $\mathcal{H}_C$, a measure of the degree of alignment of these fields:

$$\mathcal{H}_C = \frac{|\langle \mathbf{u} \cdot \mathbf{B} \rangle|}{\langle \mathbf{u}^2 \rangle^{1/2} \langle \mathbf{B}^2 \rangle^{1/2}}. \tag{10.27}$$

Ultimately, this reaches the value 1, indicating nearly perfect alignment of $\mathbf{B}$ and $\mathbf{u}$; at this stage $\mathbf{B}(\mathbf{x}) \approx \pm\mathbf{u}(\mathbf{x})$ and there is very nearly equipartition of magnetic and kinetic energy. Under this condition the inductive term $\mathbf{u} \times \mathbf{B}$ is small, allowing it to be balanced by the diffusive term $\eta\nabla \times \mathbf{B}$ in Ohm's law, even when η is very small. The associated structure of either field in the asymptotic steady state is shown in

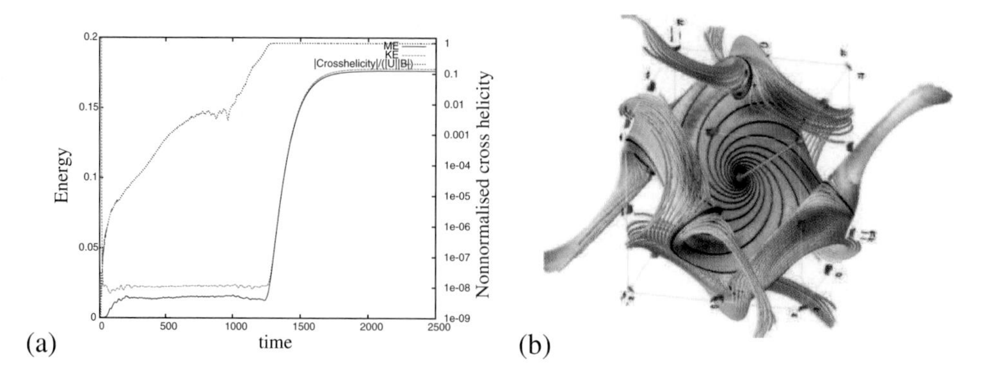

Figure 10.13 (a) Time evolution in the Archontis dynamo; magnetic energy (red), kinetic energy (green) and cross-helicity (black, dotted). (b) Field structure in the asymptotic steady state. [From Cameron & Galloway 2006 by permission of Oxford University Press on behalf of the Royal Astronomical Society.]

10.13(b). Here, the energy appears to be well distributed throughout the domain of periodicity, with no evidence of filamentary structure.

This surprising phenomenon opens up a new field of enquiry: for an arbitrary force field $\mathbf{f}(\mathbf{x})$, is there a similar tendency for an initial filamentary field generated by fast dynamo action to evolve under dynamical evolution through a 'slow-dynamo' process to a non-filamentary state of equipartition of energy? And is there a similar tendency if a time-dependent force field $\mathbf{f}(\mathbf{x}, t)$ is of natural origin, e.g. if it is the buoyancy force coupled with the Coriolis force in a rotating thermally stratified medium, as in many astrophysical contexts? We go on to consider some fundamental aspects of such dynamic problems in the following chapters.

Part III

Dynamic Aspects of Dynamo Action

11

Low-Dimensional Models of the Geodynamo

11.1 Dynamic characteristics of the segmented disc dynamo

We revert here to the segmented disc dynamo introduced in §10.4, in which the angular velocity of the disc Ω was prescribed. We now suppose that the disc is driven by a torque G, so that Ω, like the currents I and J, is time-dependent; we thus consider the coupled dynamics and electrodynamics of the system. This will serve as a useful preliminary to the fully magnetohydrodynamic situations to be considered in later chapters. We recall equations (10.12) and (10.13), which must now be coupled with the equation of motion of the disc.

We assume that the magnetic field across the disc is $B_z(s)\,\mathbf{e}_z$ in the usual cylindrical coordinates (s, φ, z), and that the current density in the disc is $j_s(s) = I/2\pi sh$, where h is the disc thickness, assumed small; the electromagnetic torque acting on the disc is then

$$\mathbf{G}_e = \int_{disc} \mathbf{x} \times (\mathbf{j} \times \mathbf{B}) \mathrm{d}V = -I\,\Phi_2/2\pi\,\mathbf{e}_z, \tag{11.1}$$

and the equation of motion (here the angular momentum equation) is

$$C\,\mathrm{d}\Omega/\mathrm{d}t = G - I\,\Phi_2/2\pi\,, \tag{11.2}$$

where C is the moment of inertia of the disc about its axis. We may regard Φ_1, Φ_2 and Ω as independent variables. Defining non-dimensional variables

$$X = \Phi_2/(2\pi GM)^{1/2},\;\; Y = (M/L)\,\Phi_1/(2\pi GM)^{1/2},\;\; Z = (M/R)\,\Omega,\;\; \tau = (R/L)\,t\,, \tag{11.3}$$

equations (10.13) and (11.2) become

$$\left.\begin{aligned} \dot{X} &= r(Y - X) \\ \dot{Y} &= mX - (1+m)Y + XZ \\ \dot{Z} &= g\left(1 + mX^2 - (1+m)X\,Y\right) \end{aligned}\right\}, \tag{11.4}$$

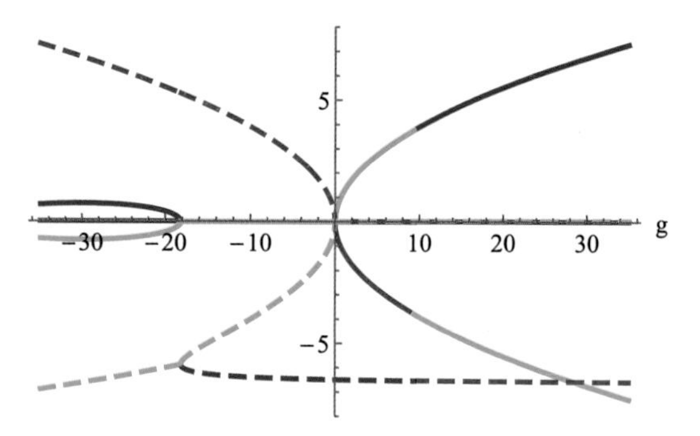

Figure 11.1 Three roots of the cubic equation (11.6) shown in red, blue and green; real parts dashed; imaginary parts solid; $m = 0.5$, $r = 5$.

where

$$m = \frac{M^2}{LL' - M^2} > 0, \quad r = \frac{R}{R'}\frac{L^2}{LL' - M^2}, \quad g = \frac{GLM}{CR^2}. \tag{11.5}$$

Note that the system (11.4) is invariant under the switch $(X, Y, Z) \to (-X, -Y, Z)$, corresponding to the underlying invariance of the MHD equations under the switch $(\mathbf{B}, \mathbf{u}) \to (-\mathbf{B}, \mathbf{u})$. The system has two fixed points at $(X, Y, Z) = (1, 1, 1)$ and $(-1, -1, 1)$.[1] The stability of either fixed points may be investigated by linear perturbation theory. Thus, substituting $X = 1 + \xi(\tau)$, $Y = 1 + \eta(\tau)$, $Z = 1 + \zeta(\tau)$ in (11.4) and linearising in the perturbations ξ, η, ζ gives three linear homogeneous equations which admit solutions of the form $(\xi, \eta, \zeta) = (\xi_0, \eta_0, \zeta_0)\mathrm{e}^{p\tau}$. The determinant condition for non-trivial solutions gives a cubic equation for p:

$$p^3 + (m + 1 + r)p^2 + (m + 1)gp + 2gr = 0. \tag{11.6}$$

The same condition is obtained for perturbations about the point $(-1, -1, 1)$, a consequence of the above symmetry.

The three roots p_1, p_2, p_3 of (11.6) (coloured red, blue and green) are shown in Figure 11.1 as functions of g for the particular choice of parameters $m = 0.5$, $r = 5$, with real parts shown dashed and imaginary parts shown solid.[2] There is a bifurcation at $g \approx -18.2$. For each $g > 0$ (and this corresponds to the applied torque being in the same sense as the sense of twist of the wire), there are two pure imaginary roots and one negative real root. The modes corresponding to the imaginary roots are (linearly) neutrally stable. However, non-linear terms are destabilising, as revealed by computation of trajectories. A sample trajectory for $0 \leq \tau < 50$ projected

[1] The system (11.4) has a superficial similarity with the Lorenz (1963) system, $\dot{x} = \sigma(y - x)$, $\dot{y} = x(\rho - z) - y$, $\dot{z} = xy - \beta z$. It is however topologically distinct from it, in that the Lorenz system has either one or three fixed points according as $\beta(\rho - 1) \lessgtr 0$, whereas the system (11.4) has two.

[2] The treatment here differs from that in Moffatt (1979), which was in error.

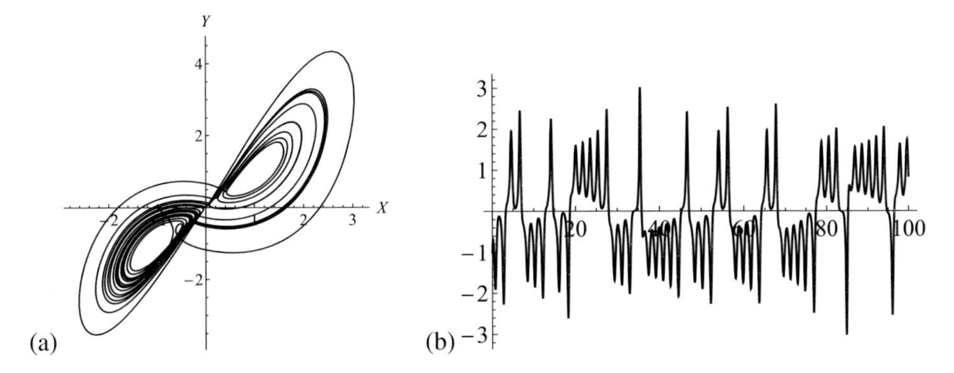

Figure 11.2 (a) Typical solution trajectory for the segmented disc dynamo ($0 < \tau < 50$), projected on (X, Y)-plane; $m = 0.5$, $r = 5$, $g = 10$. (b) Corresponding function $X(\tau)$ for $0 < \tau < 100$. The polarity reversals result from growing oscillations.

on the (X, Y)-plane (with initial point (0, 0.01, 0) and for $m = 0.5$, $r = 5$, $g = 10$) is shown in Figure 11.2a; this spirals outwards from a neighbourhood of the fixed point (1, 1) to a critical loop at which it flips over to spiral around the other fixed point $(-1, -1)$; this flipping continues randomly, as can be seen from the plot of $X(\tau)$ in Figure 11.2b; 19 such flips have occurred by time $\tau = 100$.

This is naturally of interest in the context of the geomagnetic polarity reversals described in §4.4. We may perhaps think of the dynamical system (11.4) as providing a model for the long-term variation of the geomagnetic dipole moment: the currents I and J must then be taken to represent the poloidal and toroidal currents, and Ω to represent the differential rotation that generates toroidal magnetic flux. The geometry with its right-handed twist provides the counterpart of the α-effect, and so we have here a plausible low-dimensional system representing an $\alpha\omega$-dynamo. The torque G must be taken to represent whatever drives the differential rotation, most plausibly the interaction of thermal and/or compositional convection with the Coriolis force field. The amplitude is controlled by the back-reaction of the Lorentz force, represented by the electromagnetic torque $\mathbf{G}_e$.

When $g < 0$ (i.e. $G < 0$) there is no dynamo effect. One must then ask what is the role of the positive real root p_1 evident (for negative g) in Figure 11.1. Despite the apparent instability $\sim e^{p_1 t}$ associated with this root at the linearised level, nonlinear effects take over rapidly, with the result that, provided $Z(0) < 0$, the currents I and J actually decay to zero; Ω then evolves linearly like $(G/C)\,t$, as given by (11.2). This is illustrated in Figure 11.3a. Even if $Z(0) > 0$, the dynamo fails if $-g$ is large enough to reverse the sign of $Z(\tau)$, as illustrated in Figure 11.3b for $Z(0) = 5, g = -10$; here again, the dynamo fails.[3]

[3] Of course a ‘viscous’ drag could be included in (11.2) and this would limit the linear growth of $|\Omega|$.

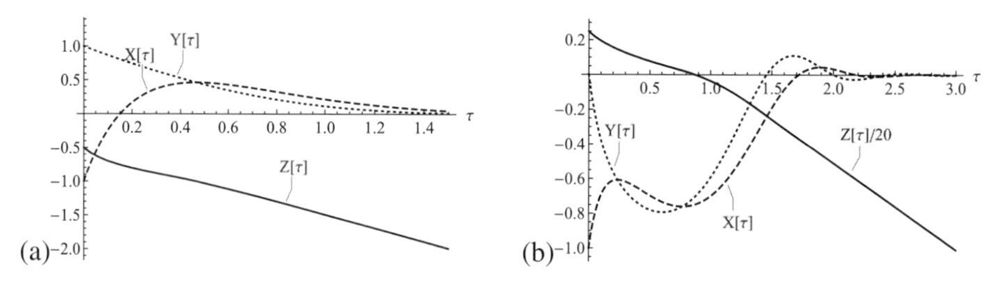

Figure 11.3 Solutions of (11.4) for $m = 0.5, r = 5$ and (a) $g = -1$, $Z(0) = -0.5$, (b) $g = -10$, $Z(0)/20 = +0.25$; in both cases, $X(\tau)$ and $Y(\tau)$ asymptote to zero, and the dynamo fails.

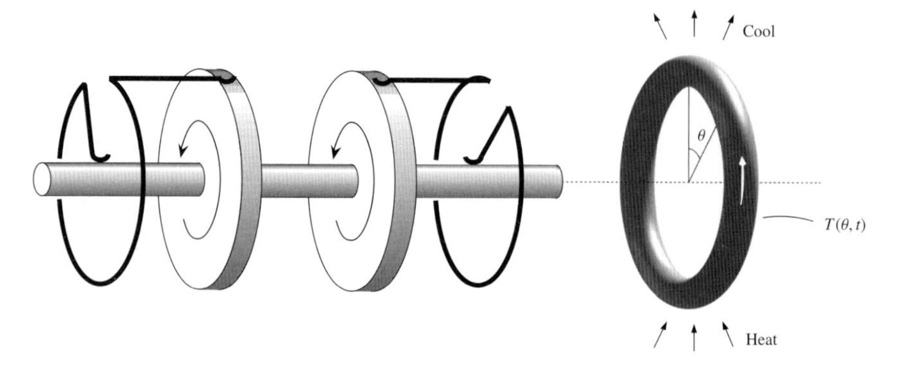

Figure 11.4 The discs on the left are assumed to be coupled with the fluid loop on the right in such a way that the fluid and the discs rotate with the same angular velocity $\Omega(t)$. [After Chui & Moffatt 1994.]

11.2 Disc dynamo driven by thermal convection

To move one step closer to the real geomagnetic situation, we now consider a model proposed by Chui & Moffatt (1994); this is still one-dimensional in character, but it involves interaction between thermal convection and magnetic self-excitation. The configuration is shown in Figure 11.4. Two discs mounted on the same shaft are shown on the left. The wires making contact with the discs are twisted in opposite senses about the axle, simulating the antisymmetric distribution of helicity relative to the Earth's equatorial plane. The magnetic fluxes are Φ_1 and Φ_2 through the two loops, and Λ_1 and Λ_2 across the two discs. The self-inductance and resistance of each loop are L and R, and the relevant mutual inductances are $M, \epsilon_1 M$ and $\epsilon_2 M$, with $M = L/\lambda$, $(\lambda > 1)$. The currents in the two circuits are I_1 and I_2; azimuthal currents in the discs are here neglected, so the model is only valid for moderate $\mathcal{R}_m$; the azimuthal currents could be included as in the previous section, but with added complication.

A circular tube containing fluid is shown on the right of Figure 11.4. When heated from below, the fluid rotates in one sense or the other, with angular velocity $\Omega(t)$, say, averaged over the tube cross section. We must imagine this loop to be coupled to the discs in such a way that they too rotate with the same angular velocity. This could in principle be achieved via strong viscous coupling, and assuming negligible disc inertia.[4]

11.2.1 The Welander loop

The behaviour of fluid in a loop heated from below was considered by Welander (1967); it is the simplest possible description of thermal convection. Welander's loop consisted of two long vertical sections joined at top and bottom. Following Chui & Moffatt (1994), we consider instead a circular loop of radius a and cross-sectional area A in a vertical plane, and we suppose that heating is provided by a distributed source $S(\theta') = -\sigma \sin\theta'$ ($\sigma > 0$), where $\theta' = \pi/2 - \theta$, with θ the 'co-latitude' as shown in Figure 11.4. Suppose the temperature of the fluid is $T_0 + T(\theta, t)$; if the fluid flows round the tube with angular velocity $\Omega(t)$, then T satisfies the heat conduction equation

$$\frac{\partial T}{\partial t} + \Omega\frac{\partial T}{\partial \theta'} = \kappa\frac{\partial^2 T}{\partial \theta'^2} - \sigma \sin\theta', \tag{11.7}$$

where κ is an effective thermal diffusivity. The forced response then has the form

$$T(\theta', t) = p(t)\cos\theta' + q(t)\sin\theta', \tag{11.8}$$

where

$$\dot{p} = -Dp - \Omega q, \quad \dot{q} = -Dq + \Omega p - \sigma. \tag{11.9}$$

In the static situation ($\Omega = 0$), we have $p = 0$ and $q = -\sigma/D = q_0$, say.

The temperature distribution (11.8) creates a density perturbation $\Delta\rho = -\alpha T$, where α is the coefficient of thermal expansion; this in turn gives rise to a force $\mathbf{F} = \Delta\rho\,\mathbf{g}$ and a net torque $\mathbf{G}$

$$\mathbf{G} = \int_V \mathbf{x} \times \mathbf{F}\,\mathrm{d}V, \tag{11.10}$$

acting on the fluid, where V is the volume occupied by the fluid in the loop. This evaluates to $\mathbf{G} = (0, 0, G)$ where

$$G = \gamma p, \quad \gamma = \tfrac{1}{2}\alpha g a V. \tag{11.11}$$

[4] A more sophisticated model might assume a thin, but hollow, conducting disc filled with conducting fluid, and heated from below; thermal instability would then cause circulating motion in the fluid, and rotation of the disc through viscous drag.

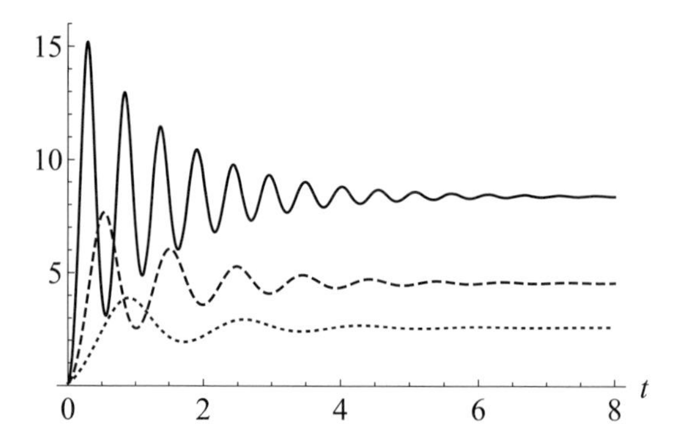

Figure 11.5 Stability of the states (11.14) for $D = k = 1, \gamma = 2$ and for $\sigma =$ 10 (dotted), 50 (dashed) and 300 (solid), illustrating how the frequency of the damped oscillations increases with increasing thermal forcing; curves computed from the non-linear equations (11.9) and (11.12) with initial conditions $p(0) =$ 0, $q(0) = -1$, $\Omega(0) = 0.1$.

It is this torque that drives the fluid motion, so that, in the absence of any electromagnetic effect, Ω satisfies the equation

$$C\dot{\Omega} = \gamma p - k\Omega, \tag{11.12}$$

where k represents the frictional resistance between fluid and tube. This equation is now coupled with the equations (11.9), forming a closed system.

In order to investigate the stability of the static equilibrium $(p, q, \Omega) = (0, q_0, 0)$, we set $q = q_0 + \hat{q}(t)$ and linearise in the variables $p(t)$, $\hat{q}(t)$ and $\Omega(t)$. Seeking solutions proportional to e^{mt}, the condition for instability $\Re\mathrm{e}\, m > 0$ is easily found to be

$$\sigma > kD^2/\gamma, \tag{11.13}$$

i.e. the situation is linearly unstable if the forcing σ is large enough to overcome the combined effects of thermal diffusion and viscous resistance. The fluid then flows in a sense determined by the sign of $\Omega(0)$.

Under the condition (11.13), equations (11.9) and (11.12) admit two steady solutions (in addition to the unstable solution $(0, q_0, 0)$),

$$\Omega = \pm\left[\frac{\gamma}{k}\left(\sigma - \frac{kD^2}{\gamma}\right)\right]^{1/2}, \quad p = \frac{k\Omega}{\gamma}, \quad q = -\frac{kD}{\gamma}. \tag{11.14}$$

The stability of these states may again be analysed with perturbations $\sim e^{mt}$; both are found to be stable, with $\Re\mathrm{e}\, m < 0$ but with $\Im\mathrm{m}\, m \neq 0$, so that perturbations oscillate as they decay. These damped oscillations are illustrated in Figure 11.5.

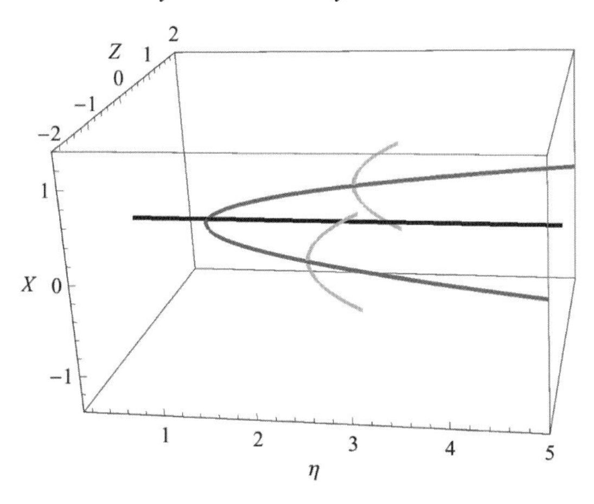

Figure 11.6 Bifurcation diagram for the thermally forced disc dynamo; the steady solutions after the first bifurcation are in the plane $X = 0$, and the steady solutions after the dynamo bifurcation are in the planes $Z = \pm Z_1$. (Here, $Z_1 = 1.2$, $\eta_1 = 2.44$.)

11.2.2 Coupling of Welander loop and Bullard disc

When the thermally driven velocity is large enough in either direction, it is clearly going to be capable of exciting a secondary dynamo instability for whichever disc/wire combination has the appropriate twist. Two additional variables, $x = \Phi_1 + \Phi_2$ and $y = \Phi_1 - \Phi_2$ are needed to describe the resulting interaction for the disc system of Figure 11.4. With shift of origin and suitable rescaling of $\{x, y, \Omega, p, q\}$ to dimensionless variables $\{X, Y, Z, P, Q\}$, the resulting fifth-order dynamical system can be expressed in the form

$$\left.\begin{aligned} \dot{X} &= c_1(-c_3\, X + c_2\, YZ) \\ \dot{Y} &= -c_3\, Y + c_2\, X Z \\ \dot{Z} &= c_4\, (P - Z - X\, Y) \\ \dot{P} &= -P + \eta Z - Q Z \\ \dot{Q} &= -Q + P Z \end{aligned}\right\}, \tag{11.15}$$

where the parameters $c_1, \ldots c_4$ (all positive) are

$$c_1 = \frac{\lambda + \varepsilon_2}{\lambda - \varepsilon_2}, \quad c_2 = \frac{(1 - \varepsilon_1^2)^{1/2}}{\lambda + \varepsilon_2}, \quad c_3 = \frac{R}{MD} \frac{1}{\lambda + \varepsilon_2}, \quad c_4 = \frac{k}{CD}, \tag{11.16}$$

and the control parameter $\eta = \sigma R / MD^2$ is the dimensionless measure of the heat source σ. As η increases from zero, the primary bifurcation occurs at $\eta = \eta_0 = 1$; at this point, rotation commences in one sense or the other. With further increase, the thermally forced rotation rate increases, and a secondary bifurcation occurs at

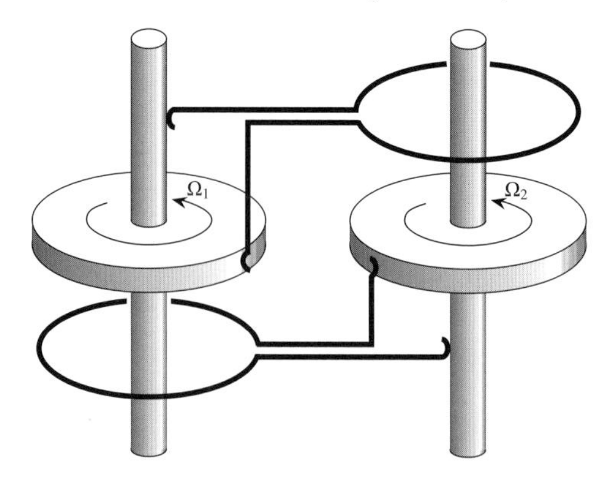

Figure 11.7 Coupled disc dynamo (after Rikitake 1958); the discs rotate (under applied torques) with angular velocities $\Omega_1(t), \Omega_2(t)$ and drive currents $I_1(t)$, $I_2(t)$ through the wires which make sliding contact with rims and axles as indicated.

$\eta = \eta_1 = 1 + c_3^2/c_2^2$. At this point, dynamo instability sets in, and for $\eta > \eta_1$, the steady states are given by $Z = \pm c_3/c_2 = \pm Z_1$, say, and

$$X = \pm Y = \left[Z_1 (\eta - \eta_1)/\eta_1\right]^{1/2}, \quad P = Q/Z_1 = Z_1(\eta/\eta_1). \tag{11.17}$$

The bifurcation diagram is shown in Figure 11.6 in the space of the variables (X, Z, η). The solution $X = Z = 0$ is the horizontal line; the steady solutions after the first bifurcation are in the plane $X = 0$, and the steady solutions after the dynamo bifurcation are in the planes $Z = \pm Z_1$.

The five-parameter five-dimensional system (11.15) evidently has a rich structure that deserves more detailed investigation than can be provided here. It serves to indicate however just how it is that dynamo action can arise when the sole input of energy is thermal. We shall encounter this sort of behaviour repeatedly in later chapters.

11.3 The Rikitake dynamo

The earliest model for reversals of the Earth's magnetic field was introduced by Rikitake (1958) and subsequently studied by Allan (1962) and Cook & Roberts (1970). This was the coupled-disc dynamo model sketched in Figure 11.7a. For this system, there are 2 degrees of freedom represented by the angular velocities $\Omega_1(t)$ and $\Omega_2(t)$ of the discs, and 2 'electrical degrees of freedom' represented by the currents $I_1(t)$ and $I_2(t)$ in the wires which must both be twisted in the same sense relative to the rotation vectors if field regeneration is to occur. Under symmetric conditions, the equations describing the evolution of this system, when a constant

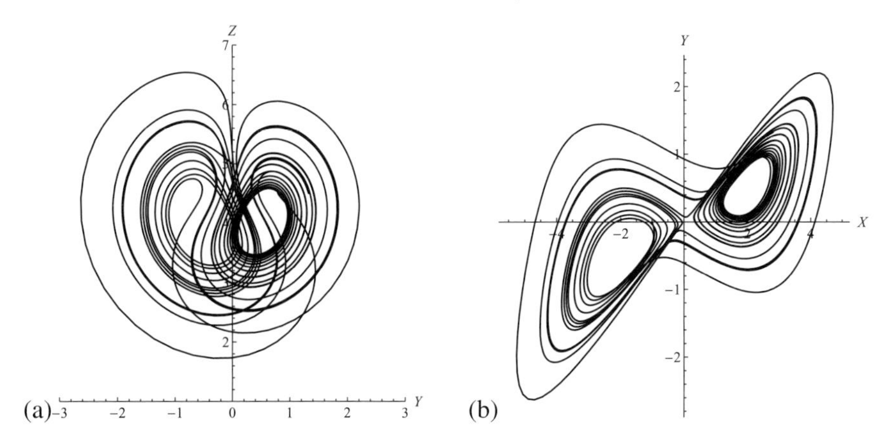

Figure 11.8 Typical trajectory for the Rikitake model in phase space (for $0 < \tau < 100$), projected (a) on the (Y, Z) plane and (b) on the (X, Y) plane (cf. Figure 11.2); for this trajectory, $\mu = 1, a = 15/4$ and $(X(0), Y(0), Z(0)) = (-1, 0, 5)$. [Adapted from Cook & Roberts 1970.]

torque G is applied to each disc, are

$$L\,\mathrm{d}I_1/\mathrm{d}t + RI_1 = M\Omega_1 I_2\,, \quad L\,\mathrm{d}I_2/\mathrm{d}t + RI_2 = M\Omega_2 I_1\,, \tag{11.18a}$$

$$C\,\mathrm{d}\Omega_1/\mathrm{d}t = C\,\mathrm{d}\Omega_2/\mathrm{d}t = G - MI_1 I_2\,, \tag{11.18b}$$

where L and R are the self-inductance and resistance of each circuit, C is the moment of inertia of each disc about its axis, and $2\pi M$ is the mutual inductance between the circuits. The terms $M\Omega_1 I_2$ and $M\Omega_2 I_1$ represent the electromotive forces arising from the rotations Ω_1 and Ω_2, while the term $-MI_1 I_2$ represents the torque associated with the Lorentz force distribution in each disc.

Here again, it must be recognised that the model is deficient in that it neglects the induced currents around the rim of each disc, so the model cannot correctly represent the high-$\mathcal{R}_\mathrm{m}$ situation. Nevertheless, the Rikitake dynamo is in some respects analogous to the geodynamo configuration and equations with structure similar to (11.18) were obtained by Robbins (1976) through truncation of a system of moment equations based on the full MHD equations ((2.134) and (12.1) below). Non-dimensionalisation of (11.18) again leads to a simple third-order dynamical system

$$\dot{X} + \mu X = ZY\,, \quad \dot{Y} + \mu Y = (Z - a)X\,, \quad \dot{Z} = 1 - XY\,, \tag{11.19}$$

where $\mu\,(> 0)$ and a are constants; this system was studied in detail by Cook & Roberts (1970). As in §11.1, there are again two fixed points that are neutrally stable on linear perturbation analysis, but unstable when non-linear effects are taken into account. The trajectories of any point $P(\tau)$ in the phase space of the variables

(X, Y, Z) define a 'velocity field'

$$\mathbf{u}_P = (\dot{X}, \dot{Y}, \dot{Z}) = (-\mu X + ZY,\ -\mu Y + (Z - a)X,\ 1 - XY)\,, \tag{11.20}$$

for which

$$\nabla \cdot \mathbf{u}_P = \frac{\partial u_1}{\partial X} + \frac{\partial u_2}{\partial Y} + \frac{\partial u_3}{\partial Z} = -2\mu\,. \tag{11.21}$$

This implies that the volume dV occupied by any element of 'phase fluid' tends to zero in a time of order μ^{-1}; any dynamical system with this property is said to be 'dissipative'.[5] This suggests that, for nearly all possible initial positions $P(0)$, the trajectory of $P(\tau)$ will ultimately lie arbitrarily close to a limit set $F(X, Y, Z) = 0$ (a 'strange attractor') with the property that

$$(-\mu X + ZY)\frac{\partial F}{\partial X} + (-\mu Y + (Z - A)X)\frac{\partial F}{\partial Y} + (1 - XY)\frac{\partial F}{\partial Z} = 0\,. \tag{11.22}$$

This behaviour was confirmed by Cook & Roberts who computed solution trajectories of (11.19), and who succeeded in describing the topological character of the limit sets for particular values of the parameters μ and a. Figure 11.8 shows the projection of a typical strange attractor (a) on the (Y, Z) plane and (b) on the (X, Y) plane, for the parameter values $\mu = 1, a = 15/4$; for these values, the fixed points of (11.19) are $(2, 1/2, 4)$ and $(-2, -1/2, 4)$, and it is evident that the trajectory circulates about these points with random switches from one to the other, but never actually approaching close to either. If P(0) is chosen very near to one of the fixed points, then the trajectory slowly spirals out from it, eventually reaching the 'critical loop' from which it switches to circulate round the other fixed point; from here on, it asymptotes to the strange attractor.

Nozières (1978, 2008), extended such models, with emphasis on the time-scales inherent in such models: a fast Alfvénic time-scale t_A characterising the oscillation frequency apparent in figures such as (4.12) or (11.2)(b), and a slower diffusive time-scale t_η characterising the duration of a reversal when it occurs. The key problem here is really to explain the large ratio between the mean duration between geomagnetic reversals ($t_d \sim 10^6$ yr) and the typical duration $t_\eta \sim 10^4$ yr of a reversal; this large ratio is not easily extracted from Nozière's analysis.

11.4 Symmetry-mode coupling

A very different description of polarity reversals has been proposed by Gissinger et al. (2010) who obtain a model system applicable to the low $\mathcal{P}_m(= \nu/\eta)$ situation pertaining to the Earth; it also seeks to describe the reversals observed in the VKS

[5] In this sense, the systems (11.4) and (11.15) are also dissipative, and the system (11.23) is dissipative if $\nu + 1 > \mu$.

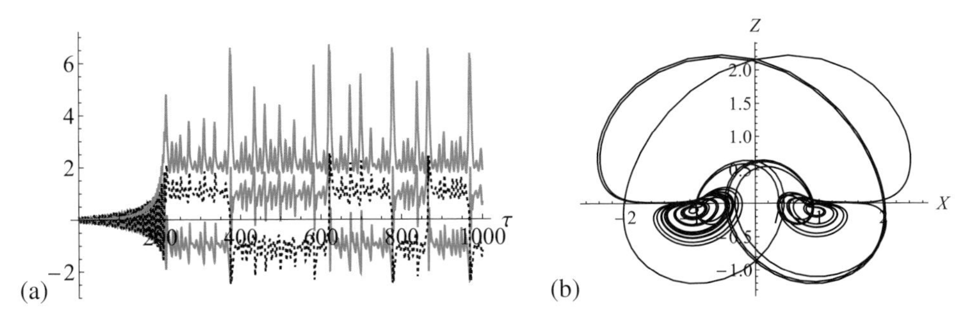

Figure 11.9 (a) Computation of $X(\tau)$ (black), $Y(\tau)$ (black dashed) and $X(\tau)^2 + Y(\tau)^2$ (black, large positive fluctuations) of the DQV system for $0 < \tau < 800$. (b) Corresponding (X, Z) projection of trajectory for $700 < \tau < 1000$ ($\mu = 0.119, \nu = 0.1, \Gamma = 0.9$, and the initial conditions for this run were $X(0), Y(0), Z(0) = (0.1, 0.01, 0.01)$). [Parameters from Gissinger et al. 2010.]

dynamo experiment (as described in §9.12). This 'DQV-system' is written in terms of the amplitudes of a dipole mode $X \equiv D(\tau)$, a quadrupole mode $Y \equiv Q(\tau)$ coupled by a 'zonal velocity mode' $Z \equiv V(\tau)$, and has the form

$$\dot{X} = \mu X - YZ, \quad \dot{Y} = -\nu Y + ZX, \quad \dot{Z} = \Gamma - Z + XY, \tag{11.23}$$

where μ, ν and Γ are parameters taken to be positive. This is rather different from the Rikitake system (11.19). The forcing term Γ in (11.23) has the effect of breaking the symmetry properties relative to a rotation R_π around an axis in the equatorial plane (as when, in the VKS experiment, the counter-rotating propellors rotate with different frequencies). So the main flow, which drives the dynamo is accounted for by positive μ (growing dipolar field). If this flow is invariant under the rotation R_π (as when, in the VKS experiment, the two impellers counter-rotate with the same angular velocity), then the evolution of the unstable dipolar mode (which reverses sign under R_π) is decoupled from that of a quadrupolar mode (invariant under R_π).[6] In the DQV model this latter mode is assumed to be linearly stable. When the symmetry in the flow is broken, these modes become coupled through the mean zonal flow V.

Figure 11.9a shows a typical evolution governed by the system (11.23) for the parameter values and initial conditions indicated in the caption. For $0 < \tau \lesssim 210$, the system evolves with growing oscillations in transient response to the initial conditions; at $\tau \approx 215$, $Z(\tau)$ changes sign for the first time, and this triggers the beginning of chaotic evolution, with random simultaneous reversals of both the dipole and quadrupole ingredients of the field. What is remarkable here is that the fluctuations of X and Y between reversals themselves appear quite chaotic (unlike the

[6] 'Dipolar' and 'quadrupolar' refer here to the decomposition of a field into a component which changes sign and one which does not change sign under the R_π rotation.

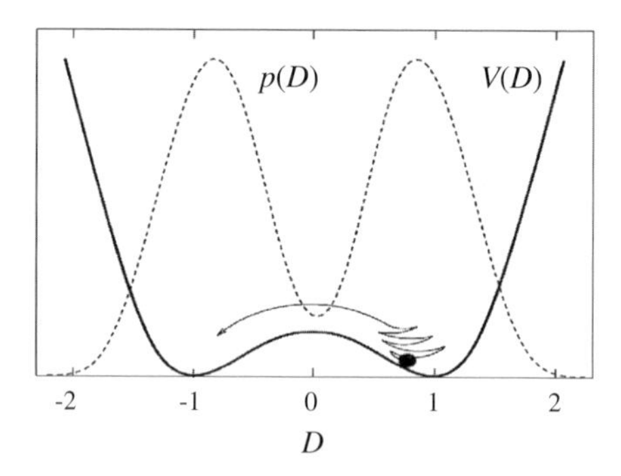

Figure 11.10 The dipole state D, represented pictorially as a particle in a potential well $V(D)$ (solid curve) with two symmetrically placed minima; random fluctuations of the α-effect are assumed as the mechanism by which the dipole can flip from one state to the other, with associated polarity reversal; the probability $p(D)$ of being in a state D is shown schematically by the dotted curve. [After Hoyng et al. 2002.]

situation for either the segmented disc dynamo or the Rikitake dynamo), and that the energy function X^2+Y^2 actually peaks during the reversals. Figure 11.9b shows the corresponding X, Z projection of the solution trajectory for $700 < \tau < 1000$; during this interval, there were four reversals, which appear as the large arcs connecting the dense 'eyes' of the phase portrait.

Gissinger et al. (2010) were in part motivated by their direct numerical simulations of the VKS experiment at the relatively low magnetic Prandtl number $\mathcal{P}_\mathrm{m} = 0.5$ (although still far from the relevant geomagnetic value estimated at $\sim 10^{-5}$) which reveal similar mode-coupling. The DQV model captures some essential qualitative features of the time series of reversals, both from paleomagnetic data and from the VKS experiment, particularly as regards the random character of the fluctuations between reversals. Their long-run solution of (11.23) also shows occasional long periods with no reversal, as for the superchrons mentioned in §4.4.

11.5 Reversals induced by turbulent fluctuations

A further model for the understanding of reversals and reversal statistics has been developed by Hoyng et al. (2002), with a view to describing as closely as possible the geomagnetic data (described as 'Sint-800') accumulated by Guyodo & Valet (1999) (for subsequent longer records, see also Valet et al. 2005 and Figure 4.12). The model is based on the idea that the geomagnetic dipole 'sits' with high probability in one of the two valleys of a bistable potential well (Figure 11.10) and that it is random fluctuations of the α-effect that can occasionally flip the dipole from one valley to the other (the 'particle-in-well' paradigm). The theory involves the Fokker–Planck equation for the probability $p(D)$ that the dipole is in state D, and is too involved to present here. The large number of adjustable parameters in the model allows for substantial, although as yet not complete, agreement with the detailed geomagnetic record.

11.5.1 Dipole-quadrupole model

A similar approach has been adopted by Pétrélis et al. (2009), who also invoke random fluctuations attributed to background turbulence to induce polarity reversals. Their model was based on the experimental observation that reversals only occurred in the VKS experiment when the two driving propellors were counter-rotated at different speeds.

The model, which preceded the DQV model, can be viewed as a simpler 'DQ model', which considers two modes with different symmetries,[7] **d** and **q**. Any field **B** can be decomposed into a symmetric and an antisymmetric component. Pétrélis et al. (2009) considered a field of the form

$$\mathbf{B}(\mathbf{x},\tau) = \mathbf{d}(\mathbf{x})\,D(\tau) + \mathbf{q}(\mathbf{x})\,Q(\tau)\,, \tag{11.24}$$

where τ is a dimensionless time. Introducing the complex variable $A(\tau) = D(\tau) + \mathrm{i}\,Q(\tau)$, an amplitude equation for A can be written. As it needs to satisfy the $A \to -A$ symmetry it contains only odd non-linearities. Writing A in the form $R\mathrm{e}^{\mathrm{i}\theta}$ and assuming that R has a shorter time-scale than the phase θ and can be eliminated, a simple evolution equation for θ can be obtained. Up to a change of the reference phase θ_0, this takes the form

$$\mathrm{d}\theta/\mathrm{d}\tau = \alpha - \sin 2\theta + \Delta\,\zeta(\tau) = -\partial V/\partial\theta + \Delta\,\zeta(\tau), \tag{11.25}$$

where $V(\theta) = -\alpha\,\theta - \frac{1}{2}\cos 2\theta$ plays the role of a 'potential function', α is a variable parameter that measures the 'DQ coupling' (i.e. the breaking of symmetry in the flow), $\zeta(\tau)$ is Gaussian white noise (representing turbulent fluctuations) and Δ is the variable amplitude of this noise. The amplitude of the modes with dipolar and quadrupolar symmetry are simply related to $\theta(\tau)$ by $D(\tau) = \cos\theta(\tau)$ and $Q(\tau) = \sin\theta(\tau)$ respectively.

The dynamics in this model can thus be represented by trajectories on a circle as shown in Figure 11.11. Non-linearities in the amplitude equation that are not retained in this model are assumed to restrict the trajectories of the system to this circle. The invariance of the system under the switch $\mathbf{B} \to -\mathbf{B}$ is preserved in this restricted phase space. If $\alpha = 0$, the modes are decoupled and there are two stable fixed points (pure dipole) at $\theta = 0$ and $\theta = \pi$, and two unstable fixed points (pure

[7] In the case of the VKS experiment, the natural symmetry is, as previously noted, the R_π rotation, while in the case of the Earth it is the symmetry about the equatorial plane (see Pétrélis et al. 2009). In the Earth situation, since the magnetic field is a pseudo-vector, it is the dipolar rather than the quadrupolar mode that is invariant under this symmetry.

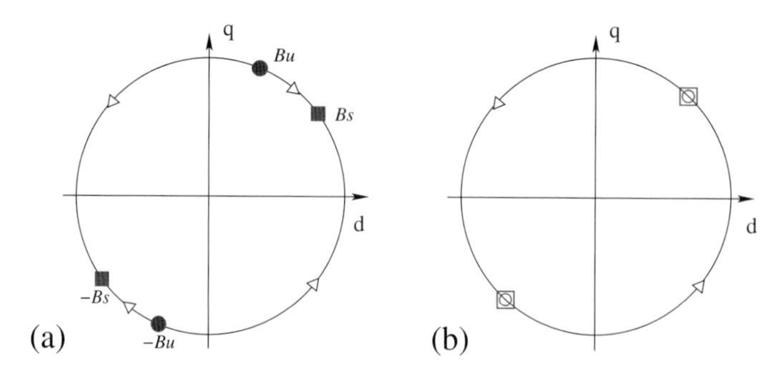

Figure 11.11 Phase space of the DQ system, displaying a saddle-node bifurcation: (a) below the onset of the saddle-node bifurcation ($\alpha < 1$) (squares: stable fixed points) (circles: unstable fixed points); (b) at $\alpha = 1$, squares and circles coincide and a saddle-node bifurcation occurs. [Reprinted figure with permission from Pétrélis et al. 2009. Copyright 2009 by the American Physical Society.]

quadrupole) at $\theta = \pi/2$ and $\theta = 3\pi/2$. For $0<\alpha<1$, the stable and unstable modes are of mixed type (Figure 11.11a). At $\alpha = 1$, the stable and unstable fixed points move into coincidence (Figure 11.11b) (a 'saddle-node bifurcation'), and for $\alpha>1$ the solution $\theta(\tau)$ describes a limit cycle. Just below $\alpha = 1$, fluctuations can drive the system from Bs to Bu (phase $Bs \to Bu$) (Figure 11.11a) and initiate a reversal (phase $Bu \to -Bs$) or an excursion (phase $Bu \to Bs$).

The potential function $V(\theta)$ is shown in Figure 11.12a for $\alpha = 0.5$ and for the 2π-interval $\pi/2<\theta<5\pi/2$, with noise represented by the high-frequency jitter near the minimum on the left. If the solution is in this 'minimal energy state' (like a particle in a potential well), it is stable, but if the amplitude of the noise increases, the 'particle' can escape over the nearby maximum and fall to the next minimum on the right. Figure 11.12b shows that, as α increases through the critical value $\alpha_c=1$, each minimum merges with its neighbouring maximum, and for $\alpha>1$, $V(\theta)$ is monotonic decreasing for all θ, i.e. there is no equilibrium solution.

When $\Delta = 0$, (11.25) may be integrated either explicitly or numerically; Figure 11.12c shows the function $\theta(\tau)$ with initial condition $\theta_0=0.3$ (the value adopted by Pétrélis et al.) and for $\alpha=0.9$ and $\alpha=1.1$, and Figure 11.12d shows the corresponding evolution of $|D|$. For $\alpha = 0.9$, θ increases from its initial value 0.3 until $\sin 2\theta = 0.9$, i.e. $\theta = \frac{1}{2}\sin^{-1} 0.9 \approx 0.56$; the solution converges to this value. For $\alpha=1.1$ on the other hand, there is no fixed point, and the solution is a superposition of a linear function and a periodic fluctuation. If α increases continuously from 0.9 to 1.1, a jump occurs as α passes through 1, from the stable 'fixed-point' solution to this 'linear-plus-periodic' solution. For $\alpha>1$, the corresponding function $|D(\tau)|$ (Figure 11.12d) is periodic with period $\widehat{\tau} = \pi(\alpha^2 - 1)^{-1/2}$, and has one gradient discontinuity in each interval of length $\widehat{\tau}$, when the dipole reverses polarity. Note

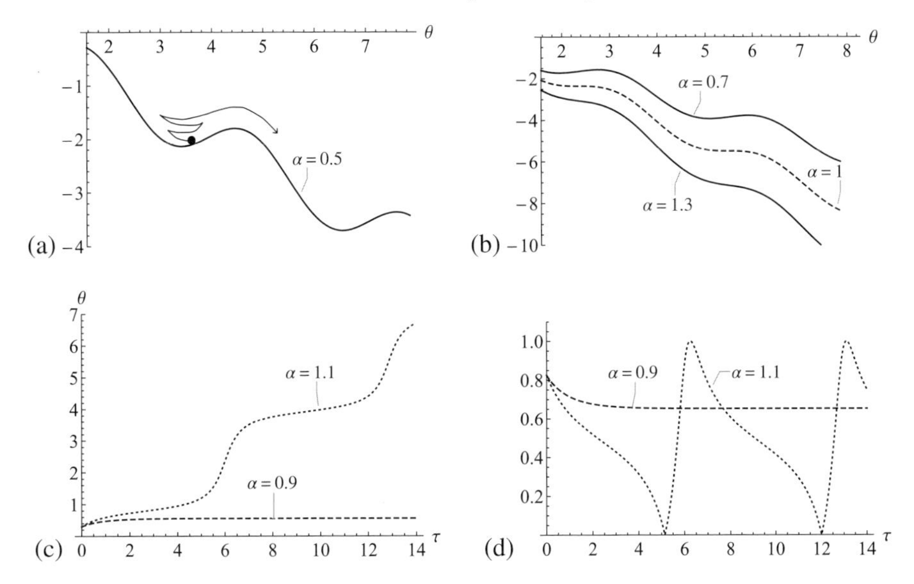

Figure 11.12 (a) The potential function $V(\theta)$, given by (11.25) for $\alpha = 0.5$ over the 2π-interval $\pi/2 < \theta < 5\pi/2$; in each such interval, there are two minima. Noise near the minimum on the left is represented by the high-frequency jitter; if the amplitude of this noise is large enough, the 'particle' can escape over the local maximum and descend to the next minimum on the right (cf. the scenario of Figure 11.10 as envisaged by Hoyng et al. 2002). (b) As α increases through the critical value $\alpha = 1$, the minima disappear and for $\alpha > 1$, $V'(\theta) < 0$ for all θ. (c) The function $\theta(\tau)$ given by (11.25) when $\Delta = 0$ for $\alpha = 0.9$ and $\alpha = 1.1$. (d) The corresponding evolution of $|D|$; when $\alpha = 1.1$, the latter is periodic with period $\widehat{\tau} = \pi(\alpha^2 - 1)^{-1/2} \approx 6.856$, and has one gradient discontinuity in each interval of length $\widehat{\tau}$.

in particular the systematic 'overshoot' of the dipole strength immediately after a reversal of polarity.

Figure 11.13a illustrates the behaviour of the component with dipolar symmetry, with noise included at the level $\Delta = 0.2$. This time series should be compared with the paleomagnetic data for the Earth's magnetic field (Figure 4.12) and the recordings of the VKS experiment (Figure 9.23). A detailed comparison of reversals is presented in Figure 11.13b (from Pétrélis et al. 2009). On the left is the result of integrating (11.25) with $\alpha = 0.9$, $\Delta = 0.2$ and $\theta_0 = 0.3$. The time-axis is scaled to give optimal fit with the geomagnetic curves on the right; the white noise is evidently responsible for the jumps, even although $\alpha < 1$. The curve for $|D(t)|$ bears comparison with the deterministic curve of (11.12)(d), and also with the curve of Figure 11.13c, which shows output from the VKS experiment, and with the curves of the right-hand panel (d), which represent the geomagnetic data of Valet et al. (2005) (see also Figure 4.12). The qualitative agreement between these sets of curves indicates that the simple model equation (11.25) with only two adjustable parameters

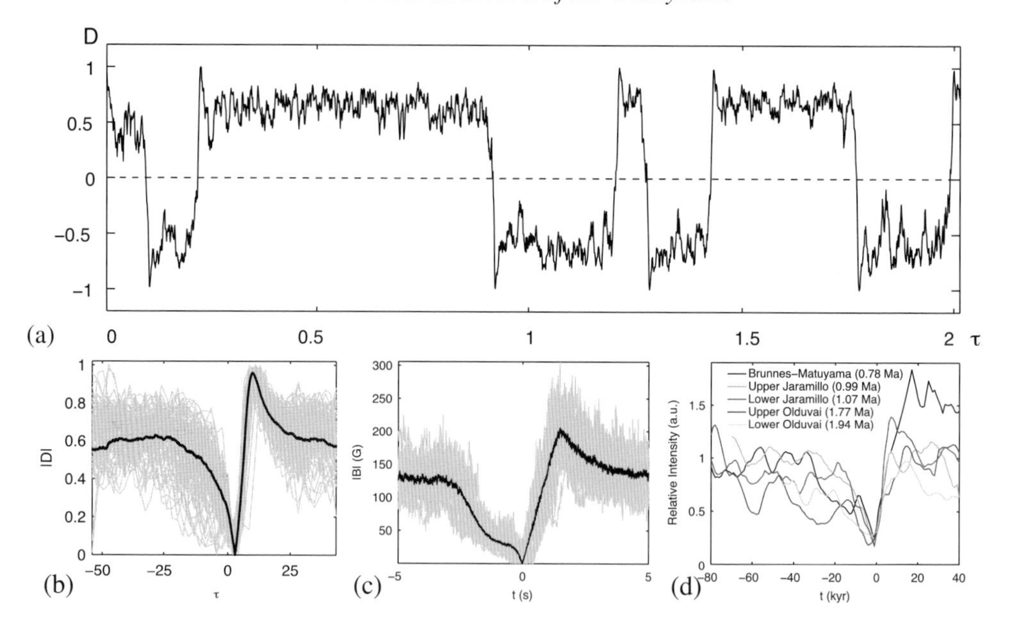

Figure 11.13 (a) Time series of the dipole amplitude for the DQ system below the bifurcation ($\alpha = 0.9$) with noise ($\Delta - 0.2$). (b)–(d) Comparison of reversals (b) in the solutions of (11.25), (c) in the VKS experiment, from Berhanu et al. (2007), and (d) in paleomagnetic data from Valet et al. (2005); black curves in (b) and (c) represent the average $\langle |D(t)| \rangle$, each realisation being represented in grey. [Reprinted figure with permission from Pétrélis et al. 2009. Copyright 2009 by the American Physical Society.]

(three if the scaling-parameter of time is included) gives a good basic description of the polarity reversal phenomenon; note particularly the steep rise and apparent 'overshoot' of the curve of $|D|$, following the reversal, properties that can also be discerned in the geomagnetic data.

Refinement of the model would involve further adjustable parameters, and it is not evident that such refinement is worthwhile; the simplicity of the model as described above is arguably its greatest virtue!

12

Dynamic Equilibration

12.1 The momentum equation and some elementary consequences

So far we have regarded the velocity field $\mathbf{u}(\mathbf{x}, t)$ as given. In this chapter, we turn to the study of dynamic effects in which the evolution of $\mathbf{u}$ is determined by the Navier–Stokes equation

$$\rho(\partial \mathbf{u}/\partial t + \mathbf{u} \cdot \nabla \mathbf{u}) = -\nabla p + \mathbf{J} \wedge \mathbf{B} + \rho\nu\nabla^2 \mathbf{u} + \rho \mathbf{F} \,. \tag{12.1}$$

Here $\mathbf{J} \wedge \mathbf{B}$ is the Lorentz force distribution, ν is the kinematic viscosity of the fluid (assumed uniform) and $\mathbf{F}$ represents any further force distribution that may be present. We limit attention to incompressible flows for which

$$\mathrm{D}\rho/\mathrm{D}t = 0 \quad \text{and} \quad \nabla \cdot \mathbf{u} = 0 \,, \tag{12.2}$$

so that bulk viscosity effects do not appear in (12.1). Although we shall be primarily concerned with motion in a rotating frame of reference, for the moment we adopt an inertial frame in which acceleration is represented adequately by the left-hand side of (12.1).

The integral equivalent of (12.1) is

$$\frac{\mathrm{d}}{\mathrm{d}t}\int_V \rho u_i \mathrm{d}V = \int_S (\sigma_{ij} + T_{ij}) n_j \mathrm{d}S + \int_V \rho F_i \mathrm{d}V \,, \tag{12.3}$$

where S is a closed surface moving with the fluid, V its interior, σ_{ij} the stress tensor given by the Newtonian relation

$$\sigma_{ij} = -p\delta_{ij} + \rho\nu\left(\partial u_i/\partial x_j + \partial u_j/\partial x_i\right) , \tag{12.4}$$

and T_{ij} the Maxwell stress tensor given by

$$T_{ij} = \mu_0^{-1}\left(B_i B_j - \tfrac{1}{2}\mathbf{B}^2 \delta_{ij}\right) . \tag{12.5}$$

With (12.3) we must clearly associate a jump condition

$$[\![(\sigma_{ij} + T_{ij}) n_j]\!] = 0 \,, \tag{12.6}$$

across any surface of discontinuity of physical properties; this may be either a fixed fluid boundary on which $\mathbf{u} = 0$ (or just $\mathbf{n} \cdot \mathbf{u} = 0$ if $\nu = 0$), or an interior surface of discontinuity moving with the fluid.

12.1.1 Alfvén waves

If ρ is uniform, it is convenient to introduce the new variable

$$\mathbf{h} = (\mu_0\rho)^{-1/2}\mathbf{B}\,, \tag{12.7}$$

in terms of which

$$\mathbf{J} \wedge \mathbf{B} = \mu_0^{-1}(\nabla \wedge \mathbf{B}) \wedge \mathbf{B} = \rho(\mathbf{h}\cdot\nabla\mathbf{h} - \tfrac{1}{2}\nabla\mathbf{h}^2)\,, \tag{12.8}$$

and (12.1) becomes

$$\frac{\partial \mathbf{u}}{\partial t} + \mathbf{u}\cdot\nabla\mathbf{u} = -\nabla P + \mathbf{h}\cdot\nabla\mathbf{h} + \nu\nabla^2\mathbf{u} + \mathbf{F}\,, \tag{12.9}$$

where $P = p/\rho + \frac{1}{2}\mathbf{h}^2$. In terms of $\mathbf{h}$, the induction equation becomes

$$\partial\mathbf{h}/\partial t + \mathbf{u}\cdot\nabla\mathbf{h} = \mathbf{h}\cdot\nabla\mathbf{u} + \eta\nabla^2\mathbf{h}\,. \tag{12.10}$$

Here $\mathbf{h}$ evidently has the dimensions of a velocity. There is a symmetry of structure in the non-linear terms of (12.9) and (12.10), which leads to some simple and important results when $\mathbf{F} = 0$ and in the limiting (non-dissipative) situation $\eta = \nu = 0$. It is a trivial matter to verify that (12.9) and (12.10) are then both satisfied by solutions of the form[1]

$$\mathbf{u} = \mathbf{f}(\mathbf{x} - \mathbf{h}_0\mathrm{t})\,, \quad \mathbf{h} = \mathbf{h}_0 - \mathbf{f}(\mathbf{x} - \mathbf{h}_0 t)\,, \tag{12.11}$$

or

$$\mathbf{u} = \mathbf{g}(\mathbf{x} + \mathbf{h}_0 t)\,, \quad \mathbf{h} = \mathbf{h}_0 + \mathbf{g}(\mathbf{x} + \mathbf{h}_0 t)\,, \tag{12.12}$$

where $\mathbf{f}$ and $\mathbf{g}$ are arbitrary functions of their arguments and $\mathbf{h}_0$ is constant. When $\mathbf{f}$ and $\mathbf{g}$ are localised (e.g. square-integrable) functions, these solutions represent waves which propagate without change of shape in the directions $\pm\mathbf{h}_0$. These are known as Alfvén waves (Alfvén 1942), and $\mathbf{h}_0 = (\mu_0\rho)^{-1/2}\mathbf{B}_0$ (with $\mathbf{B}_0$ a uniform field) is known as the Alfvén velocity. In view of the non-linearity of (12.9) and (12.10), a linear superposition of (12.11) and (12.12) does *not* in general satisfy the equations. However, if $|\mathbf{f}| \ll h_0$, $|\mathbf{g}| \ll h_0$, and if squares and products of $\mathbf{f}$ and $\mathbf{g}$ are neglected, then

$$\mathbf{u} = \mathbf{f}(\mathbf{x} - \mathbf{h}_0 t) + \mathbf{g}\,(\mathbf{x} + \mathbf{h}_0 t), \quad \mathbf{h} = \mathbf{h}_0 - \mathbf{f}(\mathbf{x} - \mathbf{h}_0 t) + \mathbf{g}(\mathbf{x} + \mathbf{h}_0 t), \tag{12.13}$$

[1] The variables $\mathbf{z}^{\pm} = \mathbf{x} \pm \mathbf{h}_0 t$ are known as Elsasser variables, after Elsasser (1946).

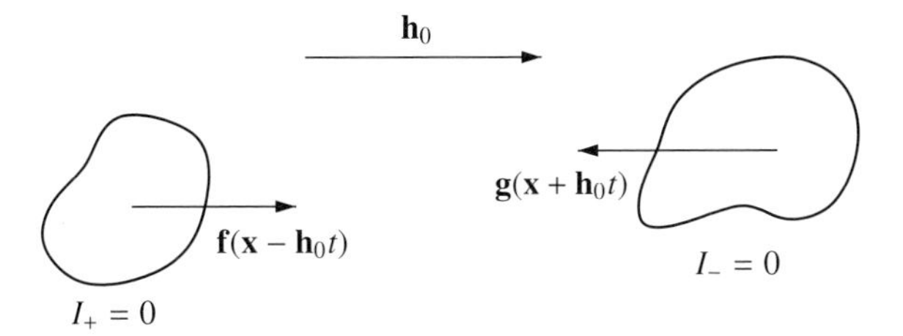

Figure 12.1 Disturbances represented by the functions $\mathbf{f}(\mathbf{x}-\mathbf{h}_0 t)$ and $\mathbf{g}(\mathbf{x}+\mathbf{h}_0 t)$ interact only while they overlap in $\mathbf{x}$-space. The time-scale characteristic of non-linear interaction of disturbances of length-scale L is evidently at most of order L/h_o.

provides a solution of the linearised equations. In general, however, solutions of the form (12.11) and (12.12) will interact in a non-linear manner when the functions $\mathbf{f}$ and $\mathbf{g}$ are overlapping (Figure 12.1).

12.1.2 Alfvén wave invariants and cross-helicity

Putting $\mathbf{h} = \mathbf{h}_0 + \mathbf{h}_1$ (with $\mathbf{h}_1 = -\mathbf{f}(\mathbf{x}-\mathbf{h}_0 t)$ in (12.11) and $+\mathbf{g}(\mathbf{x}+\mathbf{h}_0 t)$ in (12.12)) it is evident that the solutions (12.11) and (12.12) are characterised by the properties $\mathbf{u} \pm \mathbf{h}_1 = 0$ respectively. Let us define the integrals

$$I_\pm = \tfrac{1}{2}\int (\mathbf{u} \pm \mathbf{h}_1)^2 \mathrm{d}V\,, \tag{12.14}$$

on the assumption that the disturbances are sufficiently localised for these integrals to exist; here we imagine the fluid to extend to infinity in all directions, and the integral is over all space. Then $I_+ = 0$ for the solution (12.11) and $I_- = 0$ for the solution (12.12). The integrals $I_\pm$ are in fact invariants of the pair of equations (12.9) and (12.10), when $\nu = \eta = 0$. For

$$\begin{aligned}\frac{\mathrm{d}I_+}{\mathrm{d}t} &= \int (\mathbf{u}+\mathbf{h}_1)\cdot\frac{\mathrm{D}}{\mathrm{D}t}(\mathbf{u}+\mathbf{h}_1)\,\mathrm{d}V\\ &= \int (\mathbf{u}+\mathbf{h}_1)\cdot[-\nabla P + (\mathbf{h}_0+\mathbf{h}_1)\cdot\nabla(\mathbf{u}+\mathbf{h}_1)]\,\mathrm{d}V\\ &= \int \nabla\cdot\left\{-P(\mathbf{u}+\mathbf{h}_1) + \tfrac{1}{2}(\mathbf{h}_0+\mathbf{h}_1)(\mathbf{u}+\mathbf{h}_1)^2\right\}\mathrm{d}V\,,\end{aligned} \tag{12.15}$$

where we have used $\nabla\cdot\mathbf{u} = 0$, $\nabla\cdot\mathbf{h}_1 = 0$. Using the divergence theorem and the vanishing of $\mathbf{u}+\mathbf{h}_1$ at infinity ($\mathbf{u}+\mathbf{h}_1 = o(r^{-2})$ is clearly sufficient here) it follows that $dI_+/dt = 0$, and hence that $I_+ = \text{cst}$. Similarly $I_- = \text{cst}$. Note that

$$I_E = \tfrac{1}{2}(I_+ + I_-) = \tfrac{1}{2}\int (\mathbf{u}^2 + \mathbf{h}_1^2)\,\mathrm{d}V \tag{12.16}$$

is just ρ^{-1} times the total energy of the disturbance (kinetic plus magnetic), and the invariance of this quantity is of course no surprise. We have also the related invariant (Woltjer 1958)

$$I_C = \tfrac{1}{2}(I_+ - I_-) = \int \mathbf{u} \cdot \mathbf{h}_1 \, \mathrm{d}V \,. \tag{12.17}$$

The invariance of this integral admits topological interpretation (cf. the magnetic helicity integral (2.13)). In fact I_C provides a measure of the degree of linkage of the vortex lines of the **u**-field with the lines of force of the $\mathbf{h}_1$-field. To see this, consider the particular situation in which $\mathbf{h}_1 \equiv 0$ except in a single flux tube of vanishing cross section in the neighbourhood of the closed curve C, and let Φ_1 be the flux of $\mathbf{h}_1$ in the tube. Then from (12.17),

$$I_C = \Phi_1 \int_C \mathbf{u} \cdot \mathrm{d}x = \Phi_1 K \,, \tag{12.18}$$

where K is the circulation round C, or equivalently the flux of vorticity across a surface S spanning C. Hence I_C is non-zero or zero according as the $\mathbf{h}_1$-lines do or do not enclose a net flux of vorticity; I_C is the *cross-helicity* of the fields **u** and $\mathbf{h}_1$. We have already encountered this integral in §10.8 with its interpretation as the degree of alignment of the velocity and magnetic fields.

The invariance of I_+ and I_- provides some indication of the nature of the interaction of two Alfvén waves of the form (12.11) and (12.12). If **f** and **g** are both localised functions with spatial extent of order L, then the duration of the interaction (which may be thought of as a collision of two 'blobs' represented by the functions **f** and **g**) will be at most of order L/h_0. Choosing the origin of time to be during the interaction, for $t \ll -L/h_0$ (i.e. before the interaction) the solution has the form (12.13) while for $t \gg L/h_0$ (i.e. after the interaction) we must have

$$\mathbf{u} = \mathbf{f}_1(\mathbf{x} - \mathbf{h}_0 t) + \mathbf{g}_1(\mathbf{x} + \mathbf{h}_0 t), \quad \mathbf{h} = \mathbf{h}_0 - \mathbf{f}_1(\mathbf{x} - \mathbf{h}_0 t) + \mathbf{g}_1(\mathbf{x} + \mathbf{h}_0 t), \tag{12.19}$$

where $\mathbf{f}_1$ and $\mathbf{g}_1$ are related in some way to **f** and **g**. The invariance of I_+ and I_- then tells us that

$$\int \mathbf{f}_1^2 \mathrm{d}V = \int \mathbf{f}^2 \, \mathrm{d}V, \quad \int \mathbf{g}_1^2 \mathrm{d}V = \int \mathbf{g}^2 \, \mathrm{d}V \,, \tag{12.20}$$

i.e. there can be no transfer of energy between the disturbances during the interaction. The spatial structure of each disturbance will however presumably be modified by the interaction. The nature of this modification presents an intriguing problem that deserves investigation.

12.2 Lehnert waves

In a rotating mass of fluid, such as the Sun or the liquid core of the Earth, it is appropriate to refer both velocity and magnetic fields to a frame of reference rotating with the fluid. If $\mathbf{\Omega}$ is the angular velocity of this frame, then when ρ is constant the momentum equation becomes

$$\partial\mathbf{u}/\partial t + \mathbf{u}\cdot\nabla\mathbf{u} + 2\mathbf{\Omega}\wedge\mathbf{u} = -\nabla P + \mathbf{h}\cdot\nabla\mathbf{h} + \nu\nabla^2\mathbf{u} + \mathbf{F}, \qquad (12.21)$$

where P now includes the centrifugal potential $-\frac{1}{2}(\mathbf{\Omega}\wedge\mathbf{x})^2$ as well as the magnetic pressure term $\frac{1}{2}\mathbf{h}^2$. The induction equation is however invariant, i.e. we still have

$$\partial\mathbf{h}/\partial t + \mathbf{u}\cdot\nabla\mathbf{h} = \mathbf{h}\cdot\nabla\mathbf{u} + \eta\nabla^2\mathbf{h}, \qquad (12.22)$$

where $\mathbf{h}(\mathbf{x},t)$, $\mathbf{u}(\mathbf{x},t)$ are now measured relative to the rotating frame of reference. Physically, this is obvious: rigid body rotation simply rotates a magnetic field without distortion. Changes in $\mathbf{h}(\mathbf{x},t)$ in the rotating frame are therefore caused only by motion relative to the rotating frame, and by the usual process of ohmic diffusion.[2]

Suppose now that

$$\mathbf{h} = \mathbf{h}_0 + \mathbf{h}_1, \quad \mathbf{u} = \mathbf{u}_0 + \mathbf{u}_1, \qquad (12.23)$$

where $\mathbf{h}_0$ and $\mathbf{u}_0$ are uniform, and $\mathbf{h}_1$ and $\mathbf{u}_1$ represent small perturbations. Neglecting squares and products of $\mathbf{u}_1$ and $\mathbf{h}_1$, and supposing $\mathbf{F} = 0$, the linearised forms of (12.21) and (12.22) are

$$\partial\mathbf{u}_1/\partial t + \mathbf{u}_0\cdot\nabla\mathbf{u}_1 + 2\mathbf{\Omega}\wedge\mathbf{u}_1 = -\nabla P_1 + \mathbf{h}_0\cdot\nabla\mathbf{h}_1 + \nu\nabla^2\mathbf{u}_1, \qquad (12.24)$$

and

$$\partial\mathbf{h}_1/\partial t + \mathbf{u}_0\cdot\nabla\mathbf{h}_1 = \mathbf{h}_0\cdot\nabla\mathbf{u}_1 + \eta\nabla^2\mathbf{h}_1, \qquad (12.25)$$

where P_1 is the associated perturbation in P. These equations admit wave-type solutions (Lehnert 1954) of the form

$$(\mathbf{u}_1,\ \mathbf{h}_1,\ P_1) = \mathrm{Re}\left(\hat{\mathbf{u}},\ \hat{\mathbf{h}},\ \hat{P}\right)\exp\left[\mathrm{i}\left(\mathbf{k}\cdot\mathbf{x} - (\omega + \mathbf{u}_0\cdot\mathbf{k})t\right)\right], \qquad (12.26)$$

and since $\nabla\cdot\mathbf{u}_1 = \nabla\cdot\mathbf{h}_1 = 0$, it follows that

$$\mathbf{k}\cdot\hat{\mathbf{u}} = \mathbf{k}\cdot\hat{\mathbf{h}} = 0, \qquad (12.27)$$

i.e. these are transverse waves, which we may appropriately describe as 'Lehnert waves'. Substitution first in (12.26) gives the relation between $\hat{\mathbf{h}}$ and $\hat{\mathbf{u}}$

$$\hat{\mathbf{h}} = -(\omega + \mathrm{i}\eta k^2)^{-1}(\mathbf{k}\cdot\mathbf{h}_0)\,\hat{\mathbf{u}} \qquad (12.28)$$

[2] Note that this simple statement must break down at distances from the axis of rotation of order c/Ω, where c is the speed of light. At such distances, displacement currents cannot be ignored and the field inevitably lags behind the rotating frame of reference.

that we have already encountered (equation (7.74)), and substitution in (12.24) and rearrangement of the terms gives

$$-\,\mathrm{i}\sigma\hat{\mathbf{u}} + 2\boldsymbol{\Omega}\wedge\hat{\mathbf{u}} = -\mathrm{i}\mathbf{k}\hat{P}\,, \tag{12.29}$$

where

$$\sigma = \left(\omega + \mathrm{i}\nu k^2\right) - \left(\omega + \mathrm{i}\eta k^2\right)^{-1}(\mathbf{h}_0\cdot\mathbf{k})^2\,. \tag{12.30}$$

The effect of the magnetic field in (12.29) is entirely contained in the coefficient σ. This leads to an important modification of the dispersion relationship between ω and $\mathbf{k}$; but the spatial structure of the velocity field is unaffected by the presence of the magnetic field $\mathbf{h}_0$.

12.2.1 Dispersion relation and up-down symmetry breaking

To get the dispersion relationship, we first take the cross-product of (12.29) with $\mathbf{k}$; since $\mathbf{k}\cdot\hat{\mathbf{u}} = 0$, this gives

$$-\,\mathrm{i}\sigma\mathbf{k}\wedge\hat{\mathbf{u}} - 2(\mathbf{k}\cdot\boldsymbol{\Omega})\,\hat{\mathbf{u}} = 0\,. \tag{12.31}$$

Taking the cross-product again with $\mathbf{k}$ gives

$$\mathrm{i}\sigma k^2\hat{\mathbf{u}} - 2(\mathbf{k}\cdot\boldsymbol{\Omega})\mathbf{k}\wedge\hat{\mathbf{u}} = 0\,. \tag{12.32}$$

Elimination of $\hat{\mathbf{u}}$ and $\mathbf{k}\wedge\hat{\mathbf{u}}$ from (12.31) and (12.32) gives

$$\sigma^2 = 4(\mathbf{k}\cdot\boldsymbol{\Omega})^2/k^2, \ \text{ or } \sigma = \mp 2(\mathbf{k}\cdot\boldsymbol{\Omega})/k\,, \tag{12.33}$$

and from either (12.31) or (12.32) we have the corresponding simple relation between $\hat{\mathbf{u}}$ and $\mathbf{k}\wedge\hat{\mathbf{u}}$,

$$\mathrm{i}\mathbf{k}\wedge\hat{\mathbf{u}} = \pm k\,\hat{\mathbf{u}}\,. \tag{12.34}$$

Since $\hat{\omega} = \mathrm{i}\mathbf{k}\wedge\hat{\mathbf{u}}$ is the Fourier transform of the vorticity $\boldsymbol{\omega} = \nabla\wedge\mathbf{u}$ associated with the waves, it is evident from (12.34) that each constituent wave is of maximal helicity, i.e. for each wave

$$\langle\mathbf{u}\cdot\boldsymbol{\omega}\rangle = \pm k\,|\hat{\mathbf{u}}|^2\,\mathrm{e}^{-2|\omega_{\mathrm{i}}|t}\,, \tag{12.35}$$

where $\omega = \omega_{\mathrm{r}} + \mathrm{i}\omega_{\mathrm{i}}$; the decay of the waves ($\omega_{\mathrm{i}} \neq 0$) is of course associated with the processes of viscous and ohmic diffusion. Such helical flows are effective in generating an α-effect; hence their particular interest in the dynamo context.

The structure of a motion satisfying (12.34) may be easily understood by choosing axes $OXYZ$ with OX parallel to $\mathbf{k}$, so that $\mathbf{k} = (k, 0, 0)$; then $\hat{\mathbf{u}} = (0, \hat{v}, \hat{w})$ and (12.34) gives $\hat{w} = \pm\mathrm{i}\hat{v}$; in conjunction with the factor $\mathrm{e}^{\mathrm{i}\mathbf{k}\cdot\mathbf{x}} = \mathrm{e}^{\mathrm{i}kX}$ in (12.26), the motion then has a spatial structure given by

$$\mathbf{u}_1 \propto (0,\ \cos(kX + \psi),\ \pm\sin(kX + \psi))\,, \tag{12.36}$$

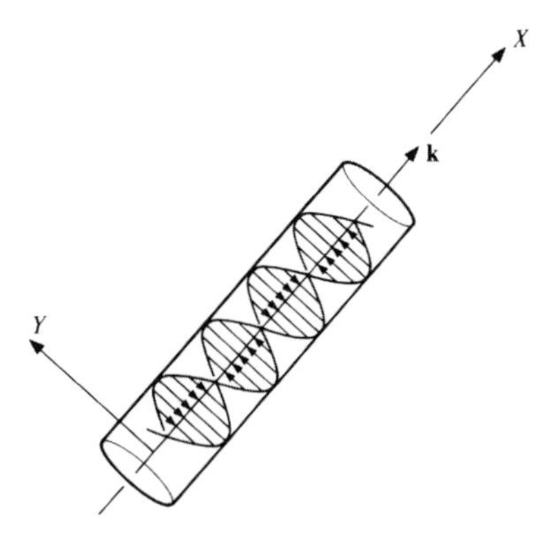

Figure 12.2 Sketch of the structure of the wave motion given by (12.36). The case illustrated, in which the velocity vector rotates in a right-handed sense as X increases, corresponds to the lower choice of sign in (12.36), with $k > 0$; the helicity of this motion as given by (12.35) is then negative.

where the phase ψ is time-dependent. This is a circularly polarised wave-motion (Figure 12.2), the velocity vector $\mathbf{u}_1$ being constant in magnitude, but rotating in direction as X increases; the sense of rotation is left-handed or right-handed according as the associated helicity $\mathbf{u}_1 \cdot \boldsymbol{\omega}_1$ is positive or negative.

When $\mathbf{h}_0 = 0$, and when viscous effects are negligible, (12.30) and (12.33) give

$$\omega = \mp 2(\mathbf{k} \cdot \boldsymbol{\Omega})/k\,, \tag{12.37}$$

the dispersion relation for pure inertial waves in a rotating fluid (see e.g. Greenspan 1968). The group velocity for such waves is given by

$$\mathbf{c}_g = \nabla_{\mathbf{k}}\omega = \mp 2k^{-3}\left(k^2\boldsymbol{\Omega} - \mathbf{k}(\mathbf{k} \cdot \boldsymbol{\Omega})\right). \tag{12.38}$$

Since $\mathrm{c}_{\mathbf{g}} \cdot \mathbf{k} = 0$, this is perpendicular to the phase velocity $\omega\mathbf{k}/k^2$. Moreover, since

$$\mathbf{c}_{\mathbf{g}} \cdot \boldsymbol{\Omega} = \mp 2k^{-1}\Omega^2 \sin^2\theta\,, \tag{12.39}$$

where θ is the angle between $\mathbf{k}$ and $\boldsymbol{\Omega}$, the value of $\mathrm{c}_{\mathbf{g}} \cdot \boldsymbol{\Omega}$ is negative or positive according as the helicity of the group is positive or negative. Loosely speaking, we may say that negative helicity is associated with upward propagation (relative to $\boldsymbol{\Omega}$) and positive helicity is associated with downward propagation. A random superposition of such waves in equal proportions would give zero net helicity but if any mechanism is present which leads to preferential excitation of upward or downward propagating waves, then the net helicity will be negative or positive respectively.

12.2.2 Inertial and magnetostrophic wave limits

Suppose now that $\mathbf{h}_0 \neq 0$, but that the Coriolis effect is still dominant in the sense that

$$|\sigma| = 2k^{-1}|\mathbf{\Omega} \cdot \mathbf{k}| \gg |\mathbf{h}_0 \cdot \mathbf{k}| . \tag{12.40}$$

(Of course if $\mathbf{h}_0$ is not parallel to $\mathbf{\Omega}$ there will always be some wave vectors perpendicular or nearly perpendicular to $\mathbf{\Omega}$ for which (12.40) is *not* satisfied.) Then, provided the diffusion effects associated with ν and η are weak, the two roots of (12.30) (regarded as a quadratic in ω) are given by

$$\omega + \mathrm{i}\nu k^2 \approx \sigma + \frac{(\mathbf{h}_0 \cdot \mathbf{k})^2}{\sigma + \mathrm{i}(\eta - \nu)k^2} , \tag{12.41}$$

and

$$\omega + \mathrm{i}\eta k^2 \approx -\sigma^{-1}(\mathbf{h}_0 \cdot \mathbf{k})^2 , \tag{12.42}$$

where σ is still given by (12.33). Clearly (12.41) still represents an inertial wave whose frequency is weakly modified by the presence of the magnetic field and which is weakly damped by both viscous and ohmic effects; in fact from (12.41) we have $\omega = \omega_r + \mathrm{i}\omega_i$, where (provided $|\eta - \nu|k^2 \ll |\sigma|$)

$$\omega_r \approx \sigma + \sigma^{-1}(\mathbf{h}_0 \cdot \mathbf{k})^2, \qquad \omega_i \approx -\nu k^2 - (\mathbf{h}_0 \cdot \mathbf{k})^2(\eta - \nu)k^2\sigma^{-2} . \tag{12.43}$$

Equation (12.42) on the other hand represents a wave which has no counterpart when $\mathbf{h}_0 = 0$. In this wave, the relation between $\hat{\mathbf{h}}$ and $\hat{\mathbf{u}}$ from (12.28) is given by

$$\hat{\mathbf{h}} = \sigma(\mathbf{k} \cdot \mathbf{h}_0)^{-1}\hat{\mathbf{u}} , \tag{12.44}$$

so that, from (12.40), $|\hat{\mathbf{h}}| \gg |\hat{\mathbf{u}}|$, i.e. the magnetic perturbation dominates the velocity perturbation. The dispersion relation (12.42) may be obtained by neglecting the contribution $\left(\partial/\partial t - \nu\nabla^2\right)\mathbf{u}_1$ to (12.24) (or equivalently the contribution $\omega + \mathrm{i}\nu k^2$ to (12.30)); setting aside the trivial effect of convection by the uniform velocity $\mathbf{u}_0$, the force balance in this wave is therefore given by

$$2\mathbf{\Omega} \wedge \mathbf{u}_1 \approx -\nabla P_1 + \mathbf{h}_0 \cdot \nabla \mathbf{h}_1 , \tag{12.45}$$

i.e. a balance between Coriolis, pressure and Lorentz forces. Such a force balance is described as *magnetostrophic* (by analogy with the term *geostrophic* used in meteorological contexts to describe balance between Coriolis and pressure forces alone). We shall describe waves of this second category with dispersion relation (12.42) as *magnetostrophic waves*. Other authors (Acheson & Hide 1973) have used the term ‘hydromagnetic inertial waves’.

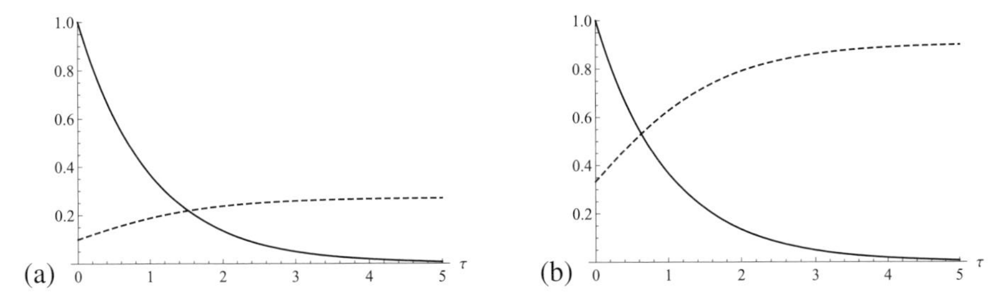

Figure 12.3 Decay of kinetic energy $E(\tau)/E(0)$ (solid) and simultaneous growth of magnetic energy $M(\tau)/E(0)$ (dashed) as given by (12.47); dimensionless time $\tau = 2\nu k_0^2 t$; $\pi K\mathcal{H}_0/2\Omega^2 = 1$; (a) from an initial level $M(0) = 0.1E(0)$; (b) $M(0) = (1/3)E(0)$; in the latter case, about half the initial kinetic energy is converted to magnetic energy by dynamo action; the magnetic field survives as a fossil field with asymptotic energy $M(0)\exp\left[\pi K\mathcal{H}_0/2\Omega^2\right]$. [From Mizerski & Moffatt 2018.]

12.3 Generation of a fossil field by decaying Lehnert waves

Lehnert waves of the inertial form (12.43) can generate an α-effect which, although decaying, is capable of generating a significant magnetic field provided $\eta \ll \nu$, a condition that holds in an extremely tenuous plasma such as that in the interstellar or intergalactic medium. To have a non-zero α-effect, conditions must be such that, as described above, the initial helicity spectrum $\hat{\mathcal{H}}(\mathbf{k})$ is non-zero. Supposing that only positive-helicity waves for which $\sigma = -2(\mathbf{k}\cdot\mathbf{\Omega})/k$ are present, Mizerski & Moffatt (2018) have shown that the electromotive force in the weak field limit is given by

$$\mathcal{E}_i = \alpha_{ij}H_j \quad \text{where} \quad \alpha_{ij} = (\nu-\eta)\int_{\mathcal{K}} \frac{k^2 k_i k_j}{8(\mathbf{k}\cdot\mathbf{\Omega})^2}\hat{\mathcal{H}}(\mathbf{k})\,\mathrm{e}^{-2\nu k^2 t}\,\mathrm{d}^3\mathbf{k}\,, \qquad (12.46)$$

and where $\mathcal{K}$ is the domain of wave number space contributing to the wave field. The exponential decay here is due solely to viscous dissipation. It is noteworthy that for this decaying wave field, ν and η have opposing effects in generating an emf.

More significantly, even when $\eta = 0$, we have here a non-zero α-effect, albeit for a limited period – the period of viscous decay. Here it is viscosity that provides the essential phase-shift between velocity and magnetic perturbations that yields a non-zero effect. Under conditions that are statistically axisymmetric about the direction of $\mathbf{\Omega}$, this provides dynamo action over this limited period. If the wave field is concentrated around a wave number k_0, then very much as described in §9.2, magnetic modes grow on a large-scale $K^{-1} \gg k_0^{-1}$; such modes have a magnetic energy $M(t)$ that grows according to the equation

$$M(t) = M(0)\exp\left[\frac{\pi K \mathcal{H}_0}{2\,\Omega^2}\left(1 - \mathrm{e}^{-2\nu k_0^2 t}\right)\right] \sim M(0)\exp\left[\frac{\pi K \mathcal{H}_0}{2\,\Omega^2}\right] \quad \text{as } t \to \infty, \tag{12.47}$$

where $\mathcal{H}_0$ is the initial mean helicity of the wave field. This growth of magnetic energy is shown in Figure 12.3. In effect, a proportion of the initial kinetic energy of the wave field is transferred by dynamo action to the large-scale magnetic field during the decay process; the magnetic field then survives as a 'fossil field' when the plasma has come to rest, because of the assumption $\eta = 0$. Of course it will then decay on a much longer time-scale $O(\eta K^2)^{-1}$; but in the intergalactic medium, this time-scale may well be much more than the age of the Universe! (Zweibel & Heiles 1997).

12.4 Quenching of the α-effect by the Lorentz force

Whenever a large-scale magnetic field grows through dynamo action, the Lorentz force ultimately reacts back upon the turbulent flow that is responsible for this growth, in such a way as to establish statistical equilibrium. It is usually a combination of mean flow and turbulence that drives the dynamo, and the Lorentz force may influence either or both of these. As far as the turbulence is concerned, the obvious saturation mechanism is through suppression of the α-effect, a mechanism now known as 'α-quenching' (Vainshtein & Cattaneo 1992, Rüdiger & Kichatinov 1993). As argued by Brandenburg (2005), this quenching is associated with the invariance of magnetic helicity which inhibits dynamo action in the limit $\eta \to 0$ (see §10.3): this means that the α-effect must somehow be quenched, in such a way as to eliminate the possibility of dynamo growth on a fast time-scale (although not necessarily, it should be noted, on a slow diffusive time-scale).

12.4.1 A simple model based on weak forcing

We consider first a very simple model that indicates the nature of the α-quenching mechanism; this is a dynamic extension of the helical-wave model introduced in §7.7. Assuming a locally uniform magnetic field $\mathbf{B}_0 = (\mu_0\rho)^{1/2}\mathbf{h}_0$, we suppose that a wave is driven by helical forcing

$$\mathbf{f}(\mathbf{x}, t) = f_0(\sin(kz - \omega t),\ \cos(kz - \omega t),\ 0) = \Re\mathrm{e}[\hat{\mathbf{f}}\,\mathrm{e}^{\mathrm{i}\varphi}], \tag{12.48}$$

where $\omega > 0, \varphi = kz - \omega t$, and where $\hat{\mathbf{f}}$ is given by

$$\hat{\mathbf{f}} = f_0(-\mathrm{i}, 1, 0), \quad \text{so that} \quad \hat{\mathbf{f}}^{\,*}\wedge \hat{\mathbf{f}} = 2\,\mathrm{i}\, f_0^2(0, 0, 1). \tag{12.49}$$

The helicity of this force field is $\langle \mathbf{f}\cdot\nabla\wedge\mathbf{f}\rangle = kf_0^2$, positive or negative according as $k \gtrless 0$. We suppose that this forcing is weak so that the amplitude of the resulting forced helical wave is small. The effect of any number of such weak forced waves is then additive – they do not interact. In particular if we superpose three forces to give 'ABC forcing' (with $A = B = C = f_0$), i.e.

$$\mathbf{f}(\mathbf{x}, t) = f_0(\cos\varphi_y + \sin\varphi_z,\ \cos\varphi_z + \sin\varphi_x,\ \cos\varphi_x + \sin\varphi_y), \tag{12.50}$$

where $\varphi_x = kx-\omega t$, $\varphi_y = ky-\omega t$, $\varphi_z = kz-\omega t$, then the mean effects are additive.

The velocity and magnetic perturbation fields generated by the forcing (12.48) have the form

$$\mathbf{u} = \hat{\mathbf{u}}\,\mathrm{e}^{\mathrm{i}\varphi}, \quad \mathbf{h} = \hat{\mathbf{h}}\,\mathrm{e}^{\mathrm{i}\varphi}, \tag{12.51}$$

(real parts understood), and as usual from the induction equation, $\hat{\mathbf{h}}$ and $\hat{\mathbf{u}}$ are related by

$$\hat{\mathbf{h}} = \mathrm{i}\,(\eta k^2 - \mathrm{i}\omega)^{-1}(\mathbf{k}\cdot\mathbf{h}_0)\,\hat{\mathbf{u}}\,, \tag{12.52}$$

from which it follows that, just as in §7.7,

$$\boldsymbol{\mathcal{E}} = \langle \mathbf{u}\wedge\mathbf{h}\rangle = \tfrac{1}{2}\Re\mathrm{e}\langle\hat{\mathbf{u}}^*\wedge\hat{\mathbf{h}}\rangle = -\frac{\eta\,|\mathbf{u}|^2k^2(\mathbf{h}_0\cdot\mathbf{k})}{\omega^2+\eta^2k^4}\,(0,\ 0,\ 1)\,. \tag{12.53}$$

But now $\mathbf{u}$ satisfies the linearised dynamical equation

$$\partial\mathbf{u}/\partial t = -\nabla P + \mathbf{h}_0\cdot\nabla\mathbf{h} + \nu\nabla^2\mathbf{u} + \mathbf{f}\,, \tag{12.54}$$

and here we may set $\nabla P = 0$ (since $\nabla\cdot\mathbf{u} = \nabla\cdot\mathbf{h} = \nabla\cdot\mathbf{f} = 0$). Hence

$$-\,\mathrm{i}\,\omega\,\hat{\mathbf{u}} = \mathrm{i}\,(\mathbf{k}\cdot\mathbf{h}_0)\,\hat{\mathbf{h}} - \nu k^2\,\hat{\mathbf{u}} + \hat{\mathbf{f}}\,, \tag{12.55}$$

and so, using (12.52) and after some simplification,

$$\hat{\mathbf{u}} = \frac{\eta k^2 - \mathrm{i}\,\omega}{\eta\nu k^4 - \omega^2 + h_0^2k^2 - \mathrm{i}\,\omega k^2(\eta+\nu)}\,\hat{\mathbf{f}}\,. \tag{12.56}$$

It follows that

$$|\mathbf{u}|^2 = \frac{\omega^2+\eta^2k^4}{\left(\eta\nu k^4 - \omega^2 + h_{0z}^2k^2\right)^2 + \omega^2k^4(\eta+\nu)^2}\,f_0^2\,, \tag{12.57}$$

and we should note here a resonant response at the Alfvén frequencies $\omega = \pm h_{0z}k$ when $\eta = \nu = 0$. Together with (12.53), this now gives

$$\boldsymbol{\mathcal{E}} = \frac{-\eta\,k^2\,\mathbf{h}_0\cdot\mathbf{k}}{\left(\eta\nu k^4 - \omega^2 + h_{0z}^2k^2\right)^2 + \omega^2k^4(\eta+\nu)^2}\,f_0^2\,(0,0,1)\,. \tag{12.58}$$

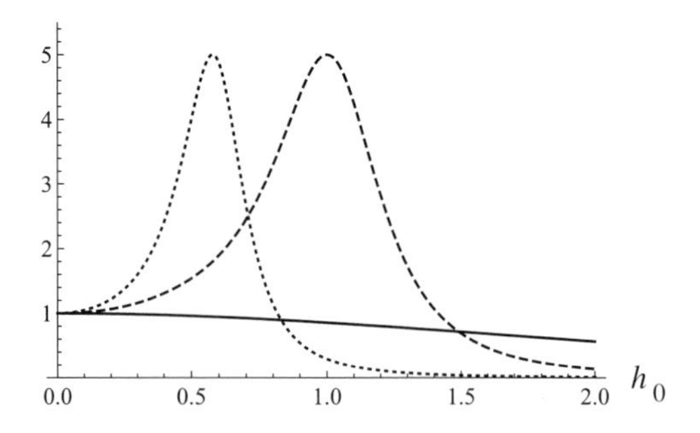

Figure 12.4 Alpha-quenching: α/α_0 as given by (12.61); $p^2 = 0.8$ and $\lambda = -0.1$ (solid), $+1$ (dashed), and $+3$ (dotted). Note the resonant response when $\lambda > 0$ at $h_0^2 = \lambda^{-1}$, and the quenching of α for $h_0^2 \gg \lambda^{-1}$.

If we now 'isotropise' by using the ABC forcing (12.50), then this results in an isotropic α-effect, $\boldsymbol{\mathcal{E}} = \alpha\, \mathbf{h}_0$, where

$$\alpha = \frac{-\eta\, k^3}{\left(\eta\nu k^4 - \omega^2 + h_0^2 k^2\right)^2 + \omega^2 k^4 (\eta + \nu)^2}\, f_0^2 \,. \tag{12.59}$$

Note that this α is negative or positive according as $k \gtrless 0$, i.e. according as the helicity of the forcing is positive or negative.

When $h_0 \to 0$ (the kinematic limit), $\alpha \to \alpha_0$ where

$$\alpha_0 = \frac{-\eta\, k^3}{\left(\eta\nu k^4 - \omega^2\right)^2 + \omega^2 k^4 (\eta + \nu)^2}\, f_0^2 \,, \tag{12.60}$$

and (12.59) may be re-expressed in the form

$$\alpha = \frac{\alpha_0}{p^2\left(1 - \lambda h_0^2\right)^2 + 1 - p^2}\,, \tag{12.61}$$

where

$$\lambda = \frac{k^2}{\omega^2 - \eta\nu k^4} \quad \text{and} \quad p^2 = \frac{\left(\eta\nu k^4 - \omega^2\right)^2}{\left(\eta\nu k^4 - \omega^2\right)^2 + \omega^2 k^4 (\eta + \nu)^2} \quad (< 1)\,. \tag{12.62}$$

Note that λ may be positive or negative, and that there is a resonance in (12.61) when $\lambda > 0$ at $h_0^2 = \lambda^{-1}$.

Figure 12.4 illustrates the behaviour for the choice $p^2 = 0.8$, and for three values of λ. When $\lambda < 0$ (the solid curve), there is a monotonic quenching of α with increasing h_0; this is the situation in particular when the forcing is steady ($\omega = 0$). In this situation, $p^2 = 1$ and

$$\alpha = \frac{\alpha_0}{\left(1 + h_0^2/\eta\nu k^4\right)^2}\,, \tag{12.63}$$

indicating strong α-quenching for $h_0^2 \gtrsim \eta\nu k^4$. Quenching at this diffusion-controlled level is described as 'catastrophic'.

When $\lambda > 0$, the situation is very different: α initially increases with increasing h_0 to a maximum at $h_0^2 = \lambda^{-1}$ (a resonant response), and is then quenched as h_0^2 increases further. If $\omega^2 \gg \eta\nu k^4$, then $\lambda \sim k^2/\omega^2$, and

$$\alpha \sim \frac{\alpha_0}{\left(1 - h_0^2 k^2/\omega^2\right)^2 + g_{\eta\nu}^2}, \tag{12.64}$$

where $g_{\eta\nu}^2 = k^4(\eta + \nu)^2/\omega^2$, now indicating strong α-quenching for $h_0^2 \gtrsim \omega^2/k^2$. The weak diffusive term $g_{\eta\nu}^2$ must be retained here, as it controls the resonance at $h_0^2 = \omega^2/k^2$.

12.4.2 *Quenching of the β-effect*

Quenching of the α-effect is presumably accompanied by a corresponding quenching of the β-effect, i.e. of turbulent diffusivity. This was first considered by Vainshtein & Cattaneo (1992) (see also Cattaneo & Vainshtein 1991), who regarded the effect as being due to suppression of the smallest scales of turbulence by the growing Lorentz force at these scales. From the point of view of mean-field theory, as for α-quenching it is equally due to modification of turbulence structures by the growing large-scale field. This is difficult to analyse (and will not be attempted here) as it requires consideration of the effect of a growing mean field $\mathbf{h}_0(\mathbf{x})$ of locally uniform gradient (rather than just a locally uniform mean field) on the forced wave field in (12.54). β-quenching was invoked by Parker (1993), although not given this name, in his study of dynamo processes near the tachocline; we defer consideration of these processes to Chapter 13.

12.5 Magnetic equilibration due to α-quenching

We now investigate how this α-quenching affects the dynamo process. For definiteness, we shall suppose that in (12.59) $k < 0$, so $\alpha > 0$. To understand the effect of the α-quenching, we may as in §9.2 consider unstable dynamo modes of force-free structure, such that $\nabla \wedge \mathbf{h}_0 = K\,\mathbf{h}_0$, where $0 < K \ll |k|$ (cf. the approach adopted by Rädler & Brandenburg 2003) and we now allow for spatial variation of $\mathbf{h}_0$ on the large-scale K^{-1}. The mean field equation for $\mathbf{h}_0$ is then simply

$$\frac{\mathrm{d}\mathbf{h}_0}{\mathrm{d}t} = (K\alpha - \eta K^2)\mathbf{h}_0 = \eta K^2 \left[\frac{q^2}{p^2\left(1 - \lambda h_0^2\right)^2 + 1 - p^2} - 1\right]\mathbf{h}_0\,, \tag{12.65}$$

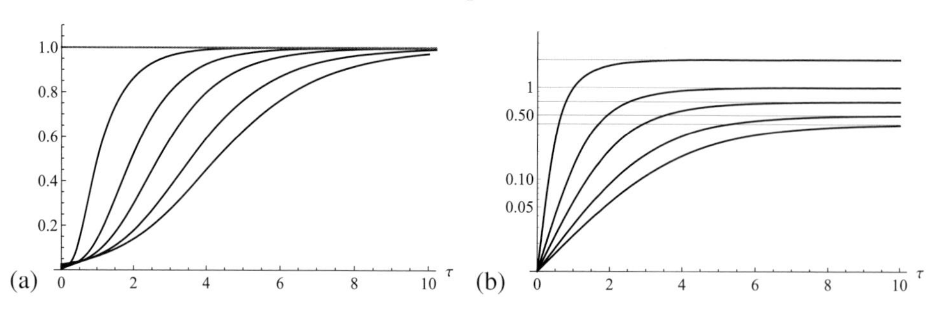

Figure 12.5 (a) Growth of $m(\tau)/(q-1)$ as determined by (12.68) with $m_0 = 0.01$, and (from left to right) $q = (\alpha_0/K\eta)^{1/2} = 3, 2, 1.7, 1.5, 1.4$. (b) Corresponding growth of $\log m(\tau)$; note the initial exponential growth (slope q^2) before saturation at $m = m_s = (q-1)$ (i.e. at $M = M_s = \frac{1}{2}\eta\nu k^2(q-1)$).

where $q^2 \equiv \alpha_0/K\eta$, and dynamo action starting from a seed field occurs provided $q^2 > 1$. We may convert (12.65) to an equation for the magnetic energy $M(t) = \frac{1}{2}\mathbf{h}_0^2$, viz.

$$\frac{\mathrm{d}M}{\mathrm{d}t} = 2\eta K^2 \left[\frac{q^2}{p^2(1-2\lambda M)^2 + 1 - p^2} - 1 \right] M. \tag{12.66}$$

We may now consider two distinct situations: steady and unsteady forcing.

12.5.1 The case of steady forcing

If the forcing is steady (i.e. $\omega = 0$ in (12.50)), then

$$p^2 = 1, \quad \text{and} \quad \lambda = -(\eta\nu k^2)^{-1}, \tag{12.67}$$

and, defining $\tau = 2\eta k^2 t$ and $m = 2M/\eta\nu k^2$, (12.66) becomes

$$\frac{\mathrm{d}m}{\mathrm{d}\tau} = \left[\frac{q^2}{(1+m)^2} - 1 \right] m. \tag{12.68}$$

This may be integrated with initial condition $m = m_0$ at $t = 0$; the result, which defines m implicitly as a function of τ, may be manipulated to the form

$$\tfrac{1}{2}q^2 \log\left[\frac{Q(m)}{Q(m_0)}\right] - \left[\log\left(\frac{m}{m_0}\right)\right] + q\left[\Theta(m) - \Theta(m_0)\right] = -\left(q^2 - 1\right)\tau, \tag{12.69}$$

where

$$\Theta(m) = \tanh^{-1}\left[(1+m)/q\right] \quad \text{and} \quad Q(m) = (m+1)^2 - q^2. \tag{12.70}$$

Interest here centres on the saturation level of m as $\tau \to \infty$, which is determined from the condition that the left-hand side of (12.69) should equal $-\infty$. Now, since

$q^2 > 1$ for dynamo action, we may choose the energy m_0 of the seed field small enough to ensure that $Q(m_0) < 0$. The roots of the quadratic $Q(m)$ are $m_\pm = -1 \pm q$; as m increases from m_0 to $m_+ = q - 1$, $Q(m)/Q(m_0)$ decreases from 1 to 0, and

$$\log\left[Q(m)/Q(m_0)\right] \to -\infty \quad \text{as} \quad m \to m_+ \,. \tag{12.71}$$

Thus $m_s \equiv m_+ = q - 1$ is the saturation level of m; and we may in fact deduce from (12.69) that

$$m \sim m_s - C \exp\left[-\left(1 - q^{-2}\right)\tau\right], \quad \text{as } \tau \to \infty, \tag{12.72}$$

where C is a positive constant. Figure 12.5a shows $m(\tau)/(q-1)$ for several values of $q > 1$, the saturation level $m_s/(q-1) = 1$ being as indicated; Figure 12.5b shows the corresponding growth of $\log m(\tau)$ with the several saturation levels again as indicated.

The corresponding saturation level of M is

$$M_s = \tfrac{1}{2}\eta\nu k^2\left[\left(\frac{\alpha_0}{K\eta}\right)^{1/2} - 1\right]. \tag{12.73}$$

Under steady forcing, we may define a characteristic rms velocity $u_0 = f_0/\nu k^2$, and a magnetic Reynolds number $\mathcal{R}_m = u_0/\eta k = f_0/\eta\nu k^3$. Then (12.73) gives

$$\frac{2M_s}{u_0^2} = \frac{1}{\mathcal{R}e}\left(\frac{k}{K}\right)^{1/2}, \tag{12.74}$$

where $\mathcal{R}e = u_0/\nu k$, the Reynolds number of the wave field. Thus for finite k/K, the saturation level is extremely low if $\mathcal{R}e \ggg 1$, as in most astrophysical contexts. However, we should note that, however large $\mathcal{R}e$ may be, equipartition of energy can still be established by dynamo action on a large enough scale K^{-1} such that $k/K = O(\mathcal{R}e^2)$. Note that it is $\mathcal{R}e$ rather than $\mathcal{R}_m$ that appears in (12.73); of course $\mathcal{R}_m = (\nu/\eta)\mathcal{R}e$, so if the magnetic Prandtl number $\mathcal{P}_m \equiv \nu/\eta$ is of order unity, then $M_s \sim \mathcal{R}_m^{-1}$ also.

12.5.2 The case of unsteady forcing with $\omega^2 \gg \eta\nu k^4$

In this situation, with $p^2 = 1 + g_{\eta\nu}^2$, the energy equation is at leading order

$$\frac{\mathrm{d}M}{\mathrm{d}t} = 2\eta K^2\left[\frac{q^2}{(1 - 2M\,k^2\omega^2)^2 + g_{\eta\nu}^2} - 1\right]M, \tag{12.75}$$

so now with $m = 2M\,k^2/\omega^2$, and $\tau = 2\eta K^2\,t$ as before, this becomes

$$\frac{\mathrm{d}m}{\mathrm{d}\tau} = \left[\frac{q^2}{(1-m)^2 + g_{\eta\nu}^2} - 1\right]m\,. \tag{12.76}$$

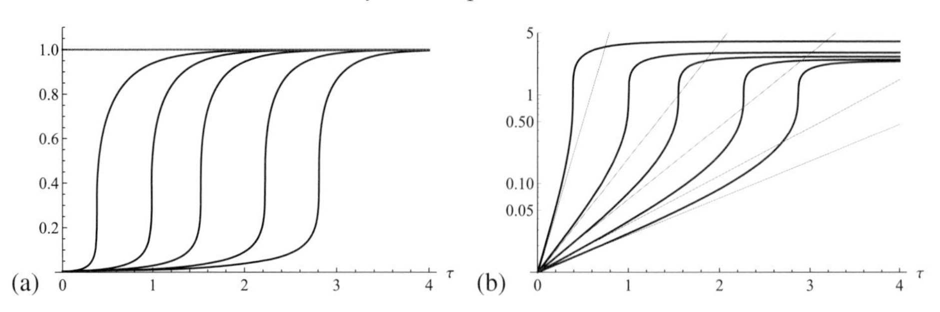

Figure 12.6 (a) Growth of $m(\tau)/(q+1)$ as determined by (12.76) with $m_0 = 0.01$, and (from left to right) $q = (\alpha_0/K\eta)^{1/2} = 3, 2, 1.7, 1.5, 1.4$. (b) Corresponding growth of $\log m(\tau)$; note the initial exponential growth (slopes q^2 as indicated), and the accelerated growth before saturation at $m = m_s = (q+1)$ [i.e. at $M = M_s = \frac{1}{2}(q+1)\omega^2/k^2$].

This may be integrated as before with initial condition $m(0) = m_0$, and in the limit $g^2_{\eta\nu} \to 0$ gives the same equation (12.69) that implicitly determines $m(\tau)$, but with the difference now that

$$Q(m) = (m-1)^2 - q^2 \quad \text{and} \quad \Theta(m) = \tanh^{-1}(m-1)/q\,. \tag{12.77}$$

Saturation now occurs when

$$m = m_s \sim q + 1 \quad \text{or equivalently} \quad M = M_s \sim \frac{1}{2}\left(\frac{\omega^2}{k^2}\right)\left[\left(\frac{\alpha_0}{K\eta}\right)^{1/2} + 1\right]. \tag{12.78}$$

The curves of $m(\tau)/(q+1)$ are shown in Figure 12.6a for the same six values of q as in Figure 12.5a; these curves were actually obtained by numerical integration of (12.76), with $g^2_{\eta\nu} = 0.001$ in order to safely pass through the singularity at $m = 1$ where the tangent to each curve is very nearly vertical; the curves are essentially unchanged as $g^2_{\eta\nu} \to 0$. Figure 12.6b shows the corresponding growth of $\log m(\tau)$, and note here the distinct change of slope in the growth curves as $m(\tau)$ passes through the resonance level.

When $\alpha_0/K\eta \gg 1$ and with $f_0^2 \sim \omega^2 u_0^2$ and $\alpha_0 \sim \eta k^3 f_0^2/\omega^4$, the result (12.78) readily translates to

$$\frac{2M_s}{u_0^2} \sim \frac{\omega^2}{kf_0}\left(\frac{k}{K}\right)^{1/2}. \tag{12.79}$$

The important thing to note here is that, unlike the previous result (12.73), the expression (12.79) does not depend on either $\mathcal{R}e$ or $\mathcal{R}_m$, but only on the characteristics of the wave field, and the scale parameter k/K. Strictly, the parameter $kf_0/\omega^2 \sim ku_0/\omega$ should be small to justify the linearised wave theory that forms the basis of the above analysis; but if kf_0/ω^2 is raised to values of order unity when the waves would be weakly interacting, the result (12.79) probably still remains reliable, at least in order of magnitude.

12.5.3 Cattaneo–Hughes saturation

Following Vainshtein & Cattaneo (1992), the saturation problem was further considered by Cattaneo & Hughes (1996) (see also Cattaneo et al. 2002; Cattaneo & Hughes 2009), who started with the Galloway–Proctor space-periodic velocity field (10.23) maintained by a corresponding force-field, and computed a quenching effect compatible (in the present notation) with

$$\alpha = \frac{\alpha_0}{1 + \mathcal{R}_m\, h_0(t)^2/u_0^2}\,; \tag{12.80}$$

here u_0 is the initial rms velocity, which remained nearly constant throughout the computation. If variables are non-dimensionalised with the box-scale k^{-1}, and the velocity scale u_0, so that in effect $k = 1$, $u_0 = 1$, then (12.80) implies quenching when $h_0^2 \sim \mathcal{R}_m^{-1}$, a potentially troublesome behaviour in the astrophysically relevant limit $\mathcal{R}_m \ggg 1$.

But consider again the unstable force-free modes of the α^2-dynamo for which $\nabla \wedge \mathbf{h}_0 = K\,\mathbf{h}_0$ with $0 < K \ll 1$. This leads just as above to an equation for the magnetic energy density $M(t)$,

$$\frac{dM}{dt} = \frac{2K\alpha_0 M}{1 + 2\mathcal{R}_m M} - \frac{2K^2 M}{\mathcal{R}_m} \approx \frac{4K^2(q^2 - M)M}{1 + 2\mathcal{R}_m M}, \tag{12.81}$$

where $q^2 = \alpha_0/2K$ and we have used $\mathcal{R}_m\alpha_0 \gg K$. This equation integrates with initial condition $M(0) = M_0$ to give

$$\left(\frac{M_0}{M}\right)\left(\frac{q^2 - M}{q^2 - M_0}\right)^{1+\mathcal{R}_m q^2} = e^{-q^2\tau}\,, \tag{12.82}$$

where $\tau = 4K^2 t$. The saturation level is evidently $M = q^2$, independent of $\mathcal{R}_m$, and this level is approached on the slow diffusive time-scale $\mathcal{O}(\mathcal{R}_m)$; in fact, for $\mathcal{R}_m q^2 \gg 1$,

$$M \sim q^2\left(1 - e^{-\tau/\mathcal{R}_m}\right). \tag{12.83}$$

Figure 12.7a shows a logarithmic plot of M/q^2 against τ as given by (12.82) for $\mathcal{R}_m = 10, 100, 1000$ and 10,000, and Figure 12.7b shows corresponding results obtained by Cattaneo & Hughes (1996) for the Galloway–Proctor flow, at $\mathcal{R}_m = \mathcal{R}e = 100$. These numerical results are qualitatively compatible with the description provided by the above simple theory. In comparing with the results

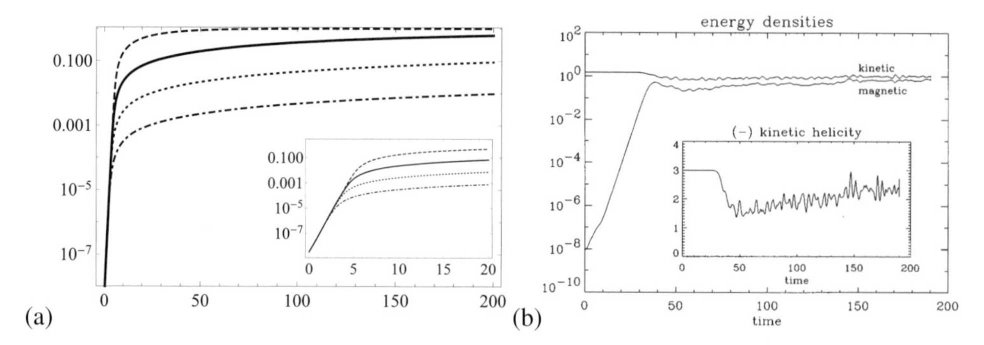

Figure 12.7 (a) Log-linear plot of M/q^2 as given by (12.82) for $\mathcal{R}_m = 10$ (dashed), 100 (solid), 1000 (dotted) and 10,000 (dash-dotted); the asymptotic level $M_s/q^2 = 1$ is the same in all cases; the time to reach this level is proportional to $\mathcal{R}_m$; the inset shows the early development up to time $\tau = 20$. (b) Computational result of Cattaneo & Hughes (1996) for the Galloway–Proctor flow (10.23) at $\mathcal{R}_m = \mathcal{R}e = 100$; the approach of magnetic energy to the saturation level occurs, as in (a), on the diffusive time-scale. [Reprinted figure with permission from Cattaneo & Hughes 1996. Copyright 1996 by the American Physical Society.]

shown in Figure 12.6b, it would appear that unsteadiness in the forcing is essential if the saturation level of M is to be independent of $\mathcal{R}_m$; moreover the change in slope of the growth curve evident in Figure 12.7b at $\tau \sim 12$ (like the change in slopes in Figure 12.6b) may well in a similar way be a consequence of an accelerated α-effect associated with an Alfvénic resonance or a range of such resonances.

In view of the above, we question the conclusion of Cattaneo & Hughes (1996) that, because of high-$\mathcal{R}_m$ quenching, "the α-effect will be ineffective and simply incapable of generating sizable mean fields". This conclusion is echoed by Charbonneau (2014) who writes in relation to the formula (12.80), "This result, which came to be known as catastrophic quenching, has profound implications for astrophysical dynamos in general, because in most astrophysically relevant situations $\mathcal{R}_m \gg 1$, implying saturation of the large-scale field at a factor $\mathcal{R}_m^{1/2}$ smaller than equipartition. In the solar interior, $\mathcal{R}_m \sim 10^{10}$ leads to a saturation at the level $|\langle B\rangle| \sim 10^{-6}$ T, which is now a long way from sunspot field strengths". In contrast to this negative diagnosis, our conclusion is that under unsteady forcing, despite the brutal α-quenching represented by formulae like (12.60) or (12.80), the magnetic energy will still saturate at approximately equipartition level, although, as for any slow dynamo, it takes a long time to get there; and we would maintain that mean-field theory does remain viable throughout the whole dynamic process of magnetic equilibration.

12.6 Quenching of the α-effect in a field of forced Lehnert waves

In similar vein, consider now the problem of α-quenching in a field of Lehnert waves in a fluid rotating with mean angular velocity $\mathbf{\Omega}$ and permeated with an initially weak mean magnetic field $\mathbf{B}_0 = (\mu_0\rho)^{-1/2}\mathbf{h}_0$. We suppose that these waves are induced by a force field $\mathbf{f}(\mathbf{x}, t)$, which may be random or space-periodic. The response $(\mathbf{u}_1, \mathbf{h}_1, P_1)$ depends on the field $\mathbf{h}_0$ as well as on $\mathbf{\Omega}$, and on the diffusivities η and ν, which may be supposed very weak. The force field $\mathbf{f}$ is considered weak enough for the linearised wave treatment of §12.2 to be applicable. We may assume[3] that $\nabla \cdot \mathbf{f} = 0$, and that $\mathbf{f}$ is reflexionally symmetric, in particular that $\langle \mathbf{f} \cdot \nabla \wedge \mathbf{f} \rangle = 0$. Any lack of reflexional symmetry in the $\mathbf{u}_1$-field then arises through the influence of rotation. In this respect, this model, first treated by Moffatt (1972), is more realistic than that treated in §12.4.

Consider first a single Fourier component $\Re\mathrm{e}\left(\hat{\mathbf{f}}\, \mathrm{e}^{\mathrm{i}(\mathbf{k}\cdot\mathbf{x}-\omega t)}\right)$ of this field, where $\mathbf{k}$ and ω are real. The result (12.28) still holds, from which we deduce a contribution to $\langle \mathbf{u}_1 \wedge \mathbf{h}_1 \rangle$ of the form

$$\tfrac{1}{2}\Re\mathrm{e}\left(\hat{\mathbf{u}}^* \wedge \hat{\mathbf{h}}\right) = \frac{\eta k^2 (\mathbf{h}_0 \cdot \mathbf{k})}{2\left(\omega^2 + \eta^2 k^4\right)} \left(\mathrm{i}\,\hat{\mathbf{u}}^* \wedge \hat{\mathbf{u}}\right) . \tag{12.84}$$

Now we must express this in terms of $\hat{\mathbf{f}}$ by means of the equation (cf.(12.29))

$$-\mathrm{i}\sigma\hat{\mathbf{u}} + 2\mathbf{\Omega} \wedge \hat{\mathbf{u}} = -\mathrm{i}\mathbf{k}\hat{P} + \hat{\mathbf{f}} , \tag{12.85}$$

where σ is still given by (12.30). When diffusion effects are weak, this has the form $\sigma = \sigma_{\mathrm{r}} + \mathrm{i}\sigma_{\mathrm{i}}$, where

$$\sigma_{\mathrm{r}} \approx \omega - \omega^{-1}(\mathbf{h}_0 \cdot \mathbf{k})^2, \quad \sigma_{\mathrm{i}} \approx k^2\left(\nu + \eta\omega^{-2}(\mathbf{h}_0 \cdot \mathbf{k})^2\right) . \tag{12.86}$$

Taking the cross-product with $\mathbf{k}$ twice, and solving for $\hat{\mathbf{u}}$, we obtain

$$\hat{\mathbf{u}} = D^{-1}\left(2(\mathbf{k} \cdot \mathbf{\Omega})\, \mathbf{k} \wedge \hat{\mathbf{f}} + \mathrm{i}\,\sigma k^2\, \hat{\mathbf{f}}\right) , \tag{12.87}$$

where

$$D = \sigma^2 k^2 - 4(\mathbf{k} \cdot \mathbf{\Omega})^2 \approx k^2(\sigma_{\mathrm{r}}^2 - \sigma_{\mathrm{i}}^2) - 4(\mathbf{k} \cdot \mathbf{\Omega})^2 + 2\mathrm{i}\, k^2 \sigma_{\mathrm{r}} \sigma_{\mathrm{i}}. \tag{12.88}$$

When ν and η are small, the term $-k^2\sigma_{\mathrm{i}}^2$ may be neglected, and there is a resonant response (i.e. $|D|^{-1}$ has a sharp maximum) when

$$\sigma_{\mathrm{r}}^2 k^2 = 4(\mathbf{k} \cdot \mathbf{\Omega})^2 , \tag{12.89}$$

i.e. when the forcing wave excites at a natural frequency–wave vector combination of the undamped system.

[3] The general $\mathbf{f}$ may be represented as the sum of solenoidal and irrotational ingredients; the latter leads only to a trivial modification of the pressure distribution (when $\nabla \cdot \mathbf{u} = 0$).

In terms of the dimensionless magnetic energy density,

$$m = (\mathbf{h}_0 \cdot \mathbf{k})^2/\omega^2\,, \tag{12.90}$$

and the *magnetic Prandtl number* $\mathcal{P}_\mathrm{m} = \nu/\eta$, the results (12.86) become

$$\sigma_\mathrm{r} = \omega(1-m), \quad \sigma_\mathrm{i} = \eta k^2(\mathcal{P}_\mathrm{m} + m), \tag{12.91}$$

and if we introduce the dimensionless Coriolis parameter

$$X^2 = 4(\mathbf{k}\cdot\boldsymbol{\Omega})^2/\omega^2 k^2\,, \tag{12.92}$$

the resonance condition (12.89) becomes

$$(1-m)^2 = X^2 \quad \text{or simply} \quad m = 1 \pm X. \tag{12.93}$$

In terms of these variables, (12.88) becomes

$$D = \omega^2 k^2 \left\{\left[(1-m)^2 - X^2\right] + \mathrm{i}\epsilon(1-m)(\mathcal{P}_\mathrm{m} + m)\right\}\,, \tag{12.94}$$

where

$$\epsilon = 2\eta k^2/\omega, \quad (0 < \epsilon \ll 1). \tag{12.95}$$

We shall suppose throughout this section that $\mathcal{P}_\mathrm{m} = \mathcal{O}(1)$.

From (12.87), $\mathrm{i}\,\hat{\mathbf{u}} \wedge \hat{\mathbf{u}}^*$ can now be calculated. Using $\mathrm{i}\,\hat{\mathbf{f}} \wedge \hat{\mathbf{f}}^* = 0$ (from the reflexional symmetry of $\mathbf{f}$), we obtain

$$\mathrm{i}\,\hat{\mathbf{u}} \wedge \hat{\mathbf{u}}^* = -4|D|^{-2}\,(\mathbf{k}\cdot\boldsymbol{\Omega})\,\mathbf{k}\,\sigma_\mathrm{r}\,k^2\,\left|\hat{\mathbf{f}}\right|^2\,. \tag{12.96}$$

In conjunction with (12.84), we then have a contribution to the tensor α_{ij} of the form

$$\hat{\alpha}_{ij}(\mathbf{k},\,\omega) = -2|D|^{-2}\sigma_\mathrm{r}(\mathbf{k}\cdot\boldsymbol{\Omega})\frac{\eta k^4}{\omega^2 + \eta^2 k^4}\left|\hat{\mathbf{f}}\right|^2 k_i k_j\,. \tag{12.97}$$

The behaviour of $\hat{\alpha}_{ij}$ as $|\mathbf{h}_0\cdot\mathbf{k}|$ increases from a low level is of central interest. First when $\mathbf{h}_0\cdot\mathbf{k} \approx 0$, we have $\sigma_\mathrm{r} = \omega$ and the expression for $\mathrm{i}\,\hat{\mathbf{u}} \wedge \hat{\mathbf{u}}^*$ (related to the helicity of the forced wave) is non-zero only when

$$\omega\,(\mathbf{k}\cdot\boldsymbol{\Omega}) \neq 0\,. \tag{12.98}$$

This condition means simply that the forcing wave must either propagate 'upwards' or 'downwards' relative to the direction of $\boldsymbol{\Omega}$; as expected from the discussion of §12.2, the velocity field will lack reflexional symmetry only if the forcing is such as to provide a net energy flux either upwards or downwards. This is the same requirement as already encountered in §12.3.

As $|\mathbf{h}_0\cdot\mathbf{k}|$ increases from zero, the variation of $\hat{\alpha}_{ij}$ is determined by the behaviour of the scalar coefficient in (12.97), $|D|^{-2}\sigma_\mathrm{r}$, σ_r being given by (12.86). To fix ideas,

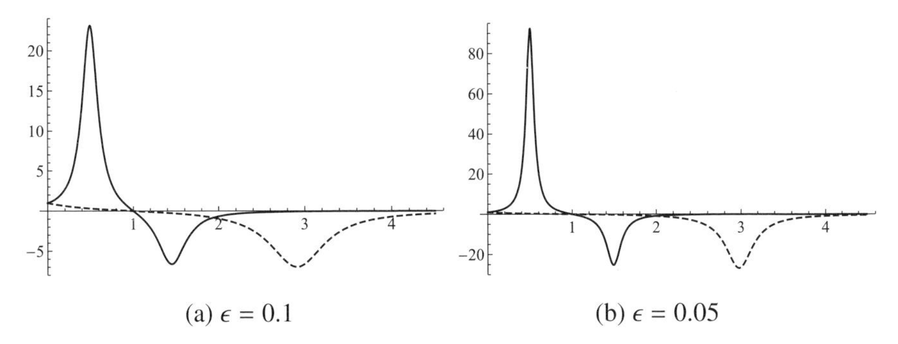

(a) $\epsilon = 0.1$ (b) $\epsilon = 0.05$

Figure 12.8 The function $F(m;\, X, \epsilon, \mathcal{P}_{\rm m}) \equiv \hat{\alpha}/\hat{\alpha}_0$ defined by (12.99) for $\mathcal{P}_{\rm m} = 1$ and $X = \frac{1}{2}$ (solid) and $X = 2$ (dashed); with decrease of ϵ, the resonances at $m = 1 \pm X$ become increasingly sharp.

suppose that $\omega > 0$ and $\mathbf{k} \cdot \mathbf{\Omega} > 0$. The resonance condition (12.93) implies that the behaviour will differ according as $X <$ or > 1.

From (12.97), we now obtain

$$\hat{\alpha} \equiv \tfrac{1}{3}\hat{\alpha}_{ii} \approx \frac{(1-m)\,[1-X^2]^2\,\hat{\alpha}_0}{\left[(1-m)^2 - X^2\right]^2 + \epsilon^2[(1-m)^2 + X^2]\,(\mathcal{P}_{\rm m}+m)^2} = F(m;\, X, \epsilon, \mathcal{P}_{\rm m})\,\hat{\alpha}_0\,, \text{ say,} \tag{12.99}$$

where $\hat{\alpha}_0 = -2\eta(k^2/3\omega^5)\, f_0^2\,(\mathbf{k} \cdot \mathbf{\Omega})[1 - X^2]^{-2}$; terms of order ϵ^3 and higher are here neglected. We have introduced the factor $[1 - X^2]^2$ in (12.99) so that $\hat{\alpha} = \hat{\alpha}_0$ when $m = 0$ and $\epsilon = 0$.

Given that $m \geq 0$, it is evident that if $X < 1$ there are two resonances[4] at $m = 1 \pm X$, while if $X > 1$ there is a single resonance at $m = 1 + X$. By way of illustration, Figure 12.8 shows the function $F(m;\, X, \epsilon, \mathcal{P}_{\rm m})\,\hat{\alpha}_0 \equiv \hat{\alpha}(m)/\hat{\alpha}_0$ for two values of ϵ, with $\mathcal{P}_{\rm m} = 1$ and $X = \frac{1}{2}$ (solid) and $X = 2$ (dashed); as expected, the resonances become sharper as ϵ decreases.

Two broad conclusions can be drawn from the above analysis. First it is clear that if a locally uniform field $\mathbf{h}_0$ grows from an initially weak level due to dynamo action, then the α-effect may intensify if the field approaches a resonance level $m = 1 \pm X$. This intensification of the α-effect is associated with the large amplitude of the response $\hat{\mathbf{u}}(\mathbf{k}, \omega)$ at resonance. Rotation by itself keeps this response at a low level, and thus acts as a constraint on the motion. The growing magnetic field can, at an appropriate level, release this constraint and can, as it were, trigger large velocity fluctuations (see Figure 12.11) which make a correspondingly large contribution to $\hat{\alpha}_{ij}$. This may seem paradoxical; analogous behaviour is, however, well known in the context of the stability of hydrodynamic systems subject to the simultaneous action of Coriolis and Lorentz forces (see e.g. Chandrasekhar 1961,

[4] The resonance at $m = 1 - X$ when $X < 1$ was noted in Moffatt (1978a); it turns out that it is only this resonance that is relevant to the process of equilibration.

Chapter 5): whereas the effects of rotation and magnetic field are separately stabilising, the two effects can work against each other in such a way that a flow that is stable under the action of rotation alone can become unstable when a magnetic field of sufficient strength is introduced. This type of behaviour has also been noted in the dynamo context by Busse (1976) who anticipated that "it is quite possible that the action of the Lorentz force releases a constraint which allows the helicity, and thereby the dynamo action, to increase". This important releasing mechanism will be encountered in the geomagnetic context in Chapter 12.

The second conclusion is perhaps less unexpected: just as in §12.4, the α-effect is ultimately quenched; as m increases, (12.99) gives the asymptotic result $\alpha \sim -m^{-3}$ (more complicated possibilities were explored in Moffatt 1972). Equilibration, and saturation of magnetic energy is then inevitable. The saturation level m_s is then the quantity of key interest.

12.7 Equilibration due to α-quenching in the Lehnert wave field

In this section, we shall suppose that the spectrum of $\mathbf{f}(\mathbf{x}, t)$ is concentrated around a single frequency–wave-number pair (ω, k), and is uniformly distributed over the hemisphere $\omega \boldsymbol{\Omega} \cdot \mathbf{k} > 0$ in wave-number space. Then, as in §12.4, we may now 'isotropise', which simply means that we re-interpret the dimensionless variables $\hat{\alpha}_0$, X^2 and m as

$$\hat{\alpha}_0 = -2\eta(k^3/3\omega^5)\, f_0^2\, \Omega \left[1 - X^2\right]^{-2}, \quad X^2 = 4\Omega^2/\omega^2, \quad m = h_0^2 k^2/\omega^2. \quad (12.100)$$

This is an approximation in that it replaces the continuum of resonances at $\omega^2 = 4(\boldsymbol{\Omega} \cdot \mathbf{k})^2/k^2 = 4\Omega^2 \cos^2\theta$, $(0 \le \theta \le \pi/2)$ by a single concentrated resonance at $\omega^2 = 4\Omega^2$, i.e. at $\theta = 0$. We make this approximation here for simplicity, in the expectation that the model still provides a correct qualitative description – see Figure 12.12.

We can now follow the same procedure as in §12.5. If again we focus on the unstable force-free mode of scale K^{-1}, the equation governing the evolution of its scaled magnetic energy density m is

$$\frac{\mathrm{d}m}{\mathrm{d}\tau} = \left(q^2 F(m; X, \epsilon, \mathcal{P}_{\mathrm{m}}) - 1\right) m, \quad (12.101)$$

where again

$$\tau = \eta K^2 t \quad \text{and} \quad q^2 = \hat{\alpha}_0/\eta K = (\Omega/\omega)(k/K)\left(2k^2 f_0^2/3\omega^4\right), \quad (12.102)$$

and the function F is defined in (12.99). We assume that $q^2 > 1$ ensuring dynamo action when m is small. To determine the saturation energy m_s when ϵ is very small,

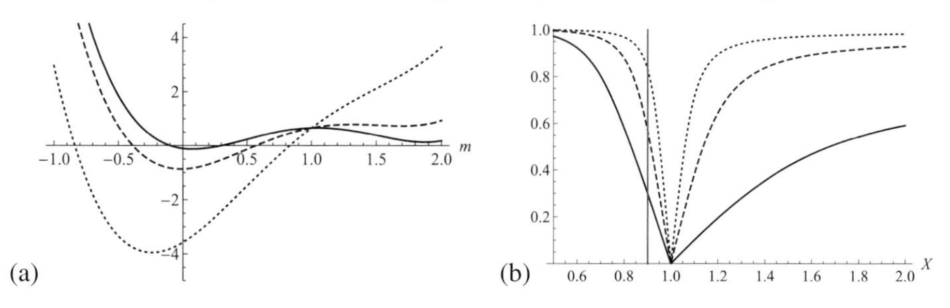

Figure 12.9 (a) The quartic $Q(m)$ defined by (12.104) for $X = 0.9$ and $q = 2$ (solid), 5 (dashed) and 10 (dotted), with the root $m = m_s$ in the range $0 < m < 1$. (b) The root $m_s(q, X)$ as a function of X for the same three values of q, the cut at $X = 0.9$ being shown by the vertical line.

we may simply set $\epsilon = 0$ in F; obviously m_s is then determined by the condition $\mathrm{d}m/\mathrm{d}\tau = 0$, i.e.

$$q^2 F(m_s;\ X, 0, \mathcal{P}_m) = 1. \tag{12.103}$$

Now as m increases from 0 to 1, $q^2 F(m;\ X, 0, \mathcal{P}_m)$ decreases monotonically from q^2 to 0. Hence the condition (12.103) is satisfied for some value $m = m_s(q, X, \mathcal{P}_m)$ where $0 < m_s < 1$; this is the saturation energy that we seek. The dependence on $\mathcal{P}_m$ disappears as $\epsilon \to 0$.

The condition $q^2 F(m; X, 0, \mathcal{P}_m) = 1$ is a quartic equation for m, viz.,

$$Q(m) \equiv \left[(m-1)^2 - X^2\right]^2 - (1-m)\, q^2\, (1-X^2)^2 = 0, \tag{12.104}$$

whose four roots may be found analytically. Two of these roots are complex and two real; one of the real roots is negative, and the other is in the range $0 \le m \le 1$. A typical situation is shown in Figure 12.9a, which shows the quartic $Q(m)$ for $X = 0.9$, and $q = 2$, 5 and 10. Figure 12.9b shows the root $m_s(q, X)$ in the range $0 < m_s < 1$ as a function of X for the same three values of q.

Figure 12.10a shows in log-linear plot the result of integrating (12.101) with $\epsilon = 0.01$, $\mathcal{P}_m = 1$ and $q = 2$, $X = 0.5$; for this value of $X\,(<1)$, the growing energy passes through the resonance at $m = 1 - X = 0.5$ resulting in the accelerated growth before saturation at $m = m_s(2, 0.5) = 0.972$, similar to that already encountered in the non-rotating situation (Figure 12.6). Figure 12.10b shows a similar plot for $q = 4$, $X = 2$ with saturation level $m_s(4, 2) = 0.890$; for this value of $X\,(>1)$, there is no acceleration in the growth rate before saturation, because the only resonance is at $m = X + 1 = 3$, outside the range $0 < m < m_s$.

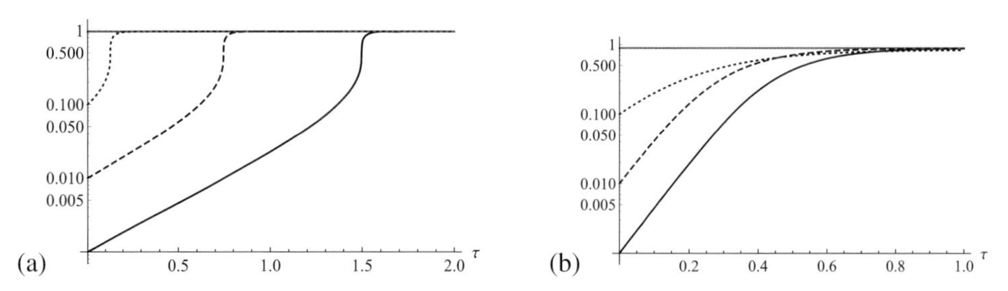

Figure 12.10 Log-linear plot of $m(\tau)$, from solution of (12.101) with initial condition $m(0) = 0.001$ (solid), 0.01 (dashed) and 0.1 (dotted); $\epsilon = 0.01$, $\mathcal{P}_m = 1$: (a) $q = 2$, $X = 0.5$, saturation level $m_s(2, 0.5) = 0.972$, as indicated by the horizontal asymptote, with accelerated growth rate before saturation; (b) $q = 4$, $X = 2$, saturation level $m_s(4, 2) = 0.890$.

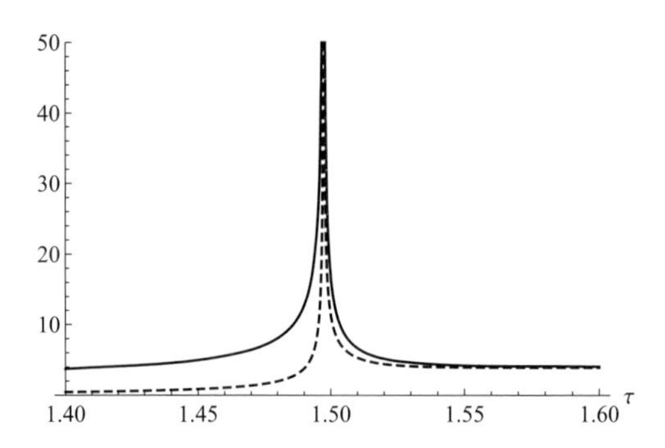

Figure 12.11 Wave-energy variables $\langle \mathbf{u}^2 \rangle / f_0^2$ (solid, the peak level is actually at ~ 9000) and $\langle \mathbf{h}^2 \rangle / f_0^2$ (dashed, peak at ~ 4500) of the forced wave field as functions of τ, showing a sharp resonance at $\tau \approx 1.497$ when $m(\tau) \approx 1 - X = 0.5$; after the resonance, equipartition of these wave energies is rapidly established.

12.7.1 Energies at resonance

The unscaled magnetic energy density is, from (12.100), $M = \frac{1}{2}h_0^2 = m\omega^2/2k^2$, with saturation level $M_s = (\omega^2/2k^2)m_s(q, X)$. It is evident from Figure 12.9b that for $q \gtrsim 10$, except in an immediate neighbourhood of $X = 1$, $m_s \sim 1$ so that $M_s = \omega^2/2k^2 = \frac{1}{2}c^2$, where c is a characteristic phase speed within the wave field.

The mean kinetic energy $\frac{1}{2}\langle \mathbf{u}^2 \rangle$ of the wave field is given, from (12.87) after a short calculation, by

$$\langle \mathbf{u}^2 \rangle = \frac{(1 - m)^2 + X^2}{\left[(1-m)^2 - X^2\right]^2 + \epsilon^2[(1-m)^2 + X^2]\,(\mathcal{P}_m + m)^2} f_0^2 , \tag{12.105}$$

and the mean energy of the associated perturbation magnetic field $\frac{1}{2}\langle \mathbf{h}^2 \rangle$ is given simply from (12.28) by

$$\langle \mathbf{h}^2 \rangle = m(\tau)\langle \mathbf{u}^2 \rangle . \tag{12.106}$$

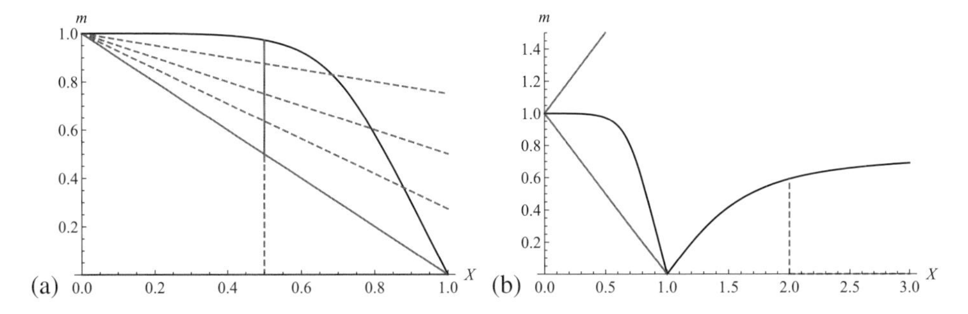

Figure 12.12 Schematics of the dynamo process in the (X, m)-plane. (a) The situation when $X < 1$; m increases through dynamo action (on the vertical line), and encounters resonance at $m = X - 1$ (on the solid line of resonance); it saturates at $m = m_s(X)$ (on the curve); strictly, the concentrated resonance at $m = X - 1$ should be replaced by a continuum of resonances in the range $X - 1 \leq m \leq m_s(X)$, as represented by the dashed lines. (b) The situation when $X > 1$; the sloping resonance lines $m = 1 \pm X$ are as shown; no resonance is encountered as m increases from 0 to $m_s(X)$.

When $X < 1$, these show a pronounced resonance as $m(\tau)$ passes through the resonance level $m = 1 - X$, as illustrated in Figure 12.11 for the same parameter values as in Figure 12.10a (with $m_0 = 0.01$). It is at this instant ($\tau \approx 0.497$ in this case) that $m(\tau)$ increases by an order of magnitude, and there is no doubt that this is caused by the transitory peak in energy of the wave field. We should perhaps caution that linear wave theory may break down during the brief passage through this resonance; however, this concentrated resonance is a consequence of the isotropisation approximation embodied in (12.100). The true situation is indicated by the schematic diagrams of Figure 12.12 and the caption to that figure.

12.8 Forcing from the boundary

An alternative means of injecting energy into the fluid system has been considered by Soward (1975). In Soward's model the fluid (supposed inviscid and highly conducting) is contained between two parallel boundaries $z = 0,\ z_0$, perpendicular to the rotation vector $\mathbf{\Omega}$; energy is injected by random mechanical excitation at $z = 0$, and is absorbed (without reflexion) at $z = z_0$. In the absence of a magnetic field, the mean helicity (averages being defined over planes $z = \text{cst.}$) is independent of z, and there is an associated non-isotropic α-effect, also independent of z. When z_0 is sufficiently large, this situation is unstable to the growth of magnetic field perturbations of the form

$$\mathbf{h}_0 = h_{00}(\sin Kz,\ \cos Kz,\ 0). \tag{12.107}$$

As soon as these reach a significant level, however, weak ohmic dissipation intervenes, leading to *spatial attenuation* of wave groups propagating from $z = 0$ to $z = z_0$, and hence the α-effect decreases in intensity with increasing z. The equilibrium level of the magnetic field at any level is determined by the local value of the pseudo-tensor α_{ij}, and so the equilibrium magnetic energy density $M = \frac{1}{2}\left\langle \mathbf{h}_0^2 \right\rangle$ is also a decreasing function of z. The scale characteristic of this spatial attenuation is found to be

$$L = \ell\, \mathcal{R}o^{-2}, \quad \mathcal{R}o = U_0/\ell\omega_0\,, \tag{12.108}$$

where U_0 is the rms velocity on $z = 0$; ℓ is a length-scale characteristic of the excitation on $z = 0$; and ω_0 is a frequency characteristic of the wave excited (assumed small compared with Ω). The treatment is based on the assumption that the *Rossby number* $\mathcal{R}o$ is small , so that, as in other two-scale approaches, $L \gg \ell$. Figure 12.13 shows solutions obtained by Soward for the two horizontal mean field components in the equilibrium situation when $z_0/L = 16.6$; a noteworthy feature of the solution is that the field rotates about the direction of $\mathbf{\Omega}$ with angular velocity approximately $0.2\left(Ro^4/Q\right)\Omega$ where $Q = \Omega\ell^2/\eta$ ($\gg 1$).

The situation in the limit $z_0/L \to \infty$ (when fluid fills the half-space $z > 0$) is rather curious. Soward argues that due to the decrease of the α-effect with height, the magnetic field must for $z \gtrsim L$ decay exponentially (the α-effect being inadequate to maintain it), and that a non-zero flux of wave energy must then propagate freely for $z \gg L$ "while the strength of the α-effect remains constant". It is difficult to accept this picture, because a non-zero α-effect always gives rise to magnetic instability when sufficient space is available, and in the situation envisaged an infinite half-space is available for the development of such instabilities. The only alternative is that the α-effect vanishes for large z, or equivalently that *all* the wave energy is dissipated in the layer $z = \mathcal{O}(L)$ of field generation, and *none* survives to propagate to $z = \infty$.

Soward draws attention to a further complication that must in general be taken into account when slow evolution of a large-scale mean magnetic field is considered. This is that, on the long time-scale associated with the evolution, non-linear interactions between the constituent background waves may also cause a systematic evolution of the wave spectra. This means that, no matter how small the perturbations $\mathbf{u}_1$ and $\mathbf{h}_1$ may be in equations (12.24) and (12.25), neglect of the non-linear terms may not be justified on the long time-scale of dynamo action associated with the mean electromotive force $\langle \mathbf{u}_1 \wedge \mathbf{h}_1 \rangle$, which is itself quadratic in small quantities.

An α that decreases with increasing $|\mathbf{B}|$ has been incorporated in a number of numerical studies of $\alpha\omega$-dynamos of the type discussed in §9.9 (e.g. Braginskii

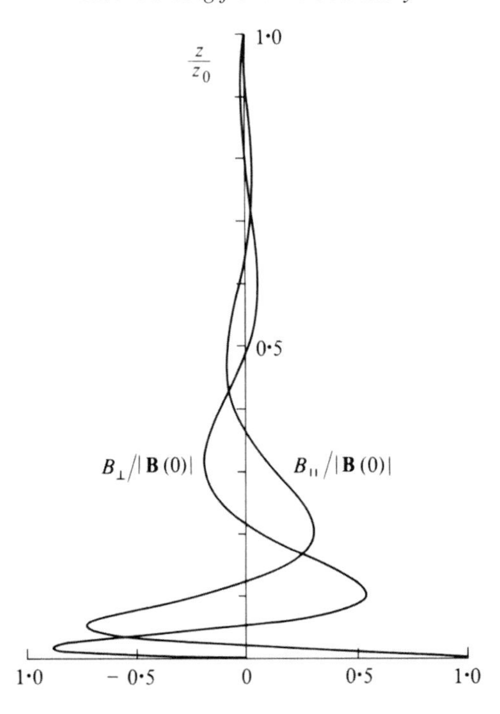

Figure 12.13 Field components parallel and perpendicular to the field at the boundary $z = 0$ associated with random excitation on $z = 0$, the boundary $z = z_0$ being perfectly absorbing; the fluid rotates about the z-axis with angular velocity Ω, and the magnetic field rotates relative to the fluid with angular velocity $0.2\,(Ro^4/Q)\,\Omega$, where Ro and Q are as defined in the text. [From Soward 1975.]

1970; Stix 1972; Jepps 1975). Stix adopted a simple 'cut-off' formula

$$\alpha = \begin{cases} \alpha_0 & (|B| < B_c) \\ 0 & (|B| > B_c) \end{cases} \tag{12.109}$$

and studied the effect on the oscillatory dipole mode excited in a slab geometry (cf. §9.6, but with $U = U(z)$). For the particular conditions adopted, the critical dynamo number for this mode was found to be $X_c = -89.0$, and the period of its oscillation was $0.993R^2/\eta$, where R is the length-scale transverse to the slab (the same scale being used in the definition of X). When $X = aX_c$ with $a > 1$, linear kinematic theory gives an oscillatory mode of exponentially increasing amplitude. The cut-off effect (12.109) limits the amplitude to a value of order B_c, and also tends to lengthen the period of the oscillation: Stix found for example that when $a = 7$, the maximum field amplitude is between $2B_c$ and $3B_c$, and the period is about $1.7R^2/\eta$. Moreover the field variation with time, although periodic, is very far from sinusoidal: bursts of poloidal field are produced by the α-effect when $|B| < B_c$, and these are followed by less pronounced bursts of toroidal field (of intensity greater than B_c) due to the effect of shear. The resulting 'spikiness' of the field (as a function of time) was also noticed by Jepps (1975) who carried out similar computations (and for a range of cut-off functions) in a spherical geometry. Charbonneau (2014) has provided an extensive review of developments in this area, with emphasis on the $\alpha\omega$-dynamo, which must somehow overcome the 'catastrophic quenching' discussed in the final paragraph of §12.5 above.

12.9 Helicity generation due to interaction of buoyancy and Coriolis forces

We have noted on several occasions that a lack of symmetry about planes perpendicular to the rotation vector $\mathbf{\Omega}$ is necessary to provide the essential lack of reflexional symmetry in random motions that can lead to an α-effect. As first pointed out by Steenbeck et al. (1966), this lack of symmetry is present when buoyancy forces $\rho'\mathbf{g}$ (with $\mathbf{g}\cdot\mathbf{\Omega} \neq 0$) act on fluid elements with density perturbation ρ' relative to the local undisturbed density ρ_0.

A simple and illuminating discussion of the mutual role of buoyancy and Coriolis forces in generating helicity was given by Hide (1976). Suppose that conditions are geostrophic and that the Boussinesq approximation (in which account is taken of density fluctuations only in the buoyancy force term of the equation of motion, and not in the inertia term) is valid. Then the equation of motion reduces to

$$2\rho_0\,\mathbf{\Omega}\wedge\mathbf{u} = -\nabla p + \rho'\mathbf{g}\,, \tag{12.110}$$

wherein for the moment we neglect any Lorentz forces. The curl of this equation (with $\mathbf{g}\wedge\nabla\rho_0 = 0$) gives

$$2\rho_0\nabla\wedge(\mathbf{\Omega}\wedge\mathbf{u}) = -\mathbf{g}\wedge\nabla\rho'\,, \tag{12.111}$$

and hence

$$2\rho_0\,(\mathbf{\Omega}\wedge\mathbf{u})\cdot\nabla\wedge(\mathbf{\Omega}\wedge\mathbf{u}) = -(\mathbf{\Omega}\wedge\mathbf{u})\cdot(\mathbf{g}\wedge\nabla\rho') = -\mathbf{\Omega}\cdot[\mathbf{u}\wedge(\mathbf{g}\wedge\nabla\rho')]. \tag{12.112}$$

Writing $\mathbf{u}_\perp = \mathbf{u} - (\mathbf{u}\cdot\mathbf{\Omega})\,\mathbf{\Omega}/\Omega^2$ for the projection of $\mathbf{u}$ on planes perpendicular to $\mathbf{\Omega}$, it may readily be verified that (12.112) may be written in the form

$$2\rho_0\,\Omega^2\,(\mathbf{u}_\perp\cdot\nabla\wedge\mathbf{u}_\perp) = -\mathbf{\Omega}\cdot[\mathbf{u}\wedge(\mathbf{g}\wedge\nabla\rho')]\,, \tag{12.113}$$

or, taking the average over horizontal planes,

$$2\rho_0\,\Omega^2\,\langle\mathbf{u}_\perp\cdot\nabla\wedge\mathbf{u}_\perp\rangle = -(\mathbf{\Omega}\cdot\mathbf{g})\,\langle\mathbf{u}\cdot\nabla\rho'\rangle + \mathbf{g}\cdot\langle\mathbf{u}\,\nabla\rho'\rangle\cdot\mathbf{\Omega}\,. \tag{12.114}$$

In the particular case where $\mathbf{g}$ is parallel to $\mathbf{\Omega}$, and if conditions are statistically homogeneous over horizontal planes, then, using

$$\nabla\cdot\mathbf{u} = \nabla_\perp\cdot\mathbf{u}_\perp + \partial w/\partial z = 0\,, \tag{12.115}$$

where w is the vertical component of $\mathbf{u}$, we have from (12.114)

$$2\rho_0\,\Omega^2\,\langle \mathbf{u}_\perp \cdot \nabla \wedge \mathbf{u}_\perp \rangle = -(\boldsymbol{\Omega}\cdot\mathbf{g})\,\langle \rho' \partial w/\partial z \rangle\,, \tag{12.116}$$

a formula that provides a direct relationship between the pseudo-scalar $\boldsymbol{\Omega}\cdot\mathbf{g}$ and at least one ingredient of the total mean helicity $\langle \mathbf{u}\cdot\nabla\wedge\mathbf{u}\rangle$. The phase relationship between ρ' and $\partial w/\partial z$ is evidently important in determining the magnitude and sign of this ingredient.

With $\mathbf{u} = \mathbf{u}_\perp + \mathbf{u}_\parallel$, the total helicity, averaged over planes $z =$ cst., is given by

$$\begin{aligned}\langle \mathbf{u}\cdot\nabla\wedge\mathbf{u}\rangle &= \langle \mathbf{u}_\perp\cdot\nabla\wedge\mathbf{u}_\perp\rangle + \langle \mathbf{u}_\parallel\cdot\nabla\wedge\mathbf{u}_\perp\rangle + \langle \mathbf{u}_\perp\cdot\nabla\wedge\mathbf{u}_\parallel\rangle \\ &= \langle \mathbf{u}_\perp\cdot\nabla\wedge\mathbf{u}_\perp\rangle + 2\langle \mathbf{u}_\parallel\cdot\nabla\wedge\mathbf{u}_\perp\rangle\,,\end{aligned} \tag{12.117}$$

using homogeneity with respect to x and y. Hence

$$\langle \mathbf{u}\cdot\nabla\wedge\mathbf{u}\rangle = \langle \mathbf{u}_\perp\cdot\nabla\wedge\mathbf{u}_\perp\rangle + 2\langle w\omega_z\rangle\,, \tag{12.118}$$

where ω_z is the component of vorticity parallel to $\boldsymbol{\Omega}$. The ingredient $2\langle w\omega_z\rangle$ is not determined (in terms of ρ') by the above argument; Hide maintains that this ingredient should be negligible compared with $\langle \mathbf{u}_\perp\cdot\nabla\wedge\mathbf{u}_\perp\rangle$ when Ω is sufficiently strong. However, order of magnitude estimates of the two contributions in (12.117) suggest that

$$\frac{\langle \mathbf{u}_\parallel\cdot\nabla\wedge\mathbf{u}_\perp\rangle}{\langle \mathbf{u}_\perp\cdot\nabla\wedge\mathbf{u}_\perp\rangle} = \mathcal{O}\left(\frac{U_\parallel/L_\perp}{U_\perp/L_\parallel}\right) = \mathcal{O}\left(\frac{L_\parallel^2}{L_\perp^2}\right), \tag{12.119}$$

where $U_\parallel$, $U_\perp$, $L_\parallel$ and $L_\perp$ are velocity and length-scales parallel and perpendicular to $\boldsymbol{\Omega}$, related (via $\nabla\cdot\mathbf{u} = 0$) by $U_\parallel/L_\parallel \sim U_\perp/L_\perp$. Since $L_\parallel/L_\perp$ is generally large in a rotation-dominated system, it seems quite possible that the second contribution to (12.118) may in fact dominate the first. Estimates such as (12.119) do however depend critically on the phase relations between the velocity components, and the matter can really only be settled within the framework of more specific models such as those treated in the following sections.

12.10 Excitation of magnetostrophic waves by unstable stratification

The following idealised problem (Figure 12.14), which has been widely studied in different forms (Braginskii 1964a, 1967, 1970; Eltayeb 1972, 1975; Roberts & Stewartson 1974, 1975), provides a basis for the detailed consideration of the effects of unstable density stratification. Suppose that fluid is contained between two horizontal planes $z = \pm z_0$, on which the temperature T is prescribed as $T = T_0 \mp \beta z_0$ respectively. In the undisturbed state, $-\beta\,(< 0)$ is then the vertical temperature gradient, and instabilities are to be expected if β is sufficiently large for the rate of release of potential energy associated with overturning of the 'top-heavy' fluid to

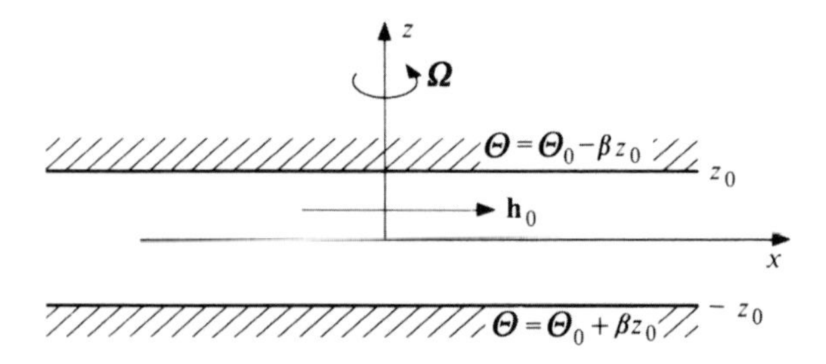

Figure 12.14 Configuration considered in §12.10. Fluid is contained between the planes $z = \pm z_0$, the lower plane being maintained at a higher temperature than the upper plane. The system rotates about the z-axis with angular velocity $\mathbf{\Omega}$, and a uniform magnetic field $(\mu_0\rho)^{1/2}\mathbf{h}_o$, with $\mathbf{h}_0$ parallel to the x-axis, is supposed present. The system is unstable when the applied temperature difference is sufficiently large.

exceed the rate of dissipation of energy due to viscous and/or ohmic diffusion. The equation for $T(\mathbf{x}, t)$ is the heat conduction equation (see §3.3),

$$\partial T/\partial t + \mathbf{u} \cdot \nabla T = \kappa \nabla^2 T \,, \tag{12.120}$$

where κ is the thermal diffusivity of the fluid, and writing $T = T_0 - \beta z + \theta(\mathbf{x}, t)$ where θ is a small perturbation, the linearised form of (12.120) (regarding $\mathbf{u} = (u, v, w)$ as small also) is

$$\partial\theta/\partial t - \beta w = \kappa \nabla^2 \theta \,. \tag{12.121}$$

The perturbation θ induces a density perturbation $\rho' = -\alpha\,\theta\rho_0$ where α is here the coefficient of thermal expansion, and the equation of motion, under the Boussinesq approximation, is

$$\partial\mathbf{u}/\partial t + \mathbf{u}\cdot\nabla\mathbf{u} + 2\mathbf{\Omega} \wedge \mathbf{u} = -\nabla P + \mathbf{h}\cdot\nabla\mathbf{h} - \alpha\theta\,\mathbf{g} + \nu\nabla^2\mathbf{u} \,, \tag{12.122}$$

where P includes a contribution due to the undisturbed density gradient. We shall restrict attention to the possibility of modes of instability having growth rate small compared with Ω; for these we have seen in §12.9 that provided $\nu/\eta \leq 1$, a legitimate approximation to (12.122) is the magnetostrophic equation

$$2\mathbf{\Omega} \wedge \mathbf{u} = -\nabla P + \mathbf{h}\cdot\nabla\mathbf{h} - \alpha\theta\,\mathbf{g} \,. \tag{12.123}$$

We suppose that in the undisturbed situation the field $\mathbf{h}_0$ is uniform and horizontal. Putting $\mathbf{h} = \mathbf{h}_0 + \mathbf{h}_1$ where $|\mathbf{h}_1| \ll h_0$, the linearised form of (12.123) is

$$2\mathbf{\Omega} \wedge \mathbf{u} = -\nabla p + \mathbf{h}_0 \cdot \nabla\mathbf{h}_1 - \alpha\theta\,\mathbf{g} \,, \tag{12.124}$$

where p is the perturbation in P, and the linearised induction equation still has the form

$$\partial \mathbf{h}_1/\partial t = \mathbf{h}_0 \cdot \nabla \mathbf{u} + \eta \nabla^2 \mathbf{h}_1 \,. \tag{12.125}$$

Equations (12.121), (12.124) and (12.125) together with $\nabla \cdot \mathbf{u} = \nabla \cdot \mathbf{h}_1 = 0$ determine the evolution of small perturbations ($\mathbf{u}$, $\mathbf{h}_1$, θ, p).

We must of course impose boundary conditions on $z = \pm z_0$; we have firstly that

$$\theta = 0, \; u_z = 0 \; \text{ on } \; z = \pm z_0 \,. \tag{12.126}$$

Secondly, neglect of all viscous effects is consistent with the adoption of the 'stress-free' conditions

$$\partial u_x/\partial z = 0, \quad \partial u_y/\partial z = 0 \; \text{ on } \; z = \pm z_0 \,. \tag{12.127}$$

Finally, the simplest conditions on $\mathbf{h}_1$ result from the assumption that the regions $|z| > z_0$ are perfectly conducting and that both magnetic and electric fields are confined to the region $|z| < z_0$; it follows that (cf. (3.147) and (3.148))

$$h_{1z} = 0, \quad \partial h_{1x}/\partial z = 0, \quad \partial h_{1y}/\partial z = 0 \; \text{ on } \; z = \pm z_0 \,. \tag{12.128}$$

The above equations and boundary conditions admit solutions of the form

$$(u,\; v,\; h_{1x},\; h_{1y},\; p) = (\hat{u},\; \hat{v},\; \hat{h}_{1x},\; \hat{h}_{1y},\; \hat{p}) \cos nz \, \mathrm{e}^{\mathrm{i}(lx+my-\omega t)} \,, \tag{12.129a}$$

$$(w,\; h_{1z},\; \theta) = (\hat{w},\; \hat{h}_{1z},\; \hat{\theta}) \sin nz \, \mathrm{e}^{\mathrm{i}(lx+my-\omega t)} \,, \tag{12.129b}$$

where nz_0/π is an integer. Substituting in (12.121), (12.124) and (12.125), and eliminating $\hat{\mathbf{h}}_1$ gives, with $\mathbf{k} = (l,\; m,\; n)$,

$$-\,\mathrm{i}\sigma\,\hat{\mathbf{u}} + 2\boldsymbol{\Omega} \wedge \hat{\mathbf{u}} = -\mathrm{i}\,(l,\; m,\; \mathrm{i}n)\,\hat{p} - \alpha\hat{\theta}\,\mathbf{g}\,, \tag{12.130a}$$

$$(-\mathrm{i}\omega + \kappa k^2)\,\hat{\theta} - \beta\hat{w} = 0\,, \tag{12.130b}$$

and, from $\nabla \cdot \mathbf{u} = 0$, we have also

$$\mathrm{i}l\,\hat{u} + \mathrm{i}m\,\hat{v} + n\,\hat{w} = 0\,. \tag{12.131}$$

Here, $\sigma = -(\omega + \mathrm{i}\eta k^2)^{-1}(\mathbf{h}_0 \cdot \mathbf{k})^2$ (cf. (12.30) in the magnetostrophic limit). We may solve (12.130a) and (12.130b) to give the velocity components in terms of $\hat{p}$ in the form

$$\frac{\hat{u}}{-l\sigma - 2\mathrm{i}\Omega m} = \frac{\hat{v}}{-m\sigma + 2\mathrm{i}\Omega l} = \frac{\hat{w}\,(\alpha\beta g + \sigma(\omega + \mathrm{i}\kappa k^2))}{\mathrm{i}n(\omega + \mathrm{i}\kappa k^2)(4\Omega^2 - \sigma^2)} = \frac{\hat{p}}{4\Omega^2 - \sigma^2}\,, \tag{12.132}$$

and (12.131) then yields the dispersion relationship for ω, which may be simplified to the form

$$Y^2(-\mathrm{i}\omega + \eta k^2)^2(-\mathrm{i}\omega + \kappa k^2) + (-\mathrm{i}\omega + \kappa k^2) - Z(-\mathrm{i}\omega + \eta k^2) = 0\,, \tag{12.133a}$$

where

$$Y = 2n\Omega/h_0^2 m^2 k, \quad Z = \alpha\beta g(l^2+m^2)/m^2k^2h_0^2\,. \tag{12.133b}$$

We are interested in the possibility of unstable modes characterised by $\omega = \omega_r + i\omega_i$ with $\omega_i > 0$. If $\omega_r = 0$, these modes are non-oscillatory, while if $\omega_r \neq 0$ they have the character of propagating magnetostrophic waves of increasing amplitude. Suppose first that we neglect dissipative effects in (12.133a), i.e. we put $\eta = \kappa = 0$. This was the ideal situation to which Braginskii (1964a, 1967) restricted attention. The roots of the cubic (12.133a) are then $\omega = 0$ and

$$\omega = \pm i(Z-1)^{1/2}/Y\,, \tag{12.134}$$

and the mode corresponding to the upper sign is clearly unstable whenever $\mathbf{k}$ is such that $Z > 1$ and $0 < Y < \infty$. Now in the geophysical or solar contexts, it is appropriate (as in §9.4) to regard the y-direction as east, so that possible values of m are restricted by the requirement of periodicity in longitude to a discrete set Nm_1 ($N = 0, \pm 1, \pm 2, \ldots$). The 'axisymmetric' mode corresponding to $N = 0$ is of no interest since, for it, $\omega = 0$. The mode that is most prone to instability is that for which $N = \pm 1$, since (other things being equal) this gives the largest value of Z. This mode would correspond to $e^{\pm i\varphi}$-dependence on the azimuth angle φ in the corresponding spherical geometry. Note that the above description in terms of the magnetostrophic approximation can only detect low frequency modes ($|\omega| \ll \Omega$) and the approximation breaks down for modes not satisfying this condition.

Neglect of dissipative effects in this problem is a dangerous simplification, for reasons spelt out by Roberts & Stewartson (1974): if $g\alpha\beta/h_0^2$ exceeds m_1^2 by any amount, no matter how small, an infinite number of modes (corresponding to large values of ℓ) apparently become unstable according to (12.134), since as $\ell \to \infty$, $Z \sim \alpha\beta g/h_0^2 m_1^2 > 1$; this result is however spurious, since $Y \to 0$ as $\ell \to \infty$ (m and n being fixed), and the condition $|\omega| \ll \Omega$ is not therefore satisfied when ℓ is very large; i.e. the contribution of $\mathbf{Du}/\mathrm{D}t$ in the equation of motion undoubtedly becomes important when ℓ is large. Also, more obviously, diffusion effects must become important for large wave number disturbances, and presumably have a stabilising effect.

Weak diffusion also has the effect of shifting the degenerate root $\omega = 0$ of (12.133a) away from the origin in the complex ω-plane. Indeed linearisation of this equation in the quantities η, κ and ω (assumed small) gives for this root the expression

$$\omega = -i\eta k^2(Z-q)/(Z-1)\,, \quad q = \kappa/\eta\,. \tag{12.135}$$

If $0 < q < Z < 1$, then the modes given by (12.134) are stable, whereas that given by (12.135) is clearly unstable with a slow growth rate determined by the weak

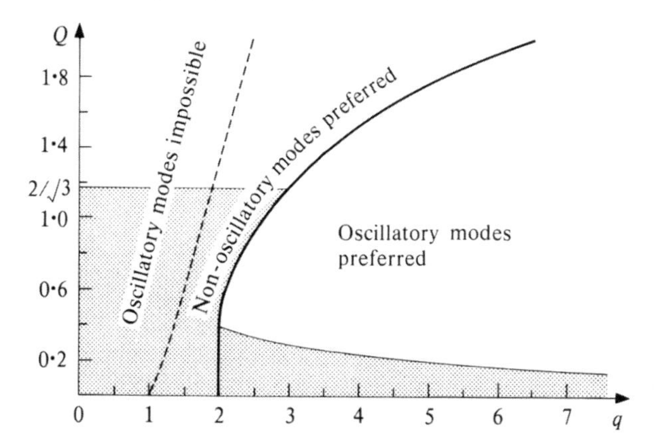

Figure 12.15 Regions of the Q–q plane in which oscillatory modes are possible and preferred. In the shaded regions, the preferred modes are oblique to the applied field (i.e. $m \neq 0$ in (12.138)), while in the unshaded regions, the preferred modes are transverse to $\mathbf{B}_0$ (i.e. $m = 0$). [Adapted from Roberts & Stewartson 1974.]

diffusion effects. The manner in which the ratio of the small diffusivities η and κ enters this criterion is noteworthy.

The behaviour of the roots of (12.133a) for varying values of k, $q = \kappa/\eta$, Y and Z was investigated by Eltayeb (1972), and the results have been summarised in §3 of Roberts & Stewartson (1974). Of particular interest is the question of whether, for given values of q, Q and $\widetilde{\mathcal{R}a}$ where

$$Q = 2\Omega\eta/h_0^2, \quad \widetilde{\mathcal{R}a} = g\alpha\beta z_0^2/\Omega\kappa\,, \tag{12.136}$$

the preferred mode of instability (i.e. that for which ω_i is maximum) is oscillatory ($\omega_\mathrm{r} \neq 0$) or non-oscillatory ($\omega_\mathrm{r} = 0$). Figure 12.15 (from Roberts & Stewartson 1974) shows the region of the (q, Q) plane in which oscillatory modes are preferred; in particular, non-oscillatory modes are always preferred (when unstable) if

$$\text{either} \quad q < 2 \quad \text{or} \quad Q \geq 3.273\,, \tag{12.137}$$

and this result is independent of the value of $\widetilde{\mathcal{R}a}$.

Let us now consider the helicity associated with unstable disturbances. With

$$\mathbf{u} = \Re\mathrm{e}(\hat{u}\cos nz,\ \hat{v}\cos nz,\ \hat{w}\sin nz)\mathrm{e}^{\mathrm{i}(l\mathbf{x}+my-\omega t)}\,, \tag{12.138}$$

it is straightforward to show that

$$\langle \mathbf{u}\cdot\nabla\wedge\mathbf{u}\rangle = \Re\mathrm{e}[\hat{w}^*(\mathrm{i}l\hat{v} - \mathrm{i}m\hat{u})]\cos nz\sin nz\,\mathrm{e}^{2\omega_\mathrm{i}t}\,, \tag{12.139}$$

the average being over horizontal planes. (Incidentally, in the notation of §12.9, the contribution $\langle \mathbf{u}_\perp\cdot\nabla\wedge\mathbf{u}_\perp\rangle$ to the total helicity is zero in this case!). Now if the disturbance is non-oscillatory, then ω and σ are both pure imaginary; putting

$$\omega = \mathrm{i}\omega_\mathrm{i}, \quad \sigma = \mathrm{i}\sigma_\mathrm{i}, \quad \sigma_\mathrm{i} = (\omega_\mathrm{i} + \eta k^2)^{-1}(\mathbf{h}_0 \cdot \mathbf{k})^2 , \tag{12.140}$$

we have from (12.132) and (12.139)

$$\langle \mathbf{u} \cdot \nabla \wedge \mathbf{u} \rangle = \frac{n\Omega(m^2 + l^2)(\omega_\mathrm{i} + \kappa k^2)|\hat{p}|^2}{(4\Omega^2 + \sigma_\mathrm{i}^2)[\alpha\beta g - \sigma_\mathrm{i}(\omega_\mathrm{i} + \kappa k^2)]} \sin 2nz\, \mathrm{e}^{2\omega_\mathrm{i} t} . \tag{12.141}$$

This is antisymmetric about the centre-plane $z = 0$; in the case[5] $nz_0 = \pi$, the helicity of unstable modes ($\omega_\mathrm{i} > 0$) is positive or negative in $0 < z < \frac{1}{2}z_0$ according as

$$\alpha\beta g(\omega_\mathrm{i} + \eta k^2) \gtrless (\mathbf{h}_0 \cdot \mathbf{k})^2(\omega_\mathrm{i} + \kappa k^2) . \tag{12.142}$$

In the critically stable case ($\omega_\mathrm{i} = 0$), the crucial importance of the ratio $q = \kappa/\eta$ is again apparent.

Further detailed consideration of this problem, and of variations involving different orientations of $\boldsymbol{\Omega}$ and $\mathbf{h}_0$ relative to the boundaries, and different boundary conditions, may be found in Eltayeb (1975).

12.11 Instability due to magnetic buoyancy

The concept of *magnetic buoyancy* was introduced by Parker (1955a) in a discussion of the process of formation of sunspots by instabilities of a subsurface solar toroidal magnetic field. Compressibility is an essential element in this type of instability, which is closely related to the instability that occurs in a stratified atmosphere when the (negative) temperature gradient is superadiabatic. Spiegel & Weiss (1982) (see also Corfield 1984) have shown how magnetic buoyancy may be incorporated within equations based on the Boussinesq approximation (to be treated in Chapter 13). Here, we can give only a very simplified description of this particular phenomenon.

Suppose that in an equilibrium situation with gravity $\mathbf{g} = (0, 0, -g)$, we have a density distribution $\rho_0(z)$ and a magnetic field distribution $\mathbf{B}_0 = (0, B_0(z), 0)$, as sketched in Figure 12.16. The equation of magnetostatic equilibrium is

$$0 = -\frac{\mathrm{d}}{\mathrm{d}z}\left(p_0(z) + \frac{1}{2\mu_0}B_0^2\right) - \rho_0(z)g , \tag{12.143}$$

where $p_0(z)$ is the pressure distribution, which may be supposed a monotonic increasing function $f(\rho_0)$ of density $\rho_0(z)$ (in the particular case of an isothermal atmosphere, $p_0(z) = c^2\rho_0(z)$ where c^2 is constant).

Suppose now that a flux tube of cross-sectional area A_1 at height $z = z_1$, with

[5] This is the second mode of instability for the domain $[-z_0, z_0]$, but the first mode if the domain is instead taken to be $[0, z_0]$.

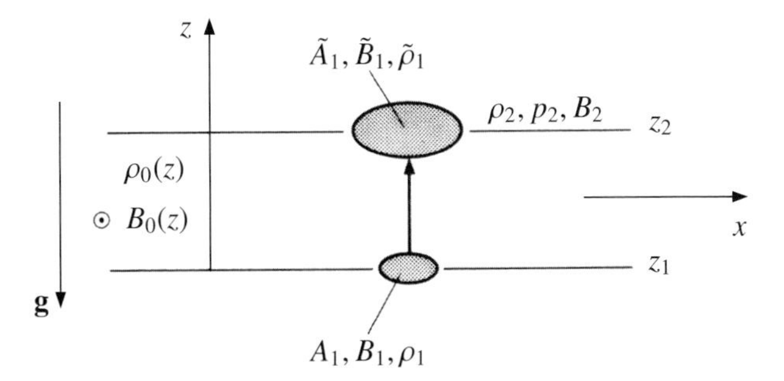

Figure 12.16 Sketch illustrating the simplest instability due to magnetic buoyancy. The shaded flux tube is displaced from the level $z = z_1$ to the level $z = z_2$; it will continue to rise if its new density $\tilde{\rho}_1$ is less than the ambient density ρ_2.

$B_0(z_1) = B_1$, $\rho(z_1) = \rho_1$, is displaced upwards to the level $z = z_2$, and that ohmic diffusion may be neglected. Then A_1, B_1 and ρ_1 will change to, say, $\tilde{A}_1, \tilde{B}_1, \tilde{\rho}_1$, and conservation of mass and of magnetic flux implies that

$$\rho_1 A_1 = \tilde{\rho}_1 \tilde{A}_1, \quad B_1 A_1 = \tilde{B}_1 \tilde{A}_1\,, \tag{12.144}$$

and in particular

$$\tilde{B}_1/\tilde{\rho}_1 = B_1/\rho_1\,. \tag{12.145}$$

During the displacement of the tube, it will tend to expand (if $\mathrm{d}\rho_1/\mathrm{d}z < 0$) in such a way as to maintain pressure equilibrium with its new environment, and if the displacement is sufficiently slow, we may suppose that this equilibrium is maintained, i.e.

$$\tilde{p}_1 + \frac{1}{2\mu_0}\tilde{B}_1^2 = p_2 + \frac{1}{2\mu_0}B_2^2\,, \tag{12.146}$$

where

$$\tilde{p}_1 = f(\tilde{\rho}_1), \quad p_2 = p(z_2), \quad B_2 = B_0(z_2)\,.$$

The tube will be in equilibrium in its new environment only if $\tilde{\rho}_1 = \rho_2$: clearly if $\tilde{\rho}_1 < \rho_2$, it will continue to rise, while if $\tilde{\rho}_1 > \rho_2$ it will tend to return to its original level. The neutral stability condition $\tilde{\rho}_1 = \rho_2$ implies $\tilde{p}_1 = p_2$, and so from (12.146), $\tilde{B}_1 = B_2$. Hence, under neutral stability conditions, (12.145) becomes

$$B_1/\rho_1 = B_2/\rho_2\,, \tag{12.147}$$

or since this must hold for every pair of levels (z_1, z_2), neutral stability requires that $\mathrm{d}(B_0/\rho_0)/\mathrm{d}z = 0$, and clearly the atmosphere is stable or unstable to the type of perturbations considered according as

$$\frac{\mathrm{d}}{\mathrm{d}z}\left(\frac{B_0}{\rho_0}\right) > \text{ or } < 0\,. \tag{12.148}$$

Defining scale-heights

$$L_B = -B_0/(\mathrm{d}B_0/\mathrm{d}z)\,, \quad L_\rho = -\rho_0/(\mathrm{d}\rho_0/\mathrm{d}z)\,, \tag{12.149}$$

the condition for instability may equally be written

$$L_B < L_\rho\,. \tag{12.150}$$

The above simple argument does not take any account of possible distortion of the field lines, and merely shows that the medium is prone to instabilities when the magnetic field strength decreases sufficiently rapidly with height. Distortion of the field lines is however of crucial importance in the problem of sunspot formation (§5.3) and it is essential to consider perturbations which *do* distort the field lines, and which are affected by Coriolis forces when account is taken of rotation. Different instability models have been studied by Gilman (1970) and by Parker (1971d). Parker studied instability modes for which the pressure perturbation δp may be neglected, i.e. modes whose time-scales are *short* compared with the time-scale associated with the passage of acoustic waves through the system. Gilman by contrast supposed that the thermal conductivity (due to radiative or other effects) is very large, and restricted attention to slow instability modes in which pressure and density remain in isothermal balance ($\delta p = c^2 \delta\rho$) and the perturbation in total pressure (fluid plus magnetic) is negligible (as in the qualitative discussion above). We shall here follow the analysis of Gilman, including the effects of rotation, but making the additional simplifying assumption that the instability growth rates are sufficiently small for the magnetostrophic approximation (neglect of $\mathrm{D}\mathbf{u}/\mathrm{D}t$ in the equation of motion) to be legitimate. This greatly simplifies the dispersion relationship, and allows some simple conclusions to be drawn.

12.11.1 The Gilman model

The equations describing small magnetostrophic perturbations about the equilibrium state of Figure 12.16 are

$$2\rho_0 \boldsymbol{\Omega} \wedge \mathbf{u} = -\nabla\chi + \mu_0^{-1}\mathbf{B}_0 \cdot \nabla\mathbf{b} + \mu_0^{-1}\mathbf{b} \cdot \nabla\mathbf{B}_0 - \rho g \mathbf{i}_z\,, \tag{12.151a}$$

$$\partial\mathbf{b}/\partial t = -\mathbf{u} \cdot \nabla\mathbf{B}_0 + \mathbf{B}_0 \cdot \nabla\mathbf{u} - \mathbf{B}_0(\nabla \cdot \mathbf{u})\,, \tag{12.151b}$$

$$\partial\rho/\partial t = -\mathbf{u} \cdot \nabla\rho_0 - \rho_0 \nabla \cdot \mathbf{u}\,, \tag{12.151c}$$

$$\nabla \cdot \mathbf{b} = 0, \quad \chi = p + \mu_0^{-1}\mathbf{B}_0 \cdot \mathbf{b}\,, \tag{12.151d}$$

and, following Gilman (1970), we assume that the pressure and density perturbations p and ρ are related by

$$p = c^2 \rho\,, \tag{12.152}$$

where c is the isothermal speed of sound. This is the situation if the effective diffusivity of heat (due to radiative transfer) is very large. In (12.151), viscous and ohmic diffusion effects are neglected, and also $\partial \mathbf{u}/\partial t$ is dropped,[6] so that we focus attention on modes with growth rate small compared with Ω. Note also that the centrifugal contribution to χ is neglected, on the reasonable assumption (in the solar context) that gravitational acceleration greatly exceeds centrifugal acceleration. If Ox, Oy, Oz are interpreted as south, east and vertically upwards with origin O in the convective zone at co-latitude θ, then the components of $\boldsymbol{\Omega}$ are given by

$$\boldsymbol{\Omega} = (-\Omega \sin\theta, 0, \Omega\cos\theta)\,. \tag{12.153}$$

Equations (12.151) and (12.152) admit solutions proportional to $e^{i(lx+my-\omega t)}$. The analysis is greatly simplified in focusing attention on disturbances for which $|l| \gg |m|$; such disturbances are particularly relevant in the sunspot context – the longitudinal separation of the two members of a typical sunspot pair gives a measure of the spatial variation in the y-direction; variation in the x-direction is unconstrained by the magnetic field $\mathbf{B}_0$, and can have a length-scale small compared with the longitudinal separation scale. If, formally, we let $l \to \infty$ in the Fourier transform of (12.151), the x-component of the equation of motion degenerates to $\hat{\chi} = 0$. Consequently, in modes for which $|l| \gg |m|$, we have equivalently

$$\chi = p + \mu_0^{-1}\mathbf{B}_0 \cdot \mathbf{b} \approx 0\,. \tag{12.154}$$

With $\mathbf{u} = (u, v, w)$, $\mathbf{b} = (b_x, b_y, b_z)$, we may then easily obtain from (12.151)-(12.154) three linear equations relating v, w and b_y, viz.

$$\begin{pmatrix} 0 & \omega q - mh_0^2 K_B & -\mathrm{i}m^2h_0^2c^2 \\ \omega q & -\mathrm{i}m^2h_0^2 & gh_0^2 \\ -\mathrm{i}m & K_\rho - K_B & -\mathrm{i}(c^2-h_0^2) \end{pmatrix} \begin{pmatrix} v \\ w \\ \omega b_y/B_0c^2 \end{pmatrix} = 0\,, \tag{12.155}$$

where[7] $q = 2\Omega\sin\theta$, $h_0 = (\mu_0\rho_0)^{-1/2}B_0$ is the local Alfvén speed, and

$$K_\rho = -\rho_0'/\rho_0 = L_\rho^{-1}, \quad K_B = -B_0'/B_0 = L_B^{-1}\,. \tag{12.156}$$

The vanishing of the determinant of coefficients in (12.155) gives a dispersion relationship in the form

$$\omega^2q^2(h_0^2+c^2) - \omega qmh_0^2(g + h_0^2K_B + c^2K_\rho) + m^2h_0^4(gK_B - m^2c^2) = 0\,. \tag{12.157}$$

If the roots of this equation are complex, then we have instability; the condition for

[6] In this respect we depart from the treatment of Gilman (1970) who obtained a more general dispersion relationship than (12.157) describing both fast and slow modes of instability. When $\Omega = 0$, Gilman found modes of instability whenever $|B_0(z)|$ decreases with height. Such modes are not helical in character, although there is little doubt that they become so when perturbed by rotation effects.

[7] Note that only the component of $\boldsymbol{\Omega}$ perpendicular to $\mathbf{g}$ influences the perturbations when $|l| \gg |m|$.

this is evidently

$$(g + h_0^2 K_B + c^2 K_\rho)^2 < 4(h_0^2 + c^2)(g K_B - m^2 c^2) . \tag{12.158}$$

Since the equilibrium condition (12.143) may (in the case of an isothermal atmosphere with c^2 = cst.) be written

$$g = h_0^2 K_B + c^2 K_\rho , \tag{12.159}$$

the conclusion from (12.158) is that the medium is unstable to perturbations whose wave number in the y (or azimuth) direction satisfies

$$m < m_c = \left(g(K_B - K_\rho)(h_0^2 + c^2)^{-1}\right)^{1/2} . \tag{12.160}$$

As anticipated in the introductory comments, a necessary condition for this type of instability is $K_B > K_\rho$ (or equivalently $L_B < L_\rho$); when this condition is satisfied, (12.160) gives the minimum scale $2\pi/m_c$ on which perturbations will grow. It is interesting that this scale does not depend on the rotation parameter q (although the growth rates given by (12.157) are proportional to q^{-1}).

With $\omega = \omega_\mathrm{r} + \mathrm{i}\omega_\mathrm{i}$, when $m < m_c$, the unstable mode ($\omega_\mathrm{i} > 0$) is given from (12.157)–(12.160) by

$$\omega_\mathrm{r}\, q = gmh_0^2 \left(h_0^2 + c^2\right)^{-1}, \quad \omega_\mathrm{i}\, q = h_0^2 \left(h_0^2 + c^2\right)^{-1/2} |m| c \left(m_c^2 - m^2\right)^{1/2} . \tag{12.161}$$

Since $\omega_\mathrm{r}\, m > 0$, these instability waves propagate in the positive y-direction, i.e. towards the east.

The helicity distribution $\langle \mathbf{u} \cdot \nabla \wedge \mathbf{u} \rangle$ associated with this type of instability is given (for large ℓ) by

$$\mathcal{H} \sim \tfrac{1}{2} \Re\mathrm{e}[\,\mathrm{i}\,\ell\,(vw^* - v^* w)] , \tag{12.162}$$

the ratio of v to w being given by (12.155). A straightforward calculation using (12.161) gives

$$v/w = -\left(h_0^2 + c^2\right)^{1/2} \left(m_c^2 - m^2\right)^{1/2} / c\,|m| , \tag{12.163}$$

and since this is real, the leading order contribution (12.162) is in fact zero. Terms of order m/l would have to be retained in the analysis throughout in order to obtain the correct expression for $\mathcal{H}$. We can however directly derive an α-effect associated with the type of instability considered. Writing

$$(\mathbf{u} \wedge \mathbf{b})_y = \tfrac{1}{2} \Re\mathrm{e} \left(u b_z^* - w b_x^*\right) = -(m/2l) \Re\mathrm{e} \left(v b_z^* - w b_y^*\right) = \alpha B_0 , \tag{12.164}$$

the coefficient α may be obtained in the form

$$\alpha = - \frac{m|m| \left(m_c^2 - m^2\right)^{1/2} h_0^2 \left[(h_0^2 - c^2) K_B + c^2 K_\rho\right] |w|^2}{2lqc \left(h_0^2 + c^2\right)^{1/2}} . \tag{12.165}$$

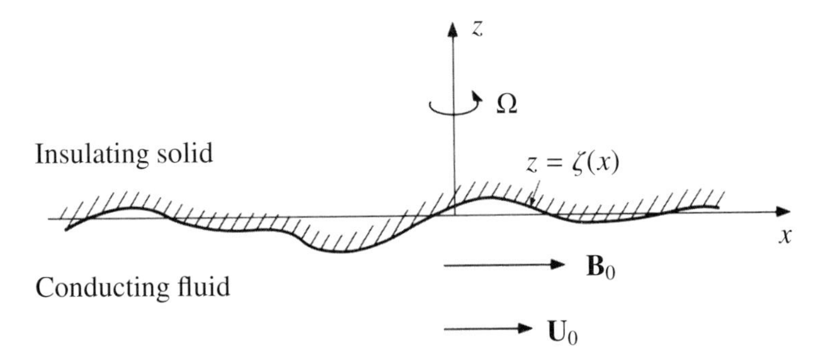

Figure 12.17 Sketch of the configuration considered in §12.12. This provides a crude model for the generation of magnetic fluctuations near the Earth's core-mantle boundary.

The fact that this is in general non-zero is of course an indication of the lack of reflexional symmetry of the motion.

12.12 Helicity generation due to flow over a bumpy surface

As a final example of a mechanism whereby helicity (and an associated α-effect) may be generated, consider the problem depicted in Figure 12.17: an insulating solid is separated from a conducting fluid by the 'bumpy' boundary $z = \zeta(x)$, and as in §12.10 we suppose that $\mathbf{\Omega} = (0,\ 0,\ \Omega)$, $\mathbf{B}_0 = (B_0, 0, 0)$. Moreover we suppose that the fluid flows over the bumps, the velocity tending to the uniform value $(U_0,\ 0, 0)$ as $z \to -\infty$. This may be regarded as a crude model of flow near the core–mantle boundary in the terrestrial context, crude because (i) the spherical geometry is replaced by a Cartesian 'equivalent', (ii) the tangential magnetic field in the terrestrial context falls to a near zero value at the core–mantle boundary because of the low conductivity of the mantle, whereas here we suppose the field to be uniform, and (iii) the bumps are regarded as two-dimensional whereas in reality they are surely three-dimensional. The model in this crude form has been considered from different points of view by Anufriev & Braginskii (1975) and by Moffatt & Dillon (1976) . These studies are relevant (i) to the tangential stress (or equivalently angular momentum) transmitted from core to mantle and (ii) to an apparent correlation between fluctuations in the Earth's surface gravitational and magnetic fields when suitably displaced in longitude relative to each other (Hide 1970; Hide & Malin 1970). Here we focus attention simply on the structure of the motion induced in the liquid and the associated helicity distribution, which is the most relevant aspect as far as large-scale dynamo effects are concerned.

If effects associated with variable density are neglected, and if we suppose that $|\zeta'(x)| \ll 1$ so that all perturbations are small, then the governing equations for the steady perturbations generated are (12.24) and (12.25) (with $\partial \mathbf{u}_1/\partial t = \partial \mathbf{h}_1/\partial t = 0$) together with $\nabla \cdot \mathbf{u}_1 = \nabla \cdot \mathbf{h}_1 = 0$. We shall also adopt the magnetostrophic approximation, which in this situation involves neglect of the convective acceleration term $\mathbf{u}_0 \cdot \nabla \mathbf{u}_1$ and of the viscous term $\nu \nabla^2 \mathbf{u}_1$ in (12.24). (The viscous term leads to a thin Ekman layer on the surface, and an associated small perturbation of the effective boundary condition (12.172a) below.) These equations admit solutions of the form

$$(\mathbf{u}_1, \mathbf{h}_1, p_1) = 2\,\Re\mathrm{e} \int_0^\infty \left(\hat{\mathbf{u}}, \hat{\mathbf{h}}, \hat{p}\right) \mathrm{e}^{\mathrm{i}\mathbf{m}\cdot\mathbf{x}} \mathrm{d}k \,, \tag{12.166}$$

where $\mathbf{m} = k(1, 0, \gamma)$ and possible values of γ are to be determined; these must satisfy $\Im\mathrm{m}\,\gamma < 0$ since the perturbations must vanish as $z \to -\infty$. Substitution in the equations and straightforward elimination of the amplitudes $\hat{\mathbf{u}}$, $\hat{\mathbf{h}}$, $\hat{p}$ leads to a cubic equation for γ^2:

$$\left(1 + \gamma^2\right) + \gamma^2 \left[Q\left(1 + \gamma^2\right) + 2\mathrm{i}A\kappa^{-1}\right]^2 = 0 \,, \tag{12.167}$$

where

$$Q = 2\Omega\eta/h_0^2, \quad A = U_0/h_0, \quad \kappa = h_0 k/\Omega \,. \tag{12.168}$$

Moreover, if the three solutions of (12.167) satisfying $\Im\mathrm{m}\,\gamma < 0$ are denoted by γ_n $(n = 1, 2, 3)$, then the ratios of the velocity and magnetic components are given in the corresponding modes by

$$\hat{\mathbf{u}}_n = a_n(k)\left(1, -\sigma_n^{-1}, -\gamma_n^{-1}\right), \quad \hat{\mathbf{h}}_n = (\mathrm{i}\sigma_n/\kappa)\,\hat{\mathbf{u}}_n \,, \tag{12.169}$$

where

$$\sigma_n^2 = -4\gamma_n^2\left(1 + \gamma_n^2\right)^{-1} . \tag{12.170}$$

In the insulating region, the field $\mathbf{B}$ is harmonic, i.e. $\mathbf{B} - \mathbf{B}_0 = (\mu_0 \rho_{\mathrm{m}})^{1/2} \nabla\Psi$ say, where ρ_{m} is the density of the solid ('mantle') region, and Ψ, being a potential function, has Fourier transform given by

$$\Psi(x, y) = 2\,\Re\mathrm{e} \int_0^\infty \hat{\psi}_0(k)\,\mathrm{e}^{-kz}\,\mathrm{e}^{\mathrm{i}kx}\,\mathrm{d}k \,. \tag{12.171}$$

The amplitudes $a_n(k)$, $\hat{\psi}_0(k)$ may be obtained in terms of the Fourier transform $\hat{\zeta}(k)$ of $\zeta(x)$ by applying the linearised boundary conditions

$$u_z = u_0 \partial\zeta/\partial x, \quad \mathbf{h} = -(\rho_{\mathrm{m}}/\rho)^{1/2} \nabla\Psi, \quad \text{on } z = 0 \,. \tag{12.172}$$

The first of these expresses the fact that the normal velocity on $z = \zeta(x)$ is zero; the second expresses continuity of both normal and tangential components of $\mathbf{B}$.

The nature of the cubic equation (12.167) makes the detailed subsequent analysis rather complicated. In the terrestrial context however both Q and A are small, and asymptotic methods may be adopted. If $Q \ll 1$ and $A\kappa^{-1} \ll Q$, the three relevant roots of (12.167) are, to leading order,

$$\gamma_1 \sim -\mathrm{i}, \quad \gamma_2 \sim -\tfrac{1}{2}(1+\mathrm{i})Q^{-1/2}, \quad \gamma_3 \sim \tfrac{1}{2}(1-\mathrm{i})Q^{-1/2}\,. \tag{12.173}$$

This means that the mode corresponding to $n = 1$ has spatial dependence $\mathrm{e}^{kz+\mathrm{i}kx}$; the corresponding velocity and magnetic perturbations are irrotational (to leading order) and penetrate a distance $\mathcal{O}(k^{-1})$ into the fluid. The helicity associated with this mode is zero. By contrast the modes corresponding to $n = 2$ and 3 have a boundary-layer character, penetrating a distance $\delta_B = \mathcal{O}\left(k^{-1}Q^{1/2}\right)$ into the fluid. Moreover these modes are strongly helical: the helicity $\mathcal{H}_n$ ($n = 2$ or 3) associated with either mode is given (Moffatt & Dillon 1976) by

$$\mathcal{H}_n(k) = \mathcal{R}e(\hat{\mathbf{u}}_n^* \cdot \mathrm{i}\mathbf{k}\wedge\hat{\mathbf{u}}_n) = 2k\left|\hat{\mathbf{u}}_n^2\right|\,\mathcal{R}e(\mathrm{i}\gamma_n/\sigma_n) = \tfrac{1}{2}k\,|\hat{\mathbf{u}}_n|^2\,Q^{-1/2}\,. \tag{12.174}$$

Since $k > 0$ in the representation (12.166) adopted, this helicity is positive for both modes, a fact that could be anticipated from the arguments of §12.2: there is clearly a *downward* flux of energy from the boundary into the fluid and the associated helicity is therefore positive. The apparent contradiction between (12.174) and the result $|\mathcal{H}(k)| \leq 2kE(k)$ bounding the helicity spectrum (§7.6) is accounted for by the fact that here we are dealing with a strongly anisotropic situation with severe attenuation of modes in the z–direction, and the argument leading to (7.54) simply does not apply. The amplitude $|\hat{\mathbf{u}}_n|^2$ in (12.174) is proportional to $|\hat{\zeta}|^2$ and also decays as $\exp\left[kz/Q^{1/2}\right]$ as $z \to -\infty$.

A recent analysis of this problem assuming a stably stratified layer below the core–mantle boundary that tends to suppress radial motion has been provided by Glane & Buffett (2018); on reasonable assumptions concerning velocity and stratification in this region, they have obtained a tangential stress distribution of sufficient magnitude to explain observed changes in the length-of-day (LOD).

In this context, Andrews & Hide (1975) have demonstrated that, in the non-dissipative limit ($\nu = \eta = \kappa = 0$), free wave motions subject to the influence of buoyancy, Coriolis and Lorentz forces can in a similar way be trapped against a solid plane boundary provided the boundary is inclined to the horizontal and provided the basic field $\mathbf{B}_0$ is parallel to the wall; the trapped modes cease to exist if $\mathbf{B}_0$ has a component perpendicular to the wall. In the study described above, a component of $\mathbf{B}_0$ perpendicular to the wall may likewise be expected to exert a strong influence on the structure of the three modes.

13

The Geodynamo: Instabilities and Bifurcations

13.1 Models for convection in the core of the Earth

Dynamo action, as we have seen, converts kinetic energy to magnetic energy. In the Earth's liquid core, this kinetic energy is most likely maintained by convective motions: light fluid elements rise through the core, converting gravitational potential energy to kinetic energy.

We have already introduced the Boussinesq approximation in §12.5 ; this is probably the simplest convection model for the Earth's core. In the spherical geometry, the temperature field has the form $T(\mathbf{x}, t) = T_s(r) + \theta(\mathbf{x}, t)$, where T_s is the steady diffusive temperature profile, and $\theta(\mathbf{x}, t)$ is the perturbation induced by the flow. With dimensionless variables based on a length-scale $L = r_o$ (the outer radius of the core) and a viscous time-scale L^2/ν, the dimensionless momentum equation (12.121) becomes

$$\frac{\mathrm{D}\mathbf{u}}{\mathrm{D}t} + \mathcal{E}^{-1}\mathbf{e}_z \times \mathbf{u} = -\nabla p + \mathcal{R}a\,\theta\,\mathbf{r} + \nabla^2\mathbf{u}\,, \quad \nabla\cdot\mathbf{u} = 0\,, \tag{13.1}$$

where $\mathbf{r} = r\,\mathbf{e}_r$, and equation (12.122) for the temperature perturbation becomes

$$\mathcal{P}r\,\frac{\mathrm{D}\theta}{\mathrm{D}t} = -\mathbf{u}\cdot\nabla T_s + \nabla^2\theta\,. \tag{13.2}$$

Here the *Ekman, Rayleigh* and *Prandtl numbers* are defined by

$$\mathcal{E} = \nu/2\Omega L^2\,, \quad \mathcal{R}a = g\alpha\beta L^5/\nu\kappa\,, \quad \mathcal{P}r = \nu/\kappa\,, \tag{13.3}$$

where α is the coefficient of thermal expansion, $\beta = -T_s'$, and g is the gravitational acceleration at $r = r_o$ (recall that for a self-gravitating body with nearly constant density, the internal gravity increases linearly with r – see Chapter 4).

Rigorous derivation of this Boussinesq model requires a double limiting process (Malkus 1964; Dormy & Soward, 2007): first, the limit of motions that are slow compared to the speed of sound, and second, the limit of a layer that is thin compared to the density scale height (i.e. the scale of density variations when the fluid is stably stratified under its own weight). The first of these limits by itself yields the so-called anelastic approximation, for which the key step involves dropping the $\partial\rho/\partial t$-term from the equation (2.135) of mass conservation to filter out sound waves (Braginsky & Roberts, 1995). This 'slow-manifold' approach is close in spirit to dropping the displacement current in Maxwell's equations to filter out electromagnetic waves.

The steady diffusive temperature profile $T_s(r)$, corresponding to the reference state of a fluid at rest, will depend on the heating mode that is considered. The two most natural modes are (i) 'internal heating', with a uniform heat source distribution motivated by the full-sphere configuration, and (ii) 'differential heating' in which a fixed temperature difference is maintained between the two concentric spheres. The former gives a temperature profile

$$T_s = -\tfrac{1}{2}\beta\, r^2 + T_0\,, \quad \nabla T_s = -\beta\, r\, \mathbf{e}_r\,, \tag{13.4}$$

whereas the latter gives

$$T_s = T_1/r + T_2\,, \quad \nabla T_s = -(T_1/r^2)\, \mathbf{e}_r\,. \tag{13.5}$$

Here, T_0, T_1 and T_2 are constants determined by the boundary temperatures.

We concentrate first on the problem of Boussinesq convection, which has been shown to provide a good model for dynamo action; the more complex anelastic formulation yields only small differences (Glatzmaier & Roberts, 1996; Raynaud et al., 2014). As we shall see, even this simple formulation leads to significant complications when applied to the geodynamo problem.

While equations (13.1–13.2) can govern either thermal convection or compositional convection (driven by an admixture of lighter elements in the core), the Prandtl number $\mathcal{P}r$ is quite different in the two situations. For thermal convection $\mathcal{P}r \sim \mathcal{O}(10^{-1})$, while for compositional convection[1] $\mathcal{P}r \sim \mathcal{O}(10^2)$ (Pozzo et al., 2012, 2014). The phenomenon of 'over-stability' (e.g. Fearn et al. 1988) can occur when both effects are considered together in 'thermo-compositional convection'.

13.2 Onset of thermal convection in a rotating spherical shell

Convection in a rotating fluid is constrained by rotation $\mathbf{\Omega}$, which tends to make the flow two-dimensional; this is a consequence of the Taylor–Proudman theorem (Greenspan 1968), which shows that $(\mathbf{\Omega}\cdot\nabla)\,\mathbf{u} = 0$ if body forces and inertia are negligible. A stronger temperature gradient is therefore needed in a rotating fluid, other things being equal, in order to drive convective motions. This has been known since the work of Chandrasekhar (1961) on the effect of rotation on convection

[1] The Prandtl number in compositional convection is frequently called the Schmidt number.

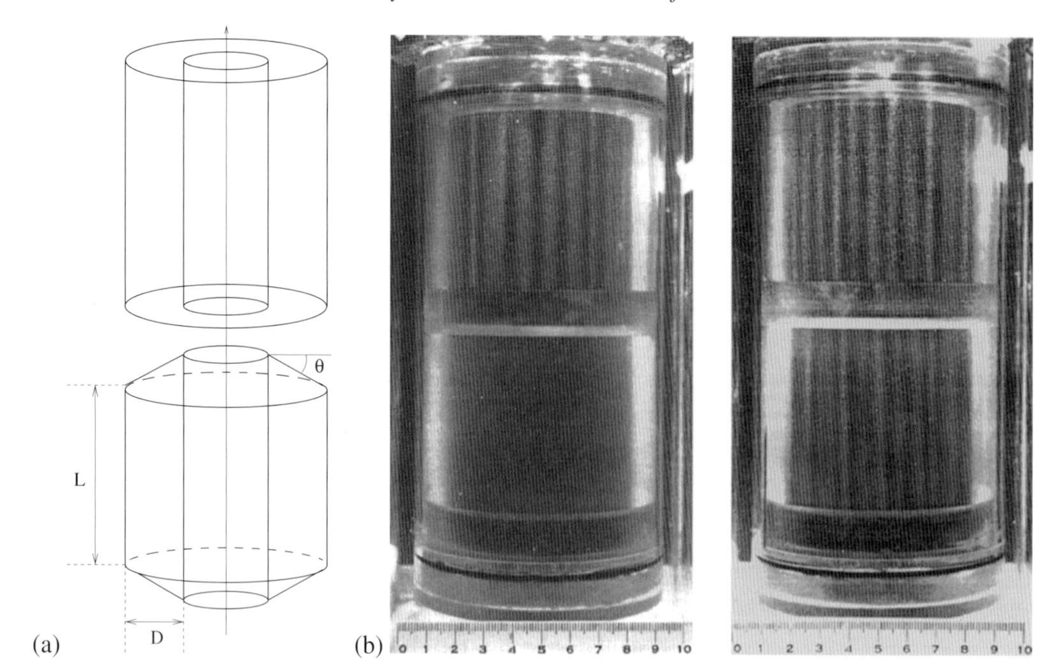

Figure 13.1 (a) Sketch of the cylindrical model of Busse (1970) providing an approximate description of the convective instability in a rapidly rotating sphere. (b) Experiment near the onset of convection between rotating cylinders (Busse & Carrigan 1974; Busse et al. 1998). Small flaky particles align themselves with the shear and make convective rolls visible. The outer cylinder is immersed in a bath of warm water, and the inner cylinder is cooled by water flowing through the axial region; centrifugal forces mimic the effect of radial gravity. The upper annular region has parallel top and bottom boundaries, whereas the lower region has conical boundaries.

between two horizontal planes. The problem of convection in a rotating spherical shell however turned out to be surprisingly difficult. The first solution in the limit of small Ekman number $\mathcal{E}$ and with internal heating was found by Roberts (1968) using the WKB approximation.

Busse (1970) subsequently identified the correct symmetry of the most unstable mode about the equatorial plane. He studied the onset of convection in the space between two finite cylinders, with a radial temperature gradient, and with ends that were inclined at an angle θ (Figure 13.1a) in order to prevent purely geostrophic convection and to mimic the stabilising effect of rotation in a sphere. Assuming $\theta \ll 1$ and $D/L \ll 1$ (the 'small gap approximation'), averaging over z led to a second-order ordinary differential equation for the pressure, and hence to determination of the critical Rayleigh number $\mathcal{R}a_c$ as a function of θ. Busse then applied these results to convection in a sphere, with uniform internal heating that gave the

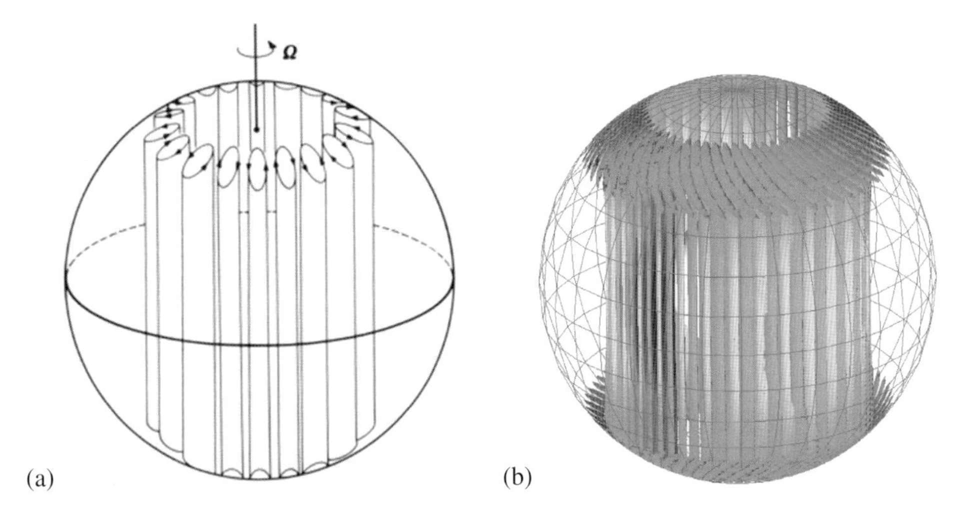

Figure 13.2 (a) Original sketch of the 'cartridge belt' geometry of the marginally unstable convective motions in an internally heated rotating sphere [from Busse 1970]. (b) Rendering of isosurfaces of the axial vorticity at the onset of convection for $\mathcal{E} = 10^{-6}$ [from Dormy 1997].

unstable temperature gradient $\beta(r) = -T_s'(r)$ proportional to r (13.4) ; the spherical geometry also defined the profile of θ as a function of distance s from the axis. The mode at the onset of convection was obtained in the form of a 'cartridge belt' of rolls (Figure 13.2a) varying slowly in the z-direction. The radial location of this cartridge belt was found by minimising $\mathcal{R}a_c$ as a function of θ, leading to $\mathcal{R}a_c \sim \theta^{4/3}$: increasing the slope θ makes it harder to drive convection. Convection starts at that finite distance from the axis where the unstable temperature gradient $\beta(s)$ is sufficient to overcome the damping effects of viscosity and thermal diffusion.

The stabilising property of rotation in conjunction with boundary conditions was demonstrated experimentally by Busse & Carrigan (1974) and Busse et al. (1998) (Figure 13.1b). Thermal convection is driven in the annular cavities between two rotating cylinders; the upper pair have horizontal boundaries at top and bottom, while the lower pair have conical boundaries. On the left, convection is suppressed in the lower cavity at a stage when convection is already well established in the upper part. On the right, at higher Rayleigh number, convection is established in both cavities.

For the most unstable mode $\sim e^{i(m\varphi-\omega t)}$ in the cylindrical geometry, Busse found the critical values

$$m_c = \left(\frac{\mathcal{P}r\,\sqrt{5/2}}{4\,\mathcal{E}\,(1+\mathcal{P}r)}\right)^{1/3}, \quad \omega_c = -(2\,\mathcal{E})^{-1}\left(\frac{10\,\mathcal{E}}{\sqrt{8}\,\mathcal{P}r\,(1+\mathcal{P}r)^2}\right)^{1/3}, \tag{13.6}$$

and the critical Rayleigh number at the onset of convection (when $\Im m\, \omega = 0$),

$$\mathcal{R}a_c = 3\left(\frac{5}{4}\right)^{5/3}\left(\frac{\mathcal{P}r}{\mathcal{E}\,(1+\mathcal{P}r)}\right)^{4/3}. \tag{13.7}$$

The non-zero frequency ω_c means that this is a wave (described as a 'quasi-geostrophic thermal Rossby wave') travelling in the φ-direction. It varies slowly in the axial direction and has small azimuthal length-scale $\mathcal{O}(\mathcal{E}^{1/3} r_o)$. Experiments in various spherical configurations confirm the elongated structure of such columns, and their drifting character (see Figure 13.3).

13.2.1 The Roberts–Busse localised asymptotic theory for small $\mathcal{E}$

The basis of this theory is the assumption that the onset of convection is in the neighbourhood of a critical cylinder $s = s_c$. Equations (13.1) and (13.2) can be linearised near this onset, giving

$$\frac{\partial \mathbf{u}}{\partial t} + \mathcal{E}^{-1}\mathbf{e}_z \times \mathbf{u} = -\nabla p + \mathcal{R}a\,\theta\,\mathbf{r} + \nabla^2\mathbf{u}, \quad \nabla\cdot\mathbf{u} = 0, \tag{13.8}$$

and

$$\mathcal{P}r\,\frac{\partial\theta}{\partial t} = \mathcal{Q}\,\mathbf{u}\cdot\mathbf{r} + \nabla^2\theta, \tag{13.9}$$

where $\mathcal{Q}(r) = 1$ for internal heating, and $\mathcal{Q}(r) = 1/r^3$ for differential (boundary) heating.

The theory adopts scalings

$$\mathcal{R}a \sim \mathcal{E}^{-4/3}, \qquad \omega \sim \mathcal{E}^{-2/3}, \qquad m \sim \mathcal{E}^{-1/3}, \tag{13.10}$$

and considers elongated structures of small horizontal scale, such that

$$\frac{1}{s}\frac{\partial}{\partial\phi} \sim \frac{\partial}{\partial s} \sim \mathcal{O}\big(\mathcal{E}^{-1/3}\big), \quad \frac{\partial}{\partial z} \sim \mathcal{O}(1), \quad \text{as } \mathcal{E} \longrightarrow 0. \tag{13.11}$$

With velocity

$$\mathbf{u} = \nabla \wedge (\psi\,\mathbf{e}_z) + w\,\mathbf{e}_z, \quad \text{and with } (\psi, w) \propto \exp[i(ks + m\varphi - \omega t)], \tag{13.12}$$

the horizontal and vertical components of (13.8) reduce at leading order to

$$\mathcal{E}\left(a^2 - i\omega\right)\left(a^2\,\psi\right) - \mathrm{d}w/\mathrm{d}z + i\,m\,\mathcal{E}\,\mathcal{R}a\,\theta = 0, \tag{13.13}$$

and

$$\mathcal{E}\left(a^2 - i\omega\right)w - \mathrm{d}\psi/\mathrm{d}z = z\,\mathcal{E}\,\mathcal{R}a\,\theta, \tag{13.14}$$

where $a^2 = k^2 + m^2/s^2$, and (13.9) gives

$$(a^2 - \mathrm{i}\,\mathcal{P}r\,\omega)\theta = \mathcal{Q}(\mathrm{i}\,m\,\psi + z\,w). \tag{13.15}$$

These equations can be combined to give a second-order ordinary differential equation for $w(z)$ governing the axial structure of the convection rolls. This is solved numerically with the condition that $\mathbf{u} \cdot \mathbf{n} = 0$ on the boundary (or equivalently that the temperature perturbation $\theta = 0$ there), leading to determination of the 'local' critical parameters.

13.2.2 The Soward–Jones global theory for the onset of spherical convection

An inconsistency in the above theory was noted by Yano (1992), in that it cannot be embedded in a WKB solution decaying exponentially in the $\pm s$-directions on both sides of the critical cylinder $s = s_c$. This is because the local approach of the Roberts–Busse theory yields a solution that does not drift as a whole, but is instead subject to 'phase-mixing' associated with the dependence of the obtained frequency ω on s. This causes a 'shearing' of the modes that decreases their length-scale in the radial direction (reminiscent of the effect of differential rotation on a transverse magnetic field, see §3.11.6).

Numerical solutions for the onset of convection (Dormy, 1997) were found to have a higher critical Rayleigh number than expected from the Roberts–Busse theory. A similar phase-mixing problem occurs for the stability of flow in the spherical shell geometry. Soward & Jones (1983) introduced a new mathematical approach, as described below, to resolve this problem, and using this approach, Jones et al. (2000) provided a 'global' asymptotic theory for the onset of convection for the full sphere problem, which determined values of the critical Rayleigh number, frequency and azimuthal wave number in agreement with the numerical results.

This theory is to be contrasted with the local Roberts–Busse theory. A WKB-asymptotic expansion of the convective mode in terms of a complex radial wave number $k(s)$ determines the frequency $\omega(s)$ of the convection mode through a dispersion relation $\omega(s, k) = \text{cst}$. According to the Roberts–Busse theory, convection sets in at the location $s = s_L$ at which $k(s_L) = 0$. It is near this point that phase mixing caused by the gradient of ω shortens the radial length-scale leading to breakdown of the WKB-approximation. Jones et al. (2000) imposed the condition that, for a point $s = s_c$ and wave number $k = k_c$, both the complex group velocity and the complex phase mixing should vanish:

$$\left(\frac{\partial \omega}{\partial k}\right)_c = 0 \quad \text{and} \quad \left(\frac{\partial \omega}{\partial s}\right)_c = 0\,. \tag{13.16}$$

They showed that, although the location s_c is complex, the wavenumber k_c is real and simply vanishes, as in the case of the local analysis. They therefore considered instead a second-order amplitude equation

$$-\frac{1}{2}\frac{\partial^2 \omega}{\partial k^2}\frac{\mathrm{d}^2 a}{\mathrm{d}s^2} + \left\{\frac{1}{2}\frac{\partial^2 \omega}{\partial s^2}(s - s_c)^2 + \mathcal{R}a_1 \frac{\partial \omega}{\partial \mathcal{R}a} - \left(\omega_1 - \frac{\partial \omega}{\partial m} m_1\right)\right\} a = 0\,. \tag{13.17}$$

Figure 13.3 (a) Side view of convection columns close to the onset of convection in a differentially heated spherical shell with aspect ratio 0.97 (narrow gap); gravity is here mimicked by centrifugal acceleration (Carrigan & Busse, 1983). (b) View from above of a full-sphere experiment close to onset, in which the outer sphere was cooled over time, thus changing the reference temperature profile (Chamberlain & Carrigan, 1986). (c) Experimental study close to onset in a hemispherical shell (aspect ratio 0.79); the rotation speed was adjusted so that the representation of true gravity by the centrifugal acceleration was as close as possible (Cordero & Busse, 1992). (d) and (e) Experiment of developed thermal convection in a rapidly rotating spherical shell with aspect ratio 1/3; the Rayleigh number here was about 50 times critical (Cardin, 1992).

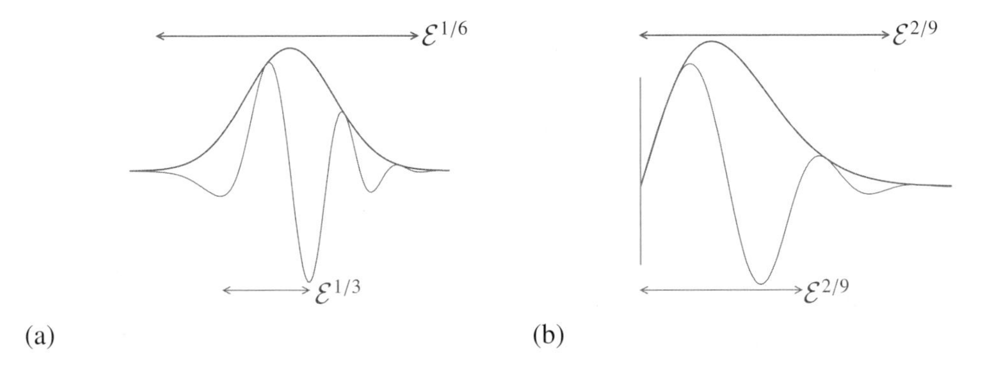

Figure 13.4 Radial structure at the onset of convection (a) for global instability and (b) for the wall-attached mode.

Instead of constructing an amplitude equation of Airy-function type, an equation with two turning points is obtained. Criticality occurs when the turning points are almost coincident at some location $s = s_c$ where phase mixing $\partial\omega/\partial s$ vanishes. This corresponds to the global instability whose structure is represented in Figure 13.4a. Remarkably, this condition is realised for a complex value of the radius s. It is nevertheless physically relevant and corresponds to the mode of instability that is actually realised.

13.2.3 Localised mode of instability in a spherical shell

In the case of a spherical annulus (or 'shell') as in the presence of an inner core, with differential heating from the boundaries, the temperature gradient increases like r^{-2} with decreasing radius, so that the onset of instability always occurs on the 'critical cylinder' tangent to the inner sphere. The unscaled amplitude equation in this case is then of 'complex Airy type'

$$-\frac{1}{2}\frac{\partial^2\omega}{\partial k^2}\frac{\mathrm{d}^2 a}{\mathrm{d}s^2} + \left\{\frac{\partial\omega}{\partial s}(s - s_c) + \mathcal{R}a_1\frac{\partial\omega}{\partial\mathcal{R}a} - \left(\omega_1 - \frac{\partial\omega}{\partial m}m_1\right)\right\} a = 0\,, \qquad (13.18)$$

with a single turning point. This has no localised solution, decaying in both directions away from the cylinder; it is however applicable to 'wall-attached' convection, which just needs to vanish at the inner boundary.

Introducing scaled variables

$$\tilde{\omega} = \omega\,\mathcal{E}^{2/3}\,, \quad \tilde{k} = k\,\mathcal{E}^{1/3}\,, \quad \tilde{m} = m\,\mathcal{E}^{1/3}\,, \quad \widetilde{\mathcal{R}} = \mathcal{R}a\,\mathcal{E}^{4/3}\,, \qquad (13.19)$$

and $x = (s - s_c)\,\mathcal{E}^{-2/9}$ (so that $\mathrm{d}/\mathrm{d}s = \mathcal{E}^{-2/9}\,\mathrm{d}/\mathrm{d}x$), (13.18) becomes

$$-\frac{1}{2}\frac{\partial^2\tilde{\omega}}{\partial\tilde{k}^2}\frac{\mathrm{d}^2 a}{\mathrm{d}x^2}\,\mathcal{E}^{-4/9} + \left\{\frac{\partial\tilde{\omega}}{\partial s}\,x\,\mathcal{E}^{-4/9} + R_1\frac{\partial\tilde{\omega}}{\partial\widetilde{\mathcal{R}}}\,\mathcal{E}^{2/3} - \left(\Omega_1 - \frac{\partial\omega}{\partial m}\,\mathcal{E}^{-1/3}\,M_1\right)\right\} a = 0\,. \qquad (13.20)$$

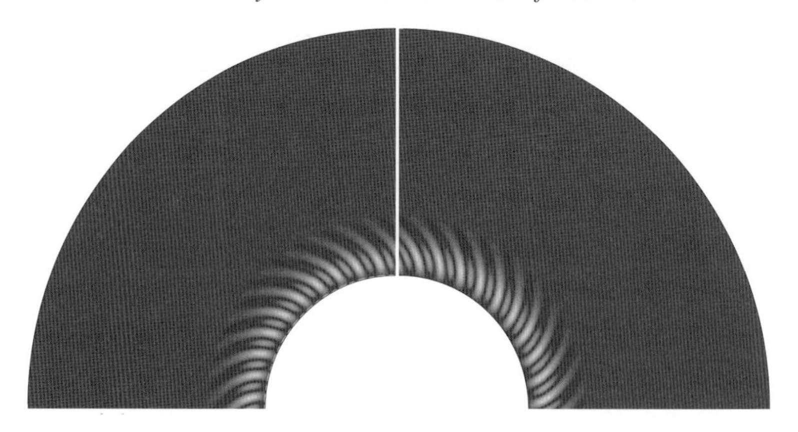

Figure 13.5 Eigenmode structure at the onset of convection with differential heating and an aspect ratio of 0.35; comparison between equatorial cross sections of the axial vorticity $\mathbf{e}_z \cdot \nabla \wedge \mathbf{u}$ in the numerical simulation at $\mathcal{E} = 10^{-7}$ (left), and the asymptotic eigenfunction for the same Ekman number (right). [From Dormy et al. 2004.]

This may be solved by noting that[2] $a(x) = \mathrm{Ai}\,(-\lambda\, x + \chi_0)$ is a solution of

$$\frac{\mathrm{d}^2 a}{\mathrm{d}x^2} + \lambda^3\, x\, a - \lambda^2\, \chi_0\, a = 0\,, \text{ where } \lambda^3 = -\frac{2\,\omega_s}{\omega_{kk}}\,. \tag{13.21}$$

It is however important to select the correct root for λ to ensure that the Airy function decays at infinity. It has to satisfy either

$$\{\Re\mathrm{e}(\lambda) < 0,\ |\mathrm{Arg}(-\lambda)\,| < \pi/3\} \quad \text{or} \quad \{\Re\mathrm{e}(\lambda) < 0,\ |\Im\mathrm{m}(\lambda)| < \sqrt{3}|\Re\mathrm{e}(\lambda)|\}. \tag{13.22}$$

Taking $\chi_0 \simeq -2.338$ to be the first zero of the Airy function, the condition $a(0) = \mathrm{Ai}(\chi_0) = 0$ is satisfied. The radial structure of the convective columns then takes the form of an Airy function for this case of wall-attached convection (see Figure 13.4b). Higher-order terms are obtained by solving

$$\tfrac{1}{2}\left(\lambda^2\, \chi_0\right)\, \omega_{kk} = R_1 \frac{\partial \omega}{\partial \mathcal{R}a}\, \mathcal{E}^{10/9} - \left(\Omega_1 - \frac{\partial \omega}{\partial m}\, \mathcal{E}^{-1/3}\, M_1\right) \mathcal{E}^{4/9}\,. \tag{13.23}$$

Here again, this 'localised' theory agrees well with direct numerical simulations performed with the same differential heating (Dormy et al. 2004; Figure 13.5).

The onset of convection is now therefore well understood, both in the case of a rapidly rotating sphere with internal heating, and in the case of a rapidly rotating spherical shell with differential boundary heating.

[2] The Airy function $\mathrm{Ai}(x)$ is defined by the integral $\mathrm{Ai}(x) = \pi^{-1} \int_0^\infty \cos\left(\frac{1}{3}t^3 + x\,t\right)\, \mathrm{d}t$; (Abramowitz & Stegun 1970, see in particular p. 448, equation (10.4.59)).

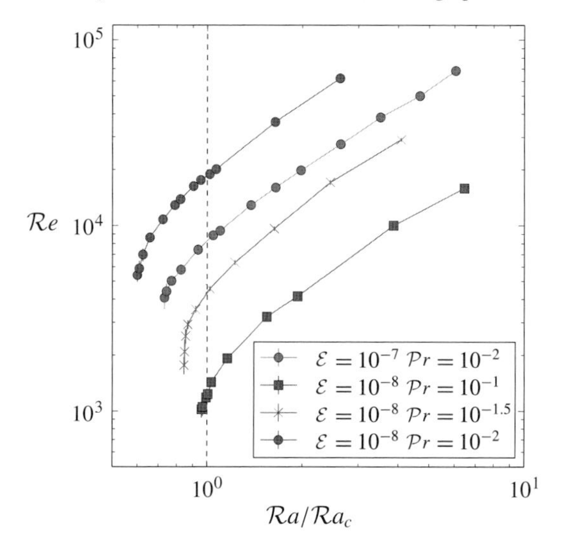

Figure 13.6 Reynolds number as a function of the departure from the onset of convection using a quasi-geostrophic model. [After Guervilly & Cardin 2016.]

13.2.4 Dynamic equilibration

Soward (1977) (see also Proctor 1994) noted that non-linearity can counteract the phase mixing of the local solution, so that convection may be sustained by non-linearity near the local threshold for instability, which is lower than the 'global' threshold for a full sphere with internal heating. This would imply 'subcritical' onset of convection in the rapid rotation limit.

This prediction is very difficult to investigate numerically. Evidence for pulsed solutions (in the form of relaxation oscillations) were reported in a modification of the cylindrical annulus model in which the slope of the boundaries was varied continuously to approach the spherical geometry (Morin & Dormy, 2009).

Using a more elaborate model, in which the heat equation was treated in three dimensions, Guervilly & Cardin (2016) did find subcritical convection, albeit through a different mechanism, and for small Prandtl number in the range $10^{-1}-10^{-2}$ (see Figure 13.6). Kaplan et al. (2017) have confirmed numerically that the onset of Boussinesq convection driven by internal heating in a rotating sphere is indeed subcritical for small enough Ekman and Prandtl numbers.

The question as to whether subcritical convection is a generic feature of the uniformly heated case in the limit of vanishing Ekman number remains open, and difficult to investigate either analytically or numerically.

13.3 Onset of dynamo action: bifurcation diagrams and numerical models

Together with the induction equation, the governing equations for convectively driven dynamos in a rotating frame of reference can be written in non-dimensional form; with $L = r_o - r_i$, L^2/η and ΔT as units of length, time and temperature respectively, and with $(\rho\mu\eta\,\Omega)^{1/2}$ as unit for the magnetic field, these equations are

$$\mathcal{E}_m\,[\partial_t\mathbf{u} + (\mathbf{u}\cdot\nabla)\mathbf{u}] = -\nabla\pi + \mathcal{E}\,\nabla^2\mathbf{u} - 2\mathbf{e}_z{\times}\mathbf{u} + \widetilde{\mathcal{R}a}\,T\,\mathbf{r} + (\nabla\times\mathbf{B}){\times}\mathbf{B}\,, \tag{13.24}$$

$$\partial_t\mathbf{B} = \nabla\times(\mathbf{u}\times\mathbf{B}) + \nabla^2\mathbf{B}\,, \qquad \partial_t T + (\mathbf{u}\cdot\nabla)T = q\,\nabla^2 T\,, \tag{13.25}$$

$$\nabla\cdot\mathbf{u} = \nabla\cdot\mathbf{B} = 0\,. \tag{13.26}$$

The system (13.24–13.26) involves four independent non-dimensional parameters: the Ekman number $\mathcal{E}$, the magnetic Ekman number $\mathcal{E}_m$, the Roberts number q and the modified Rayleigh number $\widetilde{\mathcal{R}a}$, defined by

$$\mathcal{E} = \nu/\Omega L^2\,, \quad \mathcal{E}_m = \eta/\Omega L^2\,, \quad q = \kappa/\eta\,, \quad \widetilde{\mathcal{R}a} = \alpha g \Delta T L/\eta\Omega\,. \tag{13.27}$$

In the Earth's core the first three of these parameters have order of magnitude

$$\mathcal{E} \simeq 10^{-15}\,, \quad \mathcal{E}_m \simeq 10^{-9}\,, \quad q \simeq 10^{-5}\,. \tag{13.28}$$

The fourth, $\widetilde{\mathcal{R}a}$, which controls the strength of convection, is more difficult to estimate. While the temperature difference across the core is of the order of 10^3 K, most of the heat in the core is carried along the adiabat. Only the super-adiabatic gradient is relevant in the Boussinesq framework. This deviation is only of the order of 10^{-3}K (Gubbins, 2001); the resulting rough estimate for $\widetilde{\mathcal{R}a}$ is $\widetilde{\mathcal{R}a} \simeq 10^8$.

Typical velocities in the Earth's core can be inferred from the secular variation of the magnetic field (Holme 2009), giving $U \simeq 10^{-4}\mathrm{m\,s^{-1}}$. Dimensional analysis reveals the length-scale at which inertial effects will be comparable to the effects of global rotation (the so-called Rossby radius), U/Ω, of the order of a few meters. So inertial effects are not expected to play a significant role on the global scale, on the time-scale of secular variation.

The relevant balance for the Earth's core is therefore one in which both viscous effects and inertial effects are negligible on the large scales – a state of 'magnetostrophic balance'. The limit system of equations describing this balance, omitting both viscous and inertial terms, was first considered by Taylor (1963), and has proved extremely difficult to solve numerically (see Section 13.6).

The bifurcation diagram for dynamo action in the Earth's core has been the focus of many analytical or mixed analytical-numerical studies. The key idea stems from a result of stability theory. It has been known since Chandrasekhar (1961) that whereas both rotation and an applied magnetic field are separately stabilising, the stabilising effects may in some way counterbalance each other when they

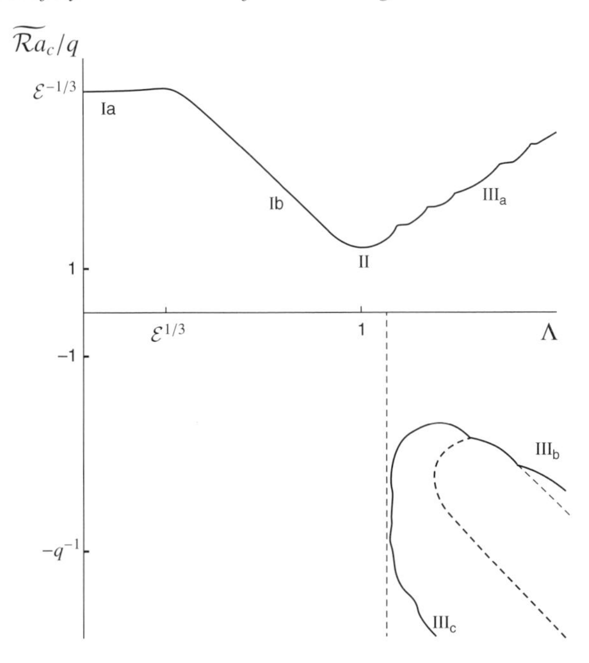

Figure 13.7 The neutral stability curve for the onset of magnetoconvection in a sphere under the influence of an azimuthal magnetic field whose strength is proportional to the distance from the rotation axis. The regions I and II represent thermal instabilities, whereas the magnetic field is the source of the energy driving the instabilities in region III. [Adapted from Fearn 1979.]

are combined. (We have already met a similar phenomenon in §12.6, where the 'releasing effect' was due to resonance between a forcing frequency and a natural frequency of the wave-supporting system.) The competing effects are here represented in terms of the *Elsasser number* $\Lambda = \sigma B^2/\rho\Omega$, where B is a measure of the field strength.

In a study of magnetoconvection in the cylindrical annulus configuration with sloping boundaries, Soward (1979) found that in general, as Λ increases from zero, the critical Rayleigh number $\widetilde{Ra}_c$ for the onset of convection increases until $\Lambda \sim \mathcal{O}(\mathcal{E}^{1/3})$, then decreases; this indicated the possible existence of a 'weak-field branch' of dynamo action, terminating when $\Lambda \sim \mathcal{O}(\mathcal{E}^{1/3})$. In a parallel study in the spherical geometry, Fearn (1979) also found that the critical $\widetilde{Ra}$ for the thermal Rossby mode described above first increases with increasing Λ, but eventually decreases to reach a minimum for $\Lambda \sim \mathcal{O}(1)$; see Figure 13.7. Regions I to III indicate different characteristics of the unstable mode: quasi-geostrophic convection occurs in region I; convection relaxes to large-scale motion with small azimuthal wave number in region II; and finally in region III the instabilities are driven by the magnetic field and can occur even for negative Rayleigh numbers. A more recent

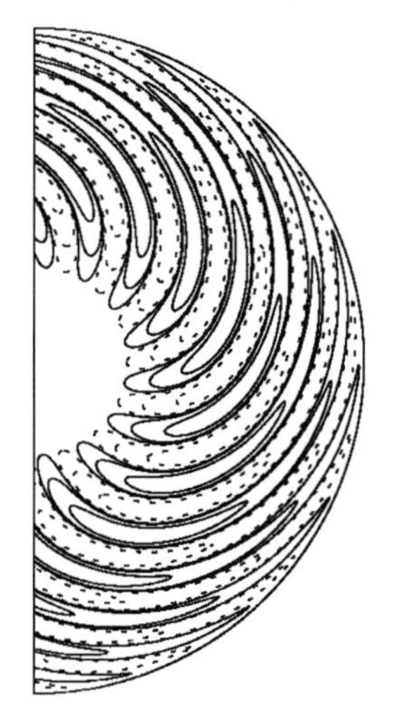

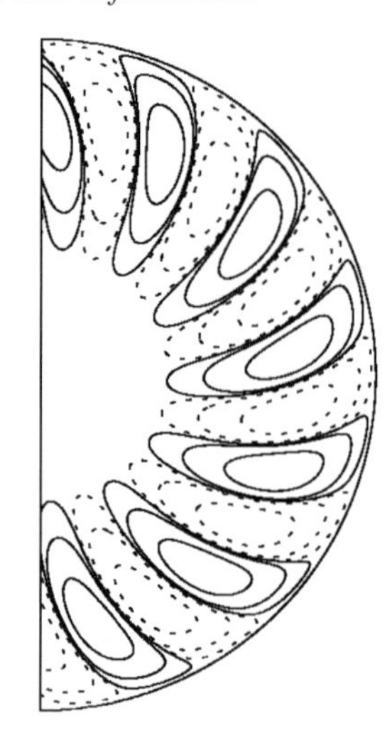

Figure 13.8 Sample structure of global convection modes occurring in a sphere with azimuthal magnetic field in the weak-field regime for $\mathcal{P}r = 0.1$, $\mathcal{P}_{\mathrm{m}} = 1$; on the left, $\lambda = 10$; on the right, $\lambda = 10^{1.5}$. [Republished with permission of the Royal Society, from Jones et al. 2003; permission conveyed through Copyright Clearance Center, Inc.]

study (Jones et al., 2003) focused on the weak-field regime and on the manner in which the convection mode changes as Λ increases. The behaviour near the onset of convection depends on a modified Elsasser number $\lambda = \mathcal{E}^{-1/3}\,\Lambda$, on $\mathcal{P}r$ and on the magnetic Prandtl number $\mathcal{P}_{\mathrm{m}} = \nu/\eta = q\,\mathcal{P}r$. Two sample diagrams are shown in Figure 13.8.

In summary, an applied magnetic field can relax the Taylor–Proudman constraint of rapid rotation; a weak applied field ($\Lambda \lesssim \mathcal{E}^{1/3}$) stabilises the flow, but $\mathcal{R}a_c$ decreases rapidly with a stronger applied field. A minimum of $\mathcal{R}a_c$ is reached when $\Lambda = \mathcal{O}(1)$.[3]

Early results on magnetoconvection led Roberts (1979) to suggest that the bifurcation diagram for the geodynamo should consist of two stable branches (Figure 13.9 in which A^2 denotes a measure of magnetic energy, e.g. the Elsasser number Λ). In the first 'weak-field' branch, the flow develops short length-scales ($\sim \mathcal{E}^{1/3}$) in directions normal to the axis of rotation, and viscosity is able to break the Taylor–Proudman constraint. As $\mathcal{R}a$ increases, the field amplitude increases and a transition occurs when the Lorentz force becomes large enough. At this stage, the weak-field solution becomes unstable, and the magnetic field experiences a runaway amplification. Saturation is achieved when the growing Lorentz force becomes comparable with the Coriolis force (i.e. $\Lambda = \mathcal{O}(1)$); this condition of 'magnetostrophic balance' characterises the resulting 'strong-field' branch.

[3] See the book by Weiss & Proctor (2009) for an in-depth analysis of this type of magnetoconvection.

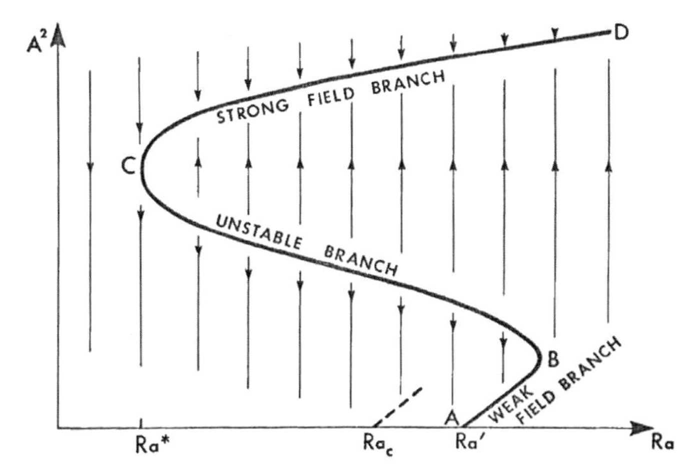

Figure 13.9 Anticipated bifurcation diagram for the geodynamo, characterised by a weak-field branch (AB) and a strong-field branch (CD); A^2 represents the square of the field amplitude, and the arrows indicate the direction of field evolution from any point. The strong field branch will always be attained from a seed field at sufficiently large Rayleigh number. [After Roberts 1979.]

13.3.1 Numerical models

Numerical simulations can retain all the terms in the governing equations and seek to approach equilibrium in different parameter ranges, particularly where the Ekman number $\mathcal{E}$ is small. This is challenging, because the limit $\mathcal{E} \to 0$ relevant to the geodynamo introduces a wide range of time-scales and length-scales. It is crucial in this limit to verify that inertial and viscous effects are indeed small in the solutions that are realised. A useful approach is to construct bifurcation diagrams for various choices of the governing dimensionless parameters, the different branches on these diagrams reflecting different dominant force balances.

Since the first numerical models of dynamo action (Zhang & Busse 1988, 1989; Glatzmaier & Roberts 1995), many regions of parameter space have been explored, allowing the construction of phase (or 'regime') diagrams. Three regimes of dynamo action have been identified (Dormy et al. 2018): weak-dipolar (WD), fluctuating-multipolar (FM) and strong-dipolar (SD); the first two of these were described by Kutzner & Christensen (2002) and Christensen & Aubert (2006) (Figure 13.10). The regime WD (circles on the figure) is characterised by a dominant axial dipole, while the regime FM (diamonds) at larger $\widetilde{\mathcal{R}a}/\widetilde{\mathcal{R}a}_c$ is multipolar, with a fluctuating dipole component.

The WD regime has been recognised since the early days of numerical modelling and is often presented as if relevant to the geodynamo; however it is now known

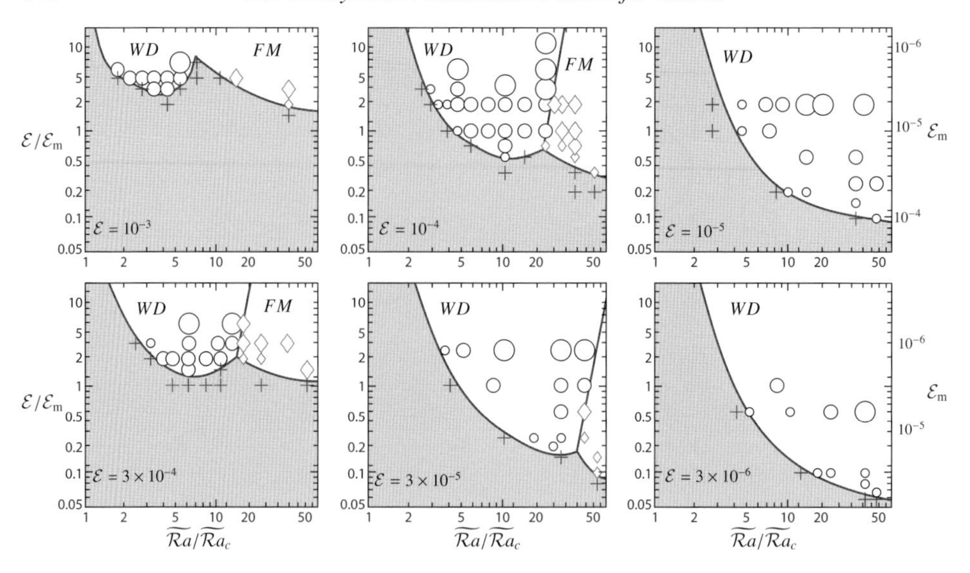

Figure 13.10 Regime diagrams resulting from three-dimensional numerical simulations in a rotating spherical shell at six different Ekman numbers as indicated; two regimes are identified: a 'weak' dipolar regime (WD) marked with circles and a fluctuating multipolar regime (FM) marked with diamonds; the dipolar regime expands as $\mathcal{E}$ decreases. [Adapted from Christensen & Aubert 2006 by permission of Oxford University Press on behalf of the Royal Astronomical Society.]

(King & Buffett 2013; Oruba & Dormy 2014a; Dormy 2016) that viscosity remains important in its force balance. Figure 13.10 indicates that, for given $\mathcal{E}$, the WD regime cannot persist if $\mathcal{P}_{\rm m} = \mathcal{E}/\mathcal{E}_{\rm m} < \mathcal{P}_{\rm mc}(\mathcal{E})$, a decreasing function of $\mathcal{E}$. In the geophysical limit, both $\mathcal{E}$ and $\mathcal{P}_{\rm m}$ are small, and for this reason numerical work usually adopts the smallest practicable values of $\mathcal{P}_{\rm m}$ for fixed small $\mathcal{E}$. From this point of view, the geophysical regime is in the lower part of the panels of Figure 13.10; but if instead (with $\mathcal{E}$ fixed) we consider the limit $\mathcal{E}_{\rm m} \to 0$ instead of $\mathcal{P}_{\rm m} \to 0$, then the geophysical regime would appear to be in the upper part (see the $\mathcal{E}_{\rm m}$-scale on the right of the rightmost panels). This indeterminacy can be resolved by adopting an appropriate 'distinguished limit' of the form $\mathcal{P}_{\rm m} \propto \mathcal{E}^{\alpha}$ $(0 < \alpha < 1)$, relating $\mathcal{P}_{\rm m}$ to $\mathcal{E}$ as both tend to zero (Dormy 2016); for example, the choice of exponent $\alpha = 2/3$ with a choice of prefactor that places $\mathcal{P}_{\rm m}$ above the shaded regions of Figure 13.10 gives $\mathcal{P}_{\rm m} = 10^{-6}$ when $\mathcal{E} = 10^{-15}$, compatible with the real geophysical situation.

When $\widetilde{\mathcal{R}a}$ increases beyond the critical value $\widetilde{\mathcal{R}a}_c$ for the onset of convection, dynamo action results from a secondary bifurcation that may be one of three types, the choice depending on $\mathcal{E}$ and $\mathcal{P}_{\rm m}$: supercritical, subcritical, or in the form of an 'isola' (Morin & Dormy 2009). Whatever the type of bifurcation, the resulting dynamo state is considered to be in the WD regime, because the convection length-

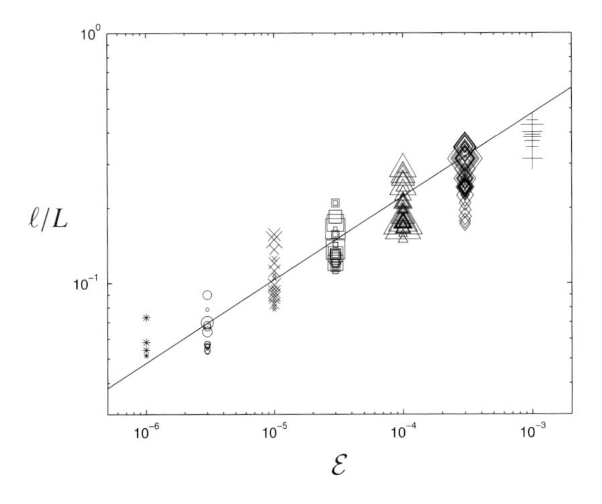

Figure 13.11 Logarithmic plot showing dependence of convective length-scale ℓ on Ekman number $\mathcal{E}$ for the weak dipole (WD) branch; the slope of the line is 1/3. [After King & Buffett 2013.]

scale still decreases like $\mathcal{E}^{1/3}$ as $\mathcal{E} \to 0$ (see Figure 13.11 from King & Buffett 2013) so that viscosity remains important in the force balance. This regime is not therefore relevant for the geodynamo.

13.3.2 Model equations for super- and subcritical bifurcations

A simple model equation for supercritical bifurcation satisfying the $B \to -B$ symmetry is

$$\dot{B} = \lambda B - B^3 , \tag{13.29}$$

where B is a typical field amplitude and $\lambda \propto \widetilde{\mathcal{R}a} - \widetilde{\mathcal{R}a}_c$. For $\lambda > 0$, this model system is characterised by three fixed points, $B = 0$ (unstable), and $B = \pm\sqrt{\lambda}$ (stable). Typical evolutions of $B(t)$ for the choice $\lambda = 1$ are shown in Figure 13.12; in these, the strong asymmetry about either state $B = \pm 1$ is a consequence of the non-linearity of the system (13.29). The corresponding bifurcation diagram is shown in Figure 13.13c.

Similarly, a subcritical bifurcation may be represented by the model equation

$$\dot{B} = \lambda B + B^3 - \alpha B^5 , \quad \alpha > 0, \tag{13.30}$$

involving a destabilising non-linearity $+B^3$, and a stabilising term $-\alpha B^5$ (see Figure 13.13b). In some cases this branch of dynamo solutions does not connect to the purely hydrodynamic mode which remains linearly stable, but consists of a detached isola of solutions (Morin, 2005; Morin & Dormy, 2009). A finite amplitude

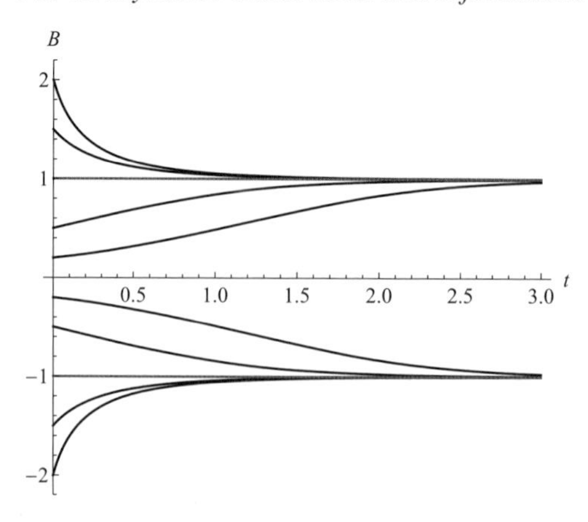

Figure 13.12 Evolution diagram for the supercritical system (13.29) with $\lambda = 1$.

initial field is needed to attain an isola state, which is bounded by turning points at two values $\mathcal{R}a_{t1}$ and $\mathcal{R}a_{t2}$ of the control parameter (Figure 13.13a). A bifurcation of this form can be represented by the model equation

$$\dot{B} = -(c + \lambda^2)\, B + B^3 - \alpha\, B^5\,, \quad (c > 0)\,, \tag{13.31}$$

like (13.30), but with the property that the state $B = 0$ is stable for all λ.

13.3.3 Three regimes, WD, FM and SD; numerical detection

The three regimes, WD, FM and SD, have been detected by numerical simulations at $\mathcal{P}r = 1$, fixed $\mathcal{E}$ and decreasing $\mathcal{E}_{\mathrm{m}}$, as shown in Figure 13.14 (Morin, 2005; Morin & Dormy, 2009; Dormy, 2016; Dormy et al., 2018). Isola diagrams similar to Figure 13.14a had been found previously with mean-field models (Soward & Jones 1983). Only the stable branches can be captured by direct time integration, but unstable branches can be found using a Newton solver (Feudel et al. 2017). The simulations that give rise to these diagrams are complex, being non-linear and three-dimensional, and the representation in terms of amplitude alone is an over-simplification. In particular, it is worth noting that for some parameter values there exist multiple solutions (Fuchs et al., 2001; Morin & Dormy, 2009).

The second (FM) regime (Kutzner & Christensen 2002; Christensen & Aubert 2006; Schrinner et al. 2012; Oruba & Dormy 2014b) corresponds to a situation in which inertia forces are so strong that the Rossby radius U/Ω is comparable with the convection length-scale ℓ. For this reason, the FM regime is also inapplicable to the geodynamo. It is possibly more relevant to stellar dynamos, and is discussed further in §14.2. (As indicated in Figure 13.10, the FM regime usually results from a tertiary bifurcation from the WD regime.)

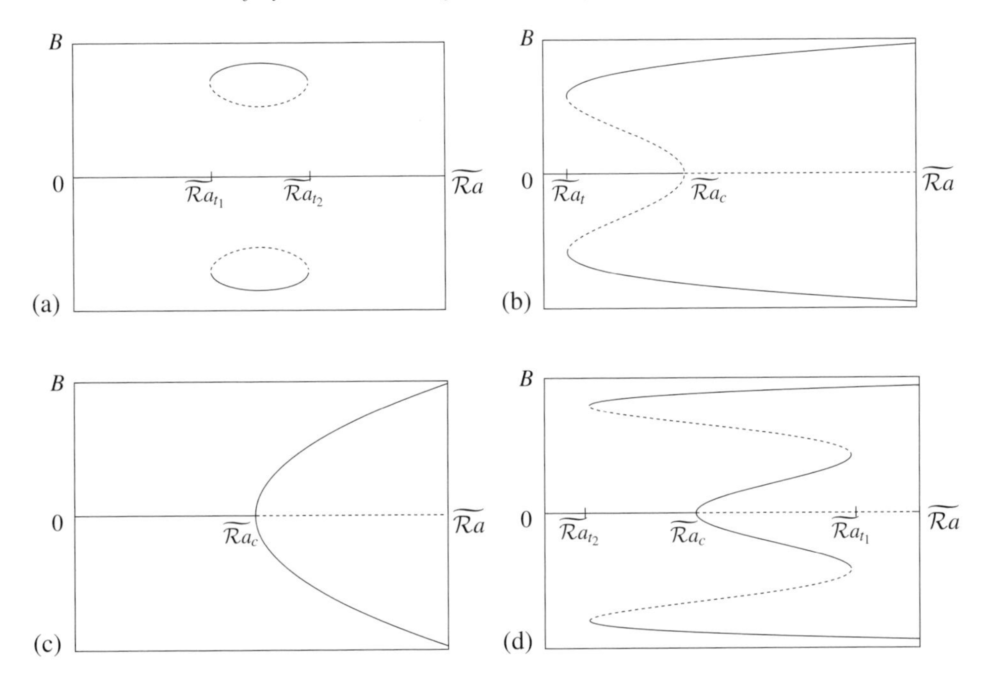

Figure 13.13 Structure of bifurcation diagrams found in numerical simulations of convectively driven spherical dynamos: (a) isola, (b) subcritical, (c) supercritical and (d) supercritical with turning point and transition from weak field to strong dipolar field; curves are solid where solutions are stable, dashed where unstable, and therefore not observed.

13.3.4 The SD regime

The third regime (SD), identified by Dormy (2016), is attainable numerically at the cost of an under-estimated magnetic diffusivity. This regime does appear to approach the relevant magnetostrophic force balance, because with increasing $\mathcal{R}a$ the Lorentz force increases faster than inertia. The first transition occurs at a turning point (a cusp catastrophe) characterised by a runaway field growth (Figure 13.16), and a hysteresis connects the weak-field and strong-field branches. The model equation

$$\dot{B} = \lambda B - B^3 + \alpha B^5 - \beta B^7, \quad \alpha > 0, \beta > 0, \tag{13.32}$$

provides a bifurcation sequence such as that of Figures 13.13d and 13.9, while still preserving the $B \to -B$ symmetry. This bifurcation sequence thus establishes a first connection between numerical models and earlier asymptotic results. In relation to Figure 13.13d, it is worth noting that the purely hydrodynamic state actually re-stabilises at larger $\mathcal{R}a$. In comparison with (13.31), this suggests a model

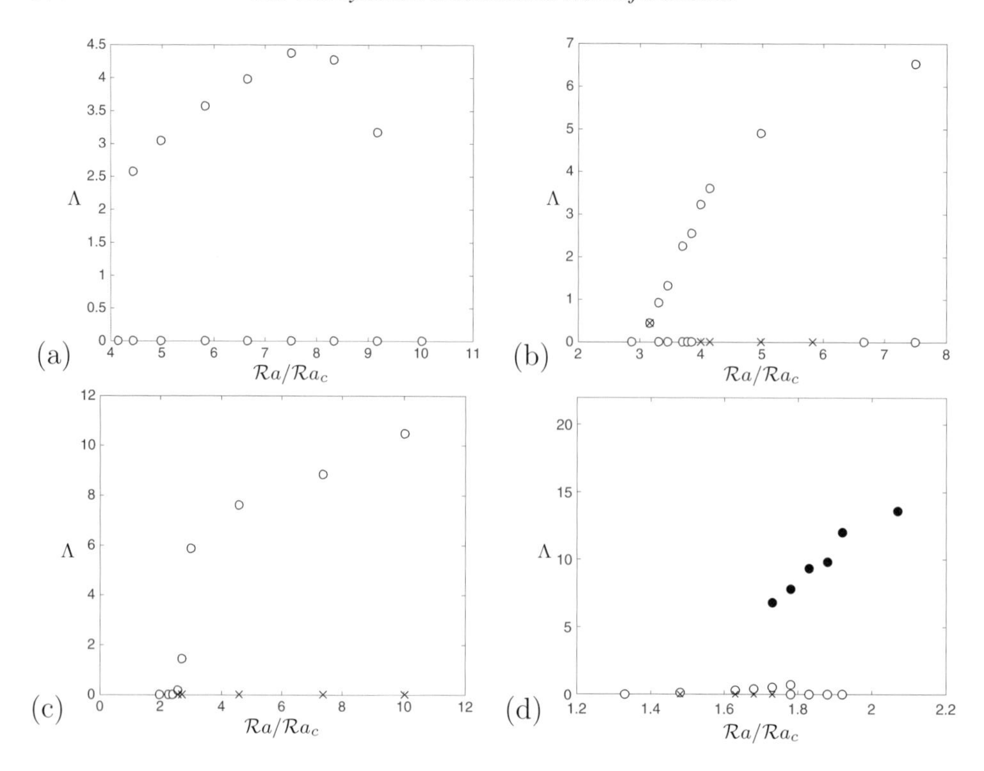

Figure 13.14 Bifurcation diagrams corresponding to those of Figure 13.13 obtained from numerical simulations. At $\mathcal{E} = 3.10^{-4}$, the bifurcation is (a) 'isola', i.e. forming an isolated branch, for $\mathcal{E}_\mathrm{m} = 2 \times 10^{-4}$, $q = 1.5$; (b) subcritical for $\mathcal{E}_\mathrm{m} = 10^{-4}$, $q = 3$; (c) supercritical for $\mathcal{E}_\mathrm{m} = 5 \times 10^{-5}$, $q = 6$; (d) supercritical with turning point and transition from weak to strong field branch for $\mathcal{E}_\mathrm{m} = 1.7 \times 10^{-5}$, $q = 18$. [After Morin & Dormy 2009; Dormy 2016.]

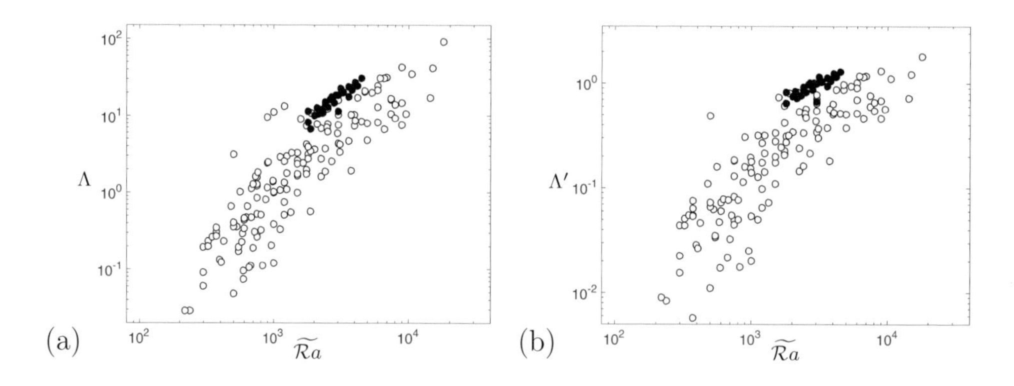

Figure 13.15 Field strength as measured by the Elsasser number Λ and the modified Elsasser number $\Lambda' = \Lambda L/(\mathcal{R}_\mathrm{m}\ell_B)$ (where ℓ_B is defined in (13.83) below) as a function of $\widetilde{\mathcal{R}a}$ for the WD and the SD regime. [After Dormy et al. 2018.]

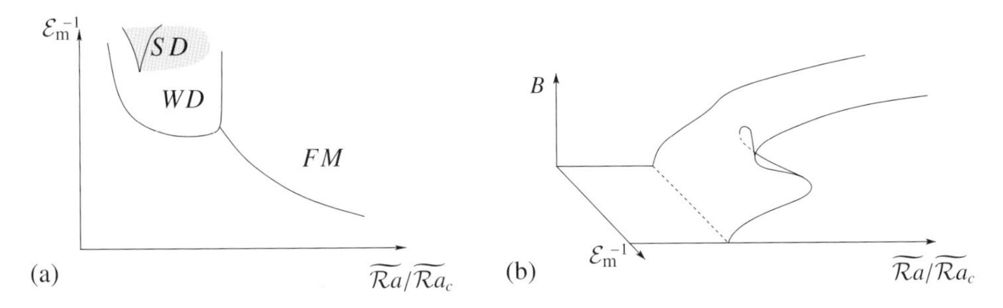

Figure 13.16 (a) Phase diagram illustrating, for a given Ekman number, the three different dynamo 'phases' observed in numerical simulations of the geodynamo, namely the weak-dipolar (WD), the fluctuating-multipolar (FM) and the strong-dipolar (SD) branch. The cusp indicates the region of bistability between the WD and the SD branch. The grey shaded region marks the strong-dipolar branch [After Dormy et al. 2018]. (b) Sketch of a tentative three-dimensional bifurcation diagram near the region of existence of the weak and strong field branches. The fold in the surface accounts for the observed two branches of solutions [Adapted from Dormy 2016].

bifurcation equation like (13.31) of the form

$$\dot{B} = (c - \lambda^2)\, B - B^3 + \alpha\, B^5 - B^7\,, \quad c > 0, \alpha > 0 \tag{13.33}$$

now giving a finite range $\lambda^2 < c$ of instability of the state $B = 0$.

The relevance of this regime for vanishing $\mathcal{E}$ and $\mathcal{E}_m$ has been further investigated by Dormy et al. (2018). Figure 13.15 shows how the WD regime (open circles) is very scattered when plotted against $\widetilde{\mathcal{R}a}$, highlighting a sensitive dependence on the small parameters $\mathcal{E}$ and $\mathcal{E}_m$, whereas the SD regime (solid circles) exhibits a much weaker dependence. The computed Elsasser number Λ varies in the range 1 to 100 for $10^3 \lesssim \widetilde{\mathcal{R}a} \lesssim 10^4$; however a finer estimate of the force balance is represented by a modified Elsasser number Λ', defined in the figure caption, which is remarkably close to unity over the same range of $\widetilde{\mathcal{R}a}$.

13.3.5 The WD/SD dichotomy

The key question is now: to what level will the magnetic energy rise when dynamo action does occur? If $\mathcal{R}a$ is just above the critical value $\mathcal{R}a_c$ for the onset of convection (the primary bifurcation), then in the absence of any magnetic effects, it is known (Malkus & Veronis 1958) that the amplitude of the motions will rise to a value of order $\varepsilon = (\mathcal{R}a - \mathcal{R}a_c)^{1/2}$. If the motion is helical, as it undoubtedly is in the rapidly rotating system, then there is an associated α-effect, in general anisotropic, but with $\alpha_0 = \mathcal{O}(\varepsilon^2)$, where α_0 is a typical ingredient of the pseudo-tensor α_{ij}. As ε increases (keeping all other parameters fixed), a critical value, ε_1 say, is reached at which dynamo action occurs (a secondary bifurcation) essentially by the α^2-mechanism (cf. Figure 11.6). This was identified in direct numerical simulations by Christensen et al. (1998). Then dynamo growth will continue until α-quenching sets in.

If the secondary bifurcation is supercritical (and if q is sufficiently small), the level of magnetic energy attained will be small compared with the kinetic energy of the cellular motion. In this situation, a perturbation approach may succeed in determining the equilibrium field structure and amplitude. It is not however certain *a priori* that such a neighbouring state exists, since, as discussed above, it is possible that a suitably oriented magnetic field may release the constraint of strong rotation, permitting more vigorous convection. In this case, a runaway situation will result; this is just the jump from the WD-regime to the SD-regime, as discussed above. The field then presumably grows to an amplitude at which the Lorentz force is comparable with the Coriolis force, i.e. an SD-regime of magnetostrophic balance will be established.

For computational reasons, the strong-field regime has so far been attained in numerical models only for $\mathcal{P}r = \mathcal{O}(1)$ and large $\mathcal{P}_{\rm m}$; it may be further studied by following a distinguished-limit hypothesis (Dormy, 2016). Several issues remain open, which numerical simulations may be expected to resolve quite soon. One such issue relates to whether dynamo action can be realised for $\mathcal{R}a < \mathcal{R}a_c$; this analytical prediction for the strong-field regime (see Figure 13.9) has not as yet been realised in numerical models. Also, $\mathcal{P}_{\rm m}$ has not been decreased significantly in the numerical work; lower values may well be needed to trigger field reversals (as described by the low-order models of Chapter 11).

13.4 The Childress–Soward convection-driven dynamo

A perturbation approach based on the idea of equilibrium at weak field strength was introduced by Childress & Soward (1972) and developed by Soward (1974) for the problem of convection in a plane layer of fluid heated from below. Both boundaries $z = 0,\ z_0$ were supposed stress-free, perfectly conducting and isothermal; as in §12.10, this leads to the simplest combination of boundary conditions for analytical solution of the stability problem.[4] In the absence of any magnetic field, the system is characterised by the Ekman, Rayleigh and Prandtl numbers, as defined in §13.2.4, but with z_0 as typical length-scale:

$$\mathcal{E} = \nu/\Omega z_0^2, \quad \mathcal{R}a = \alpha g z_0^3 \Delta\Theta/\nu\kappa, \quad \mathcal{P}_{\rm m} = \nu/\kappa\,, \qquad (13.34)$$

[4] If the stress-free condition is replaced by a no-slip condition, then, as Soward observed, Ekman layers on the two boundaries can play an important part in the long-term finite amplitude behaviour of convection cells; the system may be expected to be equally sensitive to the electrical and thermal boundary conditions.

where $\Delta\Theta$ is the temperature difference between the plates, and κ the thermal diffusivity, and it is supposed that

$$\mathcal{E} \ll 1, \quad \mathcal{P}_{\mathrm{m}} = \mathcal{O}(1)\,. \tag{13.35}$$

The critical Rayleigh number (Chandrasekhar 1961) is in these circumstances

$$\mathcal{R}a_c = \mathcal{O}(\mathcal{E}^{-4/3})\,, \tag{13.36}$$

and the horizontal wave number for which $\mathcal{R}a_c$ is minimal is

$$k_c = \mathcal{O}(\mathcal{E}^{-1/3})\, z_0^{-1}\,. \tag{13.37}$$

Soward supposed that the layer is just unstable, in the sense that

$$\frac{\mathcal{R}a - \mathcal{R}a_c}{\mathcal{R}a_c} = \mathcal{O}(\mathcal{E}^{1/3})\,, \tag{13.38}$$

so that only modes with horizontal wave vectors $\mathbf{k} = (k_1,\ k_2, 0)$ such that $|\mathbf{k}| \approx k_c$ are excited. He supposed moreover that $\mathcal{P}_{\mathrm{m}} > 1$, so that the motion at the onset of convection is steady, rather than time-periodic (Chandrasekhar 1961). Under the condition (13.38), the amplitude of the velocity in the convection cells excited has order of magnitude (Malkus & Veronis 1958)

$$u_0 = \mathcal{O}(\mathcal{E}^{1/6})\nu/z_0\,. \tag{13.39}$$

This indicates that an expansion of all mean and fluctuating field variables as power series in the small parameter $\mathcal{E}^{1/6}$ is appropriate; this was the basis of Soward's perturbation procedure.

Motions periodic in the x and y directions are determined uniquely by the vertical velocity distribution

$$w(\mathbf{x},\ t) = \mathrm{Re} \sum_{\mathbf{k}} \hat{w}(\mathbf{k}, t)\, \sin \frac{\pi z}{z_0}\, \mathrm{e}^{\mathrm{i}\mathbf{k}\cdot\mathbf{x}}\,. \tag{13.40}$$

The fact that the horizontal scale $\ell = \mathcal{O}(\mathcal{E}^{1/3})\, z_0$ is small compared with z_0 now permits the use of the methods of mean-field electrodynamics, means being defined over the horizontal plane. The magnetic Reynolds number based on u_0 and ℓ is

$$u_0\ell/\eta = \mathcal{O}\left(\mathcal{E}^{1/6}\right)(\nu/\eta)\,, \tag{13.41}$$

and this is small provided[5] ν/η is $\mathcal{O}(1)$ or less. Hence in calculating the pseudo-tensor α_{ij} (cf. §7.8) the magnetic field perturbation $\mathbf{b}$ is effectively determined (at leading order in $\mathcal{E}^{1/6}$) by

$$\eta\nabla^2\mathbf{b} = -\mathbf{B}\cdot\nabla\mathbf{u}\,, \tag{13.42}$$

[5] Soward assumed $\nu/\eta = \mathcal{O}(1)$; the fact that $\nu/\eta \ll 1$ in the core of the Earth should perhaps be incorporated in the expansion procedure to make the theory more relevant in the terrestrial context.

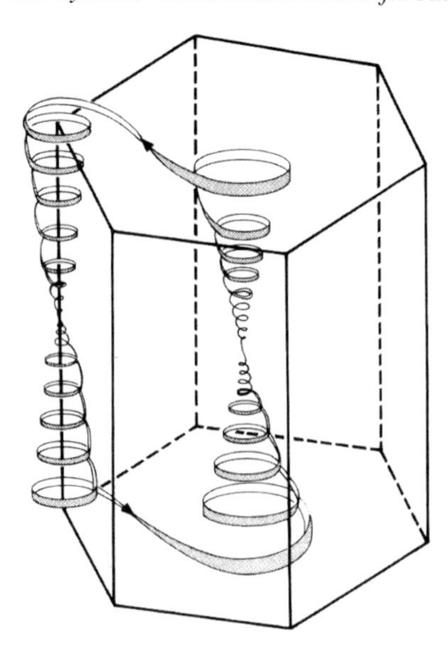

Figure 13.17 A perspective sketch of a fluid particle path which passes through the centre of a cell when the cell planform is hexagonal; the helicity associated with this type of motion is evidently antisymmetric about the centre plane $z = \frac{1}{2}z_0$, and the α-effect (given by (13.43) and (13.44)) is likewise antisymmetric about $z = \frac{1}{2}z_0$. [After Veronis 1959.]

and calculation of

$$\mathcal{E}_i = \langle \mathbf{u} \wedge \mathbf{b} \rangle_i = \alpha_{ij} B_j \tag{13.43}$$

is a relatively straightforward matter, with the result

$$\alpha_{ij} = \frac{\pi}{\mathcal{E}^{1/2} \eta z_0} \sum_{\mathbf{k}} \frac{k_i k_j}{k^6} q(\mathbf{k}, t) \sin\left(\frac{2\pi z}{z_0}\right), \tag{13.44}$$

where $q(\mathbf{k}, t) = |\hat{w}(\mathbf{k}, t)|^2$. This calculation of course requires knowledge of the phase relationships between horizontal and vertical velocity components, as determined by the linear stability analysis. In the expression (13.43), the mean field $\mathbf{B}(z, t)$ is necessarily horizontal, and it evolves according to the (now) familiar equation

$$\frac{\partial B_i}{\partial t} = \epsilon_{ijk} \frac{\partial}{\partial x_j} (\alpha_{kl}(z, t) B_l) + \eta \frac{\partial^2 B_i}{\partial z^2} \,. \tag{13.45}$$

The next stage of the calculation is to derive equations for the $q(\mathbf{k}, t)$ appearing in (13.44) by continuation of the systematic perturbation procedure to order $\mathcal{O}(\mathcal{E}^{1/2})$ at which Lorentz forces have a significant effect on the dynamics. These equations, as derived by Soward, take the form

$$\frac{\partial q(\mathbf{k},t)}{\partial t}+2\sum_{\mathbf{k}'} A(\mathbf{k},\mathbf{k}')\,q(\mathbf{k}',t)\,q(\mathbf{k},t)=\left(m(\mathbf{k},t)-\tfrac{1}{2}\sum_{\mathbf{k}'} m(\mathbf{k}',t)q(\mathbf{k}',t)\right)q(\mathbf{k},t), \tag{13.46}$$

provided all the wave vectors $\mathbf{k}$ in the velocity spectrum have equal magnitude. Here $A(\mathbf{k},\mathbf{k}')$ is a coupling coefficient representing non-linear interactions between the instability modes; $m(\mathbf{k},t)$ is a weighted average of $(\mathbf{B}\cdot\mathbf{k})^2$ over $0 \ll z \ll z_0$, and the terms involving m represent the effect of the mean magnetic field on the small-scale motions.

In order to integrate (13.45) and (13.46), it is necessary to adopt initial conditions, and in particular to specify the horizontal structure of the cellular convection pattern at time $t = 0$: if only one wave vector $\mathbf{k}_1$ is represented (i.e. $q(\mathbf{k},0) = 0$ unless $\mathbf{k} = \pm\mathbf{k}_1$), then the motion has the form of cylindrical cells with axes perpendicular to $\mathbf{k}_1$; it is almost self-evident that this motion has too simple a structure to provide dynamo maintenance of $\mathbf{B}$. If two wave vectors $\mathbf{k}_1$, $\mathbf{k}_2$ are equally represented ($q(\mathbf{k}_1,0) = q(\mathbf{k}_2,0)$ with $|\mathbf{k}_1| = |\mathbf{k}_2|$ and $\mathbf{k}_1\cdot\mathbf{k}_2 = 0$) and if the amplitudes of the various modes are equal, then the convection cells have square boundaries.[6] If three wave vectors $\mathbf{k}_1$, $\mathbf{k}_2$, $\mathbf{k}_3$, are equally represented, and if

$$|\mathbf{k}_1|^2 = |\mathbf{k}_2|^2 = |\mathbf{k}_3|^2 = -\frac{\sqrt{3}}{2}\mathbf{k}_2\cdot\mathbf{k}_3 = -\frac{\sqrt{3}}{2}\mathbf{k}_3\cdot\mathbf{k}_1 = -\frac{\sqrt{3}}{2}\mathbf{k}_1\cdot\mathbf{k}_2\,, \tag{13.47}$$

then the cell boundaries are regular hexagons; a typical particle path in this case is sketched in Figure 13.17b (Veronis 1959).

Numerical integration of (13.45) and (13.46) for the case of square-cell boundaries indicates that both the magnetic energy

$$M(t) = \frac{1}{2\mu_0}\int_0^L \mathbf{B}^2 \mathrm{d}z\,, \tag{13.48}$$

and the quantity $v(t) = q(\mathbf{k}_1,t) - q(\mathbf{k}_2,\ t)$ settle down to a time-periodic behaviour when $\eta t/L^2$ becomes large. The ratio of mean magnetic energy to mean kinetic energy in this asymptotic periodic state is of order $\mathcal{E}^{1/3}$ (i.e. small as required for validity of the 'weak-field' approach), and the fluctuations in magnetic energy are of order $\pm 3\%$ about the mean value. Soward presented these results in the form of an approach to a limit cycle in the plane of the variables $v(t)$, $M(t)$ (see Figure 13.18).

The results for hexagonal cell boundaries are naturally more complicated since now the kinetic energy is shared among three modes rather than two; nevertheless Soward's numerical integrations again indicated asymptotic weak fluctuations of the magnetic energy about its ultimate mean level, while the kinetic energy (or

[6] Cf. the velocity field (3.143) in §3.15, although in that case, four horizontal wave vectors are represented, viz. $(k_1,0,0)$, $(0,\ k_1,0)$, $2^{-1/2}(k_1,\ k_1,0)$, $2^{-1/2}(k_1,\ -k_1,0)$, with $k_1 = \pi/z_0$.

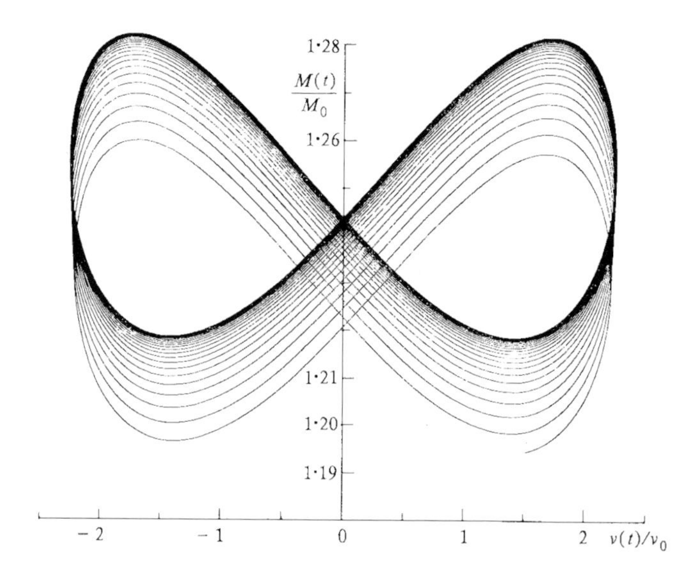

Figure 13.18 Phase plane evolution of magnetic energy $M(t)$ and the quantity $v(t)$ which represents twice the difference between the kinetic energies in the modes corresponding to wave vectors $\mathbf{k}_1$ and $\mathbf{k}_2$. For this situation, the non-linear term in (13.46) is identically zero; normalising constants M_0 and v_0 depend on the precise initial conditions of the problem. The figure shows the evolution, over 16,000 time-steps in the integration procedure, towards a limit cycle in which $M(t)$ and $v(t)$ vary periodically with time. [Republished with permission of the Royal Society, from Soward (1974); permission conveyed through Copyright Clearance Center, Inc.]

rather a substantial fraction of it) appears to 'flow' cyclically among the three modes.

It is a characteristic feature of these dynamos that both velocity and magnetic field distributions are unsteady (though ultimately periodic in time with period of order z_0^2/η).

13.4.1 Mixed asymptotic and numerical models

The Childress–Soward model has been extended by Calkins et al. (2015) through use of both asymptotic and numerical methods in order to approach the geophysically relevant parameter regime. By multi-scale techniques, these authors have derived asymptotic equations that are amenable to numerical treatment. Variables are split into slow (mean) and fast (fluctuating) components, and these are expanded in powers of Rossby number $\mathcal{R}o\,(\ll 1)$. On the small scale, this is a magnetically modified version of quasi-geostrophic convection, while on the large scale, it encapsulates 'thermal-wind' balance in the mean momentum and heat equations. In the low-$\mathcal{P}_\mathrm{m}$ limit relevant to the Earth, magnetic energy dominates over kinetic energy, and ohmic dissipation asymptotically dominates over viscous dissipation on the large scale.

The model has been extended by Calkins et al. (2016) to examine weak and intermediate field-strength situations in four convection regimes (Julien et al. 2012) (Figures 13.19a and 13.19d). In this work, as yet restricted to kinematic dynamos (i.e. neglecting the Lorentz force), the authors found that the space-dependence of the α-effect and so of the mean magnetic field generated (Figures 13.19e and 13.19f) is fairly insensitive to quite wide variation of the control parameters $\widetilde{\mathcal{R}a}$ and $\mathcal{P}r$.

Attempts to capture the strong field (SD) regime following similar procedures were initiated by Fautrelle & Childress (1982). Stellmach & Hansen (2004) were able to demonstrate the transition to the strong field branch in a plane cartesian geometry using direct numerical simulations of the full equations; but for computational reasons, as in the spherical geometry, this had to be restricted to a parameter range far from the relevant geophysical regime.

13.5 Busse's model of the geodynamo

One of the difficulties in analysing the problem of thermal convection in a rotating spherical shell (as a model of convection between the solid inner core of the Earth and the mantle) is that the radial gravity vector $\mathbf{g}$ does not make a constant angle with the rotation vector $\boldsymbol{\Omega}$. However, as we have seen, convection at small Ekman number (and when Lorentz forces are sufficiently weak) is characterised by long thin convection cells aligned with the direction of $\boldsymbol{\Omega}$. Thermal instabilities of this kind were first investigated by Busse (1970), as introduced in §13.2. Only the component $g \sin\theta$ of $\mathbf{g}$ perpendicular to $\boldsymbol{\Omega}$ is effective for cells of this structure. Since the centrifugal force $\Omega^2 r \sin\theta$ has the same θ-dependence, the effects of radial gravity can be simulated in laboratory experiments by the centrifugal force in a spherical shell rotated about a vertical axis; in such experiments, unstable stratification is provided by heating the outer sphere and cooling the inner sphere (a deliberate reversal to take account of the fact that the centrifugal force is outwards, whereas the gravitational force in the terrestrial context is inwards). Such experiments (Busse & Carrigan 1974) confirm the appearance of convection columns in the region outside the cylinder $\mathscr{C}(s_0)$ where s_0 is the radius of the inner sphere (as shown in Figure 13.3).

It was these considerations that led Busse (1975a) to consider the possibility of dynamo action due to thermal convection in the simpler cylindrical geometry of Figure 13.1a. The cylindrical annulus

$$s_0 < s < s_1\,, \quad |z| < \tfrac{1}{2}z_0 - \lambda\tilde{s}\,, \quad \tilde{s} = s - \tfrac{1}{2}(s_0 + s_1)\,, \tag{13.49}$$

has approximately the same form as the region occupied by convection cells in

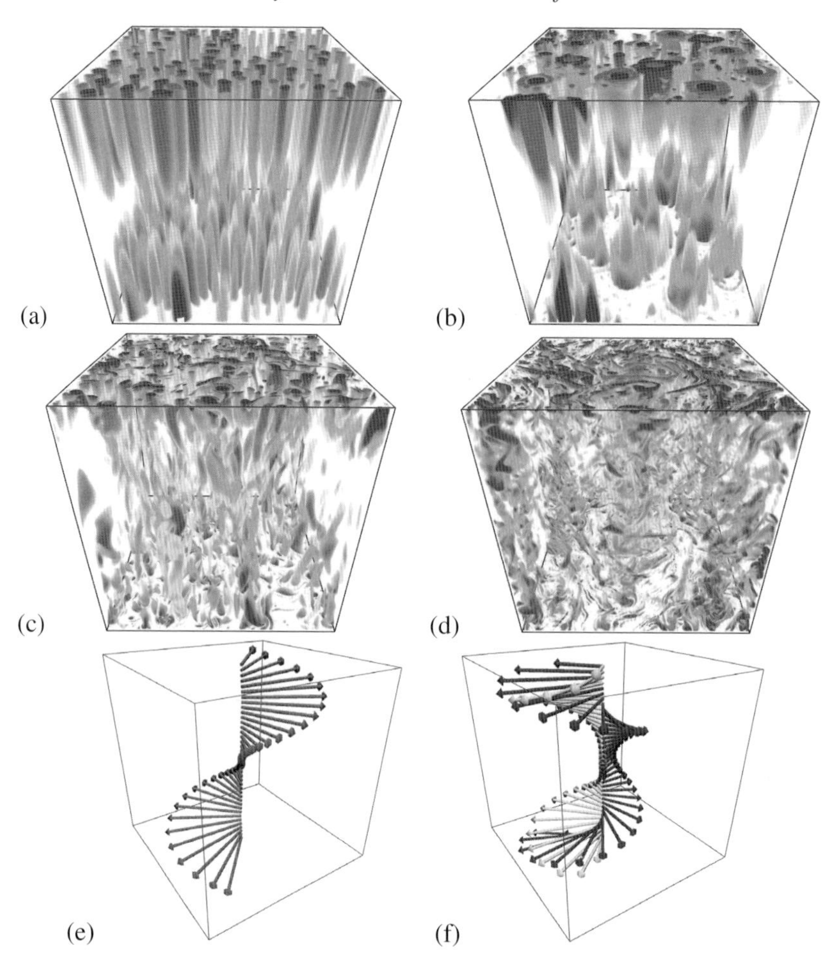

Figure 13.19 Volumetric renderings of the vertical component of the small-scale vorticity in the four convection regimes of Julien et al. (2012): (a) cellular regime $\mathcal{P}r = 1, \widetilde{\mathcal{R}a} = 10$; (b) convective Taylor column regime $\mathcal{P}r = 10, \widetilde{\mathcal{R}a} = 60$; (c) plume regime $\mathcal{P}r = 10, \widetilde{\mathcal{R}a} = 200$; (d) geostrophic turbulence regime $\mathcal{P}r = 1, \widetilde{\mathcal{R}a} = 100$; (e) mean magnetic field $\mathbf{B}$ and (f) emf $\boldsymbol{\mathcal{E}} = \alpha\mathbf{B}$ (light) and current $\mathbf{J}$ (dark) for $\mathcal{P}r = 1$ and $\widetilde{\mathcal{R}a} = 50$, at the onset of the geostrophic turbulence regime. [After Calkins et al. 2016.]

Figure 13.2a. The parameter λ represents the slope of the upper and lower surfaces of the annulus and is assumed small in the theory.[7] Moreover it was assumed that $z_0 \ll s_1 - s_0 \ll s_1$ so that (as in earlier studies) the cylindrical polar coordinates (s, φ, z) could be replaced by Cartesian coordinates (x, y, z).

[7] The perturbation procedures adopted also require that $\mathcal{E}^{1/4} \ll \lambda \ll z_0/(s_1 - s_0)$. A non-zero value for λ is of crucial importance in determining the stability properties of the system.

It is clear that simple motions of the kind represented by Figure 13.2a have zero helicity and will not in themselves be sufficient to provide an α-effect. What is needed is a superposed motion within each cell, roughly aligned with the axis of the cell, and correlated with the sense of rotation in the cell. This ingredient of the motion in Busse's theory is provided by 'Ekman suction' (Greenspan 1968) associated with the viscous Ekman layers on the upper and lower boundaries; it is of order $\mathcal{E}^{1/2}$ relative to the primary cellular motion, and provides a mean helicity linear and antisymmetric in z, and of order $\mathcal{E}^{1/2}z_0^{-1}$ relative to the mean kinetic energy. This need for an ingredient of flow parallel to two-dimensional convection cells, if dynamo action is to occur, had been recognised in an earlier study of Busse (1973) (and was already evident in the space-periodic dynamos of Roberts 1970).

Busse's calculation proceeded in three stages:

(i) The first stage consists in calculation of the critical stability conditions when there is no magnetic field present, of the structure of the critically stable disturbances, and of their amplitude when conditions are slightly supercritical; this involves the procedure of Malkus & Veronis (1958), but with the additional feature here that the small slope of the upper and lower surfaces generates weak unsteadiness in the instability modes which propagate slowly in the azimuth direction.

(ii) The second stage involves solution of the kinematic dynamo problem with the velocity field determined by stage (i); here, as in Soward (1974), the methods of mean-field electrodynamics are applied, averages being defined over the x and y variables. The dynamo mechanism is again of α^2-type, the effective value of α being linear in z (like the helicity discussed above). As in all such dynamos, the toroidal and poloidal fields are of the same order of magnitude. Busse restricts attention to fields of dipole symmetry[8] about $z = 0$, and obtained a criterion for growth of the field. Since the α-effect is proportional in intensity to $\mathcal{E}^{1/2}u_0^2$ where $u_0^2 = \langle \mathbf{u}^2 \rangle$, this criterion in effect puts a lower bound on $\mathcal{E}^{1/2}u_0^2$ in terms of other dimensionless parameters of the problem, for dynamo action to occur.

(iii) In the third stage, Busse calculated the small modification of the convection pattern due to the presence of the field excited (when $\mathcal{E}^{1/2}u_0^2$ is just large enough) as obtained in stage (ii). The field had an as yet undetermined amplitude B_0 which is assumed small. The amplitude u_0 of the modified motion and the amplitude B_0 were simultaneously determined by the condition that the fundamental magnetic mode should have zero growth rate under the α^2-action of the slightly modified convection cells. Since decrease of B_0 leads to a decrease in u_0 (other things being

[8] Busse did not actually match the field to a field in the current-free region outside the annulus. He claimed that the condition of dipole symmetry 'allows for the continuation of the meridional field towards infinity in such a way that it will decay at least as fast as a dipolar field'. The claim is plausible, but it is desirable that the actual matching should be explicitly carried out, since in the dynamo context the presence of a 'source at infinity', which might also generate a field *of dipole symmetry* about $z = 0$, must be carefully excluded.

constant), an equilibrium in which the magnetic energy density is small compared with the kinetic energy density is attained.

It is this latter fact that most distinguishes the Busse and Soward dynamos from the type of dynamo conceived by Malkus & Proctor (1975) which is characterised by magnetostrophic force balance and consequently a magnetic energy density that is the more relevant as far as the Earth's liquid core is concerned. Extension of the Busse and Soward models to strong (rather than weak) field situations (as outlined by Childress & Soward 1972) continues to offer a considerable theoretical challenge.

13.6 The Taylor constraint and torsional oscillations

We shall now consider some general aspects of the dynamics of a rotating fluid within a spherical boundary subject to a combination of Lorentz, Coriolis and buoyancy forces. As both $\mathcal{E}$ and $\mathcal{E}_{\mathrm{m}}$ are very small in the regime relevant to the geodynamo, Taylor (1963) proposed to study the limit system in which both inertia and viscosity are totally neglected in (13.24).[9] The boundary condition for $\mathbf{U}$ is then simply

$$\mathbf{U}\cdot\mathbf{n} = 0 \quad \text{on } r = R\,, \tag{13.50}$$

the no-slip condition being accommodated by an Ekman layer, thickness $\mathcal{O}(\mathcal{E}^{1/2})$ on the surface (see e.g. Greenspan 1968); and the momentum equation takes the form

$$\partial\mathbf{U}/\partial \mathrm{t} + 2\mathbf{\Omega}\wedge\mathbf{U} = -\nabla P + \mathbf{F}\,, \quad \text{with} \quad \mathbf{F} = \rho^{-1}\mathbf{J}\wedge\mathbf{B} + \alpha\Theta\,\mathbf{g}\,. \tag{13.51}$$

13.6.1 Necessary condition for a steady solution $\mathbf{U}(\mathbf{x})$

The φ-component of (13.51) is, in cylindrical polars (s,φ,z),

$$\partial U_\varphi/\partial t + 2\Omega U_s = -\partial P/\partial\varphi + F_\varphi\,. \tag{13.52}$$

The fact that P must be single-valued places an important constraint (Taylor 1963) on functions $F_\varphi(s,\varphi,z)$ for which steady solenoidal solutions of (13.51) exist. Let $\mathscr{C}(s_0)$ denote the cylindrical surface $s = s_0\,,\ |z| < (R^2 - s^2)^{1/2}$ (Figure 13.20). Since $\nabla\cdot\mathbf{U} = 0$, and $\mathbf{n}\cdot\mathbf{U} = 0$ on $r = R$, it is evident that the flux of $\mathbf{U}$ across $\mathscr{C}(s_0)$ must vanish, i.e.

$$\iint_{\mathscr{C}(s_0)} U_s(s,\varphi,z)\, s\,\mathrm{d}\varphi\,\mathrm{d}z = 0\,, \quad \text{for all } s_0\,. \tag{13.53}$$

[9] It is of course worth bearing in mind that direct numerical simulations exhibit different behaviour as the ratio $\mathcal{E}/\mathcal{E}_{\mathrm{m}}$ is varied. The assumption of a 'Taylor state' thus relies on the proposition that in a strong-field regime the field properties do not depend on this ratio of small parameters.

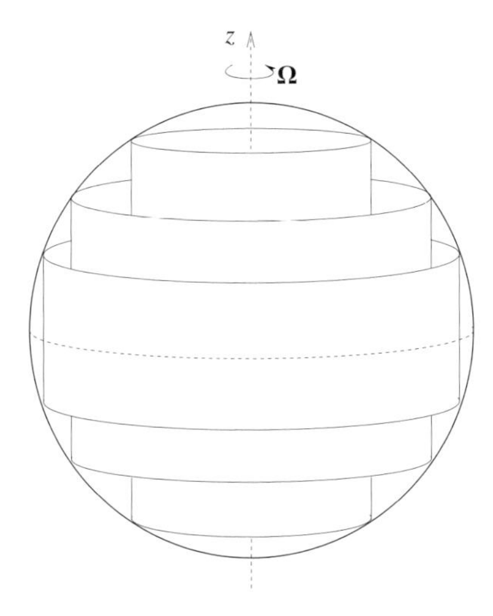

Figure 13.20 Surfaces $\mathscr{C}(s_0) : s = s_0\,,\ |z| < (R^2 - s_0^2)^{1/2}$, for various choices of s_0, corresponding to geostrophic contours in the sphere.

Hence it follows from (13.52) that if $\partial U_\varphi/\partial t = 0$, then

$$\mathcal{T}(s_0) \equiv \iint_{\mathscr{C}(s_0)} F_\varphi(s, \varphi, z) s \,\mathrm{d}\varphi\,\mathrm{d}z = 0\,, \quad \text{for all } s_0\,. \tag{13.54}$$

$\rho\mathcal{T}(s_0)\delta s_0$ is evidently the torque exerted by the Lorentz force $\mathbf{J}\wedge\mathbf{B}$ on the annular cylinder of fluid $s_0 \ll s \ll s_0 + \delta s_0$; unless this torque is identically zero, angular acceleration must result. Since $g_\varphi = 0$, the condition (13.54) may equally be written (under steady conditions $\partial\mathbf{U}/\partial t = 0$)

$$\mathcal{T}(s_0) \equiv \rho^{-1} \iint_{\mathscr{C}(s_0)} (\mathbf{J}\wedge\mathbf{B})_\varphi\, s\,\mathrm{d}\varphi\,\mathrm{d}z \equiv 0\,. \tag{13.55}$$

In this form, the condition is generally known as the *Taylor constraint.*

13.6.2 *Sufficiency of the Taylor constraint for the existence of a steady* U(x)

It is more difficult to establish that the condition $\mathcal{T}(s) \equiv 0$ is also sufficient to ensure that (13.51) can be solved for $\mathbf{U}(\mathbf{x})$. The following discussion is based on Taylor (1963), although differing in points of detail. The curl of (13.51) (with $\partial\mathbf{U}/\partial t = 0$) gives simply

$$\partial\mathbf{U}/\partial z = -(2\Omega)^{-1}\nabla\wedge\mathbf{F} = \mathbf{A}(\mathbf{x})\,,\ \text{say.} \tag{13.56}$$

Note at once that

$$\iint_{\mathscr{C}(s)} A_z s\,\mathrm{d}\varphi\mathrm{d}z = -\frac{1}{2\Omega s}\frac{\partial}{\partial s}\left(s\mathcal{T}(s)\right)\,, \tag{13.57}$$

using the definition (13.54) of $\mathcal{T}(s)$. Hence if $\mathcal{T}(s) \equiv 0$, then also

$$\iint_{\mathscr{C}(s)} A_z s \,\mathrm{d}\varphi\,\mathrm{d}z \equiv 0\,. \tag{13.58}$$

We now suppose that $\mathbf{A}(\mathbf{x})$ is a given solenoidal function satisfying (13.58), and we seek to solve (13.56) for $\mathbf{U}(\mathbf{x})$. Let $z^{\pm} = \pm(R^2 - s^2)^{1/2}$, and for any function $\psi(s, \varphi, z)$, let

$$\overline{\psi}(s, \varphi) = \frac{1}{z^+ - z^-}\int_{z^-}^{z+} \psi(s, \varphi, z)\,\mathrm{d}z\,. \tag{13.59}$$

Also let

$$\mathbf{U}_0(s, \varphi, z) = \int_{z^-}^{z} \mathbf{A}(s,\ \varphi,\ \zeta)\,\mathrm{d}\zeta\,. \tag{13.60}$$

Equation (13.56) then integrates to give

$$\mathbf{U}(s, \varphi, z) = \mathbf{U}_0(s, \varphi, z) + \mathbf{V}(s, \varphi)\,, \tag{13.61}$$

where $\mathbf{V}$ is to be found; and evidently, integrating with respect to z,

$$\mathbf{V}(s, \varphi) = \overline{\mathbf{U}}(s, \varphi) - \overline{\mathbf{U}}_0(s, \varphi)\,. \tag{13.62}$$

The s and z components of $\mathbf{V}$ are determined by satisfying $\mathbf{n} \cdot \mathbf{U} = 0$ on $r = R$, i.e. at $z = z^-,\ z^+$; so

$$sV_s + zV_z = 0 \quad \text{at } z = z^-\,, \tag{13.63a}$$

$$s(V_s + U_{0s}) + z(V_z + U_{0z}) = 0 \quad \text{at } z = z^+\,. \tag{13.63b}$$

These equations uniquely determine $V_s(s, \varphi)$ and $V_z(s,\ \varphi)$ in terms of the known function $\mathbf{U}_0$. It remains therefore to determine $V_\varphi(s, \varphi)$. Now from $\nabla \cdot \mathbf{U} = 0$,

$$\frac{1}{s}\frac{\partial}{\partial s}(sU_s) + \frac{1}{s}\frac{\partial U_\varphi}{\partial \varphi} = -\frac{\partial U_z}{\partial z} = -A_z \quad \text{from (13.56)}\,, \tag{13.64}$$

and so

$$\frac{1}{s}\frac{\partial \overline{U}_\varphi}{\partial \varphi} = -\overline{A}_z - \frac{1}{s}\frac{\partial}{\partial s}(s\overline{U}_s)\,, \tag{13.65}$$

and hence

$$\overline{U}_\varphi(\mathrm{s},\ \varphi) = -\int_0^\varphi \left(\overline{A}_z + \frac{1}{s}\frac{\partial}{\partial s} s\overline{U}_s\right) s\,\mathrm{d}\varphi + v(s)\,, \tag{13.66}$$

where $v(s)$ is an arbitrary function of integration. By virtue of (13.53) and (13.56), the function defined by (13.66) is single-valued and so $V_\varphi(s, \varphi)$ as given by (13.62) is single-valued also. This completes the proof that the condition $\mathcal{T}(s) \equiv 0$ is sufficient for the solvability of (13.56) for steady $\mathbf{U}(\mathbf{x})$.

13.6.3 The arbitrary geostrophic flow $v(s)$

The Taylor state is unique only if the function $v(s)$ in (13.66) can be uniquely determined. This follows from the requirement that $\mathcal{T}(s) \equiv 0$ for all t, so that $\partial\mathcal{T}(s)/\partial t = 0$ also. By differentiating (13.55) with respect to t and using the induction equation, we obtain (Taylor 1963)

$$\frac{\partial\mathcal{T}}{\partial t} = \frac{1}{\mu_0\rho}\iint_{\mathscr{C}(s)} \{\nabla\wedge[\nabla\wedge(\mathbf{U}\wedge\mathbf{B}) + \eta\nabla^2\mathbf{B}]\wedge\mathbf{B} + (\nabla\wedge\mathbf{B})\wedge[\nabla\wedge(\mathbf{U}\wedge\mathbf{B}) + \eta\nabla^2\mathbf{B}]\}_\varphi s\,\mathrm{d}\varphi\mathrm{d}z\,. \tag{13.67}$$

In this complicated integral, let us write

$$\mathbf{U} = \mathbf{U}_1(s,\varphi,z,t) + U_G(s,t)\,\mathbf{e}_\varphi\,, \tag{13.68}$$

where $U_G(s,t)$ is the (as yet) unknown ingredient. (Think of U_G here as a projector onto a Taylor state, just as pressure in an incompressible fluid acts a projector onto a solenoidal flow.) Now when $\mathbf{B}$ is axisymmetric,

$$\nabla\wedge(U_G\,\mathbf{e}_\varphi\wedge\mathbf{B}) = sB_s\,(\partial(U_G/s)/\partial s)\,\mathbf{e}_\varphi\,. \tag{13.69}$$

Using this result in (13.67), the condition $\partial\mathcal{T}/\partial t \equiv 0$ can be reduced to a second-order linear equation for U_G/s

$$a(s)\frac{\partial^2}{\partial s^2}\left(\frac{U_G}{s}\right) + b(s)\frac{\partial}{\partial s}\left(\frac{U_G}{s}\right) + c(s) = 0\,, \tag{13.70}$$

where

$$a(s) = \frac{1}{\mu_0\rho}\iint_{\mathscr{C}(s)} s^2B_s^2\,\mathrm{d}\varphi\,\mathrm{d}z\,, \tag{13.71a}$$

$$b(s) = \frac{1}{\mu\rho}\iint_{\mathscr{C}(s)} \left(2B_s^2 + s(\mathbf{B}\cdot\nabla)B_s\right)\,s\,\mathrm{d}\varphi\,\mathrm{d}z\,, \tag{13.71b}$$

and $c(s)$ contains all the other contributions to (13.67) which do not involve U_G.

This equation requires two boundary conditions. First the flow needs to be regular on the axis, so that U_G/s must be finite at $s = 0$. Second, if the surrounding mantle is modelled as a perfect insulator, then there is no magnetic coupling between core and mantle; the angular momentum H_z of the flow about the z-axis is then constant:

$$H_z = \int_0^{r_\mathrm{o}}\iint_{\mathscr{C}(s)} sU_G\,\mathrm{d}\varphi\,\mathrm{d}z\;\mathrm{d}s = \text{const.} \tag{13.72}$$

For prescribed H_z, the above two conditions appear sufficient to determine a unique solution of (13.70), though degeneracies may occur.

13.6.4 Deviations from the Taylor constraint

The condition $\mathcal{T}(s) \equiv 0$ may be violated if either viscous effects (which are small because $\mathcal{E} \ll 1$) or inertial effects (small because $\mathcal{E}_{\mathrm{m}} \ll 1$) are restored to the governing equations. These effects may be significant if small scales in space and/or time are present in the flow.

If just viscosity is restored, we are led to an 'Ekman state', in which flow through thin $\mathcal{E}^{1/2}$ Ekman layers compensates for the radial flow in (13.20) (Braginsky 1994; Fearn 1994, 1998). This yields an $\mathcal{E}^{1/2}$ geostrophic flow, which adjusts to maintain mass conservation; the Taylor constraint is then replaced by

$$U_G = \mathcal{E}^{-1/2} \left(1 - s^2\right)^{1/4} \iint_{\mathscr{C}(s_0)} (\mathbf{J} \wedge \mathbf{B})_\varphi \, \mathrm{d}\varphi \, \mathrm{d}z \,. \tag{13.73}$$

Alternatively, a non-vanishing magnetic torque leads to acceleration of geostrophic cylinders on short time-scales on which inertia must be restored. The leading order perturbation to the Taylor state then takes the form of an Alfvén wave in the form of a torsional oscillation propagating in the s-direction (see below).

Effects involving both small length-scales and short time-scales may be considered together. For example, inclusion of inertia in the Ekman state (Jault 1995) leads to the equation

$$\mathcal{E}_{\mathrm{m}} \frac{\partial U_G}{\partial t} = \left(1 - s^2\right)^{-1/2} \iint_{\mathscr{C}(s_0)} (\mathbf{J} \wedge \mathbf{B})_\varphi \, \mathrm{d}\varphi \, \mathrm{d}z - \mathcal{E}^{1/2} \left(1 - s^2\right)^{-3/4} U_G \,, \tag{13.74}$$

replacing (13.73).

13.6.5 Torsional oscillations when the Taylor constraint is violated

If $\mathcal{T}(s) \not\equiv 0$, then integration of (13.52) over the cylinder $\mathscr{C}(s)$ gives

$$\frac{\partial}{\partial t} \iint_{\mathscr{C}(s)} U_\varphi(s, \varphi, z) s \, \mathrm{d}\varphi \, \mathrm{d}z = \mathcal{T}(s) \,, \tag{13.75}$$

i.e. angular acceleration is inevitable. Defining

$$U_G(s) = \frac{1}{A(s)} \iint_{\mathscr{C}(s)} U_\varphi(s, \varphi, z) s \, \mathrm{d}\varphi \, \mathrm{d}z \,, \tag{13.76}$$

where $A(s) = 4\pi s(R^2 - s^2)^{1/2}$ is the area of $\mathscr{C}(s)$, (13.76) becomes

$$\frac{\partial}{\partial t} [A(s) U_G(s,t)] = \mathcal{T}(s,t) \,, \tag{13.77}$$

where we now take explicit account of time dependence. $U_G(s,t)$ is the geostrophic ingredient of the total velocity field, and it is unaffected by the rotation $\mathbf{\Omega}$. Determination of U_G is clearly equivalent to the determination of $v(s)$ in (13.66).

As described above, solving for the Taylor state amounts to solving (13.70,b) in the limit $\partial \mathcal{T}(s,t)/\partial t = 0$ (i.e. neglecting the full inertial term). Neglecting only

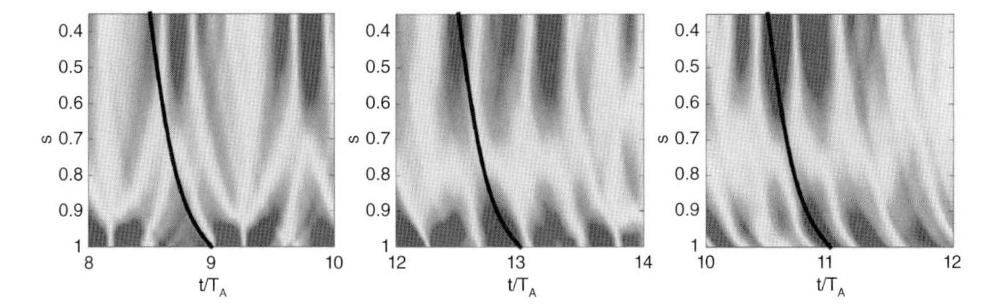

Figure 13.21 Contour plot of $U_G(t/T_A,\ s)/s$ (with Alfvén time T_A) for $Q = 0.02$ (left), 0.3 (middle) and 1.0 (right). [From Gillet et al. 2017 by permission of Oxford University Press on behalf of the Royal Astronomical Society.]

the convective acceleration, restores an acceleration term for the momentum of the cylinders, thus allowing for wave solutions.

Equations (13.77) and (13.70) clearly constitute a hyperbolic system, which may be expected to admit oscillatory solutions about the steady state for which $\mathcal{T} \equiv 0$. In such oscillatory solutions, each cylinder $\mathscr{C}(s)\,\mathrm{d}s$ of fluid rotates about its axis, coupling between the cylinders being provided by the radial field B_s via the coefficients $a(s)$ and $b(s)$. Such torsional oscillations have been studied by Braginskii (1970); the damping of the oscillations, due to Ekman layer effects on $r = R-$, has been discussed in §3 of Roberts & Soward (1972). The time-scale associated with the oscillations is determined essentially by the mean value of $h_s^2 = B_s^2/\mu_0\rho$; it is therefore of order $R/\langle h_s^2\rangle^{1/2}$, the time for an Alfvén wave to propagate on the radial field a distance of the order of the radius of the sphere.

In a situation in which the poloidal field $\mathbf{B}_P$ develops on a very long time-scale $t_\eta = \mathcal{O}\left(R^2/\eta\right)$ by dynamo action (associated with say, an α-effect), then, provided the damping time t_d for the above torsional oscillations is small compared with t_η, a quasi-static situation may be anticipated, in which $\mathcal{T} \equiv 0$ and U_G/s is the resulting steady solution of (13.70). If t_d is not small compared with t_η then torsional oscillations may be expected to persist as long as the development of the poloidal field continues.

The Taylor state, in which all inertial terms are dropped, may be considered as the result of a 'slow-manifold' approach to the full system (from which torsional waves have been filtered out). The well-posedness of this limit system raises delicate mathematical issues (Gallagher & Gérard-Varet, 2017). Solutions for axisymmetric mean-field configuration have however been achieved (Fearn & Proctor, 1987; Roberts & Wu, 2014; Wu & Roberts, 2015; Li et al., 2018).

13.6.6 Effect of mantle conductivity

The boundary condition (13.72) is not trivial, associated as it is with the double limit of an inviscid fluid and a perfectly insulating mantle. The effects of weak electrical conductivity of the lower mantle have been discussed by Jault (2003), Schaeffer & Jault (2016), and by Gillet et al. (2017) who assumed a thin layer of material of thickness δ and conductivity σ_m at the base of the mantle. Where the cylinders s =const. intersect the sphere (at $z = \pm h$), the boundary condition is then

$$B_\phi(s, \pm h) = -Q\, U_G(s)\, B_r(s, \pm h)\,, \tag{13.78}$$

where $Q = \delta\, \sigma_m\, B_0\, \sqrt{\mu_0/\rho}$ gives a dimensionless measure of the lower mantle conductance. These authors have considered torsional modes resulting from stochastic forcing in the fluid core, and have shown that when $Q = 0$, these modes consist of standing waves. If $Q > 0$, these waves propagate outwards like the torsional waves that have been detected in the Earth's core (Gillet et al. 2010 and Figure 13.21). On this basis, the low-mantle conductance was estimated to be in the range 3×10^7 to 10^8 S. Similar torsional oscillations have been extracted from numerical simulations with an imposed dipolar field by Teed et al. (2015).

13.7 Scaling laws

As we have seen, state-of-the-art numerical models of the geodynamo are still, at the time of writing, performed in a parameter regime remote from values relevant to the physics of the Earth's core.[10] In order to establish a connection between dynamo modelling and the actual geophysics, appropriate scaling laws must be developed. Such scaling laws establish the dependence of quantities such as the magnetic field strength on control parameters or computable quantities. They allow for a direct confrontation of advanced models with geophysical constraints. A realistic approach (Jones 2011) is to derive scaling laws that indicate how computable quantities are related to parameters involving viscosity, and to extrapolate these relations to conditions in the Earth's core.

Empirical scaling laws were introduced for the magnetic field strength (Christensen & Aubert, 2006) and have proved robust. They have been applied to numerical models irrespective of the parameter regime, viscous or inertial, as well as to a range of stars and planets (Christensen 2010; Schrinner et al. 2012). In these studies, the field strength is usually expressed in terms of the *Lorentz number* $\mathcal{L}o$, defined by

$$\mathcal{L}o^2 = \frac{B^2}{\mu_0\, \rho\, \Omega^2 L^2} = \Lambda\, \mathcal{E}_\mathrm{m}\,. \tag{13.79}$$

The most typical scaling laws, usually described as 'power-based', relate the field

[10] If numerical models were really approaching a magnetostrophic limit, their output should presumably be independent of viscosity. However, although $\mathcal{E}$ is small in current simulations, it is still a factor $\sim 10^6$ larger than in the Earth's liquid core, and numerical results (Figure 13.11) still exhibit a dependence on $\mathcal{E}$ indicative of viscous effects in the bulk of the flow.

strength to two computable quantities (Christensen & Aubert 2006). The first is the power per unit mass P injected by buoyancy forces, given by

$$P \propto \mathcal{R}a^{\star}\Omega^3 L^2 \quad \text{where} \quad \mathcal{R}a^{\star} \equiv \mathcal{R}a\,(\mathcal{N}u - 1)\,\mathcal{E}^3\,\mathcal{P}r^{-2}\,; \tag{13.80}$$

$\mathcal{R}a^{\star}$ is a modified Rayleigh number[11] based on the advected heat flux (represented by the *Nusselt number* $\mathcal{N}u$); P is a computable quantity (as opposed to a control parameter) whatever the thermal boundary conditions may be (Oruba 2016). The second computable quantity is the fraction of energy $f_{\rm ohm}$ dissipated by ohmic resistance,

$$f_{\rm ohm} = D_\eta/(D_\eta + D_\nu)\,, \tag{13.81}$$

where D_ν and D_η are the rates of viscous and magnetic dissipation. It is expected that $f_{\rm ohm} \to 1$ in the Earth-like limit of vanishing viscosity.

Now the time-averaged balance between energy production and dissipation implies that

$$P = D_\eta + D_\nu = D_\eta/f_{\rm ohm} = f_{\rm ohm}^{-1}\,\ell_B^{-2}\,\eta\,\mathbf{h}^2, \tag{13.82}$$

where $\mathbf{h} = \mathbf{B}/(\mu_0\rho)^{1/2}$, and ℓ_B denotes the 'magnetic-dissipation length-scale', defined by

$$\ell_B^2 = \int_V \mathbf{B}^2\,\mathrm{d}V \bigg/ \int_V (\nabla \wedge \mathbf{B})^2\,\mathrm{d}V\ . \tag{13.83}$$

Hence from (13.79), recalling that $\mathcal{E}_{\rm m} = \eta/\Omega L^2$, we obtain the scaling relation

$$\mathcal{L}o \sim f_{\rm ohm}^{1/2}\,(\ell_B/L)\,(\mathcal{R}a^{\star})^{1/2}\,(\mathcal{E}_{\rm m})^{-1/2}\,, \tag{13.84}$$

which must be satisfied by any statistically steady dynamo. Figure 13.22 demonstrates the relevance of this simple scaling law for a wide database of three-dimensional simulations (including those of Christensen & Aubert (2006)) of fully saturated dynamo action in rotating spherical shells. The simplest description (assuming that ℓ_B/L is constant) provides a good fit for all these numerical results.

The simple result (13.84) may be compared to the power-based scaling law

$$\mathbf{h}^2 \sim f_{\rm ohm}\,P^{0.68}\,\Omega^{-0.04}\,L^{0.68} \tag{13.85}$$

obtained by Christensen & Aubert (2006) as an empirical fit to numerical data. In this, the authors noted the very weak dependence on Ω, in apparent contradiction with the theory developed in Section 13.3. The scaling (13.85) can be expressed in dimensionless form as

$$\mathcal{L}o \sim f_{\rm ohm}^{1/2}\,\mathcal{R}a^{\star\,0.34}\,, \tag{13.86}$$

[11] Note that this $\mathcal{R}a^{\star}$ is denoted $\mathcal{R}a_Q^{\star}$ in Christensen & Aubert (2006) and many subsequent papers.

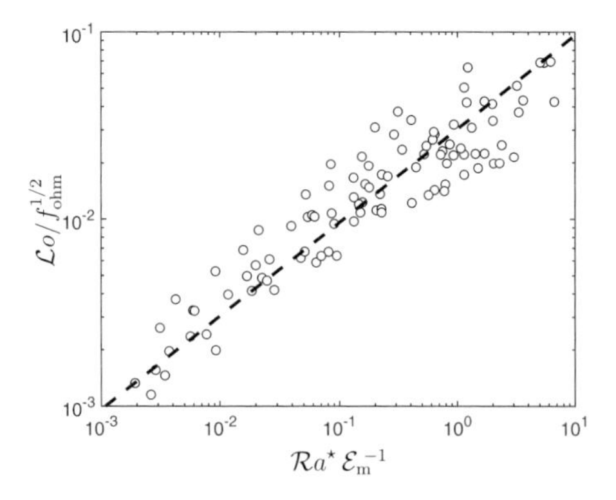

Figure 13.22 Lorentz number scaled by the fraction of ohmic dissipation vs. a combination of the flux-based Rayleigh number, and the magnetic Ekman number (13.84) ; ℓ_B/L is assumed constant. The dashed line has slope 1/2.

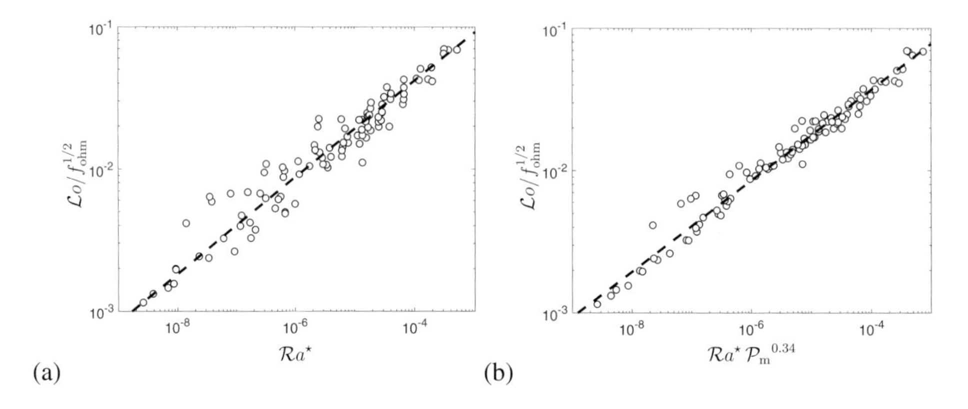

Figure 13.23 The Lorentz number corrected for the relative fraction of ohmic dissipation versus a combination of $\mathcal{R}a^\star$ and $\mathcal{P}_{\text{m}}$, as proposed by Christensen and Aubert (2006): (a) scaling law (13.86), (b) refined scaling law (13.88). (The dashed lines have slopes 0.34 and 0.32, respectively.)

as portrayed in Figure 13.23a. The authors also introduce a finer fit in the form

$$\mathbf{h}^2 \sim f_{\text{ohm}}\, P^{0.64}\, \Omega^{0.08}\, L^{0.72}\, \nu^{0.22}\, \eta^{-0.22}\,, \tag{13.87}$$

which again translates to the dimensionless form

$$\mathcal{L}o \sim f_{\text{ohm}}^{1/2}\, \mathcal{R}a^{\star\,0.32}\, \mathcal{P}_{\text{m}}^{\;0.11} \tag{13.88}$$

(see Figure 13.23b). The results (13.86) and its refined form (13.88) are empirical laws based on numerical experiments. The physical interpretation of (13.86) (Christensen & Aubert 2006) is based on an empirical scaling law for the magnetic dissipation time $\tau_\eta \sim \mathcal{R}_{\text{m}}^{-1}$, and their empirical fit $\mathcal{R}o \sim \mathcal{R}a^{\star\,0.41}$. The scalings (13.86) and (13.88) appear to provide an improvement on (13.84) (compare

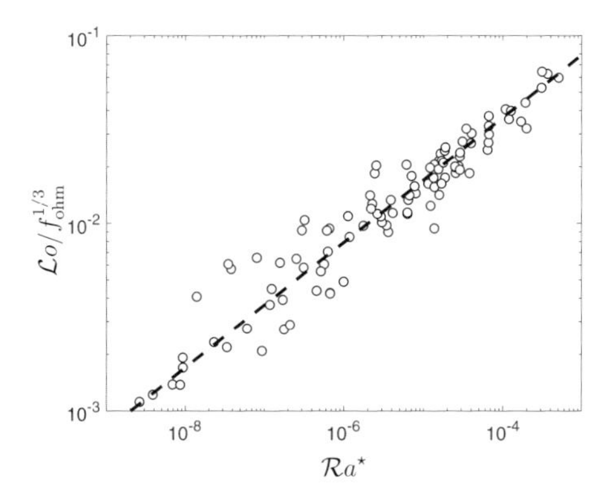

Figure 13.24 The Lorentz number corrected for the relative fraction of ohmic dissipation versus $\mathcal{R}a^{\star}$, as proposed by Davidson (2013); the dashed line has a slope $1/3$.

Figures 13.22 and 13.23) due to a finer description of ℓ_B (Oruba & Dormy, 2014a). Recall however that (13.84) simply reflects the statistical balance between energy production and dissipation, and that any improvement on this requires additional assumptions.

Davidson (2013) has introduced a modified scaling, arguing that in the asymptotic limit expected to be relevant to planetary dynamos, viscous dissipation is negligible ($f_{\rm ohm} \sim 1$), and inertial forces don't enter the dominant force balance ($\mathcal{R}o \ll 1$). Davidson's argument relies on dimensional analysis: in (13.83) with $f_{\rm ohm} = 1$, both P and ℓ_B^2/η are assumed to be independent of Ω. This implies that $\mathbf{h}^2$ depends only on L and P, and thus

$$\mathbf{h}^2 \sim P^{2/3} L^{2/3} . \tag{13.89}$$

Introducing the fraction of ohmic dissipation (which differs from unity in numerical models)

$$\mathbf{h}^2 \sim f_{\rm ohm}^{2/3} P^{2/3} L^{2/3} , \tag{13.90}$$

which can be rewritten

$$\mathcal{L}o \sim f_{\rm ohm}^{1/3} \mathcal{R}a^{\star\,1/3} , \tag{13.91}$$

a result which again conforms reasonably well with available numerical data (see Figure 13.24).

Scaling laws can also be derived from consideration of bifurcations. For numerical dynamos that are not far from the onset of dynamo action (and this is necessarily the case because of computational limitations), the amplitude equations may be used to estimate the strength of the field as a function of the departure from the critical value of the controlling parameter. Pétrélis & Fauve (2001) have derived

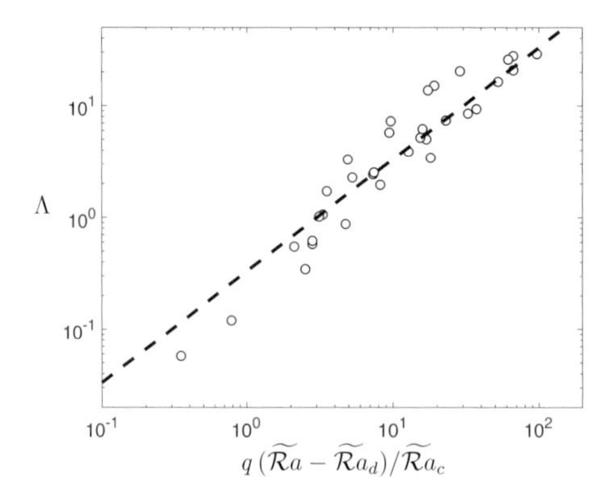

Figure 13.25 Expected scaling for a supercritical 'viscous' bifurcation as proposed by Oruba and Dormy (2014a). (The dashed line has a slope unity.)

such laws as a function of $\mathcal{R}_{\mathrm{m}} - \mathcal{R}_{\mathrm{md}}$, where $\mathcal{R}_{\mathrm{md}}$ is the critical value for dynamo action:

(i) for a laminar (viscous) dynamo

$$\mathbf{B}^2 \sim \frac{\rho\nu}{\sigma L^2}\left(\mathcal{R}_{\mathrm{m}} - \mathcal{R}_{\mathrm{md}}\right), \tag{13.92a}$$

or equivalently

$$\mathcal{L}o \sim \mathcal{E}\mathcal{P}_{\mathrm{m}}{}^{-1/2}\left(\mathcal{R}_{\mathrm{m}} - \mathcal{R}_{\mathrm{md}}\right)^{1/2} \quad \text{or} \quad \Lambda \sim \mathcal{E}\left(\mathcal{R}_{\mathrm{m}} - \mathcal{R}_{\mathrm{md}}\right); \tag{13.92b}$$

(ii) for a turbulent (inertial) dynamo

$$\mathbf{B}^2 \sim \frac{\rho}{\mu\sigma^2 L^2}\left(\mathcal{R}_{\mathrm{m}} - \mathcal{R}_{\mathrm{md}}\right), \tag{13.93a}$$

or equivalently

$$\mathcal{L}o \sim \mathcal{E}_\eta\left(\mathcal{R}_{\mathrm{m}} - \mathcal{R}_{\mathrm{md}}\right)^{1/2} \quad \text{or} \quad \Lambda \sim \mathcal{E}_\eta\left(\mathcal{R}_{\mathrm{m}} - \mathcal{R}_{\mathrm{md}}\right); \tag{13.93b}$$

and (iii) for a turbulent strong-field dynamo

$$\mathbf{B}^2 \sim \frac{\rho\Omega}{\sigma}\left(\mathcal{R}_{\mathrm{m}} - \mathcal{R}_{\mathrm{md}}\right), \tag{13.94a}$$

or equivalently

$$\mathcal{L}o \sim \mathcal{E}_\eta^{1/2}\left(\mathcal{R}_{\mathrm{m}} - \mathcal{R}_{\mathrm{md}}\right)^{1/2} \quad \text{or} \quad \Lambda \sim \left(\mathcal{R}_{\mathrm{m}} - \mathcal{R}_{\mathrm{md}}\right). \tag{13.94b}$$

The above three laws implicitly assume a supercritical bifurcation.

Oruba & Dormy (2014a) have extended (13.92b) to the case of a thermally driven dynamo, with the result

$$\mathbf{h}^2 \sim \Omega\kappa\left(\Delta T - \Delta T_d\right)/\Delta T_c\,, \tag{13.95}$$

or equivalently

$$\mathcal{L}o \sim \mathcal{E}_{\mathrm{m}}{}^{1/2}\left[\left(\widetilde{\mathcal{R}a} - \widetilde{\mathcal{R}a}_d/\widetilde{\mathcal{R}a}_c\right)\right]^{1/2}, \quad \text{or} \quad \Lambda \sim q\left[\left(\widetilde{\mathcal{R}a} - \widetilde{\mathcal{R}a}_d\right)/\widetilde{\mathcal{R}a}_c\right], \tag{13.96}$$

where $\mathcal{R}a_c$ corresponds to the onset of convection and $\mathcal{R}a_d$ to the onset of dynamo action. This expression provides a good description of current numerical models (Figure 13.25), thus highlighting yet again their viscous nature.

Finally, several points suggest that numerical simulations and the Earth's core dynamo are not in the same dynamical regime. First, the energy balance between production and dissipation provides an estimate of the flux-based Rayleigh number for the Earth's core $\mathcal{R}a^{\star}_{\text{Earth}} = 3 \times 10^{-16}$, which, when applied to scaling laws derived from numerical data, yields values of the magnetic field strength significantly weaker than expected in the core. A similar discrepancy occurs when applying the scaling law based on the magnetic Reynolds number. The last law (13.95) relies on a viscous balance. Viscous forces are indeed shown to play a major role in the bulk of the flow in numerical simulations, and not only in the vicinity of boundaries. Scaling laws derived from numerical simulations are therefore not generally applicable to the Earth's core, unless, like (13.84), they are extremely general, reflecting simply the statistical balance between energy production and dissipation.

14

Astrophysical dynamic models

14.1 A range of numerical approaches

A full description of solar and stellar dynamos requires integration of the magnetohydrodynamic equations in a rotating fluid, as in the preceding chapter. However the parameter regime relevant to stellar dynamics involves much stronger nonlinearities. The Rossby number in the Sun is estimated to be $\mathcal{O}(10^{-2})$ at the base of the convection zone, and as large as $\mathcal{O}(10^{3})$ at the photosphere (the solar surface). The magnetic Reynolds number is also significantly larger in stars than in planets, ranging from $\mathcal{O}(10^{11})$ at the base of the convective zone to $\mathcal{O}(10^{6})$ at the photosphere. It follows that direct numerical simulation (DNS) of the interplay between rotation, convection and magnetic fields in the stellar regime is far beyond current computational capacity. Stars are generally in a very turbulent state, which cannot be fully resolved on present computers. For this reason, we must rely either on full models in a simpler parameter regime remote from the real systems, or on reduced models.

A hierarchy of numerical models have been introduced in order to investigate magnetic field generation in stars. These are presented below in increasing order of complexity.

14.1.1 Low-order models

Low-order models, although simplistic like those considered in Chapter 11, are useful in that they can identify generic properties associated with symmetries. In this formalism, the 22-year solar cycle is a natural consequence of a supercritical Hopf bifurcation[1] of the magnetic field. Assuming a hydrodynamic amplitude $Z(t)$ (which represents both thermal convection and differential rotation) satisfying an

[1] A 'Hopf bifurcation' is one in which two complex conjugate growth rates (eigenvalues of the linearised perturbation system) cross the imaginary axis from left to right.

equation of the form

$$\dot{Z} = \mu - Z^2 , \tag{14.1}$$

the Hopf bifurcation for the magnetic field, taking account of the $B \to -B$ symmetry, is then described by the equation

$$\dot{B} = (\lambda + \mathrm{i}\,\omega)\, B + (a + \mathrm{i}\, b)\, Z\, B + \mathcal{O}(B\,|B|^2) , \tag{14.2}$$

where B is complex.

The linear problem corresponds to a complex eigenvalue $\lambda + \mathrm{i}\omega$ associated with a growth rate λ and frequency ω. The Hopf bifurcation results in a cyclic (i.e. time-periodic) magnetic field. Introducing $B = X + \mathrm{i}\, Y$, and retaining only leading-order (quadratic) non-linearities, we thus get

$$\left.\begin{aligned} \dot{X} &= \lambda X - \omega Y + a Z X - b Z Y \\ \dot{Y} &= \lambda Y + \omega X + a Z Y + b Z X \end{aligned}\right\} , \tag{14.3}$$

(where a and b are real parameters). This system is typical of a Hopf bifurcation. Interpreted in terms of amplitude equations for an $\alpha\omega$ mean-field dynamo, X may be interpreted as an amplitude of the toroidal field and Y of the poloidal field.

Dynamo cycles or dynamo waves are easily described through low-order models of this kind. As well as describing the 22-year sunspot cycle, such models usually seek to describe longer-term characteristics of the solar field. For example, Tobias et al. (1995) were able to describe modulations of the cycle (i.e. variations in amplitude) as a result of deterministic non-linear processes. Once the cyclic magnetic field has been established, modulations of the cycle may arise either via the interaction between fields with different symmetries (described as Type 1 modulation), or via the interaction of the magnetic field with a velocity component in a symmetry subspace (Type 2 modulation).

As already described in Chapter 5, early observations of sunspots in the seventeenth century revealed considerable activity, which was however interrupted by the Maunder minimum (see Figure 5.4). Measurements of the abundances of the cosmogenic isotopes ^{14}C (in tree rings) and ^{10}Be (in polar ice cores) indicate that magnetic cycles persisted throughout the Maunder minimum. Similar 'grand minima' have recurred aperiodically over the past 10,000 years (Usoskin 2017), indicating that the Sun switches over a long time-scale between strong modulation with deep grand minima and weaker modulation.

Tobias et al. (1995) argued that following a supercritical Hopf bifurcation leading to dynamo action with stable periodic solutions, a secondary Hopf bifurcation would provide quasi-periodic orbits and eventually chaos. In order to illustrate this,

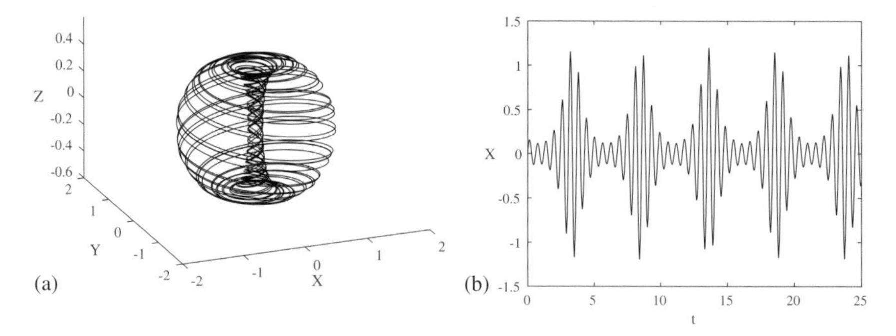

Figure 14.1 (a) A quasi-periodic trajectory in the phase-space of the variables (X, Y, Z) for the low-order model (14.4). (b) Time series for the toroidal field X showing a deterministic modulated cyclic activity. [Parameters from Tobias et al. 1995.]

they considered a modified version of (14.3) of the form

$$\left.\begin{aligned} \dot{X} &= (\lambda + a\,Z)\;X - \omega Y \\ \dot{Y} &= (\lambda + a\,Z)\;Y + \omega X \\ \dot{Z} &= \mu - Z^2 - \left(X^2 + Y^2\right) + c\,Z^3 \end{aligned}\right\}, \qquad (14.4)$$

where λ is the bifurcation parameter (and c is an additional real parameter). This deterministic dynamical system produces Type 2 modulations reminiscent of solar modulations (see Figure 14.1). Interaction between dipole and quadrupole modes leads to Type 1 modulation, as analysed by Knobloch & Landsberg (1996). These approaches were unified by Knobloch et al. (1998) through consideration of interaction between the two types of modulation. This approach can be further extended, grand minima being then interpreted as trajectories which approach a homoclinic or heteroclinic bifurcation in phase space (Weiss 2011; Weiss & Tobias 2016).

14.1.2 Mean-field models

Mean-field models offer a significant extension over low-order models in that they can incorporate spatial variations, either in one dimension (usually latitude) or in two dimensions (usually axisymmetric). Since differential rotation is so effective at creating toroidal fields, nearly all solar and stellar mean-field models are axisymmetric $\alpha\omega$-dynamos, like those already described in §9.9 and §9.10. The freedom of choice in setting the mean-field parametrisation is such that these models are capable of reproducing essential features of the large-scale solar magnetic field. For example, Beer et al. (1998) have demonstrated the existence of intensity modulations in a non-linear mean-field dynamo model that show the same route to modulation as described in low-order models. They also observed that, after a grand

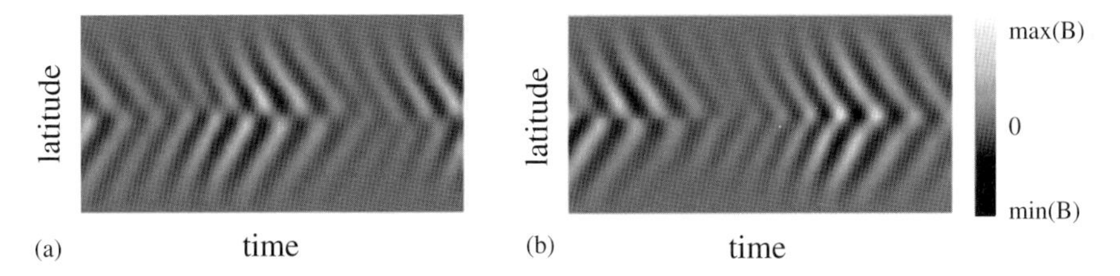

Figure 14.2 Butterfly diagrams from a mean-field model with grand minima. (a) The toroidal field is almost antisymmetric about the equator when the field is strong but is asymmetric during and just after the grand minima. In a later sequence, (b) the solution emerges from the grand minimum with quadrupolar symmetry (i.e. the toroidal field is symmetric about the equator) despite entering that minimum in a dipole (antisymmetric) state. [From Beer et al. 1998.]

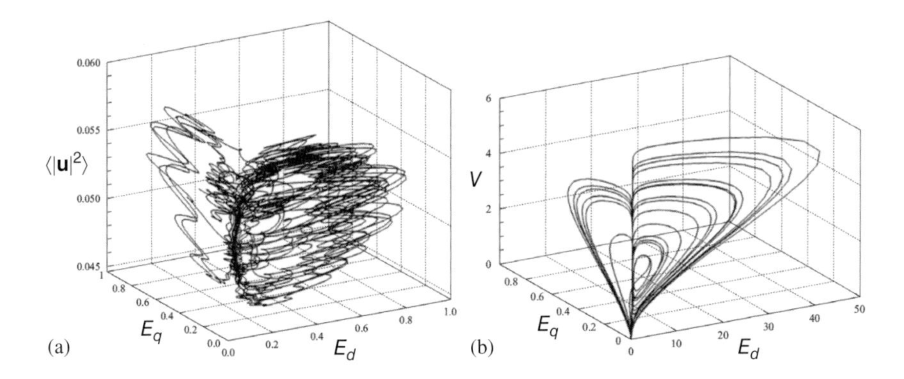

Figure 14.3 Trajectories showing flipping of symmetry during grand minima, projected onto the three-dimensional space of the dipole E_d, quadrupole E_q and mean flow $\langle |u|^2 \rangle$. (a) For the mean-field model, with the same parameters as in Figure 14.2, showing cyclic behaviour, modulation and occasional flipping from the dipole to the quadrupole symmetries. (b) For the sixth-order model of Knobloch et al. (1998) showing a similar behaviour.

minimum, the toroidal field may switch from antisymmetric to symmetric about the equatorial plane (see Figure 14.2). Similar behaviour can be observed with low-order models (as illustrated in Figure 14.3).

Although efficient in reproducing observations, mean-field models are not usually predictive. Bushby & Tobias (2007) point to the difficulties in predicting the solar cycle using mean-field models, owing to the significant modulation of the solar activity cycle due to either stochastic or deterministic processes. Deterministic chaos and stochastic fluctuations of the mean-field coefficients have been investigated (and reviewed by Ossendrijver 2003) in order to describe such spatial and temporal variability.

When the variation of mean-field coefficients with depth is taken into account, different locations for the α- and ω-effects can be incorporated. The early models of Babcock and Leighton deserve particular mention (Babcock 1961; Leighton 1969). In such models, the decay of tilted bipolar active regions regenerates the poloidal field, so that the α-effect operates at the solar surface. In this picture, bipolar magnetic regions emerge, with opposite leading/trailing polarity patterns in each hemisphere. When the magnetic regions start to decay, the leading components experience diffusive cancellation across the equator, while the trailing components have moved to higher latitudes. This process is capable of accounting for the dynamo cycle.

When only the surface differential rotation was observed, these models were very attractive. With the discovery of the tachocline, new models described as 'flux-transport dynamo models' evolved, in which the ω-effect operates in the tachocline and a meridional circulation is invoked in order to couple the surface layers to the base of the convection zone (Rempel 2006; Charbonneau 2010, 2014). This circulation acts as a 'conveyor belt' transporting poloidal flux to the tachocline and toroidal flux to the surface. In these flux-transport dynamos, it is the meridional circulation that accounts for the solar butterfly diagram and largely determines the period of the solar cycle.

Parker (1993) formulated a different model known as the 'interface dynamo'. In this model the mean toroidal field is generated by the ω-effect in the tachocline where it suppresses turbulent diffusivity, and from where it diffuses into the overlying convection zone. In contrast to models of the Babcock–Leighton type in which poloidal field is generated near the photosphere, the poloidal field in interface dynamo models is produced near the base of the convection zone, but separated by an 'interface' from the tachocline. This separation of ω- and α-effects is very like that described in the cartesian model of §9.5.

Tobias (1997) introduced a non-linear extension of Parker's cartesian model, in which the generation of the cyclic magnetic field drives a large-scale velocity field which then modulates the basic cycle (see also MacGregor & Charbonneau 1997). Mason et al. (2002) investigated both interface and surface dynamo models and found that the interface model is much more effective in explaining observed fields, with the conclusion that it makes little sense to assume that the α-effect is solely concentrated near the solar surface.

14.1.3 Direct numerical simulations

Numerical models have been developed either locally as cartesian models for shear layers near surface or tachocline, or globally for a full sphere or a spherical shell. These models are either fully compressible, or anelastic in which sound waves have been filtered out, or Boussinesq for which $D\rho/Dt = 0$ and so $\nabla \cdot \mathbf{u} = 0$.

The Boussinesq approach is free from mean-field parametrisations, but cannot describe some of the effects induced by density stratification. For this reason, many models of stellar dynamos rely on the anelastic approximation, which takes full account of the depth-dependence of the ambient density; this is at the cost of approximations that are not always easy to justify on mathematical grounds (see §13.1 and Dormy & Soward 2007). There are different anelastic formulations depending on the underlying hypotheses; in general however, as previously indicated, the output of such models does not differ much from that of Boussinesq models (Raynaud et al. 2014, 2015).

Direct numerical simulations (DNS) for a full star can be traced back to Glatzmaier & Gilman (1982). Rather than attempting to cover the vast numerical literature since then, we shall restrict attention to models that seek to establish a connection with theoretical developments, as reviewed by Miesch & Toomre (2009) and Miesch (2012).

A surprising property of direct numerical simulations is that in many cases they produce a 'reversed butterfly' diagram (perhaps 'flutter-by diagram' would be the appropriate term!), with magnetic structures drifting away from the equator (see next section). This property is incompatible with the solar dynamo. There are however some exceptions (e.g. the models of Ghizaru et al. 2010 and Käpylä et al. 2013) who have succeeded in producing solar-like butterfly diagrams.

Direct numerical simulations can also be used to constrain mean-field models. For example, simulations of turbulent compressible convection by Tobias et al. (2001) allowed the study of a buoyantly unstable region (the convection zone) overlying a stable overshoot region (the tachocline). In examining the fate of a horizontal layer of magnetic field imposed on the unstable region, they showed that the downward topological pumping of magnetic flux is robust and that the peak magnetic energy comes to reside in the lower stable region. This again militates against the models of Babcock and Leighton, in which the regeneration of poloidal field relies on toroidal flux in active regions near the solar surface. In the simulations of Tobias et al, only the strongest flux tubes were able to rise to the surface through magnetic buoyancy. The efficient downward pumping favours the lower regions of the convection zone or the overshoot region just above the tachocline as the preferred site for the α-effect and associated regeneration of poloidal field.

With a similar aim of establishing connection with mean-field models, Dikpati & Gilman (2009) and Featherstone & Miesch (2015) have investigated the importance of flux transport by meridional circulation. Raynaud & Tobias (2016) showed that the interactions between convection, differential rotation and magnetic fields can lead to modulation of the basic cycle similar to what is found with low-order models. Using fully three-dimensional anelastic simulations, they reproduced modulations previously obtained using mean-field models, including 'supermodulation' in which the system alternates between Type 1 modulation with little change of symmetry and clusters of deep minima, and Type 2 modulation involving significant changes in symmetry. The pattern of supermodulation provides evidence of deterministic chaos (Weiss & Tobias 2016).

14.2 From planets to stars

Because of their very different nature (liquid metal in one case, plasma in the other), planetary and stellar systems are studied by different communities, geophysical on the one hand, astrophysical on the other. As a practical matter however, the numerical techniques used in numerical studies of these two systems, are surprisingly similar. To a large extent this is due to the restricted parameter space accessible to present-day computers, far from conditions in the real planetary or stellar systems. As we saw in Chapter 13, the main problem for planetary dynamos concerns the extreme values of the Ekman and magnetic Ekman numbers, whereas for stellar dynamos (with no solid boundaries) it is the huge values of the Reynolds and magnetic Reynolds numbers that cause the problem.

The remarkable success of numerical models in reproducing some of the key features of the magnetic fields of both Earth and Sun led Goudard & Dormy (2008) to consider the possible importance of the aspect ratio of the dynamo region (i.e. the ratio $\mathcal{A} \equiv r_{\rm i}/r_{\rm o}$ $(0 \leq \mathcal{A} \leq 1)$ of the radii of the inner and outer spheres bounding the convecting region). In the Earth the inert solid inner core extends to some 35% of the core radius ($\mathcal{A} = 0.35$), while in the Sun the radiative zone extends to 70% of the solar radius ($\mathcal{A} = 0.70$). The convective zones of stars and planets may be expected to have all possible aspect ratios, extending even to fully convective spheres for which $\mathcal{A} = 0$.

Goudard & Dormy showed that by increasing $\mathcal{A}$, a sharp transition occurs at $\mathcal{A} \approx 0.65$ from a dipole-dominated large-scale magnetic field to a time-periodic dynamo with a weaker dipole ingredient. Above this critical value – close to that of the Sun – the strong dipole rapidly weakens, while dynamo action continues in the form of a wavy solution with quasi-periodic reversals (see Figure 14.4), thus successfully simulating some aspects of solar magnetic behaviour. This indicates that the aspect ratio $\mathcal{A}$ can play an important part in determining the type of dynamo solution that is realised.

Because of the strong symmetry of the convective flows as influenced by the rapid rotation of the planet or star, two linearly independent families of solutions can exist – those having dipole or quadrupole symmetry. Figure 14.5 shows the azimuthally averaged field from the direct numerical simulations of Goudard & Dormy (2008). The Earth-like mode is represented for aspect ratios of 0.45 (a) and 0.6 (b). The active dynamo region lies outside the tangent cylinder and is therefore

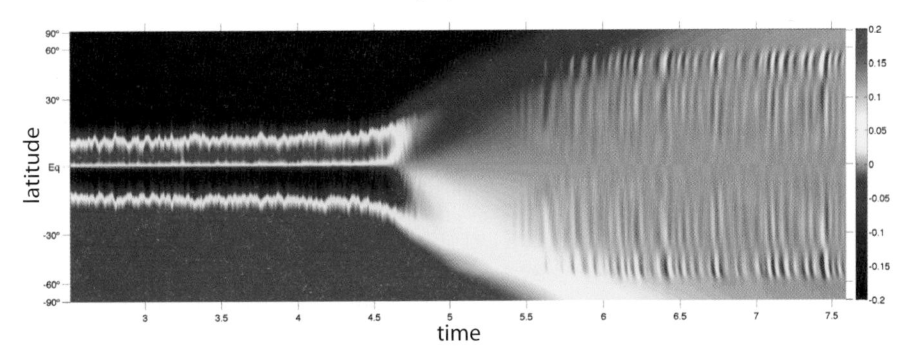

Figure 14.4 Time evolution of the radial magnetic field averaged in longitude (for an aspect ratio $\mathcal{A} = 0.65$). The initial dipole field survives for a few diffusion times and then vanishes to yield a butterfly-like diagram. [From Goudard & Dormy 2008.]

increasingly constrained as the inner sphere radius increases. The dipole eventually drops for large aspect ratio, when the volume outside the tangent cylinder becomes too small.

The weak dipolar (WD) branch, discussed in Chapter 13, is thus unstable at larger aspect ratio to an oscillatory dynamo mode. Comparison with reduced parametrised models can help interpret this transition to the solar-like mode. As a strong zonal flow develops in the multipolar mode, one may anticipate a possible transition from an α^2-dynamo to an $\alpha\omega$-dynamo as the aspect ratio $\mathcal{A}$ is increased. The situation is more complicated, as we shall see below.

14.3 Extracting dynamo mechanisms

The test-field (or 'tracer field') method, introduced by Schrinner et al. (2007, 2011), allows computation of mean-field coefficients from direct numerical simulations. These coefficients can then be used in mean-field models in order to explore underlying dynamo mechanisms. In this way, Schrinner et al. (2011) investigated the mechanism associated with the cyclic dynamos of Goudard & Dormy (2008), as illustrated in Figure 14.6. On the left is shown the azimuthally averaged radial magnetic field at the surface as a function of time, obtained by direct numerical simulation. A dynamo wave migrates from the equator until it reaches mid-latitudes where the inner-core tangent cylinder intersects the outer shell boundary (a 'flutter-by' diagram). The characteristics of the magnetic field are typical of the FM branch introduced in Chapter 13: the field is multipolar and its intensity is weak, as expressed in the case considered here by an Elsasser number $\Lambda = B_{\rm rms}^2/\mu\rho\eta\Omega = 0.13$.

The kinematically advanced tracer field (middle panel of Figure 14.6) grows

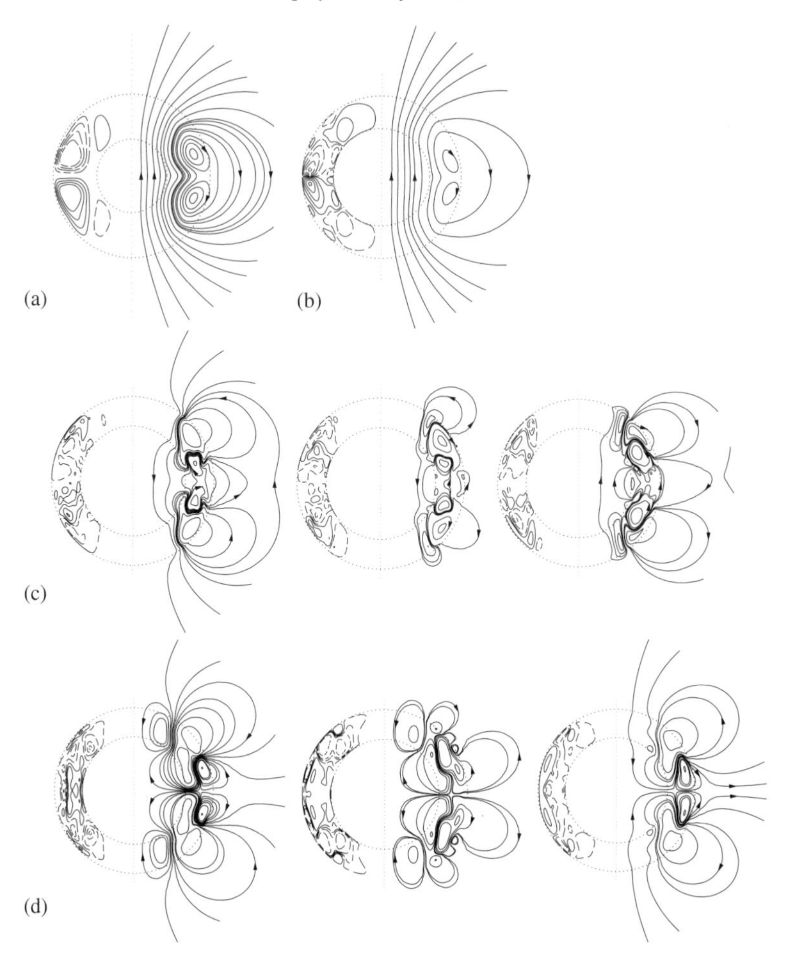

Figure 14.5 Zonal average of the magnetic field in the 3D simulations of Goudard & Dormy (2008). Contours of the toroidal part of the field are plotted in the left hemisphere and lines of force of the meridional (poloidal) part in the right hemisphere. The aspect ratio $\mathcal{A}$ is increased from 0.45 (a) to 0.6 (b) and to 0.65 (c)–(d). The sequence of dynamo waves is represented for the antisymmetric mode (c) and symmetric mode (d); this is qualitatively similar to the output of the mean-field models of P. H. Roberts (1972b) (as described in §9.9 – see Figure 9.14 on p. 264).

slowly in time, i.e. the model under consideration is kinematically unstable (see §14.5). Deviations of the tracer field from the actual field are barely detectable in the field structure. Moreover, the dynamo wave persists in the kinematic calculation (see also Goudard & Dormy 2008). Such time-dependent modes offer an interesting test for the mean-field coefficients derived from the test-field approach.

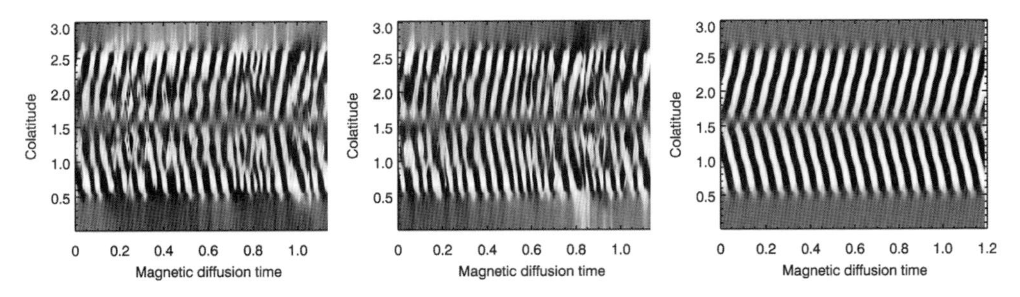

Figure 14.6 Azimuthally averaged radial magnetic field at the outer shell boundary varying with time (here a 'flutter-by' diagram) resulting from direct numerical simulation (left), kinematic calculation (middle) and mean-field calculation (right). The contour plots have been normalised by their maximum absolute value at each time step considered. [From Schrinner et al. 2011, reproduced with permission © ESO.]

The right-hand panel of Figure 14.6 shows the results of a mean-field calculation based on the dynamo coefficients derived from the test-field approach and the mean flow determined by the direct numerical simulation. The fastest growing eigenmode corresponds to a dynamo wave which compares well with the direct numerical simulation.

The influence of differential rotation can be artificially suppressed in the kinematic calculation of the tracer field without altering any other component of the flow. This results in a butterfly diagram of the kinematically advanced tracer field, as shown in the left panel of Figure 14.7. The evolution of the magnetic field is again time-periodic. The right panel of Figure 14.7 shows the butterfly diagram from the resulting kinematic mean-field α^2-dynamo. This diagram shows migration towards the equator as for the solar dynamo. This indicates that the (quasi-geostrophic) differential rotation retained in the model of Figure 14.6 differs from the solar differential rotation, as in fact already recognised from observations based on helioseismology (Figure 5.3) (see Balbus et al. 2012 for discussion of this issue).

The role of the large azimuthal shear (responsible for an ω-effect) in the dynamo cycle is therefore not obvious. Instead, the action of small-scale convection seems to be essential. Surprisingly, in this case, the corresponding $\alpha\omega$-dynamo leads to a non-oscillatory magnetic field (Schrinner et al. 2011).

14.4 Dipole breakdown and bistability

The weak dipolar mode (WD) and the fluctuating multipolar mode (FM) introduced in Chapter 13 can in general be distinguished by the ratio of a typical convective length-scale to the Rossby radius, i.e. the local Rossby number (see Schrinner et al.

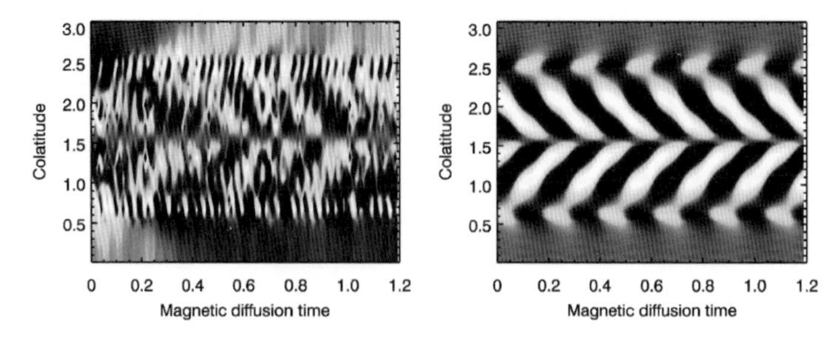

Figure 14.7 Azimuthally averaged radial magnetic field at the outer shell boundary varying with time (butterfly diagram) resulting from a kinematic calculation with suppressed ω-effect (left) and a corresponding mean-field calculation (right). The contour plots are presented as in Figure 14.6. [From Schrinner et al. 2011, reproduced with permission © ESO.]

2012 for a definition of the relevant $\mathcal{R}o_\ell$ in the case of stress-free boundaries). Models with a predominantly dipolar magnetic field are obtained if the convective scale is at least an order of magnitude larger than the Rossby radius ($\mathcal{R}o_\ell \gg 1$). In the case of stress-free boundaries (relevant to solar and stellar dynamos) the transition between these two dynamo modes can exhibit hysteresis (see Simitev & Busse 2009 and Figure 14.8). Zonal shear associated with the geostrophic zonal flow is essential for models with stress-free boundary conditions (and presumably also for no-slip boundary conditions at very small Ekman number). The abrupt transition from the WD to the FM mode is then replaced by a region of bistability in which dipolar and multipolar dynamos can co-exist.

There is a strong correlation between the topology and the time-dependence of the magnetic field in dynamo models. Sudden polarity reversals or oscillations of the magnetic field do not in general occur in dipole-dominated models, i.e. in the low Rossby number regime. Conversely, reversals and oscillations are frequent in non-dipolar models with $\mathcal{R}o_\ell > 0.1$ as well as in models with lower local Rossby numbers with stress-free boundary conditions belonging to the multipolar branch.

Whether non-dipolar models exhibit fairly coherent oscillations or irregular reversals of the magnetic field depends strongly on the magnetic Reynolds number. Dynamo models (in the non-dipolar regime) at high magnetic Reynolds number exhibit less temporal coherence. The relevance of these models to the Sun or other stars, for which the magnetic Reynolds number is very large, remains problematic.

14.5 Kinematically unstable saturated dynamos

In order to better understand how dynamo systems saturate, Cattaneo & Tobias (2009) studied the kinematic dynamo properties of velocity fields that arise from dynamos that are already saturated by the effect of the Lorentz force. They solved

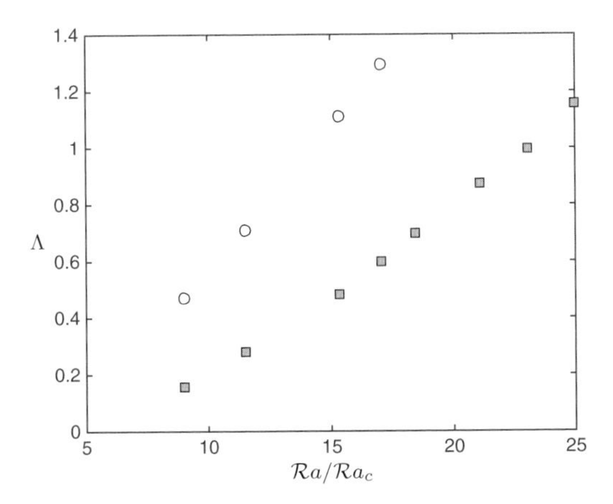

Figure 14.8 Evolution of the magnetic field strength, measured by the Elsasser number, for a Boussinesq model with stress-free boundaries. Both branches (WD and FM) are described as the Rayleigh number is varied at fixed Ekman and magnetic Prandtl numbers ($\mathcal{E} = 10^{-4}$ and $\mathcal{P}_m = 1$). Open circles denote models dominated by a dipole field (WD), and grey squares denote multipolar models (FM). [From Schrinner et al. 2012.]

two systems of equations, one for the actual (active) magnetic field and the other for the same velocity field and an independent, passive, tracer vector field, a strategy similar to the test-field approach used in §14.3. In this way, they showed that it was possible for the 'saturated flow' to act as a kinematic dynamo through a tertiary bifurcation. Schrinner et al. (2011) showed that this is the bifurcation that initiates the fluctuating multipolar (FM) regime discussed earlier, whereas the flow corresponding to a saturated dynamo on the weak dipolar branch (WD) before this bifurcation is kinematically stable.

To understand this mechanism, consider again the simplest low-order model for a supercritical bifurcation (13.29). If the unstable mode B is complex rather than real, i.e. $B = |B|e^{i\phi}$, then in order to preserve the phase of the cubic saturating term, we must write

$$\dot{B} = \lambda B - |B|^2 B = \lambda B - \bar{B} B^2 , \quad \lambda > 0. \tag{14.5}$$

This complex equation has two linearly independent unstable modes.

The saturation in the original equations occurs through the back-reaction of the quadratic Lorentz force. This may be represented by coupled model equations

$$\dot{V} = -\mu V + B^2 , \quad \dot{B} = \lambda B - \bar{B} V , \quad \mu > 0, \tag{14.6}$$

where V represents the velocity perturbation induced by the Lorentz force,

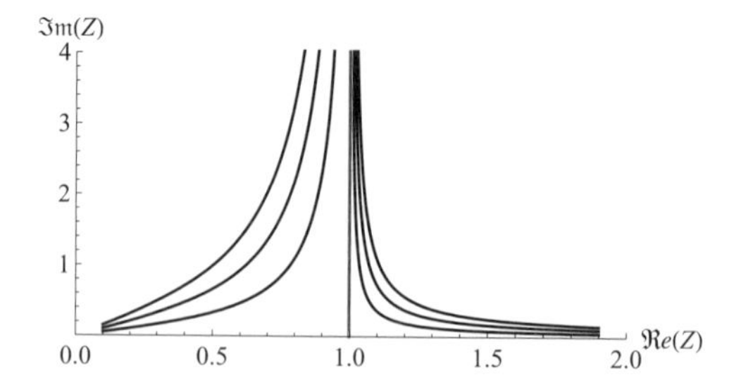

Figure 14.9 Trajectories in complex Z-plane for system (14.7), with $\dot{V} = 0$, $\lambda = \mu = 1$ and starting from the points $B = \mathcal{R}e(Z) = 0.1$, $\Im m(Z) = 0.05,\ 0.1,\ 0.15$ and $B = \mathcal{R}e(Z) = 1.9$, $\Im m(Z) = 0.05,\ 0.1,\ 0.15$.

resulting in saturation. If $\dot{V}$ is negligible (neglect of inertia!), then (14.6) provides an expanded form equivalent to (14.5) (if $\mu = 1$). For any choice of the phase, the mode governed by (14.6) is linearly unstable; and the saturation term conforms its phase to the field to provide saturation at $V = B^2/\mu$, $|B|^2 = \lambda\mu$.

The kinematically unstable saturated dynamo can then be understood by considering an additional equation for a test field $Z(t)$ in conjunction with (14.6):

$$\dot{V} = -\mu V + B^2, \quad \dot{B} = \lambda B - \bar{B} V, \quad \dot{Z} = \lambda Z - \bar{Z} V. \tag{14.7}$$

Evidently V will still saturate the growth of B, but not of Z if the phase of $Z(0)$ is different from that of $B(0)$. Typical trajectories in phase space are illustrated in Figure 14.9. The simple system (14.7) indicates precisely how a dynamo-saturated flow (here represented by V) can still act as a kinematic dynamo.

14.6 The galactic dynamo

Most of the analytical and numerical work on dynamo action in galaxies falls within the kinematic mean-field formalism, as already covered in § 9.6. It is usually assumed that the flow in the galactic disc consists of differential rotation $\omega(s)$ around the vertical axis, and small-scale cyclonic turbulence. The angular velocity $\omega(s)$ in galaxies is observed to be roughly inversely proportional to the cylindrical radius s, thus providing an ω-effect. The origin of an α-effect is more controversial. It is usually assumed that turbulence in the interstellar medium is primarily driven by supernova explosions (McCray & Snow, 1979), which often occur in clusters within the galactic disc. These explosions cause giant expanding cavities of gas known as super-bubbles, which are affected by the Coriolis forces associated with galactic rotation. This results in cyclonic motions and so helicity in the gas flow (Ferrière 1992a,b). Assuming that each explosion remnant remains spherical, Ferrière (1998) has calculated the resulting α-effect for an axisymmetric mean-field model.

Figure 14.10 Artist's illustration of the accretion disc formed by the black hole in the binary system system 4U1630-47. [NASA's Astronomy Picture of the Day, 20 November 2013; Illustration credit: NASA, CXC, M. Weiss.]

The mean-field approach can be justified only on the assumption of an adequate scale separation between the mean flow (i.e. the differential rotation) and the turbulence (Beck et al., 1996). Super-bubbles have typical scale ~ a few hundred parsecs, not dramatically less than the typical vertical scale of a galaxy (~ 1 kpc). Nevertheless, direct numerical simulations by Gissinger et al. (2009) have shown results in good agreement with earlier mean-field models.

The key difficulty with galactic dynamo models is to account for the amplification of a seed field to the current measured intensity (typically ~ μG). Most studies so far have been kinematic; however Moss & Sokoloff (2011) have discussed saturation mechanisms in the galactic context, and whether these are compatible with observations.

14.7 Accretion discs and the magnetorotational instability (MRI)

Accretion discs form when a massive star attracts ionised gas from the surrounding interstellar medium or sometimes from a neighbouring star on which it 'feeds'. The inflowing gas is generally in rotation around the star, flow towards the axis of rotation being impeded by the centrifugal force that it experiences. In cylindrical polar coordinates (s, φ, z) with origin at the centre of the star and Oz aligned along the axis of rotation, the gas flows preferentially towards the plane $z = 0$, where the accretion disc is formed.

Black holes also have accretion discs as illustrated by the NASA artist's illustration shown in Figure 14.10. Here, the black hole sucks material from the neighbouring star on the right into its accretion disc. Relativistic jets are emitted along the axis of symmetry, as described by Blandford & Payne (1982); the remarkable collimation of these jets is probably due at least in part to a strong local toroidal magnetic field generated in the immediate neighbourhood of the black hole and exerting a pinch effect (see §17.3). The fact that the jet plasma also contains atoms of heavy elements such as iron and nickel was discovered by Trigo et al. (2013), as reported in the online subtext of this image.

As an element of fluid of unit mass is drawn towards the central mass M by the gravitational force GM/s^2, its angular momentum κ is conserved if viscous effects are ignored. Its orbital velocity $v(s) = \kappa/s$ in the accretion disc thus increases as s decreases, reaching an equilibrium when the outward centrifugal force equals the inward gravitational attraction:

$$v^2/s = GM/s^2, \quad \text{or equivalently} \quad \Omega = (GM/s^3)^{1/2} = \Omega_K(s), \text{ say,} \qquad (14.8)$$

where $\Omega(s) = v(s)/s$ is the angular velocity of the element. Here, $\Omega_K(s)$ is the Keplerian angular velocity distribution, which will be most naturally established, as in planetary systems, in any accretion disc. In this state,

$$v(s) \sim s^{-1/2} \quad \text{and} \quad \kappa(s) \sim s^{1/2}. \qquad (14.9)$$

This *increase* of angular momentum κ with radius s should be particularly noted.

14.7.1 Rayleigh stability criterion

Consider a circulating flow with arbitrary angular velocity distribution $\Omega(s)$, and suppose we interchange two rings of fluid at $s = s_1, s_2, (s_2 > s_1)$, each of unit mass, their angular momenta κ_1 and κ_2 being conserved. Then the gravitational potential energy is unchanged, but the kinetic energy changes by an amount

$$\Delta E = \tfrac{1}{2}\left(\kappa_2^2 - \kappa_1^2\right)\left(\frac{1}{s_1^2} - \frac{1}{s_2^2}\right). \qquad (14.10)$$

If $\Delta E > 0$ for every pair s_1, s_2, then energy is required to effect this interchange, implying stability. This leads to Rayleigh's stability criterion (Rayleigh 1917)[2]: the

[2] It is of some historical interest to note (Moffatt 2014) that this result appeared 50 years earlier in the following problem posed, among others, by James Clerk Maxwell for the 1866 Smith's Prize Examination at the University of Cambridge:
A mass M of fluid is running round a circular groove or channel of radius a with velocity u. An equal mass is running round another channel of radius b with velocity v. The one channel is made to expand and the other to contract till their radii are exchanged. Show that the work expended in effecting the change is

$$-\frac{1}{2}\left(\frac{u^2}{b^2} - \frac{v^2}{a^2}\right)(a^2 - b^2)M.$$

Hence show that the motion of a fluid in a circular whirlpool will be stable or unstable according as the areas described by particles in equal times increase or diminish from centre to circumference.

flow is stable or unstable according as

$$\frac{d}{ds}\kappa^2(s) \gtrless 0. \tag{14.11}$$

Since $\kappa(s)$ is the circulation (as well as the angular momentum) at radius s, Rayleigh's criterion is frequently encountered in the form "the flow of an inviscid fluid is stable or unstable according as the circulation increases or decreases outwards". Viscosity is stabilising, as revealed in the seminal work of Taylor (1923), but the condition $d\kappa^2/ds > 0$ survives in a viscous fluid as a *sufficient* condition for stability of this type of flow.

This condition is evidently satisfied by Keplerian flow which is therefore stable according to Rayleigh's criterion (and we should note that it is axisymmetric perturbations of the above 'interchange' variety that are most prone to instability). And herein lies a problem for accretion discs; for if all fluid elements entering the disc have the same angular momentum κ_0, say, then they would all accumulate at the same radius κ_0^2/GM. In order to establish anything approaching the Keplerian angular momentum distribution $\kappa \sim s^{1/2}$, some mechanism must be present to break the constraint of conserved angular momentum, in other words to expel angular momentum in such a way as to allow continuing inward transport of mass and the formation of accretion discs as observed. Molecular viscosity is far too small for this purpose, and appeal must be made to some kind of turbulent viscosity; but if the basic flow is stable, how is a turbulent state to be achieved?

14.7.2 Magnetorotational instability

This impasse was overcome in the seminal pair of papers by Balbus & Hawley (1991) and Hawley & Balbus (1991), who recognised that the presence of a weak poloidal magnetic field threading the disc can lead to an instability, now known as the 'magnetorotational instability' (MRI), which can provide the required transition to turbulence. The reason is that, if two rings of fluid are now interchanged in the manner described above, the inner ring will have an excess of angular momentum relative to its new surroundings, and the outer ring will have a deficit; the resulting differential rotation will generate a toroidal field as described in §3.14. In the zero resistivity limit, the Lorentz force develops a strong φ-component, which ultimately causes each ring to rotate with its original angular velocity. The net effect, as detailed analysis confirms, is that Rayleigh's stability criterion (14.11) is now superseded by a new 'magnetorotational criterion': the flow is stable or unstable according as

$$\frac{d}{ds}\Omega^2(s) > \text{or} < 0. \tag{14.12}$$

For Keplerian flow, $\mathrm{d}\Omega^2/\mathrm{d}s < 0$, so the flow, although stable by Rayleigh's criterion, is unstable to this magnetorotational instability.

This instability was in fact discovered by Velikhov (1959), and further considered by Chandrasekhar (1960), who were mainly concerned with the stabilising effect of a strong axial magnetic field on the classical flow between differentially rotating cylinders. Velikhov did however obtain the criterion (14.12); he also provided the above interchange description and recognised that "in a sufficiently weak field this effect leads to instability of flow". However, it was only with the work of Balbus & Hawley (1991) that this instability was rediscovered in the accretion-disc context, a breakthrough that led to an explosion of research in this area. No matter how weak the poloidal field $\mathbf{B}_P$ may be, the magnetorotational mechanism leads to instability, although finite conductivity eliminates it in the strict limit $\mathbf{B}_P \to 0$.

14.7.3 Shearing-box analysis

If we adopt a local Cartesian coordinate system $Oxyz$ with moving origin at the material (Lagrangian) point $(s_0, \Omega(s_0)t, 0)$, and axes Ox in the radial direction, Oy in the local φ-direction and Oz as before, then relative to O the local flow consists of a rotation $\Omega_0\,\mathbf{e}_z \equiv \Omega(s_0)\,\mathbf{e}_z$ and a uniform shear flow

$$\mathbf{U} = (0, -\alpha x, 0) \quad \text{where} \quad \alpha = -s\mathrm{d}\Omega/\mathrm{d}s|_{s=s_0}. \tag{14.13}$$

We may assume that $\Omega_0 > 0$. For the Keplerian flow with $\Omega \sim s^{-3/2}$, $\alpha = \frac{3}{2}\Omega_0$. If we imagine a box centred on the point O and rotating with angular velocity $\Omega_0\,\mathbf{e}_z$, then the flow within this box is simply the uniform shear (14.13); hence the term 'shearing-box analysis'. A poloidal field is then represented locally by a uniform field $B_0\,\mathbf{e}_z = (\mu_0\rho)^{1/2}h_0\,\mathbf{e}_z$. This type of local representation goes back to Goldreich & Lynden-Bell (1965), and was adopted by Balbus & Hawley (1992) to provide a simplified description of the magnetorotational instability, in which differential rotation is locally represented by uniform rotation and shear combined. In what follows, we build on the review of Ogilvie (2007), and for simplicity we neglect molecular diffusion effects, i.e. we take $\eta = \nu = 0$.

Consider then perturbations of this basic state in the plane of the disc,

$$\mathbf{u} = (u_x(z,t),\, u_y(z,t)), \quad \mathbf{h} = (h_x(z,t),\, h_y(z,t)), \tag{14.14}$$

where we use the usual Alfvén scaling for the magnetic field. With this choice, all non-linearities vanish, and in the rotating frame of reference the perturbations satisfy the usual momentum and induction equations, which take the component

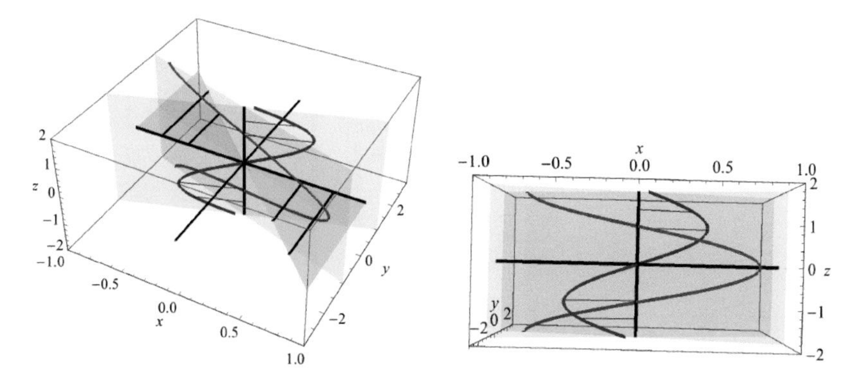

Figure 14.11 Two projections of the field structures in MRI instability of Keplerian flow ($\alpha = 3$, $h_0 = 1$): (a) the shear in the x, y-plane is shown in green; the plane $x = y$ of the velocity perturbation is shown in blue and the perpendicular plane $x = -y$ of the magnetic perturbation in red; (b) projection onto the x, y-plane; the profiles of the fields u and h are shown by the curves $3 \sin kz$ and $5 \cos kz$, respectively.

form,

$$\frac{\partial u_x}{\partial t} - 2\Omega_0 u_y = h_0 \frac{\partial h_x}{\partial z}, \quad \frac{\partial u_y}{\partial t} - \alpha u_x + 2\Omega_0 u_x = h_0 \frac{\partial h_y}{\partial z}, \tag{14.15}$$

$$\frac{\partial h_x}{\partial t} = h_0 \frac{\partial u_x}{\partial z}, \quad \frac{\partial h_y}{\partial t} + \alpha h_x = h_0 \frac{\partial u_y}{\partial z}. \tag{14.16}$$

With the further assumptions that

$$(u_x, u_y) = a\, e^{\alpha t/2} \sin kz\,(1, 1) \quad \text{and} \quad (h_x, h_y) = b\, e^{\alpha t/2} \cos kz\,(1, -1), \tag{14.17}$$

where a and b are constant amplitudes, it may be easily verified that both equations (14.16) are satisfied provided

$$h_0 k\, b = (2\Omega - \alpha/2)\, a \quad \text{and} \quad h_0 k\, a = (\alpha/2)\, b. \tag{14.18}$$

It then follows that both equations (14.15) are also satisfied provided

$$h_0^2 k^2 = (2\Omega - \alpha/2)(\alpha/2) \quad \text{and} \quad b/a = 2h_0 k/\alpha. \tag{14.19}$$

For the Keplerian $\Omega_K(s)$ (given by 14.8), this gives $(b/a)^2 = 5/3$. Thus the magnetic perturbation is (in these Alfvén units) only marginally stronger than the velocity that maintains it, for this unstable mode; the reason it is not greater, as might be expected from the shearing effect, is because the wavenumber k increases with α, so that the current $\nabla\times\mathbf{h}$ and the associated Lorentz force also increase in proportion. The instability represented by the unstable mode (14.17) (sketched in Figure 14.11) – and this is in fact the *most* unstable mode – is quite remarkable, because

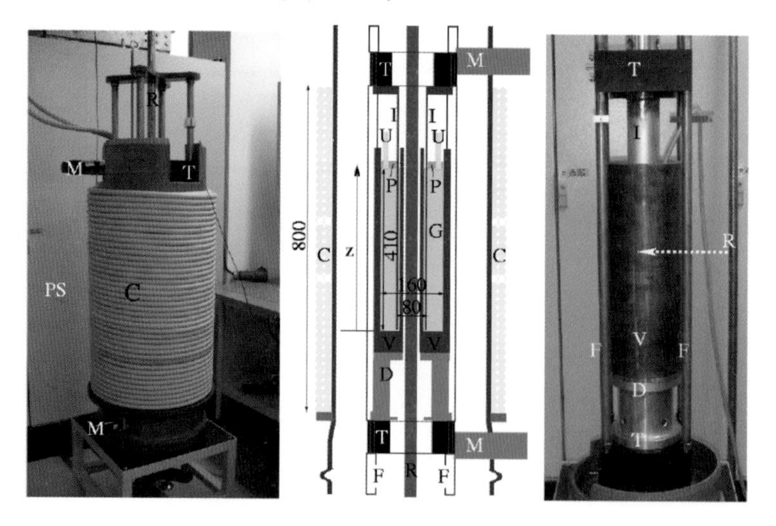

Figure 14.12 The experimental facility PROMISE of the Dresden Research Centre; the left panel shows the complete facility; the middle panel shows a schematic, in which the letters signify as follows: V = copper vessel; I = inner cylinder; G = liquid metal (GaInSn); U = ultrasonic transducers; P = Plexiglass lid; T = high precision turntables; M = motors; F = frame; C = coil; R = copper rod; PS = power supply up to 8000 A; the numbers denote dimensions in mm; the right panel shows the central module, without the coils C, and with the rod R removed from the centre. [From Stefani et al. 2007. © Deutsche Physikalische Gesellschaft, Reproduced by permission of IOP Publishing, CC-BY-NC-SA.]

it does not appear to depend on the strength of the magnetic field $h_0\,\mathbf{e}_z$, but merely requires that this should be non-zero. There are however limitations on the above simple analysis. To be relevant to an accretion disc of thickness $2z_0$, say, we obviously require that $2\pi/k \lesssim 2z_0$, i.e. that $h_0 \lesssim \alpha z_0/2\pi$, since otherwise conditions at the disc boundary need to be considered. In fact, as found by Velikhov (1959), a sufficiently strong field h_0 is stabilising through the dominant influence of the (elastic) Maxwell tension in the weakly distorted lines of force.

At the other extreme, if h_0 is very small, then k is very large, and the diffusive effects of finite viscosity and resistivity are no longer negligible. These damping effects inevitably eliminate the instability in the limit $h_0 \to 0$. The molecular diffusivities ν and η are extremely small in an accretion disc, and a very weak poloidal magnetic field is sufficient to trigger the instability.

14.7.4 Dynamo action associated with the magnetorotational instability

The question naturally arises as to whether dynamo action within an accretion disc can generate the mean poloidal field that is a crucial requirement for the magnetorotational instability This question has been addressed by Rincon et al. (2007), who have concluded that self-sustaining dynamo action can indeed occur through a cyclic process that is analogous to the process responsible for the maintenance of subcritical turbulence in shear flows. This cyclic process consists of three elements: starting with a seed field with both x and z components, the shear in the y-direction generates a y-component of field by the now conventional shearing effect – a process that can be regarded as a transient instability, like that already encountered in §3.8; the deformed field is then subject to the magnetorotational instability on vertical scales less than the scale of the initial z-component; and finally, non-linear interaction of the unstable MRI modes provides feedback capable of amplifying the original seed field.

This final stage, being essentially non-linear, means that this type of dynamo action cannot occur in a linearised model; for this reason, it must be seen as a subcritical instability of the undisturbed state. The non-linearity of the cycle means that computational exploration is required to establish the consistency and viability of the process as envisaged. Rincon et al. (2007) have carried out such computations, and have shown that dynamo action is indeed possible for magnetic Reynolds numbers $\mathcal{R}_\mathrm{m} \gtrsim 670$ and magnetic Prandtl number $\mathcal{P}_\mathrm{m}$ of order 100 and greater. The computations have been extended to lower $\mathcal{P}_\mathrm{m}$, and particularly to the range $2 < \mathcal{P}_\mathrm{m} < 10$ by Riols et al. (2013), but not as yet to small $\mathcal{P}_\mathrm{m}$. This is still an area of active current research.

14.7.5 Experimental realisation of the magnetorotational instability

Experimental detection of the magnetorotational instability under normal laboratory conditions is extremely difficult because, just as for the VKS dynamo experiment, a magnetic Reynolds number at the limit of current experimental possibility is required. An instability of possibly magnetorotational character has however been realised in the experiment of Sisan et al. (2004) at the University of Maryland, in which differential rotation was established in a spherical shell between differentially rotating spheres of radius $a = 12.5$mm and $b = 150$mm. The shell contained liquid sodium, and this was already in turbulent motion when a weak axial magnetic field B_0 was externally applied, with variable Lundquist number $S = B_0 b/\eta(\mu_0\rho)^{1/2}$ in the range $0 < S < 10$. The magnetic Prandtl number for liquid sodium is $\mathcal{P}_\mathrm{m} = 8.8 \times 10^{-6}$, and magnetic Reynolds numbers in the range $0 < \mathcal{R}_\mathrm{m} < 25$ were explored. Waves of instability were detected propagating round the axis of symmetry, which were interpreted as magnetorotational in character, but sustained dynamo action (which would have involved switching off the applied field) was not realised in this range of parameters. We note here the investigation (Petitdemange et al. 2013) of similar waves of instability in a spherical shell geometry under magnetostrophic conditions, where it was found that "an $m = 1$-mode can dominate in the non-linear regime when a sufficiently large toroidal field is present".

A less ambitious experimental approach is to artificially impose a relatively strong toroidal field in addition to the weaker axial field, in order to mimic the first 'differential-rotation' phase of the above cyclic process. This can be achieved in a cylindrical annulus, by passing a current I along the central axis. Such an experiment has been developed at the Dresden Research Centre; the apparatus (as at 2007) is illustrated in Figure 14.12 (from Stefani et al. 2007); the fluid used was the liquid metal alloy Galinstan (GaInSn). Improvements were introduced subsequently in order to minimise effects of Ekman suction at the end plates (Stefani et al. 2008, 2009). With these improvements, there is evidence of instabilities of MRI type which accord well with theoretical results. However, these experiments are conducted at relatively low $\mathcal{R}_{\mathrm{m}}$, whereas genuine magnetorotational instability, involving a strong induced toroidal field, is necessarily a high-$\mathcal{R}_{\mathrm{m}}$ phenomenon. Much however remains to be learnt from experiments of this kind, for which the technology is slowly developing.

Instabilities associated with the circular Couette-flow geometry have been explored in detail at the Princeton Plasma Physics Laboratory in a succession of papers from Ji et al. (2001) to Wei et al. (2016), with a view to experimental detection of the MRI, but again this elusive instability has not yet been convincingly detected in experiments, although the theoretical foundation appears quite secure.

15

Helical Turbulence

15.1 Effects of helicity on homogeneous turbulence

We have seen in Chapter 7 that lack of reflexional symmetry in a random 'background' velocity field $\mathbf{u}(\mathbf{x}, t)$, and in particular a non-zero value of the mean helicity $\langle \mathbf{u} \cdot \nabla \wedge \mathbf{u} \rangle$, is likely to be a crucial factor as far as the effect on large-scale magnetic field evolution is concerned. In these circumstances, it is appropriate to consider the general nature of the dynamics of a turbulent velocity field endowed with non-zero mean helicity.

In order to study turbulence with helicity it is necessary to deliberately inject lack of reflexional symmetry through appropriate control of the source of energy for the flow. The natural way to do this, as indicated by the analysis of §12.2, is to generate turbulence in a rotating fluid by some means that distinguishes between the directions $\pm\boldsymbol{\Omega}$, where $\boldsymbol{\Omega}$ is the rotation vector. For example, if a grid is rapidly drawn through a rotating fluid in the direction of $\boldsymbol{\Omega}$, the resulting random velocity field may be expected to lack reflexional symmetry.

This situation was realised in the laboratory by Ibbetson & Tritton (1975), who measured the decay of the mean-square of the three velocity components. This was followed by experiments on grid turbulence in a pipe rotating about its axis (Jacquin et al. 1990), and on turbulence in Rayleigh-Bénard convection in a cubical vessel rotating about a vertical axis and heated from below (Fernando et al. 1991), with emphasis on the Rossby-number dependence of turbulence statistics. No technique has as yet been realised for direct measurement of mean helicity;[1] however a method of detecting both writhe and twist contributions to helicity in experiments on vortex linkage and reconnection has been recently reported by Scheeler et al. (2017), and further developments in this direction arc to be expected.

[1] A possible method for making such a measurement was proposed in Appendix A3 of Léorat (1975), but has not yet been realised.

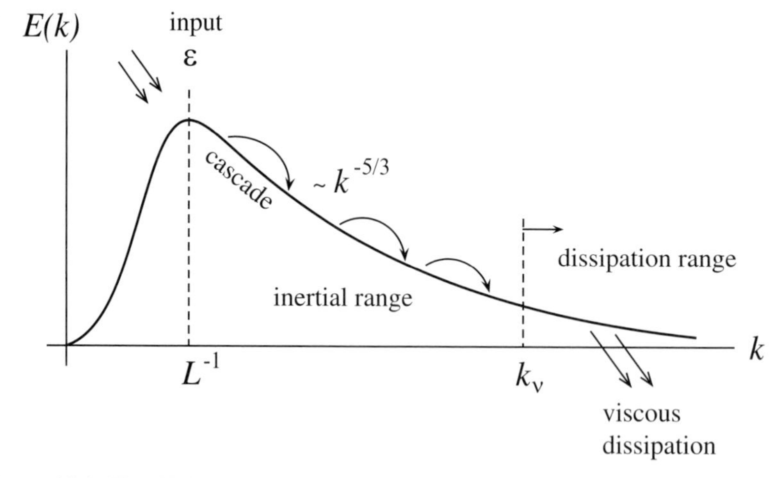

Figure 15.1 The Kolmogorov cascade: energy is injected at a rate ε at scale L, cascades to successively smaller scales and is ultimately dissipated by viscosity on the scale $\ell_\nu = k_\nu^{-1}$; conditions in the 'inertial range' $L^{-1} \ll k \ll k_\nu$ are determined solely by ε and the wave number k.

15.1.1 Energy cascade in non-helical turbulence

Before considering the effects of helicity on the dynamics of turbulence, let us first recall the essential features of high Reynolds number turbulence, as conceived by Kolmogorov (1941a) in a theory which has had great influence in a huge range of applications, and which is now generally referred to as K41. Consider a statistically steady state in which kinetic energy is generated on a length-scale L at a rate ε per unit mass. Let $u_0 = \langle \mathbf{u}^2 \rangle^{1/2}$ and suppose that the Reynolds number $\mathcal{R}e = u_0 L/\nu$ is large. Due to dynamic instability, the energy then 'cascades' in $\mathbf{k}$-space through a sequence of increasing wave number magnitudes $L^{-1} = k_0 \to 2k_0 \to 2^2 k_0 \to \cdots \to 2^n k_0 \to \cdots$ until it reaches a wave number k_ν, say, at which viscous dissipation is adequate to dissipate the energy at the rate ε (Figure 15.1). This is determined in order of magnitude by ε and ν and is therefore given (dimensionally) by

$$k_\nu \sim (\varepsilon/\nu^3)^{1/4} . \tag{15.1}$$

Moreover, on this picture, the energy level $\frac{1}{2}u_0^2$ depends on the rate of supply of energy ε to the system and on L, but not on ν, so that again on dimensional grounds

$$u_0^2 \sim (\varepsilon L)^{2/3} . \tag{15.2}$$

Eliminating ε between (15.1) and (15.2) gives

$$\ell_\nu \sim L\,\mathcal{R}e^{-3/4} . \tag{15.3}$$

In the range of wave numbers $k_0 \ll k \ll k_\nu$ (described as the 'inertial range'), the energy spectrum tensor $\Phi_{ij}(\mathbf{k})$ (see §7.8) is statistically decoupled from the energy source (which is confined to wave numbers of order k_0), and may therefore be expected to be isotropic and to be determined solely by the parameter ε which represents the rate of flow of energy across any wave number magnitude k in the inertial range. In the absence of any helicity effect, $\Phi_{ij}(\mathbf{k})$ is then given by

$$\Phi_{ij}(\mathbf{k}) = \frac{E(k)}{4\pi k^4}\left(k^2\delta_{ij} - k_i k_j\right), \tag{15.4}$$

and the energy spectrum function $E(k)$ is given, again on dimensional grounds, by

$$E(k) = C\varepsilon^{2/3}k^{-5/3} \quad (k_0 \ll k \ll k_\nu), \tag{15.5}$$

where C is a dimensionless constant of order unity. The associated vorticity spectrum function is given by

$$\Omega(k) = k^2 E(k) = C\varepsilon^{2/3}k^{1/3} \quad (k_0 \ll k \ll k_\nu). \tag{15.6}$$

For $k \gtrsim k_\nu$, both $E(k)$ and $\Omega(k)$ suffer a rapid (quasi-exponential) cut-off due to viscous dissipation. The mean-square vorticity (or 'enstrophy') $\langle\omega^2\rangle = \int_0^\infty \Omega(k)\mathrm{d}k$ is clearly dominated by contributions from the neighbourhood of the viscous cut-off wave number $k = k_\nu$, and increases as ν decreases; in fact, since the rate of dissipation of energy is $\nu\langle\omega^2\rangle$, we have the exact result

$$\langle\omega^2\rangle = \varepsilon/\nu. \tag{15.7}$$

From (15.2) and (15.7)

$$\langle\omega^2\rangle \sim \mathcal{R}e\,(u_0/L)^2, \tag{15.8}$$

or equivalently

$$\lambda = u_0/\langle\omega^2\rangle^{1/2} \sim \mathcal{R}e^{-1/2}\,L. \tag{15.9}$$

Here λ is an intermediate length-scale satisfying

$$L \gg \lambda \gg \ell_\nu. \tag{15.10}$$

One interpretation of the length λ (the Taylor micro-scale) is that it provides a measure of the mean radius of curvature of an instantaneous streamline of the flow. It is customary to use the Reynolds number based on λ, $\mathcal{R}e_\lambda \equiv \lambda u_0/\nu \sim \mathcal{R}e^{1/2}$ in reporting experimental or computational results in turbulence.

15.1.2 Intermittency

This however is not the whole story, as Kolmogorov (1962) himself recognised. He had expressed his original theory of isotropic turbulence (K41) in terms of structure

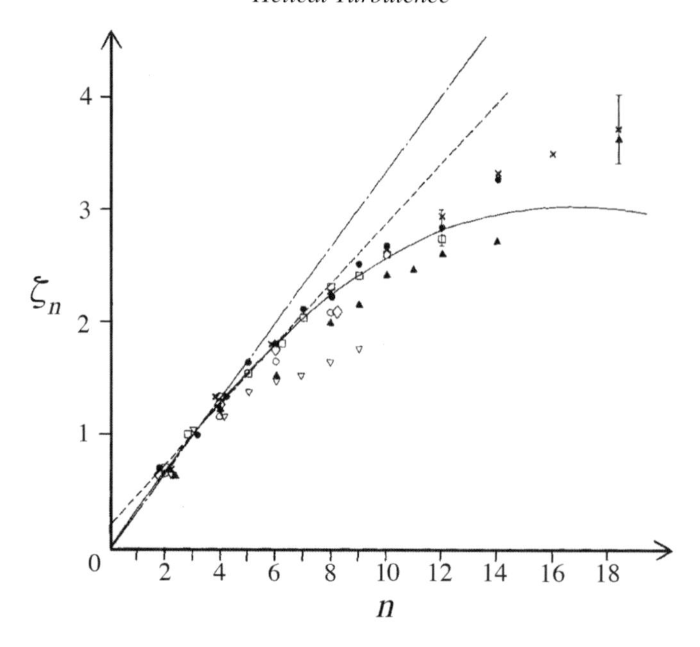

Figure 15.2 Variation of exponent ζ_n as a function of the order n. The various symbols correspond to different turbulent flows; the chain-dotted line (slope 1/3) corresponds to the K41 theory; the solid and dashed curves correspond to alternative models that take some account of intermittency. [From Anselmet et al. 1984.]

functions $S_p(r)$, defined by the spatial averages

$$S_p(r) = \langle (u(\mathbf{x}+\mathbf{r}) - u(\mathbf{x}))^p \rangle, \quad (p, 2, 3, 4, \dots), \tag{15.11}$$

where u is the component of velocity in the direction of the separation vector $\mathbf{r}$. Obviously $S_1(r) = 0$, and moments involving transverse components of velocity can be related to $S_p(r)$ through considerations of isotropy and incompressibility. According to K41, in the inertial range $\{L \gg r \gg \ell_\nu\}$, $S_p(r)$ should be determined by $r = |\mathbf{r}|$ and ε alone, and therefore, on dimensional grounds,

$$S_p(r) = C_p(\varepsilon r)^{p/3}, \tag{15.12}$$

where the constants C_p are universal, the same for any field of turbulence. The result $S_2(r) = C_2(\varepsilon r)^{2/3}$ is just the counterpart of the spectral formula (15.5), the constants C_2 and C being related. For $p = 3$, (15.12) gives

$$S_3(r) = \langle (u(\mathbf{x}+\mathbf{r}) - u(\mathbf{x}))^3 \rangle = C_3\, \varepsilon r, \tag{15.13}$$

a third order moment that is in fact determined by the process of transfer of energy from large to small scales. Kolmogorov (1941b) showed that (15.13) is in fact exact, and that $C_3 = -4/5$, a result described by Frisch (1991) as "perhaps the most important rigorous result in fully developed turbulence". There are indeed

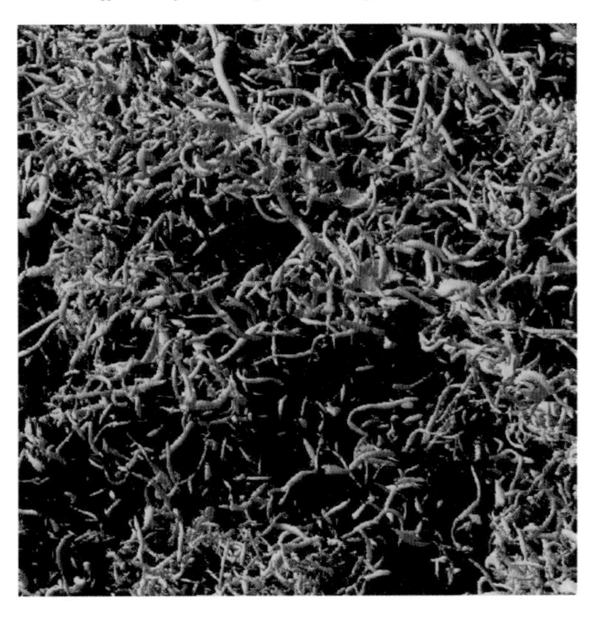

Figure 15.3 Vorticity iso-surfaces $\omega \equiv |\boldsymbol{\omega}|$=cst. in a direct numerical simulation of turbulence at $\mathcal{R}e_\lambda = 732$, showing the region where $|\omega| > \bar{\omega} + 4\sigma$; $\bar{\omega}$ and σ are, respectively, the mean and standard deviation of $|\omega|$. The size of the display domain is $(748^2 \times 1496)\ell_\nu$, this being a zoom from the computational domain of size $(5984^2 \times 1496)\ell_\nu$. [From Ishihara et al. 2007, reproduced with permission.]

very few results in the theory of turbulence that can be described in this sense as rigorous!

For $p > 3$, the result (15.12) is unfortunately not in good agreement with available experimental results. The exponents ζ_n for the modified structure functions

$$\tilde{S}_p(r) = \langle |u(\mathbf{x} + \mathbf{r}) - u(\mathbf{x})|^p \rangle \sim r^{\zeta_n} , \qquad (15.14)$$

(Anselmet et al. 1984) are shown in Figure 15.2 for $n = 2, 3, 4, \ldots 18$; these show a systematic departure from the K41 theory, which would again require that $\zeta_n \sim n/3$. This divergence from K41 can be interpreted in terms of the development of intermittency of the vorticity field as k increases, in the form of concentrated vortices which are sparsely distributed throughout the fluid. Figure 15.3 shows the vorticity field obtained in a state-of-the-art computation (Yokokawa et al. 2002; Ishihara et al. 2007); this is just a portion of the total field computed on the supercomputer 'Earth Simulator', blown up so as to reveal the intermittent character of the vorticity field; vortex tubes emerge clearly in such computations, with cross-sectional length-scale of order k_ν^{-1}. Details concerning this type of intermittency and corresponding 'anomalous scaling exponents' may be found in the review of Sreenivasan & Antonia (1997) and in the analytical treatment of Falkovich et al. (2001).

15.1.3 Effect of helicity on energy cascade

Suppose now that the source of energy on the scale L is such as to impart non-zero helicity $\langle \mathbf{u} \cdot \omega \rangle$ to the velocity field that is generated. A 'thought experiment' incorporating this behaviour has been described in §7.6. In such a situation, we may talk of 'helicity injection' as well as 'energy injection' at wave numbers of order k_0 (Brissaud et al. 1973). However the level of helicity generated is limited by a Schwarz-type inequality: defining

$$\mathcal{H}_0 = \langle \mathbf{u} \cdot \omega \rangle / \langle \mathbf{u}^2 \rangle^{1/2} \langle \omega^2 \rangle^{1/2} , \tag{15.15}$$

we must clearly have

$$|\mathcal{H}_0| \leq 1 . \tag{15.16}$$

We have already commented in §3.2 on the fact that the total helicity of a localised disturbance in an inviscid fluid is a conserved quantity (like its total energy). In homogeneous turbulence, a similar result holds:[2] the helicity density $\mathbf{u} \cdot \omega$ satisfies the equation (cf. (3.8))

$$\frac{\mathrm{D}}{\mathrm{D}t}(\mathbf{u} \cdot \omega) = -\nabla \cdot \left[\omega \left(p - \tfrac{1}{2} \rho \mathbf{u}^2 \right) \right] , \tag{15.17}$$

when $\nu = 0$, so that, taking a spatial average and using homogeneity, we have immediately

$$\mathrm{d} \langle \mathbf{u} \cdot \omega \rangle / \mathrm{d}t = 0 . \tag{15.18}$$

The helicity spectrum function $\mathcal{H}(k)$, defined (cf. 7.51) by

$$\mathcal{H}(k) = \mathrm{i} \int_{S_k} \epsilon_{ikl} k_k \Phi_{il}(\mathbf{k}) \, \mathrm{d}S , \tag{15.19}$$

is therefore presumably controlled by a process of transfer of helicity from the 'source' at wave numbers of order k_0 to the viscous sink at wave numbers of order k_ν and greater. Injection of helicity may be thought of in terms of injection of large-scale linkages in the vortex lines of the flow. These linkages survive during the (essentially inviscid) cascade process through the inertial range, but they are degraded by viscosity on the length-scale $\ell_\nu = k_\nu^{-1}$ (cf. the discussion of the twist cascade in §10.1.3).

We have seen (cf. 7.54) that $\mathcal{H}(k)$ must satisfy a 'realisability' condition

$$|\mathcal{H}(k)| \leq 2kE(k) , \tag{15.20}$$

for all k. This condition is in general much stronger than the condition (15.16); the two conditions in fact coincide only when a single wave number magnitude

[2] The possible importance of the quantity $\langle \mathbf{u} \cdot \omega \rangle$ was anticipated by Betchov 1961, but he was not aware of its invariance under Euler dynamics.

is represented in the velocity spectrum tensor. In normal turbulence with energy distributed over a wide range of length-scales, the condition (15.20) implies that $|\mathcal{H}_0| \ll 1$. To see this, suppose for example that we have maximal positive helicity at each wave number, so that $\mathcal{H}(k) = 2kE(k)$, and that $E(k)$ is given in an inertial range by the Kolmogorov spectrum (15.5). Then $\mathcal{H}(k) \propto k^{-2/3}$ in this inertial range, and so $\langle \mathbf{u} \cdot \boldsymbol{\omega} \rangle = \int_0^\infty \mathcal{H}(k)\, \mathrm{d}k$ is dominated by values of k near k_ν; this means that $\langle \mathbf{u} \cdot \boldsymbol{\omega} \rangle$ is determined by ε and ν, and so on dimensional grounds

$$\langle \mathbf{u} \cdot \boldsymbol{\omega} \rangle \sim \varepsilon^{5/4} \nu^{-7/4} . \tag{15.21}$$

Hence

$$|\mathcal{H}_0| = |\langle \mathbf{u} \cdot \boldsymbol{\omega} \rangle| / \langle \mathbf{u}^2 \rangle^{1/2} \langle \omega^2 \rangle^{1/2} \sim \mathcal{R}e^{-1/4} . \tag{15.22}$$

Even (15.22) however is an overestimate of $|\mathcal{H}_0|$ in any real turbulent situation. Suppose for example that, at time $t = 0$, we impulsively generate a random velocity field on the scale $L = k_0^{-1}$ with maximal helicity, i.e. $|\mathcal{H}_0| = \mathcal{O}(1)$, and we allow this field to evolve under the Navier–Stokes equations. Viscosity has negligible effect until $\langle \omega^2 \rangle$ has increased (by random stretching) by a factor $\mathcal{O}(\mathcal{R}e)$; during this process $\langle \mathbf{u} \cdot \boldsymbol{\omega} \rangle$ and $\langle \mathbf{u}^2 \rangle$ remain essentially constant. Hence $|\mathcal{H}_0|$ decreases by a factor $\mathcal{O}(\mathcal{R}e^{-1/2})$, i.e. at this stage

$$|\mathcal{H}_0| = \mathcal{O}(\mathcal{R}e^{-1/2}) , \tag{15.23}$$

and it will presumably remain of this order of magnitude (at most) during the subsequent decay process.

The difference between (15.22) and (15.23) provides an indication that a state of maximal helicity $\mathcal{H}(k) = 2kE(k)$ is not in fact compatible with natural evolution under the Navier–Stokes equations. Direct evidence for this is provided by the following simple argument of Kraichnan (1973). Suppose that at time $t = 0$ we have a velocity field consisting of two pure 'helicity modes', $\mathbf{u} = \mathbf{u}_1 + \mathbf{u}_2$, where

$$\mathbf{u}_1 = (\hat{\mathbf{u}}_1 + \mathrm{i}\hat{\mathbf{k}}_1 \wedge \hat{\mathbf{u}}_1) e^{\mathrm{i}\mathbf{k}_1 \cdot \mathbf{x}} + c.c., \quad \mathbf{u}_2 = (\hat{\mathbf{u}}_2 \pm \mathrm{i}\hat{\mathbf{k}}_2 \wedge \hat{\mathbf{u}}_2) \mathrm{e}^{\mathrm{i}\mathbf{k}_2 \cdot \mathbf{x}} + c.c., \tag{15.24}$$

where $c.c.$ represents the complex conjugate, and $\hat{\mathbf{k}}_n = \mathbf{k}_n / k_n \; (n = 1, 2)$. These modes satisfy

$$\omega_1 = \nabla \wedge \mathbf{u}_1 = k_1 \mathbf{u}_1, \quad \omega_2 = \nabla \wedge \mathbf{u}_2 = \pm k_2 \mathbf{u}_2 . \tag{15.25}$$

The choice of sign in $\mathbf{u}_2$ (corresponding to right-handed or left-handed circular polarisation) is retained in order to shed light on the nature of the interactions between helical modes of like or opposite polarities. Writing the Navier–Stokes equation in the form

$$\left(\frac{\partial}{\partial t} - \nu \nabla^2 \right) \mathbf{u} = -\nabla \tilde{p} + \mathbf{u} \wedge \boldsymbol{\omega} , \tag{15.26}$$

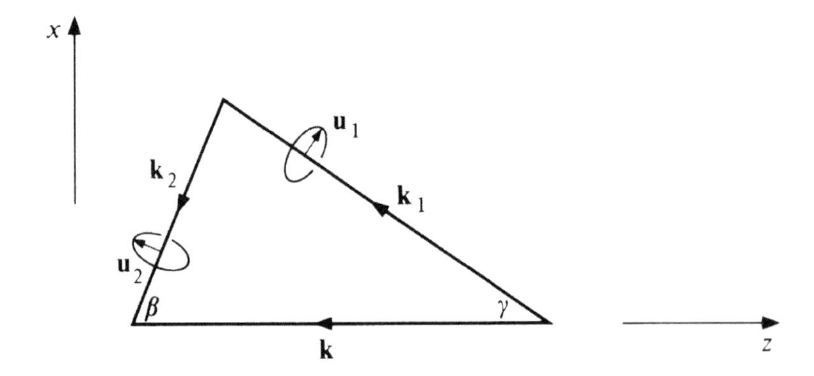

Figure 15.4 Interaction of two helical modes. Modes with wave vectors $\mathbf{k}_1$ and $\mathbf{k}_2$ are each circularly polarised (cf. Figure 12.2) and of maximum positive helicity; these interact to generate a mode of wave vector $\mathbf{k} = \mathbf{k}_1 + \mathbf{k}_2$ which is not of maximum helicity. [After Kraichnan 1973.]

where $\tilde{p} = p/\rho + \frac{1}{2}\mathbf{u}^2$, it is evident that the strength of the interaction is given by the non-linear term

$$\mathbf{u} \wedge \omega = \mathbf{u}_1 \wedge \omega_2 + \mathbf{u}_2 \wedge \omega_1 = (k_1 \mp k_2)\, \mathbf{u}_2 \wedge \mathbf{u}_1\,. \tag{15.27}$$

Since $k_1 + k_2 > |k_1 - k_2|$, it is immediately apparent that modes of opposite polarity have a tendency to interact more strongly than modes of like polarity. In fact when $k_1 = k_2$, $\mathbf{u}$ is parallel to ω when the modes are of like polarity ($\mathcal{H}_0 = 1$) and there is no non-linear interaction whatsoever!

To see that maximal helicity is not conserved, it is necessary to obtain $(\partial\mathbf{u}/\partial t)_{t=0}$ from (15.26), and this involves elimination of $\tilde{p}$ using $\nabla \cdot \mathbf{u} = 0$. The initial interaction generates modes with wave vectors $\mathbf{k}_1 \pm \mathbf{k}_2$. Let $\mathbf{k} = \mathbf{k}_1 + \mathbf{k}_2$ and choose axes (Figure 15.4) so that

$$\mathbf{k}_1 = k_1(\sin\gamma,\ 0,\ -\cos\gamma)\,, \quad \mathbf{k}_2 = k_2(-\sin\beta,\ 0,\ -\cos\beta)\,, \tag{15.28a}$$

$$\hat{\mathbf{u}}_1 = u_0(0, \tfrac{1}{2}, 0)\,, \quad \hat{\mathbf{u}}_2 = u_0(0, \tfrac{1}{2}, 0)\,, \tag{15.28b}$$

where $k_1 \sin\gamma - k_2 \sin\beta = 0$. Then, as shown by Kraichnan (1973), the initial excitation of the mode $\mathbf{u}(\mathbf{k})e^{i\mathbf{k}\cdot\mathbf{x}}$ is given (ignoring the viscous effect) by

$$\partial\mathbf{u}(\mathbf{k})/\partial t|_{t=0} = -\tfrac{1}{4}ku_0^2[\,\mp \mathrm{i}\ \sin(\beta - \gamma),\ -\sin\gamma \pm \sin\beta,\ 0]\,, \tag{15.29}$$

so that, as anticipated above, $|\partial\mathbf{u}(\mathbf{k})/\partial t|_{t=0}$ is greater when the interacting waves have opposite polarity than when they have the same polarity; moreover, since evidently $\mathrm{i}\mathbf{k} \wedge \hat{\mathbf{u}}(\mathbf{k}) \neq \pm k\hat{\mathbf{u}}(\mathbf{k})$ when $\gamma \neq \beta$, the condition of maximal helicity cannot be conserved under non-linear interactions.

The conclusion of these arguments is that, no matter how strong the level of helicity injection may be at wave numbers of order k_0, the relative level of helicity as measured by the dimensionless ratio $\mathcal{H}(k)/2kE(k)$ must grow progressively weaker with increasing k; and when k/k_0 is sufficiently large it may be conjectured (Brissaud et al. 1973) that the helicity has negligible dynamical effect, and is itself convected and diffused in much the same way as a dynamically passive scalar contaminant (Batchelor 1959).

Suppose that the rate of injection of helicity (a pseudo-scalar) at wave numbers of order k_0 is λ. This is clearly bounded by an inequality of the form $|\lambda| \leq k_0\varepsilon$ since helicity cannot be injected without simultaneous injection of energy. If helicity is injected at a maximal rate, then

$$|\lambda| \sim k_0\varepsilon \sim u_0^3/L^2 \, . \tag{15.30}$$

The helicity spectrum $\mathcal{H}(k)$ must be proportional to λ (the pseudo-scalar character of both quantities ensures this) and in the inertial range $k_0 \ll k \ll k_\nu$, the only other parameters that can serve to determine $\mathcal{H}(k)$ are ε and k; hence on dimensional grounds[3]

$$\mathcal{H}(k) = C_H\lambda\varepsilon^{-1/3}k^{-5/3}, \quad (k_0 \ll k \ll k_\nu)\, , \tag{15.31}$$

where C_H is a universal constant, analogous to the Kolmogorov constant C in (15.5). $E(k)$ is still given by (15.5), being unaffected by helicity in the inertial range. Note that (15.30), together with (15.5) and (15.31), implies that

$$|\mathcal{H}(k)| \ll 2kE(k), \quad (k_0 \ll k \ll k_\nu)\, , \tag{15.32}$$

consistent with the discussion of the previous paragraphs.

From the point of view of dynamo theory, perhaps the most significant point in the foregoing discussion is that the mean helicity in a turbulent field is given by (15.23) rather than (15.22); equivalently it is a property of the 'energy-containing eddies' of the flow (scale $\mathcal{O}(L)$) rather than of the dissipative eddies (scale $\mathcal{O}(\ell_\nu)$). A crude maximal estimate of $|\langle \mathbf{u}\cdot\boldsymbol{\omega}\rangle|$ is therefore given by u_0^2/L, independent of the Reynolds number of the turbulence. This is in fact the estimate that we adopted in previous sections (see for example the argument leading to the estimate (7.89) for α), but it is reassuring now to have some retrospective dynamical justification. A second general conclusion from the foregoing discussion is that the presence of helicity may be expected to exert a mild constraint on the energy cascade process, at any rate at wave numbers of order k_0 where the relative helicity level $|\mathcal{H}(k)|/2kE(k)$ may be quite high. The study of interacting helicity modes (Kraichnan 1973) indicates that non-linear interactions are weaker when the interacting modes have

[3] The result (15.31) was presented as one of two possibilities by Brissaud et al. (1973). The other possibility conjectured involved a 'pure helicity cascade' without any energy cascade; this possibility is incompatible with dynamical arguments, and was later abandoned (André & Lesieur 1977).

like polarity; if *all* the modes present at the initial instant have the same polarity (corresponding to a state of maximal helicity at each wave number magnitude) the net energy transfer to higher wave numbers may be expected to be inhibited, and so the decay of turbulence should be delayed. More transparently perhaps, since

$$(\mathbf{u} \cdot \omega)^2 + (\mathbf{u} \wedge \omega)^2 = \mathbf{u}^2\omega^2 , \tag{15.33}$$

maximisation of $|\langle \mathbf{u} \cdot \omega \rangle|$ may plausibly be associated with minimisation of $\langle (\mathbf{u} \wedge \omega)^2 \rangle$, and so with a weakening (on the average) of non-linear effects associated with the term $\mathbf{u} \wedge \omega$ in (15.26) (Frisch et al. 1975).

Finally, we note that André & Lesieur (1977) have numerically analysed the development of energy and helicity spectra on the basis of a closure of the infinite system of equations that gives the rate of change of nth-order correlations in terms of correlations up to order $n + 1$ (Batchelor 1953); the particular closure scheme adopted by André & Lesieur is a variant of the 'eddy-damped quasi-normal Markovian' closure (EDQNM) (Orszag 1970, 1977) by which 4th-order correlations are expressed in terms of 2nd-order correlations in a way that guarantees that the realisability conditions $E(k,t) \geq 0$, $|\mathcal{H}(k,t)| \leq 2kE(k,t)$ are satisfied at all times. Numerical integration of the resulting equations indicates (i) the development of $k^{-5/3}$ inertial ranges for both $E(k,t)$ and $\mathcal{H}(k,t)$, and (i) a significant delay in the process of energy dissipation when a high level of helicity is initially present. These results, although not conclusive (being based on a closure scheme that is exploratory in character), do support the general description of the dynamics of turbulence with helicity, as outlined in this section.

15.2 Influence of magnetic helicity conservation in energy transfer processes

Suppose now that the turbulent fluid is permeated by an equally turbulent magnetic field $(\mu_0\rho)^{1/2}\mathbf{h}(\mathbf{x},t)$, the mean value of $\mathbf{h}$ being zero. What can be said about the joint evolution of the spectra of $\mathbf{u}$ and $\mathbf{h}$? We shall suppose in the following discussion that dissipation effects (both viscous and ohmic) are weak, or equivalently that both the Reynolds number $\mathcal{R}e = u_0L/\nu$ and the magnetic Reynolds number $\mathcal{R}_\mathrm{m} = u_0L/\eta$ are large.

Consider first the new quadratic invariants that exist in the dissipationless limit $\eta = \nu = 0$. These are the total energy E_T, the magnetic helicity $\mathcal{H}_M$, and the cross-helicity[4] $\mathcal{H}_C$ (already encountered in §12.1.2):

$$E_T = \tfrac{1}{2}\langle \mathbf{u}^2 + \mathbf{h}^2 \rangle , \quad \mathcal{H}_M = \langle \mathbf{a} \cdot \mathbf{h} \rangle , \quad \mathcal{H}_C = \langle \mathbf{u} \cdot \mathbf{h} \rangle . \tag{15.34}$$

Here $\mathbf{a}$ is a vector potential for $\mathbf{h}$, i.e. $\mathbf{h} = \nabla \wedge \mathbf{a}$. Note that the value of $\mathcal{H}_M$ is

[4] In (15.34), factors involving the (uniform) density ρ and the constant μ_0 are omitted for simplicity. The term 'cross-helicity' was introduced by Frisch et al. (1975); for its topological interpretation, see Moffatt (1969).

independent of the gauge of $\mathbf{a}$. The kinetic helicity $\langle \mathbf{u} \cdot \omega \rangle$ is no longer invariant in the presence of a magnetic field distribution, because, under the influence of the rotational Lorentz force, vortex lines are no longer frozen in the fluid.

The equations of magnetohydrodynamics, (12.9) and (12.10), are invariant under the transformation $\mathbf{h}(\mathbf{x}, t) \rightarrow -\mathbf{h}(\mathbf{x}, t)$ (i.e. to every solution $(\mathbf{u}(\mathbf{x}, t),\ \mathbf{h}(\mathbf{x}, t))$, there corresponds another solution $(\mathbf{u}(\mathbf{x}, t),\ -\mathbf{h}(\mathbf{x}, t))$. If, therefore, at some initial instant $t = 0$ the statistical properties of the $(\mathbf{u},\ \mathbf{h})$ field are invariant under change of sign of $\mathbf{h}$, then they will remain so invariant for all $t > 0$ (this holds for arbitrary values of η and ν and also if there is an arbitrary forcing term in the equation of motion).[5] Under this condition, which we may describe as the condition of *magnetic sign invariance*, any statistical function of the fields $\mathbf{u}$ and $\mathbf{h}$ which appears to change sign under replacement of $\mathbf{h}$ by $-\mathbf{h}$ must in fact be permanently zero. In particular, under the condition of magnetic sign invariance,

$$\mathcal{H}_C = 0\,, \tag{15.35}$$

and similarly all moments of the form $\langle u_j u_j \ldots h_\alpha h_\beta h_\gamma \ldots \rangle$ involving an odd number of $\mathbf{h}$-factors will vanish for all t. We shall suppose, unless explicitly stated otherwise in the following discussion, that this condition of magnetic sign invariance is satisfied.

Consider now for simplicity the idealised situation in which the $\mathbf{u}$ and $\mathbf{h}$ fields are statistically invariant under rotations (but not necessarily under point reflexions). The spectrum tensors $\Phi_{ij}(\mathbf{k},\ t)$, $\Gamma_{ij}(\mathbf{k}, t)$ of $\mathbf{u}$ and $\mathbf{h}$ are then given by

$$\Phi_{ij}(\mathbf{k}, t) = \frac{E(k, t)}{4\pi k^4}(k^2\delta_{ij} - k_i k_j) + \frac{\mathrm{i}\mathcal{H}(k, t)}{8\pi k^4}\epsilon_{ijk}k_k\,, \tag{15.36a}$$

$$\Gamma_{ij}(\mathbf{k}, t) = \frac{M(k, t)}{4\pi k^4}(k^2\delta_{ij} - k_i k_j) + \frac{\mathrm{i}\mathcal{N}(k, t)}{8\pi k^2}\epsilon_{ijk}k_k\,, \tag{15.36b}$$

where

$$\tfrac{1}{2}\langle \mathbf{u}^2 \rangle = \int_0^\infty E(k, t)\,\mathrm{d}k\,, \quad \tfrac{1}{2}\langle \mathbf{h}^2 \rangle = \int_0^\infty M(k, t)\,\mathrm{d}k\,, \tag{15.37a}$$

$$\langle \mathbf{u} \cdot \omega \rangle = \int_0^\infty \mathcal{H}(k, t)\,\mathrm{d}k\,, \quad \langle \mathbf{a} \cdot \mathbf{h} \rangle = \int_0^\infty \mathcal{N}(k, t)\,\mathrm{d}k\,. \tag{15.37b}$$

Note that the spectrum tensor of the field $\mathbf{a}$ is (under the condition of isotropy) just $k^{-2}\Gamma_{ij}(\mathbf{k},\ t)$, and that $\mathcal{N}(k, t)$ is defined with respect to the invariant $\langle \mathbf{a} \cdot \mathbf{h} \rangle$ (rather than with respect to $\langle \mathbf{h} \cdot \nabla \wedge \mathbf{h} \rangle$); this leads to the factor k^{-2} in the antisymmetric term of (15.36b), as contrasted with k^{-4} in the corresponding term of (15.36a). The

[5] This result is mentioned by Pouquet et al. (1976).

realisability conditions on $\Gamma_{ij}(\mathbf{k}, t)$ are simply

$$M(k,t) \geq 0, \quad |\mathcal{N}(k,t)| \leq 2k^{-1}M(k,t)\,. \tag{15.38}$$

Let us now consider in a qualitative way how the system will respond to injection of kinetic energy and kinetic helicity at wave numbers of order k_0 (a situation studied within the framework of the EDQNM closure scheme by Pouquet et al. (1976)). Suppose that an initially weak magnetic field is present with, in particular, ingredients on scales large compared with k_0^{-1}. On the scale k_0^{-1}, these ingredients will provide an almost uniform field $\mathbf{h}_0$ say. Helical motion on scales $\lesssim k_0^{-1}$ will generate a perturbation field with magnetic helicity related to the kinetic helicity. If we represent the action of motions on scale $\ll k_0^{-1}$ by an eddy diffusivity η_e, then the magnetic fluctuations on scales $\sim k_0^{-1}$ are determined by the equation

$$\eta_e \nabla^2 \mathbf{h} \approx -(\mathbf{h}_0 \cdot \nabla)\mathbf{u}\,, \tag{15.39}$$

with Fourier transform

$$\eta_e k^2 \tilde{\mathbf{h}} \approx \mathrm{i}(\mathbf{h}_0 \cdot \mathbf{k})\tilde{\mathbf{u}}\,. \tag{15.40}$$

The corresponding relation between the spectra of $\mathbf{h}$ and $\mathbf{u}$ is

$$\Gamma_{ij}(\mathbf{k}, t) \approx \frac{(\mathbf{h}_0 \cdot \mathbf{k})^2}{\eta_e^2 k^4} \Phi_{ij}(\mathbf{k}, t) \quad (k \approx k_0)\,, \tag{15.41}$$

(a result obtained in a related context by Golitsyn 1960). The result (15.38) gives an anisotropic form for Γ_{ij} when Φ_{ij} is isotropic, due to the preferred direction of $\mathbf{h}_0$. However, if we now take account of the fact that the large-scale field $\mathbf{h}_0$ is non-uniform, all directions being equally likely, we may average over these directions[6] to obtain the isotropic relationship

$$\Gamma_{ij}(\mathbf{k}, t) \approx \frac{\langle \mathbf{h}_0^2 \rangle}{3\eta_e^2 k^2} \Phi_{ij}(\mathbf{k}, t)\,, \tag{15.42}$$

and hence, from (15.36a) and (15.36b),

$$M(k,t) \approx \frac{\langle \mathbf{h}_0^2 \rangle}{3\eta_e^2 k^2} E(k,t)\,, \quad \mathcal{N}(k,t) \approx \frac{\langle \mathbf{h}_0^2 \rangle}{3\eta_e^2 k^4} \mathcal{H}(k,t)\,. \tag{15.43}$$

The magnetic helicity generated at wave numbers of order k_0 is therefore of the same sign as the kinetic helicity.

Consider now the development of the large-scale field $\mathbf{h}_0$. We know from the general considerations of Chapter 7 that positive kinetic helicity gives rise to a negative α-effect. There is of course no *a priori* justification for a two-scale approach

[6] 'Isotropisation', as in §12.7.

when the spectrum of $\mathbf{h}(\mathbf{x}, t)$ is continuous; but *if* a two-scale approach is adopted, the evolution equation for $\mathbf{h}_0$ is

$$\frac{\partial \mathbf{h}_0}{\partial t} = \alpha \nabla \wedge \mathbf{h}_0 + \eta \nabla^2 \mathbf{h}_0 \,. \tag{15.44}$$

Writing $\mathbf{h}_0 = \nabla \wedge \mathbf{a}_0$, we have equivalently

$$\frac{\partial \mathbf{a}_0}{\partial t} = \alpha \mathbf{h}_0 - \nabla \phi_0 + \eta \nabla^2 \mathbf{a}_0 \tag{15.45}$$

for some scalar ϕ_0, and from (15.44) and (15.45), the development of the large-scale magnetic helicity is given by

$$\begin{aligned} \frac{\partial}{\partial t}\langle \mathbf{a}_0 \cdot \mathbf{h}_0 \rangle &= \alpha \langle \mathbf{a}_0 \cdot \nabla \wedge \mathbf{h}_0 \rangle + \alpha \langle \mathbf{h}_0^2 \rangle + \eta \langle \mathbf{a}_0 \cdot \nabla^2 \mathbf{h}_0 + \mathbf{h}_0 \cdot \nabla^2 \mathbf{a}_0 \rangle \\ &= 2\alpha \langle \mathbf{h}_0^2 \rangle - 2\eta \langle (\partial a_{0j}/\partial x_i)(\partial h_{0j}/\partial x_i) \rangle \,, \end{aligned} \tag{15.46}$$

using the property of homogeneity. Assuming $\langle \mathbf{u} \cdot \boldsymbol{\omega} \rangle > 0$, so that $\alpha < 0$, it is evident that the large-scale magnetic helicity generated by the α-effect will be negative (the effects of diffusion being assumed weak).

We know, moreover, from the results of §9.2 that, when dissipation is negligible, the growth of magnetic Fourier components on the scale k^{-1} due to the α-effect has a time-scale of order $(|\alpha| k)^{-1}$; hence for a given level of helicity maintained by 'injection' at wave numbers of order k_0, it is to be expected that magnetic energy (and associated magnetic helicity) will develop on progressively increasing length-scales of order $|\alpha| t$ (or equivalently at wave numbers of order $(|\alpha| t)^{-1}$) as t increases.

These qualitative considerations receive striking support from the work of Pouquet et al. (1976), who numerically integrated the four equations describing the evolution of the functions $E(k, t)$, $\mathcal{H}(k, t)$, $M(k, t)$, $\mathcal{N}(k, t)$, closed on the basis of the EDQNM scheme. Figures 15.5a and 15.5b show the development of $M(k, t)$ and $|\mathcal{N}(k, t)|$ as functions of k and for $t = 120, 240$ (the unit of time being $(k_0 u_0)^{-1}$). The system is excited by injection of kinetic energy and helicity at wave numbers of order k_0. The figures show (i) the excitation of magnetic energy at progressively larger scales as t increases, and (ii) the fact that magnetic helicity has the same sign as the injected kinetic helicity when $k/k_0 = \mathcal{O}(1)$, but the opposite sign when $k/k_0 \ll 1$. It is evident from both figures that a double-scale structure emerges in the magnetic field spectrum, the separation in the two spectral peaks becoming more marked as t increases. It must be emphasised that these results emerge from a dynamic model in which the back-reaction of the Lorentz force distribution on the velocity field is fully incorporated.

Pouquet et al. (1976) have described the above excitation of magnetic modes on ever-increasing length-scales in terms of an 'inverse cascade' of magnetic energy

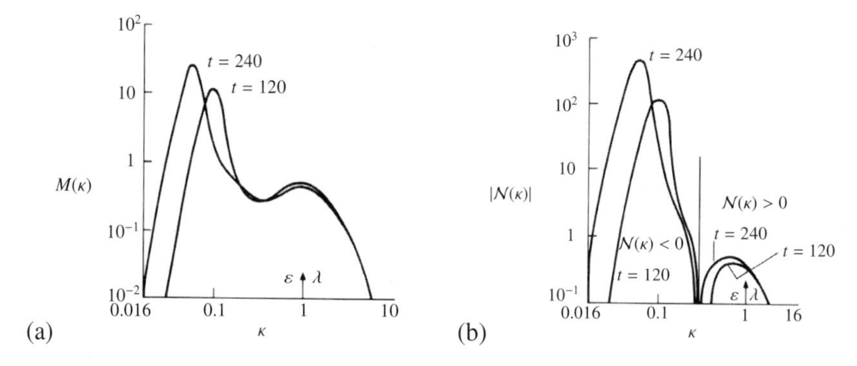

Figure 15.5 Computed development of (a) magnetic energy spectrum and (b) magnetic helicity spectrum on the basis of the EDQNM closure. Kinetic energy and helicity are injected at rates ε and λ; the injection spectra are given by $F_\varepsilon(\kappa) = \kappa^{-1}F_\lambda(\kappa) = C\varepsilon\kappa^4 e^{-2\kappa^2}$ with C chosen such that $\int_0^\infty F_\varepsilon(\kappa)\,d\kappa = \varepsilon$ and $\kappa = k/k_0$. At time $t = 0$, the (normalised) kinetic and magnetic energy and helicity spectrum functions are given by $E(\kappa) = 10M(\kappa) = C\kappa^4 e^{-2\kappa^2}$, $\mathcal{H}(\kappa) = \mathcal{N}(\kappa) = 0$. The minimum and maximum wave numbers retained in the computation were $\kappa_{\min} = 2^{-6}$, $\kappa_{\max} = 2^4$. The Reynolds number based on the initial rms velocity and the length-scale k_0^{-1} was 30, and the magnetic Prandtl number ν/η was unity. [From Pouquet, Frisch & Léorat 1976.]

and magnetic helicity. It may be merely a matter of semantics, but this terminology could be just a little misleading in the present context. The word 'cascade' suggests 'successive excitation' due to non-linear mode interactions, and 'inverse cascade' suggests successive excitations that may be represented diagrammatically in the form

$$k_0 \to 2^{-1}k_0 \to 2^{-2}k_0 \to \cdots \to 2^{-n}k_0 \to \cdots . \tag{15.47}$$

In fact in the present context, interaction of **u** and **h** fluctuations on the scale k_0^{-1} *simultaneously* generates large-scale Fourier components on all scales $k^{-1} \gg k_0^{-1}$; this is not a step-by-step process, but rather a long-range spectral process; the fact that larger scales take longer to excite does not in itself justify the use of the term 'cascade'.

A further numerical study of direct relevance to the present discussion has been carried out by Pouquet & Patterson (1978). In this work, the Fourier transforms of the coupled momentum and induction equations, appropriately truncated, were directly integrated numerically, subject to initial conditions for the Fourier components of **u** and **h** chosen from Gaussian distributions of random numbers. No attempt was made to average over different realisations of the turbulent field, and in fact the behaviour of the spectra (defined by averaging over spherical shells in wave number space) showed significant variation between different realisations.

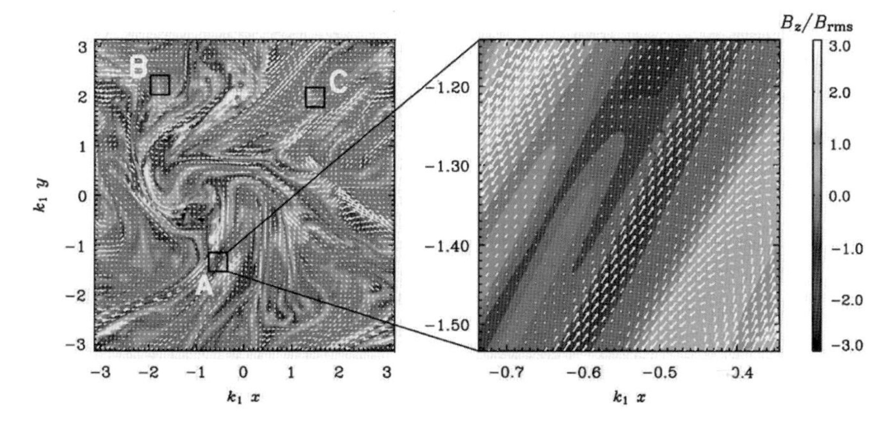

Figure 15.6 Magnetic field structures in DNS of non-helical MHD turbulence at $\mathcal{P}_m \equiv \nu/\eta = 50$; the zoom on the right shows sheet-like structures formed by the action of local straining. [From Brandenburg & Subramanian 2005.]

Nevertheless in all cases Pouquet & Patterson found a substantial net transfer of energy from kinetic to magnetic modes. They interpret this transfer in the following terms: (i) cascade of kinetic energy towards higher wave numbers due to conventional non-linear interactions; (ii) sharing of energy between kinetic and magnetic modes at small scales due to excitation of Alfvén waves propagating on the large-scale magnetic field; (iii) intensification of the large-scale field due to what is again described as an 'inverse cascade' of magnetic energy, an effect that is most noticeable in the numerical solutions when magnetic helicity (rather than kinetic helicity) is initially present. These calculations (and the EDQNM calculations of Pouquet et al. 1976) were carried out for a magnetic Prandtl number $\mathcal{P}_m = \nu/\eta$ equal to unity; calculations at $\mathcal{P}_m < 1$ (or even $\ll 1$) would be more relevant in the solar context; and at $\mathcal{P}_m \gg 1$ in the context of galaxies and galaxy clusters.

There has been an explosion of computational activity in this area, with the dramatically increasing computer power that has become available in recent decades. This has permitted direct numerical simulation (DNS) of MHD turbulence at increasingly fine resolution and ever increasing $\mathcal{R}e$ and $\mathcal{R}_m$. An extensive review of such developments has been given by Brandenburg & Subramanian (2005). The emphasis now is on seeking to understand the structures of velocity and magnetic fields that emerge from such computations. Figure 15.6 from a simulation of non-helical MHD turbulence at $\mathcal{P}_m = 50$ shows structures in the magnetic field that presumably form as the result of persistent straining as described in §3.4 et seq., the transverse scale of these structures being controlled by magnetic resistivity.[7] These magnetic structures bear comparison with the (non-magnetic) vorticity

[7] The folded striations evident in the zoom of Figure 15.6 were detected in the computations of Schekochihin

structures of Figure 15.3; they are more sheet-like, whereas the vortex structures are more tube-like. This may be because vortex sheets are prone to Kelvin–Helmholtz instability that rolls them up into vortex tubes, whereas there is no corresponding roll-up mechanism for magnetic sheets.

15.3 Modification of inertial range due to large-scale magnetic field

The presence of a strong large-scale magnetic field profoundly modifies the energy transfer process in small scales, and in particular in the inertial range, as pointed out by Kraichnan (1965). In the inertial range, the total energy content is small, relative to the total energy in scales $\gtrsim k_0^{-1}$, and dissipative processes are negligible. The large-scale velocity field $\mathbf{u}_0$ simply convects eddies on the scale k^{-1} (where $k \gg k_0$) without significant distortion, and can in effect be 'eliminated' by Galilean transformation. The large-scale magnetic field $\mathbf{h}_0$ cannot be eliminated by Galilean transformation, and in fact provides an important coupling between $\mathbf{u}$ and $\mathbf{h}$ fields on scales $k^{-1} \ll k_0^{-1}$ through the Alfvén wave mechanism discussed in §12.1. The inertial range may be pictured as a random sea of localised disturbances on scales $k^{-1}(k \gg k_0)$, propagating along the local mean field $\mathbf{h}_0$ with velocity $\pm\mathbf{h}_0$, energy transfer to smaller scales occurring due to the collision of oppositely directed disturbances. In each such disturbance magnetic and kinetic energy are equal; hence $E(k) = M(k)$ in the inertial range. Energy transfer will be maximised when there are as many disturbances travelling in the direction $+\mathbf{h}_0$ as in the direction $-\mathbf{h}_0$, i.e. when the cross-helicity $\mathcal{H}_C$ is zero. It is relevant to note here that Pouquet & Patterson (1978) have found that non-zero values of $\mathcal{H}_C$ leads to a decrease in energy transfer to small scales in numerical simulations.

The time-scale for the interaction of blobs on the scale k^{-1} is $t_k = \mathcal{O}(h_0 k)^{-1}$, and, as argued by Kraichnan (1965), the rate of energy transfer ε through the inertial range may be expected to be proportional to t_k. Since there is no dissipation in the inertial range, ε must be independent of k; moreover, if the energy cascade is local in k-space, the only other parameters that can serve to determine ε are $E(k)$ and k itself; dimensional analysis then gives

$$\varepsilon \propto h_0^{-1}(E(k))^2\, k^3\,, \tag{15.48}$$

or equivalently

$$E(k) = M(k) = A(\varepsilon h_0)^{1/2}\, k^{-3/2} \quad (k_0 \ll k \ll k_d)\,, \tag{15.49}$$

where A is a universal constant of order unity, and k_d an upper wave number for the

et al. 2004. Such structures at high $\mathcal{P}_m$ had been to some extent anticipated by Moffatt (1963); it should be noted, however, that some regions (as in the boxes B and C in Figure 15.6) contain no such striations; the folding and stretching that create them are apparently quite non-uniform.

inertial range at which dissipative effects (ohmic and/or viscous) become important. This cutoff wave number is presumably determined (in order of magnitude) by equating the time-scale of interaction of blobs $(h_0 k)^{-1}$ with the time-scale of ohmic or viscous dissipation $(\eta k^2)^{-1}$ or $(\nu k^2)^{-1}$, whichever is smaller. If, as is more usual, $\eta \gg \nu$, then evidently

$$k_d \sim h_0/\eta . \tag{15.50}$$

In (15.49) and (15.50), h_0 must be interpreted as $\langle \mathbf{h}^2 \rangle^{1/2}$, and the condition that there should exist an inertial range of the form (15.49) is $k_0 \gg k_d$, or

$$\langle \mathbf{h}^2 \rangle^{1/2} / k_0 \eta \gg 1 . \tag{15.51}$$

The spectrum (15.49) was found also by Iroshnikov (1964), and has come to be known as the Kraichnan–Iroshnikov spectrum.

15.4 Non-helical turbulent dynamo action

The generation of magnetic fields on scales of order k_0^{-1} and greater, as discussed in §15.2, is directly attributable to the presence of helicity in the smaller scale velocity and /or magnetic fields. Let us now briefly consider the problem that presents itself when both fields are assumed reflexionally symmetric, so that the α-effect mechanism for the generation of large-scale fields is no longer present.

It has long been recognised that random stretching of magnetic lines of force leads to exponential increase of magnetic energy density, as long as ohmic diffusion effects can be neglected. This random stretching is however associated with a systematic decrease in scale of the magnetic field (as in the differential rotation problem studied in §3.11), and ultimately ohmic diffusion must be taken into account. The question then arises as to whether magnetic energy generation by random stretching can still win over degradation of magnetic energy due to ohmic dissipation.

Batchelor (1950) invoked the analogy between the vorticity equation in a non-conducting fluid

$$\frac{\partial \omega}{\partial t} = \nabla \wedge (\mathbf{u} \wedge \omega) + \nu \nabla^2 \omega , \tag{15.52}$$

and the induction equation in a conducting fluid

$$\frac{\partial \mathbf{h}}{\partial t} = \nabla \wedge (\mathbf{u} \wedge \mathbf{h}) + \eta \nabla^2 \mathbf{h} , \tag{15.53}$$

(see §3.2). If, in either case, $\mathbf{u}(\mathbf{x}, t)$ is a homogeneous turbulent velocity field, we may deduce equations for $\langle \omega^2 \rangle$ and $\langle \mathbf{h}^2 \rangle$:

$$\frac{\mathrm{d}}{\mathrm{d}t}\langle \omega^2 \rangle = \left\langle \omega_i \omega_j \frac{\partial u_i}{\partial x_j} \right\rangle - \nu \left\langle \left(\frac{\partial \omega_i}{\partial x_j} \right)^2 \right\rangle , \tag{15.54a}$$

$$\frac{\mathrm{d}}{\mathrm{d}t}\langle \mathbf{h}^2 \rangle = \left\langle h_i h_j \frac{\partial u_i}{\partial x_j} \right\rangle - \eta \left\langle \left(\frac{\partial h_i}{\partial x_j} \right)^2 \right\rangle . \tag{15.54b}$$

Batchelor now argued that $\langle \omega^2 \rangle$, being a 'small-scale' property of the turbulence, is in statistical equilibrium on any time-scale (e.g. L/u_0) associated with the energy-containing eddies. Moreover the random stretching of magnetic lines of force may be expected to generate a similar statistical structure in the **h**-field as the random stretching of vortex lines generates in the ω-field. Hence if $\eta = \nu$, $\langle \mathbf{h}^2 \rangle$ may also be expected to be in statistical equilibrium, and therefore to remain constant on time-scales of order L/u_0; if $\eta < \nu$, $\langle \mathbf{h}^2 \rangle$ may be expected to grow (and Batchelor argued that this growth would continue until arrested by the back-reaction of Lorentz forces); and if $\eta > \nu$, $\langle \mathbf{h}^2 \rangle$ may be expected to decay.

The analogy with vorticity is of course not perfect, as has been pointed out in §3.2. A particular difficulty in its application to the turbulent dynamo problem is that we are really interested in the possibility of magnetic field maintenance on time-scales at least as large as the ohmic time-scale L^2/η, and this is *large* compared with L/u_0, when $\mathcal{R}_m = u_0 L/\eta \gg 1$. In the absence of 'random stirring devices', $\langle \omega^2 \rangle$ decays on the time-scale L/u_0, and if $\eta = \nu$, $\langle \mathbf{h}^2 \rangle$ may be expected to do likewise, i.e. we do not have a dynamo in the usual sense. If $\eta < \nu$, $\langle \mathbf{h}^2 \rangle$ may at first increase by a factor of order ν/η (as $\langle \omega^2 \rangle$ would do if ν in (15.54a) were suddenly decreased to a value $\nu_1 = \eta < \nu$), but may again be expected to decay in a time of order L/u_0. We can of course prevent the long-term decay of $\langle \omega^2 \rangle$ in (15.54a) by invoking a random torque distribution in (15.52); but to maintain the analogy we should then introduce the curl of a random electromotive force in (15.53) – and we are then no longer considering the problem of a self-excited dynamo.

The arguments of Batchelor (1950) were by no means universally accepted, and were the subject of widespread controversy during the 1950s and 60s. For example, Syrovatsky (1959) argued (on the basis of ideas put forward by Biermann & Schlüter 1950) that dynamo action will always occur if $\mathcal{R}_m$ is greater than some number of order unity; and Saffman (1963) argued that dynamo action might not occur under any circumstances due to the accelerating ohmic diffusion effect which occurs when the length-scale decreases, as described in §3.6. These were at the time extreme positions which failed to take account of the possible 'leak-back' to lower wave numbers which can occur and which turns out to be at the heart of the α-effect mechanism, but which may also be present in non-chiral turbulence. The possibility of such a leak-back was recognised but not quantitatively analysed by Kraichnan & Nagarajan (1967).

While the above considerations might suggest that reflexionally symmetric turbulence cannot provide sustained dynamo action, there is also the consideration (more prominent in the context of turbulence with helicity) that since the field **h** is not subject to the constraint $\boldsymbol{\omega} = \nabla \wedge \mathbf{u}$, modes of excitation may be available to **h** that are simply not available to $\boldsymbol{\omega}$ and so $\langle \mathbf{h}^2 \rangle$ may grow (until Lorentz forces intervene) even when $\eta > \nu$.

It seems likely that the questions raised by these considerations will be fully resolved only by extended numerical experimentation by methods similar to those of Pouquet & Patterson (1978). Direct numerical simulations have shown that dynamo action does in general occur in reflexionally symmetric turbulence, although the limits on the magnetic Prandtl number $\mathcal{P}_{\mathrm{m}}$ are not well established. It seems fair to say that computer capacities are still not yet adequate to yield reliable asymptotic laws in integrations over large times. Fortunately, as mentioned in the introductory chapter, the reflexionally symmetric problem, although profoundly challenging, has been largely bypassed in terrestrial and astrophysical contexts by the realisation that lack of reflexional symmetry will generally be present in turbulence generated in a rotating system, and that this lack of reflexional symmetry swamps all other purely turbulent effects as far as magnetic field generation is concerned.

15.5 Dynamo action incorporating mean flow effects

We have seen in §9.3.3 that dynamo action in a sphere can result from the α-effect alone; and that when, for example,

$$\alpha = -\alpha_0 \cos\theta\,, \tag{15.55}$$

the critical value of α_0 at which dipole modes are excited is given (numerically) by

$$R_\alpha = |\alpha_0| R/\eta = 7.64 = R_{\alpha c}\,,\ \text{say.} \tag{15.56}$$

If $R_\alpha > R_{\alpha c}$ then a mode of dipole symmetry will grow exponentially until the effects of the Lorentz force become important. We have seen in §12.4 that one manifestation of the increasing influence of the Lorentz force will be an ultimate reduction in the level of the α-effect. There is another mechanism of equilibration that may however intervene and stop the growth at an earlier stage if R_α is just a little larger than $R_{\alpha c}$, viz. the effect of the mean velocity distribution that will be driven by the large-scale Lorentz force distribution; this mean velocity will modify the structure of the growing **B**-field and in general may be expected to modify it in such a way that the rate of ohmic dissipation increases. The total energy in the

B-field may then be expected to saturate at a low level, which tends to zero as $R_\alpha - R_{\alpha c} \to 0$.

A formalism for this problem was developed by Malkus & Proctor (1975), by developing all the fields as power series in $(R_\alpha - R_{\alpha c})$ and seeking conditions for steady-state finite-amplitude magnetic field distributions. It is important to recognise that in this work the α-effect is supposed *given* by a formula such as (15.55) and unaffected by the magnetic field. Attention is focussed on the development of large-scale or 'macro' fields, and all the usual difficulties associated with the small-scale turbulent dynamics are bypassed. The equations studied are

$$\partial \mathbf{U}/\partial t + \mathbf{U} \cdot \nabla \mathbf{U} + 2\boldsymbol{\Omega} \wedge \mathbf{U} = -\nabla P + \rho^{-1} \mathbf{J} \wedge \mathbf{B} + \nu \nabla^2 \mathbf{U}\,, \tag{15.57a}$$

$$\partial \mathbf{B}/\partial t = \nabla \wedge (\alpha \mathbf{B}) + \nabla \wedge (\mathbf{U} \wedge \mathbf{B}) + \eta \nabla^2 \mathbf{B}\,, \tag{15.57b}$$

with $\alpha(\mathbf{x})$ prescribed and with initial conditions

$$\mathbf{U}(\mathbf{x}, 0) = 0\,, \quad \mathbf{B}(\mathbf{x}, 0) = \mathbf{B}_0(\mathbf{x})\,, \tag{15.58}$$

where $\mathbf{B}_0(\mathbf{x})$ is the eigenfunction of the kinematic problem (15.57b) when $\mathbf{U} = 0$ and $R_\alpha = R_{\alpha c}$ (as discussed in §9.3.3). The field $\mathbf{B}$ is of course as usual matched to an irrotational field of dipole symmetry in the region $r > R$.

When $R_\alpha = R_{\alpha c}(1 + \varepsilon)$ with $0 < \varepsilon \ll 1$, the magnetic field initially grows exponentially, and the Lorentz force (which is certainly non-zero by the general result of §2.5) generates a velocity field which develops according to (15.57a). This velocity field will continue to grow until it has a significant effect in (15.57b); and comparing the terms $\nabla \wedge (\mathbf{U} \wedge \mathbf{B})$ and $\eta \nabla^2 \mathbf{B}$, it is evident that the relevant scale for $\mathbf{U}$ at this stage[8] is $U_0 = \eta/R$.

The problem is characterised by three dimensionless numbers, the Ekman number $\mathcal{E} = \nu/\Omega R^2$, magnetic Ekman number $\mathcal{E}_{\mathrm{m}} = \eta/\Omega R^2$, and magnetic Reynolds number $\mathcal{R}_\alpha = |\alpha|_{\max} R/\eta$, and interest centres on the geophysically relevant situation

$$\mathcal{E} \ll 1, \quad \mathcal{E}_{\mathrm{m}} \ll 1\,. \tag{15.59}$$

Note that $\mathcal{E}_{\mathrm{m}}$ in fact plays the role of a Rossby number here, providing a measure of the importance of inertia forces relative to the Coriolis force in (15.57a). In the limit (15.59), the ultimate level of magnetic energy is determined by magnetostrophic balance in which Lorentz and Coriolis forces are of the same order of magnitude, i.e. the relevant scale for $\mathbf{B}$ is B_0 where

$$(\mu_0 \rho)^{-1} B_0^2 = \Omega U_0 R \sim \Omega \eta\,, \text{ i.e. } \Lambda = \mathcal{O}(1), \tag{15.60}$$

[8] Strictly $U_0 = (\eta/R) f(\varepsilon)$ where $f(\varepsilon) \to 0$ as $\varepsilon \to 0$.

and the magnetic energy may be expected to level off at a value of order $\Omega\,\eta$ (again multiplied by a function of ε which vanishes with ε).

Solutions of the above problem have been computed by Proctor (1977a), for the particular case when α is given by (15.55), and for values of $\mathcal{R}_\alpha$ in the range between the critical value 7.64 and 10.0. In each of the cases studied, Proctor found that the growth of magnetic energy was arrested by the mean flow effect, the level of equilibrium magnetic energy increasing with $R_\alpha - R_{\alpha c}$; Figure 15.7 shows the ultimate level of magnetic energy associated with the toroidal field for two cases of particular interest, (i) $\mathcal{E}_{\mathrm{m}} = 0.04,\ \mathcal{E} = 0.01$, and (ii) $\mathcal{E}_{\mathrm{m}} = 0.0025,\ \mathcal{E} = 0.005$. In case (i), the magnetic energy settled down to its ultimate level after some damped oscillations about this level. In case (ii), which was as near to the geostrophic limit ($\mathcal{E}_{\mathrm{m}} = 0,\ \mathcal{E} = 0$) as the numerical scheme would permit, there were again oscillations of small amplitude about the ultimate mean level of magnetic energy, and these oscillations showed no tendency to decay, with increasing time. The frequency of these oscillations was $\mathcal{O}(\Omega\mathcal{E}_{\mathrm{m}}{}^{1/2}) = \mathcal{O}((\Omega\,\eta)^{1/2}/R)$, and Proctor identified them with the torsional oscillations described in §13.6. The suggestion here is that when $\mathcal{E}_{\mathrm{m}}$ and $\mathcal{E}$ are sufficiently small, the steady state in which the Taylor constraint is operative is in fact unattainable, and that torsional oscillations about this state, as described in §13.6.5, are inevitable.

Greenspan (1974) studied a non-linear eigenvalue problem that presents itself in the magnetostrophic limit $\mathcal{E} = \mathcal{E}_{\mathrm{m}} = 0$, viz. suppose that $\mathbf{U} = v(s)\,\mathbf{e}_\varphi$, that $\alpha(s,z)$ is prescribed, that

$$\nabla \wedge (\mathbf{U} \wedge \mathbf{B}) + \nabla \wedge (\alpha\mathbf{B}) + \eta\nabla^2\mathbf{B} = 0 \quad (r < R)\,, \tag{15.61}$$

and finally that

$$F(s) = (\mu_0\rho)^{-1}\iint_{C(s)} [(\nabla \wedge \mathbf{B}) \wedge \mathbf{B}]_\varphi\, s\,\mathrm{d}\varphi\,\mathrm{d}z = 0 \quad (r < R)\,. \tag{15.62}$$

The matching conditions to a current-free field in $r > R$ are understood. The problem is to determine the values of the parameter $|\alpha|_{\max}R/\eta$ for which solutions $\{\mathbf{B}(\mathbf{x}), v(s)\}$ exist, and to determine the form of these solutions. Greenspan found a formal solution to this problem in terms of infinite series when $\alpha(s,z)$ is non-zero only in a thin layer on $(s^2+z^2)^{1/2} = R-$. The physical structure and significance of this solution remain to be determined.

A similar formulation to that of Malkus & Proctor (1975) was proposed by Braginskii (1975) who however considered the possibility that $|B_s|$ may be much weaker than $|B_z|$ in the liquid core of the Earth, and that in consequence the coupling of cylindrical shells due to the 'threading' of the field B_s may be so weak that the Taylor function $\mathcal{T}(s)$, though small, need not vanish identically in the steady state.

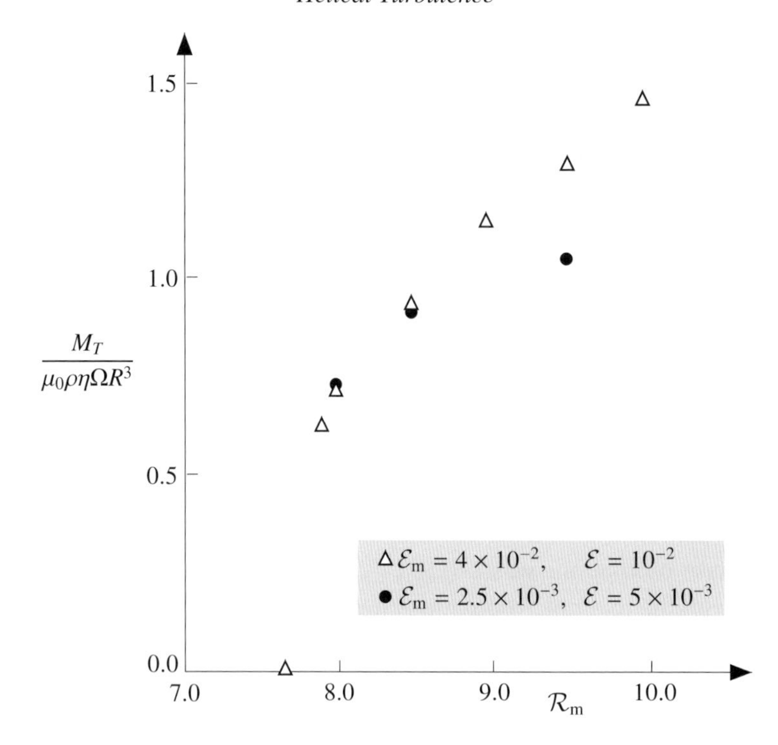

Figure 15.7 Equilibrium level of magnetic energy M_T of the toroidal field as a function of R_α for two choices of ($\mathcal{E}_m$, $\mathcal{E}$). Note that when viscosity is decreased (i.e. when $\mathcal{E}$ is decreased), the equilibrium level is lower because the Lorentz force more easily drives the mean flow, which provides the equilibration mechanism. [From Proctor 1977a.]

15.6 Chiral and magnetostrophic turbulence

Of course, in order to achieve equilibrium, energy must be supplied to the turbulence in some way, e.g. through gravitation and thermal convection modified by Coriolis forces. In such a situation, the ultimate dynamic balance is one in which Lorentz forces achieve statistical equilibrium with buoyancy and Coriolis forces, while inertial and viscous forces play a subsidiary role. Such a force balance may be described as 'magnetostrophic'. Much computational effort has been devoted to this kind of statistical equilibrium (see, for example, Brandenburg & Subramanian 2005).

The situation is most easily illustrated in the context of the Earth's liquid conducting outer core. Here, slow growth of the inner solid core leads to the release of buoyant blobs and plumes from a 'mushy zone' at the inner core boundary; these buoyant structures rise through the liquid core generating convective flow that is strongly influenced both by Coriolis effects due to the Earth's rotation and

Lorentz-force effects due to the magnetic field that is generated by dynamo action. This dynamo action in the case of the Earth involves both differential rotation (the mean-flow effect) and an anisotropic α-effect associated with a state of magnetostrophic turbulence. This type of turbulence is very different from conventional 'Kolmogorovian' turbulence, in that the dominant non-linearity is due to convection of buoyant elements rather than to fluid inertia. Nevertheless, a cascade of energy is to be expected, as argued by Moffatt (2008); see also Moffatt & Loper (1994). The discussion here parallels that of Braginsky & Meytlis (1990), who analysed the instability driven by unstable density stratification, and estimated the turbulent diffusivities associated with the resulting highly anisotropic turbulence; they did not however identify an α-effect, probably because the $\partial\mathbf{b}/\partial t$ term was neglected in their stability analysis (see below).

The depth of the liquid core is about 2300 km, but the rising elements that erupt from a mushy zone at the inner core boundary and contribute to the turbulence in the core are evidently on a much smaller scale. If this scale is of the order of 100 km or less, as seems quite likely, then the magnetic Reynolds number based on this scale is at most of order unity, and the 'low-$\mathrm{R_m}$' approximation is available at least as a first approximation for the small-scale dynamics. Moreover, this scale is still large enough for the Coriolis force to dominate over other contributions (except pressure gradient and Lorentz force) in the equation of motion. The governing equations then take the form

$$2\boldsymbol{\Omega} \times \mathbf{u} = -\nabla P + \rho^{-1}(\mathbf{B_0} \cdot \nabla)\mathbf{b} - \theta\,\mathbf{g}, \quad \nabla \cdot \mathbf{u} = 0, \tag{15.63}$$

$$\frac{\partial \mathbf{b}}{\partial t} = (\mathbf{B_0} \cdot \nabla)\mathbf{u} + \eta\nabla^2\mathbf{b}, \quad \nabla \cdot \mathbf{b} = 0, \tag{15.64}$$

$$\frac{\partial \theta}{\partial t} + \mathbf{u} \cdot \nabla\theta = S + \kappa\nabla^2\theta, \tag{15.65}$$

where ρ is now mean density, θ is buoyancy (of thermal or compositional origin), and S represents the source of buoyancy in the mushy zone, or possibly distributed throughout the liquid core. The term $\partial\mathbf{b}/\partial t$ is retained (although $\nabla \times (\mathbf{u} \times \mathbf{b})$ is discarded), because it turns out (Moffatt 2008) that the time-lag associated with this term is needed to account for the helicity that is responsible for dynamo action. Shearing of the buoyant structures by the differential rotation in the core could conceivably have a similar effect.

The important thing to note here however is that the only surviving non-linearity of this model is the term $\mathbf{u}\cdot\nabla\theta$ in (15.65). Here $\mathbf{u}$ is a linear functional of θ, $\mathbf{u} = \mathbf{u}\{\theta\}$, which can be obtained by solution of the linear equations (15.63) and (15.64), treating $-\theta\,\mathbf{g}$ as a 'source' term. Thus, equation (15.65) must be considered, not as the usual equation for a passive scalar field, but here as an equation for a very active scalar field $\theta(\mathbf{x}, t)$. We do not attempt to analyse these equations further here, but merely note that the 'magnetostrophic turbulence' that they describe is clearly very different from the usual Kolmogorovian turbulence, for which the non-linear inertia term $\mathbf{u}\cdot\nabla\mathbf{u}$ (here discarded) is responsible for the cascade of energy to small scales. The system (15.63)-(15.65) is simpler in one important respect: global regularity (in time) of solutions of such equations has been established by Friedlander & Vicol (2011). No such controlling theorem is as yet available for the Navier–Stokes equations; to establish such a theorem remains an open problem (Fefferman 2006).

16

Magnetic Relaxation under Topological Constraints

16.1 Lower bound on magnetic energy

We turn now to a process that is, in certain respects, complementary to dynamo action, i.e. the natural tendency of a magnetic field to relax to a minimum energy state compatible with any topological constraints that may apply. If for the moment we limit attention to an idealised situation in which $\mathbf{B}$ is a localised field in a fluid of infinite extent, then, from the induction equation (2.134), we may easily construct an equation for the evolution of magnetic energy $M(t) = \frac{1}{2}\int \mathbf{B}^2\,\mathrm{d}V$ (cf (6.7)) in the form

$$\frac{\mathrm{d}M}{\mathrm{d}t} = \int \mathbf{j}\cdot(\mathbf{u}\wedge\mathbf{B})\,\mathrm{d}V - \eta\int \mathbf{j}^2\,\mathrm{d}V = -\int \mathbf{u}\cdot(\mathbf{j}\wedge\mathbf{B})\,\mathrm{d}V - \eta\int \mathbf{j}^2\,\mathrm{d}V\,, \qquad (16.1)$$

in which the erosive effect of ohmic dissipation is clearly evident. The Lorentz force $\mathbf{j}\wedge\mathbf{B}$ tends to drive a flow $\mathbf{u}$ in the direction of $\mathbf{j}\wedge\mathbf{B}$, the key element of the process of magnetic relaxation. By contrast, in order to have dynamo action, we need a velocity field $\mathbf{u}$ which, at least on average, is *anti-parallel* to $\mathbf{j}\wedge\mathbf{B}$, thus working *against* the Lorentz force, and of sufficient magnitude to compensate for Joule dissipation. We have seen in earlier chapters that the helicity $\mathcal{H}$ of the flow can play a crucial role in achieving such dynamo action; and that the large-scale magnetic field generated then also has non-zero magnetic helicity $\mathcal{H}_M$.

But magnetic helicity also plays a crucial role in the complementary relaxation scenario. Here, we consider the ideal fluid limit in which $\eta = 0$, and so $\mathbf{B}$-lines are frozen in the fluid. We suppose further that the fluid is incompressible ($\nabla\cdot\mathbf{u} = 0$) (the opposite extreme of an extremely tenuous plasma will be considered in Chapter 17), and that it is confined to a finite domain $\mathcal{D}$ of maximum diameter, d say. We assume that the field $\mathbf{B}$ is also confined to this domain, and that $\mathbf{B}\cdot\mathbf{n} = 0$ on the surface $\partial\mathcal{D}$ of $\mathcal{D}$.

Since any topological structure of the magnetic field is conserved, it is perhaps

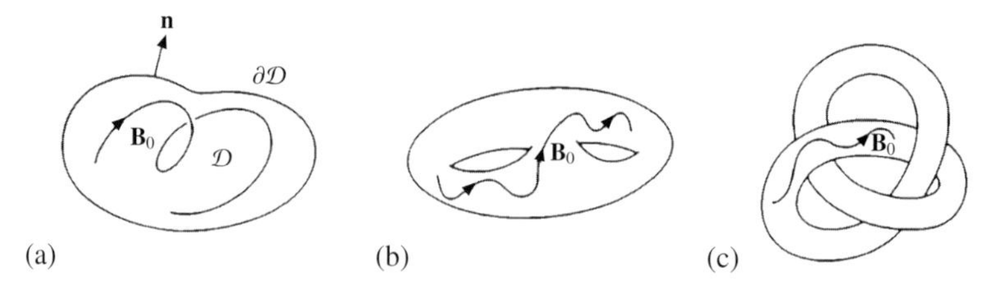

Figure 16.1 The magnetic-relaxation problem: the domain $\mathcal{D}$ may be (a) simply connected, (b) multiply connected, or even (c) multiply connected and of complex topology.

obvious that the magnetic energy must have a lower bound associated with this conserved structure. This bound, first recognised by Arnold (1974) for fields for which $\mathcal{H}_M \neq 0$, results from a combination of the Schwarz inequality

$$\int \mathbf{A}^2\,\mathrm{d}V \int \mathbf{B}^2\,\mathrm{d}V \geq \left\{\int \mathbf{A}\cdot\mathbf{B}\,\mathrm{d}V\right\}^2 = \mathcal{H}_M^2\,, \tag{16.2}$$

and the Poincaré inequality[1]

$$\int \mathbf{B}^2\mathrm{d}V \geq q^2 \int \mathbf{A}^2\mathrm{d}V\,, \tag{16.3}$$

where $q = \mathcal{O}(d^{-1})$ is a constant determined simply by the geometry and scale of the domain $\mathcal{D}$. Combining these inequalities gives immediately the result

$$\int \mathbf{B}^2\mathrm{d}V \geq q\,|\mathcal{H}_M|\,. \tag{16.4}$$

This result has been extended by Freedman (1988) to cover configurations such as the Whitehead link or the Borromean triple linkage (Figure 2.2), for which the linkage helicity is zero although the topology is certainly nontrivial. By nontrivial, we mean that there exists at least one closed **B**-line which cannot be shrunk to a point without cutting through other 'trapped' **B**-lines, such cutting being forbidden under the condition of perfect conductivity. But if the magnetic energy is to decrease to zero, this **B**-line (like every other) must shrink to a point; in so doing, the trapped **B**-lines are locally stretched without limit, the fluid being incompressible, with associated unlimited *increase* of magnetic energy. It is evident therefore that the trapping line cannot shrink to a point: the total magnetic energy must in fact be

[1] This inequality is similar to that derived in §6.2, see equation (6.12): for any non-zero solenoidal field **B** in a compact domain $\mathcal{D}$ satisfying the boundary condition $\mathbf{B}\cdot\mathbf{n} = 0$ on $\partial\mathcal{D}$, the vector potential **A** of this field is necessarily non-zero also, and **B** is determined by minimisation of the 'Rayleigh quotient' $Q = \int \mathbf{B}^2\mathrm{d}V \big/ \int \mathbf{A}^2\mathrm{d}V$, the integrals being throughout all space. This is a standard problem of the calculus of variations, from which it follows that $Q_{\min} = q^2 > 0$, where q is the lowest eigenvalue of the associated eigenvalue problem.

minimised when trapping and trapped fields have comparable energies, necessarily bounded away from zero.

Freedman's argument applies equally if the domain $\mathcal{D}$ is unbounded, the field $\mathbf{B}$ having finite energy and nontrivial topology. Suppose for example that there exists a closed tube of force $\mathcal{T}$ that is either nontrivially knotted, or that 'traps' other **B**-lines in the above sense. The volume $V_{\mathcal{T}}$ of $\mathcal{T}$ is conserved under incompressible flow, and its span d is obviously bounded for all t. The energy of the total field is greater that the energy $M_{\mathcal{T}}$ of the field $\mathbf{B}_{\mathcal{T}}$ restricted to $\mathcal{T}$, and this has a lower bound by Freedman's argument. If the helicity $\mathcal{H}_{\mathcal{T}}$ of $\mathbf{B}_{\mathcal{T}}$ is non-zero, then presumably a result of the form (16.4) still holds, with now $q \sim V_{\mathcal{T}}^{-1/3}$; it is difficult to be more precise than this.

16.2 Topological accessibility

In the domain $\mathcal{D}$, we may consider a magnetic field $\mathbf{B}(\mathbf{x}, t)$ that evolves from an initial field $\mathbf{B}_0(\mathbf{x})$ under the action of a smooth,[2] kinematically possible, transporting velocity field $\mathbf{u}(\mathbf{x}, t)$ of finite kinetic energy

$$K(t) = \tfrac{1}{2}\int_{\mathcal{D}} \rho \mathbf{u}^2 \,\mathrm{d}V < \infty \quad (0 \le t \le \infty), \tag{16.5}$$

where, as usual, ρ is the constant density of the fluid, here assumed uniform. For all finite $t > 0$, the field $\mathbf{B}(\mathbf{x}, t)$ is then also continuous and is topologically equivalent to $\mathbf{B}_0(\mathbf{x})$. However, as we shall see, as $t \to \infty$, discontinuities of $\mathbf{B}$, e.g. simple tangential discontinuities of $\mathbf{B}$ corresponding to current sheets, may develop. The limit field $\mathbf{B}_\infty(\mathbf{x})$, say, is then no longer topologically equivalent to $\mathbf{B}_0(\mathbf{x})$. It is natural however to introduce the weaker property of *topological accessibility* to describe this situation: a field $\mathbf{B}(\mathbf{x})$ may be said to be topologically accessible from a field $\mathbf{B}_0(\mathbf{x})$ if it can be obtained from $\mathbf{B}_0(\mathbf{x})$ by a process of distortion by a smooth velocity field $\mathbf{u}(\mathbf{x}, t)$ for which $K(t) < \infty$ for all $t \ge 0$, and for which, as we shall find below, it will be natural to impose a further condition of finite total dissipation of energy Φ over the whole time interval $(0, \infty)$. For a conventional Newtonian fluid with viscosity μ, this condition takes the form

$$\Phi = \mu \int_0^\infty \int_{\mathcal{D}} (\nabla \wedge \mathbf{u})^2 \,\mathrm{d}V\,\mathrm{d}t < \infty. \tag{16.6}$$

16.3 Relaxation to a minimum energy state

Suppose then that the fluid in $\mathcal{D}$ is incompressible and perfectly conducting ($\eta = 0$), but of non-zero viscosity ($\mu > 0$), thus providing a means of dissipating energy.

[2] We use the adjective 'smooth' to indicate informally 'continuously differentiable as many times as the context may require'.

Suppose that at time $t = 0$ the fluid is at rest, and permeated by an arbitrary smooth magnetic field $\mathbf{B}_0(\mathbf{x})$ of finite energy, satisfying merely $\mathbf{n} \cdot \mathbf{B}_0 = 0$ on $\partial\mathcal{D}$. The $\mathbf{B}_0$-lines may be knotted and/or linked, and they may even be chaotic in some subregions of the domain. In general, the Lorentz force $\mathbf{j}_0 \wedge \mathbf{B}_0$ will be rotational and will inevitably drive a flow $\mathbf{u}$ satisfying $\nabla \cdot \mathbf{u} = 0$ and the no-slip condition $\mathbf{u} = 0$ on ∂D; in this way, magnetic energy is converted to kinetic energy which is immediately dissipated by viscosity. By this mechanism, we may anticipate relaxation to a minimum energy state. The situation is particularly interesting if the initial field topology is non-trivial, as we shall assume in what follows; in this case, as we have seen, the magnetic energy has a positive lower bound.

The velocity field evolves according to the Navier–Stokes equation including the Lorentz force $\mathbf{j} \times \mathbf{B}$ on the right-hand side. From this, we may easily derive an equation for the kinetic energy $K(t)$:

$$\frac{\mathrm{d}K}{\mathrm{d}t} = \int \mathbf{u} \cdot (\mathbf{j} \wedge \mathbf{B})\, \mathrm{d}V - \mu \int \omega^2 \, \mathrm{d}V, \tag{16.7}$$

where $\omega = \nabla \wedge \mathbf{u}$ is the vorticity of the flow thus driven. Comparing with (16.1), it is evident that the rate of transfer of energy from magnetic to kinetic (i.e. the rate of work of the Lorentz force on the fluid) is $W(t) = \int \mathbf{u} \cdot (\mathbf{j} \wedge \mathbf{B})\, \mathrm{d}V$; and that, with $\eta = 0$,

$$\frac{\mathrm{d}}{\mathrm{d}t}(K + M) = -\mu \int \omega^2 \, \mathrm{d}V, \tag{16.8}$$

so that, as expected, $K + M$ is dissipated by viscosity. Noting now that the total energy is positive, monotonic decreasing and bounded away from zero, it must tend to a positive constant as $t \to \infty$. It follows that, in the limit,

$$\int \omega^2 \, \mathrm{d}V = 0. \tag{16.9}$$

Provided that no abnormal singularity of vorticity develops during this relaxation process,[3] controlled as it is by viscosity, it then follows that in this limit, $\omega \equiv 0$, and so $\mathbf{u} \equiv 0$ in $\mathcal{D}$ also. In this situation, let $\mathbf{B}^E$ be the limit magnetic field; with $\mathbf{u} \equiv 0$, this must evidently satisfy the magnetostatic equation

$$\mathbf{j}^E \wedge \mathbf{B}^E = \nabla p^E, \tag{16.10}$$

where $\mathbf{j}^E = \nabla \wedge \mathbf{B}^E$, and p^E is the associated pressure distribution. Here, the important thing to note is that, in the sense described in §14.2 above, the magnetostatic field $\mathbf{B}^E$ is topologically accessible from the initial field $\mathbf{B}_0$, the total energy dissipated by viscosity during the relaxation process being necessarily bounded by the initially available magnetic energy $M(0)$.

[3] It has not yet been rigorously proved that the conclusion $\omega \equiv 0$ here follows inevitably from (16.9); indeed this is still no more than a very plausible conjecture.

We emphasise that the above theory applies only in the ideal ($\eta = 0$) limit. If η is very small but non-zero, then relaxation to a nearly magnetostatic state will occur, but then slow diffusion will allow reconnection of **B**-lines, and so change of field topology. It may be conjectured that the field will continue to decay on a diffusive time-scale while being constrained by the above dynamics to remain near to a magnetostatic state. We shall encounter such behaviour in Chapter 17.

16.3.1 Alternative 'Darcy' relaxation procedure

It must be admitted that above relaxation procedure is physically artificial in that there exists no real fluid having zero resistivity but non-zero viscosity. This artificiality is not lethal however, because the procedure must be regarded merely as a means to an end, that end being demonstration of the existence of magnetostatic fields of arbitrarily prescribed topology in terms of knots and links of the **B**-lines, albeit with imbedded current sheets. Other artificial relaxation procedures may be constructed with the same objective. One that is simpler to use in numerical computations is that in which the velocity field $\mathbf{u}(\mathbf{x}, t)$ is driven directly by the Lorentz force $\mathbf{j}\wedge\mathbf{B}$ through what may be described as the 'Darcy model' (see, e.g. Whitaker 1986)

$$k\mathbf{u} = \mathbf{j} \wedge \mathbf{B} - \nabla p, \quad \nabla \cdot \mathbf{u} = 0, \quad \mathbf{n} \cdot \mathbf{u} = 0 \text{ on } \partial\mathcal{D}, \tag{16.11}$$

where k is a positive constant, and the pressure $p(\mathbf{x}, t)$ is determined by the conditions

$$\nabla \cdot \mathbf{u} = 0, \quad \mathbf{n} \cdot \mathbf{u} = 0 \text{ on } \partial\mathcal{D}, \tag{16.12}$$

or equivalently by the Neumann problem

$$\nabla^2 p = \nabla \cdot (\mathbf{j} \wedge \mathbf{B}), \quad \partial p/\partial n = \mathbf{n} \cdot (\mathbf{j} \wedge \mathbf{B}) \text{ on } \partial\mathcal{D}. \tag{16.13}$$

The first of (16.11) may be alternatively written

$$k\mathbf{u} = (\mathbf{j} \wedge \mathbf{B})_{\mathrm{S}} \tag{16.14}$$

where the right-hand-side now represents the 'solenoidal projection'[4] of $\mathbf{j}\wedge\mathbf{B}$. With this definition of **u**, the field **B** must still be supposed to evolve under the frozen field equation

$$\partial\mathbf{B}/\partial t = \nabla \wedge (\mathbf{u} \wedge \mathbf{B}). \tag{16.15}$$

[4] The Leray projection operator $\mathbb{P}[\cdot]$ may be used to define the same 'solenoidal projection', i.e. for any smooth vector field $\mathbf{X}(\mathbf{x})$,

$$\mathbb{P}[\mathbf{X}(\mathbf{x})] \equiv \mathbf{X}_{\mathrm{S}}(\mathbf{x}) = \mathbf{X}(\mathbf{x}) - \nabla[\nabla^{-2}\nabla \cdot \mathbf{X}(\mathbf{x})].$$

With these equations, an energy equation is easily obtained in the form

$$\frac{\mathrm{d}M}{\mathrm{d}t} = -k \int_{\mathcal{D}} \mathbf{u}^2 \,\mathrm{d}V, \tag{16.16}$$

with the same conclusion as before, namely that the system relaxes to a minimum-energy magnetostatic equilibrium ($\mathbf{u} \equiv 0$), i.e. $\mathbf{B}(\mathbf{x}, t) \to \mathbf{B}^E(\mathbf{x})$, where

$$\left(\mathbf{j}^E \wedge \mathbf{B}^E\right)_{\mathrm{S}} \equiv \mathbf{j}^E \wedge \mathbf{B}^E - \nabla p^E = 0, \tag{16.17}$$

the field $\mathbf{B}^E$ being topologically accessible from the initial field $\mathbf{B}_0$.

A word of warning however: if the initial field is very complex, then numerous distinct relaxed states may exist, and the particular state that is attained may depend upon the particular relaxation model adopted.

16.4 Two-dimensional relaxation

We may consider the same type of relaxation problem for two-dimensional fields $\mathbf{B} = (\partial\chi/\partial y, -\partial\chi/\partial x, 0)$, $\mathbf{u} = (\partial\psi/\partial y, -\partial\psi/\partial x, 0)$ in a two-dimensional domain $\mathcal{D}$, where $\chi(x, y, t)$ is the flux function of the field, and $\psi(x, y, t)$ the stream-function of the driven velocity. In this situation, with $\eta = 0$, the induction equation leads to equation (3.55), i.e.

$$D\chi/Dt \equiv \partial\chi/\partial t + \mathbf{u} \cdot \nabla\chi = 0. \tag{16.18}$$

Thus χ behaves like a transported scalar field. The magnetic helicity is clearly zero in this 2D situation, but the **B**-lines, χ = const., are frozen in the fluid, and this implies a strong topological constraint. Since the flow is incompressible, the area $\mathcal{A}(\chi)$ inside each such **B**-line is conserved. The function $\mathcal{A}(\chi)$ is called the 'signature' of the field,[5] which remains invariant throughout the relaxation process. This obviously places a lower bound on the magnetic energy, whatever the shape of the domain $\mathcal{D}$ may be.

Relaxation again proceeds until a magnetostatic equilibrium is established. In the two-dimensional situation considered here, the condition $\nabla \times (\mathbf{j} \wedge \mathbf{B}) = 0$ reduces to

$$\frac{\partial(\chi, \nabla^2\chi)}{\partial(x, y)} = 0, \tag{16.19}$$

implying a functional dependence, in general non-linear, between χ and $\nabla^2\chi$,

$$\nabla^2\chi = \mathcal{F}(\chi), \tag{16.20}$$

[5] If the field has a number of X-type neutral points (i.e. saddle points), then the domain $\mathcal{D}$ is the union of a set of subdomains $\mathcal{D}_i$ bounded by **B**-lines through these saddle points. There is then a corresponding set of signature functions $\mathcal{A}_i(\chi)$, one for each subdomain.

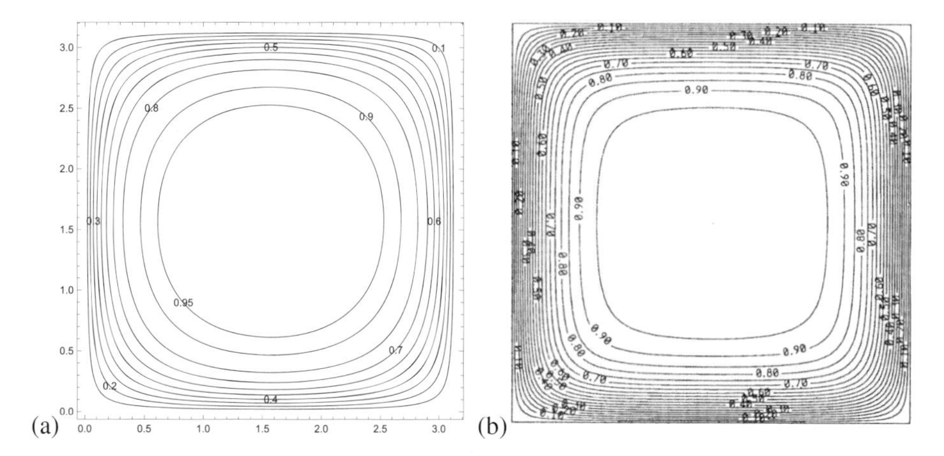

Figure 16.2 Relaxation of a field with a single O-type neutral point in a square domain: (a) initial field given by (16.21); (b) relaxed field. [From Linardatos 1993, reproduced with permission.]

where $\mathcal{F}(\chi)$ is, in principle, determined by the signature function $\mathcal{A}(\chi)$. The relationship between these two functions is not however straightforward.

In a circular domain $\mathcal{D}$, with $\chi = 0$ on $\partial\mathcal{D}$, the simplest field topology is that for which χ has a single extremum in $\mathcal{D}$, corresponding to a single elliptic ('O-type') neutral point. In this situation, as the signature $\mathcal{A}(\chi)$ is invariant during relaxation, each closed field line χ=const. attains a circular shape, this having the shortest length for prescribed area $\mathcal{A}(\chi)$. Moreover, such a configuration, being of minimum energy, is stable at least with respect to two-dimensional disturbances.

For any simply connected two-dimensional domain, for example a square domain, again the simplest field topology is that for which there is a single O-type neutral point within the domain. Now, field lines near the boundary are constrained to conform with the boundary shape, and the minimum energy state reflects a compromise for each field line between this geometrical constraint and its tendency to contract towards circular shape.

The following examples are taken from Linardatos (1993), who has treated the general situation in some detail, using the Darcy model (16.11) in the numerics. Figure 16.2 illustrates the situation when an initial field

$$\chi(x, y) = N\left[1 - \mathrm{e}^{-5\sin x \sin y}\right] \tag{16.21}$$

is given in the square domain $0 < x < \pi$, $0 < y < \pi$; here, the normalising factor $N = \left(1-\mathrm{e}^{-5}\right)^{-1} \approx 1.0068$ is chosen so that $\chi_{\max} = 1$. The relaxation of this field is quite slight, the effect being most noticeable in the interior where the nearly circular streamlines relax towards conformity with the square boundary.

A more interesting situation arises if we allow for hyperbolic ('saddle' or 'X-type') neutral points in the initial field. Figure 16.3 illustrates the situation when the initial field is given by

$$\chi(x,y) = N\left[\left(e^{5\sin\pi y} - 1\right)e^{5\sin^2 2\pi x}\sin\pi x - e^5 + 1\right], \tag{16.22}$$

where again N is chosen so that $\chi_{max}=1$ (here $N = (1-e^5)^{-2} \approx 4.602\times10^{-5}$). This field has a saddle point at $(0.5, 0.5)$, the tangents to the **B**-lines at this point being as indicated. The field is chosen to be much stronger in the regions to left and right surrounding the two O-type neutral points, so that the field rapidly relaxes to near circular form around these points; these structures press in on the tangents causing collapse of the saddle point to a tangential discontinuity (i.e. a current sheet). The relaxed field shown in Figure 16.3b illustrates this behaviour well (the process of relaxation is not quite complete here, and numerical imperfections show up near the current sheet). A 'scatterplot' of $\nabla^2\chi$ against χ is shown in Figure 16.3c for the initial field and in Figure 16.3d for the relaxed field; in the latter case, the points lie close to a curve, reflecting the functional relationship (16.20), the slight residual scatter of these points being associated with very slow relaxation in the neighbourhood of the end-points of the current sheet. The structure of the field near these end points is of particular interest. Vainshtein (1990) has shown that an equilibrium field at the end of a current sheet must have a Y-type structure in these regions, as illustrated in Figure 16.4.

For any initial field geometry within a simply connected two-dimensional domain, the numbers N_O and N_X of O- and X-type neutral points are related by Euler's identity $N_O - N_X = 1$. It seems likely that, in general, every saddle point will collapse in the above manner to a current sheet of limited extent under two-dimensional magnetic relaxation. These current sheets then become sites of diffusive reconnection of field lines, no matter how small the diffusion parameter η may be. They are also sites of intense heating in the plasma context. The formation of current sheets is therefore an important subject in its own right (Parker 1994). We shall see below that current sheets are also to be expected in fully three-dimensional relaxation.

16.5 The relaxation of knotted flux tubes

Consider the three-dimensional situation in which the initial field $\mathbf{B}_0(\mathbf{x})$ is confined to a single magnetic flux tube, whose axis has the form of a knot K. We may assume for the purpose of this discussion (i) that the tube has initially circular cross section,

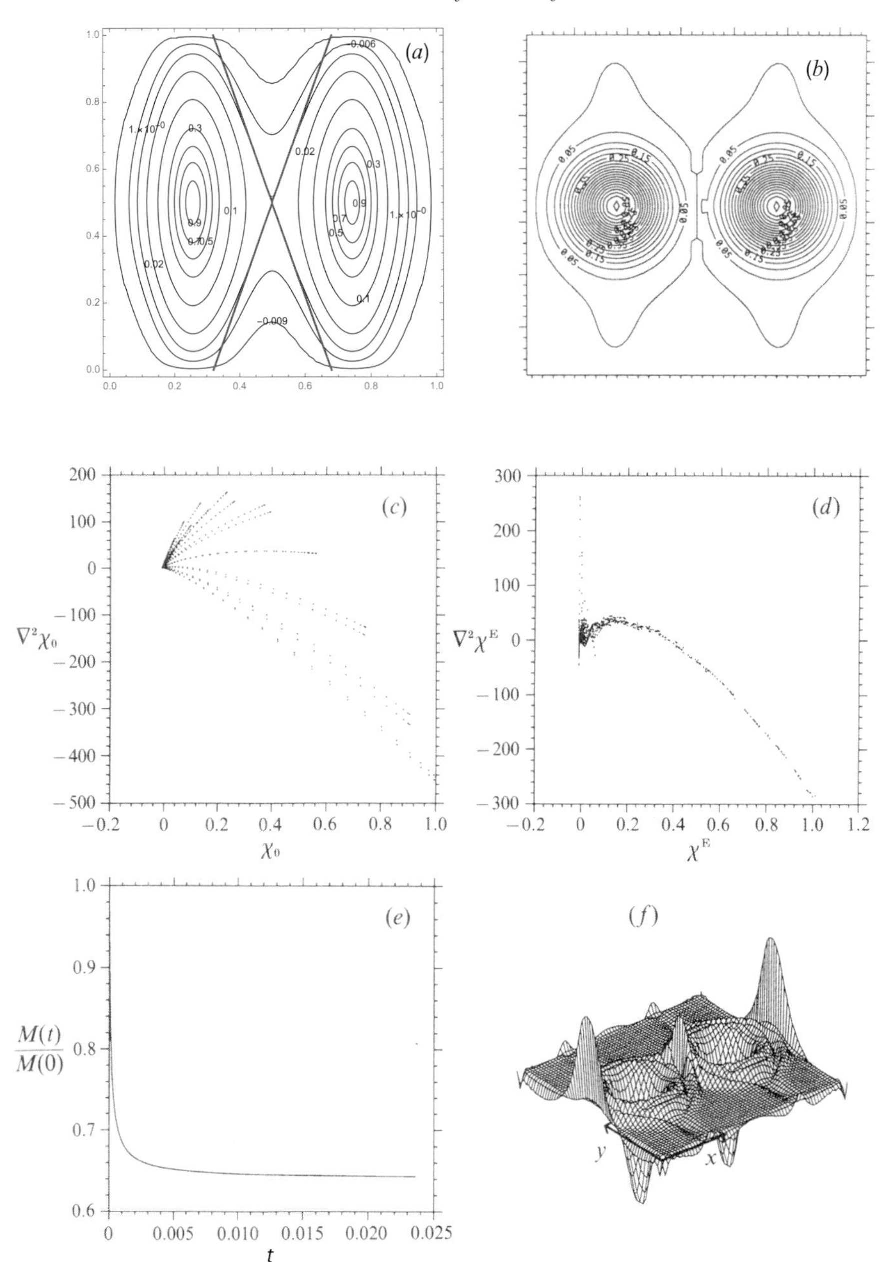

Figure 16.3 Relaxation of a field with an X-type neutral point in a square domain: (a) initial field given by (16.22); (b) relaxed field; (c) scatter plot of $\nabla^2\chi$ for initial field; (d) the same for the relaxed field providing evidence of a functional relationship $\nabla^2\chi = \mathcal{F}(\chi)$; (e) decay of magnetic energy during relaxation; (f) current distribution $j(x, y)$ in the relaxed state, showing concentrated current in the current sheet, and also at the perfectly conducting boundary. [From Linardatos 1993, reproduced with permission.]

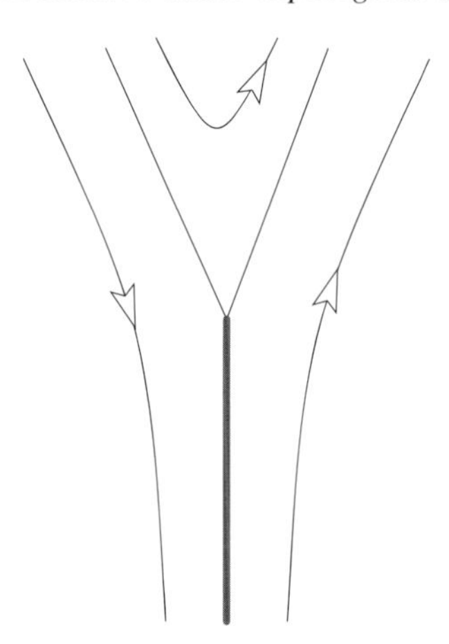

Figure 16.4 Y-type neutral point, as described by Vainshtein (1990); the thick line indicates a current sheet.

(ii) that the axial field component is continuous over the cross-section of the tube and falls smoothly to zero at the tube boundary, and (iii) that the tube is 'uniformly twisted' so that any two **B**-lines within the tube have the same linking number. As shown in §2.10.2, the magnetic helicity of such a configuration is given by

$$\mathcal{H}_M = h\Phi^2, \tag{16.23}$$

where Φ is the magnetic flux through any cross-section of the tube, $h = Wr + Tw$, and Wr and Tw are respectively the writhe and twist of the tube. Both Φ and h are conserved under deformation by a continuous velocity field $\mathbf{u}(\mathbf{x}, t)$. Moreover, if $\nabla \cdot \mathbf{u} = 0$, then the volume V of the tube is also conserved.

The Maxwell stress in such a tube consists of a magnetic pressure which is compensated by fluid pressure, and a tension within the tube that causes it to contract. As its length L decreases, its cross-sectional area A must increase to conserve the volume $V \sim A\,L$. Obviously, the contraction is impeded when the tube first makes contact with itself, a manifestation of the topological constraint (16.4). The relaxation does not cease at this stage: the knotted tube will continue to adjust itself to make contact along its full length, and its cross-section will be progressively deformed during this relaxation. Ultimately however, the tube must come to rest when the magnetic energy attains a minimum value $M_{\min}$, say. This minimum is determined by the conserved quantities Φ, V and h, and so on dimensional grounds must be expressible in the form

$$M_{\min} = m(h)\Phi^2 V^{-1/3}, \tag{16.24}$$

where $m(h)$ is a dimensionless function of the dimensionless twist parameter h; this function is an invariant property of the knot K (Moffatt 1990).

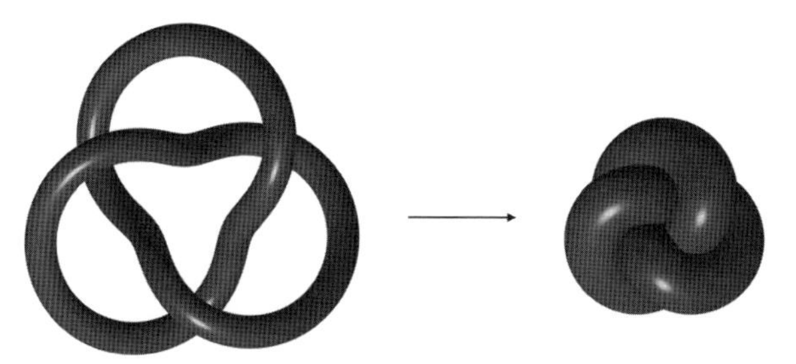

Figure 16.5 Schematic relaxation of a flux tube in the form of a (right-handed) trefoil knot.

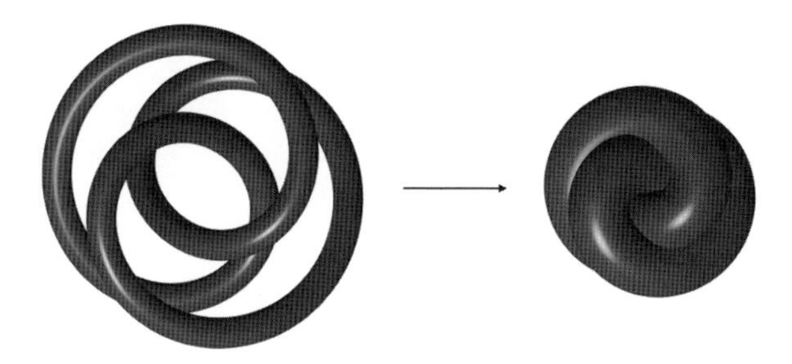

Figure 16.6 Relaxation of right-handed trefoil flux tube in the $T_{3,2}$ geometry.

16.6 Properties of relaxed state

An important feature of this process is the inevitable development of tangential discontinuities of the field **B** (i.e. of current sheets) during the final stage of relaxation. Although the initial field is zero at the tube boundary, this condition does not persist because the stronger field-lines near the tube centre can shrink by moving towards the inner part of the tube boundary (i.e. the part nearest to the local centre of curvature of the tube). The current sheets form where these inner parts come into contact in the ultimate relaxed state, as illustrated for the trefoil knot in Figure 16.5.

A second point to note is that the relaxed state for a given knot type may not be unique. A torus knot $T_{p,q}$ has p circuits the long way round the supporting torus, and q circuits the short way; and it is well-known that $T_{p,q}$ is topologically equivalent to $T_{q,p}$, in the sense that one can be continuously deformed in $\mathbb{R}^3$ into the other. Figure 16.5 shows the relaxation of $T_{2,3}$, while Figure 16.6 shows that of the topologically equivalent knot $T_{3,2}$. In the projections shown, the crossings of $T_{2,3}$ alternate in the sequence 'over, under, over, under, ... ', while the crossings for $T_{3,2}$ are in the sequence 'over, over, under, under, over, over, ... '. The two relaxed states are geometrically distinct and have different minimum-energy functions. More generally, for a given knot type K, we should anticipate a spectrum $m_i(h)$, $(i = 0, 1, 2, \dots)$

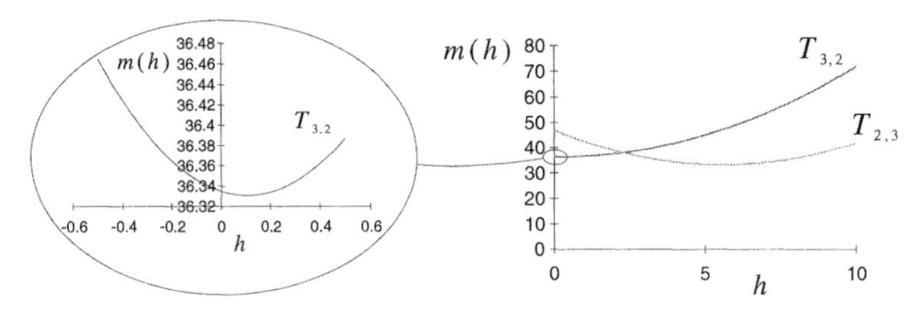

Figure 16.7 Minimum-energy functions $\bar{m}(h)$ for the trefoil knot in the configurations $T_{2,3}$ and $T_{3,2}$. [From Chui & Moffatt 1995.]

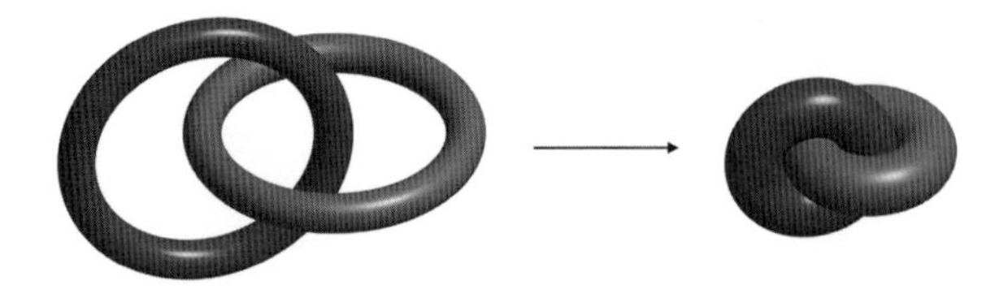

Figure 16.8 Relaxation of the Hopf link.

of minimum-energy functions, corresponding to distinct relaxed geometries, which can in general be ordered for given h: $m_0(h) < m_1(h) < m_2(h) \dots$. Interest naturally centres on the 'ground-state' of lowest energy.

Computation of these relaxed states is still beyond the power of present computation. However, progress is possible if the tube cross-section is constrained to be circular, a strategy adopted by Chui & Moffatt (1995). The constrained energy minimum is then $\bar{M}_{\min} = \bar{m}(h)\Phi^2 V^{-1/3}$; clearly $\bar{m}(h) > m(h)$ but the relative difference $\bar{m}(h)/m(h) - 1$ may be expected to be quite small. Figure 16.7 shows the computed functions $\bar{m}(h)$ for the trefoil knot in both forms $T_{2,3}$ and $T_{3,2}$. In each case, the curves are asymmetric about $h = 0$, a symptom of the chirality of the knot. Surprisingly, there is a crossover point at $h \approx 2.3$, so that for $0 < h < 2.3$ it is $T_{3,2}$ rather than $T_{2,3}$ that has the smaller energy. The absolute minima occur at $h \approx 0.11, \bar{m}_{\min} \approx 36.3$ for $T_{3,2}$, and at $h \approx 6, \bar{m}_{\min} \approx 32$ for $T_{2,3}$.

Finally, we may note that linked flux tubes must relax in a similar way to a minimum-energy state. Here however there is obviously an additional freedom corresponding to the fact that the various tubes constituting the link may have different volumes V_i, axial fluxes Φ_i, and twist parameters h_i. For example, for the Hopf link of Figure 16.8, the minimum energy must be expressible, on dimensional grounds and with obvious notation, in the form

$$M_{\min} = m\Big(h_1, h_2, \Phi_1/\Phi_2, V_1/V_2\Big)\,\Phi_1^2 V_1^{-1/3}. \tag{16.25}$$

16.7 Tight knots

The above treatment of knot relaxation is closely related to the theory of ideal (or 'tight') knots (Stasiak et al. 1998), with primary motivation from polymer physics (Deguchi & Tsurusaki 1997) and molecular biology (Simon 1996). In practical terms, one may ask, "What is the shortest length of rope with which a particular knot K may be tied?" Here it must be understood that the ends of the rope must be joined following the tying of the knot so that the axis of the rope is a closed curve. A tight knot is thus defined as a closed knotted tube of uniform circular cross section which has minimum length L for prescribed cross-sectional area $A = \pi R^2$; such a minimum is achieved by a particular geometry of the tube axis. There are in effect two constraints in determining the minimal configuration: the radius of curvature at any point of the tube axis cannot be less than R, and the tube cannot intersect itself; it does however make contact with itself along a curve or curves on the tube surface. In this respect, it differs from the magnetic relaxation problem where deformation of the flux tube cross section leads to contact over a two-dimensional surface or surfaces (i.e. current sheets). A second difference is that helicity plays no part in the construction of a tight knot, whereas prescribed helicity $h\Phi^2$ is an essential ingredient of the magnetic relaxation problem.

16.8 Structure of magnetostatic fields

As already observed, the relaxed state is an equilibrium satisfying the magnetostatic equation (16.10). This must be supplemented by a condition of continuity of total pressure $p^E + \frac{1}{2}\mathbf{B}^E \cdot \mathbf{B}^E$ across any imbedded current sheets. In regions of continuity of the field $\mathbf{B}^E$, we must evidently have

$$\mathbf{j} \cdot \nabla p^E = 0, \quad \text{and } \mathbf{B}^E \cdot \nabla p^E = 0, \tag{16.26}$$

i.e. the current lines and the field lines lie on surfaces p^E =const. This places a severe constraint on the topology of the $\mathbf{B}^E$-field. However, it may happen that p^E=const. in some subdomain $\mathcal{D}_s$, say, of the domain $\mathcal{D}$. In this case, from (16.10),

$$\mathbf{j}^E \wedge \mathbf{B}^E = 0, \quad \text{i.e.} \quad \mathbf{j}^E = \beta(\mathbf{x})\mathbf{B}^E, \quad \text{in } \mathcal{D}_s \tag{16.27}$$

for some function $\beta(\mathbf{x})$. Now however, since $\nabla \cdot \mathbf{j}^E = 0$,

$$\mathbf{B}^E \cdot \nabla\beta = 0, \quad \text{in } \mathcal{D}_s, \tag{16.28}$$

so we have still not escaped the topological constraint that $\mathbf{B}^E$-lines lie on surfaces. However, again there may be a sub-subdomain $\mathcal{D}_{ss}$ of $\mathcal{D}_s$ in which β =const. In $\mathcal{D}_{ss}$, there is no topological constraint on $\mathbf{B}^E$-lines, which are free to wander in chaotic manner throughout this sub-subdomain.

As argued in §16.3, the initial field $\mathbf{B}_0(\mathbf{x})$ in the magnetic relaxation problem may have arbitrary topology, and this topology is conserved during the ideal ($\eta = 0$) relaxation process. If the $\mathbf{B}_0$-lines are chaotic, then the field lines of the relaxed field must in some sense share the same degree of chaos. As pointed out however by Arnold (1974), the class of smooth solutions of the equation $\mathbf{j}^E = \beta \mathbf{B}^E$ with β constant is far too limited to cover every possible chaotic structure, and it follows that the chaos in the initial field must somehow be accommodated by a system of current sheets in the corresponding relaxed state. These current sheets may be densely distributed; some exotic possibilities involving a 'devil's staircase' of current sheets have been explored by Bajer (2005).

16.9 Stability of magnetostatic equilibria

A magnetostatic equilibrium that is attained via a relaxation process is evidently stable, being a state of minimum magnetic energy with respect to frozen-field perturbations (alternatively described as 'isomagnetic'). It is illuminating nevertheless to obtain a criterion (Bernstein et al. 1958) for the stability of general magnetostatic equilibria that are obtained by direct solution of the equation

$$\mathbf{j}^E \wedge \mathbf{B}^E = \nabla p^E. \tag{16.29}$$

To this end, we need to formalise what we mean by an isomagnetic perturbation. In this, we follow the procedure of Moffatt (1986).

As in §8.2, we consider a virtual displacement of the fluid resulting from a steady velocity $\mathbf{v}(\mathbf{x})$ acting over a short time-interval $(0, \tau)$. As before, we suppose that the fluid is contained in a bounded domain $\mathcal{D}$, with $\nabla \cdot \mathbf{v} = 0$ in $\mathcal{D}$ and $\mathbf{n} \cdot \mathbf{v} = 0$ on the boundary $\partial\mathcal{D}$. Let $\boldsymbol{\xi}(\mathbf{x}, t)$ be the resulting displacement of the fluid particle initially at $\mathbf{x}$. Then

$$\partial\boldsymbol{\xi}/\partial t = \mathbf{v}(\mathbf{x} + \boldsymbol{\xi}) = \mathbf{v}(\mathbf{x}) + \boldsymbol{\xi} \cdot \nabla\mathbf{v} + \cdots, \tag{16.30}$$

and so

$$\boldsymbol{\xi}(\mathbf{x}, \tau) = \tau\, \mathbf{v}(\mathbf{x}) + \tfrac{1}{2}\tau^2\, \mathbf{v} \cdot \nabla\mathbf{v} + \cdots = \boldsymbol{\eta}(\mathbf{x}) + \tfrac{1}{2}\boldsymbol{\eta} \cdot \nabla\boldsymbol{\eta} + \mathcal{O}(\eta^3), \tag{16.31}$$

or equivalently

$$\boldsymbol{\eta}(\mathbf{x}) = \boldsymbol{\xi}(\mathbf{x}, \tau) - \tfrac{1}{2}\boldsymbol{\xi} \cdot \nabla\boldsymbol{\xi} + \mathcal{O}(\xi^3), \tag{16.32}$$

where $\boldsymbol{\eta}(\mathbf{x}) = \tau\, \mathbf{v}(\mathbf{x})$ is the corresponding Eulerian displacement field, satisfying the 'admissibility conditions' $\nabla \cdot \boldsymbol{\eta} = 0$ in $\mathcal{D}$ and $\mathbf{n} \cdot \boldsymbol{\eta} = 0$ on $\partial\mathcal{D}$. The relationship between the Lagrangian displacement $\boldsymbol{\xi}$ and the corresponding Eulerian displacement $\boldsymbol{\eta}$ is indicated by the sketch of Figure 16.9.

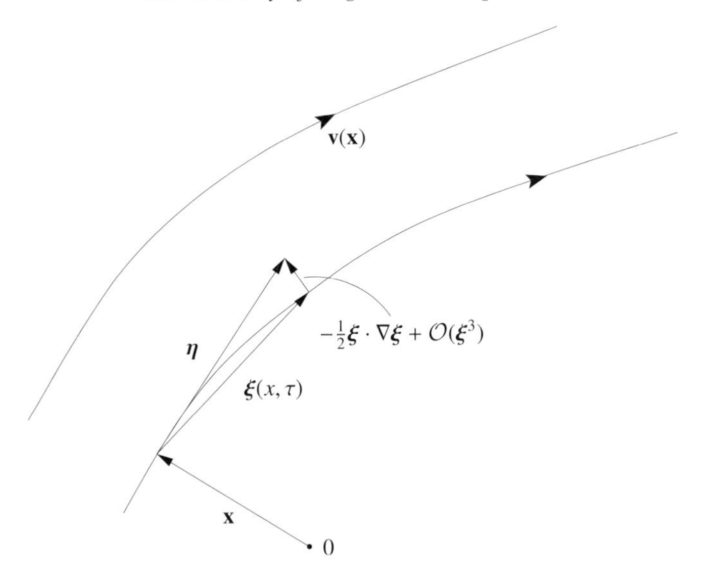

Figure 16.9 Relationship between $\boldsymbol{\xi}$ and $\boldsymbol{\eta}$, as defined by (16.31).

Now the frozen-field equation for $\mathbf{B}(\mathbf{x}, t)$ is, for small t,

$$\frac{\partial \mathbf{B}}{\partial t} = \nabla \wedge (\mathbf{v}(\mathbf{x}) \wedge \mathbf{B}) = \nabla \wedge \left\{\mathbf{v}(\mathbf{x}) \wedge \left[\mathbf{B}^E + t\,\nabla \wedge \left(\mathbf{v}(\mathbf{x}) \wedge \mathbf{B}^E\right) + \cdots\right]\right\}. \tag{16.33}$$

Hence the perturbed field at time τ is

$$\mathbf{B}(\mathbf{x}, \tau) = \mathbf{B}^E(\mathbf{x}) + \delta^1\mathbf{B} + \delta^2\mathbf{B} + \cdots, \tag{16.34}$$

where

$$\delta^1\mathbf{B} = \nabla \wedge (\boldsymbol{\eta} \wedge \mathbf{B}^E), \quad \delta^n\mathbf{B} = \frac{1}{n!}\nabla \wedge (\boldsymbol{\eta} \wedge \delta^{n-1}\mathbf{B}), \quad (n = 2, 3, \ldots). \tag{16.35}$$

The corresponding perturbed magnetic energy $M = \frac{1}{2}\int \mathbf{B}^2 \mathrm{d}V$ is

$$M = M^E + \delta^1 M + \delta^2 M + \mathcal{O}(\boldsymbol{\eta}^3), \tag{16.36}$$

where

$$\delta^1 M = \int \left[\mathbf{B}^E \cdot \delta^1\mathbf{B}\right] \mathrm{d}V, \quad \delta^2 M = \frac{1}{2}\int \left[\left(\delta^1\mathbf{B}\right)^2 + 2\,\mathbf{B}^E \cdot \delta^2\mathbf{B}\right] \mathrm{d}V. \tag{16.37}$$

It is evident that $\delta^1 M$ is a linear functional and $\delta^2 M$ a quadratic functional of $\boldsymbol{\eta}$.

Note first that any displacement field $\boldsymbol{\eta}(\mathbf{x})$ satisfying $\nabla \wedge (\boldsymbol{\eta} \wedge \mathbf{B}^E) \equiv 0$ has no effect on the field $\mathbf{B}^E$, and in particular on its energy M^E. The field is thus neutrally stable to such perturbations, which may be described as trivial as far as the stability problem is concerned. However, $\delta^1 M = 0$ even for non-trivial perturbations for which $\delta^1 \mathbf{B} = \nabla \wedge (\boldsymbol{\eta} \wedge \mathbf{B}^E) \neq 0$; for, integrating by parts and discarding terms that vanish on $\partial\mathcal{D}$, we have

$$\delta^1 M = \int_{\mathcal{D}} \left[\nabla \wedge \mathbf{B}^E \cdot (\boldsymbol{\eta} \wedge \mathbf{B}^E)\right] \mathrm{d}V = -\int_{\mathcal{D}} \boldsymbol{\eta} \cdot (\mathbf{j}^E \wedge \mathbf{B}^E)\, \mathrm{d}V, \tag{16.38}$$

and hence, by virtue of the equilibrium condition (16.29),

$$\delta^1 M = -\int_{\mathcal{D}} \boldsymbol{\eta} \cdot \nabla p^E \, \mathrm{d}V = -\int_{\partial\mathcal{D}} (\mathbf{n} \cdot \boldsymbol{\eta})\, p^E \, \mathrm{d}S = 0. \tag{16.39}$$

This is of course to be expected, since $\mathbf{B}^E(\mathbf{x})$ is a stationary state.

Similarly, the above expression for $\delta^2 M$ may be easily manipulated to the form

$$\delta^2 M = \tfrac{1}{2} \int_{\mathcal{D}} \left[\left(\nabla \wedge \left(\boldsymbol{\eta} \wedge \mathbf{B}^E\right)\right)^2 - \left(\boldsymbol{\eta} \wedge \mathbf{B}^E\right) \cdot \nabla \wedge \left(\boldsymbol{\eta} \wedge \mathbf{j}^E\right)\right] \mathrm{d}V, \tag{16.40}$$

and the equilibrium is stable (always within the zero-resistivity context) if

$$\delta^2 M \geq 0 \quad \text{for all admissible } \boldsymbol{\eta}(\mathbf{x}). \tag{16.41}$$

16.9.1 The two-dimensional situation

This criterion can be greatly simplified in the two-dimensional situation. With

$$\mathbf{B}^E = \left(\frac{\partial \chi}{\partial y}, -\frac{\partial \chi}{\partial x}, 0\right) \text{ and } \boldsymbol{\eta} = \left(\frac{\partial \psi}{\partial y}, -\frac{\partial \psi}{\partial x}, 0\right), \tag{16.42}$$

we have

$$\mathbf{j}^E = \nabla \wedge \mathbf{B}^E = (0, 0, -\nabla^2 \chi) = (0, 0, -\mathcal{F}(\chi)), \tag{16.43}$$

and

$$\boldsymbol{\eta} \wedge \mathbf{B}^E = (0, 0, G(x, y)), \text{ where } G(x, y) = \frac{\partial(\psi, \chi)}{\partial(x, y)}; \tag{16.44}$$

obviously $G = 0$ on $\partial\mathcal{D}$. It follows that

$$\int \left[\nabla \wedge \left(\boldsymbol{\eta} \wedge \mathbf{B}^E\right)\right]^2 \mathrm{d}S = \int (\nabla G)^2 \, \mathrm{d}S \geq q_0^2 \int G^2 \mathrm{d}S, \tag{16.45}$$

where q_0^2 is the lowest eigenvalue of the Dirichlet problem

$$\left(\nabla^2 + q^2\right) G = 0, \text{ in } \mathcal{D}, \; G = 0 \text{ on } \partial\mathcal{D}. \tag{16.46}$$

Combining these results, we now find from (16.40) that

$$\delta^2 M \geq \tfrac{1}{2} \int_{\mathcal{D}} \left(q_0^2 + \mathcal{F}'(\chi)\right) G^2 \mathrm{d}S\,. \tag{16.47}$$

The equilibrium is therefore stable, at least to two-dimensional disturbances, provided

$$\mathcal{F}'(\chi) \geq -q_0^2 \quad \text{throughout } \mathcal{D}. \tag{16.48}$$

Although sufficient for stability, the condition (16.48) is by no means necessary. This may be seen with reference to the example shown in Figure 16.3. For the square domain $(0<x<1,\ 0<y<1)$, the fundamental eigenfunction of the problem (16.46) is $(\sin \pi x \sin \pi y)$ and the corresponding eigenvalue is $q_0^2 = 2\pi^2 \approx 19.72$. However, the gradient $\mathcal{F}'(\chi)$ is, by inspection, of order -500 at $\chi \approx 1$, considerably less than $-q_0^2$, so the sufficient condition for stability is *not* satisfied, although the equilibrium, being the result of a relaxation process, is certainly stable. The implication is that the inequality adopted in (16.45) is very crude when, as here, the function G defined by (16.44) is remote (in a topological sense) from the above eigenfunction.

16.10 Analogous Euler flows

It is well known that there is an exact analogy between the magnetostatic equations

$$\mathbf{j} \wedge \mathbf{B} = \nabla p, \quad \mathbf{j} = \nabla \wedge \mathbf{B}, \quad \nabla \cdot \mathbf{B} = 0, \tag{16.49}$$

and the equations for steady Euler flow, written in the form

$$\mathbf{u} \wedge \omega = \nabla h, \quad \omega = \nabla \wedge \mathbf{u}, \quad \nabla \cdot \mathbf{u} = 0, \tag{16.50}$$

where $h = p/\rho + \frac{1}{2}\mathbf{u}^2$, the Bernoulli function. The analogy is given by the correspondences

$$\mathbf{u} \longleftrightarrow \mathbf{B}, \quad \omega \longleftrightarrow \mathbf{j}, \quad h \longleftrightarrow -p. \tag{16.51}$$

The minus sign in the last of these correspondences should be particularly noted. This analogy means that, to every solution $\mathbf{B}$ of the magnetostatic equations (16.49), satisfying the boundary condition $\mathbf{n} \cdot \mathbf{B} = 0$ on $\partial\mathcal{D}$, there corresponds a steady solution $\mathbf{u}(\mathbf{x}) \equiv \mathbf{B}(\mathbf{x})$ of the Euler equations satisfying $\mathbf{n} \cdot \mathbf{u} = 0$ on $\partial\mathcal{D}$.

We have shown in the foregoing sections how a magnetostatic field $\mathbf{B}^E$ of arbitrarily prescribed topology may be obtained by the process of topology-conserving magnetic relaxation. Thus we may now assert that there exists a steady solution of the Euler equations $\mathbf{u}^E (\equiv \mathbf{B}^E)$ of the same arbitrarily prescribed topology. It is important to recognise that it is the topology of $\mathbf{u}$ rather than that of ω that may be prescribed in exploiting this analogy. In *unsteady* flow of an ideal fluid, it is the topology of the *vorticity* (not the velocity) field that is conserved; but the analogy (16.51) applies only to the steady state.

Furthermore, although the relaxed magnetostatic equilibrium is, by its derivation, stable, there is no guarantee that the analogous Euler flow is stable. Perturbations to this flow are governed by the unsteady Euler equations, so the vorticity field is 'frozen in', i.e. the perturbations are necessarily 'isovortical' (Arnold 1965b). According to Arnold, the flow is stable if the kinetic energy K is either minimal or maximal in the undisturbed state with respect to such perturbations; this leads to a sufficient condition for stability of these flows that is significantly different from the condition (16.41). It has actually been shown by Rouchon (1991) that the Arnold condition is never satisfied for fully three-dimensional flows because for every choice of displacement field $\boldsymbol{\eta}_1(\mathbf{x})$ that makes $\delta^2 K$ positive, there exists another displacement field $\boldsymbol{\eta}_2(\mathbf{x})$ that makes $\delta^2 K$ negative. Thus these steady Euler flows are in effect saddle points in the function space of solenoidal velocity fields.

The analogue of a current sheet in the magnetostatic situation is a vortex sheet in the analogous Euler flow. Vortex sheets are generally subject to Kelvin-Helmholtz instability, so it is not surprising that they do not satisfy a sufficient condition for stability. The only known *smooth* fully three-dimensional Euler flow is the space-periodic 'ABC-flow'

$$\mathbf{u} = (B\cos ky + C\sin kz,\ C\cos kz + A\sin kx,\ A\cos kx + B\sin ky)\,, \tag{16.52}$$

which satisfies the Beltrami condition $\nabla \wedge \mathbf{u} = k\mathbf{u}$; the mean helicity of this flow is $\mathcal{H} = \frac{1}{2}k\left(A^2 + B^2 + C^2\right)$. According to the general result of Rouchon (1991), this flow does not satisfy Arnold's sufficient condition for stability. As argued earlier by Moffatt (1986), it seems likely that this flow is unstable to large-scale helical perturbations having the same sign of helicity as that of $\mathcal{H}$.

16.11 Cross-helicity and relaxation to steady MHD flows

The magnetic relaxation technique described in §16.3 can be adapted to describe relaxation that is not to magnetostatic equilibrium but rather to steady solutions $\{\mathbf{U}(\mathbf{x}), \mathbf{B}(\mathbf{x})\}$ of the equations of ideal magnetohydrodynamics. In this section, in which we follow the procedure of Vladimirov et al. (1999), we adopt the Poisson bracket notation

$$[\mathbf{a}, \mathbf{b}] \equiv \nabla \wedge (\mathbf{a} \wedge \mathbf{b}), \tag{16.53}$$

for any two vector fields satisfying $\nabla \cdot \mathbf{a} = \nabla \cdot \mathbf{b} = 0$. With this compact notation, and defining $\mathbf{h}(\mathbf{x}, t) = (\mu_0\rho)^{-1/2}\mathbf{B}(\mathbf{x}, t)$, the induction equation becomes

$$\mathbf{h}_t = [\mathbf{u}, \mathbf{h}], \tag{16.54}$$

and the curl of the momentum equation is

$$\omega_t = [\mathbf{u}, \omega] + [\mathbf{j}, \mathbf{h}]. \tag{16.55}$$

We have also the usual constraints $\nabla \cdot \mathbf{u} = \nabla \cdot \mathbf{h} = 0$ in $\mathcal{D}$ and $\mathbf{n} \cdot \mathbf{u} = \mathbf{n} \cdot \mathbf{h} = 0$ on the domain boundary $\partial\mathcal{D}$.

These equations conserve the total energy

$$E = \tfrac{1}{2} \int_{\mathcal{D}} \left(\mathbf{u}^2 + \mathbf{h}^2\right) \mathrm{d}V, \tag{16.56}$$

and, as we have seen in earlier sections, have two families of helicity invariants: the magnetic helicity family

$$\mathcal{H}_M = \int_{\mathcal{V}} \mathbf{a} \cdot \mathbf{h} \, \mathrm{d}V, \tag{16.57}$$

where $\mathbf{h} = \nabla \wedge \mathbf{a}$, and the cross-helicity family

$$\mathcal{H}_C = \int_{\mathcal{V}} \mathbf{u} \cdot \mathbf{h} \, \mathrm{d}V; \tag{16.58}$$

here, $\mathcal{V}$ is either $\mathcal{D}$ or any Lagrangian subdomain of $\mathcal{D}$ on whose surface $\mathbf{n} \cdot \mathbf{h} = 0$. The integral (16.57) is invariant because **h**-lines are frozen in the fluid; and the integral (16.58) is invariant because, although vortex lines are not frozen in the fluid, the flux of vorticity through any closed **h**-line *is* conserved. The family of integrals (16.58) thus provide a measure of the degree of *mutual* linkage of vorticity and magnetic fields.

16.11.1 Structure of steady states

A steady state $\mathbf{U}(\mathbf{x}), \mathbf{H}(\mathbf{x}) = (\mu_0 \rho)^{-\frac{1}{2}} \mathbf{B}(\mathbf{x})$, with $\boldsymbol{\Omega} = \nabla \wedge \mathbf{U}$ and $\mathbf{J} = \nabla \wedge \mathbf{H}$, must evidently satisfy

$$\mathbf{U} \wedge \mathbf{H} = \nabla \Phi, \qquad \mathbf{U} \wedge \boldsymbol{\Omega} + \mathbf{J} \wedge \mathbf{H} = \nabla \Psi, \tag{16.59}$$

where $\Psi = P + \frac{1}{2}\mathbf{U}^2$. As for the case of magnetostatic equilibrium, these equations impose severe constraints on the topology of the fields $\mathbf{U}(\mathbf{x})$ and $\mathbf{H}(\mathbf{x})$. First, from the first of (16.59), by arguments that will now be familiar, the **H**-lines and the **U**-lines must lie on surfaces unless $\mathbf{U}(\mathbf{x}) = \alpha \, \mathbf{H}(\mathbf{x})$, with $\alpha = $ const. But then $\boldsymbol{\Omega}(\mathbf{x}) = \alpha \, \mathbf{J}(\mathbf{x})$ also, and so the second of (16.59) becomes

$$\left(1 - \alpha^2\right) (\mathbf{U} \wedge \boldsymbol{\Omega}) = \nabla \Psi. \tag{16.60}$$

If $\alpha = \pm 1$, then we have the well-known exact solutions $\mathbf{U}(\mathbf{x}) \equiv \pm \mathbf{H}(\mathbf{x})$ for arbitrary $\mathbf{H}(\mathbf{x})$, with the interpretation that the acceleration $(\mathbf{U} \cdot \nabla)\mathbf{U}$ due to streamline curvature at any point is exactly provided by the hoop stress $(\mathbf{H} \cdot \nabla)\mathbf{H}$ of the magnetic field. If $\alpha \neq 1$, then repetition of the standard argument shows that again **H**-lines are confined to surfaces unless $\mathbf{J}(\mathbf{x}) = \beta \, \mathbf{H}$ with $\beta = $ const. In this case,

$$\alpha\beta \, \mathbf{H} = \alpha \, \mathbf{J} = \beta \, \mathbf{U} = \boldsymbol{\Omega}, \tag{16.61}$$

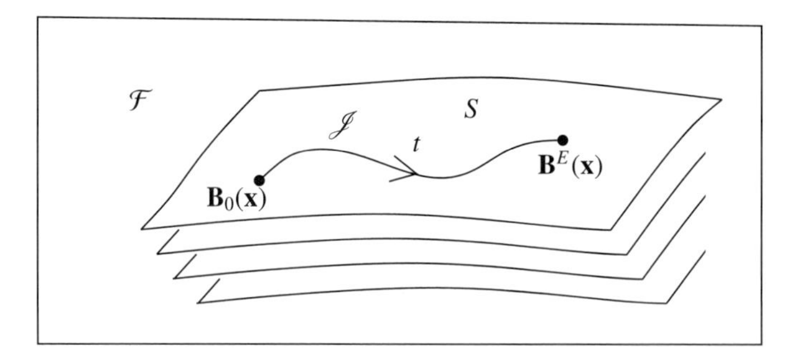

Figure 16.10 Schematic of relaxation on the imv folium.

and only under these conditions can the **H**-lines (which now coincide with the **J**-, **U**- and **Ω**-lines) wander in chaotic manner throughout the relevant subdomain.

16.11.2 The isomagnetovortical foliation

If the actual velocity field $\mathbf{u}(\mathbf{x}, t)$ and current field $\mathbf{j}(\mathbf{x}, t)$ are replaced on the right-hand sides of (16.54) and (16.55) by 'fictitious' velocity and current fields $\mathbf{v}(\mathbf{x}, t)$ and $\mathbf{c}(\mathbf{x}, t)$ satisfying $\nabla \cdot \mathbf{v} = \nabla \cdot \mathbf{c} = 0$ in $\mathcal{D}$ and $\mathbf{n} \cdot \mathbf{v} = 0$ on $\partial\mathcal{D}$, these equations become

$$\mathbf{h}_t = [\mathbf{v}, \mathbf{h}], \tag{16.62}$$

$$\omega_t = [\mathbf{v}, \omega] + [\mathbf{c}, \mathbf{h}]. \tag{16.63}$$

These equations have wider scope than (16.54) and (16.55), because, whereas $\mathbf{u}(\mathbf{x}, t)$ and $\mathbf{j}(\mathbf{x}, t)$ are constrained by the relationships $\omega = \nabla \wedge \mathbf{u}$, $\mathbf{j} = \nabla \wedge \mathbf{h}$, the fictitious fields $\mathbf{v}(\mathbf{x}, t)$ and $\mathbf{c}(\mathbf{x}, t)$ suffer no such constraints. Nevertheless, the helicity invariants (16.57) and (16.58) still survive under this unconstrained evolution. That the magnetic helicities (16.57) survive is obvious, because (16.62) implies that the **h**-lines are transported without change of topology by the **v**-field. The conservation of cross-helicity is less obvious; to prove this, it is best to write (16.62) and the 'uncurled' version of (16.63) in Lagrangian form (with $D/Dt = \partial/\partial t + \mathbf{v}\cdot\nabla$)

$$\frac{Dh_i}{Dt} = h_j \frac{\partial v_i}{\partial x_j}, \tag{16.64}$$

$$\frac{Du_i}{Dt} = v_j \frac{\partial u_j}{\partial x_i} + (\mathbf{c} \wedge \mathbf{h})_i - \frac{\partial p}{\partial x_i}. \tag{16.65}$$

Then

$$\frac{\mathrm{d}\mathcal{H}_C}{\mathrm{d}t} = \int_{\mathcal{V}} \left(\frac{D\mathbf{u}}{Dt} \cdot \mathbf{h} + \frac{D\mathbf{h}}{Dt} \cdot \mathbf{u} \right) \mathrm{d}V = \int_{\mathcal{V}} \left(h_i v_j \frac{\partial u_j}{\partial x_i} + u_i h_j \frac{\partial v_i}{\partial x_j} \right) \mathrm{d}V$$
$$= \int_{\mathcal{V}} \frac{\partial}{\partial x_i}(h_i u_j v_j)\,\mathrm{d}V = \int_{\partial\mathcal{V}} (\mathbf{n}\cdot\mathbf{h})(\mathbf{u}\cdot\mathbf{v})\,\mathrm{d}S = 0, \tag{16.66}$$

where we have used $\mathbf{n}\cdot\mathbf{h} = 0$ on $\partial\mathcal{V}$ for all t.

Equations (16.62) and (16.63) thus conserve all known topological invariants of the parent equations, but allow a much wider range of behaviour due to the freedom of choice of the auxiliary fields $\{\mathbf{v}, \mathbf{c}\}$. Under the action of this pair of auxiliary fields, evolution governed by (16.62) and (16.63) over a time interval $[t_1, t_2]$ will convert a pair $\mathcal{P}_1(\mathbf{x}) = \{\mathbf{h}_1, \omega_1\}$ to a pair $\mathcal{P}_2(\mathbf{x}) = \{\mathbf{h}_2, \omega_2\}$ having the same set of topological invariants. Conversely, we may say that two pairs $\mathcal{P}_1(\mathbf{x})$ and $\mathcal{P}_2(\mathbf{x})$ lie on the same 'isomagnetovortical (or imv) folium' of the function space $\mathcal{F}$ of such pairs if and only if there exist fields $\{\mathbf{v}(\mathbf{x}, t), \mathbf{c}(\mathbf{x}, t)\}$ that effect this conversion over a finite time interval. This requirement defines an 'imv foliation' of $\mathcal{F}$ as depicted schematically in Figure 16.10, two pairs being on the same folium if and only if they are 'accessible' one from another via (16.62) and (16.63), and therefore only if they have the same set of topological invariants. This foliation provides an appropriate generalisation of the isovortical foliation introduced by Arnold (1965b) in the context of the Euler equations.The trajectory $\mathcal{J}$ in Figure 16.10 indicates relaxation of the field $\mathbf{B}$ on the imv foilum S from its initial state $\mathbf{B}_0(\mathbf{x})$ to its asymptotic state $\mathbf{B}^E(\mathbf{x})$.

16.11.3 Relaxation to steady MHD states

Although the helicities are conserved under evolution governed by (16.62) and (16.63), the energy E given by (16.56) is not in general conserved. In fact, with repeated use of the boundary conditions $\mathbf{n}\cdot\mathbf{u} = \mathbf{n}\cdot\mathbf{v} = 0$ on $\partial\mathcal{D}$,

$$\frac{\mathrm{d}E}{\mathrm{d}t} = \int_{\mathcal{D}} \{\mathbf{h}\cdot[\mathbf{v}, \mathbf{h}] + \mathbf{u}\cdot(\mathbf{v}\wedge\omega + \mathbf{c}\wedge\mathbf{h} - \nabla P)\}\,\mathrm{d}V$$
$$= -\int_{\mathcal{D}} \{\mathbf{v}\cdot(\mathbf{u}\wedge\omega + \mathbf{j}\wedge\mathbf{h}) + \mathbf{c}\cdot(\mathbf{u}\wedge\mathbf{h})\}\,\mathrm{d}V\,. \tag{16.67}$$

We can now exploit our freedom in the choice of $\mathbf{v}$ and $\mathbf{c}$ in order to ensure monotonic decrease of E. The appropriate choice is

$$\mathbf{v} = \mathbf{u}\wedge\omega + \mathbf{j}\wedge\mathbf{h} - \nabla\phi, \quad \mathbf{c} = \mathbf{u}\wedge\mathbf{h} - \nabla\psi, \tag{16.68}$$

where ϕ and ψ are determined by the conditions $\nabla\cdot\mathbf{v} = \nabla\cdot\mathbf{c} = 0$ in $\mathcal{D}$ and $\mathbf{n}\cdot\mathbf{v} = 0$, $\psi = 0$ on $\partial\mathcal{D}$. These conditions yield a Neumann problem for ϕ and a

Dirichlet problem for ψ and so unique solutions for $\nabla\phi$ and $\nabla\psi$. Hence $\mathbf{v}$ and $\mathbf{c}$ are determined, and from (16.67) we now have

$$\frac{\mathrm{d}E}{\mathrm{d}t} = -\int_{\mathcal{D}} \left(\mathbf{v}^2 + \mathbf{c}^2\right) \mathrm{d}V\,, \tag{16.69}$$

so that E is monotonic decreasing as required.

Now, assuming $\mathcal{H}_C \neq 0$, E is obviously bounded below by the Schwarz inequality

$$E = \tfrac{1}{2}\int_{\mathcal{D}} \left(\mathbf{u}^2 + \mathbf{h}^2\right) \mathrm{d}V \geq \left|\int_{\mathcal{D}} \mathbf{u}\cdot\mathbf{h}\,\mathrm{d}V\right| = |\mathcal{H}_C|\,. \tag{16.70}$$

Hence just as for the relaxation problem treated in §16.3, E must tend to a positive limit (a minimum energy state) as $t \to \infty$. Assuming that $\mathbf{v}$ and $\mathbf{c}$ remain bounded in $\mathcal{D}$, this in fact implies from (16.69) that $\mathbf{v} \to 0$ and $\mathbf{c} \to 0$ as $t \to \infty$, and so, from (16.68), $\mathbf{u}(\mathbf{x},t) \to \mathbf{U}(\mathbf{x})$, $\mathbf{h}(\mathbf{x},t) \to \mathbf{H}(\mathbf{x})$, $\phi(\mathbf{x},t) \to \Phi(\mathbf{x})$, and $\psi(\mathbf{x},t) \to \Psi(\mathbf{x})$, where, just as in (16.59) above,

$$\mathbf{U}\wedge\mathbf{H} = \nabla\Phi, \qquad \mathbf{U}\wedge\mathbf{\Omega} + \mathbf{J}\wedge\mathbf{H} = \nabla\Psi. \tag{16.71}$$

Since all helicities are conserved, this relaxation process corresponds to evolution on an imv folium, as illustrated in Figure 16.10. The final state $\{\mathbf{U}(\mathbf{x}), \mathbf{H}(\mathbf{x})\}$ is topologically accessible from the initial state $\{\mathbf{u}(\mathbf{x},0), \mathbf{h}(\mathbf{x},0)\}$, for which the linkages of the $\mathbf{h}$-field with itself *and* with the ω-field may be arbitrarily prescribed. This implies the existence of steady states of arbitrarily complex topology; however, we should emphasise that, in this more elaborate relaxation process, current sheets and current/vortex sheets will almost inevitably develop in the limit $t \to \infty$.

17

Magnetic Relaxation in a Low-β Plasma

17.1 Relaxation in a pressureless plasma

In the previous chapter, we focussed on magnetic field relaxation in an incompressible fluid medium. We now turn to the opposite extreme of relaxation in a plasma of very low density in which the fluid pressure p is negligible in comparison with the magnetic pressure $p_M = \mathbf{B}^2/2\mu_0$: in the conventional terminology of plasma physics, this is a 'low-β' plasma: $\beta = p/p_M \ll 1$. The density field $\rho(\mathbf{x}, t)$ satisfies the equation of mass conservation

$$\frac{D\rho}{Dt} \equiv \frac{\partial \rho}{\partial t} + \mathbf{u} \cdot \nabla \rho = -\rho \nabla \cdot \mathbf{u}, \tag{17.1}$$

but as we shall see, it plays a rather passive role in the relaxation dynamics in the low-density limit.

We may get an immediate idea of the type of behaviour that may be expected through consideration of a very simple initial value problem (Bajer & Moffatt 2013). Suppose that at time $t = 0$, such a plasma is contained between two perfectly conducting boundaries $x = \pm a$, that the plasma is also perfectly conducting, and that it is at rest and permeated by a magnetic field with a single component:

$$\mathbf{B}(\mathbf{x}, 0) = (0, B_0 \sin \pi x/a, 0), \quad \text{so that} \quad p_M(\mathbf{x}, 0) = \left(B_0^2/2\mu_0\right) \sin^2 \pi x/a. \tag{17.2}$$

The initial field profile is shown in Figure 17.1a; the magnetic fluxes in the half-intervals $(-a, 0)$ and $(0, a)$ are $\pm\Phi$ respectively, where $\Phi = (2a/\pi)B_0$. The associated initial magnetic pressure distribution is shown in Figure 17.1b. This pressure distribution drives fluid down the pressure gradient towards null points of the field ($x = \pm a$ and $x = 0$) where $p_M = 0$. Thus a velocity field $\mathbf{u} = (u(x, t), 0, 0)$ is generated which carries the frozen-in field, with conservation of flux in each half interval. If some mechanism of energy loss is present, e.g. fluid viscosity, then magnetic energy decreases during this simple relaxation process, and the field must settle down to a minimum-energy state with the same fluxes $\pm\Phi$ in the two half-intervals. In

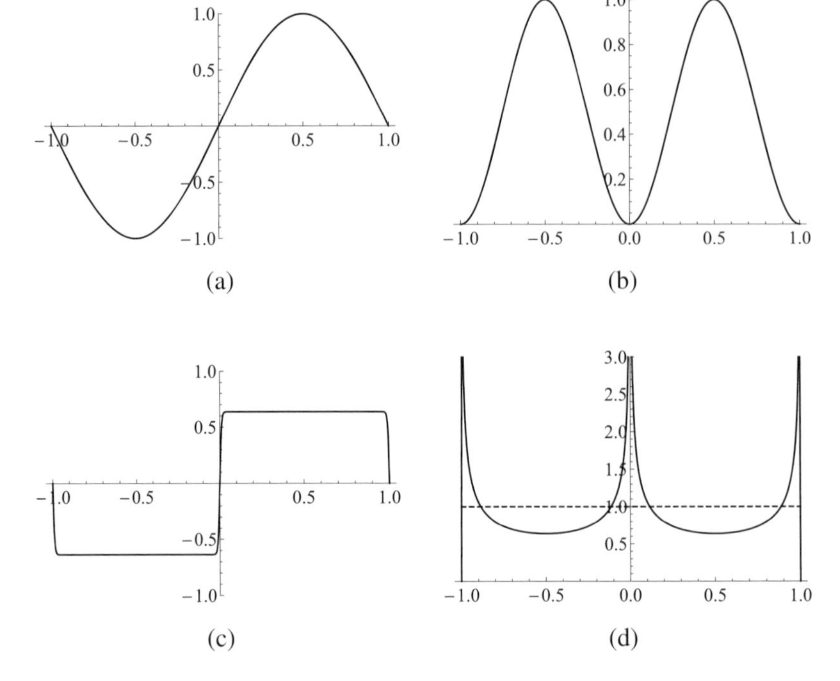

Figure 17.1 Relaxation of an initially sinusoidal field: (a) the initial field; (b) the associated initial magnetic pressure distribution; (c) the relaxed field having the same flux in each half-interval as in the initial situation; (d) the redistribution of density associated with the relaxation process.

the asymptotic state, insofar as fluid pressure p remains negligible, p_M must be constant; this asymptotic state is shown in Figure 17.1c.

For this simple geometry, with $\mathbf{B} = (0, B(x,t), 0)$, the induction equation (with $\eta = 0$) reduces (cf. 3.3) to

$$\frac{DB}{Dt} \equiv \frac{\partial B}{\partial t} + \mathbf{u} \cdot \nabla B = -B \nabla \cdot \mathbf{u}, \tag{17.3}$$

just like (17.1). These two equations imply that $D(B/\rho)/Dt = 0$, i.e. the evolution of ρ is strongly coupled to that of B. We may suppose that the initial density has the uniform value ρ_0; then the mass per unit area in the interval $(0, x)$ is $M_0(x) = \rho_0\, x$, while the magnetic flux in this same interval is $\Phi_0(x) = (a/\pi)(1 - \cos(\pi x/a))$. In the relaxation process depicted in Figure 17.1, suppose that the particle initially at position x moves to position $X(x)$; then $X(x)$ is determined by the condition that $\Phi(X) = \Phi_0(x)$; then also $M(X) = M_0(x)$, and after some calculation this determines the ultimate equilibrium density distribution $\rho^E(x)$ as

$$\rho^E(x) = \frac{\rho_0 a}{\pi\sqrt{x(a-x)}} \qquad \text{for } 0 < x < a, \tag{17.4}$$

and obviously $\rho^E(-x) = \rho^E(x)$. This density distribution is shown in Figure 17.1d. The square-root singularities at both boundaries $x = \pm a$ and at $x = 0$ are obviously unphysical; they result first because of the neglect of fluid pressure which obviously builds up ($p \sim \rho^\gamma$) as ρ increases, eventually becoming comparable with p_M; and second because of the neglect of diffusion which obviously becomes important as the discotinuities (current sheets) evident in Figure 17.1c develop. What is important to note however is that there is a strong tendency to equalise the magnetic pressure, with consequential development of these current sheets near the 'nulls' of the magnetic field and associated build-up of fluid density in these regions.

17.2 Numerical relaxation

The above behaviour has been explored in detail by Bajer & Moffatt (2013), taking account of weak diffusion, but still neglecting the build-up of fluid pressure. Diffusion near the null points leads to destruction of magnetic flux from each side and the magnetic pressure is always minimal at these points. Away from the nulls, the basic relaxation mechanism means that the 'plateau' structure of the magnetic field is maintained. The net effect is that fluid is continuously 'sucked in' to the neighbourhood of the nulls, and the level of the plateaux approaches zero (from above or below) as flux is destroyed.

The equations governing this process are the induction equation, which, for the single component magnetic field $B(x,t)$ considered here, takes the form

$$\frac{\partial B}{\partial t} = -\frac{\partial}{\partial x}(uB) + \eta \frac{\partial^2 B}{\partial x^2}, \tag{17.5}$$

and the momentum equation in the form

$$\mu \frac{\partial^2 u}{\partial x^2} = \frac{\partial}{\partial x}\left(\tfrac{1}{2}B^2\right), \tag{17.6}$$

where μ is an effective viscosity (a combination of shear and bulk viscosity);[1] here, fluid inertia is neglected as well as pressure gradient $\partial p/\partial x$, as appropriate in the low density limit. Equation (17.6) may be integrated once, with the boundary conditions $u = B = 0$ on $x = \pm a$ to give

$$\mu \frac{\partial u}{\partial x} = \tfrac{1}{2}\left(B^2 - \langle B^2 \rangle\right), \tag{17.7}$$

where the angular brackets represent an average over the range $(-a, a)$. Standard

[1] We note that, according to James Clerk Maxwell (1867) in his seminal paper *On the dynamical theory of gases*, the viscosity μ "varies as the absolute temperature, and is independent of the density". This is because momentum transport results from ion collisions, and the decrease in the frequency of these as the density decreases is compensated by the increase of the mean-free-path between collisions.

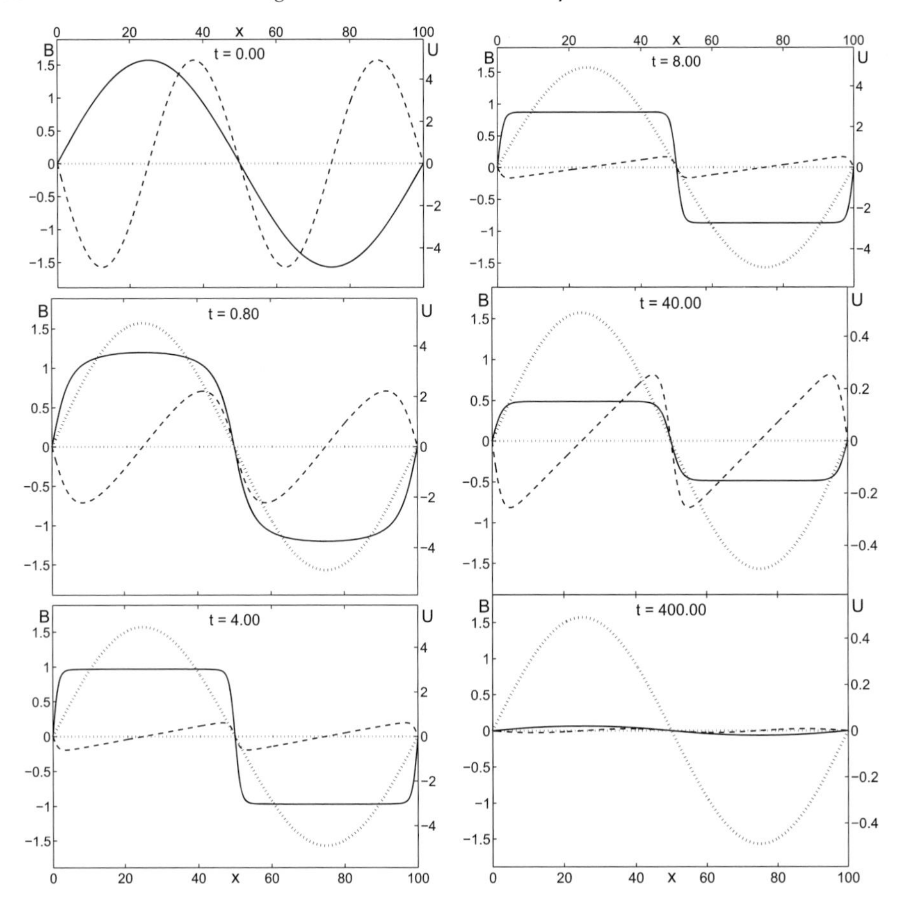

Figure 17.2 Relaxation of the initial field $B_0(x) = (\pi/2)\sin(\pi x)$, as governed by (17.5) and (17.6), with $S = aB_0/(\mu\eta)^{1/2} = 50$. Each panel shows the relaxing field $B(x,t)$ (solid), and the velocity field $u(x,t)$ (dashed); the initial field $B_0(x)$ (dotted) is also included for comparison. The left column shows the initial nearly non-diffusive relaxation on the time-scale $t_v = \mu/B_0^2$ and the formation of current sheets (cf. Figure 17.1). The right column shows the subsequent slow diffusive decay. Note the change of velocity scale (shown on the right of the panels) at $t = 40$. [Adapted from Bajer & Moffatt 2013.]

manipulation of these equations yields an energy equation in the form

$$\frac{\mathrm{d}}{\mathrm{d}t}\left\langle B^2\right\rangle = -\frac{1}{2\mu}\left\langle\left(B^2 - \left\langle B^2\right\rangle\right)^2\right\rangle - 2\eta\left\langle(\partial B/\partial x)^2\right\rangle . \tag{17.8}$$

The first term on the right indicates the tendency for B to relax to $\pm\sqrt{\langle B^2\rangle}$, according as B is positive or negative. Current sheets then inevitably form at the nulls of B, controlled only by the weak diffusive term.

Figure 17.2 shows the result of integrating equations (17.5) and (17.6), with the initial condition (17.2), and with η small. The relevant dimensionless number[2] is

$$S = aB_0/(\mu\eta)^{1/2}, \tag{17.9}$$

and this is assumed large: $S \gg 1$. The initial velocity is determined instantaneously by (17.6). The evolution proceeds as described above, with a rapid initial phase (actually on the viscous time-scale $t_v = \mu/B_0^2$) when current sheets of width $\mathcal{O}(\delta_m)$ develop, with $\delta_m/a = S^{-1}$. This is followed by a slow diffusive phase during which the level of the plateaux decreases algebraically; the width of the current sheets slowly increases during this phase from $\mathcal{O}(\delta_m)$ to $\mathcal{O}(a)$, at which stage a final period of linear decay takes over. Full details of this three-stage process are described by Bajer & Moffatt (2013), where more complex initial conditions are also considered.

17.3 The pinch effect

Suppose now that plasma is contained in the annulus between two perfectly conducting cylinders $r = \delta a$ and $r = a$, where $0 < \delta < 1$, and that a current $\mathbf{j} = (0, 0, j(r,t))$ with $j > 0$ flows in the plasma parallel to the axis. With cylindrical polar coordinates (r, θ, z), the associated magnetic field is $\mathbf{B} = (0, b(r,t), 0)$, where,[3] from (2.70),

$$j(r,t) = \frac{1}{r}\frac{\partial}{\partial r}(rb), \quad b(r,t) = \frac{1}{r}\int_{\delta a}^{r} r\, j(r,t)\mathrm{d}r. \tag{17.10}$$

The associated Lorentz force is $\mathbf{F} = (F_r, 0, 0)$ with $F_r = -j(r,t)b(r,t)$, and this will drive a radial velocity field $(u(r,t), 0, 0)$ with $u < 0$. Note that

$$F_r = -\frac{\partial p_M}{\partial r} - \frac{b^2}{r} \tag{17.11}$$

where $p_M = b^2/2$, the magnetic pressure; if $b\,\partial b/\partial r < 0$ then the magnetic pressure gradient is outwards, but under the condition $j = \partial b/\partial r + b/r > 0$, the inward 'hoop stress' $-b^2/r$ is dominant, giving $F_r < 0$.

The flux of $\mathbf{B}$ in the θ-direction per unit axial length of cylinder is

$$\Phi_\theta = \int_{\delta a}^{a} b(r,t)dr, \tag{17.12}$$

and this flux is conserved. The magnetic energy per unit axial length is

$$M = \tfrac{1}{2}\int_{\delta a}^{a} b^2 2\pi r\mathrm{d}r. \tag{17.13}$$

[2] The dimensionless number S is encountered in liquid metal MHD, where it is known as the Hartmann number.

[3] Again, for simplicity of notation, we absorb the factor μ_0 in the definition of $\mathbf{j}$.

If, as in the previous section, viscosity dissipates energy for so long as the fluid is in motion, then again it is reasonable to suppose that M is minimised subject to Φ_θ being prescribed and constant. Introducing a Lagrange multiplier λ, we must then have

$$\delta \int_{\delta a}^{a} \left(\pi r b^2 - \lambda b\right) \mathrm{d}r = \int_{\delta a}^{a} (2\pi r b - \lambda)\, \delta b \, \mathrm{d}r = 0, \tag{17.14}$$

for arbitrary δb, from which it follows that $b = \lambda/2\pi r$; and now λ is determined from the flux condition (17.12), which gives

$$\lambda = 2\pi\Phi_\theta / \log(1/\delta). \tag{17.15}$$

In this relaxed equilibrium, the current is concentrated on the inner cylinder boundary, the current density in the interior of the plasma being zero. This can be considered as the end result of the elementary principle that 'like currents attract'; they do indeed attract in this idealised situation in which the current distribution collapses inwards towards the inner boundary. The collapse is also associated with contraction of the frozen-in circular **B**-lines; hence the term 'pinch effect' (Bennett 1934).

With $b = \lambda/2\pi r$, the total axial current flowing along the inner cylinder boundary must be $I = \lambda$. The behaviour as $\delta \to 0$ is singular, in that from (17.15) $I \to 0$, but in just such a way that Φ_θ remains constant. The field b then has a $(\delta \log(1/\delta))^{-1}$ singularity near $r = \delta a$ as $\delta \to 0$. In reality however, just as in §17.1, two physical effects again conspire to prevent the development of such a singularity: (i) the build-up of fluid pressure as the plasma pinches towards $r = \delta a$, which must ultimately establish a magnetostatic equilibrium; and (ii) the effect of non-zero diffusivity of both the plasma and the inner boundary, which allows destruction of θ-flux, and which will cause this magnetostatic equilibrium to slowly evolve.

In the thermonuclear fusion context, this type of pinch, in which the current is purely in the z-direction, is known for this reason as the Z-pinch. There is an enormous literature on this topic (see for example the review of Haines 2011), concerning particularly the instabilities to which the Z-pinch may be subject. It would be inappropriate to enter into such analysis here; our principle concern will be the situation in which an axial magnetic field is also present in the cylindrical geometry, the total field being then helical in character, the situation considered in §§17.5 and 17.6 below.

17.4 Current collapse in an unbounded fluid

First however, we consider an interesting problem that presents itself when the fluid is unbounded and a current $\mathbf{j} = (0, 0, j(r,t))$ of localised radial extent flows in the axial direction. For example, we might have an initial Gaussian distribution of the form

$$j(r,0) = \frac{I}{\pi a^2} e^{-(r/a)^2}. \tag{17.16}$$

Under the action of diffusion alone, the current at time t would then be

$$j(r,t) = \frac{I}{4\pi\eta(t^* + t)} \exp\left(-\frac{r^2}{4\eta(t^* + t)}\right), \tag{17.17}$$

where $4\eta t^* = a^2$. This is just the analogue of the Lamb vortex in classical fluid dynamics; here current is the analogue of vorticity, and total current I is the analogue of circulation Γ. The magnetic field associated with the current (17.17) is $\mathbf{b} = (0, b(r,t), 0)$, where

$$rb(r,t) = \int_0^r rj\,\mathrm{d}r = \frac{I}{2\pi}\left[1 - \exp\left(-\frac{r^2}{4\eta(t^* + t)}\right)\right]. \tag{17.18}$$

Note that $rb(r,t)$ increase monotonically from 0 to $I/2\pi$ as r increases from 0 to ∞.

In a pressureless plasma however, we have a very different situation, because of the pinch effect. The Lorentz force $\mathbf{F} = (-jb, 0, 0)$ drives an inward velocity $\mathbf{u} = (u, 0, 0)$ which must now be taken into account in conjunction with diffusion. The θ component of the induction equation is then

$$\frac{\partial b}{\partial t} = -\frac{\partial}{\partial r}(ub) + \eta\frac{\partial}{\partial r}\frac{1}{r}\frac{\partial}{\partial r}(rb) = \frac{\partial}{\partial r}(-ub + \eta j). \tag{17.19}$$

The total current is $I = \int_0^\infty 2\pi r j\,\mathrm{d}r$, and this is still constant because

$$\frac{\mathrm{d}I}{\mathrm{d}t} = 2\pi\int_0^\infty \frac{\partial}{\partial t}(rj)\,\mathrm{d}r = 2\pi\left[r\frac{\partial}{\partial r}(-ub + \eta j)\right]_0^\infty = 0, \tag{17.20}$$

on the assumption that j and u are both exponentially small as $r \to \infty$. Note however that, since $b \sim I/2\pi r$ at ∞, both the flux Φ_θ and energy M per unit axial length are now infinite.

Here for simplicity, we adopt a Darcy relaxation model (16.11) for u, but now with $p = 0$, i.e.

$$0 = -jb - k^2 u, \tag{17.21}$$

so that (17.19) becomes

$$\frac{\partial b}{\partial t} = \frac{\partial}{\partial r}\left[\left(k^2 b^2 + \eta\right)\frac{1}{r}\frac{\partial}{\partial r}(rb)\right], \tag{17.22}$$

a non-linear diffusion equation, with an enhanced diffusivity $D(r,t) = \eta + k^2 b^2$.

17.4.1 Similarity solution when $\eta = 0$

We expect an initial relaxation process in which the effect of η is negligible. Let us therefore first set $\eta = 0$, and look for a similarity solution of (17.22) of the form

$$b(r,t) = \frac{J}{r} f(\zeta) \quad \text{where} \quad \zeta = \frac{\lambda r}{t^{\alpha}}, \tag{17.23}$$

and α and λ are to be determined. If we insist that rb must still increase from 0 to $J(= I/2\pi)$ as r increases from 0 to ∞, then $f(\zeta)$ must satisfy

$$f(0) = 0, \quad f(\infty) = 1. \tag{17.24}$$

Now from (17.23),

$$\frac{\partial b}{\partial t} = -\frac{J}{r} f'(\zeta)\left(\frac{\alpha\zeta}{t}\right) = -J\alpha\lambda f'(\zeta) t^{-(\alpha+1)}, \tag{17.25}$$

and also, after some simplification,

$$\frac{\partial}{\partial r}\left[\frac{b^2}{r}\frac{\partial}{\partial r}(rb)\right] = J^3 k^2 \lambda^5 \left(\frac{f^2 f'}{\zeta^3}\right)' t^{-5\alpha}. \tag{17.26}$$

Comparing powers of t in (17.25) and (17.26), we must have $\alpha+1 = 5\alpha$, i.e. $\alpha = \frac{1}{4}$, and then $f(\zeta)$ satisfies the equation

$$f'(\zeta) = -4J^2 k^2 \lambda^4 \left(\frac{f^2 f'}{\zeta^3}\right)'. \tag{17.27}$$

This integrates immediately to give

$$f = -4J^2 k^2 \lambda^4 \left(\frac{f^2 f'}{\zeta^3}\right) + 1, \tag{17.28}$$

the constant of integration being fixed by the condition $f(\infty) = 1$. If we now choose $\lambda = \frac{1}{2}(Jk)^{-1/2}$, this gives

$$f' = 4(1-f)\zeta^3/f^2, \quad \text{or equivalently} \quad \frac{f^2 \mathrm{d}f}{1-f} = 4\zeta^3 \mathrm{d}\zeta, \tag{17.29}$$

so that

$$\zeta^4 = \int_0^f \frac{f^2 \mathrm{d}f}{1-f} = -f - \tfrac{1}{2}f^2 - \log(1-f), \tag{17.30}$$

the constant of integration being now fixed by the condition $f(0) = 0$. This equation implicitly determines $f(\zeta)$; this function, plotted in Figure 17.3a, has the asymptotic behaviour

$$f(\zeta) \sim 3^{\frac{1}{3}}\zeta^{\frac{4}{3}} \text{ as } \zeta \to 0, \qquad f(\zeta) \sim 1 - e^{-\zeta^4} \text{ as } \zeta \to \infty. \tag{17.31}$$

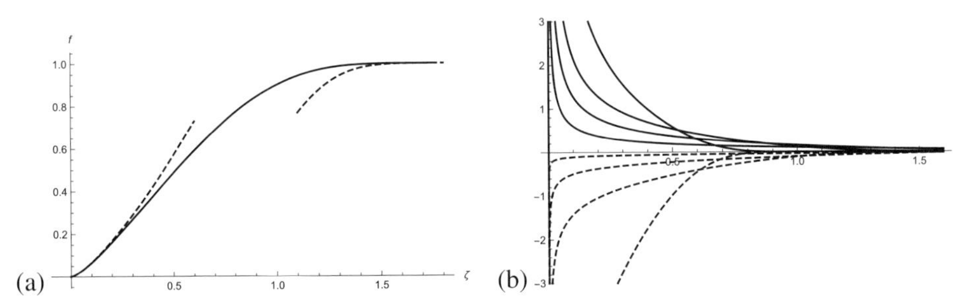

Figure 17.3 (a) The function $f(\zeta)$ defined implicitly by (17.30); the asymptotic behaviour (17.31) is shown by the dashed curves. (b) The radial distribution of current (solid) and radial velocity (dashed), as given by (17.33), at times $t =$ 0.1, 1, 3, 50.

Thus, in summary, we have

$$b(r,t) = \frac{J}{r} f(\zeta) = \left(\frac{J}{k}\right)^{1/2} \frac{f(\zeta)}{2\zeta}\, t^{-1/4} \quad \text{where } \zeta = \frac{r}{2(Jk)^{1/2}\, t^{1/4}}\,. \tag{17.32}$$

From these it follows that

$$j(r,t) = \frac{1}{k}\frac{\zeta^2(1-f)}{f^2}\, t^{-1/2} \quad \text{and} \quad u(r,t) = -\left(\frac{J^{1/2}}{2k^{7/2}}\right)\frac{\zeta(1-f)}{f}\, t^{-3/4}. \tag{17.33}$$

The resulting r-dependence of j and u at times $t = 0.1, 1.5.50$ are shown in Figure 17.3b; both are singular as $r \to 0$. Obviously the density associated with the inward fluid velocity will be singular also on the axis $r = 0$, but we expect this singularity to be controlled (as in §17.1) by the build-up of fluid pressure and/or the effect of finite resistivity.

17.5 The Taylor conjecture

Motivation for relaxation studies of the above kind, but principally for helical fields, comes partly from the magnetodynamics of the low-density astrophysical plasma, and partly from the plasma dynamics of thermonuclear fusion devices like the tokamak or the reversed-field pinch (as described, for example, by Ortolani 1989, Bodin 1990, Escande 2015). An important conjecture in the latter context was put forward by J. B. Taylor (1974) (and further developed in Taylor 1986), namely that the magnetic field in a turbulent plasma will relax to a minimum-energy state subject to the single constraint of conserved global magnetic helicity $\mathcal{H}_M$. This conjecture rests on the idea that, while magnetic field reconnections can occur on the small scales responsible for energy dissipation, the global

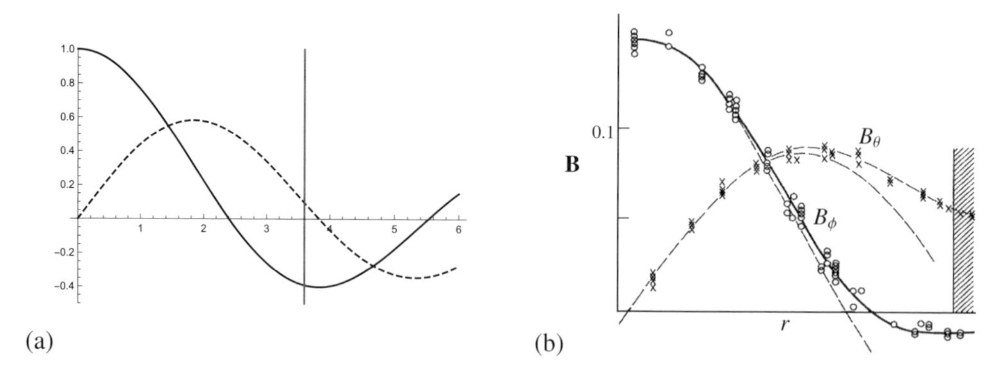

Figure 17.4 (a) Field components $B_\theta(\gamma r)/B_0$ (dashed) and $B_z(\gamma r)/B_0$ (solid) in the relaxed Taylor state, with $\mathbf{j} = \gamma\mathbf{B}$; note that $B_z(\gamma r) = 0$ at $\gamma r \approx 2.4$ and 5.6. The vertical line is placed at approximately the position of the boundary in (b). (b) Comparison of the Taylor state (dashed) with measurements in the reversed-field pinch HBTX-1A, at the Culham Laboratory, UK. [From Taylor 1986, quoting Bodin 1984.]

field topology may nevertheless be approximately conserved; equivalently that the time-scale $t_{\mathcal{H}}$ of decay of $\mathcal{H}_M$ is large compared with the relaxation time t_M of the magnetic field. This is quite plausible, because, as we have seen in the previous section, there are indeed two time-scales in the relaxation process, a short time-scale of relaxation of magnetic energy due to dissipative processes other than Joule dissipation, followed by a long slow process of diffusive decay when the global field topology (represented by fluxes through Lagrangian circuits) does slowly change.

Taylor's motivation was to explain the observed spontaneous generation of reversed axial (or 'toroidal') field in a plasma, the axial magnetic flux being maintained by currents in coils external to the plasma; this occurs in the device known for this reason as the 'reversed-field pinch', in which a significant axial current is induced to flow within the plasma, so that the poloidal and toroidal fields are of comparable magnitude. As had been shown much earlier by Woltjer (1958), minimisation of magnetic energy subject to the single constraint of conserved magnetic helicity leads to force-free fields satisfying

$$\mathbf{j} = \gamma\mathbf{B} \quad \text{with } \gamma = \text{const.} \tag{17.34}$$

In a cylindrical geometry with coordinates $\{r, \theta, z\}$ as considered by Taylor (1974), axisymmetric solutions of (17.34) are given in terms of Bessel functions (cf. §2.5)) by

$$\mathbf{B} = (0, B_\theta, B_z) = B_0(0, J_1(\gamma r), J_0(\gamma r)). \tag{17.35}$$

Here we assume $\gamma > 0$; B_θ is the poloidal field component, and B_z the axial, or toroidal, component. These components, as given by (17.35) are shown as functions of γr in Figure 17.4a. It is immediately evident that if $2.4 \lesssim \gamma a \lesssim 5.6$, as illustrated by the position of the vertical line at $\gamma a = 3.6$ in the figure, then the axial field is indeed negative near the cylinder boundary $r = a$.

Figure 17.4b shows a comparison of this 'Taylor state' with experimental measurements of both poloidal and toroidal fields (here B_θ and B_ϕ) in the original reversed-field experiment HBTX-1A, at the Culham Laboratory, UK. Here, the dashed curves represent the Taylor state, which may be compared with the best fits to the experimental points. Although the axial (toroidal) field does go negative at about the same position as the Bessel function $J_0(\gamma r)$, the axial field is quite weak in the 'reversed-field' region, and the agreement here is at best qualitative.[4] Similar results have been found in many subsequent reversed-field experiments.

Various properties of the Taylor state (17.35) are easily calculated. Since this state has the force-free property $\mathbf{j} = \gamma \mathbf{B}$, its vector potential $\mathbf{A}$ is given by a similarly simple relation $\mathbf{A} = \mathbf{B}/\gamma$. It follows that the energy and helicity per unit axial length are related by

$$\mathcal{H}_M = 2M/\gamma, \tag{17.36}$$

with

$$M = \tfrac{1}{2} B_0^2 \int_0^a \left(J_1(\gamma r)^2 + J_0(\gamma r)^2 \right) 2\pi r \, \mathrm{d}r. \tag{17.37}$$

The integral may be carried out explicitly, giving

$$M = \pi a^2 B_0^2 \left[J_0(\gamma a)^2 + J_1(\gamma a)^2 - J_0(\gamma a) J_1(\gamma a)/\gamma a \right]. \tag{17.38}$$

The total magnetic flux in the axial direction (a quantity that is determined by currents in external coils and is conserved during relaxation) is

$$\Phi_z = B_0 \int_0^a J_0(\gamma r) 2\pi r \, \mathrm{d}r = 2\pi B_0 a^2 \left[J_1(\gamma a)/\gamma a \right], \tag{17.39}$$

and, since $j_z = \gamma B_z$, the total axial current I in the plasma is simply

$$I = \gamma \Phi_z = 2\pi B_0 a J_1(\gamma a). \tag{17.40}$$

[4] The lack of detailed agreement with the theory has been attributed to the fact that the plasma resistivity is itself a function of radius, being greater near the boundary where the temperature is lower; but of course there is also the consideration that the cylinder is merely an approximation to the experimental torus, and the cross section of the torus is more nearly 'D-shaped' than circular; so qualitative agreement is the best that could reasonably be expected.

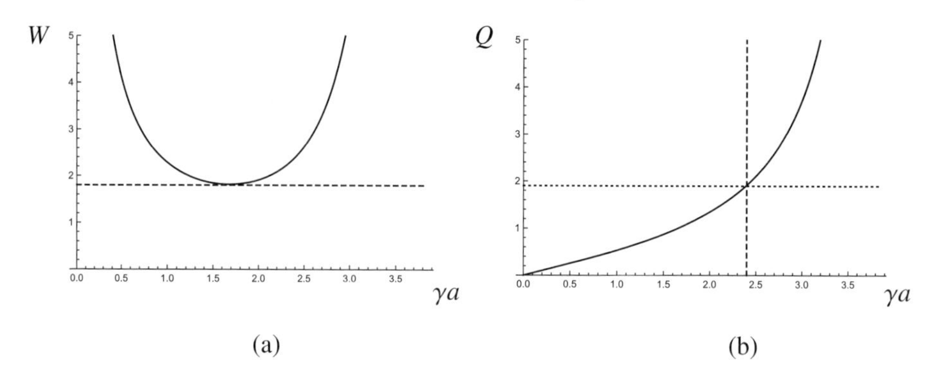

Figure 17.5 (a) The function $W(\gamma a)$ given by (17.41); note the minimum $W_{\min} \approx 1.82$ at $\gamma a \approx 1.68$. (b) The function $Q(\gamma a)$ given by (17.43); axial field reversal occurs for $\gamma a > 2.4$, where $Q(\gamma a) \gtrsim 1.8$.

From these equations, we easily find the dimensionless ratio

$$\frac{\pi a \mathcal{H}_M}{\Phi_z^2} = \gamma a \left[1 + \left(\frac{J_0(\gamma a)}{J_1(\gamma a)}\right)^2\right] - \left(\frac{J_0(\gamma a)}{J_1(\gamma a)}\right) = W(\gamma a), \text{ say.} \tag{17.41}$$

This function is shown in Figure 17.5a. Note the minimum $W_{\min} \approx 1.82$ at $\gamma a \approx 1.68$; thus there is a minimum value of the global helicity (relative to the axial flux) below which a Taylor state is not possible.

The flux of B_θ per unit axial length is

$$\Phi_\theta = B_0 \int_0^a J_1(\gamma r)\,\mathrm{d}r = B_0 a \left[(1 - J_0(\gamma a)/\gamma a\right]. \tag{17.42}$$

Note that Φ_θ and Φ_z have different dimensions, the former being a flux *per unit length*. The flux ratio is given by

$$\frac{2\pi a \Phi_\theta}{\Phi_z} = \frac{1 - J_0(\gamma a)}{J_1(\gamma a)} = Q(\gamma a), \text{ say.} \tag{17.43}$$

The function $Q(\gamma a)$ is shown in Figure 17.5b. Field reversal occurs for $\gamma a > 2.4$, and in this region, the flux ratio $Q(\gamma a) \gtrsim 1.8$: only for sufficient B_θ-flux is field reversal possible in the Taylor scenario.

The major problem that remains however is to understand the dynamical process of relaxation that somehow causes the axial field to reverse near the boundary. Taylor visualised this as an effect of vigorous turbulence in the plasma during the relaxation process, such turbulence arising from instabilities of the relaxing field. The reversal process is frequently attributed to some kind of dynamo mechanism, although of course, since an axial field is externally applied, this is not the kind of self-exciting dynamo that occurs in the absence of any external source as considered in earlier chapters; nevertheless the 'mean-field' mechanisms that gives rise to an α-effect and to a turbulent diffusivity may still be applicable. We explore this possibility in the following section.

17.6 Relaxation of a helical field

So now let us return to an initial-value problem of the type considered in §17.2. As in §17.3, we suppose that the plasma is contained in the annular domain between two perfectly conducting cylinders, $r = \delta a\,(\delta < 1)$ and $r = a$. Let $\mathbf{B} = B_0(0, b_\theta(r,t), b_z(r,t))$ be the magnetic field in this annulus. The associated current distribution is given by

$$\mathbf{j} = \nabla \times \mathbf{B} = B_0 \left(0,\ -\frac{\partial b_z}{\partial r},\ \frac{1}{r}\frac{\partial}{\partial r}(r b_\theta)\right). \tag{17.44}$$

The Lorentz force is radial and generates a velocity field $\mathbf{u} = (u(r,t), 0, 0)$, and we assume this is controlled by an effective viscosity μ. The boundary conditions are

$$u = 0, \quad \eta \mathbf{j} = 0 \quad \text{on } r = \delta a,\ a, \tag{17.45}$$

the latter arising from continuity of the tangential components of the electric field $\mathbf{E}$ across each boundary. If $\eta \neq 0$, we thus have

$$u = 0, \quad \frac{1}{r}\frac{\partial}{\partial r}(r b_\theta) = 0, \quad \frac{\partial b_z}{\partial r} = 0 \quad \text{on } r = \delta a,\ a. \tag{17.46}$$

As in §17.1, we assume that the fluid pressure gradient is negligible compared with the Lorentz force. The governing equations are then the two components of the induction equation in cylindrical polar coordinates (r, θ, z),

$$\frac{\partial b_\theta}{\partial t} = -\frac{\partial}{\partial r}(u b_\theta) + \eta \frac{\partial}{\partial r}\frac{1}{r}\frac{\partial}{\partial r}(r b_\theta), \tag{17.47}$$

$$\frac{\partial b_z}{\partial t} = -\frac{1}{r}\frac{\partial}{\partial r}(r u b_z) + \eta\, \frac{1}{r}\frac{\partial}{\partial r} r \frac{\partial b_z}{\partial r}, \tag{17.48}$$

coupled with the Navier–Stokes equation (which has only a radial component),

$$\rho\left(\frac{\partial u}{\partial t} + u\frac{\partial u}{\partial r}\right) = -B_0^2\left(\frac{1}{2}\frac{\partial}{\partial r}\left(b_\theta^2 + b_z^2\right) + \frac{b_\theta^2}{r}\right) + \mu\left(\frac{\partial^2 u}{\partial r^2} + \frac{1}{r}\frac{\partial u}{\partial r} - \frac{u}{r^2}\right), \tag{17.49}$$

and the equation of mass-conservation,

$$\frac{\partial \rho}{\partial t} = -\frac{1}{r}\frac{\partial}{\partial r}(r u \rho). \tag{17.50}$$

Note that, from (17.48) and (17.50),

$$\frac{D}{Dt}\left(\frac{b_z}{\rho}\right) \equiv \left(\frac{\partial}{\partial t} + u\frac{\partial}{\partial r}\right)\left(\frac{b_z}{\rho}\right) = \frac{\eta}{\rho}\,\frac{1}{r}\frac{\partial}{\partial r} r \frac{\partial b_z}{\partial r}, \tag{17.51}$$

so that when $\eta = 0$, the evolution of b_z is strongly coupled with that of ρ.

Equations (17.47) and (17.48), together with the boundary conditions (17.46), now have two flux invariants

$$\Phi_\theta \equiv \int_{\delta a}^{a} b_\theta \, \mathrm{d}r = \text{const.} \quad \text{and} \quad \Phi_z \equiv \int_{\delta a}^{a} b_z \, 2\pi r \, \mathrm{d}r = \text{const.}\,, \tag{17.52}$$

and of course mass per unit axial length is also conserved:

$$\mathbb{M} \equiv \int_{\delta a}^{a} \rho \, 2\pi r \, \mathrm{d}r = \text{const.} \tag{17.53}$$

The time-scale of the initial relaxation is determined by B_0 and μ, so is, on dimensional grounds $\mathcal{O}\left(\mu/B_0^2\right)$; thus, appropriate dimensionless variables are

$$\hat{r} = r/a, \quad \hat{t} = t\, B_0^2/\mu, \tag{17.54}$$

and then velocity scales as $\hat{u} = u\,\mu/B_0^2 a$. For simplicity of notation, we simply drop the 'hats' in what follows.

Following Moffatt & Mizerski (2018), we now adopt initial conditions satisfying the boundary conditions (17.46) in the form

$$u(r,0) = 0, \quad \rho(r,0) = \rho_0 = \text{const.} \quad b_z(r,0) = b_{z0} = \text{const.}, \tag{17.55}$$

and

$$r\, b_\theta(r,0) = c\left[\frac{(r-\delta)^2}{(1-\delta)^2} - \frac{2(r-\delta)^3}{3(1-\delta)^3}\right] \mathrm{e}^{-k(1-r)^2}\,. \tag{17.56}$$

Obviously the larger k, the more $b_\theta(r,0)$ is concentrated near $r = 1$, in such a way as to initiate a pinch effect. The other relevant dimensionless numbers for the problem are

$$\kappa = \eta\mu/B_0^2 a^2 \quad \text{and} \quad \varepsilon = \rho_0^2 B_0^2 a^2/\mu^2, \tag{17.57}$$

and both are assumed small: $\kappa \ll 1, \varepsilon \ll 1$.

Figure 17.6 shows a typical evolution of the field components given by numerical solution of (17.47)-(17.50), with $k = 10$, $\kappa = 10^{-4}$, $\varepsilon = 10^{-2}$. As expected, the b_θ-field tends to pinch inwards, but now carries the b_z field with it, until the inward Lorentz force associated with b_θ is balanced by the outward Lorentz force associated with b_z; from this point on, a state of slow constrained diffusion is established, in which the total Lorentz force remains close to zero.

Clearly however, this relaxation process cannot cause reversal of the axial field, and therefore cannot come anywhere near a Taylor state. This is not surprising: Taylor's (1974) scenario requires turbulence in the plasma to somehow effect this reversal. Such turbulence may give rise to an α-effect acting on the mean magnetic field, a possibility that we now investigate.

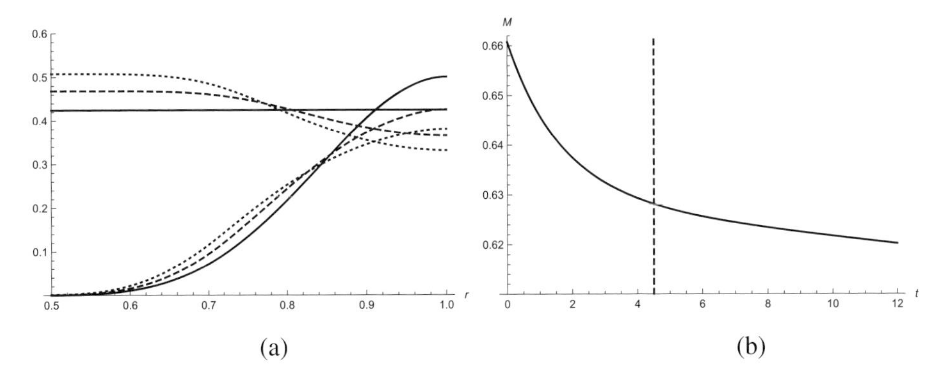

(a) (b)

Figure 17.6 (a) Early relaxation of field components $rb_\theta(r,t)$ and $b_z(r,t)$, $\kappa = 10^{-4}$; $t = 0$ (solid), 5 (dashed), 10 (dotted). (b) Associated decrease of magnetic energy; the relaxation is rapid for $t \lesssim 4.5$, and slow-diffusive for $t \gtrsim 4.5$.

17.7 Effect of plasma turbulence

Turbulence in the reversed field pinch is generally considered to arise from resistive instabilities of the relaxing field (Furth et al. 1963). This field is helical, its 'current helicity density' being represented by the pseudo-scalar $\mathbf{j} \cdot \mathbf{B}$. It seems reasonable to suppose that the turbulence resulting from instability of this field will inherit its helicity, and that the resulting α-effect will be in the usual mean electromotive force $\boldsymbol{\mathcal{E}} = \alpha\mathbf{b}$, with (still using the dimensionless variables of the previous section)

$$\alpha(r,t) = q\,\kappa\,\mathbf{j}\cdot\mathbf{b}, \tag{17.58}$$

where q is a dimensionless constant reflecting the intensity of the turbulence. The parameter κ is included here, recognising that the resistive instabilities, as the name implies, are diffusive in origin, as is also the resulting α-effect. Note that, with $\mathbf{j} = 0$ on the boundaries, $\boldsymbol{\mathcal{E}} = 0$ also, so that, with this prescription for α, the boundary conditions (17.46) are unaffected. We need only include a term

$$\nabla \times (\alpha\,\mathbf{b}) = \left(0,\ -\frac{\partial(\alpha\,b_z)}{\partial r},\ \frac{1}{r}\frac{\partial}{\partial r}(\alpha\,rb_\theta)\right) \tag{17.59}$$

in the induction equation.

The z-component here is

$$\frac{1}{r}\frac{\partial}{\partial r}(\alpha\,rb_\theta) = q\,\kappa\,\frac{1}{r}\frac{\partial}{\partial r}\left[(\mathbf{j}\cdot\mathbf{b})\,rb_\theta\right] = q\,\kappa\,\frac{1}{r}\frac{\partial}{\partial r}\left[b_z b_\theta\frac{\partial}{\partial r}(rb_\theta) - rb_\theta^2\frac{\partial b_z}{\partial r}\right] \tag{17.60}$$

and the last term here combines with the diffusion term in (17.48) to give an effective diffusion term

$$D\frac{1}{r}\frac{\partial}{\partial r}r\frac{\partial b_z}{\partial r}, \quad \text{where} \quad D = \kappa(1 - q\,b_\theta^2). \tag{17.61}$$

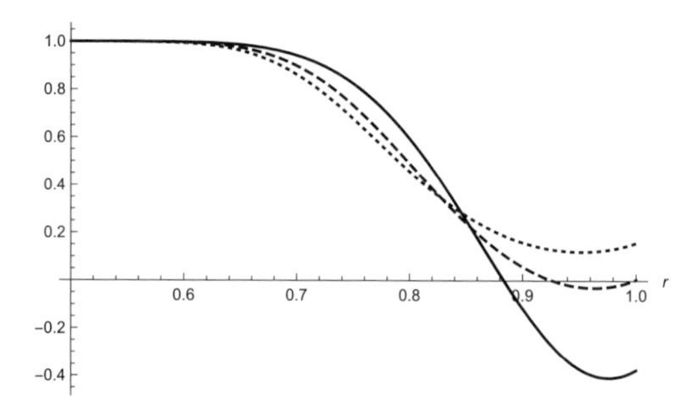

Figure 17.7 Effective diffusivity $1 - qb_\theta^2$, for $q = 5.5$, at times $t = 0$ (solid), 2 (dashed), 4 (dotted).

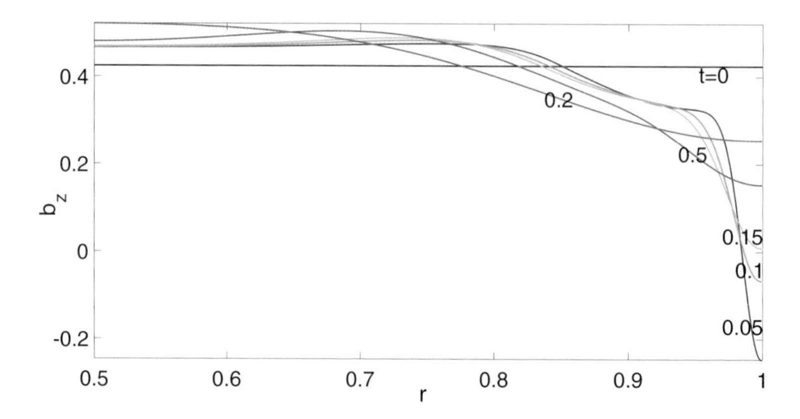

Figure 17.8 Early stage of evolution of $b_z(r,t)$ for $q = 5.5, \kappa = 0.01$. [From Moffatt & Mizerski 2018 © The Japan Society of Fluid Mechanics and IOP Publishing Ltd. Reproduced by permission of IOP publishing. All rights reserved.]

Thus, an interesting situation arises if $q\,b_\theta^2 > 1$ in any part of the range $(\delta, 1)$, for then violent instability of the mean axial field is to be expected as a result of negative diffusivity. It is further apparent, from close inspection of the equation for b_z, that b_z can become negative immediately after a time t^* in a neighbourhood of a point r^* only if $qb_\theta^2 > 1$ at (r^*, t^*).

Figure 17.7 shows $qb_\theta^2 - 1$, where $q = 5.5$ and b_θ is computed as above; this suggests that negative diffusivity effects are to be expected in the region $r \gtrsim 0.9$, as indeed found in the computation of Moffatt & Mizerski (2018)[5] shown in Figure 17.8. Here, at a very early stage, b_z does indeed go negative near $r = 1$, but this is a very transitory effect, and b_z soon recovers to positive values, as the b_θ-field pinches inwards. A very tentative conclusion is that an α-effect of the kind envisaged here may indeed play a part in generating a reversed axial field near the outer boundary.

Suppose now that in the relaxed state, the turbulence dies down and the α-effect is extinguished, and that at this stage, b_z is negative near the boundary $r = 1$.

[5] In this computation, the α-effect was included in the b_z-equation, but not in the b_θ-equation.

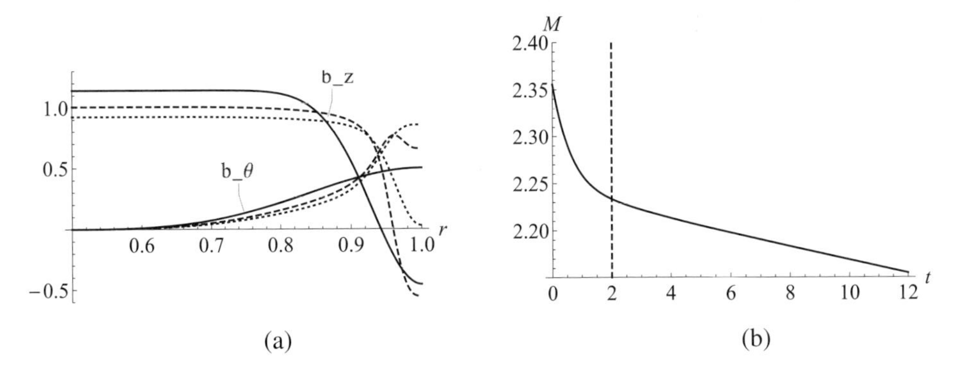

Figure 17.9 (a) Early relaxation of field components $rb_\theta(r,t)$ and $b_z(r,t)$; initial b_z given by (17.62), $\kappa = 10^{-5}$; $t = 0$ (solid), 2 (dashed), 60 (dotted). (b) Associated decrease of magnetic energy; the relaxation is rapid for $t \lesssim 2$, and slow-diffusive for $t \gtrsim 2$.

We may then adopt a new origin of time and, by way of illustration, a new initial condition

$$b_z(r,0) = C\left[1 - 1.4\exp\left(-100(1-r)^2\right)\right], \qquad (17.62)$$

where $C(= 1.14165)$ is chosen so that still $\Phi_z = 1$ (this initial condition is shown by the solid curve of Figure 17.9a); other initial conditions are unchanged. This figure (which should be compared with Figure 17.6) shows the resulting evolution, here with a reduced value of $\kappa\,(= 10^{-5})$. The relaxation is now rapid for $t \lesssim 2$, and slow-diffusive for $t \gtrsim 2$. At $t \approx 2$, b_z is still negative in the interval $0.96 < r < 1$, i.e. the negative flux near $r = 1$ is maintained during the rapid-relaxation phase; however, during the subsequent slow constrained-diffusion phase, the negative flux decreases, disappearing at $t \approx 60$, after which b_z and rb_θ continue to relax towards their asymptotic uniform positive values. Although the relaxed field is here nearly force-free in the sense that $\mathbf{j} \approx \gamma\mathbf{b}$, the coefficient $\gamma(r,t) = \mathbf{j}\cdot\mathbf{b}/\mathbf{b}^2$ is actually far from uniform throughout this evolution, which is therefore at all times far removed from a Taylor state.

17.8 Erupting flux in the solar corona

As was indicated schematically in Figure 5.5, the upwelling of magnetic flux ropes from the solar convection zone leads to eruption of such flux ropes through the chromosphere into the solar corona. Such eruptions are associated with solar flares (Heyvaerts et al. 1977) and 'coronal mass ejections' carried by the solar wind (see §5.6). This is a huge area of research (Priest 1982; Priest & Forbes 2007), peripheral to the main theme of this book. However, we touch on it here because the corona, with its exceedingly small density ($\sim 10^{-12}$ that of the solar photosphere),

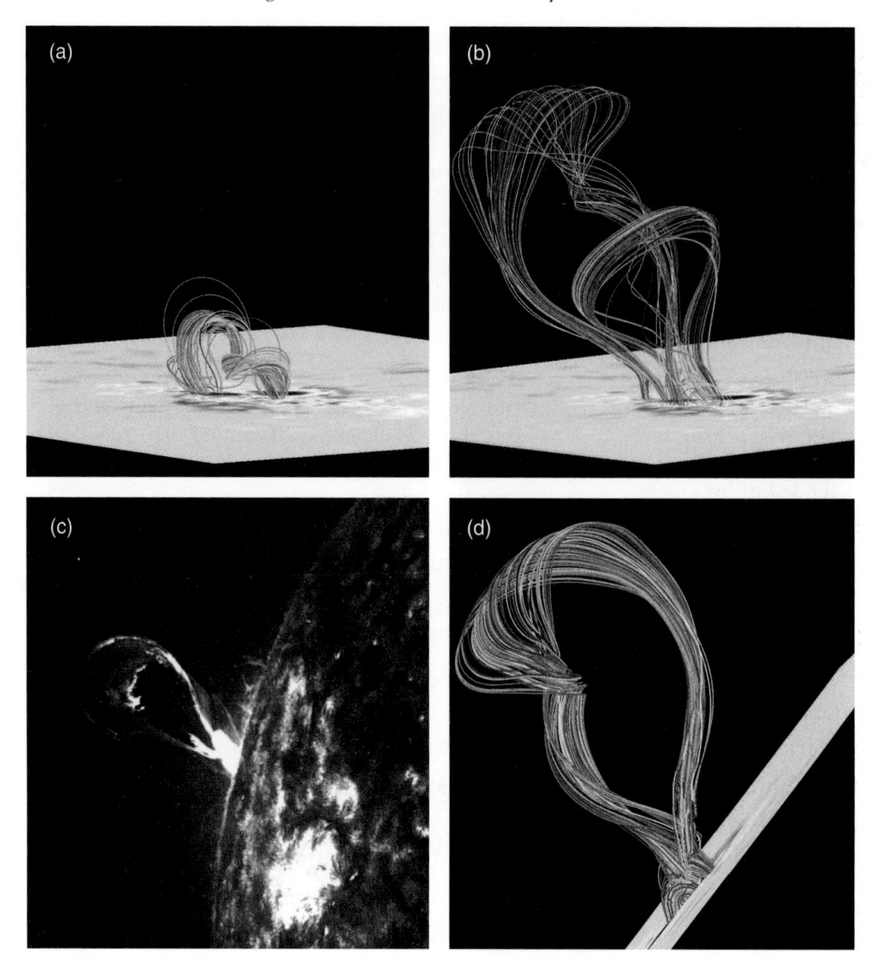

Figure 17.10 Evolution and eruption of a twisted flux rope. (a), (b) and (d) show **B**-lines just before and after the major solar eruption of 13 December 2006, computed from an MHD model involving evolution through a succession of force-free states driven by boundary condition as observed on the photosphere; (c) image of a different eruption of 14 October 2012 which shows twisting of the field that bears comparison with the twisting evident in (d). [From Amari et al. 2014; (c) image courtesy of NASA/SOO and the AIA, EVE and HMI science teams.]

provides an arena where magnetic relaxation of the erupted magnetic flux tubes most naturally occurs.

The relaxation analysis of Chapter 16 is here applicable with fluid pressure p simply set to zero. The relaxed field starting from an arbitrary initial condition must then be force-free: $\mathbf{j} \wedge \mathbf{B} \equiv 0$ (see §2.5 for general results concerning such fields), and, as we have seen in the foregoing analysis, when such fields are the result of a relaxation process, they generally exhibit imbedded current sheets or tube singularities, which in the coronal context are the site of intense coronal heating. Since a key problem in coronal physics is to explain the exceedingly high coronal temperature ($\sim 10^7$K), this heating mechanism is one of the processes that must certainly be taken into account. A second heating mechanism results from phase-mixing of Alfvén waves propagating along the non-uniform magnetic field in an erupted flux tube, which likewise results in very high field gradients.

Figure 17.10 shows the evolution and eruption of a twisted flux rope, as described in the caption to that figure (Amari et al. 2014; see also Amari et al. 2018). The computation of the field lines assumed a force-free evolution on the basis that any departure from the force-free state is immediately eliminated by magnetic relaxation. The process is driven by motion of the footpoints of the field, which are anchored on the photosphere; the observed velocity field on the photosphere provides the boundary condition for the computation. Magnetic energy is pumped into the flux tube from this lower boundary, and the dramatic eruption occurs when the energy level exceeds a critical value at which the force-free state becomes unstable, and beyond which no confined force-free state exists.

In this respect the situation is similar to that analysed by Lynden-Bell & Boily (1994) and Lynden-Bell & Moffatt (2015), who considered the force-free response of a magnetic field to a steady winding process at the lower boundary of a half-space on which the field lines are similarly anchored. The 'flashpoint' that occurs in this situation when $\eta = 0$ is analogous to the coronal mass ejections imaged in Figure 17.10.

17.9 Conclusion

For a subject of this breadth, there can be no conclusion! With every issue that is raised, new problems arise: it is a divergent process! Nevertheless, there have been great gains of understanding of the process of spontaneous magnetic field generation, and the manner in which statistically saturated states can be attained, in both planetary and astrophysical contexts. We hope that this book will give the reader a good introduction to this broad field, and at the same time take her/him to the frontiers of research in at least some of the areas that continue to attract attention, stimulated as these are by the great advances in computational power, laboratory experimentation and satellite observations of planetary and astrophysical magnetic activity. We hold to the view that there is no substitute for rigorous mathematical analysis of the underlying physical processes, a view that has guided the selection of material that we have presented. We hope that this approach will commend itself to our readers, and help to stimulate continuing research in this extremely challenging field.

Appendix
Orthogonal Curvilinear Coordinates

Let (q_1, q_2, q_3) be a system of right-handed orthogonal curvilinear coordinates. The position vector $\mathbf{x}$ may then be expressed in the form

$$\mathbf{x} = \mathbf{x}(q_1, q_2, q_3), \tag{A.1}$$

and the line element $\mathrm{d}\mathbf{x}$ is given by

$$\mathrm{d}\mathbf{x} = \frac{\partial \mathbf{x}}{\partial q_1}\mathrm{d}q_1 + \frac{\partial \mathbf{x}}{\partial q_2}\mathrm{d}q_2 + \frac{\partial \mathbf{x}}{\partial q_3}\mathrm{d}q_3 \,. \tag{A.2}$$

The *metric elements* h_1, h_2, h_3 are defined by

$$h_1 = |\partial\mathbf{x}/\partial q_1| , \quad h_2 = |\partial\mathbf{x}/\partial q_2| , \quad h_3 = |\partial\mathbf{x}/\partial q_3| , \tag{A.3}$$

and the unit triad $(\mathbf{e}_1, \mathbf{e}_2, \mathbf{e}_3)$ by

$$\mathbf{e}_1 = h_1^{-1}(\partial\mathbf{x}/\partial q_1), \quad \mathbf{e}_2 = h_2^{-1}(\partial\mathbf{x}/\partial q_2), \quad \mathbf{e}_3 = h_3^{-1}(\partial\mathbf{x}/\partial q_3) \,. \tag{A.4}$$

The assumed orthogonality implies that $\mathbf{e}_i \cdot \mathbf{e}_j = \delta_{ij}$. Thus (A.2) becomes

$$\mathrm{d}\mathbf{x} = h_1\mathbf{e}_1\mathrm{d}q_1 + h_2\mathbf{e}_2\mathrm{d}q_2 + h_3\mathbf{e}_3\mathrm{d}q_3, \tag{A.5}$$

and the square of the line element is

$$\mathrm{d}\mathbf{x}^2 = h_1^2\mathrm{d}q_1^2 + h_2^2\mathrm{d}q_2^2 + h_3^2\mathrm{d}q_3^2 \,. \tag{A.6}$$

The expressions for grad, div and curl ($\equiv$ rot) in such a coordinate system are as follows: for an arbitrary scalar field $\Phi(q_1, q_2, q_3)$,

$$\nabla\Phi \equiv \operatorname{grad}\Phi = \left(\frac{1}{h_1}\frac{\partial\Phi}{\partial q_1}, \frac{1}{h_2}\frac{\partial\Phi}{\partial q_2}, \frac{1}{h_3}\frac{\partial\Phi}{\partial q_3}\right), \tag{A.7}$$

and for an arbitrary vector field $\mathbf{V}(q_1, q_2, q_3)$,

$$\nabla\cdot\mathbf{V} \equiv \operatorname{div}\mathbf{V} = \frac{1}{h_1h_2h_3}\left\{\frac{\partial}{\partial q_1}(h_2h_3V_1) + \frac{\partial}{\partial q_2}(h_3h_1V_2) + \frac{\partial}{\partial q_3}(h_1h_2V_3)\right\}, \tag{A.8}$$

so that, in particular, the Laplacian $\nabla^2\Phi \equiv \nabla\cdot\nabla\Phi$ (commonly denoted $\Delta\Phi$) is

$$\nabla^2\Phi = \frac{1}{h_1h_2h_3}\left\{\frac{\partial}{\partial q_1}\left(\frac{h_2h_3}{h_1}\frac{\partial\Phi}{\partial q_1}\right) + \frac{\partial}{\partial q_2}\left(\frac{h_3h_1}{h_2}\frac{\partial\Phi}{\partial q_2}\right) + \frac{\partial}{\partial q_3}\left(\frac{h_1h_2}{h_3}\frac{\partial\Phi}{\partial q_3}\right)\right\}; \tag{A.9}$$

and

$$\nabla\wedge\mathbf{V} \equiv \operatorname{curl}\mathbf{V} = \frac{1}{h_1h_2h_3}\begin{vmatrix} h_1\mathbf{e}_1 & h_2\mathbf{e}_2 & h_3\mathbf{e}_3 \\ \partial/\partial q_1 & \partial/\partial q_2 & \partial/\partial q_3 \\ h_1V_1 & h_2V_2 & h_3V_3 \end{vmatrix}. \tag{A.10}$$

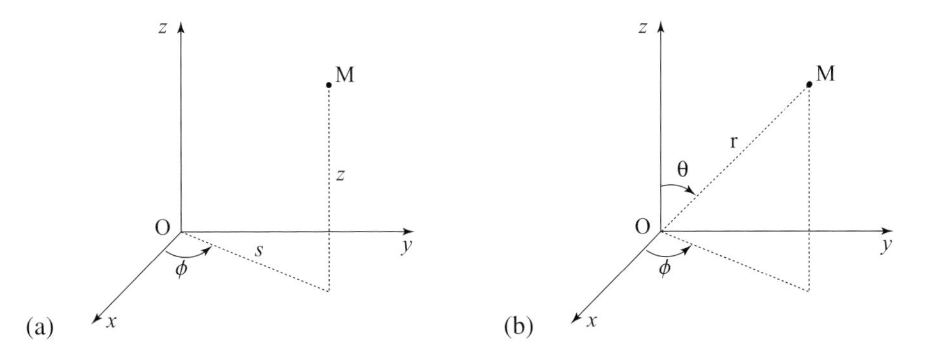

Figure A.1 (a) Cylindrical polar coordinates (s, ϕ, z). (b) Spherical polar coordinates (r, θ, ϕ).

The Laplacian of a vector may be calculated from the identity

$$\nabla^2 \mathbf{V} \equiv \nabla(\nabla \cdot \mathbf{V}) - \nabla \wedge (\nabla \wedge \mathbf{V}) . \tag{A.11}$$

Examples

The coordinate systems (other than Cartesian) most frequently used in this book are cylindrical polars (s, ϕ, z) and spherical polars (r, θ, ϕ) (Figure A.1). For these systems, the following formulae may be easily obtained using the above results.

Cylindrical polar coordinates

$$h_1 = 1,\ h_2 = s,\ h_3 = 1; \quad \mathrm{d}\mathbf{x}^2 = \mathrm{d}s^2 + s^2 \mathrm{d}\phi^2 + \mathrm{d}z^2 . \tag{A.12}$$

$$\nabla\Phi = \left(\frac{\partial \Phi}{\partial s}, \frac{1}{s}\frac{\partial \Phi}{\partial \phi}, \frac{\partial \Phi}{\partial z} \right), \tag{A.13}$$

$$\nabla \cdot \mathbf{V} = \frac{1}{s}\frac{\partial}{\partial s}(sV_s) + \frac{1}{s}\frac{\partial V_\phi}{\partial \phi} + \frac{\partial V_z}{\partial z}, \tag{A.14}$$

$$\nabla \wedge \mathbf{V} = \left(\frac{1}{s}\frac{\partial V_z}{\partial \phi} - \frac{\partial V_\phi}{\partial z}, \frac{\partial V_s}{\partial z} - \frac{\partial V_z}{\partial s}, \frac{1}{s}\frac{\partial}{\partial s}\left(s\, V_\phi\right) - \frac{1}{s}\frac{\partial V_s}{\partial \phi} \right), \tag{A.15}$$

$$\nabla^2 \Phi = \frac{\partial^2 \Phi}{\partial s^2} + \frac{1}{s}\frac{\partial \Phi}{\partial s} + \frac{1}{s^2}\frac{\partial^2 \Phi}{\partial \phi^2} + \frac{\partial^2 \Phi}{\partial z^2} . \tag{A.16}$$

Spherical polar coordinates

$$h_1 = 1,\ h_2 = r,\ h_3 = r\sin\theta; \quad \mathrm{d}\mathbf{x}^2 = \mathrm{d}r^2 + r^2\mathrm{d}\theta^2 + r^2\sin^2\theta\,\mathrm{d}\phi^2 . \tag{A.17}$$

$$\nabla\Phi = \left(\frac{\partial\Phi}{\partial r}, \frac{1}{r}\frac{\partial\Phi}{\partial\theta}, \frac{1}{r\sin\theta}\frac{\partial\Phi}{\partial\phi}\right), \tag{A.18}$$

$$\nabla\cdot\mathbf{V} = \frac{1}{r^2}\frac{\partial}{\partial r}\left(r^2 V_r\right) + \frac{1}{r\sin\theta}\frac{\partial}{\partial\theta}(\sin\theta\, V_\theta) + \frac{1}{r\sin\theta}\frac{\partial V_\phi}{\partial\phi}, \tag{A.19}$$

$$\nabla\wedge\mathbf{V} = \frac{1}{r}\left(\frac{1}{\sin\theta}\left[\frac{\partial}{\partial\theta}\left(\sin\theta\, V_\phi\right) - \frac{\partial V_\theta}{\partial r}\right], \frac{1}{\sin\theta}\frac{\partial V_r}{\partial\phi} - \frac{\partial}{\partial r}\left(rV_\phi\right), \frac{\partial}{\partial r}(rV_\theta) - \frac{\partial V_r}{\partial\theta}\right), \tag{A.20}$$

$$\nabla^2\Phi = \frac{\partial^2\Phi}{\partial r^2} + \frac{2}{r}\frac{\partial\Phi}{\partial r} + \frac{1}{r^2}\left[\frac{1}{\sin\theta}\frac{\partial}{\partial\theta}\left(\sin\theta\frac{\partial\Phi}{\partial\theta}\right) + \frac{1}{\sin^2\theta}\frac{\partial^2\Phi}{\partial\phi^2}\right]. \tag{A.21}$$

Some useful vector identities

$$\nabla\wedge(\nabla\wedge\mathbf{V}) = \nabla(\nabla\cdot\mathbf{V}) - \nabla^2\mathbf{V}, \tag{A.22}$$

$$\nabla\cdot(\Phi\mathbf{V}) = \Phi\,\nabla\cdot\mathbf{V} + \mathbf{V}\cdot\nabla\Phi, \tag{A.23}$$

$$\nabla\wedge(\Phi\mathbf{V}) = \Phi\,\nabla\wedge\mathbf{V} - \mathbf{V}\wedge\nabla\Phi, \tag{A.24}$$

$$\nabla\cdot(\mathbf{V}\wedge\mathbf{W}) = \mathbf{W}\cdot(\nabla\wedge\mathbf{V}) - \mathbf{V}\cdot(\nabla\wedge\mathbf{W}), \tag{A.25}$$

$$\nabla\wedge(\mathbf{V}\wedge\mathbf{W}) = \mathbf{V}(\nabla\cdot\mathbf{W}) - \mathbf{W}(\nabla\cdot\mathbf{V}) + (\mathbf{W}\cdot\nabla)\mathbf{V} - (\mathbf{V}\cdot\nabla)\mathbf{W}, \tag{A.26}$$

$$\nabla(\mathbf{V}\cdot\mathbf{W}) = \mathbf{V}\wedge(\nabla\wedge\mathbf{W}) + \mathbf{W}\wedge(\nabla\wedge\mathbf{V}) + (\mathbf{V}\cdot\nabla)\mathbf{W} + (\mathbf{W}\cdot\nabla)\mathbf{V}. \tag{A.27}$$

Green's formulæ (dV = *volume element;* dS = *surface element)*

$$\int_{\mathcal{D}} (\nabla\Phi\cdot\mathbf{V} + \Phi\,\nabla\cdot\mathbf{V})\,\mathrm{dV} = \int_{\partial\mathcal{D}} \Phi(\mathbf{V}\cdot\mathbf{n})\,\mathrm{dS}, \tag{A.28}$$

$$\int_{\mathcal{D}} \left(\nabla\Phi\cdot\nabla\Psi + \Phi\,\nabla^2\Psi\right)\mathrm{dV} = \int_{\partial\mathcal{D}} \Phi\,\nabla\Psi\cdot\mathbf{n}\,\mathrm{dS}, \tag{A.29}$$

$$\int_{\mathcal{D}} \left(\Phi\,\nabla^2\Psi - \Psi\,\nabla^2\Phi\right)\mathrm{d}V = \int_{\partial\mathcal{D}} (\Phi\,\nabla\Psi - \Psi\,\nabla\Phi)\cdot\mathbf{n}\,\mathrm{dS}, \tag{A.30}$$

$$\int_{\mathcal{D}} (\mathbf{V}\cdot(\nabla\wedge\mathbf{W}) - (\nabla\wedge\mathbf{V})\cdot\mathbf{W})\,\mathrm{dV} = -\int_{\partial\mathcal{D}} \mathbf{n}\cdot(\mathbf{V}\wedge\mathbf{W})\,\mathrm{dS}. \tag{A.31}$$

References

Abramowitz, M. & Stegun, I. A. 1970. *Handbook of Mathematical Functions: With Formulas, Graphs, and Mathematical Tables*. Dover.

Acheson, D. J. & Hide, R. 1973. Hydromagnetics of rotating fluids. *Rep. Prog. Phys.*, **36**, 159–171.

Alexakis, A. 2011. Searching for the fastest dynamo: Laminar ABC flows. *Phys. Rev. E*, 84, 026321.

Alfvén, H. 1942. On the existence of electromagnetic-hydrodynamic waves. *Arkiv. Mat. Astron. Fysik*, **29B**, 7 pp.

Allan, D. W. 1962. On the behaviour of systems of coupled dynamos. *Proc. Camb. Phil. Soc.*, **58**, 671–693.

Altschuler, M. D., Trotter, D. E., Newkirk, G., Jr. & Howard, R. 1974. The large-scale solar magnetic field. *Solar Phys.*, **39**, 3–17.

Alves, J., Bertout, C., Combes, F., et al. 2014. Planck 2013 results. *Astron. Astrophys.*, **571**, E1.

Amari, T., Canou, A. & Aly, J.-J. 2014. Characterizing and predicting the magnetic environment leading to solar eruptions. *Nature*, **514**, 465–469.

Amari, T., Canou, A., Aly, J.-J., Delyon, F. & Alauzet, F. 2018. Magnetic cage and rope as the key for solar eruptions. *Nature*, **554**, 211–215.

Ampère, A.-M. 1822. *Recueil d'observations electro-dynamiques*. Crochard.

Anderson, B. J., Acuña, M. H., Korth, H., et al. 2008. The structure of Mercury's magnetic field from MESSENGER's first flyby. *Science*, **321**, 82–85.

Anderson, B. J., Johnson, C. L., Korth, H., et al. 2011. The global magnetic field of Mercury from MESSENGER orbital observations. *Science*, **333**, 1859–1862.

André, J. D. & Lesieur, M. 1977. Evolution of high Reynolds number isotropic three-dimensional turbulence; influence of helicity. *J. Fluid Mech.*, **81**, 187–208.

Andrews, D. G. & Hide, R. 1975. Hydromagnetic edge waves in a rotating stratified fluid. *J. Fluid Mech.*, **72**, 593–603.

Andrews, D. G. & McIntyre, M. E. 1978. An exact theory of nonlinear waves on a Lagrangian-mean flow. *J. Fluid Mech.*, **89**, 609–646.

Angel, J. R. P. 1975. Strong magnetic fields in white dwarfs. *Ann. N. Y. Acad. Sci.*, **257**, 80–81.

Anselmet, F., Gagne, Y., Hopfinger, E. J. & Antonia, R. A. 1984. High-order velocity structure functions in turbulent shear flows. *J. Fluid Mech.*, **140**, 63–89.

Anufriev, A. P. & Braginskii, S. I. 1975. Effect of a magnetic field on the stream of a liquid rotating at a rough surface. *Magnetohydrodynamics*, **11**, 461–467.

Archontis, V. 2000. *Linear, Non-Linear and Turbulent Dynamos*. PhD thesis, University of Copenhagen, Denmark. Available online at www.astro.ku.dk/bill.

Archontis, V., Dorch, S. B. F. & Nordlund, Å. 2003. Numerical simulations of kinematic dynamo action. *Astron. Astrophys.*, **397**, 393–401.

2007. Nonlinear MHD dynamo operating at equipartition. *Astron. Astrophys.*, **472**, 715–726.

Arnold, V. I. 1965a. Sur la topologie des écoulements stationnaires des fluides parfaits. *C. R. Acad. Sci.Paris*, **261**, 17–20.

1965b. Variational principle for three-dimensional steady-state flows of an ideal fluid. *J. Appl. Math. Mech.*, **29**, 1002–1008.

1974. The asymptotic Hopf invariant and its applications. Pages 229–256 of: *Proc. Summer School in Diff. Eqs. at Dilizhan, Erevan, Armenia* [in Russian]. Armenian Academy of Sciences, Erevan. English translation: 1986 *Sel. Math. Sov.* **5**, 327–345.

Arnold, V. I. & Korkina, E. I. 1983. The growth of a magnetic field in the three-dimensional steady flow of an incompressible fluid. *Moskovskii Univ. Vestnik Seriia Mat. Mekh.*, **1**, 43–46.

Arter, W. 1985. Magnetic-flux transport by a convecting layer including dynamical effects. *Geophys. Astrophys. Fluid Dyn.*, **31**, 311–344.

Arter, W., Proctor, M. R. E. & Galloway, D. J. 1982. New results on the mechanism of magnetic flux pumping by three-dimensional convection. *Mon. Not. R. Astron. Soc.*, **201**, 57P–61P.

Babcock, H. W. 1947. Zeeman effect in stellar spectra. *Astrophys. J.*, **105**, 105–119.

1961. The topology of the Sun's magnetic field and the 22-year cycle. *Astrophys. J.*, **133**, 572.

Babcock, H. W. & Babcock, H. D. 1955. The Sun's magnetic field 1952–1954. *Astrophys. J.*, **121**, 349–366.

Backus, G. E. 1957. The axisymmetric self-excited fluid dynamo. *Astrophys. J.*, **125**, 500–524.

1958. A class of self-sustaining dissipative spherical dynamos. *Ann. Phys.*, **4**, 372–447.

Backus, G. E. & Chandrasekhar, S. 1956. On Cowling's theorem on the impossibility of self-maintained axisymmetric homogeneous dynamos. *Proc. Natl. Acad. Sci. U.S.A.*, **42**, 105–109.

Bajer, K. 2005. Abundant singularities. *Fluid. Dyn. Res.*, **36**, 301–317.

Bajer, K. & Moffatt, H. K. 1990. On a class of steady confined Stokes flows with chaotic streamlines. *J. Fluid Mech.*, **212**, 337–363.

2013. Magnetic relaxation, current sheets, and structure formation in an extremely tenuous fluid medium. *Astrophys. J.*, **779**, 169–182.

Bajer, K., Bassom, A. P. & Gilbert, A. D. 2001. Accelerated diffusion in the centre of a vortex. *J. Fluid Mech.*, **437**, 395–411.

Balbus, S. A. & Hawley, J. F. 1991. A powerful local shear instability in weakly magnetized disks. I – Linear analysis. *Astrophys. J.*, **376**, 214–222.

1992. A powerful local shear instability in weakly magnetized disks. III. Long-term evolution in a shearing sheet. *Astrophys. J.*, **400**, 595–609.

Balbus, S. A., Latter, H. & Weiss, N. 2012. Global model of differential rotation in the Sun. *Mon. Not. R. Astron. Soc.*, **420**, 2457–2466.

Barton, C. 2002. Survey tracks current position of south magnetic pole. *EOS Transactions*, **83**, 291.

Batchelor, G. K. 1950. On the spontaneous magnetic field in a conducting liquid in turbulent motion. *Proc. R. Soc. A*, **201**, 405–416.

1952. The effect of homogeneous turbulence on material lines and surfaces. *Proc. R. Soc. A*, **213**, 349–366.

1953. *The Theory of Homogeneous Turbulence*. Cambridge University Press.

1956. On steady laminar flow with closed streamlines at large Reynolds number. *J. Fluid Mech.*, **1**, 177–190.
1959. Small-scale variation of convected quantities like temperature in turbulent fluid, I. General discussion and the case of small conductivity. *J. Fluid Mech.*, **5**, 113–133.
1967. *An Introduction to Fluid Dynamics*. Cambridge University Press.
Beck, J. G. 2000. A comparison of differential rotation measurements. *Sol. Phys.*, **191**, 47–70.
Beck, R. 2012. Magnetic fields in galaxies. *Space Sci. Rev.*, **166**, 215–230.
2016. Magnetic fields in spiral galaxies. *Astron. Astrophys. Rev.*, **24**, 4.
Beck, R., Brandenburg, A., Moss, D., Shukurov, A. & Sokoloff, D. 1996. Galactic magnetism: recent developments and perspectives. *Ann Rev. Astron. Astrophys.*, **34**, 155–206.
Beer, J., Tobias, S. M. & Weiss, N. O. 1998. An active Sun throughout the Maunder minimum. *Sol. Phys.*, **181**, 237–249.
Bennett, W. H. 1934. Magnetically self-focussing streams. *Phys. Rev.*, **45**, 890–897.
Berger, M. A. & Field, G. B. 1984. The topological properties of magnetic helicity. *J. Fluid Mech.*, **147**, 133–148.
Berhanu, M., Monchaux, R., Fauve, S., et al. 2007. Magnetic field reversals in an experimental turbulent dynamo. *Europhys. Lett.*, **77**, 59001.
Berkhuijsen, E. M., Horellou, C., Krause, M., et al. 1997. Magnetic fields in the disk and halo of M51. *Astron. Astrophys.*, **318**, 700–720.
Bernstein, I. B., Frieman, E. A., Kruskal, M. D. & Kulsrud, R. M. 1958. An energy principle for hydromagnetic stability problems. *Proc. R. Soc. London*, **244**, 17–40.
Betchov, R. 1961. Semi-isotropic turbulence and helicoidal flows. *Phys. Fluids*, **4**, 925–926.
Blandford, R. D. & Payne, D. G. 1982. Hydromagnetic flows from accretion discs and the production of radio jets. *Mon. Not. R. Astron. Soc.*, **199**, 883–903.
Bloxham, J. & Jackson, A. 1992. Time-dependent mapping of the magnetic field at the core–mantle boundary. *J. Geophys. Res.*, **97**(B7), 19637–19563.
Bloxham, J., Gubbins, D. & Jackson, A. 1989. Geomagnetic secular variation. *Phil. Trans. R. Soc. London, Ser. A*, **329**, 415–502.
Bodin, H. A. B. 1984. Physics of reversed field pinch. In: *Proceedings of the 1984 International Conference on Plasma Physics*, vol. 1, pp. 417–454. EEC, Brussels.
1990. The reversed field pinch. *Nuclear Fusion*, **30**, 1717.
Bogoyavlenskij, O. 2017. Vortex knots for the spheromak fluid flow and their moduli spaces. *J. Math. Anal. Appl.*, **450**, 21–47.
Bondi, H. 1986. The rigid body dynamics of unidirectional spin. *Proc. R. Soc. London, Ser. A*, **405**, 265–274.
Bondi, H. & Gold, T. 1950. On the generation of magnetism by fluid motion. *Mon. Not. R. Astron. Soc.*, **110**, 607–611.
Boothby, W. M. 1986. *An Introduction to Differentiable Manifolds and Riemannian Geometry.* 2nd ed. Academic Press. [For Frobenius theorem, see §IV.8.]
Bourne, M., MacNiocaill, C., Thomas, A. L., Knudsen, M. F. & Henderson, G. M. 2012. Rapid directional changes associated with a 6.5 kyr-long Blake geomagnetic excursion at the Blake–Bahama outer ridge. *Earth Planet. Sci. Lett.*, **333**, 21–34.
Bouya, I. & Dormy, E. 2013. Revisiting the ABC flow dynamo. *Phys. Fluids*, **25**, 037103.
2015. Toward an asymptotic behaviour of the ABC dynamo. *EPL*, **110**, 14003.
Brachet, M. E., Meiron, D. I., Orszag, S. A., et al. 1983. Small-scale structure of the Taylor–Green vortex. *J. Fluid Mech.*, **130**, 411–452.

Braginskii, S. I. 1964a. Kinematic models of the Earth's hydromagnetic dynamo. *Geomagn. Aeron.*, **4**, 572–583.

1964b. Self excitation of a magnetic field during the motion of a highly conducting fluid. *Sov. Phys. JETP*, **20**, 726–735.

1964c. Theory of the hydromagnetic dynamo. *Sov. Phys. JETP*, **20**, 1462–1471.

1967. Magnetic waves in the Earth's core. *Geomagn. Aeron.*, **7**, 851–859.

1970. Torsional magnetohydrodynamic vibrations in the Earth's core and variations in day length. *Geomagn. Aeron.*, **10**, 1–8.

1975. An almost axially symmetric model of the hydromagnetic dynamo of the Earth, I. *Geomagn. Aeron.*, **15**, 149–156.

1993. MAC-oscillations of the hidden ocean of the core. *J. Geomagn. Geoelectr.*, **45**, 1517–1538.

1994. The nonlinear dynamo and model-Z. In: *Lectures on Solar and Planetary Dynamos*, pp. 267–304 Cambridge University Press.

1999. Dynamics of the stably stratified ocean at the top of the core. *Phys. Earth Planet. Inter.*, **111**, 21–34.

2006. Formation of the stratified ocean of the core. *Earth Planet. Sci. Lett.*, **243**, 650–656.

Braginsky, S. I. & Meytlis, V. P. 1990. Local turbulence in the Earth's core. *Geophys. Astrophys. Fluid Dyn.*, **55**, 71–87.

Braginsky, S. I. & Roberts, P. H. 1995. Equations governing convection in Earth's core and the geodynamo. *Geophys. Astrophys. Fluid Dyn.*, **79**, 1–97.

Brandenburg, A. 2005. Importance of magnetic helicity in dynamos. In: *Cosmic Magnetic Fields*, pp. 219–253. Springer.

Brandenburg, A. & Subramanian, K. 2005. Astrophysical magnetic fields and nonlinear dynamo theory. *Phys. Rep.*, **417**, 1–209.

Brandenburg, A., Moss, D. & Soward, A. M. 1998. New results for the Herzenberg dynamo: Steady and oscillatory solutions. *Proc. R. Soc. London, Ser. A*, **454**, 1283–1300.

Brandon, A. D. & Walker, R. J. 2005. The debate over core–mantle interaction. *Earth Planet. Sci. Lett.*, **232**, 211–225.

Brissaud, A., Frisch, U., Léorat, J., Lesieur, M. & Mazure, A. 1973. Helicity cascades in fully developed isotropic turbulence. *Phys. Fluids*, **16**, 1366–1367.

Brummell, N. H., Cattaneo, F. & Tobias, S. M. 2001. Linear and nonlinear dynamo properties of time-dependent ABC flows. *Fluid Dyn. Res.*, **28**, 237–265.

Brush, S. G. 1980. Discovery of the Earth's core. *Am. J. Phys.*, **48**, 705–724.

Buffett, B. A. 2002. Estimates of heat flow in the deep mantle based on the power requirements for the geodynamo. *Geophys. Res. Lett.*, **29**, GL014609.

2014. Geomagnetic fluctuations reveal stable stratification at the top of the Earth's core. *Nature*, **507**, 484–487.

Bullard, E. C. 1955. The stability of a homopolar dynamo. *Proc. Camb. Phil. Soc.*, **51**, 744–760.

1968. Reversals of the Earth's magnetic field. *Phil. Trans. R. Soc. London, Ser. A*, **263**, 481–524.

Bullard, E. C. & Gellman, H. 1954. Homogeneous dynamos and terrestrial magnetism. *Phil. Trans. R. Soc. London, Ser. A*, **247**, 213–278.

Bullard, E. C. & Gubbins, D. 1973. Generation of magnetic fields by fluid motions of global scale. *Geophys. Fluid Dyn.*, **8**, 43–56.

Bullard, E. C., Freedman, C., Gellman, H. & Nixon, J. 1950. The westward drift of the Earth's magnetic field. *Phil. Trans. R. Soc. London, Ser. A*, **243**, 67–92.

Bullen, K. E. 1946. A hypothesis on compressibility at pressures of the order of a million atmospheres. *Nature*, **157**, 405.

Burgers, J. M. 1948. A mathematical model illustrating the theory of turbulence. *Adv. Appl. Mech.*, **1**, 171–199.

Bushby, P. J. & Tobias, S. M. 2007. On predicting the solar cycle using mean-field models. *Astrophys. J.*, **661**, 1289–1296.

Busse, F. H. 1968. Steady fluid flow in a precessing spheroidal shell. *J. Fluid Mech.*, **33**, 739–751.

1970. Thermal instabilities in rapidly rotating systems. *J. Fluid Mech.*, **44**, 441–460.

1973. Generation of magnetic fields by convection. *J. Fluid Mech.*, **57**, 529–544.

1975a. A model of the geodynamo. *Geophys. J. R. Astron. Soc.*, **42**, 837–839.

1975b. A necessary condition for the geodynamo. *J. Geophys. Res.*, **80**, 278–280.

1976. Generation of planetary magnetism by convection. *Phys. Earth Planet Inter.*, **12**, 350–358.

1992. Dynamic theory of planetary magnetism and laboratory experiments. In: Friedrich, R. & Wunderlin, A. (eds.), *Evolution of Dynamical Structures in Complex Systems*, pp. 197–208. Springer Proc. Phys., vol. 69. Springer.

Busse, F. H. & Carrigan, C. R. 1974. Convection induced by centrifugal buoyancy. *J. Fluid Mech.*, **62**, 579–592.

Busse, F. H., Hartung, G., Jaletzky, M. & Sommermann, G. 1998. Experiments on thermal convection in rotating systems motivated by planetary problems. *Dyn. Atmos. Oceans*, **27**, 161–174.

Buz, J., Weiss, B. P., Tikoo, S. M., et al. 2015. Magnetism of a very young lunar glass. *J. Geophys. Res.*, **120**, 1720–1735.

Calkins, M. A., Julien, K., Tobias, S. M. & Aurnou, J. M. 2015. A multiscale dynamo model driven by quasi-geostrophic convection. *J. Fluid Mech.*, **780**, 143–166.

Calkins, M. A., Long, L., Nieves, D., Julien, K. & Tobias, S. M. 2016. Convection-driven kinematic dynamos at low Rossby and magnetic Prandtl numbers. *Phys. Rev. Fluids*, **1**, 083701.

Călugăreanu, G. 1959. L'intégrale de Gauss et l'analyse des nœuds tridimensionnels. *Rev. Math. Pure Appl.*, **4**, 5–20.

Cameron, R. & Galloway, D. J. 2006. Saturation properties of the Archontis dynamo. *Mon. Not. R. Astron. Soc.*, **365**, 735–746.

Cao, H., Dougherty, M. K., Khurana, K. K., et al. 2017. Saturn's internal magnetic field revealed by Cassini grand finale. *AGU Fall Meeting Abstracts*.

Cao, H., Russell, C. T., Christensen, U. R., Dougherty, M. K. & Burton, M. E. 2011. Saturn's very axisymmetric magnetic field: No detectable secular variation or tilt. *Earth Planet. Sci. Lett.*, **304**, 22–28.

Cardin, P. 1992. *Aspects de la convection dans la terre: Couplage des manteaux inférieur et supérieur, convection thermique du noyau liquide*. PhD thesis, Thèse Paris VI.

Carrigan, C. R. & Busse, F. H. 1983. An experimental and theoretical investigation of the onset of convection in rotating spherical shells. *J. Fluid Mech.*, **126**, 287–305.

Cattaneo, F. & Hughes, D. W. 1996. Nonlinear saturation of the turbulent α effect. *Phys. Rev. E*, **54**, R4532(R).

2009. Problems with kinematic mean field electrodynamics at high magnetic Reynolds numbers. *Mon. Not. R. Astron. Soc. Lett.*, **395**, L48–L51.

Cattaneo, F. & Tobias, S. M. 2009. Dynamo properties of the turbulent velocity field of a saturated dynamo. *J. Fluid Mech.*, **621**, 205–214.

Cattaneo, F. & Vainshtein, S. I. 1991. Suppression of turbulent transport by a weak magnetic field. *Astrophys. J.*, **376**, L21–L24.

Cattaneo, F., Hughes, D. W. & Thelen, J.-C. 2002. The nonlinear properties of a large-scale dynamo driven by helical forcing. *J. Fluid Mech.*, **456**, 219–237.

Chamberlain, J. A. & Carrigan, C. R. 1986. An experimental investigation of convection in a rotating sphere subject to time varying thermal boundary conditions. *Geophys. Astrophys. Fluid Dyn.*, **35**, 303–327.

Chandrasekhar, S. 1960. The hydrodynamic stability of inviscid flow between coaxial cylinders. *Proc. Natl. Acad. Sci. U.S.A.*, **46**, 137–141.

1961. *Hydrodynamic and Hydromagnetic Stability*. Oxford.

Chapman, S. & Bartels, J. 1940. *Geomagnetism: Vol. I, Geomagnetic and Related Phenomena; Vol. II, Analysis of the Data and physical Theories*. Clarendon.

Charbonneau, P. 2010. Dynamo models of the solar cycle. *Living Rev. Sol. Phys.*, **7**, 3.

2014. Solar dynamo theory. *Ann. Rev. Astron. Astrophys.*, **52**, 251–290.

Charbonneau, P., Christensen-Dalsgaard, J., Henning, R., et al. 1999. Helioseismic constraints on the structure of the solar tachocline. *Astrophys. J.*, **527**, 445–460.

Childress, S. 1970. New solutions of the kinematic dynamo problem. *J. Math. Phys.*, **11**, 3063–3076.

Childress, S. & Gilbert, A. D. 1995. *Stretch, Twist, Fold: The Fast Dynamo*. Lecture Notes in Physics. Springer.

Childress, S. & Soward, A. M. 1972. Convection-driven hydromagnetic dynamo. *Phys. Rev. Lett.*, **29**, 837–839.

Christensen, U. R. 2006. A deep dynamo generating Mercury's magnetic field. *Nature*, **444**, 1056–1058.

2010. Dynamo scaling laws and applications to the planets. *Space Sci. Rev.*, **152**, 565–590.

Christensen, U. R. & Aubert, J. 2006. Scaling properties of convection-driven dynamos in rotating spherical shells and application to planetary magnetic fields. *Geophys. J. Int.*, **166**, 97–114.

Christensen, U., Olson, P. & Glatzmaier, G. A. 1998. A dynamo model interpretation of geomagnetic field structures. *Geophys. Res. Lett.*, **25**, 1565–1568.

Christensen-Dalsgaard, J. 2002. Helioseismology. *Rev. Mod. Phys.*, **74**, 1073–1129.

Christensen-Dalsgaard, J., Dappen, W., Ajukov, S. V., et al. 1996. The current state of solar modeling. *Science*, **272**, 1286–1292.

Chui, A. Y. K. & Moffatt, H. K. 1994. A thermally driven disc dynamo. In: Proctor, M. R. E., Matthews, P. C. & Rucklidge, A. M. (eds.), *Solar and Planetary Dynamos*, pp. 51–58. Cambridge University Press.

1995. The energy and helicity of knotted magnetic flux tubes. *Proc. R. Soc. London, Ser. A*, **451**, 609–629.

Clarke, A., Jr. 1964. Production and dissipation of magnetic energy by differential fluid motion. *Phys. Fluids*, **7**, 1299–1305.

1965. Some exact solutions in magnetohydrodynamics with astrophysical applications. *Phys. Fluids*, **8**, 644–649.

Connerney, J. E. P. & Ness, N. F. 1988. Mercury's magnetic field and interior. In: *Mercury*, pp. 494–513. University of Arizona Press.

Connerney, J. E. P., Acuna, M. H. & Ness, N. F. 1991. The magnetic field of Neptune. *J. Geophys. Res.*, **96**, 19023–19042.

Cook, A. E. & Roberts, P. H. 1970. The Rikitake two-disc dynamo system. *Proc. Camb. Phil. Soc.*, **68**, 547–569.

Cook, A. H. 2009. *Interiors of the Planets*. Cambridge University Press.

Cordero, S. & Busse, F. H. 1992. Experiments on convection in rotating hemispherical shells: Transition to a quasi-periodic state. *Geophys. Res. Lett.*, **19**, 733–736.

Corfield, C. N. 1984. The magneto-Boussinesq approximation by scale analysis. *Geophys. Astrophys. Fluid Dyn.*, **29**, 19–28.

Courtillot, V. & Olson, P. 2007. Mantle plumes link magnetic superchrons to phanerozoic mass depletion events. *Earth Planet. Sci. Lett.*, **260**, 495–504.

Courtillot, V., Ducruix, J. & Le Mouël, J. L. 1978. Inverse methods applied to continuation problems in geophysics. In: Sabatier, P. C. (ed.), *Applied Inverse Problems*, pp. 48–82. Lecture Notes in Physics, vol. 85. Springer.

Cowling, T. G. 1934. The magnetic field of sunspots. *Not. Not. R. Astron. Soc.*, **94**, 39–48.

1957. *Magnetohydrodynamics*. Interscience.

1975a. *Magnetohydrodynamics*. Adam Hilger.

1975b. Sunspots and the solar cycle. *Nature*, **255**, 189–190.

Davidson, P. A. 2013. Scaling laws for planetary dynamos. *Geophys. J. Int.*, **195**, 67–74.

Deguchi, T. & Tsurusaki, K. 1997. Random knots and links and applications to polymer physics. *Proc. Lect. Knots*, **96**, 95–122.

Deinzer, W., von Kusserow, H. U. & Stix, M. 1974. Steady and oscillatory $\alpha\omega$-dynamos. *Astron. Astrophys.*, **36**, 69–78.

de Wijs, G. A., Kresse, G., Vocadlo, L., et al. 1998. The viscosity of liquid iron at the physical conditions of the Earth's core. *Nature*, **392**, 805–807.

Dikpati, M. & Gilman, P. A. 2009. Flux-transport solar dynamos. *Space Science Rev.*, **144**, 67–75.

Dolginov, A. Z. & Urpin, V. A. 1979. The inductive generation of the magnetic field in binary systems. *Astron. Astrophys.*, **79**, 60–69.

Dolginov, Sh., Yeroshenko, Ye. & Zhuzgov, L. 1973. Magnetic field in the very close neighbourhood of Mars according to data from the Mars 2 and Mars 3 spacecraft. *J. Geophys. Res.*, **78**, 4779–4786.

Dombre, T., Frisch, U., Greene, J. M., et al. 1986. Chaotic streamlines in the ABC flows. *J. Fluid Mech.*, **167**, 353–391.

Donati, J.-F. & Brown, S. F. 1997. Zeeman-Doppler imaging of active stars. V. Sensitivity of maximum entropy magnetic maps to field orientation. *Astron. Astrophys.*, **326**, 1135–1142.

Donati, J.-F., Collier Cameron, A., Semel, M., et al. 2003. Dynamo processes and activity cycles of the active stars AB Doradus, LQ Hydrae and HR 1099. *Month. Not. R. Soc.*, **345**, 1145–1186.

Donati, J.-F., Forveille, T., Collier Cameron, A., et al. 2006a. The large-scale axisymmetric magnetic topology of a very-low-mass fully convective star. *Science*, **311**, 633–635.

Donati, J.-F., Howarth, I. D., Jardine, M. M., et al. 2006b. The surprising magnetic topology of τ Sco: Fossil remnant or dynamo output? *Mon. Not. R. Astron. Soc.*, **370**, 629–644.

Dormy, E. 1997. *Modélisation numérique de la dynamo terrestre*. PhD thesis, Thèse IPGP, Paris.

2016. Strong-field spherical dynamos. *J. Fluid Mech.*, **789**, 500–513.

Dormy, E. & Gérard-Varet, D. 2008. Time scales separation for dynamo action. *Europhys. Lett.*, **81**, 64002.

Dormy, E. & Soward, A. M. 2007. *Mathematical Aspects of Natural Dynamos*. Chapman & Hall.

Dormy, E., Valet, J.-P. & Courtillot, V. 2000. Numerical models of the geodynamo and observational constraints. *Geochem. Geophys. Geosyst.*, **1**, 1037–1042.

Dormy, E., Soward, A. M., Jones, C. A., Jault, D. & Cardin, P. 2004. The onset of thermal convection in rotating spherical shells. *J. Fluid Mech.*, **501**, 43–70.

Dormy, E., Oruba, L. & Petitdemange, L. 2018. Three branches of dynamo action. *Fluid Dyn. Res.*, **50**, 011415.

Drobyshevski, E. M. & Yuferev, V. S. 1974. Topological pumping of magnetic flux by three-dimensional convection. *J. Fluid Mech.*, **65**, 33–44.

Dudley, M. L. & James, R. W. 1989. Time-dependent kinematic dynamos with stationary flows. *Proc. R. Soc. London, Ser. A*, **425**, 407–429.

Dungey, J. W. 1961. Interplanetary magnetic field and the Auroral zones. *Phys. Rev. Lett.*, **6**, 47–48.

Durney, B. R. 1976. On theories of solar rotation. In: *Basic Mechanisms of Solar Activity*, pp. 243–295. Intern. Astron. Union, vol. 71. Springer.

Dziewonski, A. M. & Anderson, D. L. 1981. Preliminary reference Earth model. *Phys. Earth Planet. Inter.*, **25**, 297–356.

Elsasser, W. M. 1946. Induction effects in terrestrial magnetism, I. Theory. *Phys. Rev.*, **69**, 106–116.

Eltayeb, I. A. 1972. Hydromagnetic convection in a rapidly rotating fluid layer. *Proc. R. Soc. London, Ser. A*, **326**, 229–254.

1975. Overstable hydromagnetic convection in a rotating fluid layer. *J. Fluid Mech.*, **71**, 161–179.

Escande, D. F. 2015. What is a reversed field pinch? In: Diamond, P. H., Garbet, X., Ghendrih, P. & Sarazin, Y. (eds.), *Rotation and Momentum Transport in Magnetized Plasmas*, pp. 247–286. World Scientific.

Falkovich, G., Gawędzki, K. & Vergassola, M. 2001. Particles and fields in fluid turbulence. *Rev. Mod. Phys.*, **73**, 913–975.

Faraday, M. 1832. Experimental researches in electricity. *Phil. Trans. R. Soc. London*, **122**, 125–194.

Fautrelle, Y. & Childress, S. 1982. Convective dynamos with intermediate and strong fields. *Geophys. Astrophys. Fluid Dyn.*, **22**, 255–279.

Favier, B. & Proctor, M. R. E. 2013. Growth rate degeneracies in kinematic dynamos. *Phys. Rev. E.*, **88**, 031001.

Fearn, D. R. 1979. Thermal and magnetic instabilities in a rapidly rotating fluid sphere. *Geophys. Astrophys. Fluid Dyn.*, **14**, 103–126.

1994. Nonlinear planetary dynamos. In: *Lectures on Solar and Planetary Dynamos*, pp. 219–244. Cambridge University Press.

1998. Hydromagnetic flow in planetary cores. *Rep. Prog. Phys.*, **61**, 175–235.

Fearn, D. R. & Proctor, M. R. E. 1987. Dynamically consistent magnetic fields produced by differential rotation. *J. Fluid Mech.*, **178**, 521–534.

Fearn, D. R., Roberts, P. H. & Soward, A. M. 1988. Convection, stability and the dynamo. In: Galdi, G. P. & Straughan, B. (eds.), *Energy, Stability and Convection*, pp. 60–324. Longman.

Featherstone, N. A. & Miesch, M. S. 2015. Meridional circulation in solar and stellar convection zones. *Astrophys. J.*, **804**, 67.

Fefferman, C. L. 2006. *The Millennium Prize Problems*. American Mathematical Society.

Fernando, H. J. S., Chen, R-R. & Boyer, D. L. 1991. Effects of rotation on convective turbulence. *J. Fluid Mech.*, **228**, 513–547.

Ferraro, V. C. A. 1937. The non-uniform rotation of the Sun and its magnetic field. *Mon. Not. R. Astron. Soc.*, **97**, 458–472.

Ferrière, K. 1992a. Effect of an ensemble of explosions on the galactic dynamo. I – General formulation. *Astrophys. J.*, **389**, 286–296.

1992b. Effect of the explosion of supernovae and superbubbles on the galactic dynamo. *Astrophys. J.*, **391**, 188–198.

1998. Alpha-tensor and diffusivity tensor due to supernovae and superbubbles in the galactic disk. *Astron. Astrophys.*, **335**, 488–499.

2010. The interstellar magnetic field near the galactic center. *Astron. Nachr.*, **331**, 27–33.

Feudel, F., Tuckerman, L. S., Zaks, M. & Hollerbach, R. 2017. Hysteresis of dynamos in rotating spherical shell convection. *Phys. Rev. Fluids*, **2**, 053902.

Finn, J. M. & Ott, E. 1988. Chaotic flows and fast magnetic dynamos. *Phys. Fluids*, **31**, 2992–3011.

Fletcher, A., Beck, R., Shukurov, A., Berkhuijsen, E. M. & Horellou, C. 2011. Magnetic fields and spiral arms in the galaxy M51. *Month. Not. R. Soc.*, **412**, 2396–2416.

Freedman, M. H. 1988. A note on topology and magnetic energy in incompressible and perfectly conducting fluids. *J. Fluid. Mech.*, **194**, 549–551.

Friedlander, S. & Vicol, V. 2011. Global well-posedness for an advection-diffusion equation arising in magneto-geostrophic dynamics. *Ann. Inst. H. Poincaré*, **28**, 283–301.

Frisch, U. 1991. From global scaling, à la Kolmogorov, to local multifractal scaling in fully developed turbulence. *Proc. R. Soc. London, Ser. A*, **434**, 89–99.

Frisch, U. & Villone, B. 2014. Cauchy's almost forgotten Lagrangian formulation of the Euler equation for 3D incompressible flow. *Eur. Phys. J. H.*, **39**, 325–351.

Frisch, U., Pouquet, A., Léorat, J. & Mazure, A. 1975. Possibility of an inverse cascade of magnetic helicity in magnetohydrodynamic turbulence. *J. Fluid Mech.*, **68**, 769–778.

Fuchs, H., Rädler, K. H. & Reinhardt, M. 2001. Suicidal and parthenogenetic dynamos. In: Armbuster, D. & Oprea, L (eds.), *Dynamo and Dynamics, a Mathematical Challenge*, pp. 338–346. Kluwer.

Furth, H. P., Killeen, J. & Rosenbluth, M. N. 1963. Finite-resistive instabilities of a sheet pinch. *Phys. Fluids*, **6**, 459.

Gailitis, A. 1970. Self-excitation of a magnetic field by a pair of annular vortices. *Magnetohydrodynamics*, **6**, 14–17.

1990. The helical MHD dynamo. In: Moffatt, H. K. & Tsinober, A. (eds.), *Topological Fluid Mechanics*, pp. 147–156. Cambridge University Press.

Gailitis, A. & Freiberg, Ya. 1976. Theory of a helical MHD dynamo. *Magnetohydrodynamics*, **12**, 127–130.

Gailitis, A., Lielausis, O., Dement'ev, S., et al. 2000. Detection of a flow induced magnetic field eigenmode in the Riga dynamo facility. *Phys. Rev. Lett.*, **84**, 4365.

Gailitis, A., Lielausis, O., Platacis, E., et al. 2001. Magnetic field saturation in the Riga dynamo experiment. *Phys. Rev. Lett.*, **86**, 3024.

Galileo, Galilei. 1613. *Istoria e dimostrazioni intorno alle macchie solari e loro accidenti*. Giacomo Mascardi.

Gallagher, I. & Gérard-Varet, D. 2017. *Shocks, Singularities and Oscillations in Nonlinear Optics and Fluid Mechanics*. INdAM Series, vol. 17. Springer.

Gallet, Y., Pavlov, V., Halverson, G. & Hulot, G. 2012. Toward constraining the long-term reversing behavior of the geodynamo: A new Maya superchron $\sim$ 1 billion years ago from the magnetostratigraphy of the Kartochka Formation (southwestern Siberia). *Earth Planet. Sci. Lett.*, **339**, 117–126.

Galloway, D. J. 2012. ABC flows then and now. *Geophys. Astrophys. Fluid Dyn.*, **106**, 450–467.

Galloway, D. J. & O'Brian, N. R. 1993. Numerical calculations of dynamos for ABC and related flows. In: Proctor, M. R. E., Matthews, P. C. & Rucklidge, A. M. (eds.), *Solar and Planetary Dynamos*, pp. 105–113. Cambridge University Press.

Galloway, D. J. & Proctor, M. R. E. 1983. The kinematics of hexagonal magnetoconvection. *Geophys. Astrophys. Fluid Dyn.*, **24**, 109–136.

1992. Numerical calculations of fast dynamos in smooth velocity fields with realistic diffusion. *Nature*, **356**, 691–693.

Gauss, C. F. 1832. Intensitas vis magneticae terrestris ad mensuram absolutam revocata. *Werke*, **V**, 79–118.
1838. Allgemeine Theorie des Erdemagnetismus. *Werke*, **V**, 119–193.
Gellibrand, H. 1635. *A Discourse Mathematical on the Variation of the Magneticall Needle: Together with Its Admirable Diminution Lately Discovered.* Printed by William Iones, London.
Ghizaru, M., Charbonneau, P. & Smolarkiewicz, P. K. 2010. Magnetic cycles in global large-eddy simulations of solar convection. *Astrophys. J. Lett.*, **715**, L133–L137.
Gibson, R. D. 1968a. The Herzenberg dynamo, I. *Q. J. Mech. Appl. Math.*, **21**, 243–255.
1968b. The Herzenberg dynamo, II. *Q. J. Mech. Appl. Math.*, **21**, 257–287.
Gibson, R. D. & Roberts, P. H. 1967. Some comments on the theory of homogeneous dynamo. In: Hindmarsh, W. R., Lowes, F. J., Roberts, P. R. & Runcorn, S. K. (eds.), *Magnetism and the Cosmos*, pp. 108–120. Oliver and Boyd.
1969. The Bullard-Gellman dynamo. In: Runcorn, S. K. (ed.), *The Application of Modern Physics to the Earth and Planetary Interiors*, pp. 577–601. Wiley, Interscience.
Gilbert, A. D. 1988. Fast dynamo action in the Ponomarenko dynamo. *Geophys. Astrophys. Fluid Dyn.*, **44**, 241–258.
2002. Magnetic helicity in fast dynamos. *Geophys. Astrophys. Fluid Dyn.*, **96**, 135–151.
2003. Dynamo theory. In: Friedlander, S. & Serre, D. (eds.), *Handbook of Mathematical Fluid Dynamics*, vol. 2, pp 355–441. Elsevier.
Gilbert, A. D., Mason, J. & Tobias, S. M. 2016. Flux expulsion with dynamics. *J. Fluid Mech.*, **791**, 568–588.
Gilbert, W. 1600. *De magnete magnetisque corporibus, et de magno magnete tellure*. Peter Short, London. (Translation by P. F. Mottelay 1893, republished by Dover, 1958.)
Gillet, N., Jault, D., Canet, E. & Fournier, A. 2010. Fast torsional waves and strong magnetic field within the Earth's core. *Nature*, **465**, 74–77.
Gillet, N., Jault, D. & Canet, E. 2017. Excitation of travelling torsional normal modes in an Earth's core model. *Geophys. J. Int.*, **210**, 1503–1516.
Gilman, P. A. 1970. Instability of magnetohydrostatic stellar interiors from magnetic buoyancy, I. *Astrophys. J.*, **162**, 1019–1029.
Gissinger, C., Fromang, S. & Dormy, E. 2009. Direct numerical simulations of the galactic dynamo in the kinematic growing phase. *Mon. Not. R. Astron. Soc.*, **394**, L84–L88.
Gissinger, C., Dormy, E. & Fauve, S. 2010. Morphology of field reversals in turbulent dynamos. *Europhys. Lett.*, **90**, 49001.
Glane, S. & Buffett, B. A. 2018. Enhanced core-mantle coupling due to stratification at the top of the core. *Frontiers in Earth Science*, **6**, 171–180.
Glatzmaier, G. A. & Gilman, P. A. 1982. Compressible convection in a rotating spherical shell. V - Induced differential rotation and meridional circulation. *Astrophys. J.*, **256**, 316–330.
Glatzmaier, G. A. & Roberts, P. H. 1995. A three-dimensional convective dynamo solution with rotating and finitely conducting inner core and mantle. *Phys. Earth Planet. Inter.*, **91**, 63–75.
1996. An anelastic evolutionary geodynamo simulation driven by compositional and thermal convection. *Phys. D*, **97**, 81–94.
Glatzmaier, G. A., Coe, R. S., Hongre, L. & Roberts, P. H. 1999. The role of the Earth's mantle in controlling the frequency of geomagnetic reversals. *Nature*, **401**, 885–890.
Goldreich, P. & Lynden-Bell, D. 1965. II. Spiral arms as sheared gravitational instabilities. *Mon. Not. R. Astron. Soc.*, **130**, 125–158.
Golitsyn, G. S. 1960. Fluctuations of the magnetic field and current density in a turbulent flow of a weakly conducting fluid. *Sov. Phys. Dokl.*, **5**, 536–539.

Goto, S., Fujiwara, M. & Yamato, M. 2011. Turbulence sustained in a precessing sphere and spheroids. In: *TSFP Digital Library Online*. Begell House.

Goudard, L. & Dormy, E. 2008. Relations between the dynamo region geometry and the magnetic behavior of stars and planets. *Europhys. Lett.*, **83**, 59001.

Gough, D. O. & Weiss, N. O. 1976. The calibration of stellar convection theories. *Mon. Not. R. Astron. Soc.*, **176**, 589–607.

Gourgouliatos, K. N. & Cumming, A. 2014. Hall effect in neutron star crusts: evolution, endpoint and dependence on initial conditions. *Mon. Not. R. Astron. Soc.*, **438**, 1618–1629.

Govoni, F. & Feretti, L. 2004. Magnetic fields in clusters of galaxies. *Int. J. Mod. Phys. D*, **13**, 1549–1594.

Govoni, F., Murgia, M., Feretti, L., et al. 2005. A2255: The first detection of filamentary polarized emission in a radio halo. *Astron. Astrophys.*, **430**, L5–L8.

Greenspan, H. 1968. *The Theory of Rotating Fluids*. Cambridge University Press.

1974. On α-dynamos. *Stud. Appl. Math.*, **13**, 35–43.

Gubbins, D. 1973. Numerical solutions of the kinematic dynamo problem. *Phil. Trans. R. Soc. London, Ser. A*, **274**, 493–521.

2001. The Rayleigh number for convection in the Earth's core. *Phys. Earth Planet. Inter.*, **128**, 3–12.

Guervilly, C. & Cardin, P. 2016. Subcritical convection of liquid iron metals in a rotating sphere using a quasi-geostrophic model. *J. Fluid Mech.*, **808**, 61–89.

Guyodo, Y. & Valet, J.-P. 1999. Global changes in intensity of the Earth's magnetic field during the past 800kyr. *Nature*, **399**, 249–252.

Hadamard, J. 1902. Sur les problèmes aux dérivées partielles et leur signification physique. *Bull. Univ. Princeton*, **13**, 49–52.

Haines, M. G. 2011. A review of the dense Z-pinch. *Plasma Phys. Controlled Fusion*, **53**, 093001.

Hale, G. E. 1908. On the probable existence of a magnetic field in sunspots. *Astrophys. J.*, **28**, 315–343.

Hale, G. E., Ellerman, F., Nicholson, S. B. & Joy, A. H. 1919. The magnetic polarity of sun-spots. *Astrophys. J.*, **49**, 153.

Halley, E. 1692. An account of the causes of the change of the variation of the magnetical needle; with an hypothesis of the structure of the internal parts of the Earth. *Phil. Trans. R. Soc. London*, **195**, 208–221.

Hathaway, D. H. 2015. The solar cycle. *Living Rev. Sol. Phys.*, **12**, 4.

Hawley, J. F. & Balbus, S. A. 1991. A powerful local shear instability in weakly magnetized systems. II. Nonlinear evolution. *Astrophys. J.*, **376**, 223–233.

Hénon, M. 1966. Sur la topologie des lignes de courant dans un cas particulier. *C. R. hebd. Séanc. Acad. Sci., A*, **262**, 312.

Herzenberg, A. 1958. Geomagnetic dynamos. *Phil. Trans. R. Soc. London, Ser. A*, **250**, 543–585.

Herzenberg, A. & Lowes, F. J. 1957. Electromagnetic induction in rotating conductors. *Phil. Trans. R. Soc. London, Ser. A*, **249**, 507–584.

Heyvaerts, J., Priest, E. R. & Rust, D. M. 1977. An emerging flux model for the solar flare phenomenon. *Astrophys. J.*, **216**, 123–137.

Hide, R. 1970. On the Earth's core–mantle interface. *Q. J. R. Met. Soc.*, **96**, 579–590.

1974. Jupiter and Saturn. *Proc. R. Soc. London, Ser. A*, **336**, 63–84.

1976. A note on helicity. *Geophys. Fluid Dyn.*, **7**, 157–161.

2002. Helicity, superhelicity and weighted relative potential vorticity: useful diagnostic pseudoscalars? *Q. J. R. Met. Soc.*, **128**, 1759–1762.

Hide, R. & Malin, S. R. C. 1970. Novel correlations between global features of the Earth's gravitational and magnetic fields. *Nature*, **225**, 605–609.

Higgins, G. H. & Kennedy, G. C. 1971. The adiabatic gradient and the melting point gradient in the core of the Earth. *J. Geophys. Res.*, **76**, 1870–1878.

Hiltner, W. A. 1949. Polarization of light from distant stars by interstellar medium. *Science*, **109**, 165.

Hollerbach, R. 2000. A spectral solution of the magneto-convection equations in spherical geometry. *Int. J. Numer. Meth. Fluids*, **32**, 773–797.

Hollerbach, R., Galloway, D. J. & Proctor, M. R. E. 1995. Numerical evidence of fast dynamo action in a spherical shell. *Phys. Rev. Lett.*, **74**, 3145–3148.

Holme, R. 2009. Large-scale flow in the core. In: Olson, P. & Schubert, G. (eds.), *Core Dynamics*, pp. 107–130. Treatise on Geophysics, vol. 8. Elsevier.

Holme, R. & Whaler, K. A. 2001. Steady core flow in an azimuthally drifting reference frame. *Geophys. J. Int.*, **145**, 560–569.

Howard, R. & Harvey, J. 1970. Spectroscopic determinations of solar rotation. *Sol. Phys.*, **12**, 23–25.

Howard, R. & Labonte, B. J. 1980. The sun is observed to be a torsional oscillator with a period of 11 years. *Astrophys. J. Lett.*, **239**, L33–L36.

Howe, R. 2009. Solar interior rotation and its variation. *Living Rev. Sol. Phys.*, **6**, 1.

Howe, R., Hill, F., Komm, R., et al. 2011. The torsional oscillation and the new solar cycle. Paper 012074 in: *J. Phys. Conf. Series*, vol. 271.

Howe, R., Christensen-Dalsgaard, J., Hill, F., et al. 2013. The high-latitude branch of the solar torsional oscillation in the rising phase of cycle 24. *Astrophys. J. Lett.*, **767**, L20.

Hoyng, P., Schmitt, D. & Ossendrijver, M. A. J. H. 2002. A theoretical analysis of the observed variability of the geomagnetic dipole field. *Phys. Earth Planet. Inter.*, **130**, 143–157.

Hoyt, D. V. & Schatten, K. H. 1998. Group sunspot numbers: A new solar activity reconstruction. *Sol. Phys.*, **181**, 491–512.

Hughes, D. W., Rosner, R. & Weiss, N. O. (eds.). 2007. *The Solar Tachocline*. Cambridge University Press.

Ibbetson, A. & Tritton, D. J. 1975. Experiments on turbulence in a rotating fluid. *J. Fluid Mech.*, **68**, 639–672.

Ibrahim, A. I., Swank, J. H. & Parke, W. 2003. New evidence of proton cyclotron-resonance in a magnetic strength field SGR 1806-20. *Astrophys. J. Lett.*, **584**, L17–L21.

Iroshnikov, P. S. 1964. Turbulence of a conducting fluid in a strong magnetic field. *Soviet Astron.*, **7**, 742–750.

Ishihara, T., Kaneda, Y., Yokokawa, M., Itakura, K. & Uno, A. 2007. Small-scale statistics in high-resolution direct numerical simulation of turbulence: Reynolds number dependence of one-point velocity gradient statistics. *J. Fluid Mech.*, **592**, 335–366.

Jackson, A., Jonkers, A. R. T. & Walker, M. R. 2000. Four centuries of geomagnetic secular variation from historical records. *Phil. Trans. R. Soc. London, Ser. A*, **358**, 957–990.

Jacobs, J. A. 1975. *The Earth's Core*. Academic.

1976. Reversals of the Earth's magnetic field. *Phys. Rep.*, **26**, 183–225.

1994. *Reversals of the Earth's Magnetic Field*. 2nd ed. Cambridge University Press.

Jacquin, L., Leuchter, O., Cambon, C. & Mathieu, J. 1990. Homogeneous turbulence in the presence of rotation. *J. Fluid Mech.*, **220**, 1–52.

Jault, D. 1995. Model Z by computation and Taylor's condition. *Geophys. Astrophys. Fluid Dyn.*, **79**, 99–124.

2003. Electromagnetic and topographic coupling, and LOD variations. In: Zhang, K., Soward, A. M. & Jones, C. A. (eds.), *Earth's Core and Lower Mantle*, pp. 56–76. Fluid Mech. of Astrophys. and Geophys. Taylor and Francis

Jault, D. & Le Mouël, J. L. 1989. The topographic torque associated with a tangentially geostrophic motion at the core surface and inferences on the flow inside the core. *Geophys. Astrophys. Fluid Dyn.*, **48**, 273–295.

Jault, D., Gire, C. & Le Mouël, J. L. 1988. Westward drift, core motions and exchanges of angular momentum between core and mantle. *Nature*, **333**, 353–356.

Jeffreys, H. 1926. The viscosity of the Earth (fourth paper). *Mon. Not. R. Astron. Soc. Geophys. Suppl.*, **1**, 412–424.

Jepps, S. A. 1975. Numerical models of hydromagnetic dynamos. *J. Fluid Mech.*, **67**, 625–646.

Ji, H., Goodman, J. & Kageyama, A. 2001. Magnetorotational instability in a rotating liquid metal annulus. *Mon. Not. R. Astron. Soc.*, **325**, L1–L5.

Johnson, C. L., Phillips, R. J., Purucker, M. E., et al. 2015. Low-altitude magnetic field measurements by MESSENGER reveal Mercury's ancient crustal field. *Science*, **348**, 892–895.

Jones, C. A. 2011. Planetary magnetic fields and fluid dynamos. *Ann. Rev. Fluid Mech.*, **43**, 583–614.

Jones, C. A., Soward, A. M. & Mussa, A. I. 2000. The onset of thermal convection in a rapidly rotating sphere. *J. Fluid Mech.*, **405**, 157–179.

Jones, C. A., Mussa, A. I. & Worland, S. J. 2003. Magnetoconvection in a rapidly rotating sphere: the weak–field case. *Proc. R. Soc. London, Ser. A*, **459**, 773–797.

Jones, S. E. & Gilbert, A. D. 2014. Dynamo action in the ABC flows using symmetries. *Geophys. Astrophys. Fluid Dyn.*, **108**, 83–116.

Julien, K., Rubio, A. M., Grooms, I. & Knobloch, E. 2012. Statistical and physical balances in low Rossby number Rayleigh–Bénard convection. *Geophys. Astrophys. Fluid Dyn.*, **106**, 392–428.

Kadanoff, L. P. 2000. *Statistical Physics: Statics, Dynamics and Renormalization*. World Scientific.

Kaplan, E. J., Schaeffer, N., Vidal, J. & Cardin, P. 2017. Subcritical thermal convection of liquid metals in a rapidly rotating sphere. *Phys. Rev. Lett.*, **119**, 094501.

Käpylä, P. J., Mantere, M. J., Cole, E., Warnecke, J. & Brandenburg, A. 2013. Effects of enhanced stratification on equatorward dynamo wave propagation. *Astrophys. J.*, **778**, 41.

Karman, Th. v. 1921. Über laminare und turbulente Reibung. *Z. Angew. Math. Meck.*, **1**, 244.

Khurana, K. K., Kivelson, M. G., Stevenson, D. J., et al. 1998. Induced magnetic fields as evidence for subsurface oceans in Europa and Callisto. *Nature*, **395**, 777–780.

Kida, S. 2011. Steady flow in a rapidly rotating sphere with weak precession. *J. Fluid Mech.*, **680**, 150–193.

2018. Steady flow in a rotating sphere with strong precession. *Fluid Dyn. Res.*, **50**, 021401.

Kimura, Y. & Moffatt, H. K. 2014. Reconnection of skewed vortices. *J. Fluid Mech.*, **751**, 329–345.

King, E. M. & Buffett, B. A. 2013. Flow speeds and length scales in geodynamo models: the role of viscosity. *Earth Planet. Sci. Lett.*, **371**, 156–162.

Kivelson, M. G., Khurana, K. K., Russell, C. T., et al. 1996. Discovery of Ganymede's magnetic field by the Galileo spacecraft. *Nature*, **384**, 537–541.

Kivelson, M. G., Khurana, K. K., Russell, C. T. & Walker, R. J. 2001. Magnetic signature of a polar pass over Io. *AGU Fall Meeting Abstracts*.

Klapper, I. & Young, L. S. 1995. Rigorous bounds on the fast dynamo growth rate involving topological entropy. *Commun. Math. Phys.*, **173**, 623.

Knobloch, E. & Landsberg, A. S. 1996. A new model of the solar cycle. *Mon. Not. R. Astron. Soc.*, **278**, 294–302.

Knobloch, E., Tobias, S. M. & Weiss, N. O. 1998. Modulation and symmetry changes in stellar dynamos. *Mon. Not. R. Astron. Soc.*, **297**, 1123–1138.

Kolmogorov, A. N. 1941a. The local structure of turbulence in incompressible viscous fluid for very large Reynolds numbers. *Dokl. AN SSSR*, **30**, 299–303.

1941b. Dissipation of energy in the locally isotropic turbulence. *Dokl. Akad. Nauk. SSR*, **32**, 16–18.

1962. A refinement of previous hypotheses concerning the local structure of turbulence in a viscous incompressible fluid at high Reynolds number. *J. Fluid Mech.*, **13**, 82–85.

Kovasznay, L. S. G. 1960. Plasma turbulence. *Rev. Mod. Phys.*, **32**, 815–822.

Kraichnan, R. H. 1965. Inertial-range spectrum of hydromagnetic turbulence. *Phys. Fluids*, **8**, 1385–1387.

1973. Helical turbulence and absolute equilibrium. *J. Fluid Mech.*, **59**, 745–752.

1976a. Diffusion of passive-scalar and magnetic fields by helical turbulence. *J. Fluid Mech.*, **77**, 753–768.

1976b. Diffusion of weak magnetic fields by isotropic turbulence. *J. Fluid Mech.*, **75**, 657–676.

Kraichnan, R. H. & Nagarajan, S. 1967. Growth of turbulent magnetic fields. *Phys. Fluids*, **10**, 859–870.

Krause, F. & Rädler, K-H. 1980. *Mean-Field Magnetohydrodynamics and Dynamo Theory*. Pergamon. [Republished by Elsevier 2016.]

Krause, F. & Steenbeck, M. 1967. Some simple models of magnetic field regeneration by non-mirror-symmetric turbulence. *Z. Naturforsch.*, **22**a, 671–675. See Roberts & Stix (1971), pp. 81–95.

Krause, M., Hummel, E. & Beck, R. 1989. The magnetic field structures in two nearby spiral galaxies. I – The axisymmetric spiral magnetic field in IC342. II – The bisymmetric spiral magnetic field in M81. *Astron. Astrophys.*, **217**, 4–30.

Krstulovic, G., Thorner, G., Vest, J.-P., et al. 2011. Axial dipolar dynamo action in the Taylor–Green vortex. *Phys. Rev. E*, **84**, 066318.

Kumar, S. & Roberts, P. H. 1975. A three-dimensional kinematic dynamo. *Proc. R. Soc. London, Ser. A*, **344**, 235–258.

Kutzner, C. & Christensen, U. R. 2002. From stable dipolar towards reversing numerical dynamos. *Phys. Earth Planet. Inter.*, **131**, 29–45.

Landahl, M. T. 1980. A note on an algebraic instability of inviscid parallel shear flows. *J. Fluid Mech.*, **98**(2), 243–251.

Landstreet, J. D. & Angel, J. R. P. 1974. The polarisation spectrum and magnetic field strength of the white dwarf Grw+ 70°8247. *Astrophys. J.*, **196**, 819–825.

Larmor, J. 1919. How could a rotating body such as the Sun become a magnet? In: *Rep. Brit. Assoc. Adv. Sci. 1919*, pp. 159–160. BAAS.

Lehmann, I. 1936. P'. *Publ. Bur. Centr. Séism. Internat.*, **A14**, 3–31.

Lehnert, B. 1954. Magnetohydrodynamic waves under the action of the Coriolis force. *Astrophys. J.*, **119**, 647–654.

Leighton, R. B. 1969. A magneto-kinematic model of the solar cycle. *Astrophys. J.*, **156**, 1.

Léorat, J. 1975. *La turbulence magnétohydrodynamique hélicitaire et la géneration des champs magnétohydrodynamiques a grande échelle*. PhD thesis, Univ. de Paris VII.

Levy, E. H. 1972. Effectiveness of cyclonic convection for producing the geomagnetic field. *Astrophys. J.*, **171**, 621–633.

Li, K., Jackson, A. & Livermore, P. W. 2018. Taylor state dynamos found by optimal control: axisymmetric examples. *J. Fluid Mech.*, **853**, 647–697.

Li, K., Jackason, A. & Livermore. P. W. 2018 Taylor state dynamos foundly optimal contol: antisymmetric examples. *J. Fleid Mech.* **853**, 647–697.

Liepmann, H. 1952. Aspects of the turbulence problem. *Z. Ang. Math. Phys.*, **3**, 321–342.

Lighthill, M. J. 1959. *Introduction to Fourier Analysis and Generalised Functions*. Cambridge University Press.

Lilley, F. E. M. 1970. On kinematic dynamos. *Proc. R. Soc. London, Ser. A*, **316**, 153–167.

Lin, Y., Marti, P. & Noir, J. 2015. Shear-driven parametric instability in a precessing sphere. *Phys. Fluids*, **27**, 046601.

Lin, Y., Marti, P., Noir, J. & Jackson, A. 2016. Precession-driven dynamos in a full sphere and the role of large scale cyclonic vortices. *Phys. Fluids*, **28**, 066601.

Linardatos, D. 1993. Determination of two-dimensional magnetostatic equilibria and analogous Euler flows. *J. Fluid Mech.*, **246**, 569–591.

Lister, J. R. & Buffett, B. A. 1998. Stratification of the outer core at the core–mantle boundary. *Phys. Earth Planet. Inter.*, **105**, 5–19.

Loper, D. E. 1975. Torque balance and energy budget for the precessionally driven dynamo. *Phys. Earth Planet. Inter.*, **11**, 43–60.

Lorenz, E. N. 1963. Deterministic nonperiodic flow. *J. Atmos. Sci.*, **20**, 130–141.

Lorenzani, S. & Tilgner, A. 2001. Fluid instabilities in precessing spheroidal cavities. *J. Fluid Mech.*, **447**, 111–128.

Lortz, D. 1968a. Exact solutions of the hydromagnetic dynamo problem. *Plasma Phys.*, **10**, 967.

1968b. Impossibility of steady dynamos with certain symmetries. *Phys. Fluids*, **11**, 913–915.

Lowes, F. J. 1974. Spatial power spectrum of the main geomagnetic field, and extrapolation to the core. *Geophys. J. R. Astron. Soc.*, **36**, 717–718.

2011. Why the first laboratory self-exciting dynamo was at Newcastle. *Geophys. Astrophys. Fluid Dyn.*, **105**, 213–233.

Lowes, F. J. & Wilkinson, I. 1963. Geomagnetic dynamo: A laboratory model. *Nature*, **198**, 1158–1160.

1968. Geomagnetic dynamo: An improved laboratory model. *Nature*, **219**, 717–718.

Lynden-Bell, D. & Boily, C. 1994. Self-similar solutions up to flashpoint in highly wound magnetostatis. *Mon. Not. R. Astron. Soc.*, **341**, 146–152.

Lynden-Bell, D. & Moffatt, H. K. 2015. Flashpoint. *Mon. Not. R. Astron. Soc.*, **452**, 902–909.

MacGregor, K. B. & Charbonneau, P. 1997. Solar interface dynamos I. Linear kinematic models in Cartesian geometry. *Astrophys. J.*, **486**, 484–501.

Malkus, W. V. R. 1963. Precessional torques as the cause of geomagnetism. *J. Geophys. Res.*, **68**, 2871–2886.

1964. Boussinesq equations and convection energetics. In: *Geophys. Fluid Dyn. Proc.* Woods Hole Oceanographic Institution.

1968. Precession of the Earth as the cause of geomagnetism,. *Science*, **160**, 259–264.

Malkus, W. V. R. & Proctor, M. R. E. 1975. The macrodynamics of α-effect dynamos in rotating fluids. *J. Fluid Mech.*, **67**, 417–444.

Malkus, W. V. R. & Veronis, G. 1958. Finite amplitude convection. *J. Fluid Mech.*, **4**, 225–260.

Mandea, M. & Dormy, E. 2003. Asymmetric behavior of magnetic dip poles. *Earth Planets Space*, **55**, 153–157.

Mao, S. A., Carilli, C., Gaensler, B. M., et al. 2017. Detection of microgauss coherent magnetic fields in a galaxy five billion years ago. *Nature Astron.*, **1**, 621–626.

Margot, J. L., Peale, S. J., Jurgens, R. F., Slade, M. A. & Holin, I. V. 2007. Large longitude libration of Mercury reveals a molten core. *Science*, **316**, 710–714.

Mason, J., Hughes, D. W. & Tobias, S. M. 2002. The competition in the solar dynamo between surface and deep-seated α-effects. *Astrophys. J. Lett.*, **580**, L89–L92.

Maunder, E. W. 1904. Note on the distribution of sunspots in heliographic latitude, 1874–1902. *Mon. Not. R. Astron. Soc.*, **64**, 747–761.

1913. Distribution of sunspots in heliographic latitude, 1874–1913. *Mon. Not. R. Astron. Soc.*, **74**, 112–116.

Maxwell, J. C. 1867. On the dynamical theory of gases. *Phil. Trans. R. Soc. London*, **157**, 49–88.

1873. *A Treatise on Electricity and Magnetism*. Oxford University Press.

McCray, R. & Snow, T. P., Jr. 1979. The violent interstellar medium. *Ann. Rev. Astron. Astrophys.*, **17**, 213–240.

Mestel, L. 1967. Stellar magnetism. In: Sturrock, P. (ed.), *Plasma Astrophysics*, pp. 185–228. Academic.

Miesch, M. S. 2012. The solar dynamo. *Phil. Trans. R. Soc. London, Ser. A*, **370**, 3049–3069.

Miesch, M. S. & Toomre, J. 2009. Turbulence, magnetism, and shear in stellar interiors. *Annu. Rev. Fluid Mech.*, **41**, 317–345.

Mizerski, K. A. & Moffatt, H. K. 2018. Dynamo generation of a magnetic field by decaying Lehnert waves in a highly conducting plasma. *Geophys. Astrophys. Fluid Dyn.*, **112**, 165–174.

Moffatt, H. K. 1963. Magnetic eddies in an incompressible viscous fluid of high electrical conductivity. *J. Fluid Mech.*, **17**, 225–239.

1969. The degree of knottedness of tangled vortex lines. *J. Fluid Mech.*, **35**, 117–129.

1970. Turbulent dynamo action at low magnetic Reynolds number. *J. Fluid Mech.*, **41**, 435–452.

1972. An approach to a dynamic theory of dynamo action in a rotating conducting fluid. *J. Fluid Mech.*, **53**, 385–399.

1974. The mean electromotive force generated by turbulence in the limit of perfect conductivity. *J. Fluid Mech.*, **65**, 1–10.

1978a. *Magnetic Field Generation in Electrically Conducting Fluids*. Cambridge University Press.

1978b. Topographic coupling at the core–mantle interface. *Geophys. Astrophys. Fluid Dyn.*, **9**, 279–288.

1979. A self-consistent treatment of simple dynamo systems. *Geophys. Astrophys. Fluid Dyn.*, **14**, 147–166.

1983. Transport effects associated with turbulence with particular attention to the influence of helicity. *Rep. Prog. Phys.*, **46**, 621–664.

1986. Magnetostatic equilibria and analogous Euler flows of arbitrarily complex topology. Part 2. Stability considerations. *J. Fluid Mech.*, **166**, 359–378.

1990. The energy spectrum of knots and links. *Nature*, **347**, 367–369.

1992. The Earth's magnetism – past achievements and future challenges. *IUGG Chronicle*, **22**, 1–20. (Union Lecture XXth Gen. Assembly, IUGG, Vienna 1991, published by AGU 2013.)

2008. Magnetostrophic turbulence and the geodynamo. In: Kaneda, Y. (ed.), *Computational Physics and New Perspectives in Turbulence*, pp. 339–346. Springer.

2010. Note on the suppression of transient shear-flow instability by a spanwise magnetic field. *J. Eng. Math.*, **68**, 263–268.

2014. The fluid dynamics of James Clerk Maxwell. In: Flood, R., McCartney, M. & Whitaker, A. (eds.), *James Clerk Maxwell: Perspectives on His Life and Work*, pp. 223–230. Oxford University Press.

2017. Corrigendum – the degree of knottedness of tangled vortex lines. *J. Fluid Mech.*, **830**, 821–822.

Moffatt, H. K. & Dillon, R. F. 1976. Correlation between gravitational and geomagnetic-fields caused by interaction of core fluid motion with a bumpy core–mantle interface. *Phys. Earth Planet. Int.*, **13**, 67–78.

Moffatt, H. K. & Kamkar, H. 1983. On the time-scale associated with flux expulsion. In: Soward, A. M. (ed.), *Stellar and Planetary Magnetism*, pp. 91–98. Gordon and Breach.

Moffatt, H. K. & Loper, D.E. 1994. The magnetostrophic rise of a buoyant parcel in the Earth's core. *Geophys. J. Int.*, **117**, 394–402.

Moffatt, H. K. & Mizerski, K. A. 2018. Pinch dynamics in a low-β plasma. *Fluid Dyn. Res.*, **50**, 011401.

Moffatt, H. K. & Proctor, M. R. E. 1985. Topological constraints associated with fast dynamo action. *J. Fluid Mech.*, **154**, 493–507.

Moffatt, H. K. & Ricca, R. L. 1992. Helicity and the Călugăreanu invariant. *Proc. R. Soc. London, Ser. A*, **439**, 411–429.

Moffatt, H. K. & Tokieda, T. 2008. Celt reversals: A prototype of chiral dynamics. *Proc. R. Soc. Edinb., Ser. A*, **138**, 361–368.

Monchaux, R., Berhanu, M., Bourgoin, M., et al. 2007. Generation of a magnetic field by dynamo action in a turbulent flow of liquid sodium. *Phys. Rev. Lett.*, **98**, 044502.

Monchaux, R., Berhanu, M., Aumaître, S., et al. 2009. The von Kármán sodium experiment: turbulent dynamical dynamos. *Phys. Fluids*, **21**, 035108.

Morin, J., Donati, J.-F., Petit, P., et al. 2010. Large-scale magnetic topologies of late M dwarfs. *Mon. Not. R. Astron. Soc.*, **407**, 2269–2286.

Morin, V. 2005. *Instabilités et Bifurcations associées à la Modélisation de la Géodynamo*. PhD thesis, Thèse Paris VII.

Morin, V. & Dormy, E. 2009. The dynamo bifurcation in rotating spherical shells. *Int. J. Mod. Phys. B*, **23**, 5467–5482.

Moss, D. 2006. Numerical simulation of the Gailitis dynamo. *Geophys. Astrophys. Fluid Dyn.*, **100**, 49–58.

Moss, D. & Sokoloff, D. 2011. The saturation of galactic dynamos. *Astron. Nach.*, **332**, 88–91.

Mound, J. E. & Buffet, B. A. 2005. Mechanisms of core–mantle angular momentum exchange and the observed spectral properties of torsional oscillations. *J. Geophys. Res.*, **110**, B08103.

2006. Detection of a gravitational oscillation in length-of-day. *Earth Planet. Sci. Lett.*, **243**, 383–389.

Müller, U., Stieglitz, R. & Horanyi, S. 2004. A two-scale hydromagnetic dynamo experiment. *J. Fluid Mech.*, **498**, 31–71.

Müller, U., Stieglitz, R., Busse, F. H. & Tilgner, A. 2008. The Karlsruhe two-scale dynamo experiment. *C. R. Phys.*, **9**, 729–740.

Munk, W. H. & MacDonald, G. J. F. 1960. *The Rotation of the Earth*. Cambridge University Press.

Needham, J. 1962. *Science and Civilisation in China, Vol. 4, Physics and Physical Technology*. Cambridge University Press.

Nellis, W. J. 2015. The unusual magnetic fields of Uranus and Neptune. *Modern Physics Letters B*, **29**, 1430018.

Nelson, N. J., Brown, B. P., Brun, A. S., Miesch, M. S. & Toomre, J. 2011. Buoyant magnetic loops in a global dynamo simulation of a young sun. *Astrophys. J. Lett.*, **739**, L38.

Ness, N. F. 2010. Space exploration of planetary magnetism. *Space Sci. Rev.*, **152**, 5–22.

Ness, N. F., Behannon, K. W., Lepping, R. P. & Whang, Y. C. 1975. The magnetic field of Mercury, I. *J. Geophys. Res.*, **80**, 2708–2716.

Ness, N. F., Acuna, M. H., Behannon, K. W., et al. 1986. Magnetic fields at Uranus. *Science*, **233**, 85–89.

Newitt, L. R. & Barton, C. E. 1996. The position of the North magnetic dip pole in 1994. *J. Geomagn. Geoelec.*, **48**, 221–232.

Newitt, L. R., Mandea, M., McKee, L. A. & Orgeval, J.-J. 2002. Recent acceleration of the North magnetic pole linked to magnetic jerks. *Eos Trans. AGU*, **83**, 381.

Newitt, L. R., Chulliat, A. & Orgeval, J.-J. 2009. Location of the North magnetic pole in April 2007. *Earth Planets Space*, **61**, 703–710.

Nimmo, F. 2002. Why does Venus lack a magnetic field? *Geology*, **30**, 987–990.

Nimmo, F. & Stevenson, D. J. 2000. Influence of early plate tectonics on the thermal evolution and magnetic field of Mars. *J. Geophys. Res.*, **105**, 11969–11979.

Noir, J., Brito, D., Aldridge, K. & Cardin, P. 2001a. Experimental evidence of inertial waves in a precessing spheroidal cavity. *Geophys. Res. Lett.*, **28**, 3785–3788.

Noir, J., Jault, D. & Cardin, P. 2001b. Numerical study of the motions within a slowly precessing sphere at low Ekman number. *J. Fluid Mech.*, **437**, 283–299.

Nore, C., Brachet, M. E., Politano, H. & Pouquet, A. 1997. Dynamo action in the Taylor–Green vortex near threshold. *Phys. Plasmas*, **4**, 1–3.

2001. Dynamo action in a forced Taylor–Green vortex. In: *Dynamo and Dynamics, a Mathematical Challenge*, pp. 51–58. Springer.

Nozières, P. 1978. Reversals of the Earth's magnetic field: An attempt at a relaxation model. *Phys. Earth Planet Inter.*, **17**, 55–74.

2008. Simple models and time scales in the dynamo effect. *C. R. Phys.*, **9**, 683–688.

Oersted, H. C. 1820. Experiments on the effect of a current of electricity on the magnetic needles. *Ann. Phil.*, **16**, 273.

Ogilvie, G. I. 2007. Instabilities, angular momentum transport and magnetohydrodynamic turbulence. In: Rosner, R., Hughes, D. & Weiss, N. O. (eds.), *The Solar Tachocline*, pp. 299–315. Cambridge University Press.

Oppermann, N., Junklewitz, H., Robbers, G., et al. 2012. An improved map of the galactic Faraday sky. *Astron. Astrophys.*, **542**, A93.

Orszag, S. A. 1970. Analytical theories of turbulence. *J. Fluid Mech.*, **41**, 363–386.

1977. Lectures on the statistical theory of turbulence. In: Balian, R. & Peube, J. L. (eds.), *Fluid Dynamics, Les Houches 1973*, pp. 235–374. Gordon and Breach.

Ortolani, S. 1989. Reversed field pinch confinement physics. *Plasma Phys. Controlled Fusion*, **31**, 1665.

Oruba, L. 2016. On the role of thermal boundary conditions in dynamo scaling laws. *Geophys. Astrophys. Fluid Dyn.*, **110**, 529–545.

Oruba, L. & Dormy, E. 2014a. Predictive scaling laws for spherical rotating dynamos. *Geophys. J. Int.*, **198**, 828–847.

2014b. Transition between viscous dipolar and inertial multipolar dynamos. *Geophys. Res. Lett.*, **41**, 7115–7120.

Ossendrijver, M. 2003. The solar dynamo. *Astron. Astrophys. Rev.*, **11**, 287–367.

Otani, N. F. 1993. A fast kinematic dynamo in two-dimensional time-dependent flows. *J. Fluid Mech.*, **253**, 327–340.

Parker, E. N. 1955a. The formation of sunspots from the solar toroidal field. *Astrophys. J.*, **121**, 491–507.

1955b. Hydromagnetic dynamo models. *Astrophys. J.*, **122**, 293–314.

1963. Kinematical hydromagnetic theory and its applications to the low solar photosphere. *Astrophys. J.*, **138**, 552–575.

1970. The generation of magnetic fields in astrophysical bodies. I. The dynamo equations. *Astrophys. J.*, **162**, 665–673.

1971a. The generation of magnetic fields in astrophysical bodies, II. The galactic field. *Astrophys. J.*, **163**, 255–278.

1971b. The generation of magnetic fields in astrophysical bodies, III. Turbulent diffusion of fields and efficient dynamos. *Astrophys. J.*, **163**, 279–285.

1971c. The generation of magnetic fields in astrophysical bodies, VI. Periodic modes of the galactic field. *Astrophys. J.*, **166**, 295–300.

1971d. The generation of magnetic fields in astrophysical bodies, VIII. Dynamical considerations. *Astrophys. J.*, **165**, 239–249.

1993. A solar dynamo surface wave at the interface between convection and nonuniform rotation. *Astrophys. J.*, **408**, 707–719.

1994. *Spontaneous Current Sheets in Magnetic Fields*. Oxford University Press.

Parker, R. L. 1966. Reconnexion of lines of force in rotating spheres and cylinders. *Proc. R. Soc. London, Ser. A*, **291**, 60–72.

Pavlov, V. & Gallet, Y. 2005. A third superchron during the early Paleozoic. *Episodes*, **28**, 78–84.

Pekeris, C. L. 1972. Stationary spherical vortices in a perfect fluid. *Proc. Natl. Acad. Sci.*, **69**, 2460–2462. (Corrigendum, p. 3849.)

Pekeris, C. L., Accad, Y. & Shkoller, B. 1975. Kinematic dynamos and the Earth's magnetic field. *Phil. Trans. R. Soc. London, Ser. A*, **275**, 425–461.

Petersons, H. F. & Constable, S. 1996. Global mapping of the electrically conductive lower mantle. *Geophys. Res. Lett.*, **23**, 1461–1464.

Petitdemange, L., Dormy, E. & Balbus, S. A. 2013. Axisymmetric and non-axisymmetric magnetostrophic MRI modes. *Phys. Earth Planet. Inter.*, **223**, 21–31.

Pétrélis, F. & Fauve, S. 2001. Saturation of the magnetic field above the dynamo threshold. *Eur. Phys. J. B*, **22**, 273–276.

Pétrélis, F., Fauve, S., Dormy, E. & Valet, J.-P. 2009. Simple mechanism for reversals of Earth's magnetic field. *Phys. Rev. Lett.*, **102**, 144503.

Pétrélis, F., Besse, J. & Valet, J.-P. 2011. Plate tectonics may control geomagnetic reversal frequency. *Geophys. Res. Lett.*, **38**, L19303.

Pichakhchi, L. D. 1966. Theory of the hydromagnetic dynamo. *Sov. Phys. JETP*, **23**, 542–543.

Piddington, J. H. 1972a. The origin and form of the galactic magnetic field, I. Parker's dynamo model. *Cosmic Electrodyn.*, **3**, 60–70.

1972b. The origin and form of the galactic magnetic field, II. The primordial-field model. *Cosmic Electrodyn.*, **3**, 129–146.

Poincaré, H. 1910. Sur la précession des corps déformables. *Bull. Astron., Ser. I*, **27**, 321–356.

Ponomarenko, Yu. B. 1973. Theory of the hydromagnetic generator. *J. Appl. Mech. Tech. Phys.*, **14**, 775–778.

Pouquet, A. & Patterson, G. S. 1978. Numerical simulation of helical magnetohydrodynamic turbulence. *J. Fluid Mech.*, **85**, 305–323.

Pouquet, A., Frisch, U., & Léorat, J. 1976. Strong MHD turbulence and the nonlinear dynamo effect. *J. Fluid Mech.*, **77**, 321–354.

Pozzo, M., Davies, C., Gubbins, D. & Alfè, D. 2012. Thermal and electrical conductivity of iron at Earth's core conditions. *Nature*, **485**, 355–358.

2014. Thermal and electrical conductivity of solid iron and iron-silicon mixtures at Earth's core conditions. *Earth Planet. Sci. Lett.*, **393**, 159–164.

Preston, G. W. 1967. Studies of stellar magnetism – past, present and future. In: Cameron, R. C. (ed.), *Proc. AAS-NASA Symp. The Magnetic and Related Stars*, pp. 3–28. Mono.

Priest, E. R. 1982. *Solar Magnetohydrodynamics*. Kluwer.

Priest, E. R. & Forbes, T. 2007. *Magnetic Reconnection: MHD Theory and Applications*. Cambridge University Press.

Proctor, M. R. E. 1975. *Non-linear mean field dynamo models and related topics*. PhD thesis, University of Cambridge.

1977a. Numerical solutions of the nonlinear α-effect dynamo equations. *J. Fluid Mech.*, **80**, 769–784.

1977b. The role of mean circulation in parity selection by planetary magnetic fields. *Geoph. Astrophys. Fluid Dyn.*, **8**, 311–324.

1994. Convection and magnetoconvection in a rapidly rotating sphere In: *Lectures on Solar and Planetary Dynamos*, pp. 97–114. Cambridge University Press.

Rädler, K.-H. 1969a. A new turbulent dynamo, I. *Monats. Dt. Akad. Wiss.*, **11**, 272–279. (English translation: Roberts & Stix (1971) pp. 301–308.)

1969b. On the electrodynamics of turbulent fluids under the influence of Coriolis forces. *Monats. Dt. Akad. Wiss.*, **11**, 194–201. (English translation: Roberts & Stix (1971) pp. 291–299.)

Rädler, K.-H. & Brandenburg, A. 2003. Contributions to the theory of a two-scale homogeneous dynamo experiment. *Phys. Rev. E*, **67**, 026401.

Rasskazov, A., Chertovskih, R. & Zheligovsky, V. 2018. Magnetic field generation by pointwise zero-helicity three-dimensional steady flow of an incompressible electrically conducting fluid. *Phys. Rev. E*, **97**, 043201.

Rayleigh, Lord. 1917. On the dynamics of revolving fluids. *Proc. R. Soc. London, Ser. A*, **93**, 148–154.

Raynaud, R., Petitdemange, L. & Dormy, E. 2014. Influence of the mass distribution on the magnetic field topology. *Astron. Astrophys.*, **567**, A107.

Raynaud, R., Petitdemange, L. & Dormy, E. 2015. Dipolar dynamos in stratified systems. *Mon. Not. R. Soc.*, **448**, 2055–2065.

Raynaud, R. & Tobias, S. M. 2016. Convective dynamo action in a spherical shell: symmetries and modulation. *J. Fluid Mech.*, **799**, R6.

Reiners, A. & Basri, G. 2006. Measuring magnetic fields in ultracool stars and brown dwarfs. *Astrophys. J.*, **644**, 497–509.

Reisenegger, A. 2003. Origin and evolution of neutron star magnetic fields. *arXiv:astro-ph/0307133v1*.

Rempel, M. 2006. Flux-transport dynamos with Lorentz force feedback on differential rotation and meridional flow: Saturation mechanism and torsional oscillations. *Astrophys. J.*, **647**, 662.

Rhines, P. B. & Young, W. R. 1982. Homogenization of potential vorticity in planetary gyres. *J. Fluid Mech.*, **122**, 347–367.

Rikitake, T. 1958. Oscillations of a system of disk dynamos. *Proc. Camb. Phil. Soc.*, **54**, 89–105.

Rincon, F., Ogilvie, G. I. & Proctor, M. R. E. 2007. Self-sustaining nonlinear dynamo process in Keplerian shear flows. *Phys. Rev. Lett.*, **98**, 254502.

Riols, A., Rincon, F., Cossu, C., et al. 2013. Global bifurcations to subcritical magnetorotational dynamo action in Keplerian shear flow. *J. Fluid Mech.*, **731**, 1–45.

Robbins, K. A. 1976. A moment equation description of magnetic reversals in the Earth. *Proc. Natl. Acad. Sci. U.S.A.*, **73**, 4297–4301.

Roberts, G. O. 1970. Spatially periodic dynamos. *Phil. Trans. R. Soc. London, Ser. A*, **266**, 535–558.

1972. Dynamo action of fluid motions with two-dimensional periodicity. *Phil. Trans. R. Soc. London, Ser. A*, **271**, 411–454.

Roberts, P. H. 1968. On the thermal instability of a rotating-fluid sphere containing heat sources. *Phil. Trans. R. Soc. London, Ser. A*, **263**, 93–117.

1971. Dynamo theory. In: Reid, W. H. (ed.), *Mathematical Problems in the Geophysical Sciences*, pp. 129–206. AMS.

1972a. Electromagnetic core–mantle coupling. *J. Geomagn. Geoelectr.*, **24**, 231–259.

1972b. Kinematic dynamo models. *Phil. Trans. R. Soc. London, Ser. A*, **272**, 663–698.

1979. Pure dynamo theory and geomagnetic dynamo theory. In: Cupal, I. (ed.), *Proc. First Int. Workshop on Dynamo Theory and the Generation of the Earth's Magnetic Field*, pp. 7–12. Czech. Geophys. Inst. Rep.

Roberts, P. H. & Soward, A. M. 1972. Magnetohydrodynamics of the Earth's core. *Ann. Rev. Fluid Mech.*, **4**, 117–154.

2009. The Navier–Stokes-α equations revisited. *Geophys. Astrophys. Fluid Dyn.*, **103**, 303–316.

Roberts, P. H. & Stewartson, K. 1974. On finite amplitude convection in a rotating magnetic system. *Phil. Trans. R. Soc. London, Ser. A*, **277**, 287–315.

1975. On double roll convection in a rotating magnetic system. *J. Fluid Mech.*, **68**, 447–466.

Roberts, P. H. & Stix, M. 1971. The turbulent dynamo: A translation of a series of paper by F. Krause, K.-H. Rädler and M. Steenbeck. In: *Mech. Note 60*. NCAR.

1972. α-effect dynamos, by the Bullard-Gellman formalism. *Astron. Astrophys.*, **18**, 453–466.

Roberts, P. H. & Wu, C.-C. 2014. On the modified Taylor constraint. *Geophys. Astrophys. Fluid Dyn.*, **108**, 696–715.

Rochester, M. G. 1962. Geomagnetic core–mantle coupling. *J. Geophys. Res.*, **67**, 4833–4836.

Ross, J. C. 1834. On the position of the North magnetic pole. *Phil. Trans. R. Soc. London, Series I*, **124**, 47–52.

Rouchon, P. 1991. On the Arnol'd stability criterion for steady-state flows of an ideal fluid. *Europ. J. Mech., B/Fluids*, **10**, 651–661.

Rüdiger, G. & Brandenburg, A. 1995. A solar dynamo in the overshoot layer: cycle period and butterfly diagram. *Astron. Astrophys.*, **296**, 557–566.

Rüdiger, G. & Kichatinov, L. L. 1993. Alpha-effect and alpha-quenching. *Astron. Astrophys.*, **269**, 581–588.

Russell, C. T. & Luhmann, J. G. 1997. Uranus: magnetic field and magnetosphere. In: *Encyclopedia of Planetary Science*, pp. 863–864. Springer

Saar, S. H. 1988. Improved methods for the measurement and analysis of stellar magnetic fields. *Astrophys. J.*, **324**, 441–465.

Saffman, P. G. 1963. On the fine-scale structure of vector fields convected by a turbulent fluid. *J. Fluid Mech.*, **16**, 545–572.

Salmon, R. 1988. Hamiltonian fluid mechanics. *Ann. Rev. Fluid Mech.*, **20**, 225–256.

Sarson, G. R., Jones, C. A., Zhang, K. & Schubert, G. 1997. Magnetoconvection dynamos and the magnetic fields of Io and Ganymede. *Science*, **276**, 1106–1108.

Sarson, G. R., Jones, C. A. & Zhang, K. 1999. Dynamo action in a uniform ambient field. *Phys. Earth Planet. Inter.*, **111**, 47–68.

Schaeffer, N. & Jault, D. 2016. Electrical conductivity of the lowermost mantle explains absorption of core torsional waves at the equator. *Geophys. Res. Lett.*, **43**, 4922–4928.

Scheeler, M. W., van Rees, W. M., Kedia, H., Kleckner, D. & Irvine, W. T. M. 2017. Complete measurement of helicity and its dynamics in vortex tubes. *Science*, **357**, 487–491.

Schekochihin, A. A., Cowley, S. C., Taylor, S. F., Maron, J. L. & McWilliams, J. C. 2004. Simulations of the small-scale turbulent dynamo. *Astrophys. J.*, **612**, 276.

Schlüter, A. & Biermann, L. 1950. Interstellare Magnetfelder. *Z. Naturforsch.*, **5a**, 237–251.

Schou, J., Antia, H. M., Basu, S., et al. 1998. Helioseismic studies of differential rotation in the solar envelope by the solar oscillations investigation using the Michelson Doppler imager. *Astrophys. J.*, **505**, 390–417.

Schrinner, M. 2011. Global dynamo models from direct numerical simulations and their mean-field counterparts. *Astron. Astrophys.*, **533**, A108.

Schrinner, M., Rädler, K.-H., Schmitt, D., Rheinhardt, M. & Christensen, U. R. 2007. Mean-field concept and direct numerical simulations of rotating magnetoconvection and the geodynamo. *Geophys. Astrophys. Fluid Dyn.*, **101**, 81–116.

Schrinner, M., Petitdemange, L. & Dormy, E. 2011. Oscillatory dynamos and their induction mechanisms. *Astron. Astrophys.*, **530**, A140.

Schrinner, M., Petitdemange, L. & Dormy, E. 2012. Dipole collapse and dynamo waves in global direct numerical simulations. *Astrophys. J.*, **752**, 121.

Schultz, A., Kurtz, R. D., Chave, A. D. & Jones, A. G. 1993. Conductivity discontinuities in the upper mantle beneath a stable craton. *Geophys. Res. Lett.*, **20**, 2941–2944.

Schuster, A. 1911. A critical examination of the possible causes of terrestrial magnetism. *Proc. Phys. Soc. London*, **24**, 121–137.

Schutz, B. F. 1980. *Geometrical Methods of Mathematical Physics*. Cambridge University Press.

Semel, M. 1989. Zeeman–Doppler imaging of active stars. I – Basic principles. *Astron. Astrophys.*, **225**, 456–466.

Shearer, M. J. & Roberts, P. H. 1998. The hidden ocean at the top of Earth's core. *Dyn. Atmos. Oceans*, **27**, 631–647.

Shumaylova, V., Teed, R. J. & Proctor, M. R. E. 2017. Large- to small-scale dynamo in domains of large aspect ratio: Kinematic regime. *Mon. Not. R. Astron. Soc.*, **466**, 3513–3518.

Silver, D. S. 2008. The last poem of James Clerk Maxwell. *Not. Am. Math. Soc.*, **55**, 1266–1270.

Simitev, R. D. & Busse, F. H. 2009. Bistability and hysteresis of dipolar dynamos generated by turbulent convection in rotating spherical shells. *Europhys. Lett.*, **85**, 19001.

Simon, J. 1996. Energy functions for knots: Beginning to predict physical behavior. In: Mesirov, J. p., Shulton, K. & Sumners, D. W. (eds.), *Mathematical Approaches to Biomolecular Structure and Dynamics*, pp. 39–58. Springer.

Sisan, D. R., Mujica, N., Tillotson, W. A., et al. 2004. Experimental observation and characterization of the magnetorotational instability. *Phys. Rev. Lett.*, **93**, 114502.

Smith, E., Davis, L., Jones, D. E., et al. 1974. Magnetic field of Jupiter and its interaction with the solar wind. *Science*, **183**, 305–306.

Smith, E. J., Davis, L., Jr., Jones, D. E., et al. 1980. Saturn's magnetic field and magnetosphere. *Science*, **207**, 407–410.

Snodgrass, H. B. 1983. Magnetic rotation of the solar photosphere. *Astrophys. J.*, **270**, 288–299.

1984. Separation of large-scale photospheric Doppler patterns. *Sol. Phys.*, **94**, 13–31.

Soward, A. M. 1972. A kinematic theory of large magnetic Reynolds number dynamos. *Phil. Trans. R. Soc. London, Ser. A*, **275**, 431–462.

1974. A convection-driven dynamo I. The weak field case. *Phil. Trans. R. Soc. London, Ser. A*, **275**, 611–615.

1975. Random waves and dynamo action. *J. Fluid Mech.*, **69**, 145–177.

1977. On the finite amplitude thermal instability of a rapidly rotating fluid sphere. *Geophys. Astrophys. Fluid Dyn.*, **9**, 19–74.

1978. A thin disc model of the galactic dynamo. *Astron. Nachr.*, **299**, 25–33.

1979. Convection driven dynamos. *Phys. Earth Planet. Inter.*, **20**, 134–151.

Soward, A. M. & Jones, C. A. 1983. Alpha-squared-dynamos and Taylor's constraint. *Geophys. Astrophys. Fluid Dyn.*, **27**, 87–122.

Soward, A. M. & Roberts, P. H. 2014. Eulerian and Lagrangian means in rotating, magnetohydrodynamic flows II. Braginsky's nearly axisymmetric dynamo. *Geophys. Astrophys. Fluid Dyn.*, **108**, 269–322.

Spiegel, E. A. & Weiss, N. O. 1982. Magnetic buoyancy and the Boussinesq approximation. *Geophys. Astrophys. Fluid Dyn.*, **22**, 219–234.

Spiegel, E. A. & Zahn, J.-P. 1992. The solar tachocline. *Astron. Astrophys.*, **265**, 106–114.

Spitzer, L., Jr. 1956. *Physics of Fully Ionized Gases*. Interscience.

Sreenivasan, K. R. & Antonia, R. A. 1997. The phenomenology of small-scale turbulence. *Annu. Rev. Fluid Mech.*, **29**, 435–472.

Stasiak, A., Katritch, V. & Kauffman, L. H. (eds.). 1998. *Ideal Knots*. Series on Knots and Everything, vol. 19. World Scientific.

Steenbeck, M. & Krause, F. 1966. The generation of stellar and planetary magnetic fields by turbulent dynamo action. *Z. Naturforsch.*, **21a**, 1285–1296. (English translation: Roberts & Stix (1971) pp. 49–79.)

1969a. On the dynamo theory of stellar and planetary magnetic fields, I. A. C. dynamos of solar type. *Astron. Nachr.*, **291**, 49–84. (English translation: Roberts & Stix (1971) pp. 147–220.)

1969b. On the dynamo theory of stellar and planetary magnetic fields, II. D.C. dynamos of planetary type. *Astron. Nachr.*, **291**, 271–286. (English translation: Roberts & Stix (1971) pp. 221–245.)

Steenbeck, M., Krause, F. & Rädler, K.-H. 1966. Berechnung der mittleren Lorentz-Feldstärke für ein elektrisch leitendes Medium in turbulenter, durch Coriolis-Kräfte beeinflusster Bewegung. *Z. Naturforsch.*, **21a**, 369–376. (English translation: Roberts & Stix (1971) pp. 29–47.)

Steenbeck, M., Kirko, I. M., Gailitis, A., et al. 1967. An experimental verification of the α-effect. *Monats. Dt. Akad. Wiss.*, **9**, 716–719. (English translation: Roberts & Stix (1971) pp. 97–102.)

Stefani, F., Gundrum, T., Gerbeth, G., et al. 2007. Experiments on the magnetorotational instability in helical magnetic fields. *New J. Phys.*, **9**, 295.

Stefani, F., Gailitis, A. & Gerbeth, G. 2008. Magnetohydrodynamic experiments on cosmic magnetic fields. *J. Appl. Math. Mech.*, **88**, 930–954.

Stefani, F., Gerbeth, G., Gundrum, T., et al. 2009. From PROMISE 1 to PROMISE 2: New experimental results on the helical magnetorotational instability. In: Weiss, F.-P. & Rindelhardt, U. (eds.), *Ann. Rep. 2008 Inst. of Safety Res.* Wissenschaftlich-Technische Berichte, nos. FZD–524. Forschungszentrum Dresden Rossendorf.

Stellmach, S. & Hansen, U. 2004. Cartesian convection driven dynamos at low Ekman number. *Phys. Rev.*, **70**, 056312.

Stenflo, J. O. 1976. Small-scale solar magnetic fields. In: Bumba, V. & Kleczek, J. (eds.), *Basic Mechanisms of Solar Activity, Prague 1975*. Proc. IAU Symp., vol. 71. IAU.

Stern, D. P. 2002. A millennium of geomagnetism. *Rev. Geophys.*, **40**, 1–30.

Stevenson, D. J. 1982. Reducing the non-axisymmetry of a planetary dynamo and an application to Saturn. *Geophys. Astrophys. Fluid Dyn.*, **21**, 113–127.

1983. Planetary magnetic fields. *Rep. Prog. Phys.*, **46**, 555.

Stewart, A. J., Schmidt, M. W., van Westrenen, W. & Liebske, C. 2007. Mars: A new core-crystallization regime. *Science*, **316**, 1323–1325.

Stewartson, K. & Roberts, P. H. 1963. On the motion of a liquid in a spheroidal cavity of a precessing rigid body. *J. Fluid Mech.*, **17**, 1–20.

Stix, M. 1972. Non-linear dynamo waves. *Astron. Astrophys.*, **20**, 9–12.

1973. Spherical $\alpha\omega$-dynamos, by a variational method. *Astron. Astrophys.*, **24**, 271–286.

1975. The galactic dynamo. *Astron. Astrophys.*, **42**, 85–89.

1976. Differential rotation and the solar dynamo. *Astron. Astrophys.*, **47**, 243–254.

1978. The galactic dynamo. *Astron. Astrophys.*, **68**, 459.

Størmer, C. 1937. On the trajectories of electric particles in the field of a magnetic dipole with applications to the theory of cosmic radiation. *Astrophys. Norvegica*, **2**, 193–242.

Taylor, G. I. 1921. Diffusion by continuous movements. *Proc. Lond. Math. Soc. A*, **20**, 196–211. (Collected papers, vol. 11, ed. Batchelor, G. K., Cambridge University Press, 1960, pp. 172–184.)

1923. Stability of a viscous liquid contained between two rotating cylinders. *Phil. Trans. R. Soc. London, Ser. A*, **223**, 289–343.

Taylor, G. I. & Green, A. E. 1937. Mechanism of the production of small eddies from large ones. *Proc. R. Soc. London, Ser. A*, **158**, 499–521.

Taylor, J. B. 1963. The magnetohydrodynamics of a rotating fluid and the Earth's dynamo problem. *Proc. R. Soc. London, Ser. A*, **274**, 274–283.

1974. Relaxation of toroidal plasma and generation of reverse magnetic fields. *Phys. Rev. Lett.*, **33**, 1139–1141.

1986. Relaxation and magnetic reconnection in plasmas. *Rev. Mod. Phys.*, **58**, 741.

Teed, R. J., Jones, C. A. & Tobias, S. M. 2015. The transition to Earth-like torsional oscillations in magnetoconvection simulations. *Earth Planet. Sci. Lett.*, **419**, 22–31.

Thébault, E., Finlay, C. C., Beggan, C. D., et al. 2015. International Geomagnetic Reference Field: The 12th generation. *Earth Planets Space*, **67**, 79.

Thellier, E. 1938. *Sur l'aimantation des terres cuites et ses applications géophysiques.* Thèse de doctorat, Paris.

1981. Sur la direction du champ magnétique terrestre, en France, durant les deux derniers millénaires. *Phys. Earth Planet. Inter.*, **24**, 89–132.

Thellier, E. & Thellier, O. 1959. Sur l'intensité du champ magnétique terrestre dans le passé historique et géologique. *Ann. Geophys.*, **15**, 285.

Thompson, C. & Duncan, R. C. 2001. The giant flare of 1998 August 27 from SGR 1900+14. II. Radiative mechanism and physical constraints on the source. *Astrophys. J.*, **561**, 980.

Tilgner, A. & Busse, F. H. 2002. Simulation of the bifurcation diagram of the Karlsruhe dynamo. *Magnetohydrodynamics*, **38**, 35–40.

Tobias, S. M. 1997. The solar cycle: Parity interactions and amplitude modulation. *Astron. Astrophys.*, **322**, 1007–1017.

Tobias, S. & Weiss, N. O. 2007. Stellar dynamos. In: Dormy, E. & Soward, A. M. (eds.), *Mathematical Aspects of Natural Dynamos*, pp. 281–311. The Fluid Mechanics of Astrophysics and Geophysics, vol. 13. Chapman and Hall.

Tobias, S. M., Weiss, N. O. & Kirk, V. 1995. Chaotically modulated stellar dynamos. *Mon. Not. R. Soc.*, **273**, 1150–1166.

Tobias, S. M., Brummell, N. H., Clune, T. L. & Toomre, J. 2001. Transport and storage of magnetic field by overshooting turbulent compressible convection. *Astrophys. J.*, **549**, 1183–1203.

Tobias, S. M., Rädler, K.-H. & Moffatt, H. K. (eds.). 2018. 50 years of mean-field electrodynamics. *J. Plasma Phys*. Cambridge University Press.

Toomre, A. 1966. On the coupling of the Earth's core and mantle during the 26000-year precession. In: Marsden, B. G. & Cameron, A. G. W. (eds.), *The Earth-Moon System*, pp. 33–45. Plenum.

Tough, J. G. 1967. Nearly symmetric dynamos. *Geophys. J. R. Astron. Soc.*, **13**, 393–396. (See also Corrigendum: *Geophys. J. R. Astron. Soc.* **15**, 343.)

Tough, J. G. & Gibson, R. D. 1969. The Braginskii dynamo. In: Runcorn, S. K. (ed.), *The Application of Modern Physics to the Earth and Planetary Interiors*, pp. 555–569. John Wiley.

Trigo, M. D., Miller-Jones, J. C. A., Migliari, S., Broderick, J. W. & Tzioumis, T. 2013. Baryons in the relativistic jets of the stellar-mass black-hole candidate 4U1630-47. *Nature*, **504**, 260–262.

Urey, H. C. 1952. *The Planets.* Yale University Press.

Usoskin, I. G. 2017. A history of solar activity over millennia. *Liv. Rev. Sol. Phys.*, **14**, 3.

Vainshtein, S. I. 1990. Cusp-points and current sheet dynamics. *Astron. Astrophys.*, **230**, 238–243.

Vainshtein, S. I. & Cattaneo, F. 1992. Nonlinear restrictions on dynamo action. *Astrophys. J.*, **393**, 165–171.

Vainshtein, S. I. & Ruzmaikin, A. A. 1971. Generation of the large-scale galactic magnetic field [in Russian]. *Astronom. Zh.*, **48**, 902.

1972. Generation of the large-scale galactic magnetic field. *Sov. Astron.*, **15**, 714.

Vainshtein, S. I. & Zel'dovich, Ya. B. 1972. Origin of magnetic fields in astrophysics. *Sov. Phys. Usp.*, **15**, 159–172.

Valet, J.-P., Meynadier, L. & Guyodo, Y. 2005. Geomagnetic dipole strength and reversal rate over the past two million years. *Nature*, **435**, 802–805.

Velikhov, E. P. 1959. Stability of an ideally conducting liquid flowing between cylinders rotating in a magnetic field. *Sov. Phys. JETP*, **9**, 995–998. (Translated from *Zh. Eksp. Theor. Fiz.* **36**, 1398–1404 (1959).)

Veronis, G. 1959. Cellular convection with finite amplitude in a rotating fluid. *J. Fluid Mech.*, **5**, 401–435.

Vladimirov, V. A., Moffatt, H. K. & Ilin, K. I. 1999. On general transformations and variational principles for the magnetohydrodynamics of ideal fluids. Part 4. Generalized isovorticity principle for three-dimensional flows. *J. Fluid Mech.*, **390**, 127–150.

Walker, G. T. 1896. On a dynamical top. *Q. J. Pure Appl. Math.*, **28**, 175–184.

Warwick, J. 1963. Dynamic spectra of Jupiter's decametric emission. *Astrophys. J.*, **137**, 41–60.

Wei, X., Ji, H., Goodman, J., et al. 2016. Numerical simulations of the Princeton magneto-rotational instability experiment with conducting axial boundaries. *arXiv.1612.01224v1 [physics.flu-dyn].*

Weir, A. D. 1976. Axisymmetric convection in a rotating sphere, I. Stress-free surface. *J. Fluid Mech.*, **75**, 49–79.

Weiss, B. P. & Tikoo, S. M. 2014. The lunar dynamo. *Science*, **346**, 1198.

Weiss, N. O. 1966. The expulsion of magnetic flux by eddies. *Proc. R. Soc. London, Ser. A*, **293**, 310–328.

1971. The dynamo problem. *Q. J. R. Astron. Soc.*, **12**, 432–446.

1976. The pattern of convection in the Sun. In: Bumba, V. & Kleczek, J. (eds), *Basic Mechanisms of Solar Activity, Prague 1975*. Proc. IAU Symp., vol. 71. IAU.

2011. Chaotic behaviour in low-order models of planetary and stellar dynamos. *Geophys. Astrophys. Fluid Dyn.*, **105**, 256–272.

Weiss, N. O. & Proctor, M. R. E. 2009. *Magnetoconvection*. Cambridge University Press.

Weiss, N. O. & Tobias, S. M. 2016. Supermodulation of the Sun's magnetic activity: the effects of symmetry changes. *Mon. Not. R. Astron. Soc.*, **456**, 2654–2661.

Welander, P. 1967. On the oscillatory instability of a differentially heated fluid loop. *J. Fluid Mech.*, **29**, 17–30.

Whitaker, S. 1986. Flow in porous media I: A theoretical derivation of Darcy's law. *Transp. Porous Media*, **1**, 3–25.

White, M. P. 1978. Numerical models of the galactic dynamo. *Astron. Nachr.*, **299**, 209–216.

Winch, D. E., Ivers, D. J., Turner, J. P. R. & Stening, R. J. 2005. Geomagnetism and Schmidt quasi-normalization. *Geophys. J. Int.*, **160**, 487–504.

Woltjer, L. 1958. A theorem on force-free magnetic fields. *Proc. Natl. Acad. Sci. U.S.A*, **44**, 489–491.

1975. Astrophysical evidence for strong magnetic fields. *Ann. N. Y. Acad. Sci.*, **257**, 76–79.

Wood, T. S. & Hollerbach, R. 2015. Three-dimensional simulation of the magnetic stress in a neutron star crust. *Phys. Rev. Lett.*, **114**, 191101.

Wrubel, M. H. 1952. On the decay of a primeval stellar magnetic field. *Astrophys. J.*, **116**, 291.

Wu, C.-C. & Roberts, P. H. 2009. On a dynamo driven by topographic precession. *Geophys. Astrophys. Fluid Dyn.*, **103**, 467–501.

2015. On magnetostrophic mean-field solutions of the geodynamo equations. *Geophys. Astrophys. Fluid Dyn.*, **109**, 84–110.

Yano, J.-I. 1992. Asymptotic theory of thermal convection in rapidly rotating systems. *J. Fluid Mech.*, **243**, 103–131.

Yoder, C. F., Konopliv, A. S., Yuan, D. N., Standish, E. M. & Folkner, W. M. 2003. Fluid core size of Mars from detection of the solar tide. *Science*, **300**, 299–303.

Yokokawa, M., Itakura, K., Uno, A., Ishihara, T. & Kaneda, Y. 2002. 16.4-Tflops direct numerical simulation of turbulence by a Fourier spectral method on the Earth Simulator. In: *SC '02 Proc. 2002 ACM/IEEE Conf. on Supercomputing*. ACM/IEEE.

Yoshimura, H. 1975. A model of the solar cycle driven by the dynamo action of the global convection in the solar convection zone. *Astrophys. J. Suppl.*, **29**, 467.

Zel'dovich, Ya. B. 1957. The magnetic field in the two-dimensional motion of a conducting turbulent fluid. *Sov. Phys. JETP*, **4**, 460–462.

Zeldovich, Ya. B., Ruzmaikin, A. A. & Sokoloff, D. D. 1990. *Magnetic Fields in Astrophysics*. The Fluid Mechanics of Astrophysics and Geophysics, vol. 3. Gordon and Breach.

Zhang, K.-K. & Busse, F. H. 1988. Finite amplitude convection and magnetic field generation in a rotating spherical shell. *Geophys. Astrophys. Fluid Dyn.*, **44**, 33–53.

1989. Convection driven magnetohydrodynamic dynamos in rotating spherical shells. *Geophys. Astrophys. Fluid Dyn.*, **49**, 97–116.

Zhao, J., Bogart, R. S., Kosovichev, A. G., T. L. Duvall, Jr. & Hartlep, T. 2013. Detection of equatorward meridional flow and evidence of double-cell meridional circulation inside the Sun. *Astrophys. J. Lett.*, **774**, L29.

Zweibel, E. G. & Heiles, C. 1997. Magnetic fields in galaxies and beyond. *Nature*, **385**, 131–136.

Author index

Subject index